Membrane Transport in Biology – Volume IV B

Membrane Transport in Biology

Edited by

G. Giebisch · D. C. Tosteson · H. H. Ussing

Associate Editor

M. T. Tosteson

Volume IV B

Transport Organs

Contributors

T.E. Andreoli E. Boulpaep M. Claret F.E. Curry J.H. Dirks
J.G. Forte G. Giebisch C.W. Gottschalk R.N. Khuri R. Kinne
W.E. Lassiter E.W. van Lennep T.E. Machen G. Malnic
D.W. Powell G.A. Quamme E.M. Renkin L. Reuss J.A. Schafer
S.G. Schultz I. Schulz R.A.L. Sutton K.J. Ullrich E.E. Windhager
J.A. Young

Editor

G. Giebisch

With 82 Figures and 48 Tables

Springer-Verlag Berlin · Heidelberg · New York 1979

Professor Dr. Gerhard Giebisch
Yale University, School of Medicine, Department of Physiology
333 Cedar Street, New Haven, Conn. 06510 / USA

Professor Dr. Daniel C. Tosteson, Dean
Harvard Medical School,
25 Shattuck Street, Boston, Mass. 02115 / USA

Professor Dr. Hans H. Ussing
University of Copenhagen, Institute of Biological Chemistry A
13 Universitetsparken, DK - 2100 Copenhagen

Dr. Magdalena T. Tosteson
Harvard Medical School, Department of Physiology
25 Shattuck Street, Boston, Mass. 02115 / USA

ISBN-13: 978-3-642-81258-3 e-ISBN-13: 978-3-642-81256-9
DOI: 10.1007/ 978-3-642-81256-9

Library of Congress Cataloging in Publication Data: Main entry under title: Transport organs. (Membrane transport in biology; V. 4, part A and B). Includes bibliographies and index. 1. Biological transport. I. Andreoli, Thomas E., 1935 –. II. Giebisch, G., 1927 –. III. Series, QH509.M44. vol. 4. 574.8'75s [574.1'91]. 78–23183.

Softcover reprint of the hardcover 1st edition 1979

Typesetting: Druckerei G. Appl, Wemding

2122/3020–543210

Contents Volume IV B

Contents Volume IV A XV
List of Contributors XXIII

Chapter 10 – Perfusion of Isolated Mammalian Renal Tubules
(J.A. Schafer and T.E. Andreoli) 473

A. Introduction 473

B. Experimental Techniques: Practical and Theoretical Considerations 475
I. Isolation of Renal Tubule Segments 475
II. Tubular Perfusion 476
III. Perfusing and Bathing Solutions 479
IV. Measurement of Net Volume Absorption 481
1. Collection Method 481
2. Crimped-End Method 482
V. Assessment of Hydraulic Conductivity 483
1. The Unstirred Layer Problem 483
2. The Hydraulic Conductivity Coefficient 486
VI. Electrical Measurements 488
1. Liquid Junction Potentials and Donnan Voltages 489
2. Ionic Dilution Potentials 492
3. Measurement of Transepithelial Resistance 493
VII. Tracer Flux Measurements 494
1. Unidirectional Lumen-to-Bath Fluxes 495
2. Unidirectional Bath-to-Lumen Fluxes 496
3. Evaluating P_{D_i} From Tracer Fluxes 496
4. Transmembrane Fluxes 498

C. Transport Properties of Isolated Nephron Segments 498
I. Suitability of the *In-Vitro* Preparation 499
II. The Proximal Tubule 499
1. Dissipative Transport Properties 500
2. Active Transport Processes 502
3. Mechanismus for Salt and Water Transport in the Superficial Pars Recta 505
4. Effective Luminal Hypotonicity and Isotonic Fluid Absorption 507
5. Heterogeneity of Tubular Structure and Function 509
III. The Loop of Henle 510

IV. The Distal Convoluted Tubule and Collecting Duct System 513
1. Distal Convoluted Tubule 516
2. Cortical Collecting Tubule 516
3. Water and Nonelectrolyte Permeation in the Cortical Collecting Duct: the Mechanism of ADH Action 517

References 525

Chapter 11 – Metabolic Correlates of Tubular Transport
(R. Kinne) 529

A. Introduction 529

B. Transport-Related Parameters of Metabolic Activity in Intact Cells 530
I. General Aspects 530
II. Heat Production in the Kidney 530
III. O_2 Consumption and CO_2 Production in the Kidney 531
IV. Effect of Metabolic Inhibitors on the Transepithelial Transport of Sodium 534
V. Renal Substrate Metabolism in Relation to Renal Function 540
1. Uptake and Oxidation of Substrates in Relation to Sodium Transport . 540
2. Substrate Dependence of Sodium Transport in Isolated Systems 541

C. Transport-Related Intracellular Parameters of Metabolic Activity 545
I. Determination of Enzyme Activities 545
II. Determination of ATP Content and ATP Turnover 546
III. ATP Content and α-Aminoisobutyric Acid Transport in Kidney Cortex Slices 547

D. Energetics of Transport as Studied with Isolated Renal Plasma Membranes . . . 548
I. General Aspects 548
II. Transport-Related ATP Hydrolases 549
1. Sodium 549
2. Calcium 550
a) Properties of Renal Ca^{++}-Activated ATPases 550
b) Possible Relation of Ca^{++}-ATPase to Transepithelial Ca^{++} Transport 551
3. Bicarbonate and Protons 552
4. Chloride 553
III. Sodium-Solute Cotransport Systems 554
1. General Aspects 554
2. General Characteristics of Transport by Renal Vesicles 554
3. The Sodium Gradient as the Driving Force of Intravesicular Accumulation 555
4. The Electrical Membrane Potential as the Driving Force of Intravesicular Accumulation 557
5. Energetics of Transcellular Transport in the Proximal Convoluted Tubule 558

Acknowledgements 559

References 559

Chapter 12 – Transport in Salivary and Salt Glands

(J.A. Young and E.W. van Lennep) . . . 563

Part I: Salivary Glands . . . 563

A. Introduction . . . 563

B. Anatomy and Anatomical Terminology . . . 564
I. Secretory Endpieces . . . 564
II. The Duct System . . . 565
III. Myoepithelial Cells . . . 566

C. Transport of Water and Electrolytes . . . 567
I. Theories of Secretion of Water and Electrolytes by Salivary Glands . . . 567
II. Formation of the Primary Saliva . . . 568
1. Evidence Concerning the Site or Sites of Fluid Secretion by Salivary Glands . . . 568
2. Control of Secretion and Innervation of Endpieces . . . 572
3. Composition of the Primary Secretion . . . 575
a) Osmotic Activity . . . 575
b) Electrolyte Concentrations . . . 578
i) Bicarbonate . . . 578
ii) Sodium . . . 579
iii) Potassium . . . 579
iv) Chloride . . . 580
c) Organic Solutes . . . 580
4. The Unstimulated Endpiece Cell . . . 581
5. The Stimulated Endpiece Cell . . . 590
a) Receptors and Receptor Pharmacology . . . 590
b) The Secretory Potential and Associated Ionic Fluxes . . . 592
i) Cholinergic Stimulation . . . 592
ii) Adrenergic Stimulation . . . 597
iii) Transients . . . 600
c) Stimulus-Secretion Coupling. Calcium Ions and Cyclic Nucleotides . . . 600
i) Mediation of α-Adrenergic and Cholinergic Responses . . . 601
ii) Mediation of β-Adrenergic Responses . . . 603
6. Isotonic Fluid Transport . . . 606
III. Ductal (Secondary) Modification of the Primary Secretion . . . 608
1. Flow Rate and Electrolyte Excretion Patterns in Final Saliva . . . 608
a) Parasympathetic and Parasympathomimetic Stimulation . . . 608
i) Sodium . . . 609
ii) Potassium . . . 615
iii) Bicarbonate . . . 616
b) Sympathetic and Sympathomimetic Stimulation . . . 617
c) Excretory Patterns for Calcium, Magnesium and Phosphate . . . 620
i) Calcium . . . 620
ii) Magnesium . . . 621
iii) Phosphate . . . 621
2. Micropuncture and Perfusion Studies of Salivary Duct Function . . . 622
a) Permeability Properties of Salivary Ducts . . . 625
i) Sodium . . . 625
ii) Potassium . . . 627

iii) Anions . . . 629
iv) Water and Urea . . . 632
b) Active Transport and Carrier-Mediated Passive Transport by Salivary Ducts . . . 633
i) Minimum Requirements for Maintenance of Active Transport . . . 633
ii) Sodium Reabsorption . . . 635
iii) Potassium Secretion . . . 637
iv) Transport of Bicarbonate or Protons . . . 639
c) A Transport Model for the Duct Epithelium . . . 640
3. Innervation and Autonomic Control of Ductal Transport . . . 643
4. Endocrine Control of Ductal Transport . . . 648
a) Mineralocorticoids . . . 648
b) Angiotensin . . . 653
c) Gastrointestinal Polypeptide Hormones and Related Substances . . 653
5. Relative Roles of Granular, Striated, and Excretory Ducts in Saliva Formation . . . 654

D. Transport of Proteins . . . 655
I. Transport of Albumin . . . 656
II. Transport of Immunoglobulin A (IgA) . . . 656
III. Transport of Secretory Proteins . . . 657
1. Uptake of Amino Acids into Secretory Cells . . . 659
2. Intracellular Transport . . . 659
3. Discharge of Secretion Granules . . . 661
IV. Control of Protein Secretion . . . 662
1. Control of Amylase Release . . . 662
2. Control of Mucoprotein Release . . . 663
3. Protein Secretion by Rodent Granular Duct Cells . . . 664

References . . . 665

Part II: Salt Glands . . . 675

A. Introduction . . . 675

B. Structure of Salt Glands . . . 677
I. Microanatomy . . . 677
II. Ultrastructure . . . 678

C. Adaptation to Salt Loading . . . 680

D. Neural and Hormonal Control of Salt Gland Secretion . . . 681

E. Flow Rates and Electrolyte Concentrations . . . 682

F. The Mechanism of Salt Secretion by the Tubular Endpiece . . . 683

G. The Role of the Duct System . . . 689

Acknowledgements . . . 690

References . . . 690

Chapter 13 – Gastric Secretion

(T.E. Machen and J.G. Forte) 693

A. Introduction 693

B. Morphological Features of Gastric Mucosa 694
I. Histology 694
II. Ultrastructure of the Oxyntic Cell 697

C. Ion Transport 698
I. General 698
II. H^+ Secretion 699
1. Locus of H^+ Secretion 699
2. Acid-Base Balance of Oxyntic Cells 700
3. Anion Dependence of H^+ Secretion 702
4. Cation Dependence of H^+ Secretion 703
III. Cl^- Transport 704
1. Active Transport 704
2. Exchange Diffusion 706
IV. Na^+ Absorption 708

D. Water Transport 710
I. Introduction 710
II. Diffusional Permeability to Water 711
III. Hydraulic Conductivity 711
IV. Water Flow During Secretion 713
1. Hydrostatic Pressure and Ultrafiltration 713
2. Endogenous HCl Gradients and Water Secretion 714

E. Electrophysiological Analyses of Gastric Transport 716
I. Electrogenic vs. Electroneutral H^+ and Cl^- Pumps 716
1. The Electrogenic Hypothesis 716
2. The Electroneutral Hypothesis 719
3. Conclusions 720
II. Intracellular PD's 721
III. Permeability and Conductance Pathways 722
1. Transcellular and Paracellular Conductances 722
2. Cell Membrane Permeability Characteristics 725
IV. Black Box Model of Gastric Mucosa 728

F. Biochemical Basis of Gastric HCl Secretion 729
I. Metabolic Requirements 729
II. Redox Hypotheses 731
III. ATP Utilization Hypotheses 733
IV. The Search for Gastric ATPases 734
1. HCO_3^--Stimulated ATPase 734
2. $(Na^+ + K^+)$-ATPase 735
3. K^+-Stimulated ATPase 735
V. Transport Properties of Gastric Microsomal Vesicles 737

Acknowledgements 743

References 743

Chapter 14 – Transport Across Small Intestine
(S.G. Schultz) 749

A. Introduction 749

B. The Paracellular Pathway 749
I. Permeability to Ions 750
II. Permeability to Nonelectrolytes 754
III. Electrophysiologic Implications 755

C. Active Sodium Transport 758
I. Paracellular Na Transport 759
II. Transcellular Na Transport 760

D. Relations Between Sodium Transport and the Transport of Other Solutes . . . 771
I. Na-Coupled Sugar and Aminoacid Transport 772
II. Na-Coupled Cl Transport 775

E. Summary 776

References 777

Chapter 15 – Transport in Large Intestine
(D.W. Powell) 781

A. Introduction 781

B. Water and Electrolyte Transport 782
I. Epithelial Properties 782
II. Sodium, Chloride, and Bicarbonate Transport 785
III. Potassium Transport 789
IV. Intestinal Secretion 791
V. Hormones and Electrolyte Transport 793

C. Weak Electrolyte Transport (Ammonia and Volatile Fatty Acids) 795
I. Nonionic Diffusion 796
II. Colonic Transport 798
III. Transport Models Based on the Small Intestine 800

D. Conclusion 803

Acknowledgements 804

References 804

Chapter 16 – Transport Processes in the Exocrine Pancreas
(I. Schulz and K.J. Ullrich) 811

A. Introduction 811

B. Secretion of Electrolytes and Water 812
I. Stimulatory Processes 812
II. Fluid Secretion and Ionic Requirement 815
III. Flow-Dependent Concentration Pattern in the Secreted Fluid 817
IV. Local Transport Events as Revealed by Micropuncture and Microperfusion Techniques 819

1. Electrolytes and Water 819
2. Electrical Potential Differences 822
V. Mechanism of Ion Secretion 823
1. Buffer Secretion 823
a) What Buffer Components in the Perfusate Determine Buffer and Fluid Secretion 823
b) Sidedness of the Buffer Transport 825
2. Secretion of Na^+ Ions 826
3. Possible Mechanism of Buffer Secretion 829

C. Secretion of Enzymes 831
I. Transport Processes in the Acini Involved in Stimulus-Secretion Coupling 831
1. Secretagogues of Enzyme Secretion 831
2. Electrophysiological Measurements 832
3. Role of Calcium Ions 833
II. Role of Cyclic AMP and Cyclic GMP in Stimulus-Secretion Coupling . . . 837
III. Ca^{++} and Exocytosis of Secretory Granules 839
IV. Role of Microfilaments and Microtubules 842

D. Uptake of Amino Acids Used for Enzyme Synthesis 843

Acknowledgements 845

References 845

Chapter 17 – Transport in Gallbladder
(L. Reuss) 853

A. Introduction 853

B. General Transport Properties 854
I. Morphology of the Gallbladder 854
II. Composition of Gallbladder Bile 855
III. General Characteristics of Salt and Water Transport 856
1. Salt and Water Transport: Rate, Ion Dependency, Composition of the Transported Fluid 857
2. Water Transport and Transmural Osmotic Gradients 861
3. Metabolic and Pharmacologic Aspects of Transport 862

C. Mechanisms of Ion Transport 863
I. Black-Box Electrical Properties of the Gallbladder 863
1. The Transepithelial Potential Difference 863
2. The Transepithelial Electrical Resistance 865
3. Overall Transepithelial Ionic Permeabilities 866
II. Demonstration of the Existence of a Paracellular Shunt Pathway 867
1. Electrophysiological Evidence 867
2. Morphological Evidence 869
III. Properties of the Shunt Pathway 869
1. Physicochemical Characteristics of the Paracellular Pathway 869
a) Charge of the Pathway 869
b) Mobility of the Sites 870
c) Evidence for a Free-Solution Parallel Pathway 870
d) Effects of pH and Polyvalent Cations 871
e) Permeation of Nitrogenous Cations 872
2. Role of the Lateral Intercellular Spaces in Passive Ion Permeation . . . 872

IV. Properties of the Epithelial Cell Membranes 874
1. Electrophysiological Studies 874
a) Equivalent Circuit for Gallbladder Epithelium 874
b) Ionic Permeability of the Apical Membrane 877
c) Ionic Permeability of the Basolateral Membrane 879
2. Tracer Analysis of Electrolyte Transport 879
3. Role of the Lateral Intercellular Spaces in Ion Transport 881
4. Relative Contributions of Cell Membranes and Shunt Pathway to Transepithelial Potential and Transepithelial Resistance Changes . . 882
a) Estimation of Shunt Ion Permeability Ratios from Transepithelial Measurements 883
b) Estimation of Cell Membrane Ion Permeability Ratios from Changes in Cell Membrane Potential 884
c) Effects of Transepithelial Osmotic Gradients on Potentials and Resistances 885
d) Effects of Transepithelial Current Pulses on Potentials and Resistances 887
e) Effects of Amphotericin B on Potentials and Resistances 889

D. Transport of Nonelectrolytes 890
E. Water Transport 892
1. Local Osmosis Hypotheses 892
a) Three-Compartment Hypothesis 892
b) Standing-Gradient Osmotic Flow Hypothesis 892
c) Hypertonic Interspace Mechanisms with Leaky Junctions and Distributed Solute Input 893
d) Influence of Transport-Dependent Asymmetries in Fluid Composition 894
2. Magnitude of the Hydraulic Conductivity of the Gallbladder 894
3. Route of Water Flow 895

Acknowledgements 896

References 896

Chapter 18 – Transport of Ions in Liver Cells
(M. Claret) 899

A. Introduction 899

B. Ionic Concentrations and Activities 899

C. Membrane Potential 900

D. Passive Fluxes 902
I. Distribution of Ions 902
II. Diffusion Fluxes 902
III. Facilitated Diffusion 904
1. Ouabain-Insensitive Sodium Exchange 904
2. Ouabain-Sensitive Sodium Exchange 904
3. Potassium-Facilitated Diffusion 904
IV. Factors Affecting Sodium, Potassium, and Chloride Permeabilities 905
1. Effect of Calcium on Permeability 905
a) Presumed Effect of Calcium on Permeability of Junctional Membrane 905

b) Effect of External Calcium on Permeability 906
c) Effect of Internal Calcium on Permeability 907
2. Effect of Foreign Anions on Membrane Permeability 907
3. Effect of Temperature on Permeability 907

E. Active Transport of Sodium and Potassium 908
I. Ouabain-Sensitive Fluxes 908
II. Coupling Between Sodium and Potassium Movements 909
III. Stoichiometry of the Sodium-Potassium Pump 910
1. Coupling Ratio Between Sodium and Potassium Fluxes 910
2. Cation/ATP Ratio 911
IV. Contribution of the Electrogenic Sodium Pump to Membrane Potential . . 912
V. Energy Requirement for Sodium-Potassium Transport 913
1. Measured Energy Requirement 913
2. Theoretical Energy Requirement 914

F. Regulation of Cell Volume 915

G. Calcium Transport 916

Acknowledgements 918

References 918

Subject Index 921

Contents Volume IV A

Chapter 1 – Transport of Water and Solutes Across Capillary Endothelium
(E.M. Renkin and F.E. Curry) 1

A. Morphology 1

B. Development of Capillary Pore Theory 5
I. Ultrafiltration 5
II. Diffusion 7
III. Interaction of Ultrafiltration and Diffusion 8
IV. Evaluation of Reflection Coefficients 13
V. Direct Measurement of PS by Diffusion 16

C. The Multiple Pathway Hypothesis 18
I. Theory 18
II. Application to Single Capillaries of Frog Mesentery 21
III. Application to Capillaries of Mammalian Skeletal Muscle 25

D. Characterization of the Cell Pathway 27

E. Endothelial Transport of Macromolecules 28
I. Theory 28
II. Experimental Data 30
III. Interpretation 33
IV. Endothelial Changes in Injury 35

F. Summary 36
I. Endothelial Cell Pathways 36
II. Intercellular or Transcellular Pathways 36

Appendix: Frictional Interaction of Solute and Solvent Molecules in Porous Membranes. Hydrodynamic Models 37

References 41

Chapter 2 – Electrochemistry of the Nephron

(R.N. Khuri) . . . 47

A. General . . . 47
I. Intracellular Physiology . . . 47
II. The Physical State of Intracellular Ions . . . 47
III. Electrochemical Analysis . . . 49
1. Electrochemistry . . . 49
2. Ion-Exchangers for Electrodes . . . 50
3. The Intracellular Electrochemical Technique . . . 51
IV. The Renal Tubular Epithelium . . . 57

B. Potassium . . . 59
I. Muscle Fiber Potassium . . . 59
II. Renal Cell Potassium . . . 62
1. Distal Tubule . . . 62
a) Normal Potassium Balance . . . 62
b) Changes in Potassium Balance . . . 62
c) Changes in Acid-Base Balance . . . 63
d) Effect of Aldosterone . . . 63
e) Distal Tubule: *Amphiuma* . . . 64
2. Proximal Tubule: *Necturus* . . . 65

C. Sodium . . . 69
I. Muscle Fiber Sodium . . . 69
II. Renal Cell Sodium . . . 70
1. Proximal Tubule: *Necturus* . . . 71
2. Distal Tubule . . . 75

D. Chloride . . . 77
I. Muscle-Fiber Chloride . . . 77
II. Renal Cell Chloride . . . 78
1. Proximal Tubule . . . 78
2. Distal Tubule (Rat) . . . 80

E. Bicarbonate (Cell pH) . . . 82
I. Muscle-Fiber Bicarbonate (pH) . . . 82
II. Renal-Cell Bicarbonate . . . 85

F. Ions in Combination . . . 88

G. Summary . . . 92

References . . . 93

Chapter 3 – Electrophysiology of the Kidney

(E.L. Boulpaep) . . . 97

A. Introduction . . . 97

B. Electrophysiological Methods in Renal Tubules . . . 97
I. Renal Preparations . . . 97
II. Electrode Techniques . . . 98
III. Methodological Problems in Potential Difference Measurements . . . 99
IV. Methodological Problems in Conductance Measurements . . . 99
V. Levels of Investigation and Definitions . . . 100

C. Proximal Convoluted Tubule 102
I. Transepithelial Properties of Proximal Convoluted Tubule 102
1. Transepithelial Potential Difference 102
a) Mammalian Proximal Convoluted Tubule 102
b) Amphibian Proximal Convoluted Tubule 104
2. Transepithelial Electrical Conductance 105
a) Mammalian Proximal Convoluted Tubule 105
b) Amphibian Proximal Convoluted Tubule 105
3. Passive Transepithelial Ionic Membrane Parameters 106
4. "Active" Transepithelial Ionic Membrane Parameters 110
II. Single Membrane Properties of Proximal Convoluted Tubule 111
1. Single Barrier Potential Differences 111
a) The Paracellular or Tight-Junction Barrier Potential Difference . . 111
b) The Peritubular Cell Membrane Potential Difference 111
c) The Luminal Cell Membrane Potential Difference 112
2. Single Barrier Electrical Conductance 112
3. Passive Ionic Membrane Parameters of Single Barriers 115
a) Paracellular Membrane Properties 117
b) Peritubular Cell Membrane Properties 118
c) Luminal Cell Membrane Properties 119
4. "Active" Ionic Membrane Parameters of Single Barriers 120

D. Proximal Straight Tubule 123
1. Transepithelial Potential Difference 123
2. Transepithelial Electrical Conductance 124
3. Transepithelial Ionic Membrane Parameters 124
a) Superficial Proximal Straight Tubule 124
b) Juxtamedullary Proximal Straight Tubule 125

E. Thin Descending Limb of Henle 125

F. Thin Ascending Limb of Henle 126
1. Transepithelial Potential Difference 126
2. Transepithelial Ionic Membrane Parameters 127

G. Thick Ascending Limb of Henle 127
1. Transepithelial Potential Difference 128
a) Mammalian Thick Ascending Limb 128
b) Amphibian Diluting Segments 128
2. Transepithelial Electrical Conductance 128
3. Passive Transepithelial Ionic Membrane Parameters 128
4. "Active" Transepithelial Ionic Membrane Parameters 129

H. Distal Convoluted Tubule 130
I. Transepithelial Properties of the Distal Convoluted Tubule 131
1. Transepithelial Potential Difference 131
a) Mammalian Distal Convoluted Tubule 131
b) Amphibian and Reptilian Distal Convoluted Tubule 131
2. Transepithelial Electrical Conductance 132
3. Passive Transepithelial Ionic Membrane Parameters 132
II. Single Membrane Properties of the Distal Convoluted Tubule 135
1. Single Barrier Potential Differences 135
2. Passive Ionic Membrane Parameters of Single Barriers 135
3. "Active" Ionic Membrane Parameters of Single Barriers 135

J. Cortical Collecting Tubule 136
I. Transepithelial Properties of Cortical Collecting Tubule 136
1. Transepithelial Potential Difference 136
2. Transepithelial Electrical Conductance 137
3. Passive Transepithelial Ionic Membrane Parameters 137
4. "Active" Transepithelial Ionic Membrane Parameters 138
II. Single Membrane Properties of Cortical Collecting Tubule 139

K. Medullary Collecting Duct 139
1. Transepithelial Potential Difference 139
2. Transepithelial Electrical Conductance 140
3. Passive Transepithelial Ionic Membrane Parameters 140

References 140

Chapter 4 – Sodium Chloride Transport
(E.E. Windhager) 145

A. Introduction 145

B. Proximal Tubular Transport 146
I. General Properties of the Proximal Nephron 146
II. Inhomogeneity of Proximal Tubular Morphology and Function 148
III. Net Transport Rates 151
IV. Electrophysiology of Proximal Tubules 152
V. Cotransport of Sodium with Sugars and Aminoacids 155
VI. Ion Permeabilities and Intercellular Shunt Path 158
VII. Active Transport of Sodium 160
VIII. Sodium and Chloride Entry Across the Luminal Cell Membrane 163
IX. Transport of Sodium Chloride Across the Peritubular Cell Membrane . . 167
X. Coupling of Sodium and Water Reabsorption 169
XI. Passive Reabsorption of Sodium Chloride 171
XII. Regulation of Proximal Salt Reabsorption by Passive Backflux 175

C. Loop of Henle 185

D. Distal Convoluted Tubule 190

I. General Properties 190
II. Net Movement of Sodium Chloride and Water 191
III. Electrophysiological Considerations 193
IV. Net Movement of Chloride 197
V. Water 198
VI. Interaction Between Sodium and Calcium 198
VII. Mechanisms of Transport 200
1. Peritubular Membrane 200
2. Luminal Membrane 201
3. Role of Intercellular Ionic Shunt 202
4. Cellular Mechanisms of Antidiuretic Hormone Action 203

E. Papillary Collecting Duct 203

References 205

Chapter 5 – Renal Potassium Transport
(G. Giebisch) 215

A. Introduction 215
B. General Aspects of Epithelial Transport of Potassium 216
I. Constancy of Cellular Potassium Concentration 216
II. Relationship Between Potassium and Sodium Transport 218
III. Cellular Compartmentalization of Potassium 221
IV. Cell Heterogeneity of Epithelium 222

C. General Properties and Development of Concepts of the Renal Potassium Transport System 223

D. Tubular Sites of Potassium Transport 226
I. Proximal Tubule 227
II. Loop of Henle 228
III. Distal Tubule, Collecting Tubule and Collecting Duct 231

E. Cellular Mechanisms of Renal Tubular Potassium Transport 243
I. Proximal Tubular Epithelium 243
II. Distal Tubule, Collecting Tubule and Collecting Duct 245
1. Electrical Potential Difference and Ionic Conductance of the Distal Tubule and Collecting Duct 245
a) Transepithelial Potential Profile 245
b) Properties of the Peritubular Cell Membrane 249
c) Properties of the Luminal Cell Membrane 251
2. Cellular Transport Models 254

F. Factors Affecting the Rate of Distal Tubule, Collecting Tubule, and Collecting Duct Potassium Transport 264
I. Variations in Hydration and Intake of Potassium 264
II. Distal Tubular Flow Rate, Sodium Delivery and the Effect of Diuretics . . 270
III. Acid-Base Balance 281
IV. Mineralocorticoids 288

Acknowledgements 292

References 292

Chapter 6 – Cellular Aspects of Renal Tubular Acidification
(G. Malnic and G. Giebisch) 299

A. Introduction 299

B. The HCO_3/CO_2 Buffer System 300
I. Kinetic Considerations 300
II. Equilibrium Conditions in the HCO_3/CO_2 Buffer System 302

C. Methods for the Study of Renal Acidification 305
I. Measurement of Tubular pH 305
II. Measurement of Tubular and Cellular Bicarbonate Concentrations 309
III. Measurement of Tubular pCO_2 310
IV. Measurement of Titratable Acidity 311

D. Bicarbonate Reabsorption Along the Nephron 311

E. Kinetic Aspects of Renal Tubular Acidification 314

F. Mechanism of Renal Tubular Acidification . . . 321
I. The Renal Disequilibrium pH . . . 321
II. Evidence for Active Transport of H . . . 325
III. Permeability of the Tubule to HCO_3^- Ions . . . 327
IV. Permeability of the Tubule to H Ions . . . 328
V. Participation of Proton Transport and Buffer-Base Reabsorption in Tubular Acidification . . . 328

G. Coupling of H Ion Transport to the Transport of Other Ions . . . 336
I. Na/H Exchange . . . 336
II. K/H Coupling . . . 339
III. H/Cl Coupling . . . 341

H. Control of Renal Acidification . . . 344
I. The Segmental Buffer Load . . . 344
II. The Role of Peritubular pH and pCO_2 . . . 346
III. Peritubular Base Transfer . . . 348
IV. Humoral Regulation of Acidification . . . 349

Acknowledgements . . . 350

References . . . 350

Chapter 7 – Transport of Calcium, Magnesium and Inorganic Phosphate in the Kidney
(R.A.L. Sutton, G.A. Quamme, and J.H. Dirks) . . . 357

A. Calcium . . . 357
I. Introduction . . . 357
II. Glomerulus . . . 358
III. Proximal Tubule . . . 359
1. Normal Pattern of Reabsorption . . . 359
2. Microperfusion and Microinjection Studies . . . 361
3. Free and Complexed Calcium . . . 362
4. Mechanism of Proximal Reabsorption . . . 364
IV. Loop of Henle . . . 366
V. Distal Tubule and Collecting Duct . . . 370
1. Sites of Calcium Reabsorption . . . 370
2. Hypercalcemia . . . 371
3. Parathyroid Hormone . . . 372
4. Thiazide Diuretics and Acetazolamide . . . 373
5. Vitamin D . . . 374
6. Acidosis and Alkalosis . . . 375
7. Phosphate Loading and Phosphate Deficiency . . . 375
8. Miscellaneous Factors . . . 376

B. Magnesium . . . 377
I. Introduction . . . 377
II. Glomerulus . . . 378
III. Proximal Tubule . . . 378
1. Pattern of Reabsorption . . . 378
2. Mechanism of Proximal Reabsorption . . . 381
IV. Loop of Henle . . . 381
1. Pattern of Reabsorption in the Loop . . . 381
2. Mechanism of Loop Reabsorption . . . 382

V. Distal Tubule and Collecting Duct 383
VI. Regulation of Magnesium Reabsorption by the Kidney 385
VII. Relationship of Magnesium Reabsorption to Other Cations 386
VIII. Hormonal Effects 387
1. Parathyroid Hormone 387
2. Adrenocortical Steroids 387
3. Other Hormones 388
IX. Other Factors Affecting Magnesium Reabsorption 388
1. Acidosis and Alkalosis 388
2. Alcohol 388
3. Diuretics 389

C. Phosphate 389
I. Introduction 389
II. Glomerular Filtration 391
III. Proximal Tubule 392
1. Profile of Pi Reabsorption 392
2. Relationship of Phosphate to Sodium Reabsorption 394
3. Mechanisms of Phosphate Reabsorption in the Proximal Tubule 395
4. Effect of Parathyroid Hormone (PTH) and Luminal pH 396
5. Phosphate Infusion 398
6. Phosphate Depletion 400
7. Renal Failure 401
8. Effects of Calcium Upon Phosphate Reabsorption 401
9. Other Endocrine Influences on Phosphate Transport 402
a) Vitamin D 402
b) Calcitonin and Growth Hormone 403
c) Insulin and Glucose 403
IV. Loop of Henle 404
V. Distal Tubule and Collecting Duct 405

References 407

Chapter 8 – Renal Transport of Organic Solutes
(K.J. Ullrich) 413

A. Introduction 413

B. Site of Transport of Organic Solutes within the Nephron 414

C. Methods of Evaluating Renal Transport of Organic Solutes 414
I. Transtubular Measurements 414
II. Electrical Measurements 415
III. Transport Studies with Membrane Vesicles 416
IV. Transport Studies with Reconstituted Liposomes 417

D. Theoretical Considerations on Na^+-Coupled Transport Processes 417

E. Transport of Sugars 418
I. Specificity or Structural Requirements 418
1. Sugars that are Actively Transported 419
2. Sugars that are Not Actively Transported 420
II. Sodium-Dependence of the Active Sugar Transport 421
III. Binding of Sugars and Phlorrhizin to Brush-Border Membranes 423
IV. Transport of Sugars and Na^+ Across the Brush-Border Membrane 424

1. Electrical Measurements 424
2. Transport Measurements on Brush-Border Vesicles 426
3. Transport Across Vesicles from the Contraluminal Cell Side 427
4. Transport Across Proteoliposomes 428
V. Fit of Data Gained by Different Methods 428

F. Transport of Aminoacids 429
I. Current Problems in Determining the Specificity 429
II. Sodium-Dependence of the Active Transport of Aminoacids 431
1. Measurement Across the Whole Proximal Tubule 431
2. Electrical Measurements 432
3. Transport Measurements on Membrane Vesicles 434
III. Unsolved Problems 435

G. Transport of Organic Acids and Bases 436
I. Transport Steps at Either Cell Side 437
II. Nature of the Driving Forces for the Organic Acid Transport 438
III. Physicochemical Characteristics of Substituted Phenolsulfonphthalein Dyes in Relation to Binding and Transport 439
IV. Is there Any Parallelism Between the Organic Acid Transport System in Kidney and Red Blood Cells? 440
V. Substrate-Induced Stimulation of Organic Acid Transport 440

H. Transport of Organic Buffers by Nonionic Diffusion 441

References 443

Chapter 9 – Transport of Water: Renal Concentrating Mechanism
(C.W. Gottschalk and W.E. Lassiter) 449

A. Introduction and Historical Perspective 449

B. Localization of the Concentrating Process: An Overview 451
I. The Nephron 451
II. The Vasa Recta 457

C. Countercurrent Multiplier Systems 458

D. Countercurrent Multiplication in the Inner Medulla: Function of the Thin Limbs of Henle's Loop 461

References 470

List of Contributors

Thomas E. Andreoli
Department of Internal Medicine, University of Texas Medical School
Houston, Texas, 77025 / USA

Emile Boulpaep
Department of Physiology, Yale University, School of Medicine,
333 Cedar Street, New Haven, Conn. 06510 / USA

Michel Claret
Université de Paris-Sud, Laboratoire de Physiologie Comparée,
Centre Universitaire,
Bâtiment 443, F-91405 Orsay

Fitz-Roy E. Curry
Department of Physiology, Monash University,
Clayton, Victoria 3168 / Australia

J. H. Dirks
G. F. Strang Laboratory for Medical Research, Department of Medicine,
University of British Columbia,
Vancouver, British Columbia / Canada

John G. Forte
University of California, Department of Physiology-Anatomy,
Berkeley, Calif. 94720 / USA

Gerhard Giebisch
Yale University, School of Medicine, Department of Physiology,
333 Cedar Street, New Haven, Conn. 06510 / USA

Carl W. Gottschalk
University of North Carolina, Department of Medicine,
Division of Health Affairs,
Chapel Hill, N. C. 27514 / USA

Raja N. Khuri
Department of Physiology, American University of Beirut,
Beirut/Lebanon

Rolf Kinne
Max-Planck-Institut für Biophysik, Kennedy-Allee 70,
6000 Frankfurt/Main 70 / Germany

William E. Lassiter
University of North Carolina, Department of Medicine,
Division of Health Affairs,
Chapel Hill, N. C. 27514 / USA

E. W. van Lennep
The University of Sydney, Department of Physiology, F 13,
Sydney, N. S. W. 2006 / Australia

Terry E. Machen
University of California, Department of Physiology-Anatomy,
Berkeley, Calif. 94720 / USA

Gerhard Malnic
Universidade de Sao Paulo, Instituto de Ciencias Biomedicas,
Cidade Universitaria,
Caixa Postal, 4.365, Sao Paulo / Brasil

Don W. Powell
Department of Medicine, Division of Digestive Diseases and Nutrition,
324 Clinical Sciences Building, 229 H,
Chapel Hill, N. C. 27514 / USA

Gary A. Quamme
G. F. Strang Laboratory for Medical Research, Department to Medicine,
University of British Columbia,
Vancouver, British Columbia / Canada

Eugene M. Renkin
University of California, Division of the Sciences,
Department of Human Physiology, School of Medicine,
Davis, Calif. 95616 / USA

Luis Reuss
Department of Physiology and Biophysics,
Washington University School of Medicine,
660 S. Euclid Avenue, St. Louis, MO 63110 / USA

James A. Schafer
University of Alabama, Medical Center Physiology and Biophysics,
Birmingham, Alabama 35294 / USA

Stanley G. Schultz
Department of Physiology, University of Texas Medical School,
Houston, Texas 77025 / USA

Irene Schulz
Max-Planck-Institut für Biophysik, Kennedy-Allee 70,
6000 Frankfurt/Main 70 / Germany

Roger A. L. Sutton
G. F. Strang Laboratory for Medical Research, Department of Medicine,
University of British Columbia,
Vancouver, British Columbia / Canada

Karl J. Ullrich
Max-Planck-Institut für Biophysik, Kennedy-Allee 70,
6000 Frankfurt/Main 70 / Germany

Erich E. Windhager
Department of Physiology, Cornell University Medical College,
New York City, N. Y. 10021 / USA

J. A. Young
The University of Sydney, Department of Physiology, F 13,
Sydney, N. S. W. 2006 / Australia

Chapter 10

Perfusion of Isolated Mammalian Renal Tubules

J. A. SCHAFER and T. E. ANDREOLI

A. Introduction

The purpose of this chapter is to describe the principles and techniques of microperfusion of isolated nephron segments, and to review some information, acquired by studying isolated tubules, concerning transport processes in various segments of the nephron. These processes, which effect the transformation of glomerular ultrafiltrate into urine, include, in a general sense: the transport of water and solutes between the tubular lumen and the renal interstitium; the movement of material between interstitium and peritubular capillaries; and the modulation of these processes by physiological variables such as blood pressure, blood volume, antidiuretic hormone (ADH), aldosterone, and the composition of extracellular (and possibly intercellular) fluid.

It is evident from this rather naive description that a detailed account of renal tubular function requires, at a minimum, an explicit understanding of the transport processes occurring across each segment of the renal tubular epithelium, and the integration of such information into a cohesive picture of the nephron as the final regulator of the volume and composition of the urine. It is in this context that the technique for studying transport processes across isolated renal tubular segments, first developed by BURG and his collaborators (BURG et al., 1966 and 1968), has become a formidable part of the armamentarium of the renal physiologist.

A number of methods have been applied to the study of renal tubular function. A considerable body of information concerning the location and operation of various absorption and secretion processes has been deduced from whole-kidney studies, including clearance, *in-vitro* perfusion and stop-flow methods; furthermore, better regional localization of some tubular transport processes has been obtained by *in-vitro* tissue slice and separated tubule preparations. But by far the majority of the information relating directly to tubular function has been derived from micropuncture studies on the *in-vivo* kidney.

While clearly an elegant technique, micropuncture methodology has limitations. As with all *in-vivo* techniques, alterations in systemic physiologic factors, *e. g.*, the state of extracellular fluid volume, impose a set of variable parameters that may modify tubular transport processes but may be difficult to regulate precisely during the course of an experiment. Further, because of manipulative constraints, most micropuncture studies have been confined to superficial glomeruli, proximal and distal convoluted segments of superficial nephrons, papillary collecting ducts, and the thin ascending and descending limbs of the loop of Henle. Nephron segments, such as the proximal straight tubule, and such regions of the kidney as juxtamedullary nephrons are not accessible for study by micropuncture. Manipulative requirements have also limited the ability of the investigator to regulate simultaneously the composition of both tubular fluid and peritubular capillaries. While this difficulty has been overcome in part by the technique of combined perfusion of the tubular lumen and peritubular capillaries, it should be recognized that even with this technique, the composition of fluid bathing the contraluminal side of the tubular epithelium must be inferred by assuming diffusion equilibrium between the peritubular capillary being perfused and the interstitium bathing the nephron segment under investigation. To our knowledge, this assumption has not been verified.

Many of these problems can be circumvented by the technique of isolated nephron segment perfusion. Since the tubule segments are isolated from the kidney and bathed in an *in-vitro* medium (the volume of which is large in comparision to the tubule size), the movement of material between the tubular lumen and the bathing medium can be studied independently of the requirements of interstitial fluid to capillary transport and the influence of systemic variables. The isolated perfusion method also allows the investigator near-complete freedom in choosing – and changing rapidly – the composition of the solutions perfusing and bathing the tubule segment.

As an *in-vitro* transporting epithelium, the isolated tubule offers at least one decided advantage over epithelia such as amphibian skin and urinary bladder, which have served as analogs of the renal tubular epithelium: in contrast to such epithelia, the isolated tubule consists of a single epithelial cell layer and its supporting basement membrane; thus there are no unstirred layer artifacts referable to serosal cell layers or stromal elements. Furthermore, the conditions of *in-vitro* tubule perfusion virtually eliminate bulk phase unstirred layer effects external to the epithelium (SCHAFER and ANDREOLI, 1972a), which have occasioned numerous problems in the interpretation of data from other epithelia (DAINTY and HOUSE, 1966; HAYS, 1972).

Finally, and perhaps most important as a unique contribution of the technique, *all* segments of the nephron can be isolated and perfused *in vitro,* and these segments can be obtained from superficial, intermediate or juxtamedullary nephrons. This advantage of isolated tubule perfusion has recently acquired increased significance. Micropuncture and whole kidney methods have suggested strongly that the operation of superficial and juxtamedullary nephrons might be quite different (STEIN et al., 1972; KAWAMURA et al., 1975). Using isolated tubule perfusion, KAWAMURA et al. (1975) have demonstrated quite different permeability characteristics in superficial and juxtamedullary proximal straight tubules. Not only is there heterogeneity between superficial and juxtamedullary segments, but HAMBURGER et al. (1976), using isolated perfused segments of defined regions of the proximal tubule, have shown that there is a marked variation in net fluid absorption along the length of the proximal convoluted tubule. Further, WOODHALL et al. (1976), on the basis of para-amino hippurate (PAH) transport rates and morphological studies on isolated segments of superficial and juxtamedullary nephron segments, have proposed that the proximal tubule actually contains three distinct regions, which overlap the traditional boundary of convoluted and straight portions and which have different transport functions. Indeed, it appears that even the individual segments of a given nephron population have a heterogeneous cell population: morphological studies have shown that proximal convoluted and proximal straight tubules may each contain at least three distinct cell types (SAKAGUCHI and SUZUKI, 1958; WOODHALL et al., 1976).

In the words of Lewis Carroll, the situation becomes "curioser and curioser." But it seems reasonable to argue that the study of isolated nephron segments, when integrated with data from other techniques, may be of help in unraveling the role of the nephron in homeostasis.

B. Experimental Techniques: Practical and Theoretical Considerations

This section of the chapter is intended to acquaint the reader with some general aspects of the methods used to isolate and perfuse renal tubule segments, and to describe how net volume absorption, isotopic flux, and electrical data are obtained and analyzed. In most cases, we have described the procedures used in our own laboratory (SCHAFER and ANDREOLI, 1972a and b, 1976; SCHAFER et al., 1974a–c; AL-ZAHID et al., 1976); in turn, these techniques are adapted directly from, and are quite similar to, those developed originally by BURG et al. (1966 and 1968).

I. Isolation of Renal Tubule Segments

Each perfusion experiment begins with the isolation of a specific tubule segment by freehand dissection from kidney slices; in our experience, this step is generally the rate-limiting part of the entire sequence of experimental procedures. To date, the overwhelming majority of published reports dealing with perfusion of single nephron segments have involved the rabbit renal tubule. Some preliminary results have been reported for the collecting duct and the thick ascending limb of Henle's loop from human kidney (JACOBSEN et al., 1975 and 1976). Among lower vertebrates, isolated tubule segments from the flounder (Pseudopleuronectes Americanus; BURG and WELLER, 1969), garter snake (Thamnophis spp.; DANTZLER, 1973 and 1974), frog (Rana catesbiana; STONER, 1975; IRISH and DANTZLER, 1975) have been isolated and perfused.

Isolation of tubule segments by freehand dissection was developed by BURG et al. (1966) as an outgrowth of studies with tubule segments separated from the rabbit renal parenchyma by treatment with collagenase (BURG and ORLOFF, 1962). Tubule segments separated by the collagenase method proved unsuitable for perfusion, apparently because the enzyme destroyed the necessary structural support provided by the tubular basement membrane. However, BURG et al. (1966) were also able to dissect individual tubule segments from the rabbit kidney without any pretreatment of the tissue, and virtually all subsequent reported work with the isolated perfused tubule has employed fresh untreated kidneys from which tubule segments are isolated by freehand dissection.

In our laboratory, a rabbit is maintained on water and rabbit chow *ad libitum* up until the time of the experiment, and is then rapidly decapitated. One kidney is removed, and kidney slices are placed in an appropriate medium within two minutes from the time of sacrifice. Individual tubules are stripped from one another by freehand dissection with two pairs of forceps. Care is taken to touch only the ends of each segment, which are later trimmed from the tubule, and minimal tension is applied to the tubule segment during its isolation. Orientation of the regions of the kidney is maintained during dissection, so that selective isolation of a tubule segment arising from a superficial or juxtamedullary glomerulus is possible. Precise localization of the dissected segment to the cortex or medulla can also be ensured by removal of the other tissue region before the segment is isolated[1].

The dissected segment can also be localized with respect to its distance from fixed reference points along the same nephron. For example, HAMBURGER et al. (1976) ascertain that certain of the segments they perfuse are early proximal convoluted tubules by observing the attachment of the segment to the glomerulus of origin. And GROSS et al. (1975) identify segments of early distal convoluted tubule by the attachment of the juxtaglomerular apparatus to the tubule segment.

Table 1 lists the various rabbit nephron segments which have been most widely studied to date, together with their geometric characteristics. As indicated in Table 1, workers in the field have studied 10 seemingly different segments of the rabbit nephron; and for the proximal straight tubule, segments have been obtained from two different regions of the kidney, superficial cortex and the juxtamedullary region. It should be noted that segments of varying length are obtained, depending on the nephron region under study. For example, one ordinarily obtains only a 1-mm-long segment of proximal convoluted tubule, while for the superficial *pars recta* 3- to 4-mm tubule lengths are obtained routinely.

II. Tubular Perfusion

The techniques for perfusing an isolated tubule segment involve: mounting the segment between sets of glass holding micropipets, introducing a fine inner perfusion pipet into the tubular lumen, and maintaining the tubule in a thermoregulated, appropriately gassed bathing medium. Provision is usually made

1 For further details on the methods of dissection described see: SCHAFER and ANDREOLI, 1972a; SCHAFER et al., 1974a and c.

Table 1. Average dimensions of isolated rabbit renal tubule segments

Segment	Ref.	Length mm	Inside diameter µm	Outside diameter µm	Cell height µm	Average apparent luminal area per length $cm^2\ mm^{-1} \cdot 10^3$
Proximal:						
Superficial convoluted tubule	a	1.0 (0.5–2.2)	20 (18–24)	(35–50)	(5–9)	0.630
Superficial straight tubule	b	3.2 (0.3–4.0)	22.3 (19–26)	(36–39)	(6–8)	0.701
Juxtamedullary straight tubule	c	1.4 (0.8–2.0)	20	–	–	0.630
Loop of Henle:						
Descending thin limb	d	1.2 (0.9–2.0)	(22–26)	(24–28)	~ 1.0	0.921
Ascending thin limb	e	0.4 (0.3–065)	–	–	~ 1.0	–
Ascending thick limb, cortical	f	1.3 (0.6–1.6)	(20–22)	–	–	0.660
Ascending thick limb, medullary	g	1.0 (0.6–1.6)	24.5	33.1	4.3	0.770
Distal:						
Distal convoluted tubule	h	0.5 (0.3–0.8)	–	–	–	–
Cortical collecting tubule	i	3.4 (0.5–4.5)	(20–26)	(36–38)	6.5	0.690
Outer duct medullary collecting	j	1.0 (0.5–2.0)	24	–	–	0.754
Papillary collecting duct	k	0.8 (0.5–1.4)	48 (28–80)	–	–	1.500

a: Burg and Orloff, 1968; Kokko and Rector, 1971; Kokko et al., 1971; Tune and Burg, 1971.
b: Schafer et al., 1974c.
c: Kawamura et al., 1975; Woodhall et al., 1976.
d: Kokko, 1970 and 1972b; Rocha and Kokko, 1973b.
e: Imai and Kokko, 1974b.
f: Burg and Green, 1973a; Rocha and Kokko, 1973a.
g: Rocha and Kokko, 1973a.
h: Gross et al., 1975.
i: Grantham and Burg, 1966; Grantham and Orloff, 1968; Grantham et al., 1970; Gross et al., 1975; Schafer and Andreoli, 1972a; Schafer et al., 1974a.
j: Rocha and Kokko, 1974; Schafer and Andreoli, 1972b.
k: Rocha and Kokko, 1974.

for rapid changes in the compositions of both the perfusion and the bathing solutions.

The lucite perfusion chamber has a glass coverslip bottom so that the tubule can be observed with an inverted microscope. The chamber contains ~1.2 ml of bathing solution and may be temperature-regulated, in the range 23–40° C, by means of an embedded heating coil activated by a temperature controller[2]. There is some question as to the correct temperature at which the tubule segment should be perfused to mimic *in-vivo* conditions. Although most workers using the preparation have conducted such experiments at 37–38° C, the normal body temperature of the rabbit is 39.5° C (SUTHERLAND et al., 1958). This fact should be borne in mind when attempts are made to extrapolate *in-vitro* microperfusion data to the transport characteristics of the *in-vivo* rabbit nephron.

A schematic diagram of the apparatus for tubular perfusion is shown in Figure 1, and Figure 2 is a photomicrograph of a proximal straight tubule mounted in the perfusion set-up. The proximal end of the tubule is sucked into an outer glass holding pipet with a constriction. Suction applied by a syringe draws the tubule into the pipet tip, which is ~50–100 μm long a serves to align the tubule along the cylindrical axis of the pipet, and up to a constriction which has an inside diameter equal to approximately half the outside diameter (cf. Table 1) of the given tubule segment. In order to initiate perfusion, an inner concentric pipet containing the perfusing solution is advanced through the constriction and into the tubule lumen. When the tubule lumen has opened, the inner pipet is advanced 50–100 μm beyond the tip of the outer holding pipet. The distal end of the tubule is sucked carefully into another outer glass holding pipet, referred to as the collecting pipet. No constriction or intraluminal pipet is used on this end: the inside diameter of the collecting pipet is slightly less than the outside diameter of the given tubule segment, but not so much less that the tubule lumen is constricted significantly.

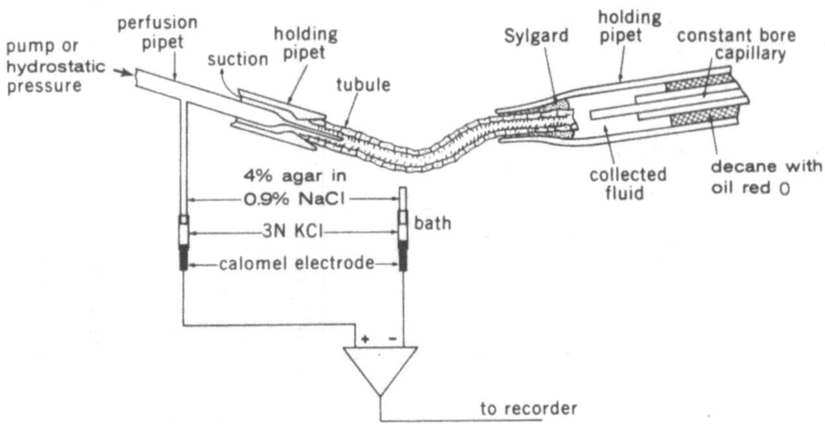

Fig. 1. Schematic diagram of isolated tubule perfusion system

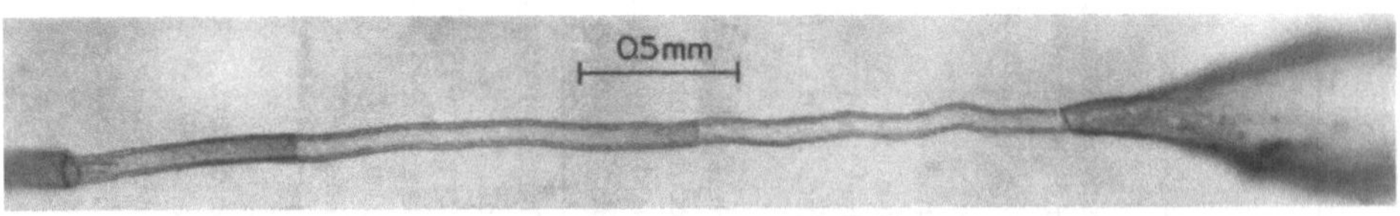

Fig. 2. Photomicrograph of perfused proximal straight tubule mounted between glass micropipets

2 By attaching a Peltier thermopile to the chamber, it is also possible to regulate chamber temperature in the range 15–40° C.

It is important, when measuring rates of fluid absorption, isotopic fluxes, transepithelial voltages, and transepithelial electrical resistances, to ensure that there is no leak pathway between the ends of the tubule and the bathing solution. Two types of seals are commonly used: mechanical (BURG et al., 1966 and 1968) and liquid silicone rubber (BURG et al., 1970). A mechanical seal is particularly effective on the perfusing end of the tubule, because the wall of the tubule is compressed between the constricted region of the outer holding pipet and the inner perfusion pipet. The seal between the tubule and the set of pipets is tested by applying suction and checking to see that this procedure produces no change in the inside diameter of the perfused tubule. Furthermore, a small amount of liquid silicone rubber (Sylgard 184) can be introduced into both perfusion and collecting pipets. This material makes a good mechanical and electrical seal by adhering to both the tissue and the glass. We find that the best test of the adequacy of the seals is the failure of any labeled impermeant volume marker added to the perfusing solution to appear in the bathing solution (see below).

Fluid contained in the inner perfusion pipet can be perfused at rates of 0.5–40 nl min^{-1}, using either a variable-speed syringe pump in series with a gas-tight syringe, or a hydrostatic pressure gradient of 5–50 cm H_2O. Fluid appearing in the collecting pipet is prevented from evaporating by a layer of decane stained with oil red 0. At desired intervals, the collected fluid is removed for analysis by timed collections into a constant volume constriction pipet (BURG, 1972a).

During the course of an experiment, the bathing solution is removed from the chamber by suction and replaced periodically to prevent evaporative losses, or to change the bath composition. The perfusing solution can also be changed rapidly, either by using fine polyethylene tubing to suck out the previous solution and add a new solution (SCHAFER et al., 1974b), or by flushing in a new solution through a concentric fine capillary sealed within the inner perfusion pipet (BURG, 1972a). The tubular epithelium can be observed and photographed by means of the inverted compound microscope; at magnifications of 100–400 X, individual tubular cells can be observed easily.

III. Perfusing and Bathing Solutions

Early studies with isolated tubule segments, particularly those involving proximal tubules, utilized rabbit serum and a rabbit serum ultrafiltrate as, respectively, bathing and perfusing solutions (KOKKO et al., 1971). However, in order to assess the effect of external medium composition on the transport characteristics of a given nephron segment, e. g., the dependence of spontaneous transepithelial voltage in the superficial proximal convoluted tubule on the presence of glucose and amino acids in the perfusate (KOKKO, 1972a and 1973), or, in both the superficial proximal convoluted tubule (KOKKO, 1973) and the superficial proximal straight tubule (SCHAFER et al., 1974c), on the Cl^- and HCO_3^- concentrations in the perfusate, a number of synthetic perfusing and bathing media have been designed.

Table 2 lists some solutions used in our laboratory to mimic conditions presumed to exist *in vivo* at points in the proximal and distal tubule. The bathing solution for proximal tubule experiments is designed to resemble rabbit serum, and contains alanine in place of the several aminoacids in serum, and acetate in place of various organic acids. The proximal tubule-perfusing solution referred to as HCO_3-KR resembles a plasma ultrafiltrate, *i. e.*, resembles the composition of tubular fluid in the early proximal tubule. The perfusing solution labeled Cl-KR has, with respect to the HCO_3-KR perfusate, a higher Cl^- concentration and a lower HCO_3^- concentration and pH, and it contains no alanine or glucose but 13.3 mM urea. Thus a Cl-KR perfusate is designed to be similar to tubular fluid found in the late mammalian proximal tubule (GOTTSCHALK, 1962/63; MALNIC et al., 1969).

Table 2. Examples of synthetic buffer solutions used to bathe and perfuse isolated rabbit tubules

	"Proximal" tubular solutions			"Distal" tubular solutions		
	Bathing solution	Perfusing solutions		Bathing solution	Perfusing solutions	
					Hypotonic	Isotonic
	HCO_3–KR	HCO_3–KR	Cl–KR			
	mM			mM		
NaCl	105.0	105.0	127.8	115.0	60.00	15 0.0
$NaHCO_3$	25.0	25.0	3.8	25.0	–	–
Na acetate	10.0	10.0	10.0	10.0	–	–
Na_2HPO_4/NaH_2PO_4	4.0	4.0	4.0	1.2	2.5	2.5
KCl	5.0	5.0	5.0	5.0	5.0	5.0
$CaCl_2$	1.8	1.8	1.8	1.0	1.0	1.0
$MgSO_4$	1.0	1.0	1.0	1.2	1.2	1.2
Glucose	8.3	8.3	0	5.5	–	–
L-alanine	5.0	5.0	0	–	–	–
Urea	–	–	13.3	–	–	–
Other additions	6% (w/v) bovine serum albumin	–	–	5% (v/v) calf serum	–	–
pH	7.4	7.4	6.6	7.4	7.4	7.4

Adapted from SCHAFER et al., 1974a, 1975, 1977; SCHAFER and ANDREOLI, 1972a.

More terminal nephron segments (cortical collecting tubule, distal convoluted tubule) are bathed in a modified Krebs-Ringer bicarbonate buffer (KRB) and are perfused with either a hypotonic or an isotonic Krebs-Ringer phosphate (KRP) buffer. The former perfusate resembles early mammalian distal tubular fluid or all distal tubule fluid in water-loaded animals; the latter perfusate resembles late cortical collecting tubule fluid in hydropenia (GOTTSCHALK, 1962–63).

IV. Measurement of Net Volume Absorption

Two major methods have been used to measure net volume transport in isolated tubule segments. In the first method, referred to as the collection method, the overall rate of loss or gain of volume by a perfused nephron segment is calculated from the difference between the measured rate of volume collection and the measured perfusion rate. In the second method, referred to as the crimped end method, the distal end of the tubule is occluded, the tubule is cannulated with an inner pipet but is not continuously perfused, and the rate of net volume movement is determined from the rate of fluid movement out of the inner pipet.

1. Collection Method

In this method (BURG et al., 1966; BURG, 1972a), the overall rate of volume loss or gain by the nephron is calculated as the difference between the perfusion and collection rates. The total collected volume of a sample may be measured with either a constant bore or a volumetric sampling pipet; the collection rate is the volume collected divided by the duration of the collection period.

The perfusion rate is generally measured from the rate of collection of a radiolabeled large molecule which does not penetrate the tubular epithelium and serves as a "volume marker." For such a molecule, the total activity collected is equal to the total activity perfused (assuming that there is no artifactual loss of the volume marker from the perfusate to the bath), and the perfusion rate is calculated as:

$$\dot{V}^{o} = \frac{\text{total cpm collected}}{\Delta t \, C^{*o}}, \tag{1}$$

where $\dot{V}^{o}$ is the perfusion rate, Δt is the duration of the collection period, and C^{*o} is the tracer concentration in the perfusing solution (*e. g.*, cpm nl^{-1}). We routinely use ^{3}H-methoxyinulin as a volume marker; in order to remove small molecular weight contaminants, the methoxyinulin is dialyzed against large volumes of water over 60 hours, using dialysis tubing with a molecular weight cut-off of 3500 daltons. Other workers have used ^{131}I- or ^{125}I-albumin (KOKKO et al., 1971), ^{125}I-iothalamate (ROCHA and KOKKO, 1973a), and ^{125}I-polyvinylpyrrolidone (ROCHA and KOKKO, 1973a).

The total net volume flow can be expressed in terms of the membrane surface area of the perfused segment. Since the apparent luminal membrane surface area is πdL, where d is the internal diameter and L is the tubule length, we have:

$$J_v \, (\text{nl sec}^{-1}\,\text{cm}^{-2}) = \frac{\dot{V}^{o} - \dot{V}^{L}}{\pi dL}, \tag{2}$$

where $\dot{V}^L$ is the collection rate. Because of a brush border luminal surface in proximal nephron segments and an irregular luminal border in others, the true surface area cannot be computed directly. Consequently, the flux per unit of true luminal membrane area is considerably lower than this calculated value.

Alternatively, the volume flow may be expressed in terms of the length of the tubule segment, which can be measured to an accuracy of at least ± 50 μm with an ocular micrometer. Using the measured length L in mm, the net volume flow, J_v, can be calculated from:

$$J_v\,(\mathrm{nl\,min^{-1}\,mm^{-1}}) = \frac{\dot{V}^o - \dot{V}^L}{L}\,. \tag{3}$$

Obviously, this method of normalizing J_v does not take into account variations in tubular diameter.

2. Crimped-End Method

This method of measuring the rate of net volume absorption was developed by GRANTHAM and his collaborators (WELLING and GRANTHAM, 1972; GRANTHAM et al., 1972) for use in experiments with proximal tubule segments or segments of tubular basement membrane denuded of epithelial cells. The tubule is mounted in the constriction pipet, and an inner concentric pipet filled with the perfusate is advanced into the tubular lumen so that the appearance of one end of the tubule is exactly the same as shown for the perfusion pipet system in Figure 1 (left). After the tubule lumen has been opened by pressure applied to the inner pipet, the distal end of the tubule is crimped and occluded by drawing a knuckle of the tubule into another glass pipet by suction. A constant pressure is then applied to the perfusion pipet. The adequacy of the preparation is assessed on the basis of three criteria (GRANTHAM et al., 1972): the absence of fluid leakage at the cannulated end of the tubule; rapid changes in the luminal diameter of the tubule segment when the pressure applied to the perfusing pipet is varied; and morphological integrity of the tissue.

The rate of net volume movement is estimated by determining the rate of delivery of fluid in the cannulating pipet. This rate may be measured by adding a radiolabeled substance to the perfusate and measuring its rate of appearance in the bathing solution; tritiated water is an ideal marker since it can be added in high concentrations to the perfusate, thereby increasing the accuracy of its measurement in the bathing solution. J_v is calculated as (GRANTHAM et al., 1972):

$$J_v\,(\mathrm{nl\,min^{-1}mm^{-1}}) = \frac{\text{THO in bath (cpm)}}{C^{*o}\cdot \Delta t \cdot L}\,, \tag{4}$$

where C^{*o} is the tracer concentration in the inner pipet, *i. e.*, the perfusing pipet. GRANTHAM et al. (1972) demonstrated the accuracy and reliability of the method of measuring J_v in proximal convoluted tubules; as expected, the intercept of a plot of J_v versus length was nearly zero.

More recently, GRANTHAM et al. (1973) have used another technique for measuring net volume transport in crimped tubules. Instead of measuring J_v from the rate of THO appearance in the bathing solution, these workers measured the rate of change in the fluid volume in the inner pipet by observing the rate of movement of a droplet of naphtha in a constant bore region of the pipet. With this method, J_v was calculated as (GRANTHAM et al., 1973):

$$J_v\,(\mathrm{nl\,min^{-1}\,mm^{-1}}) = \frac{\pi x'}{\Delta t L}\left(\frac{d'}{2}\right)^2 \tag{5}$$

where: x' is the distance moved by the naphtha droplet in the time period Δt, and d' is the inner diameter of the constant bore portion of the inner pipet. This second method is particularly valuable in the measurement of rates of net volume secretion (GRANTHAM et al., 1973 and 1974).

One reservation must be borne in mind when either crimped end method is used to measure net volume movement. If the solution introduced into the tubule segment contains a solute which is unable to cross the tubular epithelium, or which moves across the epithelium much more slowly than water, then the concentration of this substance in the tubular lumen will change with time. Consequently, the osmotic pressure exerted across the epithelium by this solute will vary with time and may alter net fluid movement. Because of this restriction, the crimped end method is unsuitable for use in tubule segments such as the cortical collecting tubule, in which the solute permeability is, in general, considerably less than the water permeability.

V. Assessment of Hydraulic Conductivity

We now consider some of the methods used for measuring hydraulic conductivity, and the application of these methods to the particular case of the isolated renal tubule. In this context, we shall discuss the equations that are particularly applicable to the calculation of hydraulic conductivities in isolated tubule segments.

1. The Unstirred Layer Problem

In order to calculate the hydraulic conductivity of any membrane or membrane system from observed rates of net volume flow in the presence of an osmotic or hydrostatic pressure gradient, *i. e.*, from the Starling equation, it is desirable that the pressure gradient across the membrane be equal to the pressure difference between opposite bulk phase solutions. This condition may not hold if there are unstirred layers, either in the bulk phases adjacent to the membrane, or, for epithelia, within the tissue itself (*e. g.*, the cytoplasm, lateral intercellular space, serosal cell layers and stromal elements). These layers may be viewed picturesquely as unmixed, diffusion-limited regions of water interposed between a membrane surface and a well-mixed solution. Defined in this manner, an unstirred layer forms a resistance for diffusion from the well-mixed solution to the membrane interface, but is a negligible constraint to laminar flow.

It now seems feasible to consider two different types of unstirred layers (SCHAFER et al., 1974a), depending on the geometric constraints to flow-diffusion processes within the latter. In the first kind, which we have termed type A, the area available for flow-diffusion processes is approximately equal to the area of the membrane with which the unstirred layer interfaces. This type of unstirred layer occurs, for example, in series with synthetic membranes (GINZBERG and KATCHALSKY, 1963; HAYS, 1968), various epithelia (HAYS, 1972; DAINTY and HOUSE, 1966; WEDNER and DIAMOND, 1969), and in most hemodialysis machines (COLTON, 1967). In certain instances, the thickness of a type-A unstirred layer can be reduced by more effective stirring of the bulk solutions (GINZBERG and KATCHALSKY, 1963; HAYS, 1972).

One approach toward assessing the effective diffusion resistance of a type A unstirred layer, *e. g.*, a bulk phase unstirred layer adjacent to a membrane, depends on varying the aqueous phase viscosity with an impermeant solute (ANDREOLI and TROUTMAN, 1971). The relationship between P_{D_i} (cm s^{-1}), the diffusional permeability coefficient of the i-th species, and the diffusion resistance of an unstirred layer, may be expressed as:

$$\frac{1}{P_{D_i}} = \frac{\Gamma}{D_i^\circ} + \frac{1}{P_{m_i}}, \tag{6}$$

where Γ is the sum of the effective thicknesses of the unstirred layers in both aqueous phases adjacent to a membrane, D_i° is the free diffusion coefficient of water or the i-th solute, and P_{m_i} is the true, or membrane, permeability coefficient for diffusion of water or the i-th solute. Equation (6) defines Γ as the unstirred layer thickness when it is assumed that the frictional constraints to diffusion in the unstirred layer are the same as in bulk solution. Thus, Γ is as an operational quantity, which does not necessarily connote the state of water in the unstirred phases. The free diffusion coefficient may be expressed in terms of the Stokes-Einstein relation-ship:

$$D_i^\circ = \frac{RT}{6\pi N a_i \eta}, \tag{7}$$

where a_i is the hydrodynamic radius of water or the i-th solute, N is Avogadro's number, R and T have their usual significance, and η is the viscosity of the solvent. The product $D_i^\circ \eta$, which will be termed β_i, is relatively constant in aqueous solutions over a wide range of temperatures (WANG, 1951; LONGSWORTH, 1963). Accordingly,

$$D_i^\circ \eta \simeq \beta_i, \tag{8}$$

and by substitution:

$$\frac{1}{P_{D_i}} = \frac{\Gamma}{\beta_i}\eta + \frac{1}{P_{m_i}}. \tag{9}$$

Equation 9 indicates that, if there are significant unstirred layers adjacent to a membrane, increasing the viscosity in the bulk aqueous phases will result in linear decrements in P_{D_i}.

We have examined this question directly in the isolated perfused cortical collecting tubule: 10-fold increases in the viscosity of both the perfusing and bathing solutions had no effect on P_{D_w}, the measured diffusional permeability coefficient for water (SCHAFER and ANDREOLI, 1972a). Thus, it seems improbable that bulk phase unstirred layer effects represent a significant factor in the isolated tubule preparation.

A membrane of a given area may also be in series with an unstirred layer having a smaller effective area within which transport processes occur. We have termed (SCHAFER et al., 1974a) such an unstirred layer type B. Clearly, for the steady-state transport of water between two aqueous phases separated by a membrane in series with a type B unstirred layer, the velocity of water flow in the latter will vary with the fractional area available for flow. Consider now the functional organization of transporting epithelia (FARQUHAR and PALADE, 1963): the apical plasma membranes are fused to form junctional complexes, and immediately adjacent to junctional complexes, the basolateral plasma membranes of adjacent cells are separated by intercellular channels which represent, with respect to apical plasma membranes, type-B unstirred layers. If transepithelial solute and water transport processes involve intercellular spaces, solute polarization within the latter may modify appreciably the solute concentration gradient across junctional complexes, and result in significant errors in estimating a transepithelial hydraulic conductivity.

WRIGHT et al. (1972), for example, recently observed that initial rates of osmotic water flow in rabbit gall bladder were considerably greater than the steady-state values. These workers deduced that the discrepancy between initial and steady-state osmotic flows was the consequence of flow-dependent changes in the concentration of solutes in the lateral intercellular spaces, and indicated the possibility that, as a result of such solute polarization, steady-state osmotic flow measurements in epithelia might result in significant underestimates of the hydraulic conductivity of the latter.

In order to consider the effects of solute polarization within unstirred layers on osmotic volume flow determinations in epithelia, we have analyzed (SCHAFER et al., 1974a) quantitatively the effects of unstirred layers and unstirred layer geometry on transient and steady-state osmotic volume flows in isolated renal tubules. In cortical collecting tubules, no difference was observed between water flow measurements made 20–30 seconds after application of the osmotic pressure difference and measurements taken 30–60 minutes thereafter (SCHAFER et al., 1974a). A theoretical analysis of these data in terms of equations for flow-diffusion processes through unstirred layers revealed that the measured values were within 10 percent of the true hydraulic conductivity. So it appears that, for the isolated mammalian cortical tubule, solute polarization within unstirred intercellular spaces has a negligible effect on hydraulic conductivity measurements.

A similar argument may apply to isolated proximal tubules. The transepithelial resistances of the latter are in the range of 5 Ω-cm^2 (LUTZ et al., 1973), *i. e.*, only 100 times more than the electrical resistance of a layer of HCO_3-KR buffer (Table 2) having the same thickness ($7.5 \cdot 10^{-4}$ cm) as the epithelium of an isolated proximal tubule. Since ion permeation in these tubules appears to involve an extracellular route (SCHAFER et al., 1974c), these considerations indicate that the diffusion resistance of intercellular spaces is, at a maximum, only 10^2 times greater than that of free solution (SCHAFER et al., 1974c). For such a diffusion resistance, and J_v values of ~1.0 nl min^{-1}mm^{-1} (*i. e.*, typical rates of isotonic fluid absorption in isolated proximal convoluted tubules [KOKKO et al., 1971], and typical rates of fluid absorption used to measure hydraulic conductivity in isolated proximal tubules [WELLING and GRANTHAM, 1972]), it may be shown that negligible solute polarization occurs within intercellular spaces (SCHAFER et al., 1975 and 1977).

2. The Hydraulic Conductivity Coefficient

Two methods can be used to measure tubular hydraulic conductivity: osmotic and hydrostatic. In the former method, the hydrostatic pressures used to perfuse isolated tubules (10–20 cm H_2O) are often far less than the osmotic pressure differences (2–5 atm) which are used to produce net volume flow. Consequently, the hydraulic conductivity can be calculated simply as $J_v/\Delta\pi$, where $\Delta\pi$ is the osmotic pressure difference between the lumen and bath, produced by a solute having a unity reflection coefficient.

If osmotic volume flow is directed from lumen to bath, the luminal osmolality will vary as a function of the distance from the tip of the perfusing pipet; for convenience, the distance x along the length of the tubule is considered to vary from $x = 0$ at the tip of the perfusion pipet to $x = L$ at the collection end. The mean osmotic pressure difference has been estimated by different investigators as the arithmetic mean of the osmolalities at $x = 0$ and $x = L$ (GRANTHAM and ORLOFF, 1968), or as a function of the logarithmic mean of the osmolalities at $x = 0$ and $x = L$ (KOKKO, 1970; SCHAFER and ANDREOLI, 1972a). Although these approximations are adequate for relatively low rates of transepithelial volume flow, they may be significantly in error for rapid flow rates.

In order to obtain an exact solution for the hydraulic conductivity under these conditions, we have solved the equations relating J_v and $\Delta\pi$ for the case where the asymmetrically distributed solute producing $\Delta\pi$ does not penetrate the epithelium, *i. e.*, has a unity reflection coefficient (AL-ZAHID et al., 1976). Under these conditions, the rate of solute flow at any point x can be expressed as:

$$\dot{V}^x C^x = \dot{V}^o C^o = \dot{V}^L C^L \tag{10}$$

where $\dot{V}^x$ is the volume flow rate at x, and C^x is the corresponding osmolality at the point x. The rate of change of volume flow with distance at x is given by the hydraulic conductivity, L_p (cm s^{-1}atm^{-1}), and the difference in osmolality between the bathing solution (C^b) and the lumen at the point is:

$$\frac{d\dot{V}}{dx} = \pi d\, L_p RT[C^x - C^b]. \tag{11}$$

Equations (10) and (11) may be solved simultaneously to give:

$$L_p = \frac{-\dot{V}^o C^o}{RTA}\left[\frac{C^o - C^L}{C^o C^L C^b} + \frac{1}{C^{b2}}\ln\frac{[C^L - C^b]\,C^o}{[C^o - C^b]\,C^L}\right], \tag{12}$$

and the average luminal osmolality, C, is:

$$C = \frac{\dot{V}^L - \dot{V}^o}{A\, L_p RT} + C^b. \tag{13}$$

L_p can also be expressed as P_f, a hydraulic permeability coefficient having the dimensions cm s^{-1}, by the relationship:

$$P_f = \frac{L_p RT}{\bar{V}_w}, \tag{14}$$

where $\bar{V}_w$ is the partial molal volume of water. P_f provides a convenient way of expressing L_p, particularly for comparison with P_{D_w}, the diffusional permeability coefficient for water; the latter, as indicated above, also has the dimensions cm s^{-1}.

A particular problem arises in attempts to measure P_f in highly permeable nephron segments such as the proximal tubule. In the latter, axial changes in solute and solvent concentrations, occurring as a consequence of radial solute and solvent fluxes, must be considered explicitly in calculating P_f. We have considered this problem in detail elsewhere (SCHAFER et al., 1978 a; ANDREOLI et al., 1978), and we summarize our approach to the problem below.

Consider an experiment conducted using a perfused proximal tubule of length L in which transepithelial osmotic volume flow is produced by perfusing the tubule segment with a hypotonic solution while the bathing solution is isotonic. Axial osmotic equilibration occurs during radial transepithelial osmosis as a consequence of two factors: loss of water from the lumen to be bath; and solute entry from bath to lumen down its concentration gradient. The diffusive solute movement is also countered by a lumen-to-bath solute flow produced as a consequence of coupling to the solvent flow. It can be seen intuitively that the absolute amount of fluid absorption ($'J_v$, cm^3/sec) occurring in the segment L will vary with the perfusion rate ($\dot{V}^o$, cm^3/sec). That is, the slower the perfusion rate, the shorter the distance x along the axis of the tubule at which osmotic equilibrium is attained, and thus the smaller will be $'J_v$. As the perfusion rate is increased, the point at which osmotic equilibrium is achieved moves farther down the length of the tubule and $'J_v$ increases. Moreover, $'J_v$ may be relatively independent of the length of the perfused segment; for example, if osmotic quilibrium at a given $\dot{V}^o$ occurs at $x = L$, then the value of $'J_v$ will be identical in all tubules having a length greater than L but perfused at the same rate.

We have utilized the perfusion rate dependence of volume absorption to estimate the hydraulic conductance (P_f) of isolated segments of superficial proximal straight and concoluted tubules (SCHAFER et al., 1978a; ANDREOLI et al., 1978). The experiments were conducted as follows. At 23° C active transport processes were inhibited so that the rate of fluid absorption was approximately zero when perfusate and bath were isosmolar. The perfusate was then exchanged for one which was 20 mOsm/l dilute with respect to the bathing solution. We then examined the dependence of the total volume flow observed in the given perfused tubule length on the perfusion rate.

In order to analyze such data, we note that, at an infinitely high perfusion rate, there would exist a uniform osmotic pressure gradient between the lumen and the bath along the entire perfused tubule length, i. e., the osmotic pressure gradient between the initial perfusate in the bathing solution would be maintained along the entire tubule length (SCHAFER et al., 1978a). Under these circumstances, P_f could be computed from the value of $'J_v$ at an infinitely high $\dot{V}^o$, subsequently termed $'J_v^{max}$, as:

$$P_f = \frac{'J_v^{max}}{\sigma_{NaCl} A \bar{V}_w \Delta C_{p \to b}}, \tag{15}$$

where $'J_v^{max}$ is expressed in cm^3/sec, A is the luminal surface area, $\bar{V}_w$ is the partial molar volume of water, σ_{NaCl} is the salt reflection coefficient, and $\sigma C_{p\rightarrow b}$ is the osmolality difference between the initial perfusate and the bathing solution. By plotting the experimental data in reciprocal fashion, i. e., $1/'J_v$ as a function of $1/\dot{V}^o$, it was possible to estimate $'J_v^{max}$ from the zero intercept on the ordinate.

The reciprocal plots gave regression coefficients in excess of 0.96 using linear regression analysis for data from both proximal straight and proximal convoluted tubules. The values of P_f computed from the zero intercept on the ordinate, as indicated above, were in the range of 5,000 μm sec^{-1} for proximal straight tubules (SCHAFER et al., 1978a) and 3,500 μm sec^{-1} for proximal convoluted tubules (ANDREOLI et al., 1978).

Finally, the "crimped-end, naphtha" method of GRANTHAM et al. (1973), *e. g.*, equation 5, can be used to compute L_p as $J_v/\Delta P$, where ΔP is an imposed hydrostatic pressure gradient. There are two major considerations particularly applicable to this method. First, as noted by WELLING and GRANTHAM (1972), appropriate correcting must be made for the drop in hydrostatic pressure occurring across the perfusing pipet. Second, as noted above, care must be taken to ensure that the luminal solute concentration, and hence the luminal fluid osmolality, remain constant during the volume flow measurement.

VI. Electrical Measurements

Three general types of electrical measurements have commonly been made in the isolated tubule preparation: transepithelial voltages which develop spontaneously in tubule segments perfused and bathed with solutions of identical salt composition; transepithelial voltages induced by perfusing and bathing with solutions of different ionic composition, thereby producing diffusion potentials of one or more ions, *i. e.*, single-salt and bi-ionic voltages; and transepithelial electrical resistance measurements. It is relevant in this regard to consider the potential sources of the electromotive forces and the resistive elements of the renal tubular epithelium. Figure 3 is a representation of the equivalent electrical circuit proposed by ROSE and SCHULTZ (1971) for an epithelial cell. The model contains three major resistive elements: R_L, the net resistance of the luminal plasma membrane, given by $1/R_L = 1/R_1 + 1/R_2$; R_b, the net resistance of the basolateral plasma membrane, given by $1/R_b = 1/R_3 + 1/R_4$; and R_e, the net resistance of the extracellular pathway, given by $1/R_e = 1/R_5 + 1/R_6$. Three electromotive forces are present: E_L and E_b are electromotive forces across, respectively, luminal and basolateral plasma membranes, which might occur as a consequence of an electrogenic active transport process across the indicated plasma membrane and/or be referable to ionic diffusion potentials between cell interior and external solutions; and E_e represents a diffusion potential across junctional complexes produced by unequal ionic concentrations across junctional complexes. Stated simply, the model illustrated in Figure 3 depicts a parallel array of cellular and extracellular elements. Thus, the pathway for passive ion permeation during transepithelial electrical resistance measurements or ionic diffusion potential measurement will depend on the relative magnitudes

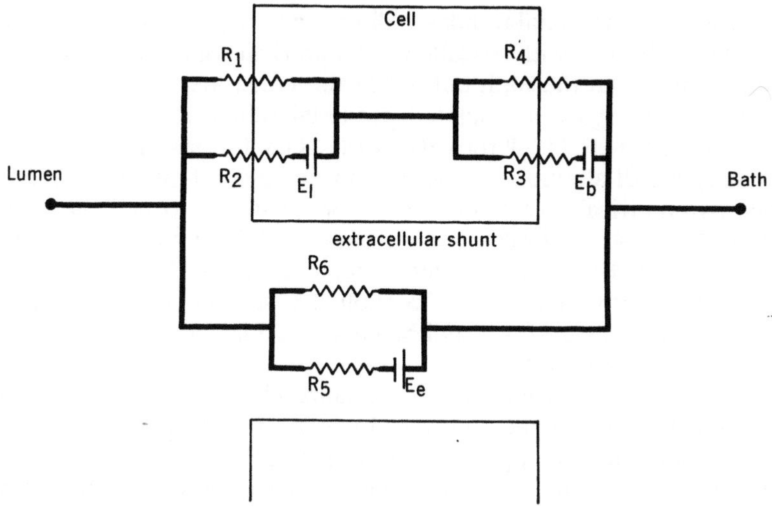

Fig. 3. Schematic diagram of electrical conductance pathways in epithelial cell layer. (Adapted from ROSE and SCHULTZ, 1971)

of R_e and $(R_L + R_b)$; or more specifically, on the relative magnitudes of R_6 and $(R_1 + R_4)$.

One important aspect of electrical measurements in isolated perfused tubule segments is that the voltage-measuring electrode in the lumen is usually the perfusing pipet itself. The transepithelial voltage recorded is, therefore, that existing between the tip of the perfusing pipet and the bathing solution. Because the transepithelial resistance of most tubule segments is comparable to the solution resistance along the tubule lumen, the length constants of proximal nephron segments are approximately 0.01 cm (HELMAN et al., 1971; LUTZ et al., 1973; BURG and GREEN, 1973a), *i. e.*, appreciably less than the lengths of perfused proximal segments (Table 1). Consequently, the recorded voltage for proximal tubule segments represents only the voltage in the vicinity of the inner pipet tip, and provides no direct information about voltages distal to this point.

1. Liquid Junction Potentials and Donnan Voltages

The transepithelial voltage (V_e, mV) of isolated tubule segments can easily be measured by incorporating electrodes into the chamber containing the bathing solution and the holder assembly for the inner perfusion pipet (Fig. 1). The electrodes contacting the solutions are salt-agar bridges connected to calomel half-cells, which in turn are connected to a high-impedence instrumentation amplifier from which the signal is passed to a recorder.

Virtually all investigators working with isolated tubules have attempted to make corrections for liquid junction potentials when reporting transepithelial voltages. However, until recently little attention has been paid to the fact that, particularly in proximal renal tubules, a diffusion voltage must exist between the perfusate and bath even with so-called "symmetrical solutions", because of the presence of albumin in the bath but not in the perfusate.

Consider in this regard an isolated proximal tubule segment perfused and bathed with isotonic HCO_3^--Krebs-Ringer (KR) solutions, the bath containing approximately 6 g/dl albumin (Table 2). In principle, a Donnan voltage arising from the asymmetrical distribution of protein should occur; yet we (e. g., SCHAFER et al., 1974c; SCHAFER and ANDREOLI, 1976) and others (e. g., BURG and ORLOFF, 1970; BURG et al., 1968; KOKKO et al., 1971; GRANTHAM et al., 1972), working with straight or convoluted segments respectively, have observed that the measured transepithelial voltage is approximately zero when active transport is abolished.

This apparent discrepancy may be rationalized by considering the conditions for measuring transepithelial voltages. In measuring transepithelial voltages, agar bridges containing an appropriate salt, generally 0.9% NaCl, are placed in the perfusing and bathing solution. Thus, in the absence of active transport processes, the observed transepithelial voltage V_e^o (mV) is:

$$V_e^o = V_{e_1 \to P} + V_D + V_{b \to e_2}, \tag{16}$$

where $V_{e_1 \to P}$ and $V_{b \to e_2}$ are the liquid junction voltages between perfusate electrode: perfusate and bath: electrode, respectively, and V_D is a transepithelial diffusion voltage. The liquid junction voltages may be computed according to the Henderson equation (MACINNES, 1961), modified in terms of ionic activities (BARRY and DIAMOND, 1970):

$$V_j = -\frac{RT}{F} \frac{\Sigma_n (u_i/Z_i)(a^b_i - a^l_i)}{\Sigma_n u_i(a^b_i - a^l_i)} \ln \frac{\Sigma_n a^l_i u_i}{\Sigma_n a^b_i u_i}, \tag{17}$$

where: u_i is the mobility of the i-th ion, a^l_i is its ionic activity in the luminal solution, a^b_i is its ionic activity in the bathing solution, z_i is its valence and V_j is the algebraic sum of the junction potentials from the perfusing to the bathing solution electrode. In the case of tubule segments with low transepithelial voltages, such as the proximal straight tubule, V_j may be of the same magnitude or greater than the total recorded voltage. Thus the absulute value of the computed transepithelial voltage after solution changes during the course of an experiment must be regarded as approximate rather than exact.

The transepithelial diffusion voltage V_D may be calculated from the expression

$$V_D = \sum_{i=1}^{n} t_i E_i, \tag{18}$$

and,

$$\sum_{i=1}^{n} t_i = l, \tag{19}$$

where E_i is the equilibrium voltage for the i-th ion, computed from the Nernst equation and the perfusate: bath-water ion concentration ratios, and t_i is the transference number for the i-th ion. It follows that, for a Donnan distribution in which the perfusate/bath water ion concentration ratios for mobile ions are the same, E_i is the same for any of the mobile ions and $V_D \simeq E_i$.

It is evident from quation 16 that V_e^o in the absence of active transport processes will be nearly zero when the algebraic sum of the electrode liquid junction potentials is approximately equal in magnitude but opposite in sign to V_D. To test this possibility, SCHAFER et al. (1978b) computed the values of these parameters when tubules were perfused with simple phosphate-buffered NaCl solutions, bathed with similar solutions containing 6g/dl albumin, and gassed with O_2. The measured bath water/perfusate concentration ratios for Na^+ and K^+ were 1.073 and 1.062, respectively; and the measured perfusate/bath water concentration ration for Cl^- was slightly higher, 1.12, due probably to the presence of unmeasured anions associated with undialyzed albumin. The value of V_e^o calculated from the algebraic sum of the electrode liquid junction potentials and V_D was 0.05 mV, *i. e.*, indistinguishable from zero.

When active transport occurs, the right-hand side of equation 16 will also include a term for V_r, the rheogenic voltage. Thus we may define an actual transepithelial voltage V_e^a as:

$$V_e^a = V_D + V_r = V_e^o - (V_{e_1 \to p} + V_{b \to e_2}). \tag{20}$$

When viewed in this context, a rheogenic voltage may represent a lumen-negative displacement from the Donnan equilibrium voltage, rather than an actual lumen-negative value. In the remaining part of this chapter, transepithelial voltages will be expressed primarily in terms of the displacement from the Donnan equilibrium voltage, *i. e.*, as V_e^o rather than V_e^a.

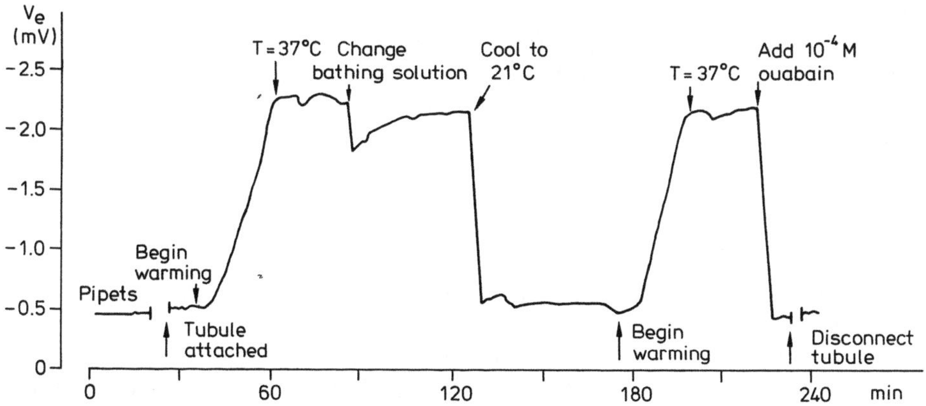

Fig. 4. Representative voltage recording from proximal straight tubule perfused and bathed with symmetrical HCO_3-KR buffers. V_e is the recorded voltage uncorrected for any diffusion potential produced by the distribution of mobile ions produced by the Donnan effect

A representative illustration of the reliability with which spontaneous voltages of small magnitudes may be measured in the isolated tubule preparation is shown in Figure 4. The voltage tracing is for an isolated proximal straight tubule exposed to symmetrical HCO_3-KR solutions. For convenience the values of V_e are expressed as V_e^o, the observed voltage; thus V_e represents the voltage displacement from a Donnan Equilibrium diffusion voltage.

The zero reference for V_e in Figure 4 was obtained by grounding both inputs of the amplifier. With no tubule attached to the pipets, a small potential of about −0.5 mV was recorded. When the tubule was attached at room temperature, the recorded potential was approximately the same as the pipet asymmetry potential, *i. e.*, the transepithelial voltage was zero. Upon warming the system to 37° C, the voltage became more lumen-negative, giving a V_e of −1.6 mV (*i. e.*, the difference between the asymmetry potential and the absolute voltage) within 10 minutes. When the tubule was cooled to 21° C or exposed to 10^{-4} M ouabain at 37° C, the voltage became indistinguishable from the pipet asymmetry potential; and when the tubule was disconnected, the latter was virtually the same as at the start of the experiment.

2. Ionic Dilution Potentials

Ionic dilution potentials, either single salt or bi-ionic, may be used to compute absolute or relative transepithelial ionic permeabilities. Clearly, since the tubular epithelium contains cellular and extracellular resistive elements in parallel, the route of passive ion permeation, and hence the interface across which ionic dilution potentials are expressed, will depend on the relationship between $(R_1 + R_4)$ and R_6 (Fig. 3).

Two sets of equations have been utilized to compute transepithelial ionic permeability coefficients, either relative or absolute, from zero-current voltages: the Nernst-Planck equations and the Goldman-Hodgkin-Katz equation. In the first approach, the zero-current voltage is computed from equations 18 and 19.

In the proximal tubule the transference number of the *i*-th ion may be related empirically to P_{D_i} (cm s^{-1}), the diffusional permeability coefficient of the *i*-th ion at zero volume flow, by the expression (Frömter et al., 1970; Schafer et al., 1975):

$$\frac{P_{D_i}}{P_{D_j}} = \frac{t_i \overline{C}_j}{t_j \overline{C}_i}, \tag{21}$$

where $\overline{C}_i$ and $\overline{C}_j$ are the mean concentrations of, respectively, the *i*-th and *j*-th ions in the two external solutions.

Alternatively, one may use the Goldman-Hodgkin-Katz equation (Goldman, 1943; Hodgkin and Katz, 1949) to express the dependence of the transepithelial voltage on the concentrations and permeabilities of the major ions. In the case of solutions containing predominantly Na^+, Cl^-, and HCO_3^-, the zero-current voltage may be expressed as:

$$V_e \simeq \frac{RT}{F} \ln \frac{P_{Na} Na^b + P_{Cl} Cl^l + P_{HCO_3} HCO_3^l}{P_{Na} Na^l + P_{Cl} Cl^b + P_{HCO_3} HCO_3^b} . \tag{22}$$

The limitations of equation 22 have been discussed in detail elsewhere (FRÖMTER et al., 1970; SCHAFER et al., 1975). In the present context, the major issue worth noting is that equation 22 is most appropriate in instances where ions permeate *via* a common, extracellular pathway, *e. g.*, in the proximal tubule.

We know of no *a-priori* reasons why Eq. (18) or Eq. (22) should be the more appropriate formulation for expressing zero-current transepithelial voltages referable to ion gradients across a renal tubule. As it happens, the issue may not be a major one, at least for isolated proximal tubules, since the results with either equation are approximately the same, so long as V_e is $\leqq$ 15 mV. For example, when proximal straight tubules are perfused and bathed at 21° C with, respectively, Cl-KR and HCO_3-KR solutions, V_e is +3.7 mV (SCHAFER et al., 1974c). And for this tubule segment, P_{Na} and P_{Cl}, estimated from tracer fluxes at zero volume flow, are $0.23 \cdot 10^{-4}$ and $0.73 \cdot 10^{-1}$ cm s^{-1}, respectively (SCHAFER et al., 1974c). From these data, one can compute a P_{HCO_3} value of $0.03 \cdot 10^{-4}$ cm s^{-1}, using Eq. (18), and a P_{HCO_3} value of $0.03 \cdot 10^{-4}$ cm s^{-1}, using Eq. (22).

3. Measurement of Transepithelial Resistance

Transepithelial resistance can be measured by passing a constant current pulse across the tubular epithelium and observing the resultant change in the transepithelial voltage. Several pipet arrangements have been used to accommodate the dual functions of passing current and measuring voltage at the perfusing pipet. One method provides a third concentric pipet within the inner perfusion pipet for measuring V_e, while using the inner perfusion pipet as the current electrode (BURG et al., 1968; LUTZ et al., 1973). A second method employs a layer of platinum black applied to the inner perfusion pipet for passing current, while the inner perfusion pipet serves as the electrode for measuring V_e (LUTZ et al., 1973). Finally, a single electrode, the inner perfusion pipet, may be used to pass current and measure V_e (HELMAN et al., 1971). Under these circumstances, the voltage drop across the tip of the perfusion pipet must first be nulled using a bridge circuit or a negative-resistance external circuit; the resulting V_e across the tubule can then be measured. This latter method can be applied only to tubule segments having relatively high transepithelial electrical resistances.

When the current pulse is passed at one end of the tubule, *e. g.*, from the perfusing pipet, the resultant ΔV_e is maximal at the perfusion end of the segment and is attenuated with distance away from this point, due to current flow across the tubular epithelium. Thus cable analysis must be applied to the measurement of transepithelial resistances in perfused tubules. HELMAN et al. (1971) have applied standard cable equations (TAYLOR, 1963) to the particular case presented by the tubule, i. e., a cylinder of finite length, L, having a presumably homogenous wall and luminal cross-sectional area. The equations used by HELman et al. (1971), and subsequently LUTZ et al. (1973), for transepithelial resistance measurements in cortical collecting tubules and proximal tubule segments are:

$$\frac{L}{\lambda} = \cosh^{-1}(V_e^o / V_e^L), \tag{23}$$

$$R_c = \frac{V_e^o \tanh(L/\lambda)}{I_o \lambda}, \tag{24}$$

$$R_e = \frac{2\pi r V_e^o \lambda}{I_o} \tanh(L/\lambda), \tag{25}$$

where R_c is the resistance of the luminal solution (Ω cm^{-1}), R_e is the epithelial resistance (Σ cm^2), I_o is the current passed, λ is the length constant (TAYLOR, 1963), and V_e^o and V_e^L are the recorded voltages at the perfusing and collecting end of the segment, respectively.

Since λ for isolated renal tubules is approximately 0.1 mm for proximal tubule segments (LUTZ et al., 1973), while most isolated proximal tubule lengths studied are greater than 0.5 mm (Table 1), a proximal segment cannot be voltage-clamped accurately to a desired value along the entire tubule length when the clamping current is passed only from the perfusing pipet. Rather, voltage clamping of an entire segment requires the use of an axial electrode from the perfusion to the collection end. Although such a technique has been devised for short-circuit current measurements in the Necturus proximal tubule *in vivo* (SPRING and PAGNELLI, 1972), the much smaller luminal diameter of mammalian tubules has thus far precluded application of this technique to isolated rabbit nephron segments.

VII. Tracer Flux Measurements

The measurement of solute permeability coefficients and net solute fluxes in isolated perfused tubules should obviously include attention to the problems attendant to flux measurements in all epithelia; and also to problems which are unique to the isolated tubule preparation. Among the latter, three factors are particularly noteworthy. First, while the composition of the bath may be changed rapidly (*i. e.*, in less than 2 seconds [SCHAFER et al., 1974a]), no method currently available permits replacement of the perfusing solution within the same time scale. Second, the volume of the bath, approximately 1.2 ml, is quite large with respect to either the volume of the cellular compartment, approximately 1–4 nl, or the volume of the luminal compartment, approximately 0.5 nl. Consequently, in the case of bath-to-lumen tracer fluxes with a relatively permeable species, attention must be directed to the possibility of tracer equilibration between the bath and the lumen. Finally, because the volume of the perfusate is relatively small, and because luminal fluid moves in an axial direction, the luminal concentration of a tracer which is transported from lumen to bath will vary along the length of the tubule segment; thus the tracer concentration profile in the luminal solution must be computed for each particular set of experimental conditions.

The general equation describing J_i^{net}, the net flux of the i-th species from lumen to bath, is:

$$J_i^{net} = \frac{\dot{V}^o C_i^o - \dot{V}^L C_i^L}{A}, \tag{26}$$

where C_i^o and C_i^L are the concentrations of the i-th species in the perfusate and collected fluid, respectively; $\dot{V}^o$ and $\dot{V}^L$ are the perfusion and collection rates, respectively; and A is the luminal surface area. Evidently, C_i^o and C_i^L can be measured either by chemical methods or with tracers. In the latter case, the isotope-specific activity in the perfusate must remain constant. However, it is frequently desirable, and manipulatively more convenient, to measure fluxes and permeability coefficients from unidirectional tracer fluxes.

1. Unidirectional Lumen-to-Bath Fluxes

The unidirectional flux of a solute from lumen to bath can be measured by adding tracer-labeled solute to the perfusing solution and measuring its rate of appearance in the bathing solution. For convenience, the isotope concentration in the bathing solution C^{*b} should be considerably less than that in the perfusing solution C^{*0} and the collected fluid C^{*L}, a condition which is usually easy to satisfy due to the relatively large volume of the bath and the absence of significant bulk phase unstirred layers (SCHAFER and ANDREOLI, 1972a). During fluid transport, an apparent luminal efflux coefficient, K_i^{lb} (cm s^{-1}) for the i-th species, can be computed from the rate of tracer disappearance from the luminal solution by using the expression (SCHAFER and ANDREOLI, 1972a):

$$K_i^{lb} = \frac{\dot{V}^0 - \dot{V}^L}{A} \left[\frac{\ln (C_i^{*o}/C_i^{*L})}{\ln (\dot{V}^o/\dot{V}^L)} + \left(\frac{1+\sigma_i}{2} \right) \right], \tag{27}$$

where σ_i is the reflection coefficient of the i-th solute. At zero volume flow, equation 27 reduces to (GRANTHAM and BURG, 1966):

$$K_i^{lb} = \frac{\dot{V}^o}{A} \ln \frac{C^{*o}}{C^{*L}}. \tag{28}$$

Evidently, the K_i^{lb} term in Eqs. (27) and (28) includes, in principle, terms for active and passive solute transport, and for exchange diffusion. The lumen-to-bath flux of the i-th species can then be computes as:

$$J_i^{lb} = K_i^{lb} \overline{C}_i^l, \tag{29}$$

where $\overline{C}_i^l$ is, to a sufficient approximation (KOKKO et al., 1971), the arithmetic mean of C_i^o and C_i^L, *i. e.* the mean luminal concentration.

If the rate of efflux of the tagged species from the luminal solution is small, the difference between C^{*o} and C^{*L} becomes small, and Eqs. (27) and (28) become indeterminate. Under these conditions, J_i^{lb} may be computed from the steady-state rate of tracer appearance in the bath according to the expression (BURG and GREEN, 1973a):

$$J_i^{lb} = \frac{-\dot{V}^o C_i^o}{L} \ln \left[1 - \frac{N_b^* t_o}{N_o^* t_b}\right], \tag{30}$$

where N_b^* is the number of counts accumulating in the bath over time t_b, and N_o^* is the total counts perfused in the time period t_o. The use of equation 30 assumes that both $\dot{V}^x$ and C^x are constants along the tubule length.

2. Unidirectional Bath-to-Lumen Fluxes

The most convenient way of measuring bath-to-lumen fluxes is to keep the tracer concentration in the lumen negligibly low with respect to the tracer concentration in the bath. For tracer species having relatively high rates of transport from bath to lumen, this condition can often be satisfied by using appropriately short tubule lengths and appropriately high perfusion rates (SCHAFER et al., 1974c). For these circumstances, J_i^{bl}, the bath-to-lumen flux of the i-th solute, can be computed from (FRINDT and BURG, 1972):

$$J_i^{bl} = \frac{\dot{V}^L C_i^b C^{*L}}{C_i^{*b} A} \tag{31}$$

where C_i^b and C_i^{*b} are, respectively, the unlabeled and labeled concentrations of the i-th solute in the bath.

When tracer accumulation in the luminal fluid is appreciable and a lumen-to-bath back-flux occurs, Eq. (31) is not suitable. Rather a correction term must be provided which accounts for lumen-to-bath back-flux. IMAI and KOKKO (1974b) have derived an equation for these circumstances:

$$K_i^{bl} = \frac{C^{*L}}{C^{*b}} \left[\frac{K_i^{lb}}{1\text{-}\exp(-AK_i^{lb}/\overline{\dot{V}})}\right], \tag{32}$$

where K_i^{bl} is an influx coefficient (cm s^{-1}) and $\overline{\dot{V}}$ is the arithmetic mean of $\dot{V}^o$ and $\dot{V}^L$. Evidently, the calculation of K_i^{bl} from Eq. (32) also requires assessment of K_i^{lb} (Eqs. 27 and 28).

3. Evaluation of P_{D_i} From Tracer Fluxes

In order to assess the significance of K_i^{lb} and K_i^{bl}, with particular reference to the relationship between these coefficients and P_{D_i} (cm s^{-1}), the diffusional permeability of the i-th species at zero volume flow, a number of other factors must be evaluated. These include, at a minimum: the symmetry of K_i^{lb} and K_i^{bl}; and the possible effect of active transport processes, exchange diffusion, and entrainment of solute and solvent flows on K_i^{lb} and K_i^{bl}. In short, if it can be established that a flux coefficient, either for efflux or influx, depends on no other driving forces except those for a diffusion process, then that flux coefficient may be equated with P_{D_i}.

For example, SCHAFER et al. (1974c) assessed unidirectional $^{22}Na^+$ fluxes in isolated superficial proximal straight tubule lengths exposed to symmetrical HCO_3-KR solutions. At 37° C, *i. e.*, during isotonic fluid absorption coupled to active transport processes, K_{Na}^{lb} was $0.36 \cdot 10^{-4}$ cm s^{-1}; and when 10^{-4} M ouabain was added to the bath and net fluid absorption and the sponatneous transepithelial voltage was zero, K_{Na}^{lb} fell to $0.28 \cdot 10^{-4}$ cm s^{-1}. This last value was not significantly different from $0.24 \cdot 10^{-4}$ cm s^{-1}, the value of K_{Na}^{bl} computed from J_{Na}^{bl} measurements at 37° C (*i. e.*, during isotonic fluid absorption) from Eq. (31) and either the Goldman-Hodgkin-Katz equation:

$$K_{Na}^{bl} = -\frac{J_{Na}^{bl} RT}{V_e z_{Na}^2 F} \left[\frac{1 - \exp \dfrac{z_{Na} F V_e}{RT}}{C_{Na}^b} \right] \tag{33}$$

or Fick's first law:

$$K_{Na}^{bl} = -\frac{J_{Na}^{bl}}{C_{Na}^b}, \tag{34}$$

where it should be noted that: for the case of $V_e = 0$, Eq. (33) may be reduced to Eq. (34) by L'Hospital's rule; and for V_e values in the range ± 5 mV (the spontaneous V_e in these tubules exposed to symmetrical HCO_3-KR solutions at 37° C is approximately -1.0 mV [SCHAFER and ANDREOLI, 1974c]), Eqs. (33) and (34) give results which vary by less than 3 percent. Since K_{Na}^{bl} measured during isotonic fluid absorption from lumen to bath was not significantly different from K_{Na}^{lb} measured at zero volume flow, one may conclude that K_{Na}^{bl} was not affected significantly by fluid absorption coupled to active salt absorption, or by entrainment of solute and solvent flows. Based on these considerations, one may conclude that K_{Na}^{bl} for these conditions may be equated with $P_{D_{Na}}$.

We argue that similar assessments should be made in any consideration of P_{D_i} values from unidirectional tracer fluxes in isolated renal tubules. Indeed, in the case of other nephron segments, the determination of P_{D_i} values from unidirectional tracer fluxes may be even more complex. Consider, for example, the distal convoluted tubule or the cortical collecting tubule. In such nephron segments, the magnitude of V_e is appreciably greater than in the proximal tubule and there is, at present, no way of defining precisely the change in the luminal voltage in an axial direction. Furthermore, the magnitude of passive ion flux, for example, Na^+, in the lumen to bath direction is appreciably less than the active component of Na^+ flux in that direction. Thus, for such nephron segments, the assignment of P_{D_i} values from unidirectional tracer fluxes may be more difficult than in the case of proximal renal tubules.

4. Transmembrane Fluxes

Finally, one can measure the steady-state concentration of a tagged solute within renal tubular cells. The basic technique is to remove the perfused tubule segment rapidly from the perfusion system after the tubular cells have had ample time to establish a steady-state concentration of the solute in question. The tubule segment is extracted in a small droplet of trichloroacetic acid (TUNE and BURG, 1971) to remove the accumulated solute and ions, then extracted in chloroform and acetone to remove oil adhering to the tubule, and then air-dried. The dried tubule segment can be weighed on a quartz fiber balance, and the intracellular water can be computed from the dry weight and the premeasured relation between the volume of distribution of tritiated water and a volume marker such as ^{14}C-Inulin (TUNE and BURG, 1971). From the extracted solute activity minus the extracellular activity divided by the intracellular water, the intracellular concentration may be measured.

A particularly useful technique for the simultaneous measurement of opposing transepithelial and transmembrane fluxes using double labeling was described by TUNE and BURG (1971). These investigators perfused proximal tubular segments with a solution containing a known concentration (C^{*o}) of ^{14}C-glucose and unlabeled glucose (C^{o}). The bath contained known concentrations of ^{3}H-glucose (C^{**b}) and unlabeled glucose (C^{b}). The concentration of glucose collected could be estimated from the relationship developed by KOKKO et al. (1971):

$$C^{L} = \frac{C^{**L}\,C^{o}}{C^{**o}} + \frac{C^{*L}\,C^{b}}{C^{**b}}\,. \tag{35}$$

Similarly, the cell glucose concentration was given as:

$$C^{c} = \frac{C^{**c}\,C^{o}}{C^{**o}} + \frac{C^{*c}\,C^{b}}{C^{*b}}\,. \tag{36}$$

C. Transport Properties of Isolated Nephron Segments

As indicated in the preceding section, *in-vitro* perfusion of isolated nephron segments has provided a useful means for evaluating the transport characteristics, both active and passive, of the different segments of the nephron and of the same segment of the nephron from different regions of the kidney. This portion of the chapter will consider some of these transport characteristics, with particular emphasis on results obtained from rabbit renal tubules.

I. Suitability of the *In-Vitro* Preparation

It is evident that the suitability of the *in-vitro* isolated tubule preparation depends both on the viability of the preparation and on the ability of the *in-vitro* preparation to yield results comparable to those observed in, or at least expected from, the *in-vivo* mammalian nephron. A number of such correlations will be indicated in succeeding sections of this chapter. In the present context, three observations are particularly germane. First, single *in-vitro* renal tubules maintain intracellular-to-extracallular ionic distribution ratios comparable to those expected for the *in-vivo* condition. In isolated proximal convoluted tubules exposed to aqueous solutions similar to the HCO_3-KR perfusing and bathing media listed in Table 2, BURG et al. (1966) obtained the following values for intracellular electrolyte content, all expressed as mEq l^{-1} of tissue water: Na^+, 69.7; K^+, 110; and Cl^-, 55.7. The extracellular ionic concentrations in these experiments were: Na^+, 151.2 mEq l^{-1}; K^+, 5.0 mEq l^{-1}, and Cl^-, 122 mEq l^{-1}.

Second, DELLASEGA and GRANTHAM (1973) have found that isolated segments of proximal convoluted, proximal straight, and cortical collecting tubules are able to regulate cell volume when suspended in hypotonic media, possibly by a ouabain-insensitive cellular salt extrusion process. These workers found that, when such tubules were exposed to hypotonic media at 37° C, there was a rapid cellular swelling. This osmometric swelling phase was followed by a rapid restoration of cell volume, *i. e.,* by isosmotic intracellular volume regulation. The volume regulatory process in hypotonic media was unaffected by ouabain, but was inhibited either by hypothermia or by elevated K^+ concentrations in extracellular fluid. On the basis of these observations, DELLASEGA and GRANTHAM (1973) concluded that cells of isolated renal tubule segments possess a volume-regulatory mechanism dependent on solute efflux, possibly KCl, from the cells in response to a reduction in extracellular fluid osmolality.

Finally, BURG and ORLOFF (1968) found that the rate of isotonic fluid absorption coupled to active salt transport was, for isolated proximal convoluted tubules, in the range of 1.2 nl min^{-1} mm^{-1}, a result subsequently confirmed by a number of other investigators (*e. g.,* KOKKO et al., 1971; IMAI and KOKKO, 1972 and 1974a; GRANTHAM et al., 1972). Recently, CHONKO et al. (1975a and b), using micropuncture techniques, obtained a value of 1.9 nl min^{-1} mm^{-1} for the rate of tubular fluid absorption from the *in-vivo* rabbit proximal tubule. It seems clear from these observations that, at least in the case of isotonic fluid absorption in proximal convoluted tubules, there is a reasonable agreement between *in-vitro* and *in-vivo* experimental results.

II. The Proximal Tubule

Traditionally, the proximal tubule has been considered in terms of two morphologic components, the proximal convolution and the pars recta, or proximal straight tubule. In the interval between 1966–1974, most studies with isolated proximal tubules were aimed primarily at obtaining either convoluted or straight segments. No particular reference was made, in the case of proximal convolutions, to whether the tubule segment was close to or remote from the glomerulus of origin. Likewise, uniform attention was not directed to whether a proximal tubule, either convoluted or straight, was derived from a glomerulus having an origin in either the superficial or juxtamedullary regions of rabbit renal cortex, although it should be recognized that, in most such studies (cf. below), tubule segments were in fact obtained from superficial cortical regions.

However, it now seems likely that there are dimensions of complexity to the structural and functional heterogeneity of rabbit proximal tubules beyond those envisioned by classification into convoluted and straight portions, depending on the distance of the tubular epithelium from the glomerulus of origin (SEELY, 1973; WOODHALL et al., 1976; HAMBURGER et al., 1976; JACOBSON and KOKKO, 1976), and whether the tubule segment arises from glomeruli situated in superficial or juxtamedullary regions of the cortex (KAWAMURA et al., 1975). For these reasons, we will first summarize the major observations on transport processes in proximal convoluted or proximal straight tubules, citing whenever possible more explicit characterization of the tubule segment under study, and then report the available information addressed particularly to the issue of structural and functional heterogeneity of proximal tubule segments.

1. Dissipative Transport Properties

Table 3 compares the dissipative permeability characteristics of proximal convoluted and proximal straight tubules, the latter including primarily superficial tubule segments. Both proximal convoluted and proximal straight tubules have remarkably low transepithelial electrical resistances, in the range of 7–8.2 Ω cm^2. Other electrically leaky epithelia, such as the Necturus proximal convoluted tubule (BOULPAEP, 1967), rabbit ileum (FRIZZELL and SCHULTZ, 1972), and rabbit gallbladder (BARRY et al., 1971), have transepithelial resistances in the range of 40–80 Ω cm^2. Thus it is clear that, even among electrically leaky epithelia, the total ionic conductance of mammalian proximal renal tubules is quite high.

Although the total ionic conductances of the convoluted and straight portions of the proximal tubule are quite similar, the ionic permeability coefficients in these two nephron segments vary. Table 3 indicates that, in the superficial proximal straight tubule, the ionic selectivity sequence, estimated either by tracer fluxes and/or electrical methods, is $Cl^- > Na^+ > HCO_3^-$ (SCHAFER et al., 1974c and 1975). In the proximal convoluted tubule the P_{Na}/P_{Cl} ratio, measured either during *in-vivo* micropuncture (BOULPAEP, 1967; FRÖMTER et al., 1970; BOULPAEP and SEELY, 1971) or *in-vitro* microperfusion (IMAI and KOKKO, 1971), is slightly in excess of unity. As indicated in Table 3, P_{Na} in superficial proximal straight tubules is approximately fourfold less than in proximal convoluted tubules whereas P_{Cl} is nearly the same in the two segments; thus it follows that the P_{Na}/P_{Cl} ratio is lower in straight tubules than in convoluted tubules, primarily because Na^+ is less permeant in the former than in the latter.

Table 3. Permeability properties of isolated proximal tubule segments

	Proximal convoluted tubule	Proximal straight tubule*
Ionic Permeability Coefficients (P_i, μm s^{-1}):		
Na^+	0.40–1.19[b,c,f]	0.23–0.26[d,j]
Cl^-	0.19–0.65[b]	0.56–0.73[d,j]
HCO_3^-**	–	0.04–0.20[k,m]
Acetate	–	0.14[h]
Isethionate	–	0.02[k]
Phosphate	0.054[a]	0.051[a]
Nonelectrolyte Permeability Coefficients (P_i, μm s^{-1}):		
Urea	0.53–0.63[c,e]	0.45[i]
Glucose	0.72[l]	–
Mannitol	–	0.250[i]
Sucrose	0.069[c]	0.078[i]
Hydraulic Conductivity Coefficients (P_f, μm sec^{-1}):		
Intact Epithelium	~ 3500[h]	~ 500[n]
Reflection Coefficients (σ_i):		
NaCl	0.68[f]	–
Na^+	–	0.9–1.0[k]
Cl^-	–	0.78–0.95[k]
HCO_3^-	–	0.97[k]
Urea	0.91[f]	–
Electrical Resistance Properties:		
Resistance (Ω cm^2)	7.0[g]	8.2[g]
Length Constant (cm)	0.01[g]	0.01[g]

[a] Dennis et al., 1976
[b] Imai and Kokko, 1971
[c] Imai and Kokko, 1972
[d] Kawamura et al., 1975
[e] Kokko, 1972
[f] Kokko et al., 1971
[g] Lutz et al., 1973
[h] Schafer and Andreoli, 1976
[i] Schafer and Andreoli, unpubl. observations
[j] Schafer et al., 1974c
[k] Schafer et al., 1975
[l] Tune and Burg, 1971
[m] Warnock and Burg, 1977
[n] Andreoli et al., 1978; Schafer et al., 1978a.
* Most of the data listed were derived exclusively from experiments with superficial proximal straight segments. However, the proximal straight segments from which the resistance data of Lutz *et al.* (1973) were derived were not identified by the location of the nephron.
** Estimated from NaCl-$NaHCO_3$ bionic potentials.

Three other considerations are noteworthy in this context. First, it is widely believed that passive ion transport in epithelia involves an extracellular route (USSING and WINDHAGER, 1964; FRÖMTER and DIAMOND, 1972; FRIZZELL and SCHULTZ, 1972). In order to evaluate this possibility in superficial proximal straight tubules, SCHAFER et al. (1974c) measured zero-current NaCl dilution potentials at zero volume flow (*i. e.*, 21° C) by varying the NaCl concentrations in the HCO_3-KR (Table 2) perfusing or bathing solutions by isosmotic mannitol replacement. The salt dilution voltages were equal in magnitude but opposite in sign for a given reduction in the NaCl concentration in either the perfusing and bathing solutions. Thus, it seemed probable that the salt dilution potentials were expressed across a single symmetrical interface having the same P_{Na}/P_{Cl} ratio on either side. Since the ionic permeability properties of proximal tubular luminal and peritubular membranes differ in some species (BOULPAEP, 1967; GIEBISCH, 1968), it was concluded that the salt dilution potentials were expressed across junctional complexes. It follows from this argument that the low transepithelial resistance of these tubules is referable to a high ionic conductance of junctional complexes, *i. e.*, $(R_5 + R_6)$ in Figure 3.

Second, both in gallbladder (BARRY and DIAMOND, 1971) and in rabbit ileum (FRIZZELL and SCHULTZ, 1972), it has been suggested that ion permeation involves neutral or weakly polar sites. In isolated proximal straight tubules, SCHAFER et al. (1974c) found that a P_{Na}/P_{Cl} ratio of 0.3, estimated from tracer fluxes in HCO_3-KR media, adequately described zero-current NaCl dilution potentials over a concentration range of 35–150 mM NaCl. In charged membranes, the P_{Na}/P_{Cl} ratio varies with the external salt concentration, because of the Donnan distribution of ions between the external solutions and the membrane exchange sites (BAKER et al., 1964). Since the P_{Na}/P_{Cl} ratio in proximal straight tubules was independent of external NaCl concentrations, SCHAFER et al. (1974c) argued that, in these tubules, the ion permeation pathways contained a negligibly small quantity of charged sites.

Third, there are important functional differences between the ion permeation pathways in superficial and deep proximal straight tubules. Thus, in contrast to the superficial proximal straight tubule, where P_{Cl} exceeds P_{Na}, KAWAMURA et al. (1975) found that proximal straight tubules from juxtamedullary regions were more permeable to Na^+ than to Cl^-. The relevance of this functional heterogeneity to transport processes occurring in superficial and deep nephrons requires clarification.

2. Active Transport Processes

Table 4 summarizes certain of the active and passive transport characteristics of isolated proximal tubules. The data for spontaneous net volume absorption, spontaneous transepithelial voltages, and net Na^+ fluxes are for tubules exposed either to symmetrical HCO_3-KR solutions (Table 2) or comparable buffers, or to rabbit serum (bath) and rabbit serum ultrafiltrate (perfusate). Under these conditions at 37° C, both the convoluted tubule and the superficial proximal straight tubule carry out isotonic fluid absorption and develop spontaneous

Table 4. Active or net transport processes in proximal tubule segments exposed to symmetrical solutions

	Proximal convoluted tubule	Proximal straight tubule*
Net volume absorption (J_v, nl min^{-1} mm^{-1}):		
21–25° C, or with ouabain	0–0.4[a,d]	0–0.1[d,l,m]
37° C	0.84–1.40[a,d,e,i]	0.31–0.47[a,d,f,l,m]
Transepithelial voltage† (V_e, lumen to bath):		
21–25° C, or with ouabain	–1.2[h]	0[l,m]
37° C	–3.8 to –5.8[b,g,h,j]	–1.1 to –2.0[j,l,m]
Na^+ fluxes, 37° C (J_{Na}, nEq s^{-1} cm^{-2}):		
lumen to bath	20[i]	4.86[l]
bath to lumen	15.5[i]	3.34[l]
net	4.5[i]	1.52[l]
$[Na^+]$ perfused/$[Na^+]$ collected	~ 1.0[i]	~ 1.0[l]
Cl^- fluxes, 37° C (J_{cl}, nEq s^{-1} cm^{-2}):		
lumen to bath	–	9.5[k]
bath to lumen	–	8.3[k]
net	–	1.2[k]
Phosphate fluxes, 37° C (J_{PO_4}, nmoles s^{-1} cm^{-2}):		
lumen to bath	0.150[c]	0.051[c]
bath to lumen	0.010[c]	0.007[c]
net	0.140[c]	0.044[c]
Glucose fluxes, 37° C (J_{glu}, nmoles s^{-1} cm^{-2}):		
lumen to bath*	2.66[n]	–
bath to lumen*	0.49[n]	–
net absorption (at T_{max})**	2.17[n]	0.14[n]
PAH fluxes, 37° C (J_{PAH}, nEq s^{-1} cm^{-2}):		
lumen to bath*	–	0.0005[o]
bath to lumen*	–	–
net secretion (T_{max})*	0.0062[o]	0.0231[o]
Acetate fluxes, 37° C (J_{Ac}, nEq s^{-1} cm^{-2}):		
lumen to bath	–	0.379[k]
bath to lumen	–	0.140[k]
net	–	0.291[k]

a BURG and ORLOFF, 1968
b BURG and ORLOFF, 1970
c DENNIS et al., 1976
d GRANTHAM et al., 1972
e IMAI and KOKKO, 1972
f KAWAMURA et al., 1975
g KOKKO, 1973
h KOKKO and RECTOR, 1971
i KOKKO et al., 1971
j LUTZ et al., 1973
k SCHAFER and ANDREOLI, 1976
l SCHAFER et al., 1974c
m SCHAFER et al., 1975
n TUNE and BURG, 1971
o TUNE et al., 1969

† As discussed in connection with equations 16–19, the values of V_e are the observed transepithelial voltages and hence represent displacements from the Donnan equilibrium voltage.

* Data given for net volume absorption, transepithelial voltage, Na^+, Cl^- and acetate fluxes are from superficial proximal straight tubules. Segments perfused in the glucose and PAH transport studies were not localized to superficial or juxtamedullary origin.

** Approximated from data given in the reference indicated.

lumen-negative transepithelial voltages, but the values for both J_v and V_e are appreciably greater in convoluted than straight segments; and as indicated in Table 4, J_v and V_e are inhibited in both segments, either by cooling or by ouabain.

In the isolated proximal convoluted tubule, certain other factors merit particular consideration. During net volume absorption at 37° C, KOKKO et al. (1971) observed that there was no change in the luminal Na^+ concentration between the perfusate and the collected fluid. However, when the perfusate contained the relatively impermeant solute, raffinose, in place of a portion of the NaCl, net volume absorption resulted in a further decrease in the luminal NaCl concentration, such that collected fluid to bathing solution NaCl concentration ratios as low as 0.76 were observed (KOKKO et al., 1971). Thus, as in the rat micropuncture experiments of WINDHAGER and GIEBISCH (1961), net volume absorption, in the presence of a nonabsorbable solute, could establish a transepithelial NaCl gradient. Net Na^+ flux in the presence of identical perfusing and bathing solutions was ~75 percent of the unidirectional Na^+ flux from lumen to bath, and the Na^+ flux ratio (l→b/b→l) of 1.3 was considerably greater than that predicted by the Ussing flux ratio equation (KOKKO et al., 1971). These data, together with the observations that either ouabain or cooling rapidly inhibited both J_v and V_e, strongly suggested that net volume absorption in the proximal convoluted tubule was driven by the active transport of Na^+ from lumen to bath (KOKKO et al., 1971), much as had been concluded from the micropuncture experiments of WINDHAGER and GIEBISCH (1961).

There was also little doubt that the isolated convoluted tubule established a negative transepithelial voltage, at least in the region of the tip of the perfusing pipet, when the perfusate and bath had essentially identical ionic compositions (BURG et al., 1968). Since the voltage-measuring electrode was the inner perfusing pipet, which could be localized visually in the tubular lumen, voltage measurements from isolated tubule experiments were of considerable importance in supporting the reassessment of the micropuncture measurements of proximal tubular transepithelial voltage initiated by FRÖMTER and HEGEL (1966).

KOKKO and RECTOR (1971) and KOKKO (1973) have also observed that both the magnitude and sign of V_e depend on the luminal solution composition: when alanine, glucose, and HCO_3^- were omitted from the perfusate, V_e became lumen-positive rather than lumen-negative (KOKKO and RECTOR, 1971; KOKKO, 1973). On the basis of these observations, it was suggested that both fluid absorption and the transepithelial voltage might depend upon co-transport of glucose and alanine with Na^+, and on the coupled transport of Na^+ with HCO_3^- (KOKKO, 1973). But while CARDINAL et al. (1975) have been able to confirm the effects of varying luminal composition on V_e, these workers could not establish a systematic relationship between V_e and J_v. Evidently, the question warrants further consideration.

DENNIS et al. (1976) have measured unidirectional and net phosphate fluxes using ^{32}P and ^{33}P in both proximal convoluted and straight tubules. Their results, shown in Table 3 and 4, are consistent with net phosphate absorption in both segments but with a much higher lumen-to-bath flux in the convoluted than in the straight segment. Ouabain (10^{-5} M) was observed to inhibit net phosphate absorption completely. Bath to lumen phosphate fluxes were much lower than the lumen to bath fluxes and were unaffected by ouabain. The latter results indicate that the passive phosphate permeability of both segments is low, on the order of 0.05 $\mu m\ s^{-1}$.

Finally, TUNE and BURG (1971) observed that glucose was actively absorbed in both proximal convoluted and straight tubules, although the process was considerably more rapid in the former segment. Since the glucose concentration in the cells of proximal convoluted tubule segments was more than two times that in either the perfusate or the bathing solution, TUNE and BURG attributed the active step in the transport to the luminal plasma membrane of the tubular cells. In contrast to glucose, PAH was secreted by both convoluted and straight segments but more rapidly in the straight segment, and the active mechanism appeared to be localized to the peritubular plasma membrane of the tubular cells (TUNE et al., 1969; BURG and ORLOFF, 1969).

In the superficial proximal straight tubule exposed to symmetrical HCO_3-KR solutions, SCHAFER et al. (1974c) found (Table 4) that the net lumen-to-bath Na^+ flux at 37° C accounted quantitatively for isotonic fluid absorption, that J_v, V_e and net Na^+ flux became indistinguishable from zero, either at 23° C or at 37° C in the presence of ouabain, and that the flux ratio for unidirectional Na^+ fluxes was considerably greater than that expected solely for passive ion transport. SCHAFER and ANDREOLI (1976) found that passive Cl^- transport accounted for approximately 28% of net Na^+ absorption, and that, during fluid absorption in tubules exposed to symmetrical HCO_3-KR solutions, the tubular fluid Cl^- concentration rose from 116.6 to 127.6 mEq l^{-1}; the other counter-ions for Na^+ transport were: acetazolamide-sensitive HCO_3^- transport, ~50 percent, and active acetate absorption, ~25 percent.

3. Mechanisms for Salt and Water Transport in the Superficial Pars Recta

Two issues relating to the characteristics of salt and water absorption warrant particular consideration: the relative contributions of active and passive transport processes to NaCl absorption; and the mechanism of isotonic fluid absorption. In this context, SCHAFER et al. (1975 and 1977) evaluated salt and water transport in superficial proximal tubules perfused and bathed at 37° C with a Cl-KR and an HCO_3-KR solution, respectively (Table 2), i. e., by exposing these tubules to luminal and peritubular solutions similar to those which might be seen by this nephron segment *in vivo* (GOTTSCHALK 1962–63; MALNIC et al., 1969; BENNETT et al., 1968).

The major experimental results are shown in Table 5. The rates of fluid absorption were comparable to those seen with symmetrical HCO_3-KR solutions (*e. g.*, Table 4), but the spontaneous transepithelial voltage was lumen-positive rather than lumen-negative. Moreover, when the preparation was cooled to 21° C, neither fluid absorption nor the transepithelial voltage became zero, as occurs with symmetrical HCO_3-KR solutions (Table 4). Rather, Table 5 indicates that, at 21° C, J_v was approximately one-third, and V_e slightly greater, than the comparable values at 37° C (essentially identical results obtain when

Table 5. Salt and water absorption in the superficial pars recta exposed to asymmetrical solutions

T^o	J_v	V_e	$J_{Cl}^{net}/J_{Na}^{net}$
°C	nl min^{-1} mm^{-1}	mV	%
37	0.33	+3.2	85%
21	0.13	+3.7	100%

The perfusate contained a Cl-KR buffer and the bath an HCO_3-KR buffer (Table 2). Adapted from SCHAFER et al., 1975, 1977.

ouabain is used as a transport inhibitor at 37° C [SCHAFER et al., 1975 and 1977]). Table 5 also indicates that, at 37° C and 21° C, Cl^- accounted for 85 percent and 100 percent, respectively, of the anion component in an isotonic salt absorbate. Thus these observations were in accord with an hypothesis first proposed by RECTOR et al. (1966): as a consequence of preferential absorption of HCO_3^- and the attendant rise in tubular fluid Cl^- concentrations (cf. above), an electrochemical gradient driving passive NaCl absorption might develop.

A quantitative assessment of these data depended on an evaluation of the diffusion constraints within intercellular spaces. Since passive ion permeation in these tubules probably involves an extracellular route (SCHAFER et al., 1974c), approximate limits for these constraints could be deduced by comparing the transepithelial resistance of a proximal straight tubule with that of a layer of HCO_3-KR buffer having a thickness equivalent to the epithelium, *i. e.*, $7.5 \cdot 10^{-4}$ cm (Table 1). Table 4 indicates that the transepithelial resistance of proximal straight tubules is approximately 8.2 Ω-cm^{-2}; and a layer of 150 mM Krebs-Ringer buffer having a thickness of $7.5 \cdot 10^{-4}$ cm has a computed resistance of ~0.05 Ω-cm^2. Thus, for the case of passive ion transport through an extracellular route, the diffusion resistance of intercellular spaces, exclusive of junctional complexes, could be, at a maximum, 163 times greater than that of free solution. On the basis of these considerations, and the fact that the Cl^-/HCO_3^- selectivity ratio in these tubules (Table 3) was far greater than in free solution, SCHAFER et al. (1974c) defined a dimensionless parameter α, which expressed the ratio of the diffusion resistance of intercellular spaces, exclusive of junctional complexes, to that of free solution, and which had a range of:

$$1 \leq \alpha \leq 100.$$

Evidently, for $\alpha = 1$, all of the diffusion resistance to passive ion permeation is assigned to junctional complexes; and for $\alpha = 100$, the bulk of the diffusion resistance is assigned to intercellular spaces.

An analysis of the flow-diffusion equations for intercellular spaces in terms of the variable parameter α led to the following general conclusions (SCHAFER et al., 1975 and 1977). First, for $1 \leq \alpha \leq 100$ and the experimental conditions listed in Table 5, either at 21° C (SCHAFER et al., 1975) or at 37° C (SCHAFER et al., 1977), the intercellular spaces were in virtual diffusion equilibrium with the bathing solutions, even when all active transport sites, for the 37° C case, were situated at the apical ends of intercellular spaces. Stated in another way, because of the remarkably low transepithelial electrical resistance of these tubules, and hence the low diffusion resistance of intercellular spaces, it was not possible to account for the observed rates of fluid absorption, either at 37° C or at 21° C (Table 5), in terms of the standing gradient mechanism of DIAMOND and BOSSERT (1967). But it was possible to rationalize the rates of fluid absorption from asymmetrical solutions, either at 21° C or at 37° C, in terms of the chemical gradient for water transport from lumen to bath referable to the unequal Cl^- and HCO_3^- concentrations in luminal and bathing solutions, and the unequal reflection coefficients for these two anions (Table 3). Second, the results in Table 5 are consistent with the view that approximately one-third of NaCl absorption from superficial proximal straight tubules exposed to asymmetrical solutions is due to passive driving forces, the latter referable to the lumen-positive voltage, the lumen-to-bath Cl^- concentration gradient, and possibly entrainment of solute and solvent flows. In accord with this possibility, the theoretical calculations of SCHAFER et al. (1977) indicated that, for $1 \leq \alpha \leq 100$, 33.5 percent of NaCl absorption at 37° C could be accounted for in terms of passive processes, primarily diffusional.

4. Effective Luminal Hypotonicity and Isotonic fluid Absorption

We have developed a model to explain fluid absorption coupled to solute absorption in the isolated mammalian proximal tubule. The model is based on the assumptions that passive ion flows in this epithelium occur paracellularly (*e. g.*, FRÖMTER and DIAMOND, 1972; FRÖMTER et al., 1970; SCHAFER et al., 1974c; USSING and WINDHAGER, 1964), and as deduced from the low electrical resistance of the epithelium (cf. above), that the intercellular spaces are in virtual diffusion equilibrium with the bathing solution. If this is the case, then intercellular spaces cannot serve as a central hypertonic compartment and external osmotic gradients must be identified. However, in view of the remarkably high transepithelial hydraulic conductance observed either in straight (SCHAFER et al., 1978a) or convoluted (ANDREOLI et al., 1978) rabbit proximal tubules, relatively small effective osmotic gradients would be required to account for the rates of isotonic fluid absorption observed in these segments. If the intercellular spaces are in diffusion equilibrium with peritubular media, then the osmotic driving force between the luminal and bathing solutions may be expressed in terms of external solution composition as:

$$\frac{J_v}{P_f \bar{V}_w} = \sum_{i=1}^{n} \sigma_i (\bar{C}_{l_i} - C_{b_i}), \tag{37}$$

where i refers to the *i*-th solute, C_{b_i} is the osmolar concentration in the bathing solution and $\bar{C}_{l_i}$ is the integrated average of the luminal osmolar concentration. The relation betwenn J_v and the effective osmotic driving force, *e. g.*, the right-hand side of quation 37, indicates that for a P_f value of 3,500 μm sec^{-1} (Table 3), one requires effective transepithelial osmolality differences in the range of

1.3–3.2 mOsm/l to provide an adequate driving force for the spontaneous rates of volume absorption listed in Table 4.

Thus as a consequence of the high hydraulic conductance of the proximal nephron, one may account for isotonic volume absorption on the basis of very small effective osmolality gradients between the luminal and bathing solutions without invoking a hypertonic intercellular space. Even if one considers rates of fluid absorption observed for the in vivo proximal convoluted tubule of the rat (RECTOR et al., 1966) and rabbit (CHONKO, 1975a), approximately 2.0 nl min^{-1} mm^{-1}, relatively small effective osmolality gradients, approximately 6 mOsm/l, are adequate to account for such rates of volume flow.

We (SCHAFER et al., 1975; ANREOLI and SCHAFER, 1978) have considered two separate but not mutually exclusive mechanisms to account for the development of such effective transepithelial osmolality gradients:

1) Changes in the anion composition of the tubular fluid occur along the length of the proximal nephron as a consequence of tubular fluid acidification; this results in the preferential absorption of bicarbonate and an increase in the tubular fluid to bath Cl^- concentration ratio (SCHAFER et al., 1976; 1974c; WARNOCK and BURG, 1977). The generation of such axial anion asymmetry produces an osmotic driving force for fluid absorption if the reflection coefficient for bicarbonate exceeds that for chloride, even when the cryoscopic osmolalities of the luminal and bathing solution are identical.

In this context, it is important to recognize that, even in those tubule segments which are more permeable to Na^+ than to Cl^-, the development of axial anion asymmetry, *i. e.*, preferential HCO_3^- absorption, will produce fluid absorption. Thus when the cryoscopic osmolality of the luminal solution remains nearly constant, the major driving force for fluid absorption under such circumstances will depend upon the difference in Cl^- and HCO_3^- concentrations between the luminal and peritubular solution, and the difference between the reflection coefficients for HCO_3^- and Cl^-. This fact is particularly relevant to juxtamedullary nephrons, which are more permeable to Na^+ than to either Cl^- or HCO_3^- (BERRY et al., 1978). However, because these nephron segments are considerably more permeable to Cl^- than to HCO_3^- (BERRY et al., 1978), net volume absorption will result if these tubules acidify the luminal fluid, thereby producing the axial development of anion asymmetry between the external solutions.

It should be noted that the principle of the axial development of asymmetry between the luminal and peritubular solutions is not intended to be restricted to Cl^- and HCO_3^-. Any solute which is absorbed preferentially in the early regions of the proximal nephron can produce an effective osmotic force if the reflection coefficient of the solute in question is relatively high with respect to solutes remaining in the luminal solution, and if the proximal tubular fluid remains nearly isotonic.

2) ANDREOLI and SCHAFER (1978) have argued that the development of luminal hypotonicity might procude fluid absorption in the presence of symmetrical NaCl solutions in the luminal and bathing solution, since preferential solute absorption cannot occur. Luminal hypotonicity would develop as a consequence of net solute efflux, driven primarily by active Na^+ and passive Cl^- absorption. Thus a steady-state disequilibrium is achieved between luminal and bathing

solutions in the presence of isotonic fluid absorption, *i. e.*, when the absorbate is isosmotic with respect to the steady-state luminal solution. And because of the high hydraulic conductance of proximal renal tubules, the degree of absolute luminal hypotonicity needed to account for approximately 0.5 nl min^{-1} mm^{-1} of volume absorption would require the development of a transepithelial NaCl gradient of only 0.8 mM.

5. Heterogeneity of Tubular Structure and Function

As indicated at the beginning of this section, it has now become evident that considerable heterogeneity of proximal tubular structure and function exists, not only with respect to distance from the glomerulus of origin, but also, for a given nephron segment, depending on glomerular location, *i. e.*, whether superficial or juxtamedullary. Table 6 illustrates some of these heterogeneity characteristics. HAMBURGER et al. (1976) found that proximal convoluted tubules dissected from close to the glomerulus of origin had rates of fluid absorption that were twice those of more distal proximal convolutions; these proximal convolutions, in turn, had fluid absorption rates more nearly comparable to proximal straight tubules. Similarly, JACOBSON and KOKKO (1976) and KAWAMURA et al. (1975) found that the magnitude of the lumen-negative transepithelial voltage (in symmetrical solutions) was greater in early proximal convolutions than in late convolutions. These workers also found that the P_{Na}/P_{Cl} ratio declined progressively along the length of superficial isolated proximal tubules, and that the P_{Na}/P_{Cl} ratio was uniformly higher, for a given nephron segment, in juxtamedullary than in superficial tubules.

Finally, it has traditionally been held that organic acid transport occurs primarily in the proximal straight tubule. In accordance with this view, TUNE et

Table 6. Comparison of discrete regions of superficial and juxtamedullary proximal tubules

	Superficial			Juxtamedullary		
	Early PCT	Late PCT	PST	Early PCT	Late PCT	PST
Net volume absorption at 37 °C (J_v, nl min^{-1} mm^{-1})	1.38[a]	0.62[a]	0.47–0.52[a,c]	–	–	0.56[c]
Transepithelial voltage at 37 °C (mV, lumen to bath)	–4.3[b]	–3.1[b]	–2.1[c]	–4.7 ± 0.6[b]	–3.8 ± 0.1[b]	–1.8[c]
P_{Na}/P_{Cl}	1.56[b]	0.3[b]	0.3–0.5[c,d]	2[b]	2[b]	2[c]
Net PAH secretion rate (pmol mm^{-1} min^{-1})	0.32[e]	1.44[e]	1.04*[e]	0.32[e]	1.44[e]	0.25[e]

[a] HAMBURGER et al., 1976
[b] JACOBSON and KOKKO, 1976
[c] KAWAMURA et al., 1975
[d] SCHAFER et al., 1974
[e] WOODHALL et al., 1976
* Approximated from data given in the reference indicated. The J_v and V_e data are for symmetrical HCO_3-KR solutions.

al. (1969) found that the maximal rate of PAH secretion was greater in proximal straight tubules than in proximal convolutions (Table 4); GRANTHAM et al. (1973) found that aryl acid secretion occurred in proximal straight, but not proximal convoluted, tubules; and CHONKO et al. (1975) demonstrated net uric acid secretion in proximal straight tubules, but not proximal convolutions. But it now appears that this issue is more complex. WOODHALL et al. (1976) have presented preliminary data from the rabbit proximal tubule which they interpret to demonstrate three morphologically distinct regions along the proximal tubule, referred to as S_1, S_2 and S_3: S_1 includes the early proximal convoluted tubule of both superficial and juxtamedullary nephrons; S_2 includes both convoluted and straight portions of superficial nephrons, but is confined entirely to the convoluted portion of juxtamedullary nephrons; and S_3 comprises the entire proximal straight portion of juxtamedullary nephrons but only the lower third of superficial proximal straight tubules. As seen in Table 6, the rates of PAH section measured by *in-vitro* perfusion varied widely, depending on the segment perfused. It is evident from these considerations that the important issue of proximal tubular heterogeneity warrants further study.

III. The Loop of Henle

The technique of perfusing isolated tubule segments has allowed considerable advances in our knowledge of the operation of urinary concentrating mechanisms, *i. e.*, the counter-current multiplication events in the renal medulla. Until the advent of this technique, knowledge about these processes was inferred almost entirely from micropuncture analyses of fluid samples from accessible portions of the collecting duct, distal tubules and proximal tubules, and short papillary sections of thin ascending and descending limbs of loops of Henle from juxtamedullary nephrons. Using *in-vitro* perfusion techniques, all regions of the loop of Henle and collecting system have now been studied.

The basic passive permeability properties of the component portions of the loop of Henle are given in Table 7. KOKKO (1970) demonstrated that the thin descending limb of the loop of Henle has an osmotic water permeability at least as high as that of the proximal tubule (Table 3), and higher than that of any other segment in the rabbit kidney. However, the solute permeability of this segment is low: the reflection coefficients for NaCl and urea are in excess of 0.95, and the permeability coefficients for Na^+ and urea are 0.19 and 0.15 μm s^{-1}, respectively (ROCHA and KOKKO, 1973b; KOKKO, 1972b). IMAI and KOKKO (1974b) have demonstrated that the ascending thin limb of the loop has a very low water permeability, but a very high NaCl permeability and a moderate urea permeability (Table 7); and these workers were unable to demonstrate any active ion transport, spontaneous transepithelial voltage, or net volume absorption in the thin ascending limb.

Tables 7 and 8 summarize the transport characteristics of the thick ascending limb. Both BURG and GREEN (1972b and 1973a) and ROCHA and KOKKO (1973a) reported that the thick ascending limb exhibited a spontaneous trans-

Table 7. Passive permeability characteristics of component segments of the loop of Henle

	Thin Descending Limb	Thin Ascending Limb	Thick Ascending Limb
Osmotic water permeability (P_f, μm s^{-1})	2414[d]	~0[c]	1–11[a,b,f]
Diffusional water permeability (P_{Dw}, μm s^{-1})	–	5.27[c]	–
P_{Na} (μm s^{-1})	0.19[g]	2.55[c]	0.28–0.63[b,f]
P_K (μm s^{-1})	0.25[g]	–	–
P_{Cl} (μm s^{-1})	–	11.70[c]	0.11–0.14[b,f]
P_{urea} (μm s^{-1})	0.15[e]	0.70[c]	0.09[h]
σ_{NaCl}	0.96[d]	–	–
σ_{urea}	0.95[e]	–	–
Electrical resistance (Ω cm^2)	–	–	21–25[b]
Length constant (cm)	–	–	0.01[b]

[a] Burg 1972b
[b] Burg and Green, 1973a
[c] Imai and Kokko, 1974b
[d] Kokko, 1970
[e] Kokko, 1972b
[f] Rocha and Kokko, 1973a
[g] Rocha and Kokko, 1973b
[h] Rocha and Kokko, 1974

Table 8. Active transport characteristics of the thick ascending limb of the loop of Henle*

Spontaneous V_e (mV, lumen to bath)	+6.7 to +7.0
Net volume absorption	~0
Maximum collected/perfused osmolality ratio at slow flow	0.68–0.70
Sodium fluxes** (J_{Na}, mEq cm^{-2} s^{-1})	
lumen to bath	5.9–15.6
bath to lumen	3.8–9.3
net	2.1–6.3

* Data taken from Rocha and Kokko (1973a) and Burg and Green (1973a) at 37° C with symmetrical perfusing and bathing solutions.

** Recalculated from data given in units normalized per apparent tubule inner surface area assuming a mean tubule inside diameter of 20 μm (see Burg and Green, 1973a), or $6.29 \cdot 10^{-3}$ cm^2 cm^{-1} length.

epithelial voltage of +6 to +7 mV (lumen-positive) at 37° C, which they demonstrated was due to active lumen-to-bath Cl^- transport. Since the water permeability of the thick ascending limb is also virtually zero, NaCl absorption driven by active Cl^- transport results in dilution of tubular fluid. Further, it is probable that the thick ascending limb may be the site of action of several important diuretic agents. Mersalyl (Burg and Green, 1973b), ethacrynic acid (Burg and Green, 1973c) and furosemide (Burg et al., 1973) all inhibit active

Cl^- transport in the thick ascending limb of the loop of Henle, as does the cardiac glycoside ouabain, which is usually persumed to be a specific inhibitor of Na^+, K^+-ATPase (BURG and GREEN, 1973a).

KOKKO and RECTOR (1972) have integrated these data into a model which provides two spatially distinct sites for renal countercurrent multiplication mechanisms: an active step in the outer medulla and a passive step in the inner medulla. The salient features of this model can be qualitatively described in the following way. The first multiplication step depends on active Cl^- transport producing NaCl efflux from water-impermeable thick ascending limbs; thus fluid entering the distal tubule is both hypotonic and low in salt. During antidiuresis, ADH-enhanced water abstraction from urea-impermeant cortical and outer medullary collecting ducts results in accumulation of urea in fluid entering papillary collecting ducts. Since the latter are urea-permeable, passive urea transport down a chemical gradient from tubular fluid to medullary interstitium contributes to medullary hypertonicity, thereby providing a second, but in this case passive, multiplication step. Simultaneously, osmotic equilibration of papillary collecting duct fluid with the medullary interstitium results in the formation of hypertonic urine.

KOKKO and RECTOR (1972) have rationalized the progressive concentration and dilution of tubular fluid in, respectively, descending and ascending thin limbs, entirely in terms of passive flows. Consider, for example, a medulla whose osmolality ranges from 300 mOsm/kg at the cortico-medullary junction to 1400 mOsm kg^{-1} at the papillary tip. In keeping with tissue analyses (VALTIN, 1966), approximately half of the medullary hypertonicity is assigned to NaCl and the remainder to urea. Isotonic fluid containing 280 mOsm/kg NaCl entering the highly water-permeable but urea and Na^+-impermeable descending thin limb is concentrated almost entirely by water abstraction, so that fluid entering the ascending thin limb has a higher NaCl concentration and a lower urea concentration than the medullary interstitium. These passive driving forces between lumen and interstitium, coupled with the fact that the thin ascending limb is more permeable to NaCl than to urea, poise the system for fluid dilution. As fluid moves up the water-impermeable ascending thin limb, passive NaCl efflux from lumen to interstitium exceeds passive urea influx from interstitium to tubular fluid, resulting in both progressive dilution of ascending limb fluid and, concomitantly, urea recycling from papillary collecting ducts through the interstitium to ascending thin limbs. Finally, the process begins again by active NaCl transport from the thick ascending limb.

There is theoretical support for this view. STEPHENSON (1972) considers *vasa recta* loops to be sufficiently permeable to water and solutes for them to be treated as a single tube, or "central core," which is open at the cortical end, closed at the papillary end, and aligned in parallel with the loop of Henle and the collecting duct. His calculations indicate that, for nephron segments having relative permeabilities for urea, salt, and water similar to those listed in Tables 7 and 8, the combination of passive salt efflux from ascending thin limbs and passive urea efflux from medullary collecting ducts could raise the osmolality of the central core, thereby providing the driving force for osmotic water abstraction from descending thin limbs and collecting ducts. Stephenson's equations, like the model of KOKKO and RECTOR (1972), link active salt transport in the outer medulla and cortex and urea impermeability in cortical and outer medullary collecting ducts into a mechanism for urea recycling in the inner medulla.

Not all workers agree with such models for counter-current multiplication, and there are two major reasons for disagreement. First, KOKKO and RECTOR (1972) have proposed that 96 percent of osmotic equilibration of descending thin limb fluid is due to water abstraction and only 4 percent to urea entry; thus, NaCl contributes more than 90 percent to the osmolality of fluid reaching the papillary bend of Henle's loop. In contrast, PENNELL et al. (1974) have found that, in the rat, urea addition contributes 40 percent to the osmolality increase in ascending thin limb fluid. These workers therefore argue that, if urea in tubular fluid at the tip of Henle's loop equilibrates with the papillary interstitium, there exists little or no driving force for passive urea recycling from collecting ducts and interstitium to thin ascending limbs. Second, the hypothesis of KOKKO and RECTOR (1972) requires passive NaCl transport down a concentration gradient from ascending thin limb to medullary interstitium. But MARSH and AZEN (1975) have found the Na^+ concentration in the thin ascending limb of the hamster to be either the same as or less than that in the ascending *vasa recta;* assuming that the latter was in equilibrium with interstitial fluid, these workers conclude that active Na^+ transport contributes at least in part to inner medullary hypertonicity. In short, the counter-current multiplier model of KOKKO and RECTOR (1972), which derives largely from the *in-vitro* characteristics of isolated rabbit nephron segments, is not entirely supported by *in-vivo* micropuncture results in the rat (PENNELL et al., 1974) and hamster (MARSH and AZEN, 1975), and requires further evaluation.

IV. The Distal Convoluted Tubule and Collecting Duct System

Morphologically, the distal nephron of the rabbit has been described as consisting of four regions (BURG and STONER, 1974): 1) The diluting segment, including medullary and cortical segments of the thick ascending limb of Henle's loop up to the *macula densa.* The transport functions of this segment have been described above in the context of the loop of Henle. 2) The distal convoluted tubule, considered to extend from the *macula densa* to the junction of two or more distal convoluted tubules to form a cortical collecting tubule. The distal convoluted tubule in the rabbit is quite short. 3) The cortical collecting tubule, formed from the confluence of two or more distal convoluted tubules; the collecting tubule then runs straight and unbranched through the cortex and most of the outer medulla. For convenience, this region of the collecting duct can be further divided into cortical and outer medullary segments; as will be described in this section, cortical and outer medullary collecting ducts have rather similar transport characteristics, but these differ appreciably from those of papillary, or inner medullary collecting ducts. 4) The final segment is the inner medullary or papillary collecting duct which has multiple junctions of tubules (BURG and STONER, 1974). Table 9 summarizes the major transport characteristics of isolated segments of these different regions of the rabbit nephron.

Table 9. Transport properties of distal nephron segments

	Distal convoluted tubule	Cortical collecting tubule	Outer medullary collecting duct	Papillary collecting duct
Diffusional water permeability ($P_{D_w}^{l\to b}$, μm s^{-1}, 25° C)				
−ADH	~0[f]	3.8–4.7[e,i]	–	4.0[h]
+ADH	~0[f]	9.7–14.2[e,i]	–	5.7[h]
Osmotic water permeability ($P_f^{l\to b}$, μm s^{-1}, 25° C)				
−ADH	–	20[l]	–	~0[h]
+ADH	–	186	–	68[h]
Osmotic water permeability 6$P_f^{b\to l}$, μm s^{-1}, 25° C)				
−ADH	–	88[l]	–	–
Osmotic water permeability ($P_f^{l\to b}$, μm s^{-1}, 37° C)				
−ADH	–	56[a]	–	–
+ADH	–	440	–	–
Nonelectrolyte permeability coefficients (P_{D_i}, 25° C)				
urea (l→b, −ADH)	–	0.009–0.030[c,j]	0.031[h]	0.22[h]
urea (l→b, +ADH)	–	0.009–0.020[c,j]	0.034[h]	0.24[h]
urea (l→b, with net b→l water flow produced by luminal urea, −ADH)	–	0.216	–	–
urea (b→l, with net b→l water flow produced by luminal urea, −ADH)	–	0.033[l]	–	–
acetamide (l→b, −ADH)	–	0.034[c]	–	–
acetamide (l→b, +ADH)	–	0.036[c]	–	–
thiourea (l→b, −ADH)	–	0.024[c]	–	–
thiourea (l→b, +ADH)	–	0.026[c]	–	–
Transepithelial voltage (V_e, mV, lumen to bath)				
37° C	−40[f]	+ 4 to − 35[f,m]	–	–
25° C	–	−10 to − 45[d,g]	–	–

Table 9 (continued)

	Distal convoluted tubule	Cortical collecting tubule	Outer medullary collecting duct	Papillary collecting duct
Net ion fluxes (J_i^{net}, nEq cm^{-2} s^{-1})				
Na^+, 25° C	–	0.41[d]	–	–
Na^+, 37° C	–	1.23[b]	–	–
K^+, 37° C	–	−0.30[b]	–	–
Cl^-, 37° C	–	0.93[b]	–	–
Ionic permeability coefficients (P_{D_i}, μm s^{-1})				
Na^+, 25° C (from b→l flux)	–	0.006[d]	–	–
Na^+, 37° C (from b→l flux)	–	0.008[b]	–	–
K^+, 37° C (from l→b flux)	–	0.100[b]	–	–
Cl^-, 37° C (from b→l flux)	–	0.470[b]	–	–
Electrical resistance (Ω cm^2), 25° C length constant (cm)	–	867[g]	–	–
Reflection coefficients (σ_i, +ADH)				
Urea	–	1.0[j]	1.0[j]	–
NaCl	–	1.0[j]	–	–
Sucrose	–	1.0[j]	–	–

[a] Al-Zahid et al., 1976
[b] Burg and Stoner 1974
[c] Burg et al., 1970
[d] Frindt and Burg, 1972
[e] Grantham and Burg, 1966
[f] Gross et al., 1975
[g] Helman et al., 1971
[h] Rocha an Kokko, 1974
[i] Schafer and Andreoli, 1972a
[j] Schafer and Andreoli, 1972b
[k] Schafer et al., 1974a
[l] Schafer et al., 1974b
[m] Stoner et al., 1974

1. Distal Convoluted Tubule

Gross et al. (1975) found that distal convoluted tubule, when perfused and bathed in media similar to, respectively, an ultrafiltrate of rabbit serum and serum, developed a spontaneous transepithelial voltage of -40.4 mV, lumen-negative. Significantly, the spontaneous transepithelial voltage in this nephron segment seemed to be independent of mineralocortical activity (at least as judged by the lack of difference in the spontaneous voltage between tubules obtained from rabbits maintained on a regular diet without exogenous mineralocorticoid, and those obtained from animals maintained on a high-K^+ diet and supplemented with parenteral desosycorticortisone acetate), and was unaffected by the addition of vasopressin to the bathing media. But the addition of either ouabain to the bath, or amiloride to the luminal solution, inhibited V_e. Finally, these workers showed that isolated distal convoluted tubules had a negligibly low diffusional water permeability coefficient, which was unaffected by ADH.

2. Cortical Collecting Tubule

As shown in Table 9, the cortical collecting tubule has a high transepithelial resistance, $\sim 870\ \Omega\ cm^2$, especially in comparison to either proximal tubules or thick ascending limbs. Even at 25° C, the cortical collecting tubule exhibits several characteristics of an actively transporting epithelium. When the tubule is perfused with an isotonic Krebs-Ringer phosphate buffer (pH 7.4) and bathed in a Krebs-Ringer bicarbonate buffer (pH 7.4) gassed with 5% CO_2 – 95% O_2, the transepithelial voltage varies in the range $+10$ to -70 mV (Frindt and Burg, 1972). Frindt and Burg (1972) found that cortical collecting tubules isolated from rabbits maintained on a high-K^+, low-Na^+ diet exhibited initial V_e's in the range -20 to -70 mV and generally reached relatively stable V_e's of -40 to -50 mV after 150 minutes' perfusion. On the other hand, cortical collecting tubules isolated from rabbits maintained on a low-K^+, high-Na^+ diet had initial V_e's of 0 to $+10$ mV, which became stable at -2 to -10 mV after about 200 minutes of perfusion (Frindt and Burg, 1972). It seems reasonable to assume that the variation of spontaneous transepithelial voltage with dietary K^+ and Na^+ depends on variations in the *in-vivo* rates of endogenous mineralocorticoid secretion, a conclusion which is also supported by the results of Gross et al. (1975), who found that cortical collecting tubules obtained from rabbits maintained on a regular diet exhibited spontaneous voltages of $+3.7$ mV, lumen-positive, while tubules obtained from rabbits on a high-K^+ diet supplemented with exogenous mineralocorticoid had spontaneous transepithelial voltages of -33.8 mV, lumen-negative.

Grantham et al. (1970) have investigated the mechanisms of cation transport in the cortical collecting tubule at 25° C. When the perfusing and bathing solutions both contained 150 mM Na^+ and 5 mM K^+, the Na^+ concentration decreased along the length of the tubule, and the K^+ concentration increased by an approximately equal amount. The transepithelial voltage in these experiments was between −20 and −70 mV, and at the lowest perfusion rates used, the luminal Na^+ concentration fell to 60 mM and the K^+ concentration rose to over 100–125 mM. The final lumen-to-bath distribution of both Na^+ and K^+ developed against the electrochemical gradients for these ions; therefore the results were interpreted to indicate active absorption of Na^+ and active secretion of K^+, even at 25° C (GRANTHAM et al., 1970). The active transport of both cations was inhibited by ouabain, removal of Na^+ from the perfusate, or the removal of K^+ from the bathing solution (GRANTHAM et al., 1970). FRINDT and BURG (1972) measured the lumen-to-bath Na^+ flux at 25° C, and found it to be in the order of 0.4 nEq cm^{-2} s^{-1}.

Similar studies have also been conducted in the cortical collecting tubule at 37° C by STONER et al. (1974). Although the results at 37° C were similar to those at 25° C, Table 9 shows that the net Na^+ fluxes were significantly greater at 37° C than at 25° C. The rate of Na^+ transport was approximately one-half that reported for the ascending thick limb (BURG and GREEN, 1973a), and one-fourth that reported for proximal convoluted tubules (Table 4). STONER et al. (1974) also demonstrated that at 37° C, 10^{-4} amiloride added to the perfusate, but not to the bathing solution, inhibited Na^+ and K^+ transport and changed V_e from lumen-negative to lumen-positive. These authors attributed the positive V_e to urinary acidification. In keeping with this view, when acetazolamide was added to the bathing solution or when CO_2 was eliminated (both in the presence of luminal amiloride), V_e decreased toward zero. Acetazolamide also increased the magnitude of the lumen-negative potential in the absence of amiloride, again suggesting the presence of an electrogenic H^+ secretory mechanism.

Finally, it is well recognized that, in amphibian epithelia such as toad urinary bladder or frog skin, vasopressin has dramatic effects on the spontaneous transepithelial voltage and active Na^+ transport. In the mammalian cortical collecting tubule, these effects are less evident: FRINDT and BURG (1972) found that the addition of vasopressin to the bath produced a transient decrease in the magnitude of V_e and an increase in net lumen-to-bath Na^+ transport; however, both effects lasted less than 30 minutes and were followed by a return to control levels or below. Similarly, GROSS et al. (1975) found that the hormone had minimal effect, if any, on the spontaneous transepithelial voltage in this tubule segment.

3. Water and Nonelectrolyte Permeation in the Cortical Collecting Duct: the Mechanism of ADH Action

Table 9 summarizes the major effects of antidiuretic hormone on water and nonelectrolyte permeation in cortical collecting tubules (it should also be noted from Table 9 that in the instances studied to date, ADH produces virtually the same effect on transport processes in outer medullary collecting ducts): there is a ten-fold increase in P_f and approximately a four-fold increase in P_{D_w}, so that P_f/P_{D_w}, the ratio of osmotic to diffusional water permetability coefficients, rises from approximately four, in the absence of hormone, to approximately 13, in the presence of hormone; the hormone has virtually no effect on the transepithelial permeation rates for small hydrophilic amides, so that in the presence of ADH, P_{D_i} for such solutes is vanishingly low with respect to either P_{D_w} or P_f; and the ADH-dependent value of σ_i is unity for NaCl, urea, and sucrose.

GRANTHAM and BURG (1966), GANOTE et al. (1968), and SCHAFER and ANDREOLI (1972a) showed that, when cortical collecting tubules were perfused and bathed with, respectively, hypotonic and isotonic Krebs-Ringer buffers (in which NaCl accounts for the majority of solution osmolality) in the absence of ADH, the tubular epithelium gradually became flattened, and the lateral intercellular spaces were progressively less visible. When ADH was added to the bath, the cells swelled, their luminal borders bulged, and the lateral intercellular spaces became distinctly visible. In other words, luminal surfaces (*i. e.,* apical plasma membranes and/or junctional complexes) constitute the ADH-sensitive interface for water transport, and, in the absence of ADH, the rate-limiting site for water permeation. Further, both GRANTHAM et al. (1969) and SCHAFER et al. (1974b) provided evidence consistent with the view that the primary route for transepithelial lumen-to-bath osmosis was cellular, rather than paracellular; and SCHAFER et al. (1974a) demonstrated that there were negligible osmotic flow transients in cortical collecting tubules, *i. e.,* that solute polarization within intracellular spaces during osmosis was relatively minor and resulted in a less than 10 percent error in underestimating P_f.

It is likely that luminal surfaces are also the rate-limiting site for the transport of hydrophilic nonelectrolytes across cortical collecting tubules, both in the presence and absence of ADH. SCHAFER and ANDREOLI (1972b) found that when cortical collecting tubules were perfused and bathed with isosmotic Krebs-Ringer solutions, replacement of a portion of NaCl in the perfusing solutions with urea had no effect on tubular morphology, but when urea replaced a portion of the NaCl in the bathing solutions, the tubular cells swelled rapidly. It can be concluded from these observations that even in the presence of ADH, luminal surfaces of cortical collecting tubules are rather impermeable to urea, especially in comparison to basolateral membranes.

In principle, P_f will equal P_{D_w} when net water transport across a given membrane occurs exclusively by a solubility-diffusion process. But in the vast majority of natural and synthetic membranes, P_f exceeds P_{D_w}, i. e., the P_f/P_{D_w} ratio exceeds unity. Two classes of explanations have generally been proposed to account for such disparities. First, KOEFOED-JOHNSEN and USSING (1953) and PAPPENHEIMER (1953) reasoned that membranes might contain aqueous pores sufficiently large to permit laminar net volume flow. In aqueous cylinders, laminar flow is proportional to r^4, where r is the tube radius, while diffusion is proportional to r^2. Accordingly, in a porous membrane, P_f will exceed P_{D_w}. By combining Poiseuille's law for laminar flow with Fick's first law of diffusion, the relationship between P_f and P_{D_w} for a porous membrane may be expressed as:

$$\frac{P_f}{P_{D_w}} = 1 + \frac{RT}{8\eta D_w^o \bar{V}_w} r^2, \tag{38}$$

where pore radius r is given in Angstrom units (10^{-8} cm), η is aqueous viscosity (poise), and D_w^o is the free diffusion coefficient of water in water ($cm^2\ s^{-1}$). At

25° C, the term $RT/8\eta\ D_w^o\ \bar{V}_w$ for relatively dilute aqueous solutions can be expressed as:

$$\frac{RT}{8\eta D_w^o \bar{V}_w} \simeq 0.08$$

Unstirred layers adjacent to a membrane may also account for P_f/P_{D_w} ratios in excess of unity: during net fluid transport, volume flow to the membrane interface is not impeded by unstirred layers; but at zero volume flow, tracer water flux is hindered by the total diffusion resistance of the membrane plus unstirred layers. Since $1/P_{D_w}$ is the total resistance to tracer water diffusion, DAINTY (1963) proposed that, for the case of a membrane in series with unstirred layers, P_f and P_{D_w} might be related by an equation for series resistances:

$$\frac{1}{P_{D_w}} = \frac{1}{P_f} + \frac{\alpha}{D_w^o}, \tag{39}$$

where α is the effective thickness (cm) of the unstirred layers and D_w^o is the free diffusion coefficient of water in water. Equation (39) indicates that, for entirely diffusional net water transport across a membrane, P_f will exceed P_{D_w} significantly when the α/D_w^o term becomes large with respect to $1/P_f$.

The results in Table 9 indicate that ADH increases P_f, P_{D_w} and the P_f/P_{D_w} ratio in cortical collecting tubules, where luminal plasma membranes are the rate-limiting site for water permeation (cf. above). If the P_f/P_{D_w} discrepancy in these tubules were due to laminar water flow through membrane pores, the ADH-independent and ADH-dependent P_f/P_{D_w} ratios predict, in terms of equation 39, that the hormone increases the radii of luminal membrane pores from 6.4 Å to 12.1 Å. But the ADH-dependent reflection coefficient for urea is unity, and $P_{D_{urea}}$ is unaffected by the hormone (Table 9). Since the effective hydrodynamic radius of urea is approximately 1.8 Å, one expects on the basis of molecular sieving theory that for 12.1 Å pores, the ADH-dependent reflection coefficient for urea would be less than unity and the hormone would have increased $P_{D_{urea}}$ appreciably. Thus, pore theory *per se* is not adequate to account for the effects of ADH on water transport across luminal membranes of cortical collecting tubules.

However, unstirred layer effects, due either to bulk phase unstirred layers or to cellular constraints to diffusion, might account for the disparity between P_f and P_{D_w}. Taking D_w^o to be $2.36 \cdot 10^{-5}$ cm² s⁻¹ (WANG et al., 1953), equation 39 predicts an α value of $153 \cdot 10^{-5}$ cm to rationalize the ADH-dependent disparity between P_f and P_{D_w}. *i. e.,* to argue that the route of ADH-dependent osmotic water flow across luminal membranes is entirely diffusional in nature, and that P_{D_w} is less that P_f because of diffusion resistances in series with luminal membranes. The internal diameter of cortical collecting tubules is only $22\text{–}25 \cdot 10^{-4}$ cm (Table 1); and tenfold increments in the viscosity of perfusing and bathing solutions, a maneuver which predictably increases the diffusion resistance of bulk phase unstirred layers in series with synthetic bilayer membranes (ANDREOLI and TROUTMAN, 1971), have no effect on the ADH-dependent values of P_{D_w} in cortical collecting tubules (SCHAFER and ANDREOLI,

1972a). Thus it seems improbable that bulk phase unstirred layers contribute to the disparity between P_f and P_{D_w} in these tubules.

Alternatively, the epithelial cell layer itself, exclusive of luminal membranes, might act as a diffusion resistance sufficient to account for the ADH-dependent P_f/P_{D_w} discrepancy. But the thickness of cortical collecting tubule epithelium is $6–7 \cdot 10^{-4}$ cm s^{-1}, and the value of α required from equation 39 to account for the difference between the ADH-dependent values of P_f and P_{D_w}, is, as indicated above, approximately $153 \cdot 10^{-4}$ cm. In other words, for cellular diffusion constraints to account for the ADH-dependent P_f/P_{D_w} ratio, the resistance to water diffusion in the epithelial cell layer must be 21–25 times greater than in free solution.

To test this possibility, Schafer and Andreoli (1972a) measured the permeability coefficients of highly lipophilic solutes, such as butanol, 5-hydroxyindole, and pyridine, in cortical collecting tubules. Each of these solutes has an oil:water partition coefficient (β_{oil}) in excess of 0.5, and hence might be expected to permeate readily the hydrophobic regions of luminal membranes; and solutes such as butanol have a zero reflection coefficient in cortical collecting tubules (Schafer and Andreoli, 1972a). On the basis of these considerations, Schafer and Andreoli (1972a) reasoned that $1/P_{D_i}$ (R_{D_i}, s cm^{-1}) for these lipophilic solutes would provide an index to the diffusion resistance of the epithelial cell layer of cortical collecting tubules, exclusive of luminal membranes.

Table 10 summarizes the P_{D_i} values for butanol, 5-hydroxyindole, and pyridine in cortical collecting tubules, and also lists both the observed R_{D_i} values (*i. e.*, $1/P_{D_i}$) and the values of R_{D_i} predicted for the diffusion of these solutes through a $6 \cdot 10^{-4}$ cm layer of water, *i. e.*, a layer of water having a thickness equivalent to the cell layer of cortical collecting tubules. ADH had no effect on the P_{D_i} values for these solutes; and the observed R_{D_i} values were 12–26 times greater than the R_{D_i} values predicted for diffusion in free solution. Thus, to the extent that the resistance of the epithelium to the diffusion of these solutes provides an index ot the constraints imposed by the cell layer, exclusive of luminal plasma membranes, to transepithelial water diffusion, these data imply that the mode of water transport across luminal membranes during osmosis is

Table 10. The diffusional permeability coefficients of lipophilic solutes in cortical collecting tubules

Solute	ADH	P_{D_i}	R_{D_i} Observed	R_{D_i} Predicted
	μU ml^{-1}	cm $s^{-1} \cdot 10^4$	s cm^{-1}	
pyridine	0	12.6	794	68.2
pyridine	250	13.2	—	
butanol	0	12.2	820	67.8
butanol	250	12.4	—	
5-hydroxyindole	0	4.03	2480	92.8
5-hydroxyindole	250	3.94	—	—

Adapted from Schafer and Andreoli, 1972a.

diffusional. According to this view, P_f rather than P_{D_w} provides a more accurate index to the rate of water diffusion across luminal membranes; and ADH enhances the rate of water diffusion across luminal membranes.

Two other factors are noteworthy in this context. First, SCHAFER et al. (1974a) provided evidence that the cellular constraints to water diffusion in cortical collecting tubules were referable to a 25-fold reduction in the fractional area available for water transport in the epithelial cell layer. Second, the ADH-dependent value of P_f in toad urinary bladder is approximately 200–240 · 10^{-4} cm s^{-1} (HAYS and LEAF, 1962), and HAYS and FRANKI (1970) have estimated the ADH-dependent value of P_{D_w} in the epithelial cell layer of that tissue, exclusive of supporting stroma, to be 11 · 10^{-4} cm s^{-1}. There is close accord between these data and the comparable values in cortical collecting tubules (Table 9). Moreover, PARISI and PICCINI (1973) have recently provided direct evidence for significant hindrances to water diffusion in the epithelial cell layer of the toad bladder. Thus it may be that, both in cortical collecting tubules and toad urinary bladder, osmotic water flow across apical membranes is diffusional in nature; and in both epithelia, comparable molecular events may account for ADH-dependent increments in water transport.

Recently, PIETRAS and WRIGHT (1975) and LEVINE et al. (1976) showed that ADH not only increased water permeation, but also increased the permeabilities of several moderately lipophilic solutes in the toad bladder. AL-ZAHID et al. (1976) examined the same issue in the isolated cortical collecting tubule: the permeability coefficients of three moderately lipophilic solutes, butyramide, isobutyramide and antipyrine (all labeled with ^{14}C), were measured at 25° C. Table 11 shows that ADH produced, at best, a two-fold increase in P_{D_i}

Table 11. The effect of ADH on the diffusional permeability coefficients for moderately lipophilic solutes in cortical collecting tubules

Solute	β_{oil}	bath[ADH]	P_{D_i}	P
		μU ml^{-1}	cm s^{-1} · 10^4	
Isobutyramide	0.0095	0	0.12 ± 0.02	–
Isobutyramide	0.0095	250	0.26 ± 0.05	–
	Mean paired difference		0.14 ± 0.04	<0.01
n-butyramide	0.0095	0	0.14 ± 0.04	–
n-butyramide	0.0095	250	0.28 ± 0.02	–
	Mean paired difference		0.14 ± 0.02	<0.01
antipyrine	0.032	0	0.25 ± 0.07	
antipyrine	0.032	250	0.40 ± 0.06	–
	Mean paired difference		0.16 ± 0.02	<0.01

The data are expressed as mean values ± SEM. Adapted from AL-ZAHID et al., 1976.

for these solutes, *i. e.,* appreciably less than the effect of the hormone on P_f (Table 11).

In order to arrive at some tentative conclusions about the molecular nature of water and solute permeation pathways in cortical collecting tubules, AL-ZAHID et al. (1976) analyzed water and solute permeation across cortical collecting tubules in terms of comparable processes in synthetic lipid bilayer membranes. It seems clear from the current evidence (FINKELSTEIN and CASS, 1967; ANDREOLI and TROUTMAN, 1971) that osmotic water flow across synthetic bilayer membranes occurs by way of a solubility-diffusion process; in other words, P_f represents the true permeability coefficient for water diffusion across synthetic bilayer membranes. Thus, given the assumption that osmotic water flow across luminal membranes of the collecting tubule (cf. above) is diffusional in nature, it is relevant to evaluate water and nonelectrolyte discrimination in the latter with respect to synthetic bilayer membranes. Such a comparison is

Table 12. A comparison of P_f and P_{D_i} in cortical collecting tubules and synthetic lipid bilayer membranes

System	Solute	Solute β_{oil}	P_{D_i}	P_f	P_f/P_{D_i}	References
			($cm\ s^{-1} \cdot 10^4$)			
Cortical collecting tubules						
−ADH	urea	0.00014	0.03	20	666	a
−ADH	acetamide	0.0008	0.034	20	588	a
−ADH	butyramide	0.01	0.14	20	142	b
+ADH	urea	0.00014	0.02	197	9850	a
+ADH	acetamide	0.0008	0.036	197	5472	a
+ADH	butyramide	0.01	0.32	197	615	b
Lipid bilayer membranes						
Sheep red cell lipid/decane (planar)	acetamide	0.0008	0.83	22.9	22.7	c
lecithin/decane (planar)	urea	0.00014	0.036	19	527	d
	acetamide	0.0008	1.43	19	13	d
	thiourea	0.0012	0.046	19	413	d
lecithin/chloroform-decane (spherical)	acetamide	0.0008	0.24	13	54	e
	valeramide	0.023	1.83	13	7.1	e

The solute oil:water partition coefficients are from COLLANDER (1949) and COLLANDER and BÄRLUND (1933). The values of P_f and P_{D_i} are for 22–25° C. Adapted from AL-ZAHID et al. (1976).

[a] BURG et al., 1970 GRANTHAM and BURG, 1966; SCHAFER and ANDREOLI, 1972a and b; SCHAFER et al., 1974a and b

[b] AL-ZAHID et al., 1976

[c] ANDREOLI et al., 1969

[d] GALLUCI et al., 1971; HANAI and HAYDON, 1966; LIPPE, 1969; VREEMAN, 1966

[e] POZNANSKY et al., 1976

presented in Table 12. It is evident from Table 12 that the ADH-independent P_f in collecting tubules is the same as in bilayer membranes, while the ADH-dependent P_f is approximately ten-fold greater than in bilayer membranes. Moreover, for the solutes listed in Table 12, the ADH-dependent P_f/P_{D_i} ratio in cortical collecting tubules is at least ten times that in bilayer membranes.

In other words, using synthetic bilayer membranes as a frame of reference, it is difficult to envision ADH-mediated effects on the hydrophobic regions of luminal membranes which permit, simultaneously, at least ten-fold higher rates of water diffusion but appreciably lower permeation rates for nonelectrolytes having β_{oil} values $\geqq 0.0008$, than in synthetic bilayers. Rather, AL-ZAHID et al. (1976) have proposed that ADH-mediated water and nonelectrolyte diffusion across luminal membranes may involve parallel pathways: one for water and the other for moderately lipophilic solutes. The water diffusion pathway, although appreciably more permeable to water than synthetic bilayer membranes, virtually excludes small, relatively hydrophilic solutes such as urea, thiourea and acetamide. The regions for nonelectrolyte diffusion are less permeable than synthetic bilayer membranes to solutes having β_{oil} values $\geqq 0.0008$.

Additional support for this view may be derived from a comparison of the activation energies for water and nonelectrolyte diffusion through synthetic lipid bilayer membranes and those in cortical collecting tubules. In bilayer membranes, the activation energy for water diffusion (termed E_A^w) can be expressed as the sum of two terms: the energy required for a water molecule to break the four hydrogen bonds formed with neighboring water molecules (termed $E_A^{H_w}$); and a nonhydrogen bond term (termed $E_A^{D_w}$), *i. e.*, the energy for water diffusion within the membrane (PRICE and THOMPSON, 1969; COHEN, 1975). It is also probable that the activation energy for nonelectrolyte diffusion (termed E_A^s) across a bilayer membrane can be rationalized in terms of the same process, with $E_A^{H_s}$ and $E_A^{D_s}$ representing the components of E_A^s due, respectively, to rupture of hydrogen bonds in solution and diffusion within the membrane. COHEN (1975) computed 1.8 kcal mol^{-1} as the average activation energy per hydrogen bond in aqueous solution, either for water or for solutes. And in liposomes, COHEN (1975) found that $E_A^{D_w}$ and $E_A^{D_s}$ varied depending either on liposome composition or, for liposomes of a particular composition, on whether the liposomes were above or below their transition temperature; but importantly, for liposomes of a given composition at the same temperature, $E_A^{D_s}$ (for all solutes tested) and $E_A^{D_w}$ were the same.

In cortical collecting tubules, AL-ZAHID et al. (1976) found that the ADH-dependent activation energies for water and moderately lipophilic solute (*e. g.*, Table 11) permeation were, respectively, 8.9 and 16.6–19.6 kcal mol^{-1}; and that these energies were unaffected by ADH. Taking four, three and two as the number of aqueous hydrogen bonds for, respectively, water, isobutyramide or butyramide, and antipyrine, these E_A^w and E_A^s data yield, following the analysis applied to synthetic bilayer membranes, apparent values of 0.9 and 10.6–15.6 kcal mol^{-1} for, respectively, $E_A^{D_w}$ and $E_A^{D_s}$ for luminal membranes of cortical collecting tubules.

Two alternate explanations may account for this disparity: either $E_A^{D_w}$ and $E_A^{D_s}$ differ within the hydrophobic regions of luminal membranes; or, water traverses a different pathway than moderately lipophilic species in these membranes. Thus, to the extent that activation energy measurements in liposomes provide a frame of reference for assessing transport processes in the hydrophobic regions of luminal membranes, we may infer that, in the latter, water and moderately lipophilic solutes traverse parallel permeation pathways.

Since ADH increased the permeation rates for moderately lipophilic solutes (Table 11) without affecting the activation energy for solute permeation (cf. above), it seems improbable that ADH produced a generalized increase in the fluidity of the hydrocarbon matrix in luminal membranes, since, in comparison with synthetic bilayers, fluidity increases should have been accompanied by a fall in E_A^s due to a reduction in $E_A^{D_s}$. Alternatively, AL-ZAHID et al. (1976) have suggested that nonelectrolyte diffusion across cortical collecting tubules may proceed *via* discrete hydrophobic pathways in luminal membranes, and that ADH increases the number of such pathways; a corollary of this hypothesis is that, in luminal membranes, the hydrophobic region itself may be a mosaic structure containing hydrophobic pathways for butyramide, isobutyramide, and antipyrine permeation, whose number may be increased by ADH, and more ordered hydrophobic regions unaffected by ADH, which have lower permeation rates and appreciably higher activation energies for transport of moderately lipophilic solutes.

The pathway for water diffusion in luminal membranes differs in at least two major respects from water diffusion in synthetic bilayers: the ADH-dependent P_f in luminal membranes exceeds reported P_f values in synthetic bilayers (Table 12); and, in luminal membranes with or without ADH, the apparent $E_A^{D_s}$ for nonelectrolytes exceeds the apparent $E_A^{D_w}$ (cf. above), while in liposomes of a given composition, $E_A^{D_w}$ equals $E_A^{D_s}$ (COHEN, 1975). AL-ZAHID et al. (1976) have rationalized these data by suggesting that water might diffuse through luminal membranes via small aqueous channels. In that case, water molecules might not need to be fully dehydrated in order to diffuse through these channels, and $E_A^{H_w}$ might then be appreciably less than 8 kcal mol^{-1}. For example, in bilayer membranes, gramicidin A dimers form aqueous channels having radii of approximately 2 Å (URRY et al., 1971) and water/urea selectivity ratios in excess of 10^3 (FINKELSTEIN, 1974). Thus, a 2 Å radius channel in luminal membranes, similar to the gramicidin A dimer in bilayer membranes, might account for both the high ADH-dependent P_f/P_{D_i} ratios for urea, acetamide, and thiourea (Table 12), and the lower activation energy for water with respect to nonelectrolyte permeation (cf. above).

Such a channel, if present n luminal membrane, would need to have an appreciably lower Na^+ conductance than a gramicidin-A channel. FINKELSTEIN (1974) found that the water permeability of a gramicidin-A channel was $1.2 \cdot 10^{-14}$ cm^3 s^{-1}. Hence, an ADH-dependent P_f of $186 \cdot 10^{-4}$ cm s^{-1} (Table 9), requires $1.6 \cdot 10^{12}$ gramicidin-A channels per cm^2 of luminal membrane; since the Na^+ conductance of a gramicidin-A channel is $6 \cdot 10^{-12}$ Ω^{-1} (FINKELSTEIN, 1974) this number of channels would give a luminal membrane resistance of 0.1 Ω cm^2. But the transepithelial resistance of cortical collecting tubules, with or without ADH, is in excess of 800 Ω cm^2 (Table 9), and HELMAN (1973) has found that luminal plasma membranes have appreciably higher electrical resistances than either junctional complexes or basolateral membranes. Accordingly, if ≃2 Å radius aqueous channels having the same water permeability as gramicidin-A channels constituted the route for ADH-mediated water diffusion in luminal plasma membranes, the Na^+ conductance of such a putative channel would need to be approximately 10^{-4} of the Na^+ conductance of a gramidicin-A channel.

In summary, on the basis of present experimental data, it may be argued (AL-ZAHID et al., 1976), that in cortical collecting tubules, and possibly other ADH-sensitive tissues, water and moderately lipophilic solutes traverse luminal plasma membranes by parallel diffusion pathways. The water permeation pathways are appreciably more permeable to water than are synthetic bilayer membranes, and virtually exclude small hydrophilic solutes such as urea; the pathways for diffusion of moderately lipophilic solutes occur in parallel with the water permeation pathway, and require a greater free energy change per molecule (*i. e.*, a higher activation energy) for permeation of such solutes than do the water pathways for water permeation. ADH may increase water and moderately lipophilic solute permeation through apical membranes by increasing the number (or frequency distribution) of both of these pathways in apical plasma membranes.

References

AL-ZAHID, G., SCHAFER, J. A., TROUTMAN, S. L., ANDREOLI, T. E.: J. Membrane Biol. 31: 103 (1977).

ANDREOLI, T. E., TROUTMAN, S. L.: J. gen. Physiol. **57**, 464 (1971).

ANDREOLI, T. E., SCHAFER, J. A.: Am. J. Physiol. **234**, F349 (1978).

ANDREOLI, T. E., DENNIS, V. W., WEIGL, A. M.: J. gen. Physiol. **53**, 133 (1969).

ANDREOLI, T. E., SCHAFER, J. A., TROUTMAN, S. L.: Kidney International **14**, 263 (1978).

BAKER, P. F., HODGKIN, A. L., MEVES, H.: J. Physiol. (Lond.) **170**, 541 (1964).

BARRY, P. H., DIAMOND, J. M.: J. Membrane Biol. **3**, 93 (1970).

BARRY, P. H., DIAMOND, J. M.: J. Membrane Biol. **4**, 295 (1971).

BARRY, P. H., DIAMOND, J. M., WRIGHT, E. M.: J. Membrane Biol. **4**, 358 (1971).

BENNETT, C. M., BRENNER, B. M., BERLINER, R. W.: J. clin. Invest. **47**, 203 (1968).

BERRY, C. A., WARNOCK, D. G., RECTOR, F. C.: Am. J. Physiol. **235**, F234–F245 (1978).

BOULPAEP, E. L.: In: Transport and Funktion intercellulärer Elektrolyte (F. Krück, Ed.). Munich: Urban and Schwarzenberg 1967, p. 98.

BOULPAEP, E. L., SEELY, J. F.: Amer. J. Physiol. **221**, 1084 (1971).

BURG, M. B.: Yale J. Biol. Med. **45**, 321 (1972a).

BURG, M. B.: In: Proceedings of the Fifth International Congress of Nephrology, Vol. 2 (H. Villareal, Ed.). Basel: Karger 1972b, p. 112.

BURG, M. B., GREEN, N.: Amer. J. Physiol. **224**, 659 (1973a).

BURG, M. B., GREEN, N.: Kidney International **4**, 245 (1973b).

BURG, M. B., ORLOFF, J.: Amer. J. Physiol. **203**, 327 (1962).

BURG, M. B., ORLOFF, J.: J. Clin. Invest. **47**, 2016 (1968).

BURG, M. B., ORLOFF, J.: Amer. J. Physiol. **217**, 1064 (1969).

BURG, M. B., ORLOFF, J.: Amer. J. Physiol. **219**, 1714 (1970).

BURG, M. B., WELLER, P. F.: Amer. J. Physiol. **217**, 1053 (1969).

BURG, M. B., GRANTHAM, J. J., ABRAMOW, M., ORLOFF, J.: Amer. J. Physiol. **210**, 1293 (1966).
BURG, M. B., ISAACSON, L., GRANTHAM, J., ORLOFF, J.: Amer. J. Physiol. **215**, 788 (1968).
BURG, M. B., HELMAN, S., GRANTHAM, J., ORLOFF, J.: In: Urea and the Kidney (B. Schmidt-Nielsen, Ed.). Amsterdam: Excerpta Medica Foundation 1970, p. 193.
BURG, M. B., STONER, L., CARDINAL, J., GREEN, N.: Amer. J. Physiol. **225**, 119 (1973).
BURG, M., STONER, L.: Fed. Proc. **33**, 31 (1974).
CARDINAL, J., LUTZ, M. D., BURG, M. B., ORLOFF, J.: Kidney International **7**, 94 (1975).
CHONKO, A. M., OSGOOD, R. W., NICKEL, A. E., FERRIS, T. F., STEIN, J. H.: J. clin. Invest. **56**, 232 (1975a).
CHONKO, A. M., LOWE, C. M., GRANTHAM, J. J.: Clin. Res. **23**, 358A (1975b).
COHEN, B. E.: J. Membrane Biol. **20**, 205 (1975).
COLLANDER, R.: Acta chem. scand. **3**, 717 (1949).
COLLANDER, R., BÄRLUND, H.: Acta bot. fenn. **11**, 1 (1933).
COLTON, C. K.: Federal Clearinghouse Accession No. PB 182–281. Public Health Service, Artificial Kidney-Chronic Uremic Program, National Institute of Arthritis and Metabolic Disease: 1967.
DAINTY, J.: Advanc. Bot. Res. **1**, 279 (1963).
DAINTY, J., HOUSE, C. R.: J. Physiol. (Lond.). **185**, 172 (1966).
DANTZLER, W. H.: Amer. J. Physiol. **224**, 445–453 (1973).
DANTZLER, W. H.: Amer. J. Physiol. **226**, 634 (1974).
DELLASEGA, M., GRANTHAM, J. J.: Amer. J. Physiol. **224**, 1288 (1973).
DENNIS, V. W., WOODHALL, P. B., ROBINSON, R. R.: Amer. J. Physiol. **231**, 979 (1976).
DIAMOND, J. M., BOSSERT, W. H.: J. gen. Physiol. **50**, 2061 (1967).
FARQUHAR, M. G., PALADE, G. E.: J. cell. Biol. **17**, 375 (1963).
FINKELSTEIN, A.: In: Drugs and Transport Processes (B. A. Callingham, Ed.). Baltimore, Md: Univ. Park Press 1974, p. 241.
FINKELSTEIN, A., CASS, A.: Nature **216**, 717 (1967).
FRINDT, G., BURG, M. B.: Kidney International **1**, 224 (1972).
FRIZZELL, R. A., SCHULTZ, S. G.: J. gen. Physiol. **59**, 318 (1972).
FRÖMTER, E., DIAMOND, J.: Nature New Biol. **235**, 9 (1972).
FRÖMTER, E., HEGEL, U.: Pflügers Arch. **291**, 107 (1966).
FRÖMTER, E., MULLER, C. W., WICK, T.: In: Electrophysiology of Epithelial Cells (G. Giebisch, Ed.). Stuttgart: Shattauer 1970, p. 119.
GALLUCCI, E., MICELLI, S., LIPPE, C.: Arch. Internat. Physiol. Biochem. **79**, 881 (1971).
GANOTE, C. E., GRANTHAM, J. J., MOSES, H. L., BURG, M. B., ORLOFF, J.: J. Cell Biol. **36**, 355 (1968).
GIEBISCH, G.: J. gen. Physiol. **51**, 315a (1968).
GINZBERG, B. Z., KATCHALSKY, A.: J. gen. Physiol. **47**, 403 (1963).
GOLDMAN, D. E.: J. gen. Physiol. **27**, 37 (1943).
GOTTSCHALK, C. W.: Harvey Lectures Ser. B. **58**, 99 (1962/63).
GRANTHAM, J. J., BURG, M. B.: Amer. J. Physiol. **211**, 255 (1966).
GRANTHAM, J. J., ORLOFF, J.: J. clin. Invest. **47**, 1154 (1968).
GRANTHAM, J. J., GANOTE, C. E., BURG, M. B., ORLOFF, J.: J. Cell Biol. **41**, 562 (1969).
GRANTHAM, J. J., BURG, M. B., ORLOFF, J.: J. clin. Invest. **49**, 1815 (1970).
GRANTHAM, J. J., QUALIZZA, P. B., WELLING, L. W.: Kidney International **2**, 66 (1972).
GRANTHAM, J. J., IRWIN, R. L., QUALIZZA, P. B., TUCKER, D. R., WHITTIER, F. C.: J. clin. Invest. **52**, 2441 (1973).
GRANTHAM, J. J., QUALIZZA, P. B., IRWIN, R. L.: Amer. J. Physiol. **226**, 191 (1974).
GROSS, J. B., IMAI, M., KOKKO, J. P.: J. clin. Invest. **55**, 1284 (1975).
HAMBURGER, R. J., LAWSON, N. L., SCHWARTZ, J. H.: Amer. J. Physiol. **230**, 286 (1976).
HANAI, T., HAYDON, D. A.: J. theoret. Biol. **11**, 370 (1966).
HAYS, R. M.: J. gen. Physiol. **51**, 385 (1968).
HAYS, R. M.: In: Current topics in Membranes and Transport (F. Bronner, A. Kleinzeller, Eds). New York; Academic 1972, p. 339.
HAYS, R. M., FRANKI, N.: J. Membrane Biol. **2**, 263 (1970).
HAYS, R. M., LEAF, A.: J. gen. Physiol. **45**, 905 (1962).
HELMAN, S. I.: Amer. Soc. Nephrol. Abs. **6**, 49 (1973).
HELMAN, S. I., GRANTHAM, J. J., BURG, M. B.: Amer. J. Physiol. **220**, 1825 (1971).
HODGKIN, A. L., KATZ, B.: J. Physiol. **108**, 37 (1949).

IMAI, M., KOKKO, J. P.: Amer. Soc. Nephrol. Abs. **5**, 34 (1971).
IMAI, M., KOKKO, J. P.: J. clin. Invest. **51**, 314 (1972).
IMAI, M., KOKKO, J. P.: Kidney International **6**, 138 (1974a).
IMAI, M., KOKKO, J. P.: J. clin. Invest. **53**, 393 (1974b).
IRISH, J. McC. III, DANTZLER, W. H.: Fed. Proc. **34**, 392 (1975).
JACOBSON, H. R., KOKKO, J. P.: J. clin. Invest. **57**, 818 (1976).
JACOBSON, H. R., GROSS, J. B., KAWAMURA, S., KOKKO, J. P.: Kidney International **8**, 480 (1975).
JACOBSON, H. R., KAWAMURA, S., GROSS, J. B., KOKKO, J. P.: Clin. Res. **24**, 468A (1976).
KAWAMURA, S., MASAHI, I., SELDIN, D. W., KOKKO, J. P.: J. clin. Invest. **55**, 1269 (1975).
KOEFOED-JOHNSEN, V., USSING, H. H.: Acta physiol. scand. **28**, 60 (1953).
KOKKO, J. P.: J. clin. Invest. **49**, 1835 (1970).
KOKKO, J. P.: Yale J. Biol. Med. **45**, 332 (1972a).
KOKKO, J. P.: J. clin. Invest. **51**, 1999 (1972b).
KOKKO, J. P.: J. clin. Invest. **52**, 1352 (1973).
KOKKO, J. P.: Fed. Proc. **33**, 25 (1974).
KOKKO, J. P., RECTOR, F. C.: J. clin. Invest. **50**, 2745 (1971).
KOKKO, J. P., RECTOR, F. C., Jr: Kidney International **2**, 214 (1972).
KOKKO, J. P., BURG, M. B., ORLOFF, J.: J. clin. Invest. **50**, 69 (1971).
LEVINE, S. D., FRANKI, N., EINHORN, R., HAYS, R. M.: Kidney International **9**, 30 (1976).
LIPPE, C.: J. molec. Biol. **29**, 669 (1969).
LONGSWORTH, L. G.: J. phys. Chem. **67**, 689 (1963).
LUTZ, M. D., CARDINAL, J., BURG, M. B.: Amer. J. Physiol. **225**, 729 (1973).
MacINNES, D. A.: The Principles of Electrochemistry. New York: Dover Publications 1961, p. 232.
MALNIC, G., ENOKIBARA, H., AIRES, M. M., VIEIRA, F. L.: Pflügers Archiv **309**, 21 (1969).
MARSH, D. J., AZEN, S. P.: Amer. J. Physiol. **228**, 71–79 (1975).
PAPPENHEIMER, J. R.: Physiol. Rev. **33**, 387 (1953).
PARISI, M., PICCINI, Z. F.: J. Membrane Biol. **12**, 227 (1973).
PENNELL, J. P., LACY, F. B., JAMISON, R. L.: Kidney International **5**, 337 (1976).
PIETRAS, R. J., WRIGHT, E. M.: J. Membrane Biol. **22**, 107 (1975).
POZNANSKY, M., TONG, S., WHITE, D. C., MILGRAM, J. M., SOLOMON, A. K.: J. gen. Physiol. **67**, 45 (1976).
PRICE, H. D., THOMPSON, T. E.: J. molec. Biol. **41**, 443 (1969).
RECTOR, F. C., MARTINEZ-MALDONADO, M., BRUNNER, F. P., SELDIN, D. W.: J. clin. Invest. **45**, 1060 (1966).
ROCHA, A. S., KOKKO, J. P.: J. clin. Invest. **52**, 612 (1973a).
ROCHA, A. S., KOKKO, J. P.: Kidney International **4**, 326 (1973b).
ROCHA, A. S., KOKKO, J. P.: Kidney International **6**, 379 (1974).
ROSE, R.C., SCHULTZ, S. G.: J. gen. Physiol. **57**, 639 (1971).
SAKAGUCHI, H., SUZUKI, Y.: Keio J. Med. **7**, 17 (1958).
SCHAFER, J. A., ANDREOLI, T. E.: J. clin. Invest. **51**, 1264 (1972a).
SCHAFER, J. A., ANDREOLI, T. E.: J. clin. Invest. **51**, 1279 (1972b).
SCHAFER, J. A., ANDREOLI, T. E.: J. clin. Invest. **58**, 500 (1976).
SCHAFER, J. A., PATLAK, C. S., ANDREOLI, T. E.: J. gen. Physiol. **64**, 201–227 (1974a).
SCHAFER, J. A., TROUTMAN, S. L., ANDREOLI, T. E.: J. gen. Physiol. **64**, 228 (1974b).
SCHAFER, J. A., TROUTMAN, S. L., ANDREOLI, T. E.: J. gen. Physiol. **64**, 582 (1974c).
SCHAFER, J. A., PATLAK, C. S., ANDREOLI, T. E.: **66**, 445 (1975).
SCHAFER, J. A., PATLAK, C. S., ANDREOLI, T. E.: Amer. J. Physiol. **232**, F154 (1977).
SCHAFER, J. A., PATLAK, C. S., TROUTMAN, S. L., ANDREOLI, T. E.: Am. J. Physiol. **234**, F340 (1978a).
SCHAFER, J. A., TROUTMAN, S. L., WATKINS, M. L., ANDREOLI, T. E.: Am. J. Physiol. **234**, F332, (1978b).
SEELY, J. F.: Amer. J. Physiol. **225**, 48 (1973).
SPRING, K. R., PAGNELLI, C. V.: J. gen. Physiol. **60**, 181 (1972).
STEIN, J. H., CONGBALAY, R. C., KARSH, D. L., OSGOOD, R. W., FERRIS, T. F.: J. clin. Invest. **51**, 1709 (1972).
STEPHENSON, J. L.: Kidney International **2**, 85 (1972).
STONER, L. C.: Fed. Proc. **34**, 392 (1975).

STONER, L. C., BURG, M. B., ORLOFF, J.: Amer. J. Physiol. **227**, 453 (1974).
SUTHERLAND, G. B., TRAPANI, I. L., CAMPBELL, D. H.: J. appl. Physiol. **12**, 367 (1958).
TAYLOR, R. E.: Phys. Tech. Biol. Res. **6**, 219 (1963).
TUNE, B. M., BURG, M. B.: Amer. J. Physiol. **221**, 580 (1971).
TUNE, B. M., BURG, M. B., PATLAK, C. S.: Amer. J. Physiol. **217**, 1057 (1969).
URRY, D. W., GOODALL, M. C., GLICKSON, J. D., MAYERS, D. F.: Proc. nat. Acad. Sci. (Wash.) **68**, 1907 (1971).
USSING, H. H., WINDHAGER, E. E.: Acta physiol. scand. **61**, 484 (1964).
VALTIN, H.: J. clin. Invest. **45**, 337 (1966).
VREEMAN, H. J.: Kon. Ned. Akad. Wetensch. Proc. Ser. B. Phys. Sci. **69**, 542 (1966).
WANG, J. H.: J. Amer. Chem. Soc. **73**, 510 (1951).
WANG, J. H., ROBINSON, C. V., EDELMAN, I. S.: J. Amer. Chem. Soc. **75**, 466 (1953).
WARNOCK, D. G., BURG, M. B.: Amer. J. Physiol.: Kidney. **231**, F20 (1977).
WEDNER, H. J., DIAMOND, J. M.: J. Membrane Biol. **1**, 92 (1969).
WELLING, L. W., GRANTHAM, J. J.: J. clin. Invest. **51**, 1063 (1972).
WINDHAGER, E. E., GIEBISCH, G.: Amer. J. Physiol. **200**, 581 (1961).
WOODHALL, P. B., TISHER, C. C., SIMONTON, ROBINSON, R. R.: Clin. Res. **24**, 417 A (1976).
WRIGHT, E. M., SMULDERS, A. P., TORMEY, J. M.: J. Membrane Biol. **7**, 198 (1972).

Chapter 11

Metabolic Correlates of Tubular Transport

R. KINNE

To Evamaria and Daniel
for their support and patience

A. Introduction

The net transport of solutes across a biological membrane can, in a linear range of the system, be described as the sum of active and passive transport components. Active transport is that portion of the net transport that is observed in the absence of any electrochemical potential differences for all components across the epithelium. This active transport can be defined thermodynamically by $J_i^{act} = L_{iA'} \cdot A'$ where $L_{iA'}$ is the phenomenological coefficient that couples the flow of i to the affinity of a chemical reaction A', an equivalent to the thermodynamic force conjugated to the reaction flow (SAUER, 1973). Therefore it is this process where coupling between metablism and transport has to be expected. Accordingly, inhibition of active transport is observed when metabolic inhibitors are applied to transporting epithelia.

In analyzing metabolic correlations of epithelial transport, three main questions may be asked. First: what is the magnitude of L_{iA} and A', and how are these parameters altered if the active transport of i is changed by inhibitors or stimulators. This analysis by linear nonequilibrium thermodynamics has been performed extensively by ESSIG and his associates (ESSIG, 1975). Secondly, one would like to know which biochemical pathways produce the energy that is used ultimately for the active transport, and thirdly, what molecular mechanisms are involved in the energy coupling during active transport.

In the kidney the last two questions can be answered with some certainty only for the proximal tubule, where the different components of solute transfer have been analyzed and where measurements of solute transport during local alterations of the metabolism could be performed. In addition, membrane vesicles could be isolated from proximal tubular epithelial cells, which can be used to elucidate the mechanisms and driving forces involved in transport of solutes across the luminal and contraluminal membranes. For all other parts of the tubule, conclusions have to be drawn mainly from studies with the whole kidney, where due to the functional and biochemical inhomogeneity of the organ and to intrarenal compensation phenomena, interpretation of the results is very difficult.

It has to be kept in mind, however, that in addition to the coupling between active transport and metabolism — which is the topic of this review — the passive transport properties of the epithelia are also influenced by the metabolic activity of the cells. Maintenance of the functional organization of the cell and the cellular organelles requires energy, and energy-dependent regulatory processes such as membrane phosphorylation (WALTON et al., 1975) may play an important role in transporting epithelia.

B. Transport-Related Parameters of Metabolic Activity in Intact Cells

I. General Aspects

In this chapter, transport-related parameters of metabolic activity that can be observed without destruction of the cells will be discussed. The discussion will start with the most general parameters, such as heat production, O_2 uptake, and CO_2 production of the kidney, and then turn to the more defined uptake and utilization of substrates and to the action of metabolic inhibitors on the tubular transport. It will turn out that only a few studies performed so far allow reasonably valid conclusions on the metabolic pathways supporting tubular transport. Further studies in which active transport of solutes and metabolic activity of individual tubular segments can be correlated are definitely needed. One possible way for such experiments might be to combine the perfusion of isolated tubules with the microspectrophotometric determination of the redox state of pyridine nucleotides and cytochromes. Such a preparation may allow direct study of the interrelation between transepithelial transport and cellular metabolism under a variety of experimental conditions.

II. Heat Production in the Kidney

A general indicator of metabolic activity and work performance is heat production by the cells. As early as 1928, REIN and JANNSEN determined the heat production of the kidney by direct calorimetry. In continuation of these studies (GRUPP and JANNSEN, 1958) it was found that the heat production similar to the oxygen consumption was independent of the blood flow of the kidney above $2\ ml \cdot g^{-1} \cdot min^{-1}$. A further development in this area was the introduction of the heat accumulation technique, by which heat production during the first seconds after occlusion of the renal artery can be measured in defined areas of the kidney (AUKLAND et al., 1969).

Under various conditions oxygen consumption and heat accumulation agreed quite well (HIERHOLZER, 1957) and thus it was concluded that this technique seemed suitable for the determination of the renal metabolic rate (GRUPP and JANNSEN, 1958; AUKLAND et al., 1969). The last authors observed that the initial temperature rise after occlusion of the renal artery is higher in the cortex than in the outer and inner medulla. Furthermore, using this methodology,

AUKLAND et al. (1969) and LIE et al. (1973) tried to determine the site of action of diuretics in dog kidney. Mersalyl reduced heat production mainly in the outer medulla, which was interpreted as an indication that sodium transport in the loops of HENLE was inhibited. Chlorothiazide acted predominantly on cortical tubular segments, whereas ouabain reduced the metabolic rate both in the outer medulla and, although to a lesser extent, in the cortex (LIE et al, 1973). Similarly, ethacrynic acid and furosemide acted mainly in the loops of Henle (AUKLAND et al., 1969). The conclusions drawn from the heat accumulation studies on the site of action of diuretics were confirmed by several other authors using microperfusion of isolated tubules (BURG and GREEN, 1973 a and b; BURG et al., 1973), micropuncture (HOLZGREVE et al., 1964) and quantitative histochemistry (SCHMIDT and DUBACH, 1970).

In the heat accumulation studies, however, information on the actual ion transport performed by the tubular segments located in the area of the kidney investigated is lacking. Therefore, a direct correlation between the metabolic rate, as determined by heat production, and solute transport cannot be established. Furthermore, the spatial resolution of the method is limited — several tubular segments contribute to the signal — and the signal itself is too unspecific to allow conclusions on the mode of coupling between tubular transport and metabolism.

III. O_2 Consumption and CO_2 Production in the Kidney

Following the finding of ZEHRAN (1956) that oxygen consumption and active sodium transport across the short-circuited frog skin are linearly correlated, several authors tried to find a similar correlation for the kidney. DEETJEN and KRAMER (1961) and THURAU (1961) were the first to find experimental conditions under which renal net sodium reabsorption could be altered without a change in renal blood flow and GFR[1]. Under the circumstances of increased sodium load, salyrgan poisoning, sodium sulfate infusion, and glucose and mannitol diuresis, a relation of 7.1 mEq sodium transported per 1 mEq O_2 consumed was found (see Fig. 1 and Table 1). A similar ratio was found by KILL et al. (1961), LASSEN et al. (1961), WOLF et al. (1969), FÜLGRAFF et al. (1969) and FUJIMOTO et al. (1964). SCHUREK et al. (1975) made the same observation in the isolated perfused rat kidney. These results demonstrate that sodium reabsorption in the mammalian kidney involves active transport processes that are energized by oxidative metabolism. This point had been in dispute in the early sixties.

In further pursuit of the question of a fixed proportion between sodium transport and respiration, however, evidence for uncoupling of the two processes was found. For example, infusion of mannitol (KILL et al., 1961), dichlorphenamide (KESSLER et al., 1964), ethacrynic acid (WOLF et al., 1969) and

1 glomerular filtration rate.

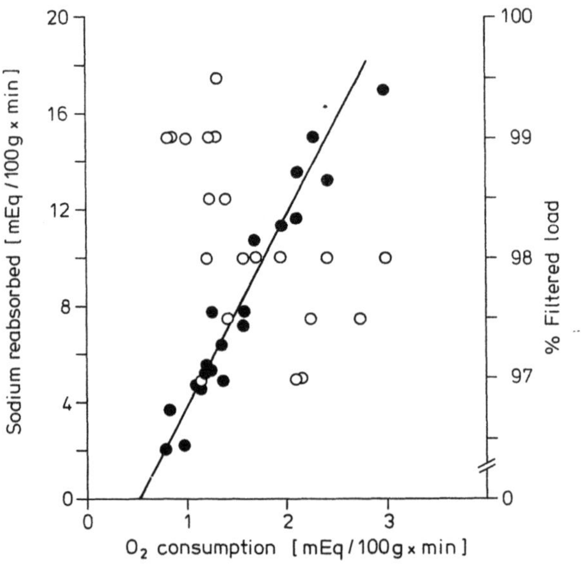

Fig. 1. Correlation between O_2 consumption and sodium reabsorption in dog kidney *in vivo*. Data represent spontaneous variations of the two parameters in normal dogs and are taken from DEETJEN and KRAMER (1961). Figure indicates very close correlation between O_2 consumption and net sodium reabsorption when expressed as amount of sodium reabsorbed in mEq/100 g · min (●), and lack of correlation between O_2 consumption and sodium reabsorption when expressed as percent of filtered load (○) (see section C II)

furosemide (FÜLGRAFF et al., 1969) into the renal artery of dogs reduced net reabsorption of sodium, either without a change in or with a proportionally lesser reduction of oxygen consumption. For these instances several possible explanations must be considered.

The net transport of sodium (or any electrolyte) across an epithelium is the sum of the passive transport by solvent drag and by electrochemical potential difference and of the active transport. This is expressed in Eq. (1):

$$J_i = (1 - \sigma_i)\, \bar{c}_i\, J_v - P_i \left(\Delta c_i + \frac{zF}{RT}\, \Delta \varphi \bar{c}_i\right) + J_{act}, \qquad (1)$$

where P represents the permeability coefficient, σ the reflection coefficient, Δc the concentration difference, $\Delta\varphi$ the electrical potential difference, $\bar{c}$ the mean concentration, J_v the volume flow, and J_{act} the active transport component. This active transport component is by definition the transport observed when all physical driving forces are zero, and it is coupled to a chemical reaction (SAUER, 1973; ULLRICH, 1973). In the proximal tubule of rat kidney the active transport component for Na^+ ions comprises approximately 35–40 percent of the total Na^+ transport, 60 percent being due to passive driving forces (FRÖMTER et al., 1973). Eq. (1) clearly demonstrates that changes in net transport of sodium across the epithelium can occur, for example, by a change in the solvent drag or diffusion component, without alteration of the active transport component.

From Eq. (1) it is also obvious that the ratio net sodium transported/oxygen consumed depends on the proportion between active and passive transport components. Since water permeability, sodium permeability, and electrical potential difference are different in the various segments of the nephron (see ULLRICH, 1973), the ratio between net sodium transport and oxygen consumption can be expected to vary along the nephron. In the distal convolution, for example, the water permeability (L_p) and sodium permeability are low and a lumen-negative transepithelial potential is observed. Therefore a smaller contribution of passive driving forces to the net transport can be expected and the ratio of net sodium transport to O_2 consumption should therefore be lower than in the proximal tubule.

Furthermore, according to studies of BURG and GREEN (1973c) sodium transport in the ascending limb of the loop of Henle is postulated to be passive, and furosemide and ethacrynic acid are thought to inhibit active chloride transport and not primarily sodium transport. The inhibition of chloride transport, however, secondarily also affects the sodium transport because the transepithelial electrical potential (inside positive) is reduced, which can act as a driving force for sodium reabsorption. Thus O_2 consumption in the thick ascending limb might be correlated in the first place to active chloride transport and not to sodium transport.

Therefore an important factor that influences the results in experiments on the metabolic requirements of sodium transport in the intact kidney is the intrarenal compensation (MENG, 1972). HOWARDS et al. (1968) observed, for example, that in the dog a depression of fractional sodium reabsorption in the proximal tubule can be compensated by the more distal parts of the nephron and thus does not necessarily lead to an increased net excretion of sodium. Such shifts in sodium reabsorption from the proximal parts of the tubule (with a high ratio of net sodium transport / O_2 consumption) to more distal parts (with a lower ratio) could explain the higher oxygen requirement for the sodium reabsorption observed in the experiments mentioned above (KILL et al., 1961; KESSLER et al., 1964; WOLF et al., 1969; FÜLGRAFF et al., 1969).

Another interesting observation concerning the apparent uncoupling of Na^+ reabsorption and O_2 consumption during ethacrynic acid application was made by SCHUREK et al. (1972). In studies with isolated perfused rat kidneys they demonstrated that ethacrynic acid, in the presence of NAD-dependent substrates such as glucose, pyruvate and oxalacetate, inhibits O_2 consumption and sodium reabsorption of the kidney; after addition of succinate (FAD-dependent), however, O_2 consumption increased, but sodium reabsorption decreased. The authors concluded that ethacrynic acid inhibits the oxidation of NAD-dependent substrates (with a P/O ratio of 3) and favors the oxidation of FAD-dependent substrates (P/O ratio of 2). Because of the lower energy production in the respiratory chain and the higher amount of oxygen required per ATP formed, inhibition of sodium transport without a decrease in oxygen consumption can occur.

Irrespective of the above mentioned ambiguities there was for a long time general agreement that in the proximal tubule (since 70% of sodium are reabsorbed in this part of the nephron also the major part of oxygen consumption

Table 1. Energy requirements for active sodium transport

Species	Tissue	Na/O_2[a]	Na/ATP[b]	Na/ATP (corrected for passive components)
Dog	Kidney	27[c]	4.5	2.7[g]
Rabbit	Kidney	29[d]	4.8	2.9
Frog	Skin	16–20[e]	3.0	3.0
Toad	Bladder	16.5[f]	2.8	2.8
Man	Red blood cell			2.8–3.0[h]

[a] mol sodium transported per mol O_2
[b] mol sodium transported per mol ATP calculated on the assumption that 6 mol ATP are produced per mol O_2 (P/O Ratio of 3)
[c] Mean value derived from a table given by COHEN and BARAC-NIETO (1973)
[d] TORELLI et al. (1966)
[e] ZEHRAN (1956)
[f] LEAF et al. (1959)
[g] According to FRÖMTER et al. (1973), approximately 60% of the net sodium transport in the proximal tubule of rat kidney is passive
[h] From comparison of ATP hydrolysis to sodium movement (SEN and POST, 1964; GARRAHAN and GLYNN, 1967)

was attributed to this segment) an unusually high ratio of net sodium transport to oxygen consumption exists. Compared to other sodium-transporting epithelia, such as frog skin (ZEHRAN, 1956) and toad bladder (LEAF et al., 1959), in the tubule apparently 50 percent more sodium was transported per mol of O_2 and, assuming identical P/O ratios for the three tissues, also per mol ATP. However, taking into account that in the proximal tubule only 40 percent of the net sodium transport is active (FRÖMTER et al., 1973) (see Table 1, fifth column) the Na/ATP ratio is similar for all three ion-transporting epithelia.

It should also be pointed out that this ratio is identical for cells performing transepithelial transport and for cells performing transport across only one cell membrane. Thus in the erythrocyte also three sodium ions are transported per ATP hydrolyzed (SEN and POST, 1964; GARRAHAN and GLYNN, 1967). This finding suggests that the basic mechanisms involved in transepithelial and transmembrane sodium transport are identical. Transepithelial transport would then depend on a specific localization of the sodium transporting system within the epithelial cell, which indeed has been found recently (KINNE et al., 1971; HEIDRICH et al., 1972; SCHMIDT and DUBACH, 1971).

IV. Effect of Metabolic Inhibitors on the Transepithelial Transport of Sodium

The strict dependence of the oxygen consumption of the kidney on the amount of sodium reabsorbed suggests that it is mainly oxidative metabolism that supplies the energy for the active sodium transport. This idea was further supported by results obtained with metabolic inhibitors, which are known to interact with the respiratory chain, ATP synthesis, and substrate oxidation. The results of these experiments are shown in Table 2. In the mitochondria Amytal and Q_o are thought to inhibit electron transfer from NADH-dehydrogenase to coenzyme Q, antimycin A inhibits electron flow between cytochrome b and c_1, and electron flow between cytochrome $(a + a_3)$ and O_2 can be blocked by cyanide and azide. Oligomycin is thought to inhibit ATP synthesis by interaction with the mitochondrial ATPase, DNP[2] and CFCCP[3] uncouple the oxidative phosphorylation, and iodoacetamide inhibits mainly glycolysis; monofluoroacetate, methoxyindolyl carbonic acid, and malonate reduce the activity of the tricarboxylic acid cycle, whereas 4-pentenoic acid inhibits the β-oxidation of long-chain fatty acids.

In the intact kidney, when inhibitors were applied by infusion into the renal artery, only cyanide consistently reduced the net sodium reabsorption (STRICKLER and KESSLER, 1963; FUJIMOTO et al., 1964; ROSS et al., 1973). In one case Antimycin A was also found inhibitory, whereas azide and Amytal were found to be ineffective (STRICKLER and KESSLER, 1963). The lack of action of Amytal might indicate that electron flow from succinate *via* the succinate dehydrogenase system to coenzyme Q is sufficient to support sodium transport. However, in the same studies malonate did not reduce net sodium reabsorption; hence the succinate dehydrogenase reaction seems not to be essential for the sodium transport. The lack of inhibition of transport by azide could not be explained. Another puzzling feature in the studies with metabolic inhibitors in the whole kidney was that DNP did not inhibit sodium transport although it increased the O_2 consumption of the kidney (FUJIMOTO et al., 1964). DNP is known to uncouple respiration and phosphorylation in isolated mitochondria, and this enhances the electron flow through the respiratory chain. This finding was interpreted as an indication that energy-rich compounds other than ATP are capable of supporting the sodium transport (STRICKLER and KESSLER, 1963; FUJIMOTO et al., 1964).

Although almost all authors investigating the effect of metabolic inhibitors on the kidney were aware of the hazardous nature of assuming the same action of the inhibitor in the isolated mitochondria as in the intact cell, little attention was paid to the problems of the inhibitor concentration present within the tubular cells during the experiments and to the possibility of intrarenal compensation of events occurring in the proximal tubule.

Various workers attempted to avoid these difficulties by micropuncture and microperfusion studies. The most extensive study in this respect was performed by GYÖRY and KINNE (1971). They demonstrated that the capability of a metabolic inhibitor to affect volume (sodium) reabsorption in the proximal tubule is not only dependent on its concentration, but depends also on the route of application. For example, cyanide and CFCCP did not inhibit isotonic volume transport when applied intratubularly but only when applied peritubularly. The opposite was found for Antimycin A and oligomycin. In this instance it was in

2 Dinitrophenol.
3 Carbonyl cyanide p-trifluoromethoxyphenylhydrazone.

Table 2. Effect of metabolic inhibitors on sodium transport in the kidney

Author	Preparation	Kind and concentration of inhibitor	Effect on sodium transport
MUDGE and TAGGART (1950)	Dog kidney *in vivo*	Injection (i. v.) of dinitrophenol (10 ng kg^{-1})	No effect on sodium excretion
STRICKLER and KESSLER (1963)	Dog kidney *in vivo*	Infusion into the renal artery of:	
		cyanide (~77 μM/20 min)	Diuresis
		Antimycin A (~1 μM/20 min)	Diuresis
		iodoacetamide (1.6 mM/20 min)	Diuresis
		dinitrophenol (1.1 mM/20 min)	No effect
		azide (1.8 mM/20 min)	No effect
		Amytal (0.2 mM/20 min)	No effect
		malonate (8.7 mM/20 min)	No effect
WEINSTEIN and KESSLER (1963)	Dog kidney *in vivo*	Infusion into the renal artery of:	
		Coenzyme Q_0 (1 μM/20 min)	Inhibits sodium reabsorption by 17%
FUJIMOTO et al. (1964)	Dog kidney *in vivo*	Infusion into the renal artery of:	
		cyanide (0.1 mM/10–40 min)	10% reduction in sodium reabsorption
		Coenzyme Q_0 (0.1 mM/10–40 min)	No change
		dinitrophenol (~1 mM/10–40 min)	No change
ROSS et al. (1974)	Isolated perfused rat kidney	Cyanide (10 μM),	Inhibits sodium reabsorption by 48%
		cyanide (10 μM), and iodoacetamide (0.1 mM)	Inhibits sodium reabsorption 96%
HOHENEGGER et al. (1974)	Dog kidney *in vivo*	4-Pentenoic acid (6.6 μM)	Natriuresis
KLEINMANN et al. (1973)	Dog kidney *in vivo*	4-Pentenoic acid (6.6 μM)	Natriuresis
WINDHAGER and GIEBISCH (1961)	Rat kidney *in vivo*, proximal tubule, Gertz method	Luminal application of cyanide (4 mM)	95% Inhibition of sodium reabsorption
CHERTOK et al. (1966)	Rat kidney *in vivo*, proximal tubule, Gertz method	Luminal application of:	% Of initial perfusate remaining after 20 s
		cyanide (0.1 mM)	78.2
		dinitrophenol (0.1 mM)	55.1
		Amytal (0.1 mM)	71.8
		control	36.1

Table 2 (continued)

Author	Preparation	Kind and concentration of inhibitor	Change in sodium transport
Hernandez et al. (1969)	Rat kidney *in vivo*, proximal tubule, Gertz method	Luminal application of:	Half of the initial perfusate reabsorbed after:
		dinitrophenol (0.1 mM)	11.9 s
		Oligomycin (2 γ ml^{-1})	
		single doses	11.3 s
		cumulative doses	14.1 s
		FCCP (1 mM)	23.2 s
		control	9 s
Weinstein (1970)	Rat kidney *in vivo*, proximal tubule, Gertz method	Luminal application of:	Half of the initial perfusate reabsorbed after:
		cyanide (0.1 mM)	20.3 s
		dinitrophenol (0.1 mM)	9.0 s
		control	9.0 s
Györy and Kinne (1971)	Rat kidney *in vivo*, proximal tubule, Gertz method	cyanide (5 mM)	
		peritubular	100% inhibition
		intratubular	No inhibition
		CFCCP (1 mM)	
		peritubular	100% inhibition
		intratubular	No inhibition
		Oligomycin (0.1 mM)	
		peritubular	Inconsistent
		intratubular (with 0.5% albumin)	80% inhibition
		Antimycin (0.1 mM)	
		peritubular	Inconsistent
		intratubular (with 0.5% albumin)	82% inhibition
Maude (1970)	Rat kidney slices *in vitro*, proximal tubule, Gertz method	Present in the incubation medium	Half of the initial perfusate reabsorbed after:
		No substrate	39.6 s
		α-Ketoglutarate (3 mM)	19.1 s
		α-Ketoglutarate and monofluoroacetate (6 mM)	22.0 s
		α-Ketoglutarate and methoxyindolcarboxylic acid (0.5 mM)	32.2 s
		α-Ketoglutarate and malonate (15 mM)	31.2 s
		Succinate (10 mM)	20.7 s
		Succinate and monofluoroacetate (6 mM)	44.4 s

Table 2 (continued)

Author	Preparation	Kind and concentration of inhibitor	Change in sodium transport
		Succinate and methoxy-indolcarboxylic acid (0.5 mM)	21.6 s
		Oxalacetate (3 mM)	18.0 s
		oxalacetate and mono-fluoroacetate (6 mM)	40.2 s
		oxalacetate and methoxy-indolcarboxylic acid (0.5 mM)	30.0 s
		oxalacetate and malonate (15 mM)	26.4 s

addition necessary that albumin was present in the tubular lumen, together with the inhibitors. The authors concluded from their results with cyanide and CFCCP that during exclusive intratubular application the inhibitor concentration in the mitochondria is probably not high enough to cause inhibition of sodium transport because the inhibitors rapidly permeate the epithelium and are lost continuously into the blood of the peritubular capillaries. The very high permeability of the proximal tubule to cyanide and dinitrophenol was demonstrated directly by SILBERNAGL (1973), who studied the resorption of metabolic inhibitors in micropuncture experiments. The intratubular route of application of cyanide used by other authors probably explains the failure to inhibit sodium transport completely (CHERTOK et al., 1966; HERNANDEZ et al., 1969; WEINSTEIN, 1970, 1974). The effect of albumin, together with the lag time of about 4 minutes observed for the inhibition of volume flow by intratubular application of antimycin A and oligomycin, was interpreted to indicate that these inhibitors do not penetrate the apical cell membrane but are taken up into the cell *via* pinocytosis and are released into the cytoplasm when the permeability characteristics of the pinocytic vesicles change during their transformation into secondary lysosomes.

A low permeability of the luminal membrane to oligomycin is already evident from the studies of HERNANDEZ et al. (1969), in which oligomycin inhibited isotonic volume transport only after repeated application.

The studies of GYÖRY and KINNE (1971) not only demonstrated the importance of the route of application for metabolic inhibitors, but also suggested that ATP generated by oxidative phosphorylation is the major direct energy source for transepithelial transport of sodium in the proximal tubule. However, the question as to whether it is exclusively ATP could not be answered with certainty, because the inhibitors applied from the lumen inhibited sodium transport by only 80 percent, regardless of the concentration used. The discrepancy between the findings with cyanide and CFCCP and the results with antimycin A and oligomycin might be explained in several ways.

First of all, it is known that cyanide in the concentration used inhibits other systems, such as carbonic anhydrase, in addition to oxidative phosphorylation, and CFCCP is reported to increase the proton conductivity of renal plasma membranes (MURER et al., 1976). Secondly, an effect of cyanide and CFCCP on the passive permeability of the tubular epithelium has to be considered. As shown by WEINSTEIN (1972), DNP, for example, increases the permeability of the tubular epithelium for mannitol, sucrose, and raffinose. Since the net sodium reabsorption determined in these studies is composed of active and passive components (see above), such a change may lead to a decrease in net transport that is not related to an inhibition of the active sodium transport. It is also possible that the amount of antimycin and oligomycin that enters the cell is limited because these inhibitors are taken up into the cell by pinocytosis, which itself is energy-dependent. Furthermore, since in the experiments with oligomycin and antimycin only the tubular lumen was perfused, the remaining transport might have been maintained by glycolysis. However, this seems unlikely, since SATO (1975) demonstrated that during perfusion of the peritubular capillaries the inhibition of volume transport by cyanide was not abolished if glucose was present in the peritubular perfusate.

Thus in all probability more than 80 percent, if not 100 percent of the active sodium transport in the proximal tubule seems to be driven by ATP and not by other hypothetical energy-rich intermediates of the oxidative phosphorylation. In this connection it should be mentioned that in view of modern concepts of oxidative phosphorylation (for reviews see MITCHELL, 1969; REID, 1974), the existence of high energy-rich intermediates that are able to leave the mitochondria and support transport processes at the cell membrane is very unlikely. There is increasing evidence that in oxidative phosphorylation ATP is generated by the dissipation of a proton gradient that is locally built up during mitochondrial respiration. Hence, the "coupling factor" is the proton gradient, and correspondingly, classic uncouplers have been shown to increase the proton permeability of artificial and mitochondrial membranes (PRESSMAN, 1968).

The reducing equivalents that are oxidized in the respiratory chain are – according to studies on the action of monofluoroacetate, methoxyindolcarboxylic acid, and malonate on the tubular fluid reabsorption in leached rat kidney cortex slices (MAUDE, 1970) – most probably produced in the tricarboxylic acid cycle. Thereby, in some instances, apparently sufficient ATP can be produced without the participation of the whole cycle. As shown in Table 2, methoxyindolcarboxylic acid, for example, which inhibits the lipoic acid-dependent oxidation of α-ketoglutarate, reduces fluid reabsorption in the presence of α-ketoglutarate but not in the presence of succinate. Similarly, monofluoroacetate inhibits the isotonic volume transport if oxaloacetate or succinate is used as substrate but is ineffective when α-ketoglutarate is present. However, high concentrations of malonate inhibit the transport regardless of what substrate is provided as metabolic fuel.

The acetyl-CoA converted to CO_2 in the tricarboxylic acid cycle is derived at least partly from long-chain fatty acids. This can be concluded from the inhibitory effect of 4-pentenoic acid on the sodium reabsorption in dog kidney (KLEINMANN et al., 1973; HOHENEGGER, 1974) and from the observation that in dog kidney (PITTS, 1975) 16–21 percent and in the isolated perfused rat kidney (BERTERMANN et al., 1973) 25 percent of total renal CO_2 are derived from long-chain fatty acids. Earlier investigations have also demonstrated a significant renal uptake of nonesterified fatty acids (DIES et al., 1970a; NIETH and SCHOLLMEYER, 1966).

V. Renal Substrate Metabolism in Relation to Renal Function

1. Uptake and Oxidation of Substrates in Relation to Sodium Transport

On the basis of the interdependence of renal O_2 consumption and sodium transport, it would be anticipated that also a linear relationship exists between the uptake of the metabolic fuel and sodium transport. Of special interest in this respect was the question as to whether the transport was supported by any specific substrate or whether the energy for reabsorption was derived from a common pool supplied by various interchangeable fuels.

The first studies on the possible linkage between substrate uptake and renal work were performed by BARAC-NIETO and COHEN (1968), who observed that in dog kidney *in vivo* the spontaneous rates of glomerular filtration and sodium reabsorption were correlated inversely with the simultaneously determined rates of renal uptake of free fatty acids. Experiments with chlorothiazide, however, showed a positive correlation between sodium reabsorption and free fatty acid uptake which was highly significant when the decrease in sodium reabsorption was compared to the decrease in renal free fatty acid uptake. DIES et al. (1970b), studying the effect of furosemide on sodium reabsorption and free fatty acid uptake, also observed a parallel reduction in these two parameters. They found in addition a positive relation between lactate uptake and net sodium transport by the kidneys. When the ureteral pressure was increased in order to reduce sodium reabsorption a significant correlation was observed only between lactate uptake and sodium reabsorption, and not with the corresponding free fatty acid uptake (DIES et al., 1970b).

Thus the results obtained on the role of lactate and free fatty acid as metabolic fuel for sodium reabsorption were not unequivocal; for the latter substrate, depending on the experimental conditions, positive or negative correlations could be found.

The whole situation became even more complicated when it was realized that the substrates taken up by the kidney were not completely oxidized but partly incorporated into cellular constituents or interconverted into other metabolites. Thus it was found by WEIDEMANN and KREBS (1969) that *in vitro* less than 25 percent of the oleate consumed by rat kidney cortex slices was converted into respiratory carbon dioxide, whereas about two-thirds was incorporated into the tissue lipids. An even lower proportion was found by BARAC-NIETO and COHEN (1971) for palmitate oxidation by the dog kidney *in vivo*. For lactate the proportion of substrate oxidized is far greater than for the free fatty acids. In the dog kidney *in vivo* approximately 50 percent of the lactate taken up is oxidized, the rest being converted into glucose and other products (COHEN, 1975).

These observations prompted the reinvestigation of the link between the – now more precise – oxidation of substrates and the net sodium reabsorption. In the intact dog, lactate oxidation (BRAND et al., 1974) and glucose oxidation (GARZA-QUINTERO et al., 1975) were found to be independent of net Na^+ reabsorption when the reabsorption was reduced by elevation of ureteral pressure. The authors concluded that "either lactate (and glucose) oxidation is not supplying energy for active sodium transport or is coupled to a moiety of sodium transport which is unaffected by raising ureteral pressure". However, in the isolated rat kidney (FRANKE et al., 1975) perfused with a mixture of substrates in their physiological concentrations, acetate, pyruvate, and lactate oxidation to CO_2 parallels sodium transport when it is reduced by ouabain, by decreased perfusion pressure, or by decreased sodium concentration in the perfusate (Fig. 2).

Thus the results on the relation between oxidation of a substrate and the sodium reabsorption are also conflicting. The most probable reasons for this are a) that in the kidney several substrates can replace each other in their metabolic role (see also ROSS et al., 1973), b) that the net Na^+ transport is composed of several components (see Eq. (1).

2. Substrate Dependence of Sodium Transport in Isolated Systems

The isolated perfused kidney offers the possibility of studying the effect of individual substrates on the sodium reabsorption. As shown in Table 3, during prolonged perfusion of isolated dog kidney WAUGH and KUBO (1969) observed that in the presence of glucose alone natriuresis developed, which could be

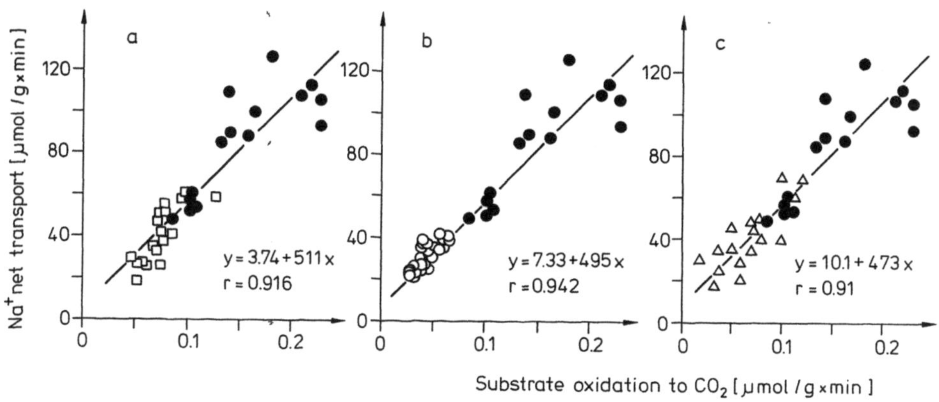

Fig. 2a–c. Correlation between oxidation of pyruvate and sodium transport in isolated perfused rat kidney. Data, redrawn from FRANKE et al. (1975), represent results of experiments in which sodium transport was reduced. (a) by addition of 0.5–1 mg ml^{-1} ouabain to the perfusion medium (□); (b) by reduction of sodium concentration in perfusion medium from 140 mM to 100 mM (○); (c) by lowering perfusion pressure to 80 Torr (△). Control periods indicated by closed circles

Table 3. Substrate dependence of sodium transport and the effect of transport inhibition on renal substrate uptake

Substrate	Change in uptake or oxidation during reduction of sodium transport	Effect on sodium reabsorption
D-glucose	Unaltered oxidation: during increased ureteral pressure (BRAND et al., 1974) during ouabain infusion (FRANKE et al., 1975)	No improvement in the proximal tubule during cyanide inhibition (SATO, 1975) Improved distal function (ROSS et al., 1973)
Pyruvate	Reduced oxidation: during ouabain infusion, reduced perfusion pressure, and reduced sodium load (FRANKE et al., 1975)	No improvement[a] (WAUGH and KUBO, 1969) Improved reabsorption (SCHUREK et al., 1975)
Lactate	Reduced uptake: during furosemide infusion and increased ureteral pressure (DIES et al., 1970b) Unaltered oxidation: during increased ureteral pressure (GARZA-QUINTERO et al., 1975) Reduced oxidation: during ouabain infusion, reduced perfusion pressure, and reduced sodium load (FRANKE et al., 1975)	Improved reabsorption (SCHUREK et al., 1975; COHEN, 1975) No improvement[a] (WAUGH and KUBO, 1969)
Acetate	Reduced oxidation: during ouabain infusion, reduced perfusion pressure, and reduced sodium load (FRANKE et al., 1975)	Improved transport in the proximal tubule of rat kidney slices (MAUDE, 1970) Supports isotonic volume flow in the proximal tubule (RUMRICH and ULLRICH, 1968)
Butyrate		Improved reabsorption[b] (ROSS et al., 1973)
Free fatty acids	Reduced uptake: during chlorothiazide infusion (BARAC-NIETO and COHEN, 1968) during furosemide infusion (DIES et al., 1970b) Unaltered uptake: during increased ureteral pressure (DIES et al., 1970b)	
Citrate		Improved reabsorption[c] (WAUGH and KUBO, 1969) No improvement of transport in the proximal tubule of rat kidney slices (MAUDE, 1970)

[a] lactate and pyruvate compared to glucose
[b] compared to endogenous substrates
[c] compared to glucose, pyruvate, and lactate

Table 3 (continued)

Substrate	Change in uptake or oxidation during reduction of sodium transport	Effect on sodium reabsorption
α-Ketoglutarate		Improved transport in the proximal tubule of rat kidney slices (MAUDE, 1970)
Glutamate		Improved reabsorption (SCHUREK et al., 1975)
Proline		Improved reabsorption (WAUGH and KUBO, 1969)
Succinate		Improved transport in the proximal tubule of rat kidney slices (MAUDE, 1970)
Fumarate		No improvement of transport in the proximal tubule of rat kidney slices (MAUDE, 1970)
Malate		Partial improvement of transport in the proximal tubule of rat kidney slices (MAUDE, 1970)
Oxalacetate		Improved transport in the proximal tubule of rat kidney slices (MAUDE, 1970) Improved reabsorption (SCHUREK et al., 1975)

prevented when lactate, pyruvate, and citrate or lactate, pyruvate and proline were added but not by lactate and pyruvate alone. These authors attributed the effect of proline to its rapid permeation into mitochondria and conversion to intramitochondrial intermediates of the tricaboxylic acid cycle.

The investigations on the basic requirements for the function of the isolated cell-free perfused rat kidney performed by SCHUREK et al. (1975) also indicate the importance of a supply of tricarboxylic acid cycle intermediates or their immediate precursors for the sodium reabsorption.

The relationship between sodium reabsorption and substrate oxidation in the isolated perfused rat kidney was investigated by COHEN (1975) and by ROSS et al. (1973). COHEN (1975) observed that when lactate oxidation was increased from 0.2 to 1 mol min^{-1} g^{-1} by raising the lactate concentration in the perfusate to 10 mM, the fractional sodium reabsorption increased from 77 percent to approximately 90 percent. The author concluded that although the effect of lactate on sodium reabsorption was relatively small it shows that lactate supports sodium transport in the proximal tubule.

ROSS et al. (1973) demonstrated that in the isolated perfused rat kidney, fatty acids could double the net sodium reabsorption by the kidney, and a further increase was observed after the addition of glucose to the perfusion medium containing fatty acids. Since the latter effect was not observed with 3-O-methylglucose, a nonmetabolizable sugar, which like glucose is transported in a sodium-dependent manner, the authors excluded that the glucose sodium cotransport can explain the stimulation of transport by glucose. When the metabolic fate of glucose was studied under these conditions the following results were obtained. Glucose (5 mM) as single substrate was removed at a high rate from the perfusion medium, approximately 17 percent was converted to lactate, whereas the remainder of the glucose was probably oxidized. In the presence of a fatty acid both glucose removal and lactate formation were significantly depressed; over 85 percent of the glucose removed was recovered as lactate. There was, however, no inhibition of sodium reabsorption under these conditions. The observations again point to the fact that depending on the presence of substrates in the perfusion fluid, the substrates used by the kidney are interchangeable.

It should be noted in this context that the addition of substrates to the perfusion medium of isolated kidneys does not improve only the sodium transport, but that other functions of the kidney also seem to benefit from this addition. Thus, for example, SCHUREK et al. (1975) observed an increase in glucose reabsorption and in potassium excretion. The latter was also found by Ross et al. (1973) during the action of glucose on the kidney. This fact led the authors to conclude that glucose improves mainly the function of the distal parts of the nephron where potassium secretion takes place.

Substrate dependence of the sodium transport in the proximal tubule was investigated extensively by MAUDE (1970) in micropuncture experiments performed in kidney slices. MAUDE compared the rate of isotonic volume transport of proximal tubules in substrate-depleted kidney slices with the transport rate after addition of various substrates and metabolic inhibitors. As shown in Table 3, acetate, α-ketoglutarate, succinate, malate and oxalacetate were capable of acting as substrates for the isotonic volume transport. Together with the results on the action of metabolic inhibitors in these studies, this clearly indicates that the oxidation of citric acid cycle intermediates provides the energy for sodium transport in the proximal tubule. Accordingly, acetate supports isotonic volume transport in proximal tubules during perfusion of the peritubular capillaries (RUMRICH and ULLRICH, 1968).

Table 4 reflects an attempt to correlate the findings obtained from functional studies *in vivo* (the relation of substrate uptake and oxidation to sodium transport) to results obtained *in vitro* in the kidney slice (no transepithelial sodium transport) with respect to the substrate requirement of the different nephron segments. Although for an integration of the results from both kinds of studies into this cheme several assumptions hat to be made (*i. e.* concerning the site and mechanism of action of the maneuvers used to reduce sodium reabsorption) the following picture emerges: a) there seems to be no difference in the substrate requirements of tubular cells for their basic cellular work and their transport work; and b) there is a variety of substrates which are able to replace each other, depending on their relative concentration. These observations seem to lead to the conclusion that the special function of the tubular cell (or epithelial cell in general) is due to a specific arrangement of membrane-bound transport systems t(he development of a polarity of the cell, as discussed below) and not caused by the presence of specific metabolic pathways that produce energy exlusively used for transport purposes. That the energy produced by the cell is equivalent for membrane transport processes and for other cellular functions

Table 4. Comparison of the metabolic fuels used by the various nephron segments for transepithelial transport and for cellular metabolism in the absence of transepithelial transport

Nephron segment	Substrate used according to studies on relation to sodium transport *in vivo*	Substrate used according to studies on O_2 consumption of kidney slices *in vitro* (no transepithelial transport)
Proximal tubule	Glucose (BRAND et al., 1974) Lactate (DIES et al., 1970b; GARZA-QUINTERO et al., 1975) Pyruvate, α-ketoglutarate, and oxalacetate (MAUDE, 1970) Acetate (RUMRICH and ULLRICH, 1968) Free fatty acids (BARAC-NIETO and COHEN, 1968; DIES et al., 1970b)	Glucose[a], acetoacetate, butyrate, oleate, lactate (WEIDEMANN and KREBS, 1969)
Thick ascending limb of Henle's loop	Free fatty acids (BARAC-NIETO and COHEN, 1968; DIES et al., 1970b)	Free fatty acids and glucose (HOHENEGGER, 1975)

In this table it is assumed that increased ureteral pressure affects mainly distal tubular transport, that chlorothiazide acts mainly on the cortical nephron segments, and that furosemide acts primarily on the thick ascending limb of Henle's loop. In all instances the active transepithelial transport is thought to be affected.

[a] From the studies of GUDER and SCHMIDT (1976) it is evident that the activity of glycolytic enzymes is very low in the proximal tubule. Therefore, in the *in-vitro* studies glucose is probably metabolized mainly by the distal portions of the nephron, which are present in the kidney slices together with the proximal segments.

such as gluconeogenesis is also indicated by the stimulatory action of ouabain (which inhibits the "sodium pump") on gluconeogenesis in rat kidney cortex slices (FRIDERICHS and SCHONER, 1973).

C. Transport-Related Intracellular Parameters of Metabolic Activity

In essence two kinds of intracellular parameters were investigated in studies on the relation between renal metabolism and renal function, the activity of enzymes and the concentration and turnover of the adenine nucleotide pool.

I. Determination of Enzyme Activities

This topic has been reviewed recently by SCHMIDT and GUDER (1976) and will be discussed only briefly. In microdissected tubular segments the activity of those enzymes was measured that catalyze irreversible steps in glycolysis, the pentose phosphate pathway, gluconeogenesis, and the citric acid cycle (HÖHMANN et al.,

1969; Schmidt and Guder, 1976). From the activity of these enzymes the maximal capacity of the various metabolic pathways can be roughly estimated and their contribution to the cellular energy production can be determined. The exact rates are difficult to calculate since the relevant concentrations of the substrates, products, cofactors, and regulators are not known. According to these studies the different portions of the nephron exhibit the following metabolic activity. The proximal tubule has a very low glycolytic activity and high citric acid cycle activity, and is the only segment of the (rat) kidney where gluconeogenesis takes place. In the thick ascending limb of Henle's loop and in the distal convolution, high levels of glycolytic enzymes and citric acid cycle enzymes have been found, whereas in the collecting duct glycolysis and probably glycogenolysis are the main routes of energy production (Guder and Schmidt, 1976).

II. Determination of ATP Content and ATP Turnover

Studies of this kind have severe limitations, not only in the kidney but also in other tissues. First of all there is the technical problem of maintaining the *in vivo* level of the metabolites during fixation and extraction of the tissue. This is only possible if rapid freezing of the tissue is used for fixation, since a transient anoxia reduces the ATP levels drastically in the highly metabolically active kidney cortex (Kessler, 1970; Gerlach et al., 1961). Even if one manages to measure the "real" levels of the metabolites, these values are of limited value for studies on the coupling between active transport and metabolism. Owing to intracellular compartmentalization, the actual concentration of ATP at the site of the cell where the transport takes place is not known and the metabolite concentrations reflect steady-state levels; these only very indirectly indicate the turnover of the adenine nucleotide pool, which is the parameter one would like to know.

The levels of the energy-rich adenosinephosphates were first determined by Gerlach et al. (1961) in the rat kidney and by Kessler (1970) in dog kidney. In both species, cortical ATP decreases rapidly during one minute of ischemia, whereas medullary ATP is slightly reduced or not altered. In the dog, in contrast to the rat, adenylate kinase, an enzyme that catalyzes the reaction 2 ADP $\rightleftharpoons$AMP + ATP, is very active in the cortex, and during ischemia the ADP levels therefore remain almost unchanged, whereas the AMP level increases in proportion to the decrease in ATP level.

The results obtained by Gerlach et al. seem to be reliable since the freeze-clamping technique was used, which preserves the *in vivo* state of the cell reasonably well (see also Zwiebel et al., 1970). It has to be mentioned, however, that even with this technique quite different results can be obtained, according to whether the kidney has been manipulated in any way before the clamping is performed.

Studies on the action of inhibitor compounds on adenine nucleotides of renal cortex and sodium excretion in dog kidney *in vivo* have been performed by

URBAITIS and KESSLER (1971). The interpretation of their data is limited by the fact that the adenine nucleotide levels were determined with a technique in which the kidney became anoxic before it was frozen in liquid nitrogen. Therefore the control values of ATP content in nontreated animals dropped from 1.2 $\mu mol\,g^{-1}$ wet weight, as determined by the freeze clamping method, to 0.5 $\mu mol\,g^{-1}$ and the levels of nucleotides no longer reflected the status of the cell *in vivo*. Furthermore, fractional sodium reabsorption (percentage of filtered load) was used for the correlation between sodium transport and the energy state of the cell. As discussed above, however, the transport parameter related directly to the metabolic activity of the kidney is not fractional sodium reabsorption but the active component of the net sodium transport (see also Fig. 1).

Similar reservations apply to the studies on the relation between ATP turnover and sodium reabsorption in dog kidney performed by KESSLER et al. (1968). In these studies there is the additional difficulty that it is not quite clear what factors influence the labeling of ATP with ^{32}P in addition to the cellular metabolism. One might envisage that the phosphate permeability of the plasma membranes is also one of the rate-limiting steps for the incorporation.

Therefore the studies on the ATP content and the ATP turnover reported above do not provide a reliable basis for conclusions on the ATP dependence of the renal sodium transport.

III. ATP Content and α-Aminoisobutyric Acid Transport in Kidney Cortex Slices

Another study on the interdependence of transport properties of the cell and the adenine nucleotide system was performed by REA and SEGAL (1972). These authors studied the effect of substrates, metabolic inhibitors, anaerobiosis and storage at 4° C on ATP level and on α-aminoisobutyric acid uptake into rat renal cortex slices. Uptake was determined as accumulation above the medium concentration after 60 minutes. At this time the concentration of AIB in the renal cell is determined by the relation between influx and efflux. The active energy-requiring step in this steady state is the uptake into the cell, whereas the efflux is governed solely by passive forces. The essential feature of the findings of REA and SEGAL is that a constant high accumulation (distribution ratio) of about 3–3.2 is achieved regardless of the ATP content of the cells in the range of 0.5–1.5 mM. The accumulation seems to be reduced only if the cellular ATP level is below ~ 0.5 mM.

From studies on kidney slices it is known that the uptake of α-aminoisobutyric acid is stimulated by the presence of sodium ions (FOX et al., 1964; THIER et al., 1967), and experiments with isolated brush-border vesicles (EVERS et al., 1976) have indicated that the intravesicular accumulation of neutral aminoacids is dependent on the electrochemical potential difference of sodium across the membrane. Therefore a correlation between the electrochemical gradient of sodium and the intracellular accumulation should be expected. Indeed, reduction of the concentration of sodium in the medium (ULLRICH et al., 1974b) or increase of the intracellular sodium concentration after inhibition of the sodium-pumping system by ouabain has been found to be associated with an impaired accumulation and decreased transepithelial transport (ULLRICH et al., 1977). Anaerobiosis, dinitrophenol, oligomycin and probably also maleic acid are known to inhibit the sodium pump by reducing predominantly the ATP production of the cell. All these situations are accompanied with low ATP levels and low AIB accumulation. Ouabain, in contrast, increases the ATP level slightly (FRIDERICHS and SCHONER, 1973), reduces the transmembranal electrochemical potential difference of sodium across the plasma membrane, and suppresses AIB uptake. On the other hand, slices stored at 4° C for 48 hours (and probably also those stored for 24 h and 72 h) establish their normal sodium gradient during incubation at 37° C and accumulate AIB although the intracellular ATP level is reduced by 30–50 percent. Thus the correlative factor for AIB accumulation seems to be the electrochemical potential difference of sodium and not the intracellular ATP concentration. Unfortunately, no determinations of intracellular sodium concentrations were made by REA and SEGAL in their experiments; these might have provided further support for the assumption presented above. Such determinations might also answer the question as to whether – as would be the consequence of these considerations – even at an ATP content of 0.4 $\mu mol\, g^{-1}$ the normal distribution of ions across the cell membrane can be maintained.

D. Energetics of Transport as Studied with Isolated Renal Plasma Membranes

I. General Aspects

The recent development of techniques to isolate plasma membranes from the renal proximal tubule (for a recent review see KINNE, 1976) and to study their transport properties as vesicles (KINNE et al., 1975) offers a new possibility for investigation of the problem of the energetics of transport. In the isolated plasma membrane vesicles the composition of the intravesicular and the extravesicular medium can be altered at will, and therefore detailed studies on the driving forces of the transport of molecules across the cell membranes are possible. Since such studies can be performed on the apical and on the basolateral membranes of the proximal tubule separately, they should also provide an understanding of the energetics of transepithelial transport, where these two membranes are arrayed in series.

Two transport-related phenomena observed with isolated plasma membranes will be discussed below. The first comprises the observation that several ions stimulate the ATPase activity of the membranes. The existence of such ion-dependent ATPases is taken as evidence that in the intact cell the transport of the ion across the respective cell border can be active since energy can be provided by ATP hydrolysis for the movement of the ions across the membrane. The second includes the transport studies performed with membrane vesicles in the absence of ATP, where the existence of sodium-cotransport systems and the role of the electrochemical potential difference of sodium as a driving force for the intravesicular accumulation and for the transcellular transport of solutes is discussed.

For studies on ATP-dependent transport processes in vesicles it is preferable to have inside out-oriented vesicles. This orientation would allow ATP present at the outside of the vesicle to reach a former cytoplasmic face of the membrane. In the brush-border membrane fraction the orientation is almost exclucively outside out, so that it has so far not been possible to study ATP-dependent reactions. The baso-lateral membrane fraction is composed of outside-out and inside-out vesicles, and it has therefore recently been possible to observe ATP-driven calcium accumulation in this fraction (GMAJ et al., 1977).

II. Transport-Related ATP Hydrolases

1. Sodium

Evidence for the presence of an active transport system for sodium in the plasma membranes of renal epithelial cells has been obtained from the observation that the hydrolysis of ATP by these membranes in the presence of magnesium is stimulated by the addition of sodium and potassium (BLOND and WHITTAM, 1964). The presence of this Na^+-K^+-ATPase in kidney membranes leads to the conclusion that this enzyme, similarly to the erythrocyte, is involved in the active extrusion of sodium out of the cell against the electrochemical potential gradient which, due to the inside negative cell potential and the low intracellular sodium concentration, is present across the plasma membrane.

Evidence for the involvement of Na^+-K^+-ATPase in the transepithelial sodium reabsorption in the proximal tubule came from several observations. Firstly, it was found by quantitative histochemistry (SCHMIDT and DUBACH, 1971) and histochemistry (ERNST, 1975), and in membrane fractionation studies (KINNE et al., 1971; HEIDRICH et al., 1972) that the Na^+-K^+-ATPase of the proximal tubular epithelial cell is present exclusively in the baso-lateral plasma membranes and not in the brush-border membranes. This cellular distribution corroborated physiological considerations of a passive entry of sodium into the cell at the luminal border and an active transfer at the contraluminal cell side. Secondly, the enzymatic activity rises in an adaptive way in rat kidneys when the demand for sodium reabsorption is chronically increased (KATZ and EPSTEIN, 1967). Furthermore, ouabain and other cardiac glycosides as specific inhibitors of Na^+-K^+-ATPase produce a sodium diuresis during infusion into the renal artery (RUEDAS and WEISS, 1967; NELSON, 1970; NECHAY et al., 1970; TORETTI et al., 1972), and inhibit sodium transport in single surface tubules of rat kidney slices (MAUDE, 1969), in micropunctured rat (GYÖRY et al., 1972), golden hamster (ULLRICH et al., 1977), and *Necturus* tubules (SCHATZMANN, 1958), and in isolated rabbit tubules (GRANTHAM, 1970).

Interestingly, the sodium transport is inhibited differently in various species. A very sensitive tissue preparation is the rabbit proximal tubule where the sodium transport-related potential difference is brought to zero by the addition of 10^{-5} M ouabain to the bathing medium (BURG and ORLOFF, 1970). In rat proximal tubules, when the peritubular capillaries are perfused with an even higher concentration of ouabain ($5 \cdot 10^{-3}$ M), isotonic solute transfer is inhibited by a maximum of 44 percent (GYÖRY et al., 1972); in the isolated perfused rat kidney also, approximately 50 percent of sodium is reabsorbed in the presence of $1.2–2.4 \cdot 10^{-2}$ M ouabain (ROSS et al., 1974). In the golden hamster 1 mM ouabain inhibits isotonic sodium reabsorption in the proximal tubule by 80 percent (ULLRICH et al., 1977).

The difference observed in the sensitivity of the transport systems to ouabain correlates with the well-known differences in the ouabain sensitivity of the Na^+-K^+-ATPase. Therefore, it is reasonable to assume that the ouabain-sensitive component of the sodium reabsorption involves the sodium-potassium-activated ATPase and thus is linked to the cellular metabolism *via* ATP. The different magnitude of maximum inhibition is most probably due to the composed nature of the net sodium transport across the epithelium. As already discussed above, sodium reabsorption in the proximal tubule of the rat involves active transport (~ 40%) and passive transport *via* solvent drag (~ 30%) and coupled to electrical driving forces (~30%). Application of ouabain to the early proximal rat tubule results in an increase in the transepithelial inside positive electrical potential (FRÖMTER, 1975); thus, although the active transport of Na^+ might be completely inhibited, the passive driving forces might be able to maintain the residual sodium transport during ouabain poisoning. The ouabain-insensitive transport system, which generates the inside positive potential, is probably an electrogenic proton/HCO_3^- pump (FRÖMTER, 1975; FRÖMTER et al., 1976). This movement of sodium together with bicarbonate can be inhibited by azetazolamide; accordingly, for example, in the golden hamster a 93 percent inhibition of sodium reabsorption was observed when 0.2 mM azetazolamide was imposed on 1 mM ouabain (ULLRICH et al., 1977). Thus, one might hypothesize that two active transport processes are involved in proximal tubular sodium reabsorption, one for Na^+ (Cl^-), mediated by the Na^+-K^+-ATPase, and one for (Na^+) HCO_3^-, mediated by the electrogenic proton transport system — the ion in brackets accompanies the respective actively transported ion passively. The different magnitude of ouabain inhibition in the various species thus might be explained by a different proportion of the two transport processes.

2. Calcium

a) Properties of Renal Ca^{++}-Activated ATPases

In analogy to the Na^+-K^+-ATPase, a calcium-stimulated Na^+-K^+-ATPase has been postulated to be involved in active calcium transfer across the erythrocyte membrane (SCHATZMANN and ROSSI, 1971). In renal plasma membranes two types of Ca^{++}-ATPase have been detected: one enzyme requires magnesium ATP as substrate and is – in a way similar to the red cell enzyme – maximally stimulated by 0.1 mM calcium (high-affinity ATPase) (MOORE et al., 1974); the other ATPase exhibits optimal activity in the absence of magnesium at a calcium concentration of 5–10 mM (low-affinity ATPase) (PARKINSON and RADDE, 1971; KINNE-SAFFRAN and KINNE, 1974a). The activity of the low-affinity enzyme is about 20 times higher than the activity of the high-affinity ATPase. Both ATPases can be distinguished from the mitochondrial ATPase because they are insensitive to azide and oligomycin.

In two investigations the calcium pumping or binding properties of renal membranes have been compared to the activity of the high-affinity Ca^{++}-ATPase system. In rabbit kidney plasma membranes PALMER and POSEY (1970) found no correlation between enzyme activity and calcium uptake when nucleotide specificity, calcium affinity and sensitivity to mercurials were tested. They could, however, correlate the phosphorylation of the membranes to the amount of calcium taken up. Thus they concluded that under the experimental conditions employed, calcium binding to the membranes rather than transport into vesicles was studied. This assumption was supported by the fact that oxalate did not stimulate the uptake of calcium by the membranes. On the other hand, stimulation of uptake by oxalate was demonstrated by MOORE et al. (1974), who studied the calcium pump activity in plasma membranes and microsomes isolated from rat kidney. In this study also, no correlation between the amount of ATP hydrolyzed and the amount of calcium transported could be demonstrated. In both investigations the stimulation of ATP-hydrolysis by the addition of calcium was about 60 times higher than the amount of calcium taken up by the vesicles. However, in regard to the probably low degree of vesiculation of the membrane fractions and the lack of information on the orientation of the vesicles, as also mentioned by MOORE et al. (1974), close coupling is not to be expected.

b) Possible Relation of Ca^{++}-ATPase to Transepithelial Ca^{++} Transport

The high-affinity ATPase has several properties that make it attractive to speculate that this enzyme is involved in the transtubular calcium transport. Its high affinity for calcium would enable the enzyme to work effectively at the low intracellular level of free calcium. The indication that it might be capable of catalyzing active calcium accumulation in membrane vesicles is also compatible with this postulated role. The only fact that would contradict this assumption is that the enzyme appears to be a constituent of the endoplasmic reticulum and not of the plasma membranes (MOORE et al., 1974). However, certainly these localization studies have to be repeated with the improved separation techniques now available.

The low-affinity Ca^{++}-ATPase shows one important feature in respect of transepithelial calcium transport: it is localized in the peritubular membrane, where an active step in the transepithelial transport is assumed. At this side of the cell a high electrochemical potential difference has to be overcome during the extrusion of calcium. The low affinity of the enzyme for calcium does not *per se* exclude the possibility that it is involved in the calcium transfer; it might be compensated by the high capacity of the system. In connection with the low-affinity Ca^{++}-ATPase, one always has to be aware of the possibility that this enzyme might be related to the presence of actin-myosin filaments in the plasma membranes. Such filaments have been demonstrated in the basal part of proximal tubular cells (ROSTGAARD et al., 1972).

Thus, at the moment the function of the Ca^{++}-ATPase and its relation to transepithelial Ca^{++} transport is far from being understood. Micropuncture studies revealed a close relationship between transtubular sodium and calcium transport (FRICK, 1965) and showed that the calcium transport is dependent on the ambient sodium concentration and is inhibited by ouabain (ULLRICH et al., 1976). Therefore, it is also possible that active calcium transfer at the peritubular side involves a sodium/calcium exchange, as demonstrated in the squid axon (BLAUSTEIN, 1974) and recently also in kidney membranes (GMAJ, MURER, KINNE, unpubl.).

3. Bicarbonate and Protons

A bicarbonate-stimulated ATPase has been implied to represent a primary active transport system for HCO_3^-/H^+ in gastric mucosa, pancreas, submaxillary duct, and renal proximal tubule (KASBEKAR and DURBIN, 1965; SACHS et al., 1972; SIMON et al., 1972a and b; WIZEMANN et al., 1974; WAIS and KNAUF, 1975; KINNE-SAFFRAN and KINNE, 1974b).

Although the first investigations of KATZ and EPSTEIN (1971) indicated that the bicarbonate-stimulated ATPase activity found in renal plasma membrane fractions was due to mitochondrial contamination, it has recently been demonstrated that there is a parallel distribution and enrichment of alkaline phosphatase, a marker enzyme of the microvillous membrane, and HCO_3^- ATPase during free flow electrophoretic separation of luminal and contraluminal plasma membranes (KINNE-SAFFRAN and KINNE, 1974b). At the same time, succinic dehydrogenase, a mitochondrial marker enzyme, was not enriched in the isolated brush-border fraction. This indicated that a nonmitochondrial ATPase is present in renal plasma membranes and that this enzyme is confined to the luminal plasma membrane of the proximal tubular cell, where the active step in the transepithelial H^+/HCO_3^- transport is thought to be localized. The apparent K_m of the enzyme for bicarbonate (16 mM l^{-1}) also compares well with the recently determined HCO_3^- dependence of tubular H^+/HCO_3^- transport (FRÖMTER et al., 1976).

An electrogenic transfer of protons (or bicarbonate) across the luminal membrane has been postulated by FRÖMTER (1975) on the basis of electrophysiological experiments. Unfortunately it has not yet been possible to demonstrate HCO_3^--stimulated, ATP-driven proton transport in isolated brush-border membrane vesicles (KINNE and MURER, unpubl.). This is probably due to the fact that the brush-border membrane vesicles studied were oriented outside out and therefore ATP had no access to the catalytic site of the proton-translocating system. As discussed below, only electroneutral sodium-dependent proton movements could be found in these vesicles (MURER et al., 1976). The relation between this exchange system and the ATPase is still not clear; although the ATPase activity seems not to be influenced by sodium, in the absence of ATP an Na^+/H^+ exchange reaction might be catalyzed by the enzyme system.

A bicarbonate-stimulated ATPase has also been described in the luminal membranes of collecting duct epithelial cells (SCHWARTZ et al., 1974). According to ULLRICH and EIGLER (1958), marked acidification of the urine occurs in the collecting ducts; therefore again, a link between the luminal HCO_3^--ATPase and active proton secretion at the apical cell pole might be postulated.

4. Chloride

In a variety of biological systems, thiocyanate inhibits anion (especially chloride) transport across epithelial surfaces (DAVENPORT, 1940; WOLFF, 1964; SACHS et al., 1972; EPSTEIN et al., 1973). In the kidney, active transport of chloride has been described in the thick ascending limb of Henle's loop (BURG and GREEN, 1973a), and it was therefore of interest to elucidate the possible role of the thiocyanate-inhibited ATPase in the transport of this structure. As shown by SOLOMON et al. (1975), a thiocyanate-sensitive Mg^{++}-ATPase is present in the mitochondrial and in the microsomal fraction of rat kidney cortex and rat kidney outer medulla (the part of the kidney where the thick ascending limb of Henle's loop is located), and the activity in the microsomal fraction is probably related to the existence of a nonmitochondrial (plasma membrane-bound) thiocyanate-sensitive ATPase. The specific activity of this enzyme in microsomes of medullary tissue was, however, only 58 percent of the activity in comparable membrane fragments from renal cortex. The authors therefore concluded that if the thiocyanate-inhibited ATPase were connected with the chloride transport one might anticipate a higher activity in the plasma membranes of the outer medulla than in the cortex. Because this was not found, a role of thiocyanate-inhibited ATPase in chloride transport analogous to that of Na^+/K^+-ATPase for sodium and potassium seems unlikely.

This conclusion cannot be regarded as definitive, because it still might be possible that the luminal membranes of the cells in the thick ascending limb have a very high activity of the enzyme, which might not have been detected in the above-mentioned experiments because no separation of the luminal and contraluminal plasma membranes of the cells in the thick ascending limb of Henle's loop was achieved. There is additionally some uncertainty as to whether the thiocyanate-inhibited ATPase is an individual ATPase or whether several ATPases (including the HCO_3^--ATPase; see above) are inhibited by thiocyanate.

III. Sodium-Solute Cotransport Systems

1. General Aspects

In the elucidation of the molecular mechanism of transport processes in epithelia, it becomes more and more evident that the sodium gradient established across the plasma membrane by the action of the Na^{+}-K^{+}-ATPase is capable of providing driving force for active transepithelial transport of solutes. The pioneering ideas on the sodium gradient-coupled movement of sugars and aminoacids put forward by CRANE (1962) and EDDY (1968) have been supported by studies on intact cells (ULLRICH, this Volume, Chapter 10, ULLRICH et al., 1974a and b), and have been proven by studies on isolated plasma membrane vesicles. The evidence that has accumulated over recent years from these last-mentioned studies suggests that essentially all solutes that have been shown to be transported in a sodium-dependent manner by epithelia undergo co- or countertransport with sodium at one cell side or the other. It has also become clear for electroneutral solutes such as sugars and neutral aminoacids that the electrochemical potential difference of sodium, and not simply the concentration difference, is the factor promoting the transfer across the cell membrane (MURER and HOPFER, 1974; KINNE et al., 1975; GECK and HEINZ, 1976). Therefore, the determination of the electrical membrane potential in studies on the energetics of sodium-dependent transport becomes necessary.

In the next section, the studies on the transport properties of plasma membrane vesicles can be presented only in a very condensed form. The reader is referred to a recent review on membrane molecular aspects of tubular function for further information (KINNE, 1976).

2. General Characteristics of Transport by Renal Vesicles

Under appropriate conditions, renal plasma membranes vesiculate and enclose an intravesicular space into which transport of solutes occurs. The vesicles are osmotically reactive; they shrink when the osmolality of the extravesicular space is increased by the addition of nonpermeant solutes, and they burst if they are suspended in hypo-osmotic media (osmotic shock). This behavior is one of the criteria that makes it possible to distinguish between transport into the vesicles and binding to or incorporation into the membranes (HOPFER et al., 1973; KINNE et al., 1975; EVERS et al., 1976; HOFFMANN et al., 1976; BERNER and KINNE, 1976).

In general, the passive uptake of solutes across the membranes may occur *via* simple diffusion or by carrier-mediated, so-called facilitated diffusion systems. The last-named transport systems show saturability, substrate specificity and inhibition by specific inhibitors. For a given membrane vesicle the relative contribution of the two pathways to the uptake varies, depending on the substrate and on its concentration. In the renal brush border, in the presence of a NaCl gradient, the transport by facilitated diffusion exceeds transport *via* simple diffusion by a factor of ~10 for phosphate, of ~ 3 for D-glucose and of ~ 2 for phenylalanine (all substances in 1 mM concentration) (HOFFMANN et al., 1976; KINNE et al., 1975; EVERS et al., 1976).

Furthermore, it has been established that the transport properties of the luminal and contraluminal (baso-lateral) plasma membranes are different. Sodium-dependent transport systems for glucose (KINNE et al., 1975), phenylalanine (EVERS et al., 1976), phosphate (HOFFMANN et al., 1976), and protons (MURER et al., 1976) have been observed only in the brush-border membrane. The transport systems for these solutes located in the baso-lateral plasma membranes seem to be carrier-mediated but sodium-independent.

3. The Sodium Gradient as the Driving Force of Intravesicular Accumulation

In isolated plasma membrane vesicles the intravesicular sodium can be completely removed, and therefore the effect of a sodium gradient on substrate uptake can easily be studied. On the other hand it is also possible to equilibrate the intra- and extravesicular space with respect to cations and anions and to study transport under these conditions. Figure 3 shows the uptake of D-glucose by brush-border vesicles under the two different conditions, and for comparison in the presence of potassium. It is evident that in the presence of a sodium gradient across the membrane the initial uptake of D-glucose is stimulated and

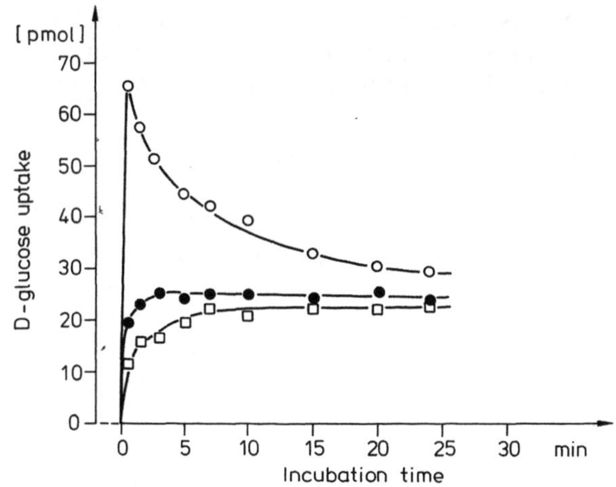

Fig. 3. Uptake of D-glucose into brush-border membrane vesicles in presence of sodium gradient. Isolated brush-border vesicles containing 100 mM mannitol and 1 mM Tris-HEPES buffer, pH 7.4, incubated at 25° C in solution containing in addition 1 mM ^{3}H-glucose and 100 mM NaCl (○—○) or 100 mM KCl (□—□). At time indicated, amount of D-glucose taken up by vesicles was determined by a rapid filtration method (HOPFER et al., 1973; KINNE et al., 1975). The closed circles (●—●) represent uptake of glucose in absence of sodium gradient across brush-border membrane; vesicles had been pre-equilibrated with 100 mM mannitol, 1 mM Tris-HEPES and 100 mM NaCl, and were incubated in the same medium containing 1 mM ^{3}H-D-glucose in addition. (Data taken from KINNE, 1976)

that furthermore the uptake curve shows an overshoot, *i. e.* the amount of D-glucose found in the vesicles transiently exceeds the intravesicular concentration observed at equilibrium; at equilibrium the concentration inside and outside the vesicle can be considered to be identical. The overshoot most probably indicates – such drastic osmotic changes of the vesicle volume seem to be excluded – that for a short time the concentration of D-glucose inside the vesicle is higher than outside, *i. e.* that a transient accumulation of glucose by the vesicles *via* cotransport is observed. The overshoot is absent if there is no sodium gradient across the membrane (see Fig. 3). These findings indicate a coupling of sodium flux to glucose flux, an assumption which is further supported by the fact that even in the absence of a cation gradient, the uptake of glucose is faster in the presence of sodium than in the presence of potassium. Such a stimulation points to a direct interaction of sodium with the transport system, one of the prerequisites for a molecular flux coupling.

Similarly, sodium gradient-supported transient intravesicular accumulations have been observed for phenylalanine (EVERS et al., 1976) and phosphate (HOFFMANN et al., 1976) in renal brush-border vesicles.

Thermodynamically such coupling processes can be described by Eq. (2)

$$J_i = L_{ii}\Delta\eta_i + L_{i\mathrm{Na}^+}\Delta\eta_{\mathrm{Na}^+}, \tag{2}$$

which links the flux of the solute i (J_i) to the electrochemical potential difference of i ($\Delta\eta_i$) and to the flux of sodium, which is dependent on the electrochemical potential difference of sodium ($\Delta\eta_{\mathrm{Na}^+}$) and is coupled *via* the cross-coefficient $L_{i\mathrm{Na}^+}$ to the transport system of i; L_{ii} is the straight coefficient, which describes the properties of the transport system for i. For electroneutral solutes such as D-glucose the equation can be simplified by replacing $\Delta\eta_{\mathrm{Glu}}$ by Δc_{Glu}:

$$J_{\mathrm{Glu}} = L_{\mathrm{GluGlu}}\Delta c_{\mathrm{Glu}} + L_{\mathrm{GluNa}^+}\Delta\eta_{\mathrm{Na}^+}$$

From this equation the occurrence of an overshoot can be explained as follows: at the beginning of the experiment, Δc_{Glu} and $\Delta\eta_{\mathrm{Na}^+}$ are positive and uptake of glucose is observed. As soon as the intravesicular glucose concentration due to the persisting $\Delta\eta_{\mathrm{Na}^+}$ of the sodium gradient becomes higher than the extravesicular concentration, Δc_{Glu} becomes negative and J_{Glu} approaches zero when $\Delta\eta_{\mathrm{Na}^+}$ and $-\Delta c_{\mathrm{Glu}}$ cancel each other (peak of the overshoot). Since $\Delta\eta_{\mathrm{Na}^+}$ is not constant but is continuously diminished by a rapid influx of sodium into the vesicle, D-glucose then leaves the vesicles following the negative Δc_{Glu} until all gradients $\Delta\eta_{\mathrm{Na}^+}$ and Δc_{Glu} are zero and equilibrium is reached. When no sodium gradient is present, $\Delta\eta_{\mathrm{Na}^+}$ is zero and only equilibration of the intra- and extravesicular space is observed.

These considerations also demonstrate that the occurence of an overshoot depends on the relative rates of uptake of substrate and of sodium; if the rate of sodium entry is higher than that of the substrate, no overshoot can be observed in the isolated vesicles (see for example BERNER et al., 1976). This lack of overshoot *in vitro*, however, does not exclude the possibility that the sodium gradient provides the energy for the transport in the intact cell, because *in vivo* the sodium gradient across the membrane is maintained by the action of the Na^+/K^+-ATPase. The sodium-dependent phosphate transport system in the small intestine seems to belong to this category (BERNER et al., 1976).

4. The Electrical Membrane Potential as the Driving Force of Intravesicular Accumulation

As indicated in Eq. (2) given above, the electrochemical gradient ($\Delta\eta_{Na^+}$) of sodium is one of the driving forces for the uptake of substrates by sodium coupled transport systems. Since $\Delta\eta_{Na^+}$ is composed of the chemical term and an electrical term, the membrane potential, even in the absence of a chemical gradient of sodium, should be able to support intravesicular accumulation. Such an experiment is shown in Figure 4. Here the membrane potential across the brush-border membrane has been altered by the addition of KSCN to the incubation medium (EVERS et al., 1976). Since the anion SCN^- has a higher permeability than the cation K^+, a diffusion potential is generated across the membrane, which renders the intravesicular space negative and thus increases $\Delta\eta_{Na^+}$; correspondingly a transient accumulation of L-phenylalanine in the brush-border vesicles is observed.

According to these results obtained *in vitro*, changes of the electrical potential across the epithelium of the proximal tubule and across the brush-border membrane have been observed when glucose and aminoacids are applied to the tubular lumen in the presence of sodium (FRÖMTER and GESSNER, 1974; MARU-

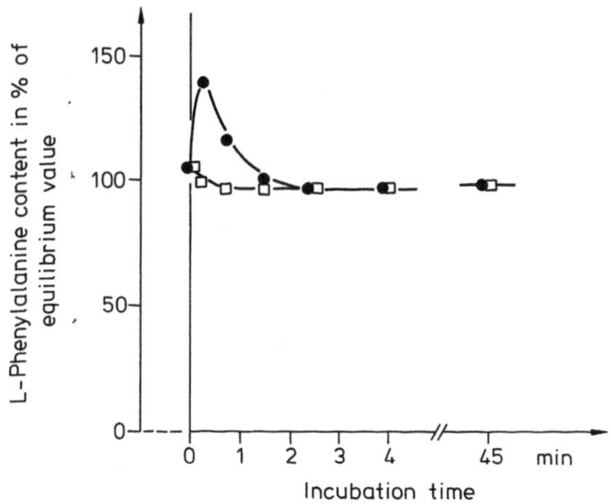

Fig. 4. Accumulation of L-phenylalanine in isolated brush-border membrane vesicles induced by changes of membrane potential. Isolated brush-border vesicles incubated for 1 h at 25° C in solution containing 100 mM mannitol, 50 mM Na_2SO_4, 5 mM Tris-HEPES, pH 7.4, and 1 mM ^{3}H-L-phenylalanine. At zero time 10 µl of 1.6 M KSCN (●——●) or 0.8 M K_2SO_4 (□——□) added to 150 µl of membrane suspension. Values expressed as percentages of new equilibrium reached after 45 min of incubation, and taken from EVERS et al. (1976). Addition of KSCN renders inside of brush-border vesicles negative because of high permeability of membrane to SCN and low permeability to potassium

YAMA and HOSHI, 1972; KOKKO, 1972; SAMARZIJA and FRÖMTER, 1979a). The substrate-induced depolarization of the cell is dependent on the substrate concentration and can be explained by an increased sodium entry across the brush border into the cell. Thus, the change in membrane potential can be considered as a cellular correlate of sodium-dependent tubular transport processes. This holds, however, only for solutes where transfer across the brush border as a solute/sodium complex is associated with the transfer of positive charges. Phosphate, for example, as shown by HOFFMANN et al. (1976), is transferred (at pH 7.4) across the brush border predominantly in an electroneutral manner, *i. e.* the transfer of negative charges by phosphate is compensated by the simultaneous movement of sodium ions. This process is electrically silent (SAMARZIJA and FRÖMTER, 1979b). The H^+/Na^+ exchange (or OH^-/Na^+ cotransport) in the brush border also seems to be electroneutral (MURER et al., 1976) and therefore cannot be detected by potential measurements.

5. Energetics of Transcellular Transport in the Proximal Convoluted Tubule

From the results obtained with isolated plasma membranes on the localization and properties of transport systems, which are summarized in Figure 5, the kind

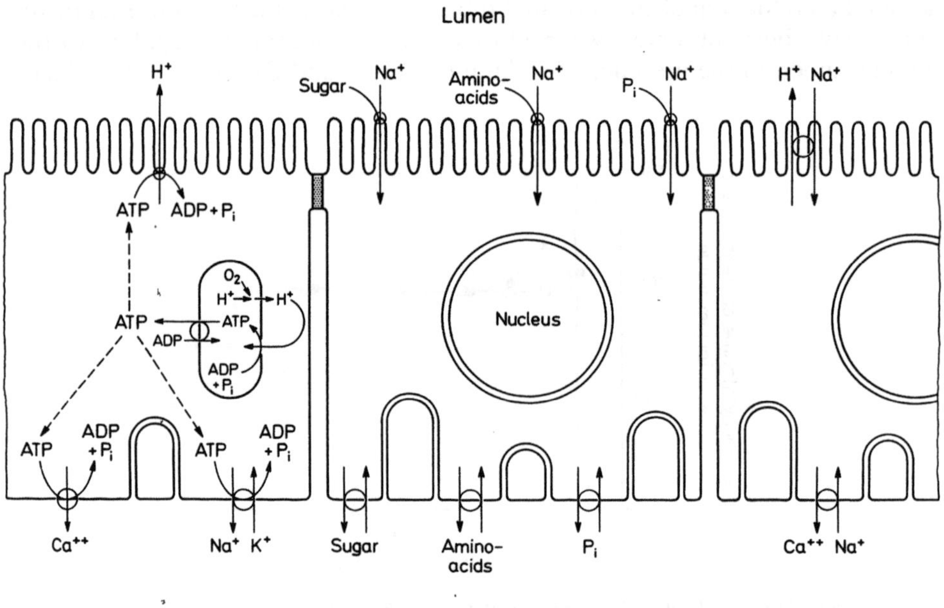

Fig. 5. Localization of active and sodium solute cotransport systems in the proximal tubule epithelial cell. In cell on left-hand side all active transport systems are shown. Assumed that Na^+-K^+-stimulated ATPase, Ca^{++}-stimulated ATPase, and HCO_3^--stimulated ATPase represent active transport systems for sodium, calcium, and protons respectively. In middle, sodium-dependent cotransport systems for sugars, neutral aminoacids, and phosphate, and sodium-independent transport systems for these solutes compiled. In cell on right, countertransport systems are assembled

of driving forces in the active reabsorption of the various substances can be deduced. For sodium, an active transport system is located in the baso-lateral plasma membranes of the tubular cell. The active transepithelial transfer of this ion is thus linked directly to the ATP-generating metabolism. Sugars (D-glucose), neutral aminoacids (phenylalanine), and phosphate are transferred across the brush-border membrane by sodium-cotransport systems, the electrochemical potential difference of sodium thereby providing the energy for the intracellular accumulation against an electrochemical gradient. They leave the cell *via* sodium-independent transport systems located at the basal pole of the cell. The driving force for these transport processes – the electrochemical potential difference of sodium – is secondarily linked to the metabolism, because the concentration gradient of sodium is maintained by the action of the sodium/potassium-stimulated ATPase. For the bicarbonate-buffer system and calcium, probably both a primary active ATPase and an exchange system with sodium have to be considered. Therefore direct as well as indirect coupling to metabolism might exist.

In this connection, it should be pointed out that the active transport systems for sodium and calcium, as well as the sodium-dependent transport systems for aminoacids and phosphate, are not only found in polarized epithelial cells but also in nonpolarized cells, such as nucleated erythrocytes and Ehrlich's ascites tumor cells (EDDY, 1968; BROWN and LAMB, 1975). Transepithelial transport thus seems to require different composition of the luminal and contraluminal cell border, whereby transport systems are arranged in such a way that the substrate transported reacts successively with the different transport systems during its passage through the cell.

Acknowledgements

The author wishes to express his gratitude to *Drs. K. J. Ullrich, W. Guder, H. Murer,* and *E. Kinne-Saffran* for the stimulating discussions which contributed to the final form of the manuscript.

References

AUKLAND, K., JOHANNESEN, J., KIIL, F.: Scand. J. clin. Lab. Invest. **23**, 317 (1969).
BARAC-NIETO, M., COHEN, J. J.: Amer. J. Physiol. **215**, 98 (1968).
BARAC-NIETO, M., COHEN, J. J.: Nephron **8**, 488 (1971).
BERNER, W., KINNE, R.: Pflügers Arch. **361**, 269 (1976).
BERNER, W., KINNE, R., MURER, H.: Biochem. J. **160**, 467 (1976).
BERTERMANN, H., FRANKE, H., HULAND, H., WEISS, C.: Res. exp. Med. **161**, 1 (1973).
BLAUSTEIN, M. P.: Rev. Physiol. Biochem. Pharmacol. **70**, 33 (1974).
BLOND, D. M., WHITTAM, R.: Biochem. J. **92**, 158 (1964).
BRAND, P. H., COHEN, J. J., BIGNALL, M. C.: Amer. J. Physiol. **227**, 1255 (1974).

BROWN, K. D., LAMB, J. F.: J. Physiol. (Lond.) **251**, 58P (1975).
BURG, M. B., GREEN, N.: Amer. J. Physiol. **224**, 659 (1973a).
BURG, M. B., GREEN, N.: Kidney International **4**, 245 (1973b).
BURG, M. B., GREEN, N.: Kidney International **4**, 301 (1973c).
BURG, M. B., ORLOFF, J.: Amer. J. Physiol. **219**, 1714 (1970).
BURG, M. B., STONER, L., CARDINAL, J., GREEN, N.: Amer. J. Physiol. **225**, 119 (1973).
CHERTOK, R. J., HULET, W. H., EPSTEIN, B.: Amer. J. Physiol. **211**, 1379 (1966).
COHEN, J. J., BARAC-NIETO, M.: In: Handbook of Physiology, Section 8 (J. Orloff and R. W. Berliner, Eds.). Baltimore, Md: Williams and Wilkins 1973, p. 909.
COHEN, J. J.: Medical Clinics of North America **59**, 523 (1975).
CRANE, R. K.: Fed. Proc. **21**, 891 (1962).
DAVENPORT, H. W.: Amer. J. Physiol. **129**, 505 (1940).
DEETJEN, P., KRAMER, K.: Pflügers Arch. **273**, 636 (1961).
DIES, F., HERRERA, J., MATOS, M., AVELAR, E., RAMOS, G.: Amer. J. Physiol. **218**, 405 (1970a).
DIES, F., RAMOS, G., AVELAR, E., MATOS, M.: Amer. J. Physiol. **218**, 411 (1970b).
EDDY, A. A.: Biochem. J. **108**, 195 (1968).
EPSTEIN, F. H., MAETZ, J., de RENZIS, G.: Amer. J. Physiol. **224**, 1295 (1973).
ERNST, A.: J. Cell Biol. **66**, 586 (1975).
ESSIG, A.: Biophys. J. **15**, 651 (1975).
EVERS, J., MURER, H., KINNE, R.: Biochim. biophys. Acta **426**, 598 (1976).
FOX, M., THIER, S., ROSENBERG, L., SEGAL, S.: Biochim. biophys. Acta **79**, 167 (1964).
FRANKE, H., MALYUSZ, M., WEISS, C.: Current Problems in Clinical Biochemistry, Vol. 4; Biochemical Aspects of Renal Function (S. Angielski and U. C. Dubach, Eds). Berne: Huber 1975, p. 169.
FRICK, A., RUMRICH, G., ULLRICH, K. J., LASSITER, W. E.: Pflügers Arch. **286**, 109 (1965).
FRIDERICHS, D., SCHONER, W.: Biochim. biophys. Acta **304**, 142 (1973).
FRÖMTER, E.: Fortschritte der Zoologie **23**, 248 (1975).
FRÖMTER, E., GESSNER, K.: Pflügers Arch. **351**, 85 (1974).
FRÖMTER, E., RUMRICH, G., ULLRICH, K. J.: Pflügers Arch. **343**, 189 (1973).
FRÖMTER, E., SATO, K., GESSNER, K.: In: Advances in Nephrology (Proc. 6th Int. Congress of Nephrology, Florence, 1975; S. Giovannetti, V. Bonomini and G. D'Amico, Eds). Basel: Karger 1976, p. 108.
FUJIMOTO, M., NASH, F. C., KESSLER, R. H.: Amer. J. Physiol. **206**, 1327 (1964).
FÜLGRAFF, G., WOLF, K., ADELMANN, J., KRIEGER, A.-K.: Arch. Pharmak. exp. Path. **263**, 485 (1969).
GARRAHAN, P. J,, GLYNN, I. M.: J. Physiol. **192**, 217 (1967).
GARZA-QUINTERO, R., COHEN, J. J., BRAND, P. H., YOUNG, J. K.: Amer. J. Physiol. **228**, 549 (1975).
GECK, P., HEINZ, E.: Biochim. biophys. Acta **443**, 49 (1976).
GERLACH, E., BADER, W., SCHWORER, W.: Pflügers Arch. **272**, 407 (1961).
GMAJ, R., MURER, H., KINNE, R.: Pflügers Arch. **368**, R21 (1977).
GRANTHAM, J. J., BURG, M., ORLOFF, J.: J. clin. Invest. **49**, 1815 (1970).
GRUPP, G., JANNSEN, S.: Pflügers Arch. **267**, 58 (1958).
GUDER, W. G., SCHMIDT, U.: Kidney International **10**, S-32 (1976).
GYÖRY, A. Z., KINNE, R.: Pflügers Arch. **327**, 234 (1971).
GYÖRY, A. Z., BRENDEL, U., KINNE, R.: Pflügers Arch. **335**, 287 (1972).
HEIDRICH, H. G., KINNE, R., KINNE-SAFFRAN, E., HANNIG, K.: J. Cell Biol. **54**, 232 (1972).
HERNANDEZ, J., CAPEK, K., HELLER, J., NORAKOVA, A.: Experientia **25**, 125 (1969).
HIERHOLZER, K.: Naunyn-Schmiedebergs Arch. exp. Path. **232**, 255 (1957).
HOFFMANN, N., THEES, M., KINNE, R.: Pflügers Arch. **362**, 147 (1976).
HOHENEGGER, M.: In: Current Problems in Clinical Biochemistry, Vol. 4: Biochemical Aspects of Renal Function (S. Angielski and U. C. Dubach, Eds). Berne, Stuttgart, Vienna: Huber 1975, p. 150.
HOHENEGGER, M., BRECHTELSBAUER, H., FINSTERER, U., PRUCKSUNAND, P.: Pflügers Arch. **351**, 231 (1974).
HÖHMANN, B., ZWIEBEL, R., YAMAGATA, A., KINNE, R.: Pflügers Arch. **312**, 110 (1969).
HOLZGREVE, H., FRICK, A., RUMRICH, G., WIEDERHOLT, M., ULLRICH, K. J.: 3. Symposion der Gesellschaft für Nephrologie, 1964: Normale und pathologische Funktionen des Nierentubulus (K. J. Ullrich and K. Hierholzer, Eds). Berne, Stuttgart: Huber 1965, p. 147.

Hopfer, U., Nelson, K., Perrotto, J., Isselbacher, K. J.: J. biol. Chem. **248**, 25 (1973).
Howards, S. S., Davis, B. B., Knox, F. G., Wright, F. S., Berliner, R. W.: J. clin. Invest. **47**, 1561 (1968).
Janssen, S., Rein, H.: Ber. Ges. Physiol. **42**, 567 (1928).
Kasbekar, D. K., Durbin, R. P.: Biochim. biophys. Acta **105**, 472 (1965).
Katz, A. I., Epstein, F. H.: J. clin. Invest. **46**, 1999 (1967).
Katz, A. I., Epstein, F. H.: Enzyme **12**, 499 (1971).
Kessler, R. H.: Proc. Soc. exp. Biol. **134**, 1091 (1970).
Kessler, R. H., Weinstein, S. W., Nash, F. D., Fujimoto, M.: Nephron **1**, 221 (1964).
Kessler, R. H., Landwehr, D., Quintanilla, A., Wesely, S. A., Kaufmann, W., Arcila, H., Urbaitis, B. K.: Nephron **5**, 474 (1968).
Kiil, F., Aukland, K., Refsum, H. E.: Amer. J. Physiol. **201**, 511 (1961).
Kinne, R.: Medical Clinics of North America **59**, 615 (1975).
Kinne, R.: In: Current Topics in Membranes and Transport, Vol. 8 (F. Bronner and A. Kleinzeller, Eds). New York: Academic 1976, p. 209.
Kinne, R.: In: Kidney and Urinary Tract Physiology, II Vol. 11 (K. Thurau, Ed.). (MTP International Review of Science.) London: Butterworths; Baltimore, Md: University Park Press 1976, p. 169.
Kinne, R., Schmitz, J. E., Kinne-Saffran, E.: Pflügers Arch. **329**, 191 (1971).
Kinne, R., Murer, H., Kinne-Saffran, E., Thees, M., Sachs, G.: J. Membrane Biol. **21**, 375 (1975).
Kinne-Saffran, E., Kinne, R.: J. Membrane Biol. **17**, 263 (1974a).
Kinne-Saffran, E., Kinne, R.: Proc. Soc. exp. Biol. **146**, 751 (1974b).
Kleinmann, J. G., Mandelbaum, J., Levin, M. L.: Amer. J. Physiol. **224**, 95 (1973).
Kokko, J. P.: Clin. Res. **20**, 598 (1972).
Lassen, N. A., Munck, O., Thaysen, J. H.: Acta physiol. scand. **51**, 371 (1961).
Leaf, A., Page, L. B., Anderson, J.: J. biol. Chem. **234**, 1625 (1959).
Lie, M., Johannesen, J., Kiil, F.: Amer. J. Physiol. **225**, 1181 (1973).
Maruyama, T., Hoshi, T.: Biochim. biophys. Acta **282**, 214 (1972).
Maude, D. L.: Amer. J. Physiol. **216**, 1199 (1969).
Maude, D. L.: Biochim. biophys. Acta **215**, 216 (1970).
Meng, K.: Arzneim.-Forsch. (Drug Res.) **22**, 796 (1972).
Mitchell, P.: In: The Molecular Basis of Membrane Function. (D. C. Tosteson, Ed.). Englewood Cliffs, N. J.: Prentice-Hall 1969, p. 483.
Moore, L., Fitzpatrick, D. F., Chen, T. S., Landon, E. J.: Biochim. biophys. Acta **345**, 405 (1974).
Mudge, G., Taggart, J. V.: Amer. J. Physiol. **161**, 173 (1950).
Murer, H., Hopfer, U.: Proc. nat. Acad. Sci. (Wash.) **71**, 484 (1974).
Murer, H., Hopfer, U., Kinne, R.: Biochem. J. **154**, 597 (1976).
Nechay, B. R., Palmer, R. F., Chinoy, D. A., Posey, V. A.: J. Pharmacol. exp. Ther. **157**, 599 (1967).
Nelson, J. A., Nechay, B. R.: J. Pharmacol. exp. Ther. **175**, 727 (1970).
Nieth, H., Schollmeyer, P.: Nature **209**, 1244 (1966).
Palmer, R. F., Posey, V. A.: J. gen. Physiol. **55**, 89 (1970).
Parkinson, D. K., Radde, I. C.: Biochim. biophys. Acta **242**, 238 (1971).
Pitts, R. F.: Medical Clinics of North America **59**, 507 (1975).
Pressman, P.: Proc. Fed. Amer. Soc. exp. Biol. **27**, 1283 (1968).
Rea, C., Segal, S.: Kidney International **2**, 101 (1972).
Reid, R. A.: Biochem. Soc. Spec. Publ. **4**, 113 (1974).
Rein, H., Janssen, S.: Ber. Ges. Physiol. **42**, 565 (1928).
Ross, B. D., Epstein, F. H., Leaf, A.: Amer. J. Physiol. **225**, 1165 (1973).
Ross, B. D., Leaf, A., Silva, P., Epstein, F. H.: Amer. J. Physiol. **226**, 624 (1974).
Rostgaard, J., Kristensen, B. I., Nielsen, L. E.: Z. Zellforsch. **132**, 497 (1972).
Ruedas, G., Weiss, C.: Pflügers Arch. **298**, 12 (1967).
Rumrich, G., Ullrich, K. J.: J. Physiol. (Lond.) **197**, 69P (1968).
Sachs, G., Wiebelhaus, V. D., Blum, A. L., Hirschowitz, B. I.: In: Gastric Secretion. (G. Sachs, E. Heinz and K. J. Ullrich, Eds). New York, London: Academic 1972, p. 321.
Samarzija, I., Frömter, E.: (in preparation) (1979a).

SAMARZIJA, I., FRÖMTER, E.: (in preparation) (1979b).
SATO, K.: In: Current Problems in Clinical Biochemistry, Vol. 4: Biochemical Aspects of Renal Function (S. Angielski and U. C. Dubach, Eds). Berne: Huber 1975, p. 175.
SAUER, F.: In: Handbook of Physiology, Section 8: Renal Physiology (J. Orloff and R. W. Berliner, Eds). Baltimore, Md: Williams and Wilkins 1973, p. 399.
SCHATZMANN, H. J., ROSSI, G. L.: Biochim. biophys. Acta **241**, 379 (1971).
SCHATZMANN, H. J., WINDHAGER, E. E., SOLOMON, A. K.: Amer. J. Physiol. **195**, 570 (1958).
SCHMIDT, U., DUBACH, U. C.: Nephron **7**, 447 (1970).
SCHMIDT, U., DUBACH, U. C.: Pflügers Arch. **330**, 265 (1971).
SCHMIDT, U., GUDER, W. G.: Kidney International **9**, 233 (1976).
SCHUREK, H. J., BRECHT, J. P., BRANDT, P.: In: Proceedings of 8th Symposium of the Gesellschaft für Nephrologie (R. Heintz and H. Holzhuter, Eds). Aachen: Heintz & Holzhuter 1972, p. 89.
SCHUREK, H. J., BRECHT, J. P., LOHFERT, H., HIERHOLZER, K.: Pflügers Arch. **354**, 349 (1975).
SCHWARTZ, I. L., SHLATZ, L. J., KINNE-SAFFRAN, E., KINNE, R.: Proc. nat. Acad. Sci. (Wash.) **71**, 2595 (1974).
SEN, A. K., POST, R. L.: J. biol. Chem. **239**, 345 (1964).
SILBERNAGL, S.: In: Nieren- und Hochdruckkrankheiten **3**, 18 (1974).
SIMON, B., KINNE, R., KNAUF, H.: Pflügers Arch. **337**, 177 (1972a).
SIMON, B., KINNE, R., SACHS, G.: Biochim. biophys. Acta **282**, 293 (1972b).
SOLOMON, R. J., SILVA, P., BEND, J. R., EPSTEIN, F. H.: Amer. J. Physiol. **229**, 801 (1975).
STRICKLER, J. C., KESSLER, R. H.: Amer. J. Physiol. **205**, 117 (1963).
THIER, S., BLAIR, A., FOX, M., SEGAL, S.: Biochim. biophys. Acta **135**, 300 (1967).
THURAU, K.: Proc. Soc. exp. Biol. **106**, 714 (1961).
TORELLI, G., MILLA, E., FAELLI, A., CONSTANTINE, S.: Amer. J. Physiol. **211**, 576 (1966).
TORETTI, J., HENDLER, E., WEINSTEIN, E., LONGNECKER, R. E., EPSTEIN, F. H.: Amer. J. Physiol. **222**, 1398 (1972).
ULLRICH, K. J.: In: Handbook of Physiology, Section 8: Renal Physiology (J. Orloff and R. W. Berliner, Eds). Baltimore, Md: Williams and Wilkins 1973, p. 377.
ULLRICH, K. J., EIGLER, F. W.: Pflügers Arch. **267**, 491 (1958).
ULLRICH, K. J., RUMRICH, G., KLÖSS, S.: Pflügers Arch. **351**, 35 (1974a).
ULLRICH, K. J., RUMRICH, G., KLÖSS, S.: Pflügers Arch. **351**, 49 (1974b).
ULLRICH, K. J., RUMRICH, G., KLÖSS, S.: Pflügers Arch. **364**, 223 (1976).
ULLRICH, K. J., CAPASSO, G., RUMRICH, G., PAPAVASSILIOU, F., KLÖSS, S.: Pflügers Arch. **368**, 245 (1977).
URBAITIS, B. K., KESSLER, R. H.: Amer. J. Physiol. **220**, 1116 (1971).
WAIS, U., KNAUF, H.: Pflügers Arch. **361**, 61 (1975).
WALTON, K. G., DELORENZO, R. J., CURRAN, P. F., GREENGARD, P.: J. gen. Physiol. **65**, 153 (1975).
WAUGH, W. H., KUBO, T.: Amer. J. Physiol. **217**, 277 (1969).
WEIDEMANN, M. J., KREBS, H. A.: Biochem. J. **112**, 149 (1969).
WEINSTEIN, S. W., KESSLER, R. H.: Amer. J. Physiol. **205**, 123 (1963).
WEINSTEIN, S. W.: Amer. J. Physiol. **219**, 978 (1970).
WEINSTEIN, S. W.: Amer. J. Physiol. **223**, 583 (1972).
WEINSTEIN, S. W.: In: Kidney and Urinary Tract Physiology, Vol. 6/1 (K. Thurau, Ed.). (MTP International Review of Science.) London: Butterworths; Baltimore: University Park Press 1974.
WINDHAGER, E. E., GIEBISCH, G.: Nature (Lond.) **191**, 1205 (1961).
WIZEMANN, V., CHRISTIAN, A.-L., WIECHMANN, J., SCHULZ, I.: Pflügers Arch. **347**, 39 (1974).
WOLF, K., BIEG, A., FÜLGRAFF, G.: Europ. J. Pharmacol. **7**, 342 (1969).
WOLFF, J.: Physiol. Rev. **44**, 45 (1964).
ZEHRAN, K.: Acta physiol. scand. **36**, 300 (1956).
ZWIEBEL, R., WIECHMANN, J., HÖHMANN, B., KINNE, R.: Hoppe-Seylers Z. physiol. Chem. **351**, 854 (1970).

Chapter 12

Transport in Salivary and Salt Glands

J.A. YOUNG and E.W. VAN LENNEP

Part I: Salivary Glands

To Leon Schneyer in memoriam

A. Introduction

Properly speaking, the term, salivary gland should be taken as including any tissue that normally discharges a secretory product into the oral cavity. The functions of such secretions will include: 1. moistening of the buccal mucosa with the implied role of supporting oral hygiene and appetite; 2. providing lubrication to aid in the swallowing of solid and semisolid foodstuff; 3. providing enzymes destined to play a role in digestion; 4. providing an electrolyte-containing medium suitable for the solution and activation of salivary enzymes. Of course salivary glands may have other functions, such as the synthesis and secretion of hormones (ELLISON, 1967; ITO, 1960; LEVI-MONTALCINI and ANGELETTI, 1961, 1964; LEVI-MONTALCINI and COHEN, 1960; COHEN, 1962; COHEN and ELLIOTT, 1963; COHEN et al., 1972; NAUGHTON et al., 1969; WEIMAR and HARAGUCHI, 1975), and the glands may be specialized to secrete venom (PEARSON, 1942, 1950), but such functions will not be considered within the scope of the present article, which will be restricted to a consideration of trans-epithelial transport leading to the production of saliva subserving the normal functions as outlined above.

Although salivary glands, as defined here, occur in many species, including invertebrates, our attention will be restricted to the glands of mammalian vertebrates, particularly to the three major salivary glands, the parotid, sublingual and the mandibular (or submandibular in humans) glands. This restriction is dictated largely by the paucity of studies on the function of nonmammalian salivary glands. (Insect glands have been investigated extensively (BERRIDGE and OSCHMAN, 1972) and in many ways are found to be remarkably similar to mammalian salivary glands, but the separate evolutionary origin of these glands justifies our considering them as unrelated to mammalian salivary glands although subserving similar functions.)

The literature on the subject of transport in salivary glands is very extensive. This is partly because of the ready accessibility of the glands to the interested investigator, but also reflects the enormous structural diversity of the glands, a diversity that contrasts strikingly with the structural homogeneity of the pancreas as seen in a wide range of mammalian species. This structural heterogeneity of salivary glands is associated with a comparable physiological heterogeneity that necessitates performance of many comparative physiological investigations before generalizations concerning salivary functions can be advanced with much confidence. The extent of this diversity of structure and function will become apparent in the following pages where differences among species will be emphasized in order to discourage development of simplistic (as distinct from simple) hypotheses.

B. Anatomy and Anatomical Terminology

Salivary glands are compound glands (*i. e.*, have branched ducts) with specialized secretory endpieces; both ducts and endpieces may contain more than one cell type. The following terminology, described in detail elsewhere (YOUNG and VAN LENNEP, 1978), will be used in this monograph. No attempt will be made to describe glandular anatomy in detail – the interested reader is referred to relevant review articles (YOUNG and VAN LENNEP, 1978; SCHAFFER, 1927; ZIMMERMANN, 1927; FAHRENHOLZ, 1967; KURTZ, 1964).

I. Secretory Endpieces

The term secretory endpiece, or more simply endpiece, is taken to refer to the specialized terminal parts of the blind ductal tree, from which it is believed the fluid and most of the characteristic organic components of saliva are derived. Endpieces vary greatly in shape among different glands. They may resemble berries on a stalk, when they are called acini (*e. g.* human parotid gland), or they may be cylindrical in form, when they are called tubules (human sublingual gland); intermediate forms called tubulo-acini are also recognized (*e. g.* rat sublingual gland). The shape of the endpieces may be of more than anatomical significance. On the one hand, the form of the endpiece lumen and the nature of its junction with intercalated ducts could influence flow and pressure patterns and, on the other, if the endpieces were functionally heterocrine, there would be opportunity for cells near the intercalated ducts to modify the secretion formed by cells near the blind end. No physiological information is available to illuminate these ideas, however.

Homocrine and heterocrine denote, respectively, that the secretory endpiece contains one, or more than one, distinct cell type. A mixed gland has endpieces of more than one shape (acinus, tubule or tubulo-acinus) – such glands are classed as heterocrine, since the different endpieces are composed of different cells although individual endpieces may contain cells of only one type.

Cell types may be *serous, seromucous* or *mucous.* Although many workers have tried to give rigorous histochemical definitions for these terms, no attempt has yet proven entirely satisfactory. They will be used here in their classical sense, *i. e.* based on cell morphological characteristics without prejudice to histochemical characterization of the secretion products. Thus serous cells, viewed under the light microscope, contain small, discrete refractile granules that are usually eosinophilic; under the electron microscope they are homogeneous and electron-dense. Mucous cells contain large granules, closely packed and often ill-defined. They stain poorly with haematoxylin and eosin in paraffin sections but appear quite homogeneous in araldite sections stained with toluidine blue; electron-microscopically, the granules often appear fused with an electron-translucent matrix. All these cell types have conspicuous stacks of rough endoplasmic reticulum in the basal region and several perinuclear Golgi complexes.

Heterocrine endpieces usually consist of short mucous tubules with serous or seromucous demilunar caps (*e. g.* dog mandibular gland); occasionally, the pattern is reversed and serous or seromucous tubules with mucous demilunes are seen (mandibular gland of *Erinaceus europaeus*). In other cases (*e. g.* rabbit mandibular gland) short serous or seromucous tubules terminating in a number of morphologically distinct, seromucous acini are observed.

In fresh tissue, the endpiece lumen can be discerned with the aid of a dissecting microscope but only if retrograde injection of a coloured solution is attempted. As a consequence, attempts to micropuncture endpieces directly have been largely unsuccessful. When visualized, the lumen often extends between the endpiece epithelial cells as small intercellular secretory canaliculi. Such canaliculi are often seen between neighbouring serous (or seromucous) cells but only rarely between mucous cells (however see human labial glands for a striking exception; Tandler et al., 1969); they are especially conspicuous between the neighbouring cells of serous demilunes capping mucous tubulo-acini.

II. The Duct System

Three duct types will be distinguished: intercalated, striated and excretory ducts.

The term intercalated duct needs little explanation. Such ducts, when present at all, are narrow structures, joining endpieces to the next ductal order, generally the striated ducts. Usually the epithelium is cuboidal or flattened and embraces only a very narrow lumen. Normally there are two orders of intercalated ducts, the second of which shares a common embryological origin, at least in the rat mandibular gland, with the cells of the secretory endpiece (Jacoby and Leeson, 1959; Leeson and Jacoby, 1959). In some species, the intercalated duct cells retain their primitive ability to synthesize and secrete organic material, but this function does not persist in many species and seems unlikely to be of great importance, even when it does.

Intercalated ducts terminate in striated ducts. These are lined by a simple, or less commonly, by a pseudostratified, cuboidal or low-columnar epithelium and have, as a characteristic feature, a radial, eosinophilic striation of extreme density, covering the basal third of the cell. The extent to which the striated ducts are developed varies among species and among glands within a single species. In general they are most conspicuous in parotid and mandibular glands and inconspicuous in sublingual and minor salivary glands (LEESON, 1967); as a rule, they are inconspicuous whenever a gland is predominantly mucus-secreting, although the buccal gland of *Macaca mulatta* is a remarkable exception (YOUNG and VAN LENNEP, 1978). Studies with the electron microscope reveal that the basal striations are due largely to the presence of mitochondria, aligned in cytoplasmic processes formed by extensive infolding of the basal and basolateral cell membranes.

Greatly modified striated ducts, called granular ducts, are found in the mandibular glands of many rodents, in the mandibular and parotid glands of the hedgehog (*Erinaceus europaeus*), the mandibular gland of the tree shrew (*Blarina brevicauda*), and the parotid gland of the vampire bat (*Desmodus rotundus*) (HENRIKSEN, 1972; JUNQUEIRA et al., 1965; TANDLER and MACCALLUM, 1974). These ducts have few basal striations and their apical cytoplasm contains numerous secretion granules; in some species, the extent of the granulation is much greater in males (CARAMIA, 1966a and b; LACASSAGNE, 1940a and b).

The striated (and granular) ducts drain into two or more generations of excretory ducts, which unite, in monostomatic glands, to form a single main excretory duct. In most animals, such as the rat (TAMARIN and SREEBNY, 1965), the epithelium of these ducts consists of a layer of tall columnar cells with apically located nuclei and conspicuous basal striations; light- and dark-staining cells can be distinguished, as well as scattered basal cells that have no contact with the lumen.

III. Myoepithelial Cells

Salivary endpieces are usually supported by a basket-like trellis composed of one or more cells, thought to be contractile in nature and to contain filaments of actomyosin; since the cells lie within the basement lamina that envelops the endpiece, they should be considered part of the endpiece (YOUNG and VAN LENNEP, 1977, 1978). They are usually called myoepithelial cells. Cells of similar appearance, but spindle-shaped, invest the intercalated ducts. Although such cells may extend onto the ends of the striated ducts, myoepithelial cells do not normally surround these structures, nor are they seen surrounding excretory ducts. However, they can be seen investing striated ducts in the sheep parotid (VAN LENNEP et al., 1977). The morphology and physiology of salivary myoepithelial cells have recently been the subject of an extensive review (YOUNG and VAN LENNEP, 1977) and they will not be discussed further in this paper. However, the authors wish to draw attention to the fact that the cells have an important, albeit indirect role to play in salivary transport, by providing support for the endpieces that would otherwise be distended during the secretory process and, in special cases, by providing a propulsive force to aid in the expulsion of especially viscous secretions.

C. Transport of Water and Electrolytes

I. Theories for Secretion of Water and Electrolytes by Salivary Glands

Some salivary glands, when deprived of all recognized secretory stimuli, are still observed to secrete spontaneously at a low rate (BURGEN and EMMELIN, 1961), while others are not observed to secrete at all under such conditions. Vigorous secretion is normally produced by parasympathetic nerve activity but, in many glands, sympathetic nerve activity is also effective. Although an adequate blood supply is obviously necessary to support sustained secretory activity, it has been known since early in the last century that salivary glands can be made to secrete by nerve stimulation, even after death of the animal, or against outflow resistances in excess of arterial pressure (LUDWIG, 1851; see also BRETTEL, 1869). Hence it follows that the secretory process triggered off by nerve stimulation cannot involve simple pressure ultrafiltration of plasma but must involve some active energy-consuming process. It has always been assumed that a component of the secretion originated in the endpieces but it is only in the case of organic constituents of saliva (see below) that we can say with certainty that the endpieces play a major role. Of course it can scarcely be doubted that some fluid and electrolytes must enter the saliva from the endpieces, since a medium for solution of secreted proteins and for their carriage to the exterior is obviously essential, but some consideration needs to be given to the possibility that a significant component of salivary fluid enters beyond the endpiece, *e. g.* in the intercalated or granular ducts (see SCHNEYER and SCHNEYER, 1961; HOLZGREVE et al., 1966; YOUNG and VAN LENNEP, 1978). Within the endpiece, little is known about the fluid secretory capacity of the various cell types, serous, seromucous or mucous. Since homocrine glands can all produce liquid secretions (even mucous glands can have vigorous flow rates) it is clear that all cell types can transport some fluid but the relative contribution of the different cell types comprising a heterocrine endpiece is a matter for speculation.

Up until about 1950, theories of salivary secretion ascribed separate secretory roles to endpieces and striated ducts (intercalated ducts and excretory ducts were either ignored or treated as inert conduits) although it should be pointed out that attention was focussed largely on secretion of organic products (BABKIN, 1950). With development of a better understanding of mechanisms of transmembrane transport of salt and water and an appreciation of the thermodynamic problems involved in postulating formation of a hypotonic secretion, theories on secretion of the salt and water components of saliva began to be developed in more detail. It had been known since the time of HEIDENHAIN (1878) that the salt content of saliva was lower than in mammalian plasma and that the osmolality, lowest in saliva collected at low secretory rates, rose asymptotically towards a constant value as secretory rate reached a maximum. THAYSEN and his colleagues studied this phenomenon and first enunciated their 2-stage secretory hypothesis to explain it in 1954 (THAYSEN et al., 1954). Studying human parotid saliva, they found that its osmotic activity was due principally to Na^+, K^+, Cl^- and HCO_3^- and that the concentration of these ions changed with salivary flow rate in such a way that total osmotic activity, normally low, rose to a maximum either equal to or lower than in plasma. Subsequently, THAYSEN (1960) enunciated more explicitly his hypothesis that salt and water were first secreted into saliva as an isotonic primary secretion with plasma-like electrolyte composition and that the fluid was modified subsequently by reabsorption of Na^+ during passage of the primary saliva along the duct system to the exterior. Both THAYSEN and subsequent workers ascribed formation of the primary secretion to the endpieces and sodium reabsorption to the striated ducts. BURGEN, observing that saliva was often excessively rich in potassium, inclined to believe that the endpiece secretion was potassium-rich and that it first re-equilibrated with plasma in the intercalated ducts before being rendered hypotonic in the striated ducts (BURGEN and SEEMAN, 1958).

The essential points of the two-stage hypothesis, still the basis of all secretory theories, were:

i. That endpieces produce an isotonic primary secretion, rich in sodium and that it is the major, or sole, site of fluid secretion by the gland.
ii. That the primary fluid, during its passage from the endpiece to the mouth, is rendered hypotonic by a process of sodium reabsorption in excess of water across an epithelium having a relatively low hydraulic conductivity.

In the following pages an attempt will be made to examine what is known about transport processes in endpiece and ductal cells with the object of testing and modifying the two-stage secretory hypothesis as necessary.

II. Formation of the Primary Saliva

1. Evidence Concerning the Site or Sites of Fluid Secretion by Salivary Glands

As pointed out above, the Thaysen two-stage hypothesis states that saliva is produced initially by secretion of a plasma-like primary fluid. Since at least some fluid must enter the saliva at the endpiece in order to dissolve and flush out the exportable proteins, one must term this the primary fluid. However, the question then arises as to whether the endpiece is the only site of fluid secretion or whether fluid is also secreted at some more distal point (in the gland duct system) in a manner analogous to the admixture process thought by many to be involved in the formation of pancreatic juice (SEWELL and YOUNG, 1975). Surprisingly, the answer to this question is based mainly on indirect evidence.

One approach to this problem would be to study the rate of formation of endpiece secretions separately from the rate of excretion of the final saliva. This is no easy matter, since the primary fluid is not formed by filtration and no substance is available that is analogous to inulin in its role as a marker of glomerular filtration. Attempts have been made to use salivary enzymes, such as amylase, as markers of primary fluid, but these have not been very successful (a) because one cannot assume that the enzyme concentration in primary fluid remains constant during sustained stimulation, and (b) because no single enzyme could be used for all salivary glands. Indeed, not only do some glands, such as the sheep parotid and the rat mandibular gland (CHAUNCY and QUINTARELLI, 1961; ELLISON, 1967; JUNQUEIRA and FAVA-de-MORAES, 1965), secrete very little amylase but, in some glands it is not certain whether the enzyme is secreted by cells of the endpiece or by cells of the ducts (CHRÉTIEN and ZAJDELA, 1965; SMITH et al., 1971; SMITH and FROMMER, 1972a and b). Studies on the amylase content of primary and final saliva from the parotid glands of rat, mouse, and rabbit (MANGOS et al., 1973b–d) show almost no change in enzyme concentration along the duct system during either cholinergic or adrenergic stimulation. This would seem to provide excellent evidence that the endpiece-intercalated duct region is the only important site of fluid secretion at least in these closely related parotid glands (see HOLZGREVE et al., 1966; YOUNG and VAN LENNEP, 1978).

Urea, also, has been tried as a marker of endpiece secretory rate, on the assumption that it enters the endpiece secretion easily and does not cross the ductal epithelium readily in either direction (MANGOS and MCSHERRY, 1970). The assumption has proven wrong on both counts. Thus, in resting glands, although the urea concentration in endpiece fluid is nearly the same as in plasma, it is lower after stimulation. Similarly, microperfusion experiments show that although the main duct is relatively impermeable to urea (see Section C. III. 2. a. iv) there is sufficient transductal efflux to invalidate calculations of water movement at low gland secretory rates, when urea concentration in saliva can rise to nearly three times the plasma concentration (rat mandibular gland, unpublished observation). It should also be mentioned that, in most glands so far studied, with the exception of the cat sublingual and the sheep parotid, the endpiece lumen is too narrow to be micropunctured, so that samples of "primary" secretion have to be collected from the intercalated ducts (YOUNG and MARTIN, 1971); in any case, even when single endpieces have been punctured, it has not proven possible to estimate their individual rates of secretion.

Another approach to the study of the formation of the primary fluid has been to attempt to destroy duct function by retrograde injection of toxins such as mercuric salts (BURGEN, 1967), ouabain (MANGOS and BRAUN, 1966; MARTINEZ, 1971) and poly-L-lysine (MANGOS et al., 1973c and d; MANGOS and MCSHERRY, 1968; SEWELL and YOUNG, 1973). Such experiments are based on the assumption that toxins can be applied selectively to the duct system by retrograde injection and that ducts so treated will then act as passive conduits, conveying the primary secretion unchanged to the exterior. Figure 1 shows results of such experiments for rat parotid and mandibular gland, injected retrograde with poly-L-lysine in doses just lower than those needed to cause secretion to stop altogether. In the rat parotid, the results are quite striking – treatment caused only a slight reduction in secretory rate in response to parasympathomimetic stimulation but salivary sodium concentrations became nearly plasma-like. This behaviour can be interpreted as indicating that poly-L-lysine inhibited ductal Na^+ reabsorption but had no effect on secretion of primary fluid, all of which must normally have been secreted at a point beyond the site of action of poly-L-lysine. While this interpretation might indeed be correct, it is worth pointing out

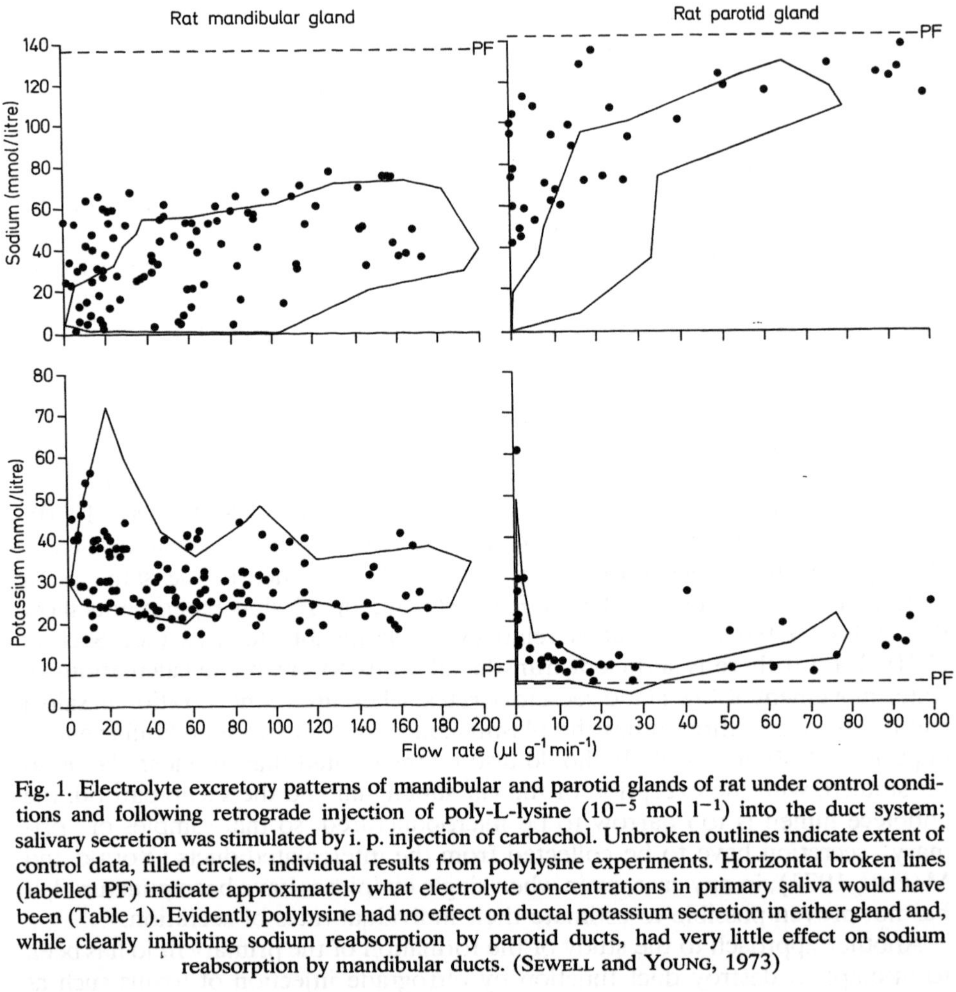

Fig. 1. Electrolyte excretory patterns of mandibular and parotid glands of rat under control conditions and following retrograde injection of poly-L-lysine (10^{-5} mol l^{-1}) into the duct system; salivary secretion was stimulated by i. p. injection of carbachol. Unbroken outlines indicate extent of control data, filled circles, individual results from polylysine experiments. Horizontal broken lines (labelled PF) indicate approximately what electrolyte concentrations in primary saliva would have been (Table 1). Evidently polylysine had no effect on ductal potassium secretion in either gland and, while clearly inhibiting sodium reabsorption by parotid ducts, had very little effect on sodium reabsorption by mandibular ducts. (SEWELL and YOUNG, 1973)

that the same experiment is much less convincing when performed in the rat mandibular gland (Fig. 1; SEWELL and YOUNG, 1973). Furthermore, in all cases, the volume of the polylysine solution that had to be injected exceeded by far the possible volume of the entire duct tree and must clearly have been forced through the epithelium at some point — hence one cannot say with certainty at what site in the gland the drug acted[1]. It is also noteworthy that even if the drugs, when injected retrogradely, did act only on duct cells, no one has succeeded in determining whether the inhibitory action was on all three orders of

1 It is well known that many drugs acting on salivary gland cells, including stromal cells such as intraglandular nerve ganglion cells, can be administered by retrograde duct injection (EMMELIN et al., 1973) *e. g.* ductal application of nicotine can provoke salivation and the effect can be blocked by the administration of hexamethonium intravenously (EMMELIN et al., 1954).

salivary ducts (intercalated, striated, and excretory) or whether inhibition was localized to a single duct segment.

A similar approach to the above, involving activation of salivary myoepithelial cells, has been tried in the sheep parotid (BLAIR-WEST et al, 1969). In this gland it has been alleged that there are no secreto-motor sympathetic fibres and that sympathetic nerve stimulation acts only to cause myoepithelial cells to contract (BLAIR-WEST et al., 1969) although there is evidence to show that the gland posesses some β-adrenergic secreto-motor receptors (PATTERSON and TITCHEN, 1975). This led WRIGHT and his colleagues (BLAIR-WEST et al., 1969) to argue that stimulation of myoepithelial cell contraction would cause preformed endpiece secretion to be expelled from the duct system, unchanged by ductal fluid and electrolyte transport processes. Micropuncture studies (COMPTON and YOUNG, 1976; COMPTON et al., 1979), as well as measurements of ductal electrolyte transport rates, have shown that the assumption is false, a fact that might have been guessed from experience of stop-flow analysis in renal physiology, where tubular flow rates can be an order of magnitude greater than in salivary glands and yet the nephron has ample time in which to modify the composition of tubular fluid during the expulsion phase of the experiment.

Failing a satisfactory means for determining the rate of formation of endpiece fluid, one might attempt to answer the question, is there a second site of fluid secretion, by studying the function of each duct segment separately. In the case of the excretory ducts, micropuncture and microperfusion studies indicate quite clearly that these structures reabsorb fluid rather than secrete it (YOUNG et al., 1967; SCHNEYER, 1968; SCHNEYER et al., 1972; YOUNG, 1973) or, in the case of the main duct of some glands, are inert and allow equilibration of saliva with the interstitial fluid (*e. g.* KNAUF and FRÖMTER, 1970a). In the case of striated and granular ducts the position is less clear, since no means are yet available for individual perfusion of the ducts. Micropuncture of striated ducts has occasionally proven possible in the rat sublingual gland (MARTIN and YOUNG, 1971a), and has shown their contents to be hypotonic. Retrograde catheterization of second-order excretory ducts in the rat mandibular gland indicates that fluid entering the excretory ducts is already hypotonic, profoundly so in the stimulated gland (YOUNG and SCHÖGEL, 1966); this suggests strongly that the striated and granular ducts reabsorb salt to render the primary saliva hypotonic rather than form a secretion of their own. On the other hand, it is surprising to note that micropuncture of the surface structures of the rat and mouse mandibular glands has not yielded any hypotonic samples, since granular ducts are quite frequently found close to the gland surface and should therefore be readily accessible to a micropuncture pipette (YOUNG and MARTIN, 1971). The ultrastructure of striated ducts (YOUNG and VAN LENNEP, 1978) resembles that of reabsorptive epithelia such as kidney convoluted tubules and gallbladder mucosa (BERRIDGE and OSCHMAN, 1972), but it must be said that the granular ducts seen in many rodent mandibular glands do not have such conspicuous absorptive characteristics, although basal infoldings are often readily seen (TAMARIN and SREEBNY, 1965). Since the ducts clearly secrete organic material (DOREY and BHOOLA, 1972; CUTLER and CHAUDRY, 1973; MATTHEWS, 1974a and b), it would hardly be surprising if such structures were found to have some capacity to secrete fluid. In mice, castration causes the protein-secretory function of these ducts to involute and, interestingly, the basolateral labyrinth then becomes more conspicuous (CHRÉTIEN, 1972a–c; DROZ et al., 1973).

While a good case can be made out to suggest that striated ducts are reabsorptive and do not secrete fluid, there is really no suggestive evidence to justify such a conclusion in relation to intercalated ducts. Micropuncture studies have not proven helpful, since endpieces themselves have not been punctured (except in the cat sublingual gland – see above), so there is no way of telling whether the "acinar-intercalated duct fluid" referred to in micropuncture experiments comes from endpieces alone, or from both endpieces and ducts. The relatively simple ultrastructural appearance of intercalated duct cells in most, but not all animals (YOUNG and VAN LENNEP, 1978), and the fact that they are not present in some salivary glands, has encouraged the view that these ducts are inert (GARRETT, 1976; TANDLER, 1972). However, it must be emphasized that there is no ultrastructural feature that can be identified as characteristic of cells secreting water and electrolytes – the protein-synthesizing apparatus of secretory cells is associated only with secretion of organic material, often in a solid or semisolid state. Intercalated duct cells are by no means to be described as having an "inert-looking" ultrastructure, and among duct cells in general, they are the ones most frequently described as having a close-contact, hypolemmal innervation (ALM et al., 1973; BOLANDE and TOWLER, 1973; COWLEY and SHACKLEFORD, 1970; GARRETT, 1972, 1974; YOHRO, 1971). From an embryological point of view, it would not be at all surprising if intercalated ducts proved to be capable of secreting fluid since at least second-order ducts share a common embryological origin with endpiece cells, both cell types developing from cells of the primitive terminal tubule (YOUNG and VAN LENNEP, 1978; see also Section B. II). Furthermore, the terminal tubules themselves are known to be capable of secreting fluid, sometimes quite vigorously (SCHNEYER and SCHNEYER, 1961; HOLZGREVE et al., 1966; SCHNEYER and HALL, 1968), as are the embryonic cells of intermediate ductal segments (YOUNG and VAN LENNEP, 1978). It should also be mentioned that in salivary glands in which endpiece atrophy (described histologically as complete or nearly complete) has been induced by chronic main duct obstruction, there is good evidence that some degree of fluid secretion, often quite copious, can still be provoked by parasympathetic or parasympathomimetic stimulation (EMMELIN et al., 1974; SCHNEYER and SCHNEYER, 1961). Such experiments could indicate ductal secretion, perhaps from intercalated ducts, but incomplete atrophy of endpieces in unexamined regions of the gland could just as well account for the results, since atrophy following duct ligation is known to be patchy and of unpredictable severity (HARRISON and GARRETT, 1975, 1976a and b). It would be interesting to repeat these experiments and study serial thick sections of the entire salivary gland.

2. Control of Secretion and Innervation of Endpieces

From the foregoing, it would seem reasonable to conclude that fluid secretion in salivary glands takes place only in the endpiece-intercalated duct region, even if a clear separation of the roles of the intercalated ducts and of the endpieces (and of the different cell types within a heterocrine endpiece) cannot be made. It seems appropriate to examine what factors are known to control production of this composite primary secretion.

First it should be pointed out that some glands, such as the sheep parotid and the cat and dog sublingual gland, secrete spontaneously at a slow but steady rate, even when totally denervated and deprived of other known stimuli (EMMELIN, 1967). (Sheep parotid glands continue secreting for 10–15 minutes after death; BRETTEL, 1869.) Spontaneous secretion has been attributed also to the rat sublingual and the rabbit mandibular glands (OHLIN and PEREC, 1965; NORDENFELT and OHLIN, 1957) although in these cases it is easy to abolish secretion with deep anaesthesia or lowered blood pressure. Most other salivary glands are said not to secrete spontaneously but by no means is it excluded that, in these cases, a small flow of spontaneous secretion is produced and subsequently reabsorbed fully in the salivary ducts.

Increase in the rate of salivary secretion above the basal level is achieved naturally by autonomic nerve activity, always by parasympathetic activity, and sometimes also by sympathetic activity. In the case of the sympathetic innervation, the postganglionic fibres arise in the superior cervical ganglion (where they can easily be stimulated electrically) and pass into the gland with the major blood vessels. The ganglia of the parasympathetic system are more diffuse. The parotid gland is supplied largely by postganglionic fibres of the auriculotemporal nerve, arising from the otic ganglion, but mandibular and sublingual glands are usually supplied with preganglionic fibres that synapse with ganglion cells lying along the main excretory ducts or within the gland substance. As a consequence, although direct electrical stimulation of postganglionic fibres is possible for parotid glands, it is seldom possible for mandibular or sublingual glands. Only in the Australian marsupial possum, *Trichosurus vulpecula*, is a postganglionic nerve trunk clearly discernible, running along the main excretory duct of the mandibular gland: it has been succesfully stimulated electrically (CREED and McDONALD, 1970)[2].

Electron microscopy shows that, within the gland substance, the postganglionic, nonmyelinated fibres are composed of narrow cylindrical sections, alternating with periodically placed varicosities. The varicosities contain a few large-diameter (80–120 nm) vesicles with dense cores and a large number of small vesicles (40–50 nm), which appear electron-lucent in cholinergic terminals but contain dense cores (after permanganate fixation) in adrenergic fibres; both types of fibre run side-by-side within the same Schwann cell sheath (ALM et al., 1973; BOGART and de LEMOS, 1973; ENEROTH et al., 1969; GARRETT, 1972; HAND, 1972a). Innervation sites consist simply of varicose regions bare of Schwann cell covering, at which vesicles have accumulated. Two types of parenchymal neuroeffector junction are distinguished, depending on whether the

2 It should be noted that not all nerve fibres entering the possum mandibular gland are postganglionic since some small ganglia have been found within the substance of this gland (BLOOD et al., 1977).

nerve terminal penetrates the basement membrane (hypolemmal) or remains outside it (epilemmal) (GARRETT, 1975, 1976). In the case of epilemmal junctions, a distance of at least 100 nm intervenes between the nerve membrane and the membrane of the parenchymal cell; in the case of hypolemmal junctions the distance is only about 20 nm. There may be a number of neuroeffector junctions along the course of an axon, and sometimes the fibres are seen running between adjacent epithelial cells. No ultrastructural specialization has been observed on the parenchymal side of a neuroeffector junction apart from a conspicuous subsurface cistern of endoplasmic reticulum (HAND, 1970). The biological importance of hypolemmal as against epilemmal innervation is a matter only of speculation (GARRETT, 1972) but, clearly, either mode of innervation suffices to permit secretomotor activity and it is certainly not necessary to demonstrate hypolemmal innervation of a particular cell type in order to claim that the cell is activated by secretomotor nerves. Obviously, single-shock nerve stimulation would be more likely to produce an electrophysiological and a secretory response when hypolemmal innervation was present (CREED and WILSON, 1969; FRITZ and BOTELHO, 1969b; LUNDBERG, 1955) although this does not seem to be a universal rule (FRITZ, 1971; KAGAYAMA and NISHIYAMA, 1972; NORDENFELT and OHLIN, 1957).

In most glands the density of innervation is greater around the endpiece than around other glandular structures, although the pattern is extremely variable (GARRETT, 1972). In glands receiving a dual secretomotor innervation, both types of autonomic fibre enmesh the endpiece although cholinergic nerves predominate among the hypolemmal fibres. The frequency with which hypolemmal fibres occur is very variable, even within a single species (GARRETT, 1972), and seems to diminish with increasing age of an animal (FRIEDMANN and HODGES, 1975; KAGAYAMA and NISHIYAMA, 1972; YOHRO, 1971).

If it can be assumed that primary fluid formed in the endpiece-intercalated duct region is subject only to nett fluid reabsorption during its passage to the exterior, it follows that the salivary secretory rate must always be less than the rate of formation of primary fluid and that the two rates will tend to the same value as gland stimulation becomes maximal, *i. e.* at high secretory rates the rate of saliva excretion may be assumed equal to the rate of primary fluid production. In response to parasympathomimetic stimulation, the maximum secretory rates achieved range from 100 to 400 μl per minute per g of wet gland tissue, depending on the gland and species being studied. In general, salivary glands secrete far more vigorously in response to parasympathetic (and parasympathomimetic) than to sympathetic (and sympathomimetic) stimulation. However many glands respond quite vigorously to either mode of stimulation [*e. g.* rat mandibular gland: YOSHIDA et al. (1967), YOUNG and MARTIN (1971), MARTINEZ et al. (1975a and b)] and occasionally, sympathomimetic responses can be as large as parasympathomimetic (KALADELFOS and YOUNG, 1974). The size of a secretory response will be determined not only by the intrinsic properties of the secretory cells, but also by the extent and effectiveness of their innervation and by the capacity of the glandular blood supply to adapt so as to maintain an increased secretory rate. Since pharmacological stimulation, usually slightly more effective than direct nerve stimulation, reveals the same difference between parasympathetic and sympathetic responses as does nerve stimulation (SCHNEYER et al., 1972), it seems unlikely that the difference can be ascribed to the degree of effectiveness of the two types of innervation. On the other hand, there is little reason to doubt that vasoconstriction during sympathetic nerve stimulation is a factor contributing to the limited salivary secretory response (EMMELIN, 1967) although use of sympathomimetics with no vasoconstrictor action still does not normally result in production of secretory rates equal to those induced by parasympathomimetics. It seems reasonably safe to conclude that the intrinsic capacity of salivary secretory cells to respond to stimulation is greater for parasympathomimetic than for sympathomimetic agents.

In the above discussion, salivary flow rates were related to the wet weight of glandular tissue. This practice is widely followed (a) to allow for differences in the sizes of different glands among different species and (b) to allow for differing gland size among different animals belonging to a single species. The practice is based on the assumption that secretory mass bears a fixed relation to total gland mass, an assumption that is in fact untrue. Thus, within the one species, there is evidence that body sodium status is an important determinant of the relative masses of ducts and endpieces comprising a given gland (COMPTON et al., 1975; COMPTON, 1976) and, furthermore, among different species there is variation in the fraction of the total gland mass made up of ducts (YOUNG and VAN LENNEP, 1978). Of course one could express flow rates as a function of endpiece mass (based on a morphometric analysis of the glands), but it remains to be shown that secretory rate is related to this mass – at least one cause of a huge increase in endpiece mass, chronic treatment with isoproterenol (SCHNEYER, 1972), has little or no effect on gland secretory rate (MANGOS et al., 1969a and b). Failure to appreciate the importance of the assumption underlying the correction of flow rate by gland mass can lead to erroneous conclusions (MANGOS et al., 1969a and b).

3. Composition of the Primary Secretion

a) Osmotic Activity

Since the driving force for the formation of the primary secretion is thought to be active transport of solute into the endpiece lumen with a consequent generation of an osmotic gradient (SCHNEYER et al., 1972), it is to be expected that samples of primary fluid would be isotonic or slightly hypertonic at low secretory rates and might become more hypertonic at high secretory rates. However, if secretion were to take place *via* secretory canaliculi (see Section B, above), so that standing osmotic gradients could be established, one might observe that the primary secretion remained isotonic even when secretory rates changed by up to three orders of magnitude (DIAMOND and BOSSERT, 1967; OSCHMAN and BERRIDGE, 1970; YOUNG, 1973).

Table 1 summarizes the information available on the osmotic activity of primary saliva. It appears that the primary secretion of the mandibular and parotid glands of rat, mouse and rabbit are close to isotonic (rabbit mandibular and

Table 1. Electrolyte composition and osmotic activity of primary saliva in various glands and species. Concentrations ($mmol \cdot l^{-1}$) and osmolality ($mOsm \cdot kg^{-1}$) are mean values ± S.D.

	$[Na^+]$	$[K^+]$	$[Cl^-]$	Osmolality	Source
Rat					
Mandibular gland					
Resting					
immature	160.0±12.6	–	120.0± 3.1	354.0±24.2	a
normal	146.0± 4.5	–	121.0±10.1	307.0± 9.5	b
normal	135.3±12.6	6.7± 2.7	–	308.0±14.7	c
normal	136.2±19.2	8.4± 2.9	105.3±17.7	–	d
Stimulated					
pilocarpine	141.6±18.6	5.1± 3.4	–	299.0± 4.6	c
carbachol	139.0±10.1	4.0± 1.6	99.7±14.8	–	d
isoproterenol	136.4±12.5	5.2± 3.6	91.5±12.5	–	d
Sublingual gland					
Resting					
normal	126.1±10.2	11.9± 7.2	99.3±13.0	–	e
Stimulated					
carbachol	131.9± 6.0	8.9± 6.0	95.7±11.3	–	e
Parotid gland					
Resting					
normal	141.9± 6.7	5.2± 1.5	–	302.5±11.7	f
Stimulated					
acetylcholine	141.6± 7.1	5.1± 0.9	112.7± 5.9	–	g
isoproterenol	134.2± 5.2	13.2± 2.9	76.3± 4.6	–	g
pilocarpine					
– normal	141.4± 8.5	4.8± 0.9	–	291.1± 9.3	f, h
– adrenalectomized	131.3± 5.7	6.9± 0.8	–	–	h
Mouse					
Mandibular gland					
Resting					
normal	151.6±10.4	5.1± 0.6	124.6± 7.0	297.4± 7.8	i
Stimulated					
pilocarpine	157.2±12.1	4.9± 1.6	122.9±12.6	300.6± 9.6	i
Parotid gland					
Resting					
normal	141.6± 4.3	5.6± 1.1	112.6± 4.7	283.1± 8.7	i
Stimulated					
pilocarpine	143.5± 6.3	5.2± 0.8	114.2± 5.6	287.0±13.6	i
Rabbit					
Mandibular gland					
Resting					
normal	153.0±20.4	11.5± 9.0	128.1±19.8	307.6±39.2	j
normal	85.6±37.6	13.2±10.6	81.2±14.8	263.7±57.4	k
sodium deprived	74.1±23.9	11.6± 7.2	–	–	k
DOCA	109.4±35.9	11.8± 6.0	–	–	k
dexamethasone	96.9±42.5	10.7±10.8	–	–	k

Table 1 (continued)

	$[Na^+]$	$[K^+]$	$[Cl^-]$	Osmolality	Source
Stimulated					
pilocarpine and carbachol					
– normal	82.2±35.4	6.9± 4.3	77.5±14.7	294.8±30.5	k
– sodium-deprived	78.5±36.2	10.9± 9.6	–	–	k
– DOCA	133.4±28.5	9.3± 6.2	–	–	k
– dexamethasone	101.9±35.0	7.7± 5.8	–	–	k
		Parotid gland			
Resting					
normal	159.7±27.9	3.1± 4.6	131.6±25.8	311.4±40.7	j
		Cat			
		Mandibular gland			
Resting					
normal	158.0±29.0	10.7± 5.2	127.0±21.0	–	l
normal	142.0±13.0	10.9± 4.8	117.0±11.0	–	l
Stimulated					
carbachol	159.0±26.0	14.9± 7.6	121.0±28.0	–	l
isoproterenol	135.0±14.0	11.7± 5.2	116.0± 8.0	–	l
		Sublingual gland			
Resting					
normal	137.0±19.0	8.4± 5.6	128.0±18.0	–	m
Stimulated					
carbachol	145.0±11.0	10.1± 6.5	122.0±18.0	–	m
		Possum *(Trichosurus vulpecula)*			
		Mandibular gland			
Resting					
normal	122.4±24.5	10.4± 7.2	–	–	n
Stimulated					
acetylcholine or carbachol	106.9±34.0	6.9± 4.4	–	–	n
		Sheep			
		Parotid gland			
Resting					
sodium replete	120.2±23	11.5± 9	53.3±15	312±20	p
sodium deprived	81.8±26	12.6±13	58.5±24	308±36	p
Stimulated					
sodium replete	132.6±22	9.0± 6	49.8±15	296±25	p
sodium deprived	82.5±34	11.4±13	53.5±21	288±29	p

Source:
a Holzgreve et al. (1966)
b Martinez et al. (1966)
c Young and Schögel (1966)
d Young and Martin (1971)
e Martin and Young (1971a)
f Mangos et al. (1966)
g Mangos et al. (1973a)
h Mangos and McSherry (1969)
i Mangos et al. (1973b)
j Mangos et al. (1973c)
k Compton and Young (1977)
l Kaladelfos and Young (1974)
m Kaladelfos and Young (1973)
n Compton (1976)
p Compton et al. (1979)

parotid primary saliva was slightly hypertonic [MANGOS et al., 1973c] as was mouse mandibular primary saliva [MANGOS et al., 1973b], whereas primary saliva from mouse parotid [MANGOS et al., 1973b] and from rat mandibular and parotid was isotonic [MARTINEZ et al., 1966; YOUNG and SCHÖGEL, 1966; MANGOS et al., 1966]; primary fluid from immature rat mandibular glands was rather more hypertonic [HOLZGREVE et al., 1966]). Stimulation had little or no effect on primary fluid osmolality. It is tempting to attribute the near-isotonic osmotic activities of the primary fluids to the action of the secretory canaliculi since they are quite conspicuous in mature salivary endpieces. One might predict that the canaliculi would be less well developed in those glands that have slightly hypertonic primary secretions, but no morphometric data are available to test this prediction. Certainly in the immature rat mandibular gland, where endpieces have not developed at all, secretory canaliculi are not present (LEESON and JACOBY, 1959), so that it need occasion no surprise that the secretion of the terminal tubule is hypertonic.

Attention should be drawn here to the fact that the monovalent electrolyte content of sheep parotid primary fluid, especially in sodium-deprived animals, indicates that the primary fluids might have been hypotonic. However, when the melting points of micropuncture samples were measured directly, it was found that this secretion, too, was isotonic (COMPTON and YOUNG, 1976; COMPTON et al., 1979).

b) Electrolyte Concentrations

Table 1 also summarizes the available data on the Na^+, K^+, and Cl^- concentrations of primary saliva. The normal volume of primary saliva obtainable by micropuncture (about 10^{-10} litres) is too small to permit direct determination of HCO_3^- concentration and, although determination of pH might be feasible, no attempt had been made to measure it in primary fluid samples. Similarly, no attempt had been made to measure divalent cations and anions such as Ca^{2+} and $HPO_4^{2-}/H_2PO_4^-$. In the sheep parotid, however, the volume of saliva obtainable by micropuncture is relatively large ($\sim 10^{-8}$ l) and both pH and phosphate content have been measured (COMPTON et al., 1979). According to the sodium status of the sheep, the pH of the parotid primary saliva (at a pCO_2 of 25–30 mm Hg) ranged from 7.47 to 8.24 and the total inorganic phosphate concentration from 12.2 to 14.3 mol l^{-1}; from the pH it can be calculated that the bicarbonate concentration, at equilibrium, ranged from 20 to over 40 mmol l^{-1}. It needs to be stressed, however, that the final saliva secreted by the parotids of all Bovidae is exceptionally rich in phosphate and bicarbonate and the primary saliva might be expected to reflect this fact.

i) *Bicarbonate.* The lack of an easy direct method for determination of primary fluid HCO_3^- concentration led MANGOS and his colleagues to attempt an indirect approach (MANGOS et al., 1973c and d; MANGOS and MCSHERRY, 1968, 1971) by injecting the basic polyaminoacid, poly-L-lysine, retrogradely up the duct system of various exocrine glands with the intention of destroying ductal transport activity without interfering with the rate of formation of primary fluid by the endpieces. They hoped that polylysine, by "*eliminating transductal fluxes of ions*" (MANGOS and MCSHERRY, 1971), would allow the endpiece secretion to pass to the exterior unchanged in composition, allowing them to obtain samples of primary fluid in almost unlimited volumes, thereby facilitating biochemical analysis. Unfortunately, this approach has proven misleading rather than helpful (see Section C. II. 2). Furthermore, the polylysine-treated duct becomes quite leaky to all ions, so that any substance present in primary fluid in a concentration different from plasma will tend to re-equilibrate with plasma while within the duct lumen (*cf.* toad bladder; MAMELAK et al., 1969).

ii) *Sodium.* From an inspection of Table 1, it is apparent that the primary secretion of the salivary glands of rat, mouse and cat have sodium concentrations close to those seen in plasma. Small falls in sodium concentration, produced by isoproterenol stimulation, can be attributed to the greater capacity of the sympathomimetics to induce secretion of proteins – these polyanions would be expected to occupy volume in solution and might well be secreted associated with other cations such as Ca^{2+}, H^+ and K^+. In the glands of rabbit, sheep and possum *(Trichosurus vulpecula)*, however, the sodium concentration is often vastly different from that of plasma and seems to be dependent on body sodium status and steroid activity. Since the fluctations in Na^+ concentration were not matched by reciprocal changes in K^+ concentration in any animal, one might be tempted to infer that the sodium-poor primary fluid must have been hypotonic. However, studies on the sheep parotid show that the osmotic activity, determined from sample melting point, remains isotonic or slightly hypertonic. What the additional osmotically active solute could be is a matter for speculation, but $HPO_4^{2-}/H_2PO_4^-$ is one possible candidate, since phosphates are known to be present in sheep parotid saliva in high concentrations (COATS and WRIGHT, 1957); it could, of course, prove in part to be a nonionic solute.

It is interesting to note that, while sodium deprivation caused a lowering in the primary fluid sodium concentration in both rabbit and sheep (Table 1), treatment, in the rabbit with either the mineral-active steroid, desoxycorticosterone acetate (DOCA), or the glucocorticoid, dexamethasone, caused the sodium concentration to increase. (In the rat parotid, adrenalectomy lowered sodium concentration but no more than could be explained by a concomitant fall in plasma Na^+ concentration [MANGOS and MCSHERRY, 1969].) Available studies on the effects of mineral-active steroids on transepithelial transport indicate that they stimulate Na^+ reabsorption, but existing information does not allow us to predict what effect they might have on a sodium-secreting epithelium, although one might have expected Na^+ deprivation to have the same effect as treatment with DOCA.

iii) *Potassium.* The pattern of potassium concentration in primary fluid is rather more uniform than the pattern of sodium concentration (Table 1). In most animals, the primary potassium concentrations are greater than the concentration in plasma and fall somewhat after cholinergic stimulation. The primary potassium concentrations are particularly high in the sheep parotid (12–16 mmol l^{-1}) and in the rabbit mandibular gland (6.5–13.2 mmol l^{-1}) but show no consistent pattern of change in response to the marked changes in primary sodium concentration seen in these animals in response to sodium deprivation or steroid treatment. The slight fall in potassium concentration seen in most animals during cholinergic stimulation probably reflects a lowering of the intracellular potassium concentration (LUNDBERG, 1958) consequent on a stimulation-induced potassium efflux from the secretory cells to the interstitium (see Section C. II. 5.b).

iv) *Chloride.* Fewer studies on the chloride content of primary fluid are available, but in most cases the concentration has been found to be plasma-like or slightly higher than in plasma; it has always been found to be lower than the primary sodium concentration. In the rat mandibular gland, a linear relation between the chloride concentration in the primary fluid and in the serum has been demonstrated and interpreted as indicating passive equilibration of chloride between the interstitium and the endpiece lumen (YOUNG and MARTIN, 1971). In those glands where primary fluid chloride concentration is higher than in the interstitium, it may be that the transepithelial potential difference is positive.

c) *Organic Solutes*

As mentioned earlier, attempts have been made to measure the concentration of various organic substances in primary saliva in an effort to find a substance that would serve as a marker for the rate of formation of the primary secretion. MANGOS and MCSHERRY (1970) reported that the saliva/plasma ratio for urea in the primary saliva of the rat parotid was 0.95 ± 0.20 (S. D.) at low rates of secretion and 0.58 ± 0.18 at high rates of secretion. This can be interpreted as indicating that urea is at equilibrium between the primary saliva and the interstitial fluid under resting conditions but that its diffusion coefficient is not large enough for equilibrium to be maintained when the gland is stimulated. An attempt has also been made to determine the glucose concentration in rat parotid primary fluid (MANGOS et al., 1973d). The values reported – between 9 percent and 20 percent of the corresponding blood level at low rates of secretion – suggest that the endpiece-intercalated duct epithelium is rather less permeable to glucose than to urea. An old study of HEBB and STAVRAKY (1936) suggests that cat mandibular endpieces may be even less permeable to glucose but that intravenous injections of adrenalin increase the permeability markedly.

Surprisingly, only two attempts seem to have been made to determine the protein content of primary saliva. In the rat parotid, MANGOS et al. (1973d) report that amylase in the primary fluid has a concentration of 286 U ml^{-1} during acetylcholine stimulation and 41 800 U ml^{-1} during isoproterenol stimulation; total protein concentrations, respectively, were 1.1 and 9.6 mg ml^{-1}. The difference between the effects of the two autonomic agonists accords with their known potencies in stimulating amylase secretion (see below, Section C. II. 5.c). In this study, MANGOS also observed that the enzyme and protein concentrations in samples of primary fluid and simultaneously collected final saliva were the same, allowing the conclusion that additional fluid neither enters nor leaves the primary saliva in appreciable quantities during its passage along the striated and excretory ducts. In another study in the mouse parotid (MANGOS et al., 1973b), MANGOS observed a pattern similar to that for the rat but made the additional observation that amylase concentration in the primary fluid was not constant at different secretory rates, suggesting that processes for secretion of fluid and enzymes are at least partially independent of one another.

4. The Unstimulated Endpiece Cell

Many research workers have attempted to measure the cell membrane potentials of salivary parenchymal cells (Table 2). In all cases, the smallness of the cells being investigated has dictated the use of glass microelectrodes filled with a concentrated salt solution, usually 3M KC1 (LING and GERARD, 1949; PETERSEN, 1974a). Normally, in such experiments one would expect to impale only parenchymal cells, which might include the secretory cells of the endpiece, myoepithelial cells, and the cells of the intercalated and striated ducts (and of granular ducts when they occur). Since the microelectrode is inserted from the gland surface, it is inevitable that endpiece cells should make up the bulk of the units from which recordings are obtained, although, in rat and mouse mandibular glands, the cells of the granular ducts ought also to form a significant proportion of the units (YOUNG and MARTIN, 1971). In heterocrine glands, where endpiece cells are of two or more distinct types, *e. g.* endpieces of dog, possum, and rabbit mandibular glands and of all three major salivary glands of the cat (YOUNG and VAN LENNEP, 1978), any of the constituent endpiece cell types might be impaled. Not unreasonably, most workers seem to discount the possibility that they may have impaled myoepithelial cells, presumably on account of their small dimensions.

Some investigators have made a deliberate attempt to record from units lying deeper in the gland substance (*e. g.* deeper than 560 μm), in the hope of recording from duct cells, which ought there to make up a small but definite proportion of the units actually impaled (LUNDBERG, 1955, 1958; ICHIOKA et al., 1965). Unfortunately, identification of the site of impalement depends largely on a statistical evaluation of the results (ICHIOKA et al., 1965; SCHNEYER and SCHNEYER, 1965a; SCHNEYER et al., 1972; FRITZ and BOTELHO, 1969a and b) or on a preconceived notion of what the outcome ought to be (HENRIQUES and SPERLING, 1966), since attempts to label the impaled cells for subsequent histological investigation (HENRIQUES and SPERLING, 1966) have not proven very popular.

It has proven extremely difficult to set up rigid objective criteria defining what constitutes a successful impalement. It has always been recognized that use of glass microelectrodes tends to yield potential values that are somewhat too low, presumably because the punctured cell membrane may fail to seal completely around the walls of the electrode, with the consequent development of a short-circuiting current path in parallel with the recording electrode (LING and GERARD, 1949; FRÖMTER et al., 1971; PETERSEN, 1976c). Hence, it is inevitable that electrophysiologists apply an element of selection in choosing what criteria to follow when deciding whether an impalement record is satisfactory (and acceptable) or not. In general, a successful impalement is one in which, following advancement of the electrode tip into the tissue, there is a sudden, sharp shift in the recorded potential that approaches a new stable level monophasi-

Table 2. Potential difference (P. D.) recorded across the basal membrane of parenchymal cells of salivary glands in various species. The symbols de, hy and bi denote, respectively, depolarization, hyperpolarisation and a biphasic (de-hy) response. *Ach* and *Mech* indicate acetylcholine and methylcholine, *Nad* and *Adr* indicate noradrenaline and adrenaline, *dbcAMP* indicates dibutryl cyclic AMP and *pn* and *sn* indicate parasympathetic and sympathetic nerve stimulation, respectively. All values are in mV. Where possible mean values ± S. D. are given, recalculated, if necessary, from the original data. The morphology of each type of secretory endpiece is based on description of YOUNG and VAN LENNEP (1978)

Resting P. D. (mV)	Change in P. D. (mV) caused by stimulation Parasympathetic	 Sympathetic	Source
		Rat	
		Mandibular gland (homocrine seromucous acini)	
40.1± 1.9	de, hy or bi *(pn)*		a
27.8± 1.3 I	(0–20)	–	a
68.7± 3.2 III			a
28.0±10.9	–	–	b
23.0± 6.6 I	de, hy, bi, or no change *(pn)*	de, hy, bi, or no change *(sn)*	b,d
41.0± 2.6 III	(1–20)	(1–20)	b,d
immature rats:			
20.0± 6.5 I	–	–	c
55.0±10.3 III	–	–	c
		Parotid gland (homocrine serous acini)	
20.0± 7.6	–	–	b
		Golden Hamster	
		Mandibular gland (homocrine seromucous acini)	
15–30 I	–	–	
30–42 II	–	–	g
42–82 III	–	–	
		Mouse	
		Mandibular gland (homocrine seromucous acini)	
56.9± 9.3	de, hy or bi *(Ach)*	–	e*
66.9± 6.9	de (9.8) *(Ach)*	–	e*
50.9± 7.5	hy (14.8) *(Ach)*	–	e*
60.4± 8.5	bi (+4.9, −10.0) *(Ach)*	–	e*
		Parotid gland (homocrine serous acini)	
67.6± 1.3 (SEM)	–	hy (12) *(Adr)*	f*, v*
70.6± 0.4 (SEM)	hy (12.3) *(Ach)*	de (9.4) (Isoproterenol)	f, u*
		hy (3.5) *d.b.c.*AMP	u*

* Indicates studies *in vitro*. I, II, III refer to various types of unit as described originally by LUNDBERG (1955)

Table 2 (continued)

Resting P. D. (mV)	Change in P. D. (mV) caused by stimulation Parasympathetic	Sympathetic	Source
Rabbit			
Mandibular gland (central serous tubules with terminal and peripheral seromucous acini)			
59.2± 9.1	de, hy or bi *(Ach)*	–	e*
47.6± 2.4 (SEM)			h
46.9± 2.4 "	hy (12.8) *(pn)*	–	h
48.6± 1.5 "	bi (+8.4, −7.5) *(pn)*	–	h
12–13 (*n*=66)	hy (7.7) *(Mech)*	No response *(sn)*	i
Parotid gland (homocrine serous acini)			
15–29 (*n*=58)	hy (8.0) *(Mech)*	hy (4.2) *(sn)*	i
Cat			
Mandibular gland (mucous tubules with seromucous demilunes)			
22 (*n*=100) I	hy (35) *(pn)*	hy *(sn)*	j
33 II	hy (20) *(pn)*	de *(sn)*	j
80 III	de (60) *(pn)*	de *(sn)*	j
20.7± 5.4	–	–	k
38.2± 9.4	de (16.9) *(Ach)*		e*
31.7± 1.2 (SEM)			h
29.0± 1.1 "	hy (25.4) *(pn)*	–	h
39.8± 2.1 "	bi (+5.2, −15.2) *(pn)*	–	h
32.7± 0.9 "			h
29.9± 1.0 "	–	hy (24.7) *(sn)*	h
35.7± 1.2 "	–	bi (+3.5, −19.4) *(sn)*	h
25.0± 7.0	hy (10) *(pn, Ach)*	bi *(pn, Nad, Adr)*	l, t
Sublingual gland (mucous tubulo-acini with serous demilunes)			
33 (*n*=100)	hy (35) *(pn)*	hy (20) *(sn)*	m
Parotid gland (seromucous acini with occasional mucous cells)			
29.7±16.4			w
22.2± 8.0 I	hy (6) *(pn)*	No change *(sn)*	n, p
55.7± 9.0 III	No change *(pn)*	No change *(sn)*	n, p
Dog			
Mandibular gland (mucous tubules with seromucous demilunes)			
42.0± 7.0	hy (15) *(pn, Ach)*	–	q, r

* Indicates studies *in vitro*. I, II, III refer to various types of unit as described originally by LUNDBERG (1955)

Table 2 (continued)

Resting P. D. (mV)	Change in P. D. (mV) caused by stimulation Parasympathetic	Sympathetic	Source
	Possum *(Trichosurus vulpecula)*		
	Mandibular gland (serous tubules capped with seromucous acini)		
23.5± 6.0	hy *(pn, Ach)*	No effect *(sn)*	s
		bi *(Nad, Adr)*	s
28.0± 8.0	–	–	t

Source:

a Ichioka et al. (1965)
b Schneyer and Schneyer (1965a)
c Schneyer et al. (1968)
d Schneyer and Yoshida (1969)
e Nishiyama and Petersen (1974)*
f Pedersen and Petersen (1973)*
g Henriques and Sperling (1966)
h Kagayama and Nishiyama (1974)
i Fritz (1971)
j Lundberg (1955)
k Petersen (1970, 1972)
l Creed and Wilson (1969), Creed (1972)
m Lundberg (1957a)
n Fritz and Botelho (1969a)
p Fritz and Botelho (1969b)
q Imai (1965a)
r Yoshimura and Imai (1967)
s Creed and McDonald (1970)
t Creed and McDonald (1973)
u Petersen (1976a)*
v Petersen and Pedersen (1974)*
w Fritz (1974)

* Indicates studies *in vitro*. I, II, III refer to various types of unit as described originally by Lundberg (1955)

cally and is maintained at the new level for a long period (observed on a minute rather than a second time scale since resealing after impalement seems to take about one minute [Petersen, 1976c]); following withdrawal of the electrode, there should be an equally sharp reversion of the recorded potential to the pre-insertion level. In practice, the potential recorded on impalement may often shift with a very slow time course, may rise to a high value initially and then fall to a stable value, may never stabilize but rather fall slowly towards zero, or may disappear suddenly after a few seconds. Petersen (1974a) has examined the selection criteria closely and concludes that stability is the most reliable criterion of successful impalement. He accepts recordings in which the potential reaches a stable level monophasically and remains steady for some minutes. He rejects recordings in which the potential has an early falling phase. To his criteria should be added a strong warning that one should not prejudge what constitutes a successful impalement only from the magnitude of the recorded value.

Since glandular parenchymal cells form part of an epithelium, it follows that after a successful impalement, further advance of the electrode should give rise to a second potential shift if the luminal cell membrane can be penetrated. Such two-step potential profiles can be obtained in many epithelia, such as thyroid follicles (Woodbury and Woodbury, 1963) and kidney tubules (Frömter et al., 1971). In mammalian salivary glands these profiles have definitely been obtained from cat sublingual endpieces (LUNDBERG, 1957a, 1958) and perhaps also from dog mandibular endpieces (IMAI, 1965a), but from no other gland. In some cases, where insertion of the electrode into the endpiece lumen proved difficult, workers have attempted to circumvent the problem by placing one recording electrode in an endpiece cell and the other in the lumen of a large duct (*e.g.* IMAI, 1974). Such attempts must always lead to erroneous conclusions since one may not ignore the electrical cable properties of a small tubular structure filled with an electrolyte solution of finite resistivity (HEGEL et al., 1967) — in fact, the electrical recordings obtained in such studies reflect mainly the transepithelial potential difference across the duct wall in the immediate vicinity of the electrode tip. Even in so large a salivary duct as the main excretory duct of the human submandibular gland, the electrical length constant is only 6.9 mm (KNAUF and FRÖMTER, 1970c) and in the main duct of the rabbit mandibular gland it is 0.38 mm (FRÖMTER et al., 1974).

Table 2 summarizes available information on resting potentials across the basal plasma membrane of salivary parenchymal cells. It is immediately apparent that there is a wide variability of results, ranging from as low as 12–13 mV (FRITZ, 1971: rabbit mandibular gland) to as high as around 71 mV (PEDERSEN and PETERSEN, 1973: mouse parotid, *in vitro*) and 80 mV (LUNDBERG, 1955: cat mandibular gland, deeply placed units). Part of this scatter undoubtedly reflects real interspecies differences, but it seems reasonable also to conclude that recording *in vivo* resulted in values lower than are obtained by recording *in vitro*. Thus in the rabbit mandibular gland, the basal membrane potential of resting endpiece cells was reported as 50.2 ± 3.7 mV and 46.9 ± 2.4 mV when recorded *in vivo* (KAGAYAMA and NISHIYAMA, 1974) and 59.2 ± 9.1 mV, *in vitro* (NISHIYAMA and PETERSEN, 1974); similarly potentials of 31.7 ± 1.2 mV for cat mandibular gland studied *in vivo* and 38.2 ± 9.4 for cat mandibular gland studied *in vitro* have been reported (Table 2). PETERSEN (1974a) believes that the higher potentials obtained *in vitro* reflect the greater ease of obtaining stable recordings when vascular pulsations and transmitted muscular movements are eliminated by use of an *in-vitro* preparation. While this conclusion seems very reasonable, it must be borne in mind that the membrane potential is determined by the relative conductances of the membrane to those ions not in equilibrium across the membrane, conductances which themselves depend on the composition of the fluid bathing the membrane (*e.g.* the Ca^{2+} content of the bathing fluid: NISHIYAMA and PETERSEN [1974]). Hence it must always be remembered that the stable recording obtained *in vitro* might not prove to be identical to the values actually existing *in vivo*, if these could be recorded with equal facility. There is always a temptation in electrophysiology to treat larger values as more acceptable than smaller ones.

If one confines the comparison among species only to those studied *in vivo*, the variability is seen to become somewhat less. Thus the mandibular glands of rat, hamster, cat, and possum appear to have many units with potentials of 20–30 mV, while rabbit and dog mainly have units with potentials of 30–40 mV. Mouse mandibular glands appear from the *in-vitro* studies to have units with even higher potentials, of 60–70 mV. Fewer studies on parotid glands are available, but the data suggest that parotid and mandibular glands have units with similar potentials for each species investigated. Only one investigator has studied a sublingual gland (in the cat), so generalization is not possible (LUNDBERG, 1957a–c, 1958).

When LUNDBERG (1955) performed his first electrophysiological study on the cat mandibular gland, he identified three different types of unit from which he could record, partly on the basis of the site of the units within the gland tissue and partly on the basis of the size of the resting membrane potential and the changes in it that could be induced by sympathetic and parasympathetic nerve stimulation (see Section C. II. 5. b. i). The most frequently encountered unit (Type I), which was found from direct observation to come from endpiece cells, had an average basal membrane potential of −22 mV (cytoplasm-negative). A second type of unit, less frequently encountered (Type II), came from the superficial region of the gland but could not be localized visually. Type II resting potentials were higher on average (−33 mV). A third unit (Type III), recorded only when the electrode was inserted deeper into the gland, had an average resting potential of −80 mV. LUNDBERG (1955, 1957a) was inclined to ascribe Type I and II units, respectively, to the mucous tubular cells and the seromucous demilunar cells of the endpiece, and Type III units to the striated duct cells. Although his evidence was indirect and not conclusive, his arguments in support of these assumptions still appear plausible. More recent studies of cat mandibular gland, *in vivo* (KAGAYAMA und NISHIYAMA, 1974) and *in vitro* (NISHIYAMA and PETERSEN, 1974) have confirmed the existence of two types of superficial unit, but no further information is available to indicate if they correspond to two anatomically distinct cell types.

The possibility of recording from different types of unit in other salivary glands has been examined for the mandibular glands of rat (SCHNEYER and SCHNEYER, 1965a; SCHNEYER et al., 1968; ICHIOKA et al., 1965), golden hamster (HENRIQUES and SPERLING, 1966), mouse (NISHIYAMA and PETERSEN, 1974), and rabbit (KAGAYAMA and NISHIYAMA, 1974), and also for cat parotid (FRITZ and BOTELHO, 1969a). Although most of these workers have observed two or three types of responses to stimulation (hyperpolarization, depolarization, and a biphasic response consisting of depolarization followed by hyperpolarization), none has succeeded in ascribing a particular type of response to a histologically identifiable cell type. In the rat mandibular gland, where granular ducts are superficial, it seems inevitable that ductal cells must sometimes have been impaled instead of the seromucous cells of the homocrine endpieces (YOUNG and VAN LENNEP, 1978). Hence the conclusion, reached on statistical grounds by SCHNEYER and SCHNEYER (1965a) and by ICHIOKA et al. (1965), that there are two types of superficial units in the rat mandibular gland, one with resting potentials of 20–30 mV (presumed endpiece cells) and one with potentials of 40–70 mV (presumed granular duct cells), seems highly plausible, even though the two types of unit could not be distinguished from one another on the basis of their responses to nerve stimulation. In contrast, NISHIYAMA and PETERSEN (1974) could not distinguish two types of unit in the mouse mandibular gland despite the fact that it too possesses granular ducts in abundance; it is perhaps unfortunate that NISHIYAMA and PETERSEN (1974) seem only to have used female mice, since the granular ducts are much more conspicuous in male mice

(CARAMIA, 1966a and b; LACASSAGNE, 1940a and b; YOUNG and VAN LENNEP, 1978).

Regardless of any possible anatomical significance that the different types of glandular unit may have, it seems clear that the resting membrane potentials of most types of salivary unit yet investigated are lower, sometimes considerably lower, than the resting membrane potentials of skeletal muscle cells (~ −90 mV), where the membrane behaviour approximates closely to the behaviour of an ideal potassium-selective electrode (HODGKIN and HOROWICZ, 1959). The question arises of the extent to which the lower potential differences found in salivary units reflect inadequate recording techniques, a different intracellular ionic composition, or different permeability and pump properties of the salivary cell membrane. Although PETERSEN (1973, 1974a, 1976c) reasonably holds the opinion that inadequate recording technology is a contributory factor, it seems unlikely to be the only one, since lower membrane potentials appear to be characteristic not only of salivary parenchymal cells but of nonexcitable cells in general (WILLIAMS, 1970; MATTHEWS, 1974). It is a pity that electrophysiological studies in mouse salivary glands *in vivo* are not available, since this would resolve the question concerning the extent to which high membrane potentials in that animal are merely an expression of a species differences (PEDERSEN and PETERSEN, 1973).

Table 3 summarizes evidence tending to exclude the possibility that salivary cell ionic content is notably different from that of muscle cells. The table shows the intracellular electrolyte content of the five glands (in three species) that have so far been studied. These data are based on estimates of total tissue ionic content, corrected for extracellular fluid contamination by use of inulin-space determination and an assumed knowledge of the ionic content of salivary extracellular fluid. Such estimates, which are based on inevitably inaccurate measurements, involve also making the assumption that different gland cell types have similar ionic composition and that the intracellular ions are dispersed homogeneously throughout the cell or, at least, that the ionic activities of the fluid bathing the cytoplasmic surface of the cell membrane bear a fixed relation to the total intracellular ionic content. In the case of salivary glands the estimates are particularly prone to error, since inulin-space determinations do not include an estimate of the volume of saliva already formed and present in the gland duct system, and the estimates do not allow for the fact that, if saliva does not have a plasma-like composition, neither can the salivary gland interstitial fluid from which ultimately the saliva is formed. Nevertheless, despite these potential sources of error, it is apparent that, like most other mammalian cells, salivary parenchymal cells are relatively rich in potassium and poor in sodium and chloride.

Table 3 also shows the calculated electrochemical potentials of Na^+, K^+, and Cl^- ions in the cell cytoplasm, relative to the interstitial fluid. It seems apparent that Na^+ and K^+ are maintained at electrochemical potentials displaced (in opposite directions) far from equilibrium, whereas Cl^- is quite close to its equilibrium point (especially when one bears in mind that the concentrations of Na^+ and Cl^- shown in Table 3 are probably overestimates and the cytoplasmic concentrations of K^+, underestimates). Membrane-bound, Na^+-, K^+-activated,

Table 3. Summary of available information concerning the basal membrane potential differences, $\Delta\psi$ (mV), the intracellular electrolyte concentrations $[Na_i^+]$, $[K_i^+]$ and $[Cl_i^-]$ (mmol $kg_{H_2O}^{-1}$), and the associated ionic electrochemical potentials, η^{Na}, η^{K} and η^{Cl} (kJ mol^{-1}), for salivary endpiece cells of various species; + and − signs indicate polarity of cytoplasm with respect to interstitium. Values given are averages for all parenchymal cells with no distinction made between various types of endpiece and duct cell. Inulin space was used to estimate the extracellular fluid (*e. c. f.*) volume

	$\Delta\psi$	$[Na_i^+]$	η^{Na}	$[K_i^+]$	η^{K}	$[Cl_i^-]$	η^{Cl}	Source
Rat								
Mandibular gland								
resting	−28	9.8	−9.55	157.0	+7.14	35.8	−0.19	a
Parotid gland								
resting	−20	21.6	−6.76	142.0	+7.62	48.3	−0.19	a
e.c.f.		143.1		3.4		110.8		a, b
Cat								
Mandibular gland								
resting	−22	47.0	−5.21	115.0	+6.47	72.0	+0.68	c, d
Sublingual gland								
resting	−33	53.0	−5.98	147.0	+6.03	80.0	+2.03	e
stimulated	−56	78.0	−7.24	125.0	+3.38	—	—	e
e.c.f.		158.0		4.0		125.0		e
Dog								
Mandibular gland								
resting	−42	3.0	−14.09	117.0	+4.54	62.0	+1.83	f, g
stimulated	−57	34.0	− 9.26	89.0	+2.41	60.0	+3.18	f, g
e.c.f.		150.0		4.0		148.0		f, g

Source:
a SCHNEYER and SCHNEYER (1960)
b Extracellular fluid concentrations for the rat taken from MARTIN and YOUNG (1971a)
c LUNDBERG (1955)
d WILLS and FENN, cited by BURGEN and EMMELIN (1961) p. 205
e LUNDBERG (1957a, 1958)
f IMAI (1965a and b)
g YOSHIMURA and IMAI (1967)

ouabain-sensitive, ATPases are abundant in salivary glands (SCHWARTZ et al., 1963; SCHWARTZ and MOORE, 1968; BUNDGAARD et al., 1977), and all available evidence indicates that these enzymes are responsible for maintaining the transmembrane gradients for Na^+ and K^+ (SCHNEYER et al., 1972). The high electrochemical potential for chloride in the dog mandibular gland (YOSHIMURA and IMAI, 1967) reflects the higher values recorded for membrane potentials; if PETERSEN (1974a) is correct in his assertion that most resting membrane potentials that have been reported are too low it could very well be that chloride is concentrated in all salivary cells against an electrochemical gradient. This would explain why replacement of extracellular chloride with sulphate in mouse parotid gland increases the resting membrane potential (NISHIYAMA and PETERSEN, 1974); a sustained increase should not occur if chloride is distributed passively across the membrane (HODGKIN and HOROWICZ, 1959). (See however, Section C. II. 5. b. i)

From the foregoing it seems reasonable to conclude that the lower membrane potentials of salivary parenchymal cells are not due to any marked difference in electrolyte content between salivary and other mammalian cells. Hence it follows that the difference must be due to different permeability and pump characteristics in the salivary basal cell membrane. WILLIAMS (1970) and MATTHEWS (1974) have reviewed the topic of membrane potentials and permeability properties in nonexcitable cells, including some salivary cells, and conclude that while nonexcitable cell membranes show the same differential permeability to sodium and potassium as does muscle cell membrane, the ratio of the permeabilities of Na^+ and K^+ (P_{Na}/P_K) is much larger in nonexcitable cells. WILLIAMS (1970) calculates that while amphibian skeletal muscle has a value of above 0.01 for P_{Na}/P_K and a membrane potential of -90 mV, most mammalian cells, both excitable and nonexcitable, have ratios of 0.04 to 0.40 and membrane potentials only of -40 to -60 mV. PETERSEN (1976c), however, stresses that the differences between gland cell membranes and muscle membranes are not very great in many cases.

In salivary glands the relation between the resting membrane potential and the bathing fluid potassium concentration, $[K_o^+]$, has been studied in a number of glands and found to be substantially less than 61 mV per decade change in $[K_o^+]$, the Nernst prediction for an ideal potassium-selective membrane. Thus IMAI (1965b) and YOSHIMURA and IMAI (1967) reported a slope of 15 mV/decade for the dog mandibular gland and PETERSEN and POULSEN (1967b) reported a slope of 7 mV/decade for the cat mandibular gland (see also PETERSEN, 1971c). CREED (1974) reports failing to observe any relation between membrane potential and $[K_o^+]$ in either cat or possum mandibular glands, even when $[K_o^+]$ was increased to 70 mmol · l^{-1}. More recently, however, much higher slopes of around 50 mV/decade have been reported for mouse mandibular and parotid glands studied *in vitro* (PEDERSEN and PETERSEN, 1973; NISHIYAMA and PETERSEN, 1974); this would be expected, given the high resting membrane potentials observed in mouse salivary glands.

Thus, we may conclude that the origin of the resting membrane potential in salivary parenchymal cells is similar to its origin in other mammalian cells, excitable and nonexcitable. The cell membranes differ, at least quantitatively, from those of nerve and muscle tissue in having a greater resting sodium permeability; the magnitude of this sodium conductance depends inversely on the external calcium concentration, as it does in excitable tissues (NISHIYAMA and PETERSEN, 1974). In addition, like other secretory cell membranes that have been studied (NISHIYAMA and PETERSEN, 1975; PETERSEN, 1975), the salivary cell basal membrane seems to be electrically nonexcitable, *i. e.* the current-voltage relation is linear, at least when sodium conductance is reduced by a high external calcium concentration (NISHIYAMA and PETERSEN, 1974). In the resting state, therefore, the membrane potential is generated largely as a potassium diffusion potential, short-circuited by a sodium diffusion potential of opposite polarity; chloride ions, if they really are at equilibrium across the membrane, would make no contribution to the potential. There is some evidence that the membrane sodium-potassium pump responsible for setting up the electrochemical gradients for sodium and potassium may itself be electrogenic and therefore capable of influencing the membrane potential (PEDERSEN and PETERSEN, 1973), but whether this holds normally or only under extreme conditions (such as the transitions from $[K_o^+]=0$ to $[K_o^+]=4.7\ \mathrm{mmol \cdot l^{-1}}$) is unknown (see PETERSEN, 1976c).

At the present time it seems simplest to suppose that, at rest, sodium and potassium cross the cell membrane *via* protein ionophores with binding sites for positive ions that, at rest, are occupied largely by calcium ions that are in diffusional equilibrium with extracellular and intracellular calcium pools. The presence of calcium determines the "pore size" to sodium and potassium following an inverse relation, so that withdrawing calcium from the medium increases membrane conductance to sodium and potassium, and adding calcium decreases conductance. Some of these pores may be specialized to interact with neurotransmitter substances ("excitable pores").

The above discussion has been concerned largely with the basal membrane of the resting salivary cell. Very little information is available concerning the potential across the luminal membrane, largely because of the technical difficulties associated with driving the recording electrode through the cell and controlling the placement of its tip in the endpiece lumen. LUNDBERG (1957a, 1958) was able to achieve this in cat sublingual gland after first distending the endpiece by duct obstruction; he was unable to do the same in the cat mandibular gland (LUNDBERG, 1955). IMAI (1965a) appears to have been successful in the dog mandibular gland. In the cat sublingual gland the lumen was about 2 mV positive to the interstitium, and in the dog mandibular about 2 mV negative. The significance of these values is discussed later (Section C. II. 5.b).

5. The Stimulated Endpiece Cell

a) Receptors and Receptor Pharmacology

So far as is known, mammalian salivary glands normally secrete only in response to nerve stimulation, although some naturally occurring polypeptides have been shown to be sialogogic (ERSPAMER and ERSPAMER, 1962; BERTACCINI et al., 1965a and b; BERTACCINI and DE CARO, 1965; LEEMAN and HAMMERSCHLAG, 1967; LEMBECK et al., 1968), so the possibility of the existence of a yet undiscovered sialogogic polypeptide local hormone cannot be excluded. Adrenalin, of course, is also sialogogic, but not at normally circulating concentrations. The autonomic pharmacology of salivary secretion has been exhaustively reviewed quite recently (EMMELIN et al., 1973) and will be considered outside the scope of the present work, except for the pharmacology of events taking place at the neuroeffector junctions and within the secretory cells themselves.

All salivary glands so far studied receive a parasympathetic innervation and secrete in response to intravenously injected acetylcholine. The effect of acetylcholine is mimicked by muscarine (FRASER, 1957; WASER, 1958) and is blocked by the administration of atropine, justifying the conclusion that salivary endpiece cells possess muscarinic cholinergic receptors; these receptors can also be activated by a wide range of other parasympathomimetic drugs with known muscarinic activity (EMMELIN et al., 1973). Activation of these muscarinic receptors by whatever means causes production of copious quantities of saliva, usually hypotonic, with moderately low concentrations of enzymes such as amylase and, in mucous-secreting glands, variable quantities of glycoproteins.

In contrast to the parasympathetic innervation, the sympathetic innervation of salivary glands is extremely variable. Some glands seem to receive a rich innervation and others, even in the same species, receive few or no sympathetic secretomotor fibres at all; of those that are innervated, some seem to have predominantly α receptors, some β and some both α and β. In the rat, sympathetic nerve stimulation causes a vigorous response in the mandibular gland (EMMELIN et al., 1965; OHLIN, 1966; YOSHIDA et al., 1967). The response is predominantly of β type, modified slightly by α-receptor activation (see Section C. III. 1. b). A pure β response to isoproterenol can also be elicited; one obtains a smaller volume of saliva of striking composition (SCHNEYER, 1962; YOUNG and MARTIN, 1971). The response to the α-agonist, phenylephrine, is vigorous and resembles the response to parasympathomimetic agents closely (MARTINEZ et al., 1975; see Fig. 6). The rat parotid gland also responds vigorously to sympathetic nerve stimulation (EMMELIN, 1967; GARRETT and THULIN, 1975); this effect is mediated, in part at least, by β receptors, since isoproterenol causes formation of moderate volumes of a potassium- and bicarbonate-rich fluid (MANGOS et al., 1973d) not dissimilar in composition to the isoproterenol-induced saliva of the rat mandibular gland. The response to a pure α agonist has not been published. The rat sublingual gland gives little or no response to sympathetic nerve stimulation or to the administration of α- or β-adrenergic agonists (MARTIN and YOUNG, 1971a; EMMELIN et al., 1965). The rabbit parotid responds vigorously to sympathetic stimulation (NORDENFELT and OHLEN, 1957) but the mandibular gland gives almost no response to nerve or drug stimulation (EMMELIN, 1967; COMPTON, 1976; FRITZ, 1971). The cat mandibular gland has long been known to give a particularly brisk secretory response to sympathetic nerve stimulation and to the administration of α-adrenergic agonists (LANGLEY, 1878; EMMELIN, 1955; EMMELIN and ENGSTRÖM, 1960; DOUGLAS and POISNER, 1963; KAHN et al., 1969) but, in some cats similarly brisk responses can be produced by the β-agonist, isoproterenol (KALADELFOS and YOUNG, 1974). The cat sublingual gland gives a definite, albeit sluggish, flow response to isoproterenol, but the electrolyte content of the saliva remains unchanged (Fig. 4; KALADELFOS, 1971). The cat parotid gives little or no response (EMMELIN, 1967; FRITZ and BOTELHO, 1969a and b). In contrast to the cat mandibular gland, the dog mandibular gives a sluggish response to sympathetic nerve stimulation, which appears to be mediated exclusively *via* β-receptors (EMMELIN et al., 1969a and b; EMMELIN and GJÖRSTRUP, 1973, 1974). The dog parotid responds even less well (EMMELIN, 1967). The ductal application of adrenalin provokes secretion from the human submandibular but not the parotid gland (EMMELIN and STRÖMBLAD, 1954), while intravenous administration of isoproterenol stimulates both glands to secrete (KATZ and MANDEL, 1968). Observations on the sheep parotid gland (BLAIR-WEST et al., 1969; COMPTON, 1976) and the mandibular gland of the Australian brush-tailed possum, *Trichosurus vulpecula*

(CREED and McDONALD, 1970; COMPTON, 1976) suggest that these organs do not secrete in response to sympathetic stimulation. However, it would appear that COMPTON (1976) and BLAIR-WEST et al., (1969) overlooked a small adrenergic secreto-motor response of exclusively β type (PATTERSON and TITCHEN, 1975).

While it is possible, allowing for a few exceptions, to make the generalization that parasympathetically-evoked saliva is hypotonic and relatively poor in organic constituents, it is more difficult to generalize about sympathetic saliva, partly because fewer systematic studies are available, and partly because of the variability of the response in different glands. In general it is justified to conclude that saliva collected following sympathetic nerve stimulation is smaller in volume and richer in organic material. The organic material may largely be amylases, as is the case for the parotid glands of the mouse and the rat (SCHNEYER and HALL, 1965; MANGOS et al., 1973b and d; GARRETT and THULIN, 1975), or specialized glycoproteins, as for the cat mandibular gland (KAHN et al., 1969; DISCHE et al., 1970). In addition, sympathetic saliva is usually richer in potassium and bicarbonate and may, but need not, be poorer in sodium and chloride (SCHNEYER et al., 1972; EMMELIN et al., 1973).

b) The Secretory Potential and Associated Ionic Fluxes

i) *Cholinergic Stimulation.* Stimulation of the secretomotor nerves supplying any salivary gland causes a shift in the basal membrane potential (Table 2). In the cat mandibular gland, the gland in which LUNDBERG (1955) made his first observations, this potential shift was usually found to be a marked hyperpolarization of up to 35 mV, regardless of whether the stimulus was parasympathetic or sympathetic. Units giving such a response (Type I) had a low resting potential of −22 mV, were activated by single-shock parasympathetic nerve stimulation (duration of responses <1 s) or multiple-shock sympathetic nerve stimulation, and had a long latency of around 350 ms between the start of stimulation and the onset of the hyperpolarizing response ("secretory potential"). LUNDBERG recognized a second group of superficial units (Type II), which were characterized by a higher resting membrane potential (−33 mV), a more sustained hyperpolarizing response (~7 s) resulting from single shock parasympathetic nerve stimulation, and a depolarizing response to sympathetic stimulation. He considered both Type I and Type II responses to come from endpiece cells. Subsequently, PETERSEN (1970b, 1972) and CREED (CREED and WILSON, 1969; CREED, 1974; CREED and McDONALD, 1973) confirmed LUNDBERG's observations concerning Type I potentials. CREED and WILSON (1969) discussed the long latency of onset of the hyperpolarizing response, and calculated that a maximum of 60 ms of a total of 350 ms could be accounted for by nerve conduction times and delays at the ganglionic junction. (Subsequent studies in the Australian marsupial possum, *Trichosurus vulpecula,* where ganglionic delays could be avoided, still revealed a long latent period that could not be fully accounted for [CREED and McDONALD, 1970].) It seems that a significant period elapses between the arrival of transmitter at the postjunctional membrane and initiation of the secretory potential.

More recently, KAGAYAMA and NISHIYAMA (1974) have observed somewhat higher resting potentials in cat mandibular units, which they subdivided into two groups on the basis of their response to both parasympathetic and sympathetic nerve stimulation. One group, with a mean resting potential of 29–30 mV, responded to nerve stimulation with a hyperpolarization of about 25 mV after a latent period of 290 ms. The other group, with higher resting potentials of 36–40 mV, responded biphasically to stimulation, first with a depolarization of 3–5 mV, and then with a hyperpolarization of 15–20 mV. In this group, the latency of onset of depolarization was much shorter, about 170 ms. The same authors also identified similar groups of units in the rabbit mandibular gland (Table 2), where again the high-resting-potential group gave a biphasic secretory response and had a much shorter latency (168 ms) than the low-resting-potential group, which gave a hyperpolarizing response (550 ms). In the hyperpolarizing group of units (for rabbit but not cat mandibular gland), the authors made the important observation that a decrease in membrane resistance, with an onset latency of about 150 ms, preceded the hyperpolarizing response by up to 400 ms. They concluded that the real onset latency was about the same for both groups, the only difference being that, in units with potentials below a critical reversal value, no depolarizing response is seen; rather there is a "silent" increase in membrane conductance. Nevertheless, a response latency of 150 ms is still large and, on the basis of CREED and WILSON's (1969) estimate, indicates that a delay of about 90 ms follows release of transmitter at the neuroeffector junction before a response of any kind is seen in the postjunctional membrane. If the neuroeffector junctions were of hypolemmal type and, consequently, only about 20 nm wide (GARRETT, 1972, 1974; YOUNG and VAN LENNEP, 1978), it would be most unlikely that diffusion of acetylcholine across the junction could account for more than a few milliseconds of the delay (*cf.* amphibian neuromuscular junctions: KATZ and MILEDI, 1965), but if the junctions were of epilemmal type, when the junction is greater than 100 nm in width and the basement membrane separates the nerve terminal from the secretory cell, the delay due to acetylcholine diffusion would obviously be much greater. In neither the rabbit nor the Australian possum mandibular gland are hypolemmal junctions seen in relation to the seromucous acinar cells of the endpiece, although they are seen near the granular serous tubular cells at the base of the endpiece (KAGAYAMA and NISHIYAMA, 1972; GARRETT, 1974; BLOOD et al., 1977; YOUNG and VAN LENNEP, 1978); in the cat mandibular gland the intensity of hypolemmal cholinergic innervation is extremely variable, so that in some animals junctions are seen frequently, while in others they are not seen at all (GARRETT, 1972, 1974; KAGAYAMA and NISHIYAMA, 1972; YOUNG and VAN LENNEP, 1978). Hence it cannot be excluded that a substantial component of the onset latency observed by KAGAYAMA and NISHIYAMA (1974) and many other authors arises from the nature of the neuroeffector junctions involved, although it seems a little unlikely to be the sole explanation, since presumably some units that have been studied would have had hypolemmal innervation and should therefore have shown much shorter latencies.

Other studies, carried out *in vitro* on mandibular and parotid glands of the mouse and on mandibular glands of the rabbit, have also shown variable types of secretory response, hyperpolarization, depolarization or a biphasic response (PEDERSEN and PETERSEN, 1973; NISHIYAMA and PETERSEN, 1974). In the mouse mandibular gland, the hyperpolarizing group had the lowest resting membrane potential (51 mV), the biphasically responding group had an intermediate resting potential (60 mV), and the depolarizing group had the highest resting potential (67 mV). Interestingly, however, the mouse parotid gland (PEDERSEN and PETERSEN, 1973; PETERSEN and PEDERSEN, 1974; PETERSEN, 1976a) and the cat mandibular gland (NISHIYAMA and PETERSEN, 1974), when studied *in vitro*, gave only hyperpolarizing responses after exposure to acetylcholine despite having high resting membrane potentials.

Variable secretory potentials have also proven to be a feature of the rat mandibular gland, although there does not seem to be any association with the level of the resting membrane voltage. ICHIOKA et al. (1965) recorded resting potentials from three depths in the rat mandibular gland, superficial (0–280 μm), intermediate (280–560 μm), and deep (> 650 μm). At all three levels they were able to record units that, on statistical probability grounds, could be subdivided into a group having low resting potentials (28 mV) and a group having high potentials (69 mV). The relative frequency of these two potentials did not differ among the units recorded at each level within the gland. Nerve stimulation, activated reflexly by oral stimulation, resulted in depolarization in 39 percent of all units, hyperpolarization in 29 percent, and mixed and irregular biphasic responses in the remainder. The type of response obtained was unrelated to the distance of the unit below the gland surface or to the size of the resting membrane potential. SCHNEYER and SCHNEYER (1965a) also observed units with high and low resting potentials, although they separated their groups on the basis of the relative frequencies of endpiece and granular duct cells rather than on the basis of a distribution probability. Subsequently SCHNEYER and YOSHIDA (1969), who studied only units having low resting potentials, showed that direct stimulation of parasympathetic and sympathetic secretory nerves could cause either a hyperpolarizing or a depolarizing response. In these studies, particularly that of ICHIOKA et al. (1965), it is clear that even if one studied only units that have potentials fulfilling PETERSEN's (1974a) requirements for stability, one would still have observed all possible response types.

Before attempting to summarize what is known of the origin of the secretory potential, it is appropriate to comment on the effect of stimulation on the luminal membrane potential. As mentioned earlier, only two workers have succeeded in recording from the endpiece lumen, LUNDBERG (1957a and b) in the cat sublingual, and IMAI (1965b) in the dog mandibular gland. Both investigators recorded the potential difference across the entire epithelium and obtained the luminal membrane potential by subtracting the basal membrane potential from it. LUNDBERG reported that the endpiece lumen was only about 2–3 mV positive to the interstitium while IMAI found that the lumen was slightly negative; both workers concluded that the apical and basal membrane potentials were of similar magnitude but of opposite polarity. Both authors also observed that parasympathetic nerve stimulation, which caused hyperpolarization of the basal membrane, also caused hyperpolarization of the luminal membrane; LUNDBERG (1957a), in addition, found that the hyperpolarization of the two membranes followed different time courses. However, the significance of all these observations is obscure since endpiece epithelia almost certainly belong to the so-called "leaky" class, in which low-resistance shunt pathways for current flow exist between cells across the *zonulae occludentes* (*cf.* FRÖMTER and DIAMOND, 1972). In the presence of such shunts, the significance of the size and polarity of the luminal membrane potential at rest, and the influence of nerve stimulation of it, cannot be assessed (*cf.* SCHULTZ, 1973) without some knowledge of the conductance properties of the luminal and basal membranes and of the shunt pathway. LUNDBERG (1957b) attempted to obtain such information by means of a current-injection technique, but in order to calculate membrane resistances he assumed that current spread from cell to cell *via* low-resistance connecting pathways would be negligible; since such intercellular connections are now known to occur between most if not all neighbouring epithelial cells, his calculated membrane resistances must be considered as impossibly low (PETERSEN, 1976c). In the pancreas, where there is evidence to suggest that all the cells comprising a single acinus are electrically coupled, PETERSEN (1976c) now calculates that the specific resistance of the basal membrane of the acinar cells is about 2000 Ω cm^2; it seems highly likely that the corresponding value for salivary endpiece cells will be similar. From LUNDBERG's (1957b) original studies it is clear that the transepithelial resistance is about 100 times smaller than this, so it can be inferred that the shunt resistance also is very small. However, no inference can be drawn about the size of the resistance of the luminal membrane. Until such data are forthcoming it is profitless to speculate on the origin of the potential difference across the luminal cell membrane, at rest or during stimulation.

In essence, one of two possible mechanisms could account for the salivary secretory potentials. A change in membrane ionic conductances might take place, giving rise to various nett ionic fluxes as ion concentrations in the cell cytoplasm approach their equilibrium concentrations, or there might be a change in the current carried by an electrogenic membrane-bound ion pump. LUNDBERG (1957c, 1958) favoured the latter view and, on the basis of perfusion experiments in which he replaced chloride with nitrate in the perfusion fluid, proposed that the secretory potential arose from activation of a basally located chloride pump. However, since he had noticed a decrease in membrane resistance during stimulation, LUNDBERG (1957b) also entertained the idea that the secretory potential might be associated with an increase in potassium conductance, leading to a shift in the membrane potential from its low resting value towards the potassium equilibrium potential (LUNDBERG, 1958). The transient efflux of potassium into the venous effluent of salivary glands that occurs immediately following the onset of stimulation (BURGEN, 1956) obviously accorded more readily with the proposal that stimulation increased P_K.

Since 1957 the accumulated evidence in favour of a change in P_K rather than an increase in the rate of an electrogenic chloride pump has become overwhelming (see reviews of SCHNEYER et al., 1972; PETERSEN, 1973, 1975, 1976c). Thus, it is now realized that secretory potentials need not be hyperpolarizing, as would be required by the chloride pump theory, but can be depolarizing or biphasic (Table 2) and, regardless of their form, they are always associated with a decrease in the resistance of the basal cell membrane (LUNDBERG, 1957b; IMAI, 1965a; YOSHIMURA and IMAI, 1967; KAGAYAMA and NISHIYAMA, 1974; NISHIYAMA and PETERSEN, 1974; PETERSEN, 1974b, 1975, 1976a). This decrease in resistance is associated with a sharp increase in potassium efflux from the gland into the interstitium (BURGEN, 1956; BURGEN and SEEMAN, 1958; BURGEN and TERROUX, 1962; SCHNEYER and SCHNEYER, 1964; PETERSEN, 1970a, 1971a–d, 1973, 1975; POULSEN, 1973a, 1974). In addition, not only does the basal membrane at rest behave like a partly short-circuited potassium electrode (IMAI, 1965b; YOSHIMURA and IMAI, 1967; PETERSEN and POULSEN, 1967b; PETERSEN, 1971c; PEDERSEN and PETERSEN, 1973; NISHIYAMA and PETERSEN, 1974), but the size of the secretory potential itself (or, rather, its hyperpolarizing component) is inversely related to the potassium concentration in the external solution (YOSHIMURA and IMAI, 1967; PETERSEN, 1970b, 1971c and d; NISHIYAMA and PETERSEN, 1974). Another argument against the chloride pump hypothesis comes from perfusion experiments in which chloride ions were removed from the extracellular fluids. Although replacement of chloride by nitrate abolished both secretion and the secretory potential (LUNDBERG, 1957b, 1958; IMAI, 1965a)[3], replacement of chloride with sulphate had no marked effect on the secretory potential, despite the fact that it caused an immediate reduction in secretory rate (IMAI, 1965b; PETERSEN and POULSEN, 1968); a similar result was obtained by replacing chloride salts with sucrose (PETERSEN, 1971a). In experiments with chloride-containing perfusion fluid, replacement of sodium with tetraethylammonium, or of potassium with sodium, actually increased the secretory potential although secretion was abolished (PETERSEN, 1970b). Finally, PETERSEN (1970a and c) has shown that the secretory potential is not abolished by concentrations of dinitrophenol that inhibit secretion and cell K^+ uptake, suggesting that its development is not an energy-consuming process.

From the above it is clear that stimulation, both cholinergic and adrenergic, causes an increase in membrane conductance associated with an efflux of potassium ions from the cell (BURGEN, 1956; PETERSEN, 1970a; BATZRI et al., 1973) and a decrease in intracellular potassium concentration (Table 3). Since even the resting membrane seems to have a high sodium conductance (see above), and since it is known that salivary glands show an enhanced uptake of sodium when stimulated (BURGEN, 1956; SCHNEYER, 1967; YOSHIMURA and IMAI, 1967; PETERSEN, 1971c; POULSEN, 1974; see also Table 3), the question arises of how the secretory potential is modulated by sodium fluxes and changes in membrane sodium conductance. PETERSEN (1970b) has shown for the cat mandibular gland that replacement of sodium in the bathing fluid with tetraethylammonium enhances the size of the secretory potential and increases the slope of the line relating secretory potential to external potassium concentration to over 40 mV/decade. Plausibly he interpretes these data as indicating that sodium influx, occurring concomitantly with potassium efflux, normally reduces the size of the secretory hyperpolarization considerably. Similar findings have also been

3 However, PETERSEN and POULSEN (1969) were unable to confirm that replacement of chloride with nitrate could abolish the secretory potential.

reported for the mouse mandibular gland (NISHIYAMA and PETERSEN, 1974) and the cat parotid (FRITZ, 1974).

A change in sodium conductance may also serve to explain the initial depolarization seen when biphasic secretory potentials are observed. Perfusion studies performed by POULSEN (1974) on the cat mandibular gland show quite clearly that stimulated glands take up sodium concomitantly with the release of potassium. POULSEN (1973a) has also shown that K^+ is released with a latency of 540 ms ± 180 S.D., although it is doubtful whether the technique is accurate enough to permit one to say whether one flux might not have preceded the other by an interval as small as 200 ms. However, although conclusive direct evidence is lacking, indirect evidence in support of the idea that the depolarization phase of the secretory potential is due to an increase in sodium conductance is quite strong. Biphasic potentials are not seen when the external sodium concentration is reduced to zero (NISHIYAMA and PETERSEN, 1974) nor are they seen when the resting membrane potential is low (KAGAYAMA und NISHIYAMA, 1974; PETERSEN, 1973, 1975). Of particular importance is the finding that the membrane resistance decreases concomitantly with the onset of depolarization (KAGAYAMA and NISHIYAMA, 1974; NISHIYAMA and PETERSEN, 1974). These data are best explained by postulating that the initial conductance change is a nonselective increase in both sodium and potassium conductance, the direction of the resultant potential change being determined by the level of the resting membrane potential—sodium influx is predominant at membrane potentials above −50 mV (in mouse mandibular gland) and potassium efflux is predominant at lower potentials. The only alternatives that one might consider would be a conductance change leading to a calcium influx or a chloride efflux. Although calcium influx almost certainly does take place during the depolarizing phase, it seems unlikely to be responsible for the potential change, since biphasic potentials have been seen in glands bathed in calcium-free media. A chloride efflux consequent on an increase in chloride conductance cannot be ruled out so easily since it is difficult to exclude the possibility that the intracellular electrochemical potential for chloride might not be slightly positive (Table 4; see also Section C. II. 4). If this were the case, then the failure to see depolarization at low resting potentials would simply indicate that, in units of this kind, the membrane potential was near the equilibrium value for chloride because the membrane had an especially high chloride conductance.

Following the depolarization (or silent) phase in which sodium and potassium conductances appear to increase, there appears to be a further sharp increase in potassium conductance responsible for the delayed hyperpolarization phase seen in almost all endpiece cells (discussed above). NISHIYAMA and PETERSEN (1974) speculate that this latter phase may be triggered off by calcium ions that enter the cell along with sodium during the depolarization phase. There is no convincing evidence for or against this, however.

ii) *Adrenergic Stimulation.* Table 2 summarizes all available information on the secretory potentials produced by sympathetic nerve stimulation or administration of sympathomimetic agonists. LUNDBERG (1955) was the first to study the effect of sympathetic nerve stimulation on the salivary cell resting potential; he found that units of Type I hyperpolarize and those of Type II and III depolarize. The response of Type II units contrasted with their response to parasympathetic stimulation, which caused hyperpolarization. CREED and WILSON (1969) also studied the cat mandibular gland. They saw only hyperpolarizing units but, in contrast to the effect of parasympathetic stimulation, they observed that the secretory potential lasted only about 2 s, even when stimulation was maintained. They also observed that the latency of response was very long, averaging 662 ms ± 178 S. D. Other early studies on the rat mandibular gland (SCHNEYER and YOSHIDA, 1969), the rabbit parotid (FRITZ, 1971) and the cat sublingual gland (LUNDBERG, 1957a, 1958) revealed only units that hyperpolarized in response to sympathetic nerve stimulation; such units all had relatively low resting membrane potentials. Studies on the rabbit mandibular gland (FRITZ, 1971), the cat parotid gland (FRITZ and BOTELHO, 1969a and b; FRITZ, 1974) and the mandibular gland of *Trichosurus vulpecula* (CREED and MCDONALD, 1973) revealed no response to sympathetic activation (although *Trichosurus* mandibular gland gave a small biphasic response to administration of adrenalin and noradrenalin).

More recent studies on the cat mandibular gland and the mouse parotid have shown that secretory potentials are somewhat more complex than the above might have indicated. KAGAYAMA and NISHIYAMA (1974), studying the cat mandibular gland *in vitro*, observed two types of units; one type, with resting potentials of about −30 mV, showed a hyperpolarizing response of 25 mV to sympathetic nerve stimulation with a latency of 500 ms, while the other type, with resting potentials of −36 mV, gave a biphasic response, first depolarizing by 3.5 mV and then hyperpolarizing by 19.4 mV. The amplitude of the initial depolarization was dependent on the magnitude of the resting potential. The latency of onset was 534 ms. Both types of unit had much longer latencies than were observed for preganglionic cholinergic stimulation (291 and 176 ms).

PETERSEN has pursued these observations in greater detail using an *in-vitro* preparation of the mouse parotid gland (PEDERSEN and PETERSEN, 1973; PETERSEN and PEDERSEN, 1974; PETERSEN, 1976a and b). Adrenalin (10^{-5}M) caused a transient hyperpolarization of the membrane potential from a resting value of −68 mV to a peak value of −80 mV. Higher concentrations were not more effective; a small effect was detectable at 10^{-8}M but not 10^{-9}M. However, in the presence of the α-receptor blocking agent, phentolamine (10^{-5}M), adrenalin always caused a transient depolarization of the resting potential (around 7 mV). Use of the β-receptor agonists isoproterenol and metaproterenol also caused depolarizations (of 13 and 7 mV) and the α-agonist synephrine always caused hyperpolarization. The hyperpolarizing effect of adrenalin persisted in the presence of the β-blocker, propranolol, whereas the effect of isoproterenol was blocked. Interestingly, the response to adrenalin was sustained rather than transient in the presence of propranolol. Neither the α- nor the β-agonist effects were blocked when the gland was perfused with a calcium-free solution containing the chelating agent EGTA. In contrast, calcium ions appear to be essential for the development of cholinergic secretory potentials in cat and possum mandibular glands (CREED and MCDONALD, 1973) but not for mouse mandibular gland (NISHIYAMA and PETERSEN, 1974). The response to adrenalin was enhanced greatly by withdrawal of potassium from the bathing medium and severely reduced when the potassium concentration was increased to 20 mmol l^{-1}; ouabain did not block the responses. Unfortunately data are not available on the effect of bath potassium concentration on *β*-agonist responses. In his most recent study, PETERSEN (1976b) shows that both the α- and the β-agonist responses are associated with a drop in membrane input resistance of about 1–2 MΩ. Surprisingly, he also reported that dibutyryl cyclic AMP (2 mmol l^{-1}) did not produce the same response as the β-agonist isoproterenol; rather it caused a membrane hyperpolarization.

It is interesting to note that sympathetic stimulation induces biphasic and depolarizing responses as well as hyperpolarizing responses. Since the responses are associated with decreases in input resistance that are attributable to increased ionic conductances of the basal membrane (see above), and since α-adrenergic stimulation is known to cause potassium efflux from parotid glands (PETERSEN, 1970a; BATZRI et al., 1973), it is tempting to conclude that sympathetically activated secretory potentials are produced by changes in sodium and potassium conductances that are qualitatively similar to those produced in response to cholinergic stimulation. Since the data available for rat parotid and rat and cat mandibular gland (see above) suggest that α-adrenergic and cholinergic stimulants produce similar secretory responses (*i. e.* production of large volumes of a low-protein secretion) it is interesting to note that both modes of stimulation produce quantitatively similar secretory potentials. On the other hand, since β-adrenergic stimulation is thought to induce mainly protein secretion, with a minimal volume of salt and water, it is perhaps not surprising that it is associated with a qualitatively distinctive secretory response. Nerve stimulation and adrenalin administration, which activate both *α*-and *β*-receptors, could thus be expected to produce a composite potential response, either biphasic or transient and cancelling.

It is surprising that withdrawal of extracellular calcium, which severely reduces the α-adrenergic secretory response (PETERSEN, 1976a and b), does not block the development of the secretory potential; similarly, administration of dibutryl cyclic AMP, which simulates the β-adrenergic secretory response (BATZRI and SELINGER, 1973), did not produce the same type of secretory potential as did isoproterenol. Explanation of these anomalies has not yet advanced beyond the realm of speculation. It is clearly necessary to make systematic measurements of membrane conductance, both before and during the onset of the secretory potential, and to study the effects of changing electrolyte concentrations using both α- and β-agonists on the one preparation.

The fact that α- and β-agonists produce qualitatively different secretory potentials need not imply that their membrane actions are qualitatively different. Both modes of stimulation may affect sodium and potassium conductance simultaneously but in different proportions and with different time courses.

A comment should also be made on the extremely long response latencies (~ 500–600 ms) observed by KAGAYAMA and NISHIYAMA (1974) for the cat mandibular gland following sympathetic nerve stimulation. These latencies are twice those seen for parasympathetic nerve stimulation, for which, in contrast, preganglionic rather than postganglionic fibres have to be stimulated. Provided subsequent studies do not reveal the existence of a "silent" conductance change preceding the onset of the secretory potential by 300–400 ms (*cf.* parasympathetic response), the most likely explanation for the long latency would appear to be the great width of the neuroeffector junction (~ 100–200 nm, with the basement membrane intervening between nerve terminal and gland cell), since the cat mandibular gland is said to have only epilemmal sympathetic junctions (GARRETT, 1966b; KAGAYAMA and NISHIYAMA, 1972; YOUNG and VAN LENNEP, 1978).

iii) *Transients.* BURGEN (1956) and BURGEN and TERROUX (1962) originally noticed that abrupt changes in the intensity of salivary secretomotor nerve stimulation caused fluctuations in the sodium and potassium concentrations of saliva and salivary venous blood that persisted for several minutes, until new steady-state concentrations were achieved. In the case of potassium, onset of stimulation or a shift in stimulation frequency caused sharp, more or less simultaneous increases in potassium concentration in saliva and venous blood, associated with a concomitant decrease in the potassium concentration in the salivary tissue. At the same time, there is a transient decrease in venous sodium concentration (POULSEN, 1974) and a transient increase in salivary sodium concentration (BURGEN, 1964; PETERSEN and POULSEN, 1967a); this saliva sodium transient, which is delayed in relation to the potassium transient, is accompanied by a secondary transient fall in saliva potassium concentration (PETERSEN and POULSEN, 1968). Electrolyte transients, including those involving other ions such as iodide, which is actively transported in some salivary glands, are discussed in detail elsewhere (SCHNEYER et al., 1972).

With the realization that secretomotor nerve stimulation causes changes in cell membrane ion conductance, the cause of salivary venous transients becomes clear — they simply represent the transition stages of secretory cells from one steady state of cellular ion concentration to another. Since such transients have been discussed extensively elsewhere (BURGEN and EMMELIN, 1961; SCHNEYER et al., 1972) we need not consider them further. In contrast to the venous transients, however, the salivary transients are not so easily explained. It has usually been assumed that transients arise from endpiece cells, but although the truth of this is undeniable in the case of venous transients (since the bulk of the glandular mass is composed of endpiece cells), it is not improbable that saliva transients are of ductal origin, since it is now clear that the ducts play a major role in determining saliva composition and are known to be under autonomic nervous control (see Section C. III. 3). Micropuncture studies have not provided any evidence to indicate whether saliva transients originate in endpieces or ducts and, while duct perfusion studies might be able to settle the question, no evidence is yet available. For the present, the origin of saliva (as distinct from venous) transients must be considered as a completely open question.

c) Stimulus-Secretion Coupling. Calcium Ions and Cyclic Nucleotides

Since the neurotransmitter substances, acetylcholine and noradrenalin, do not enter the cell but interact with receptors located in the basolateral membrane, intracellular "second messengers" are needed in order to bring about secretion, an event that must ultimately involve the apical membrane. Since there seem to be two separate secretory events in salivary glands, secretion of water and electrolytes on the one hand, and discharge of synthesized organic products on the other, it seems not unreasonable to anticipate that there should be two separate systems of intracellular messengers, but since secretion of protein is scarcely possible without secretion of some carrier fluid, it should also occasion no surprise if the two messenger systems interact.

To date, only three substances have been implicated with certainty in the messenger system, calcium ions and the two cyclic nucleotides, cyclic adenosine 3′, 5′monophosphate (cyclic AMP), and cyclic guanosine 3′, 5′monophosphate (cyclic GMP). The nature of the receptor-agonist interaction determines which messenger substances are released into the cytoplasm and the nature of the cells' own biochemical processes then determines what the secretory outcome will be. Similar messenger systems are involved in the control of secretion by the pancreas and, doubtless, other exocrine glands, but the individual properties of each glandular system differ, so it is not possible to produce a single generalized description of stimulus-secretion coupling for all exocrine glands (*e. g.* in salivary glands acetylcholine causes secretion predominantly of salt and water, whereas in the pancreas it acts mainly to cause enzyme secretion). The present discussion will be confined strictly to what is known about salivary glands. Available information on the pancreas is summarized elsewhere in this volume (ULLRICH and SCHULZ), and a number of excellent general reviews on secretion and second messengers are also readily available (BERRIDGE, 1975; RUBIN, 1970; ROBISON et al., 1971). Since the secretion of salt and water, on the one hand, and of protein, on the other, do not occur alone, regardless of the stimulus, it is appropriate to consider the topic of stimulus-secretion coupling in relation both to protein and to salt and water in the following sections, although all other aspects of protein secretion will be considered elsewhere in this review (see Section D).

i) *Mediation of* α-*Adrenergic and Cholinergic Responses.* It is thought that acetylcholine and α-adrenergic agonists have a similar action on salivary glands, at least in some animals. This conclusion is based on studies of the secretions of intact salivary glands, particularly the rat and cat mandibular gland and the rat parotid (see Sections C. II. 5. a and C. III. 1. b), and on studies on *in-vitro* preparations of rat and mouse parotid gland. In rat parotid and mandibular slices, both α-adrenergic and cholinergic agonists cause relatively little amylase release but do cause marked potassium release[4] (BATZRI et al., 1973; SCHRAMM and SELINGER, 1974; MARTINEZ et al., 1976; MARTINEZ and QUISSELL, 1976); ultra-

4 This release of potassium ions into the bath reflects a change in the basal membrane conductance to potassium. It has been interpreted as direct evidence for secretion of ions and water into the endpiece lumen (SCHRAMM and SELINGER, 1976) in terms of a discarded theory that primary fluid formation is dependent on active secretion of potassium. Although it now seems clear that the prime driving force for production of primary fluid is sodium transport (see Section C. II. 3), the use of potassium release into the bath as a marker of an agonist-receptor interaction that ultimately leads to fluid secretion remains valid, provided always that one realizes that it may be possible to uncouple the two events *i. e.* that potassium release might persist when secretion of fluid has been blocked or *vice versa*.

structurally, this response is reflected in the formation of vacuoles in the endpiece cells (BATZRI et al., 1973). In the mouse parotid, both agonists cause similar electrical changes (see Table 2). Cyclic AMP does not seem to be involved in the mediation of either type of response (BATZRI et al., 1971).

DOUGLAS and POISNER (1963) were the first to notice that omission of calcium from the fluid perfusing the cat mandibular gland resulted in a reduced secretory response to noradrenalin and acetylcholine. The calcium requirement is not absolute, but sustained secretion is greatly hindered and the secretion of protein, in response to these stimulants, is blocked. After prolonged exposure to Ca^{2+}-free solutions, although fluid secretion is greatly reduced, secretory potentials and associated potassium release and re-uptake are hardly affected (NIELSEN and PETERSEN, 1972; NISHIYAMA and PETERSEN, 1974). In the rat parotid and mandibular glands, α receptor-stimulated vacuole formation and potassium release are inhibited in the absence of calcium in the bathing medium (SELINGER et al., 1973; MARTINEZ et al., 1976; MARTINEZ and QUISSEL, 1976). On the basis of these and various electrophysiological studies (see above), PETERSEN (1976a) advances the hypothesis that the agonist-receptor interaction causes initially an increase in membrane Na^+ and K^+ conductance associated with or caused by release of membrane-bound calcium (*cf.* pancreas: PETERSEN and UEDA, 1976b). Removal of extracellular calcium presumably acts by altering the level of membrane calcium – it has been established that the membrane sodium conductance is controlled by extracellular calcium (NISHIYAMA and PETERSEN, 1974).

Although most authors agree that one of the effects of cholinergic and α-adrenergic stimulation is to raise the concentration of calcium in the cytoplasm, there are differences of opinion as to how and from where this calcium enters. SELINGER et al. (1974) argue that it is transferred across the basal membrane, but recent studies in the pancreas make it appear likely that calcium originates in surface and intracellular membranes (PETERSEN and UEDA, 1976b). Regardless of where it normally comes from, studies with the calcium ionophore A23187 make it clear that calcium carried from the exterior on the ionophore can mimic the potassium-releasing response to cholinergic and α-adrenergic stimulation (SELINGER et al., 1974; BUTCHER, 1975). It is not known whether fluid secretion was stimulated under these conditions although this is the case in insect salivary glands (PRINCE et al., 1973).

While calcium is obviously important for maintenance of secretion, it appears that cyclic GMP is also involved in the cholinergic and α-adrenergic responses. Increases in tissue concentration of this cyclic nucleotide have now been reported associated with acetylcholine treatment of many tissues including myocardium (GEORGE et al., 1973), isolated neutrophils (IGNARRO, 1974), cerebral cortex (LEE et al., 1972), ileum (LEE et al., 1972), thyroid (YAMASHITA and FIELD, 1972), pancreas (ALBANO et al., 1975; ROBBERECHT, 1976), rat mandibular glands and *ductus deferens* (SCHULTZ et al., 1973). As in other systems, the increase in salivary gland cyclic GMP is dependent on external calcium, and the two substances together must be regarded as regulators of fluid secretion. How they operate is a matter of speculation – in some way they control the apical pump mechanism for salt and water transport. This mechanism cannot be stimulated merely by raising intracellular sodium (POULSEN, 1973b; LAUGESEN et al., 1976), although obviously this will be a co-requisite for secretion.

Although the concentration of protein in saliva formed in response to cholinergic or α-adrenergic stimulation is much lower than in saliva formed during β-adrenergic stimulation (see Section C. II. 5 a), the total output of protein is by no means negligible since salivary flow rates are much greater (see MARTINEZ et al., 1975b). The mechanism of this type of protein secretion differs from that provoked by β-agonists in its dependence on extracellular calcium (PETERSEN and UEDA, 1976a; PETERSEN, 1976a). Thus, in the absence of extracellular calcium, cholinergic and α-adrenergic stimulants can only cause a small and transient secretion of amylase by the mouse parotid; subsequent readmission of calcium to the bathing medium then gives rise to a sudden, sustained burst of amylase secretion. Similarly, addition of manganese ions (5 mM) to the medium blocks the amylase release caused by acetylcholine. These findings, which indicate that acetylcholine or α-adrenergic agonists can only cause amylase release by causing an influx of calcium into the cell from the exterior, contrast with findings related to β-adrenergic stimulation (see below), where external calcium seems to be unimportant for amylase secretion, but mobilization of intracellular stores is essential.

ii) *Mediation of β-Adrenergic Responses.* As mentioned above, in many salivary glands, sympathetic nerve stimulation (or administration of adrenalin or noradrenalin) causes a vigorous secretion of protein, usually dissolved in a small volume of fluid; this ability to stimulate protein release independently of the secretion of large volumes of fluid, is mediated *via* β-adrenergic receptors (Section C. II. 5.a). Although it had been known since 1958 that cyclic AMP was involved in the mediation of many hormone responses, it was not until 1965 that it was implicated in the protein secretory response of exocrine glands to neurotransmitter substances. In that year BDOLAH and SCHRAMM (1965) found that dibutyryl-cyclic AMP (db-cAMP) or the phosphodiesterase inhibitor theophylline could induce amylase release by rat parotid gland tissue incubated *in vitro* (see also BABAD et al., 1967; MALMUD, 1972)[5]. Subsequently it was shown that adrenergic stimulation of the rat parotid could cause a marked increase in tissue levels of cyclic AMP (RASMUSSEN and TENENHOUSE, 1968; GUIDOTTI et al., 1972) and fractionation of parotid gland homogenates has revealed the presence of a membrane-bound, catecholamine-sensitive adenylate cyclase

5 The technique for studying amylase release by rat parotid *in vitro* involves agitation of small fragments of rat parotid in a temperature-controlled incubation medium. Amylase secretion is monitored by ascertaining the amount released into the incubation medium and is expressed as a percentage of the total amylase present in the system, *i. e.* medium and tissue, at the end of the experiment. It should be pointed out that secretion rates so calculated do not represent rates of discharge of amylase from the pool of the enzyme existing at the start of the experiment, since synthesis continues and is increased by autonomic agonist stimulation, provided there is a sufficient intracellular or extracellular pool of free amino acids (KUIJPER-LENSTRA et al., 1975; KUIJPER-LENSTRA and KRAMER, 1975). It should also be mentioned that agitation of small fragments of parotid tissue might result in amylase release where secretion *in vivo* might not have been observed, *e. g.*, if insufficient volumes of carrier fluid were not secreted concomitantly. Limitations of this technique for studying salt and water secretion are discussed in footnote No. 4.

(SCHRAMM and NAIM, 1970). In this last-mentioned study, SCHRAMM and NAIM showed that catecholamine stimulation of the enzyme could be blocked by propranolol but that its activity was independent of the presence of calcium. It is tempting to postulate that the β-agonist receptor is associated with adenylate cyclase in the basal cell membrane but histochemical studies of rat parotid and sublingual glands localize the enzyme only on the apical membrane of serous cells and not at all in mucous cells (KIM and HAN, 1975; KIM, 1976).

The interacting roles of calcium and cyclic AMP in enzyme secretion are complex and not easily understood. It has been reported that parotid amylase release is blocked after prolonged incubation of the gland tissue in calcium-free media containing EGTA (SELINGER and NAIM, 1970) but on the other hand, HARFIELD and TENENHOUSE (1973) report near normal amylase secretions after prolonged periods of stimulation in the absence of extracellular calcium. The calcium ionophore A23187 can induce a small secretion of amylase but the effect is much less pronounced than that induced by isoproterenol (SELINGER et al., 1974). It can probably be concluded that while there is a requirement for calcium, amylase release is largely independent of extracellular supplies of the ion, and that calcium requirements can be met from intracellular stores.

The study of intracellular calcium is rather difficult. The rat parotid gland contains about 10 times the amount of calcium found in rat liver and, within the gland cells, there seem to be several pools that take up $^{45}Ca^{2+}$ to different extents and at different rates (FEINSTEIN and SCHRAMM, 1970; WALLACH and SCHRAMM, 1971). The zymogen granule fraction had the highest calcium concentration but a low binding capacity, whereas the microsomal fraction had a lower calcium content but a higher binding ability. The zymogen fraction of cell calcium is tightly bound to amylase and may serve to promote granule stability. A microsomal fraction of the rat parotid exhibits an ATP-dependent calcium uptake (SELINGER et al., 1970) as does a similar fraction of dog mandibular gland (WATSON et al., 1974). The rat microsomal fraction also possesses a specific cyclic AMP-binding site (SALOMON and SCHRAMM, 1970), which makes it tempting to hypothesize that the cyclic nucleotide in some fashion regulates the ATPase that controls this calcium uptake.

BERRIDGE (1975), on the basis of the available information, has advanced a general hypothesis to account for salivary stimulus-secretion coupling (Fig. 2). He proposes that calcium is essential both for fluid and for enzyme secretion. Acetylcholine, he proposes, provokes a large influx of sodium and calcium through the ionophore-receptor and this calcium then elicits a potassium conductance rise at the basal membrane and activates a fluid secretory mechanism at the apical membrane (perhaps in conjunction with cyclic GMP). Noradrenalin, he envisages, acts by increasing the intracellular level of cyclic AMP (involving stimulation of a basal membrane-bound adenylate cyclase) which, in turn, releases calcium from an intracellular reservoir. This released calcium may serve to trigger fluid secretion at the apical membrane and, in conjunction with cyclic AMP, may stimulate zymogen granule exocytosis. As RASMUSSEN (1970) has proposed, cyclic AMP may phosphorylate granule membranes *via* a protein kinase, thereby priming them for interaction with calcium.

SCHRAMM et al. (1972) and SELINGER et al. (1974) have shown that isolated

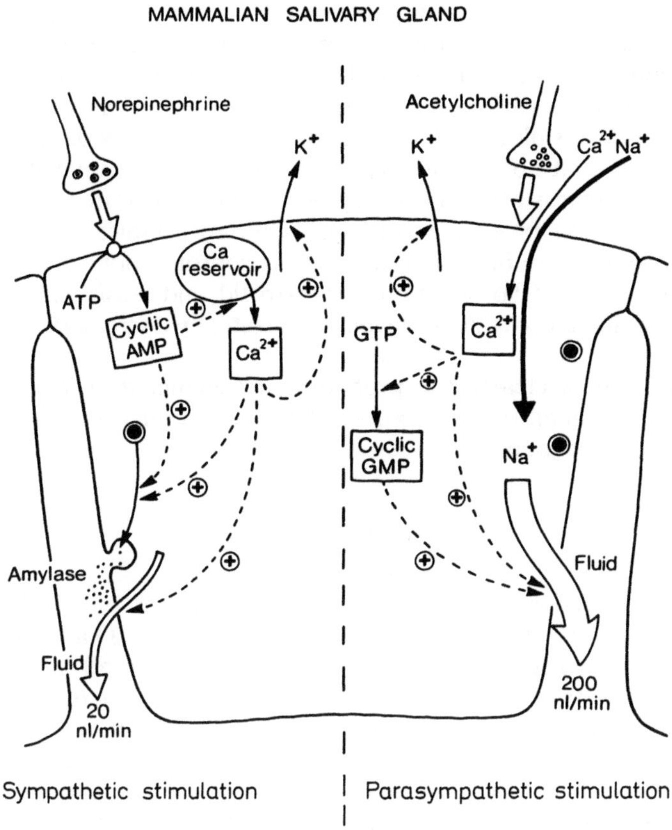

Fig. 2. Model to account for action of noradrenalin and acetylcholine on endpiece cells of rat parotid gland. Noradrenalin, released from sympathetic nerve terminals, interacts with β-adrenergic receptors in basal plasma membrane and stimulates synthesis of cyclic AMP, which then initiates release of amylase at luminal plasma membrane. One action of cyclic AMP may be to release calcium ions from intracellular reservoirs which, in turn, may play a role in release of amylase but may also initiate an increase in potassium conductance at the basal membrane and provoke secretion of some fluid at the luminal membrane. Acetylcholine, released from parasympathetic nerve terminals, interacts with basally located muscarinic receptors and induces an increase in membrane sodium and calcium conductance, which leads to an influx of these ions into the cell. There is a concomitant or subsequent increase in potassium conductance of the basal membrane and efflux of potassium from cell to interstitium. These events are associated in some way with increased intracellular cyclic GMP, which, with calcium ions, seems to be responsible for causing vigorous secretion of fluid across the luminal membrane. For simplicity, interaction of noradrenalin with an α-adrenergic receptor, which mimics the effect of acetylcholine on a muscarinic receptor, is not shown. (From BERRIDGE, 1975)

secretion granules from the rat parotid form pseudopodia when incubated in a medium containing ATP, Mg^{2+}, cyclic AMP, theophylline, creatine phosphate, creatine kinase and a protein kinase fraction from rat parotid tissue. These pseudopodia were similar to those seen in electron micrographs of stimulated parotid glands. Since omission of cyclic AMP or ATP reduced pseudopodium formation, the authors suggested that the role of cyclic AMP was to promote membrane changes that facilitate granule fusion with the apical cell membrane.

The interpretation of the experiments described above is complicated by the fact that there seem to be no "pure" α- or β-agonists. Each drug type exhibits some overlapping with the other as well as some nonspecific effects. Thus the α-agonist phenylephrine can cause a vigorous secretion of amylase by parotid tissue and the effect is blocked fully by propranolol and only partially by phentolamine. *In vitro*, but not *in vivo*, phenylephrine stimulates parotid adenylate cyclase, an effect that is also blocked by propranolol (BUTCHER et al., 1975). It thus appears that this prototype α-agonist exerted a typical β-agonist effect. Similarly a β-agonist can have α effects. Thus, isoproterenol has been reported as stimulating cyclic GMP production in rat parotid, and causing amylase release at a time when tissue cyclic AMP levels were actually decreased (DURHAM and BUTCHER, 1974).

At this stage it is clearly too premature to enunciate a simple, unifying hypothesis to explain all the phenomena described in relation to salivary gland stimulus-secretion coupling.

6. Isotonic Fluid Transport

As discussed in Section C. II. 3. the primary secretions of all glands so far studied appear to be isotonic or slightly hypertonic (Table 1) and, although the sodium content is somewhat variable, the secretions can all be described as sodium-rich and potassium-poor. In all cases (except for the mandibular gland of the possum, *Trichosurus vulpecula*), cholinergic stimulation had little or no effect on the electrolyte composition of the primary fluid (Table 1). Despite the obvious differences in composition of the primary fluid among the various glands studied, the similarities seem great enough to encourage the belief that the underlying secretory mechanisms are similar, especially since the electrophysiological correlates of secretion (Section C. II. 5.b) among the various glands studied are also very similar.

Our information concerning the electrochemical gradients against which the ions in the primary secretion must be secreted is most nearly complete in the case of the cat sublingual gland, where data are available concerning the electrolyte composition of the primary fluid, cell cytoplasm and interstitial fluid, both at rest and during cholinergic stimulation, together with data for the potential differences across the basal membrane and across the endpiece epithelium (KALADELFOS and YOUNG, 1973; LUNDBERG, 1958). Table 4 shows the calculated electrochemical potential difference for Na^+, K^+ and Cl^- between the cytoplasm and the interstitium and between the endpiece lumen and the interstitium. (Values cannot be calculated for the difference between lumen and cytoplasm since the value of the electrical potential difference from lumen to cell has not been measured and it cannot be calculated from the two potential differences available without a knowledge of the intercellular shunt resistance.) Considering first the gradients from interstitium to lumen, it is apparent that, in the resting state, sodium and chloride are close to equilibrium but potassium is evidently concentrated in the lumen against a concentration gradient. After stimulation it appears that chloride, too, has entered the lumen against an

Table 4. The electrochemical potential differences (kJ mol^{-1}) for Na^+, K^+ and Cl^- across the basal cell membrane (η_c-η_o) and across the entire epithelium (η_i-η_o) of the cat sublingual gland. The symbols, c, o and i indicate, respectively, cytoplasm, interstitium and lumen. Values are calculated from LUNDBERG (1958) and KALADELFOS and YOUNG (1973a) according to standard formulae *e.g.*

$$(\eta_c^{Na}-\eta_o^{Na}) = RT \ln \frac{Na_c}{Na_a} + zF(\psi_c-\psi_o)$$

	Na		K		Cl	
	Resting	Stimulated	Resting	Stimulated	Resting	Stimulated
$\eta_c-\eta_o$	−5.98	−7.24	+6.03	+3.38	+2.03	–
$\eta_i-\eta_o$	+0.29	−0.80	+1.83	+1.06	−0.13	+1.16

electrochemical gradient. When one considers the gradient from interstitium to cytoplasm, however, it becomes apparent that at the basal membrane both sodium and potassium must have been subject to active transport; during stimulation the conclusion still holds, although the electrochemical gradients are reduced. Chloride, too, appears to be concentrated in the cell against a gradient. Although estimates of the gradients from cytoplasm to lumen are not available, it seems highly probable that potassium could have entered the lumen from the cell down an electrochemical gradient and that sodium must have been transported into the lumen against a gradient.

Thus, to explain potassium entry into the primary saliva, one need not postulate more than the existence of a basally located potassium pump that builds up a high potassium electrochemical potential in the cytoplasm, whence the ion enters the lumen passively. The existence of basally located, membrane-bound, ouabain-sensitive Na^+, K^+-ATPases in salivary glands is very well documented (SCHWARTZ et al., 1963; SCHNEYER and SCHNEYER, 1965b; SCHWARTZ and MOORE, 1968; BUNDGAARD et al., 1976). To account for sodium entry into the lumen, it is now generally held that a sodium extrusion pump is located on the apical cell surface. Direct evidence for the existence of this pump, comparable to the evidence supporting the existence of the basal membrane Na^+-, K^+-ATPase, is not available, although PETERSEN (1971b, 1972) has published some data to suggest that there is an apical sodium pump that is sensitive to ethacrynic acid but not to ouabain. Given the existence of some such pump, the entry of chloride into the lumen down a favourable electrochemical gradient would then be expected – the current pathway could as readily be *via* a transcellular route as *via* the intercellular junctions. The low transepithelial potential seen in the resting endpiece suggest that the epithelium is of the "leaky" type, associated with vigorous water transport (FRÖMTER and DIAMOND, 1972), across which chloride would be expected to pass mainly *via* the tight junctions.

Although, as mentioned earlier (Section C. II. 5.b), it is no longer believed that an electrogenic chloride pump is responsible for the basal membrane secretory potential, it cannot yet be said that a basally located anion pump does not exist. Chloride appears to be concentrated against a small electrochemical potential in the cytoplasm of both the dog mandibular gland and the cat sublingual gland (Table 3). While one cannot place too much reliance on these data, which are based on estimates of inulin space and of total tissue chloride content, the fact that removing extracellular chloride causes an increase in salivary resting membrane potentials (NISHIYAMA and PETERSEN, 1974) should be considered as important corroborative evidence for active chloride uptake. If proof that an active chloride uptake mechanism were to emerge, it would be necessary to re-evaluate the evidence for an apical sodium pump.

The demonstration that the primary fluid is isotonic at rest, and remains so during stimulation, when secretory rates can increase by more than two orders of magnitude, has prompted the search for secretory channels equivalent to the tortuous intercellular spaces described in gallbladder and other absorbing epithelia, which DIAMOND (DIAMOND and BOSSERT, 1967) has shown could explain isotonic fluid transport. Overlooked altogether in early ultrastructural studies, or mistaken for basolateral infoldings, these secretory canaliculi are tortuous luminal projections, like the hollow fingers of a glove, that extend between cells but are separated from the basolateral intercellular spaces by tight junctions or terminal bars. They have been demonstrated clearly in most serous and seromucous glands but are rare in pure mucous glands (YOUNG and VAN LENNEP, 1978). However attempts to relate their dimensions to the functional state of the tissue at the time of fixation have not met with much success, so it cannot be taken as established that isotonic secretion is mediated *via* these channels, although it seems highly likely (see BERRIDGE and OSCHMAN, 1972, for a comparable picture in insect glands).

III. Ductal (Secondary) Modification of the Primary Secretion

1. Flow Rate and Electrolyte Excretion Patterns in Final Saliva

Since the first studies of HEIDENHAIN in 1878, electrolyte excretory patterns in parasympathetically evoked saliva obtained from many different salivary glands have been described. However, because of the morphological complexity of the glands, only recently has it become possible to examine models proposed to explain the complex excretory patterns that have been observed; use of the techniques of micropuncture and microperfusion has allowed segments of the salivary duct system to be studied in isolation from the rest of the gland.

Current models of salivary secretion are based on the two-stage hypothesis proposed first by THAYSEN (THAYSEN et al., 1954; THAYSEN, 1960). However the salivary electrolyte excretory patterns differ vastly from one another in various glands, among different species, and within a single species, and it is not possible to apply the Thaysen hypothesis without modifying it considerably to accommodate these variations.

a) Parasympathetic and Parasympathomimetic Stimulation

The electrolyte excretory patterns for Na^+, K^+ and HCO_3^- for the salivary glands of ten species are summarized in Table 5, read in conjunction with Fig. 3.

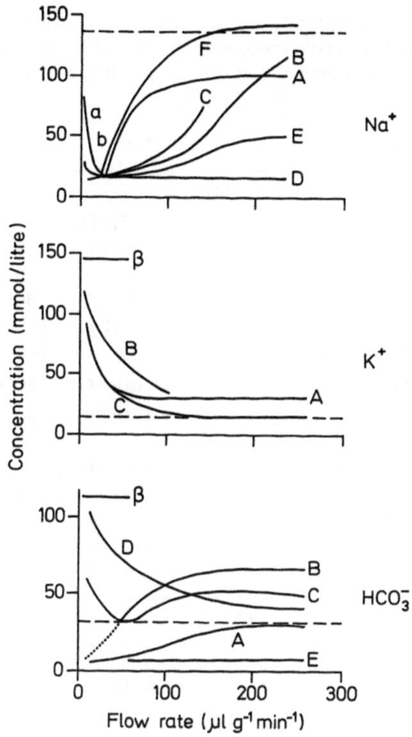

Fig. 3. Electrolyte excretory patterns in mammalian salivary glands. Salivary electrolyte concentrations, plotted schematically, as functions of salivary secretory rates. Each pattern shown is described for one or more salivary glands (see text and Table 5 for details). Horizontal broken lines indicate what the concentrations of each ion in primary saliva might be (see Table 1)

For each ion referred to in Fig. 3, a broken horizontal line is used to indicate a probable primary saliva concentration, although it should be appreciated that these concentrations vary considerably among different glands (Table 1); furthermore, the bicarbonate concentration has not been measured directly and can only be inferred from a knowledge of the concentrations of other electrolytes.

i) *Sodium.* In general terms, the sodium excretory pattern produced in response to cholinergic stimulation is one in which sodium concentrations are low at low secretory rates and rise with increasing secretory rates. The variation among different glands is considerable, however. One can distinguish four distinct patterns (with additional subvarieties). In one group (pattern F in Fig. 3), the sodium concentration rises with increasing secretory rate, following a sigmoid configuration to reach a plateau when it is equal to or even greater than the concentration in primary saliva and in plasma. Such patterns have only been reported with certainty in the rat parotid (SCHNEYER and HALL, 1965) and the sheep parotid (COATS and WRIGHT, 1957; THAYSEN and TARDING, 1974; COMPTON, 1976; see also Fig. 20); similar patterns may also be developed by the sublingual glands of cat (KALADELFOS and YOUNG, 1973; see also Fig. 4) and dog

Table 5. Summary of electrolyte excretory patterns exhibited by salivary glands of various species. For each electrolyte, the concentration observed at maximum rates of secretion is given in mmol l^{-1}, together with symbols in brackets that indicate the shape of the electrolyte excretory pattern as defined in Fig. 3*

Gland	Stimulus	Na^+	K^+	HCO_3^-	Source
	Rat				
Mandibular gland	Pilocarpine	5(bD)	35(A)	–	1
	Carbachol	40(bE)	29(A)	42(C)	2,3
	Parasympathetic nerve	10	46	28	5
	Phenylephrine	20(E)	40(B)	–	3
	Isoproterenol	6(b)	150(β)	129(β)	2,3,4
	Sympathetic nerve	21	149	91	5,39
Sublingual gland	Carbachol	60(bA)	30(B)	45(B)	6
Parotid gland	Parasympathetic nerve	170(F)	20(A)	–	7
	Parasympathetic nerve (immature)	80–120(B)	–	–	45
	Acetylcholine	30(E)	15(B)	15(E)	8
	Carbachol	120(A)	15(A)	–	Fig. 1
	Pilocarpine	115(aB)	18(A)	20(A?)	9,10,40
	Isoproterenol	50(C)	60(B)	105(β)	8
	Mouse				
Mandibular gland	Pilocarpine	5(bD)	40(A)	12(E)	11
Parotid gland	Pilocarpine	130(A)	15(A)	25(A)	11
	Pilocarpine	60(aE)	–	–	41
	Rabbit				
Mandibular gland	Carbachol + pilocarpine	50(A)	15(C)	25(D)	12
	Carbachol + pilocarpine	100(B)	10(C)	–	13
	Parsympathetic nerve	30(aD)	40(C)	–	14
Parotid gland	Pilocarpine	110(A)	10(A)	12(A)	15
	Carbachol + pilocarpine	140(A)	10(C)	30(A)	12
	Sheep				
Mandibular gland	Reflex (nerve)	20(bC)	7(A)	23(A)	16
Parotid gland	Parasympathetic nerve	175(F)	9(A)	140(B)	17
	Acetylcholine and methylcholine sodium depleted	80(A)	40(A)	95(B)	13,18,44
	sodium replete	160(F)	10(C)	113(B)	13,18,44
	Possum *(Trichosurus vulpecula)*				
Mandibular gland	Acetylcholine	46(E)	30(A?)	–	13
	Cat				
Mandibular gland	Parasympathetic nerve	40(bC)	10(C)	–	19,20
	Carbachol	51(bC)	9(C)	29(C)	21
	Sympathetic nerve	48	17	–	22
	Isoproterenol	26(bE)	26(B)	35(D)	21
Sublingual gland	Carbachol	141(F?)	11(C)	0	24
	Isoproterenol	130(F?)	9(C)	0	23

Table 5 (continued)

Gland	Stimulus	Na^+	K^+	HCO_3^-	Source
		Dog			
Mandibular gland	Parasympathetic nerve	70(bA)	15(A)	–	25,43
	Pilocarpine	100 (A or B)	12(A)	10–30 (A or E)	15,26,27
Sublingual gland	Pilocarpine	150(F?)	12(?)	5(E)	15
Parotid gland	Parasympathetic nerve	110(B)	14(A)	50(B?)	28
	Pilocarpine	80(A)	6(C)	50(B)	15
		Monkey *(Macaca mulatta)*			
Mandibular gland	Pilocarpine	90(A)	40(A)	80(B)	29
Parotid gland	Pilocarpine	110(A)	20(A)	110(B)	29
		Baboon *(Papio anubis)*			
Parotid gland	Methylcholine	105(A)	25(A)	115(B?)	30
		Man			
Submandibular gland	Pilocarpine	70(B)	15(A)	–	31
	Reflex (nerve)	7(C)	15(A)	6(E)	32
	Reflex (nerve)	55(C)	14(A?)	36(B)	37
Sublingual gland	Pilocarpine	50(B)	5(C)	–	31
Parotid gland	Reflex (nerve)	80(B)	28(A)	40(B)	33,38
	Reflex (nerve)	20(aB)	8(C)	53(B)	42
	Pilocarpine	80(B)	12(A)	60(B)	34
	Pilocarpine	70(aB)	15(A)	–	31
	Methylcholine	90(A)	19(A)	60(B)	35
Labial glands	Pilocarpine	30(bE)	12(A)	–	36

Source:
1 Young and Schögel (1966)
2 Young and Martin (1971)
3 Martinez et al. (1975a)
4 Schneyer (1962)
5 Yoshida et al. (1967)
6 Martin and Young (1971a)
7 Schneyer and Hall (1965)
8 Mangos et al. (1973d)
9 Mangos and Braun (1966)
10 Mangos and McSherry (1969)
11 Mangos et al. (1973b)
12 Mangos et al. (1973c)
13 Compton (1976)
14 Smaje (1973)
15 Yoshimura et al. (1959)
16 Kay (1960)
17 Coats and Wright (1957)
18 Thaysen and Tarding (1974)
19 Petersen and Poulsen (1967a)
20 Langstroth et al. (1938)
21 Kaladelfos and Young (1974)
22 Kahn et al. (1969)
23 Kaladelfos (1971)
24 Kaladelfos and Young (1973)
25 Imai et al. (1970)
26 Gregerson and Ingalls (1931)
27 Martinez and Martinez (1972a)
28 Brusilow and Cooke (1959)
29 Bruckman and Mangos (1970)
30 Siegel (1972)
31 Knauf and Frömter (1970a)
32 Blomfield et al. (1973)
33 Hildes and Fergusen (1955)
34 Kreusser et al. (1972)
35 Thaysen et al. (1954)
36 Wiesmann et al. (1972)
37 Dawes (1974)
38 Dawes (1969)
39 Schneyer (1975)
40 Sommer et al. (1975)
41 Kaiser et al. (1974a)
42 Kaiser et al. (1974b)
43 Siegel (1976)
44 Compton et al. (1979)
45 Schneyer and Hall (1968b)

* See addendum to this table on p 664.

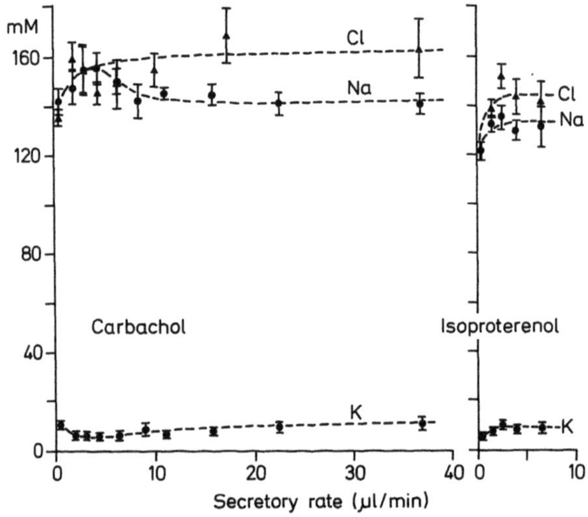

Fig. 4. Electrolyte excretory patterns in cat sublingual gland undergoing stimulation with carbachol or isoproterenol. Each value is the mean (± S.E.M.) of determinations on 6–40 samples from glands of 5–11 cats. Electrolyte concentrations are largely independent of flow rate, but a hint of ductal electrolyte transport activity is present at extremely low secretory rates. Isoproterenol induced lower maximum secretory rates than carbachol, but electrolyte excretory patterns for these two drugs were similar. (KALADELFOS and YOUNG, 1973; KALADELFOS, 1971)

(YOSHIMURA et al., 1959), although in these glands the sodium concentrations seen at low flow rates are never as low. A much commoner excretory pattern (lines A and E in Fig. 3) is similar to pattern F except that the plateau concentration achieved at high secretory rates is much lower than that in plasma or primary fluid. Such patterns have been reported for almost all salivary glands listed in Table 5 (*e. g.* see Fig. 5), although they are unusual in human salivary glands. A third excretory pattern (lines B and C in Fig. 3) exists, in which sodium concentrations rise with increasing flow rate but do not appear to reach a plateau. These are less commonly seen but have been reported in the mandibular glands of rabbits (COMPTON, 1976; see also Fig. 18), sheep (Kay, 1960), cats (KALADELFOS and YOUNG, 1974) and dogs (IMAI et al., 1970) as well as in all three major salivary glands and the labial glands of man (KNAUF and FRÖMTER, 1970a; WIESMANN et al., 1972; KREUSSER et al., 1972). Finally, in some salivary glands the sodium concentration remains low at all rates of salivary secretion (line D in Fig. 3). This has been reported for the mandibular glands of rat (YOUNG and SCHÖGEL, 1966), mouse (MANGOS et al., 1973b), and rabbit (SMAJE, 1973).

From the foregoing it is clear that the sodium excretory patterns vary greatly among species and among different glands in a single species. In addition, however, the pattern varies for a single gland, particularly when different modes of stimulation are used or the animal's sodium balance is altered. Thus, in the rat parotid, SCHNEYER and HALL (1965), who used direct stimulation of the parasympathetic nerves to elicit saliva, reported that sodium excretion followed pattern F, so that saliva was nearly isotonic at all except the lowest flow rates. In

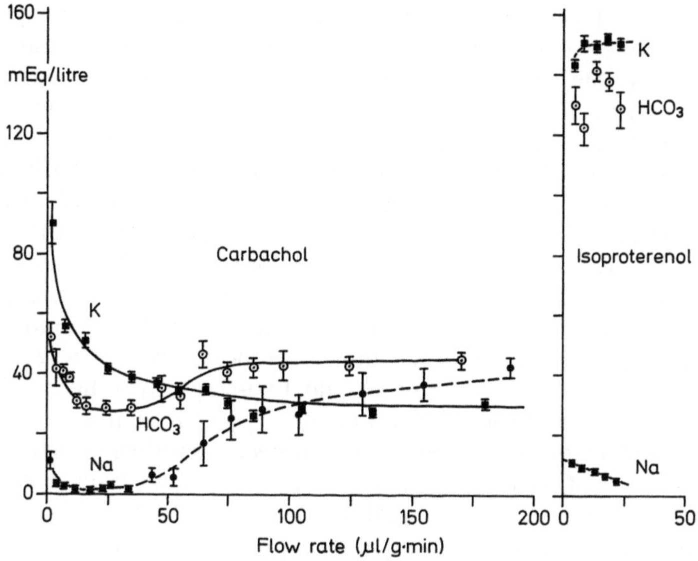

Fig. 5. Electrolyte excretory patterns for the rat mandibular gland undergoing stimulation with carbachol or isoproterenol. (From YOUNG, 1973)

contrast, MANGOS, using pilocarpine or acetylcholine as the stimulant, reported pattern B (pilocarpine) or E (acetylcholine) and concluded that the saliva is markedly hypotonic at all flow rates (MANGOS and BRAUN, 1966; MANGOS and McSHERRY, 1969; MANGOS et al., 1973d). Using carbachol as stimulant, SEWELL and YOUNG (1973) observed pattern A (Fig. 1). Similar variability, attributable to alterations in sodium status, has been reported for the sheep parotid and the rabbit mandibular gland (see section C. III.4.c).

In many glands the sodium concentration also shows a sharp rise at extremely low flow rates (patterns a and b in Fig. 3), so that the overall excretory curve takes on a U shape instead of being sigmoid (*e. g.* pattern aB instead of B). This trend has been reported for the mandibular glands of rat (YOUNG and SCHÖGEL, 1966; YOUNG and MARTIN, 1971; see Fig. 5), mouse (MANGOS et al., 1973b), rabbit (SMAJE, 1973); cat (PETERSEN and POULSEN, 1967a; KALADELFOS and YOUNG, 1974), and sheep (KAY, 1960), for the parotid gland of mouse (KAISER et al., 1974a), rat (MANGOS and BRAUN, 1966; SOMMER et al., 1975), and man (KNAUF and FRÖMTER, 1970a; KAISER et al., 1974b), and for the rat sublingual gland (MARTIN and YOUNG, 1971a). Its occurrence in extremely variable, however, and on occasion the same investigator has seen it in one study and failed to see it in another on the same gland (*cf.* MANGOS and BRAUN [1966] with MANGOS and McSHERRY [1969]). This rise is difficult to interpret in terms of the simple two-stage hypothesis, since one would expect the salivary sodium concentration to approach a minimum steady-state concentration at very low flow rates, when the primary fluid would be expected to remain in contact with the sodium-reabsorbing ductal epithelium for an extremely long time. Various explanations may be considered (YOUNG and SCHÖGEL, 1966). It is possible, but perhaps unlikely, that ductal sodium reabsorption ceases in the absence of any secretory stimulus. More likely is the existence of a short ductal segment near the tip of the collection catheter in which sodium reabsorption does not occur and re-equilibration takes place. Such a segment might be artifactual, induced traumatically by the catheter tip, or it might occur naturally – for instance, the lower end of the main duct of the mandibular gland frequently has an inactive histological appearance reminiscent of buccal mucosa (YOUNG and VAN LENNEP, 1978). Electrical studies on the human parotid main duct suggest that a segment near the oral opening is inactive (KNAUF and FRÖMTER, 1970b and c).

Setting aside the variable rise in sodium concentration seen at very low flow rates, the questions that arise from an examination of the sodium excretion patterns are (1), what accounts for the differences, and (2) whether all the patterns can be explained in terms of the two-stage hypothesis. Clearly an important determinant of the sodium excretory pattern seen in final saliva will be the functional mass of the secretory endpieces (producing primary saliva) relative to the mass of ductal tissue concerned with sodium reabsorption. Thus a gland with relatively little sodium reabsorptive capacity would be expected to produce saliva with a high sodium content (pattern F), whereas a gland with relatively large absorptive capacity would produce a sodium-poor saliva (pattern D). Intermediate patterns (*e. g.* line B) are thus easy to explain.

It is less simple to explain why some glands apparently reach plateaus substantially below the primary sodium concentration (line A), since a simple decrease in contact time (when primary fluid is in the duct lumen) should result in an exponential rise in the salivary sodium concentration, approaching the primary fluid concentration asymptotically. In cases where a lower asymptote is approached one must conclude that the ducts have responded to an increased load of primary fluid by increasing their rate of reabsorption, *i. e.* that some kind of ductulo-endpiece balance exists. One possible mechanism for this could be that ducts, like endpieces, are under autonomic control. Evidence for this hypothesis is presented below (Section C. III. 3).

Variability in the secretory response of a single gland type in one species may have several causes. If, as foreshadowed above, ducts and endpieces both respond to autonomic agonists, then different modes of stimulation could be expected to produce different excretory patterns if the endpieces and ducts can respond independently of one another. This certainly seems to be the case (see Section C. III. I. a). Another important cause of variability in glandular response is a variability in the sodium status of the experimental animals. The influence of this factor on salivary sodium content is clearly established in the case of the sheep parotid and the rabbit mandibular glands (Section c. III. 4. c) and may be present in other glands as well. Although no definite evidence is available, it seems not unlikely that hormonal agents other than those involved in regulating salt balance could also modify salivary excretory patterns (Section C. III. 4. c).

Finally, one needs to comment on the variability in sodium excretory patterns seen in a relatively homogeneous population of salivary glands. It is well known to anyone who has studied salivary excretory patterns that glands from two similarly handled animals can give quite distinct secretory responses (*e. g.*

PETERSEN and POULSEN, 1967a; KNAUF and FRÖMTER, 1970a); similarly, a repeat stimulation in a single gland can elicit a different response (MARTIN and YOUNG, 1971a). In such cases one is loath to invoke hormonal causes, although these cannot be excluded. However, when it is realized that the salivary secretory response depends not only on the co-ordinated response of numerous endpieces and ducts, but also on the vasodilator response of blood vessels supplying these parts independently, it may perhaps appear more remarkable that the patterns resemble one another as closely as they do and the variability actually observed need occasion no surprise.

ii) *Potassium.* In contrast to the excretory curves for sodium, the curves for potassium are relatively uniform among the various glands studied (Table 5, Fig. 3). In response to cholinergic stimulation, the potassium concentrations in saliva fall from a relatively high concentration at low flow rates towards a lower plateau concentration at high flow rates. The shape of the curve is angled and more nearly resembles a rectangular hyperbola than an exponential[6]. Three forms of the curve can be recognized. In pattern A (Fig. 3), which is by far the commonest (*e. g.* see Figs. 1 and 5), the curve attains a plateau concentration at high flow rates, the value of which is unmistakably greater than in the primary fluid or in the plasma. In pattern C (Fig. 3), which is seen in some parotid and sublingual glands but is uncommon in mandibular glands, the plateau concentration reached is equal to that in primary fluid. Pattern B (Fig. 3) is intermediate in form and a true plateau value is not attained—hence it is not possible to say whether the asymptote will be greater than the concentration in the primary fluid or not. In all three cases there is some variability in the size of the concentration seen at low flow rates but, in general, the concentration is higher than 50 mmol l^{-1} and failure to obtain such values (*e. g.* BRUSILOW and COOKE, 1959) usually reflects a poor technique of sample collection. However in those cases where the sodium concentration rises at extremely low flow rates (pattern a or b), there is a reciprocal fall in potassium concentrations (*e. g.* YOUNG and SCHÖGEL, 1966).

The concentration to which the salivary potassium concentration falls asymptotically at high secretory rates varies considerably from species to species although in general it is around 20 mmol l^{-1}. Values as high as 30–40 mmol l^{-1} are normal for the mandibular glands of the rat (YOUNG and SCHÖGEL, 1966) and the rhesus monkey (BRUCKMAN and MANGOS, 1970), whereas values as low as 6–7 mmol l^{-1} are reported for the mandibular glands of the sheep (KAY, 1960) and the dog (YOSHIMURA et al., 1959). In some cases the potassium concentration rises slightly at very high flow rates (*e. g.* PETERSEN and POULSEN, 1967a; IMAI et al., 1970; SIEGEL, 1972; COMPTON, 1976) so that no true asymptote can be defined. This is probably due to persistence of salivary potassium transients (Section C. II. 5. c) for longer than anticipated by the investigator (*cf.* Fig. 8.6 of BURGEN and EMMELIN, 1961); normally, sample collection only begins at a time when the investigator believes the transient has subsided.

The two-stage hypothesis postulates that the salivary ducts secrete potassium

6 An equation of the form $x(y-b) = c$ can be employed to describe the data. The fit obtained is always better than can be obtained from a semilogarithmic transformation.

at a constant rate into the primary saliva during its passage along the ducts. Such a model can probably account for pattern C of the three potassium excretory curves (Fig. 3). At low flow rates, contact time is long enough to allow the ducts to establish a high steady-state potassium concentration and, as flow rate increases and contact time diminishes, the concentration approaches the primary level asymptotically. In those cases where pattern B is observed, one need only postulate that the activity of the ducts relative to the endpieces is greater, a conclusion that is borne out by the observation in these glands that sodium absorption is more extensive than in the C pattern group. Similar reasoning can also be extended to the A pattern group of glands, but in this case it must be stressed that the concentration does not appear to be approaching the primary level. This implies that the rate of ductal potassium secretion increases as gland secretory rate increases, *i. e.* there is ductulo-endpiece balance. How this comes about is unknown, although an independent action of autonomic agonists on ductal potassium secretory rate seems likely to be an important factor (see Section C. III. 3). It may be that the curve merely reflects the kinetics of the potassium secretory mechanism. As in the case of the sodium excretory curve, the roles of body salt balance and of adrenocortical hormones are certainly of some importance although, except in the rather spectacular case of the sheep parotid, salivary potassium concentrations are rather less affected by changes in mineral metabolism than are sodium concentrations.

iii) *Bicarbonate.* As in the case of sodium, the excretion curves for bicarbonate are extremely variable (Table 5). At one extreme (pattern E in Fig. 3) there are salivas in which bicarbonate concentrations are extremely low (0–15 mmol l^{-1}) and remain so at all secretory rates *e. g.* rat parotid (MANGOS et al., 1973d) and dog sublingual (YOSHIMURA et al., 1959). More commonly, the bicarbonate concentrations are low and rise slightly with increasing secretory rates but never to values greater than in plasma or the probable value in the primary fluid (line A in Fig. 3). In some glands the pattern resembles line A except that the bicarbonate concentration rises with increasing flow rate to values well in excess of probable primary levels (pattern B). Such situations arise in saliva from the rat sublingual gland (YOUNG and MARTIN, 1971), the sheep parotid (COATS and WRIGHT, 1957) and the dog parotid (YOSHIMURA et al., 1959) as well as the parotid and mandibular glands of many primates (Table 5). In one gland, the rabbit mandibular (MANGOS et al., 1973c), the bicarbonate concentration is high at low flow rates and falls with increasing flow to a plateau value that is greater than or equal to the plasma concentration (line D in Fig. 3). In two other glands, the cat mandibular (KALADELFOS and YOUNG, 1974) and the rat mandibular (YOUNG and MARTIN, 1971; see also Fig. 5), the bicarbonate concentration is high at low flow rates, falls to plasma-like values with increasing flow rate, and then rises again at maximum flow rates (line C in Fig. 3). In the rat parotid the bicarbonate concentration is high at low secretory rates but falls to very low levels as flow rate increases and then conforms to pattern A of Fig. 3 (SOMMER et al., 1975).

Without definite knowledge of the bicarbonate concentration in the primary fluid, it is difficult to put forward a model that will account for the various bicarbonate excretory curves convincingly. As mentioned earlier, the small volume of primary saliva obtainable by micropuncture has discouraged attempts to measure primary bicarbonate concentrations directly. However, from a knowledge of the primary fluid concentrations of Na^+, K^+ and Cl^- (Table 1), one can infer that the primary bicarbonate concentration is unlikely to be greater than 40 mmol l^{-1} in the salivary glands of rat, mouse and rabbit. In the cat mandibular gland the corresponding concentration would be 40–50 mmol l^{-1} and in the cat sublingual gland it would be 20–30 mmol l^{-1}. Data on which to base estimates for other glands are not available.

In the case of the mandibular glands of the rat, rabbit, and cat, and the sublingual glands of the rat, where the bicarbonate concentrations in final saliva exceed the probable maximum primary levels, one is drawn to the hypothesis that the ducts secrete bicarbonate. The increasing concentration of this ion sometimes seen as flow rate increases (lines B and C) would then be interpreted as the result of stimulated ductal secretion (see Section C. III. 3). In other glands, such as the rat parotid and the cat sublingual, where the bicarbonate concentrations in final saliva are extremely low, one might be tempted to postulate ductal reabsorption of bicarbonate (or secretion of hydrogen ions), but the position must be considered as uncertain since we do not know how low the bicarbonate concentration in the primary saliva could be. BRUCKMAN and MANGOS (1970) were struck by the extremely high bicarbonate concentrations (line B) observed in monkey mandibular and parotid saliva (*cf.* the parotid saliva of the baboon: SIEGEL, 1972); they proposed that the primary saliva must have had a bicarbonate concentration in excess of 80–110 mmol l^{-1} and that the ducts secrete hydrogen ions by a mechanism having a maximum transport capacity. Since the primary fluid composition in the case of monkey salivary glands is not available, this hypothesis cannot be excluded; it would seem simpler, however, to suppose that the ducts secrete bicarbonate by a stimulable mechanism into a low-bicarbonate primary fluid, as they clearly must do in the rat mandibular and sublingual glands. Ductal secretion of bicarbonate has been demonstrated in a recent micropuncture study on the sheep parotid, another gland that is known to produce a saliva that is extremely rich in bicarbonate (COMPTON et al., 1979; see Section C. II. 2. b).

In summary, one can conclude that in many glands the primary fluid contains relatively little bicarbonate and that the ducts secrete the ion, sometimes by a stimulable mechanism (rat mandibular gland) and sometimes by a transport maximum-limited process (rabbit mandibular gland). Some glands, such as the cat sublingual, have a bicarbonate-poor primary fluid and ducts that either secrete hydrogen ion or at least do not secrete much bicarbonate.

b) Sympathetic and Sympathomimetic Stimulation

In Table 5 and Figure 3, the electrolyte excretory patterns associated with catecholamine stimulation are also shown. Relatively few such studies are available, presumably because the lower salivary flow rates obtained make analysis more difficult. In addition, many salivary glands lack sympathetic secretomotor innervation and do not secrete in response to administration of catecholamines (Section C. II. 5. a).

The electrolyte composition of rat mandibular saliva has been studied during nerve stimulation (YOSHIDA et al., 1967; SCHNEYER, 1975) and during administration of the α-adrenergic agonist phenylephrine (MARTINEZ et al., 1975a) and the β-adrenergic agonist isoproterenol (SCHNEYER, 1962; YOUNG and MARTIN, 1971; MARTINEZ et al., 1975a and b). The patterns observed following α-adrenergic stimulation (Fig. 6) resemble very closely those obtained with parasympathomimetic stimulation (YOUNG and MARTIN, 1971; YOSHIDA et al., 1967; see also Fig. 5), lending support to the concept that activation of muscarinic and α-adrenergic receptors produces the same sequence of intracellular responses (Section C II. 5. c). The patterns produced in response to isoproterenol are vastly different (pattern β in Fig. 3; see also Fig. 5). Potassium and bicarbonate concentrations are extremely high, 150 and 129 mmol l^{-1} respectively, and independent of flow rate; sodium concentrations are very low (see Fig. 5) and fall slightly with increasing flow rate. Stimulation simultaneously with iso-

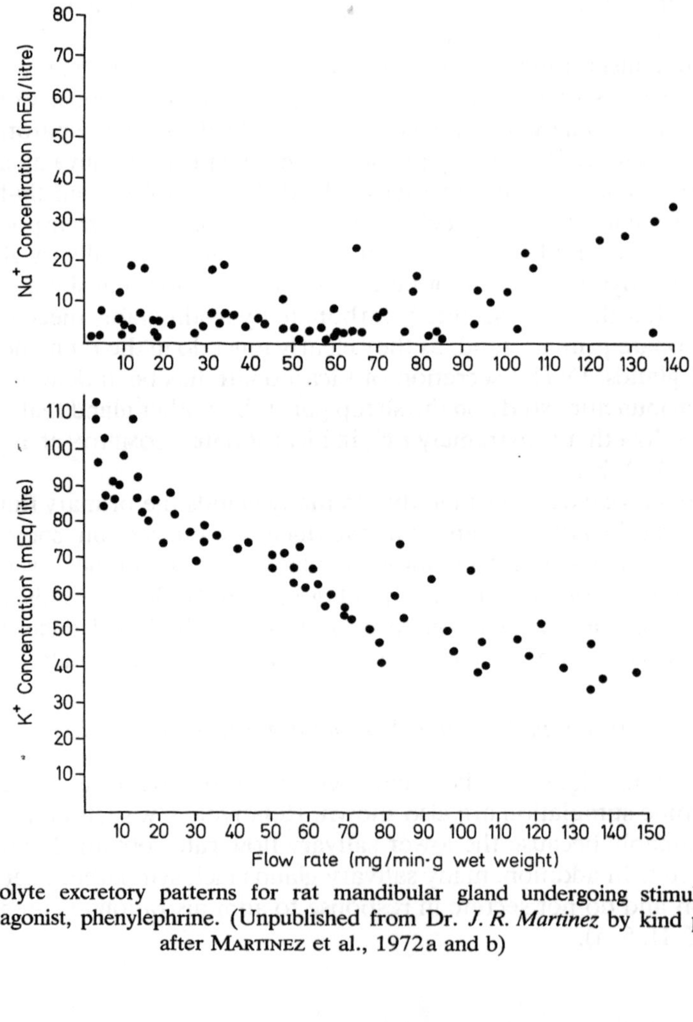

Fig. 6. Electrolyte excretory patterns for rat mandibular gland undergoing stimulation with α-adrenergic agonist, phenylephrine. (Unpublished from Dr. *J. R. Martinez* by kind permission; after MARTINEZ et al., 1972a and b)

proterenol and carbachol produces an additive response with respect to electrolyte but not fluid excretion (YOUNG and MARTIN, 1971). Electrical stimulation of the sympathetic nerves produces a response resembling β-adrenergic activation modified slightly by α-adrenergic stimulation (YOSHIDA et al., 1967; SCHNEYER, 1975).

The action of isoproterenol on the rat parotid gland has been investigated by MANGOS et al. (1973d). Its effect on bicarbonate is similar to the effect produced on the rat mandibular gland but its effects on potassium and on sodium are rather different; the excretory curves resemble parasympathetic pattern C in the case of both ions. It appears to provoke a greater excretion of potassium than does acetylcholine, but the difference is much less marked than in the mandibular gland.

The only other animal in which sympathetic excretory patterns have been studied is the cat. Isoproterenol produces a vigorous response in the cat mandibular gland (KALADELFOS and YOUNG, 1974) with sodium excretion conforming to pattern bE (Fig. 3), potassium to pattern B, and bicarbonate to pattern D. Sympathetic nerve stimulation gives rise to higher sodium and lower potassium concentrations in the saliva (KAHN et al., 1969). Unfortunately, although many studies are available on the flow responses to sympathetic nerve stimulation and α-adrenergic drug administration (EMMELIN, 1967), few authors have interested themselves in the electrolyte content of the saliva. In the cat sublingual gland isoproterenol produces patterns similar to those induced by carbachol although maximum flow rate is much lower (Fig. 4; KALADELFOS, 1971).

Since the electrolyte composition of sympathetic saliva, however evoked, always differed from that of parasympathetic saliva collected at equivalent flow rates, it is clear that the two-stage hypothesis cannot account for the composition of both types of saliva unless we postulate that parasympathetic and sympathetic stimulation have different effects on the endpieces and/or ducts. So far as the endpieces are concerned, it is clear from inspection of Table 1 that sympathetic and parasympathetic stimulants do not evoke primary salivas of vastly different composition although β-adrenergic agonists usually produce lower flow rates than do cholinergic (EMMELIN, 1967; YOUNG and MARTIN, 1971; MANGOS et al., 1973d; KALADELFOS, 1971; KALADELFOS and YOUNG, 1974) or α-adrenergic agonists (MARTINEZ et al., 1975a; EMMELIN, 1967). From this it follows that the two types of stimulant must have disproportionate effects on ductal water and electrolyte transfer processes, a conclusion that has been established directly in microperfusion experiments (see Section C. II. 3). In the case of the cat salivary glands, the differences between the effects of isoproterenol and cholinergic drugs on ductal transfer of Na^+, K^+ and HCO_3^- is evidently slight, but in the case of the rat mandibular and parotid glands the differences are very substantial *e. g.* isoproterenol-evoked mandibular saliva has a potassium concentration of 150 mmol l^{-1} and a bicarbonate concentration of 129 mmol l^{-1}, whereas the concentrations of these ions in carbachol-evoked saliva are only 29 and 42 mmol l^{-1}, respectively (Table 5). YOUNG and MARTIN (1971) have concluded for the rat mandibular gland that carbachol and isoproterenol both stimulate ductal secretion of K^+ and HCO_3^- to the same extent but that, since isoproterenol causes formation of primary secretion at a much slower rate than carbachol, the final salivary concentrations of the two ions are very much greater in isoproterenol-evoked saliva. A similar explanation may apply to the parotid gland.

c) Excretory Patterns for Calcium, Magnesium and Phosphate

i) *Calcium*. In the early literature there are many studies reporting the calcium content of saliva (often mixed saliva) collected at rest and during stimulation (BURGEN and EMMELIN, 1961; SCHNEYER and SCHNEYER, 1967), but it is only more recently that excretory patterns of individual glands have been studied. In interpreting data on salivary calcium content, it is important to remember that not all calcium in saliva or in serum is ionized. Thus, in rat serum, the total calcium concentration is 5.4 mEq l^{-1}, of which only 3.6 mEq l^{-1} is ultrafilterable (DREISBACH, 1959); of this, only 82 percent would be ionized, so that $[Ca^{2+}]$ would only be about 3 mEq l^{-1}. Similarly, at least 20 percent of salivary calcium is protein-bound and, again, only around 80 percent of the remainder would be in the ionic form (DREISBACH, 1960). Data dealing with these questions are difficult to obtain and most studies refer only to total salivary calcium, expressed, however, in mEq l^{-1}.

In human parotid saliva, the total calcium concentration varies between 1.5 and 3.0 mEq l^{-1}, showing a tendency to rise with increasing secretory rates (KREUSSER et al., 1972; DAWES, 1969; KAISER et al., 1974b). VOGEL et al. (1965) claim that only half this calcium is ionized, so that, even at maximum secretory rates, it appears that the ionized calcium concentrations are lower than in serum (2.5 to 3.0 mEq l^{-1}). Nevertheless, parotid salivary calcium levels seem to be correlated positively with serum levels (BURGEN and EMMELIN, 1961). Studies on the human submandibular gland reveal similarly low concentrations of total salivary calcium, which exhibit either no flow dependency (BLOMFIELD et al., 1973) or only a slight rise with increasing flow rate (DAWES, 1974). Interestingly, the calcium content of human labial saliva (WIESMANN et al., 1972) was much higher (4–9 mEq l^{-1}) and showed a distinct flow dependency which had the same form as a potassium excretory curve (Fig. 3).

Flow rate-dependent calcium excretory curves of this form have also been reported for the mandibular gland of cat (NIELSEN and PETERSEN, 1970) and rat (MARTINEZ et al., 1975a and b). In the cat, the plateau concentration was 1 mEq l^{-1} and in the rat about 2.5 mEq l^{-1}. Interestingly, MARTINEZ et al. (1975a and b) report that the β-adrenergic agonist isoproterenol and the α-agonist phenylephrine produce excretory curves for total calcium that do not differ noticeably (allowance being made for flow rate) from those produced by the cholinergic agonist carbachol. This is surprising, since it has been reported that sympathetic saliva is rich in calcium (BURGEN and EMMELIN, 1961). This may represent a difference between parotid and mandibular saliva (NIELSEN and PETERSEN, 1970).

It is tempting to invoke the same mechanisms to account for the calcium and the similarly shaped potassium excretory curves, *i.e.* ductal secretion into a primary fluid that is relatively poor in calcium. However since salivary ionized calcium concentrations seem to be lower than those of plasma and since nothing is known of the primary fluid calcium content, such speculation is premature. NIELSEN and PETERSEN (1970) rightly point out that calcium in the primary fluid, particularly if protein-bound, would be expected to become concentrated in mandibular saliva, which is produced at low secretory rates as a secondary consequence of ductal water absorption. This seems to us to be the most probable explanation and we are reluctant to invoke any ductal transport mechanisms at present. Since water reabsorption at low flow rates seems to be more conspicuous in mandibular than in parotid glands, one might anticipate seeing flow-dependency more readily in mandibular saliva.

ii) *Magnesium.* Magnesium levels in saliva evoked by parasympathetic stimulation are always lower than those in serum, although the total magnesium concentration may exceed that of the serum in sympathetically evoked saliva (BURGEN and EMMELIN, 1961). Excretory patterns for magnesium have been studied in human parotid (KREUSSER et al., 1972), submandibular (BLOMFIELD et al., 1973; DAWES, 1974), and labial (WIESMANN et al., 1972) salivary glands and in the mandibular and sublingual glands of cat (NIELSEN and PETERSEN, 1970). In all cases, the total concentrations observed were below the levels of free magnesium in the serum and, except for the cat sublingual gland (where data were too few to be assessed), the excretory patterns resembled potassium excretory curves (Fig. 3), with relatively high concentrations at low secretory rates, falling hyperbolically towards a plateau value as flow rate increased. Although KREUSSER et al. (1972) favoured the concept of ductal secretion of magnesium, we incline to agree with NIELSEN and PETERSEN (1970) that the excretory curves are best explained by assuming that magnesium enters the saliva in the primary fluid and becomes secondarily concentrated in the ducts in the same manner as calcium (see above).

iii) *Phosphate.* Salivary phosphate content is usually about two to three times that of plasma and up to 80 percent is present as the inorganic form (BURGEN and EMMELIN, 1961; SCHNEYER and SCHNEYER, 1967). Studies on excretory patterns are complicated by the fact that inorganic phosphate exists as both the monovalent and the divalent ion and that the relative proportions of the two forms change with the pH of the saliva. Figure 7, taken from DAWES (1969), shows how phosphate concentration, expressed both as mEq l^{-1} and as mmol l^{-1}, changes in human parotid saliva with changing secretory rate. The figure also shows how pH changes and how the proportions of HPO_4^{2-} and $H_2PO_4^-$ change; similar patterns have also been reported for the human parotid by KREUSSER et al. (1972) and by HILDES and FERGUSON (1955). In the resting state, the total inorganic phosphate concentration is about 10 mmol l^{-1} and, as the salivary flow rate increases, the concentration falls to a plateau value of about 3 mmol l^{-1}. As the salivary pH increased from about 5.6 to 7.8, the part of the total concentration of inorganic phosphate represented by the divalent form increased in concentration from 1 to 3 mmol l^{-1}, whereas, of course, the concentration of the monovalent form decreased to less than 1 mmol l^{-1}. Similar excretory curves have been described in other glands such as the sheep mandibular (KAY, 1960) and parotid (COATS and WRIGHT, 1957) glands, the human submandibular (BLOMFIELD et al., 1973; DAWES, 1974) and the cat mandibular (NIELSEN and PETERSEN, 1970). Cat mandibular saliva is unusual in that the total inorganic

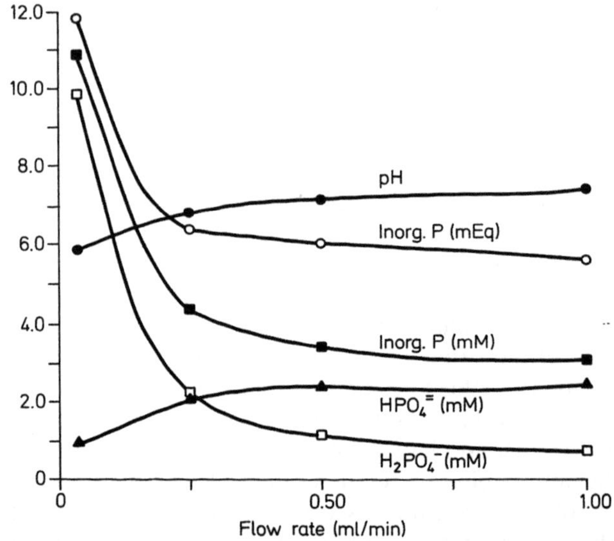

Fig. 7. Phosphate content and pH of human parotid saliva plotted as a function of salivary secretory rate. Total inorganic phosphate, expressed both in mmol l^{-1} and in mEq l^{-1}, is shown. Since pH increases as secretory rate increases, concentrations of $H_2PO_4^-$ and HPO_4^{2-} change in reciprocal fashion. (From DAWES, 1969)

phosphate content is lower than that of plasma. At the opposite extreme, the secretions of the mandibular and parotid glands of sheep and other Bovidae are remarkable for their high phosphate concentrations. Thus, in sheep mandibular saliva, the phosphate content can range from 80 mmol l^{-1} at low secretory rates to 3 mmol l^{-1} at maximum secretory rates (KAY, 1960), and in sheep parotid saliva, the concentrations range from 20 mmol l^{-1} in resting saliva to 12–16 mmol l^{-1} in saliva collected at high secretory rates (COMPTON et al., 1978).

As was the case for calcium and magnesium, it has proven difficult to put forward a secretory model for phosphate ions because of the lack of information about the phosphate content of primary fluid. Recent micropuncture studies on the sheep parotid, which reveal that the primary fluid phosphate concentration is 12–14 mmol l^{-1} regardless of secretory rate (COMPTON et al., 1979), make it clear that the bulk of the salivary phosphate, at least in the case of the sheep parotid, enters the saliva at the endpiece; no significant additional secretion appears to take place in the duct system. In view of the extremely high phosphate content of bovid saliva, however, some caution is needed in extending these conclusions to other mammals.

2. Micropuncture and Perfusion Studies of Salivary Duct Function

In the light of the available information on the composition of the primary secretion, and from an analysis of electrolyte excretory patterns, it is clear that the salivary duct system has a major role to play in the regulation of salivary composition. For any detailed study of the function of the ducts, direct access to them is essential, so that their contents can be sampled by micropuncture and their transport properties assessed by microperfusion. Unfortunately, the structure of salivary glands has defeated most attempts at micropuncture of striated ducts, which lie at the centre of each lobule surrounded on all sides by endpieces. Although puncture of these ducts has occasionally proven possible (*e. g.* MARTIN and YOUNG, 1971 a), it must be emphasized that our knowledge of their function is still largely conjectural, being based mainly on an appraisal of the composition of the primary fluid presumed to be entering them and of the fluid in the excretory ducts presumed to have just emerged from them. The other ductal structures, granular ducts and excretory ducts, are quite accessible.

MARTINEZ was the first worker to puncture excretory ducts (MARTINEZ et al., 1966; HOLZGREVE et al., 1966). He demonstrated that the sodium concentration and osmotic activity of rat mandibular saliva decreased between the endpieces and the lobar ducts but, in addition, he made the important observation that the sodium concentration declined even further along the main excretory duct. YOUNG and SCHÖGEL (1966) extended these studies to show that the salivary potassium concentration increased progressively along the duct system. They also demonstrated that while the main excretory duct could alter salivary electrolyte concentrations in the resting gland, the major sites at which such changes took place in the stimulated gland were the striated and smaller excretory ducts. MANGOS subsequently repeated these experiments in the rat parotid and in the parotid and mandibular glands of mouse and rabbit, and obtained essentially similar results, although in each species, the excretory ducts of the parotid were found to be less active than the excretory ducts of the mandibular gland (MANGOS et al., 1966, 1973b and c).

Taking advantage of the fact that the main excretory duct of the rat mandibular gland seemed to function in the same way as the striated ducts, YOUNG et al. (1967) developed a technique for studying its electrolyte transport properties by microperfusion *in vivo*. The technique that they employed is illustrated in Fig. 8a. The main duct is opened near the hilum and a polyethylene cannula of about 60 μm diameter is inserted 1–2 mm downstream and tied in place; a second cannula for the collection of perfusion samples is threaded a similar distance down the duct through the oral opening. Microperfusion is carried out from hilum to mouth at physiological rates and, since the duct wall is impermeable to inulin (FIELD and YOUNG, 1973; MARTIN et al., 1973), this substance can be added to the perfusion fluid as a marker of nett water movement. By attaching a side arm to the perfusion system, it is possible to record the transepithelial potential difference at the tip of the perfusion catheter, using the fluid-filled catheter itself as the recording electrode. In like manner one may record potential differences from the collection catheter.

This *in-vivo* perfusion technique has proven valuable for characterizing ductal electrolyte transport properties, but is limited by the fact that the composition of the periductal fluid cannot be varied at will. Furthermore, where drug actions on the duct have to be assessed, it proves difficult to distinguish between direct local effects and indirect effects mediated *via* peripheral nerves or by changes in systemic blood pressure. Accordingly, KNAUF (1972a) developed a technique

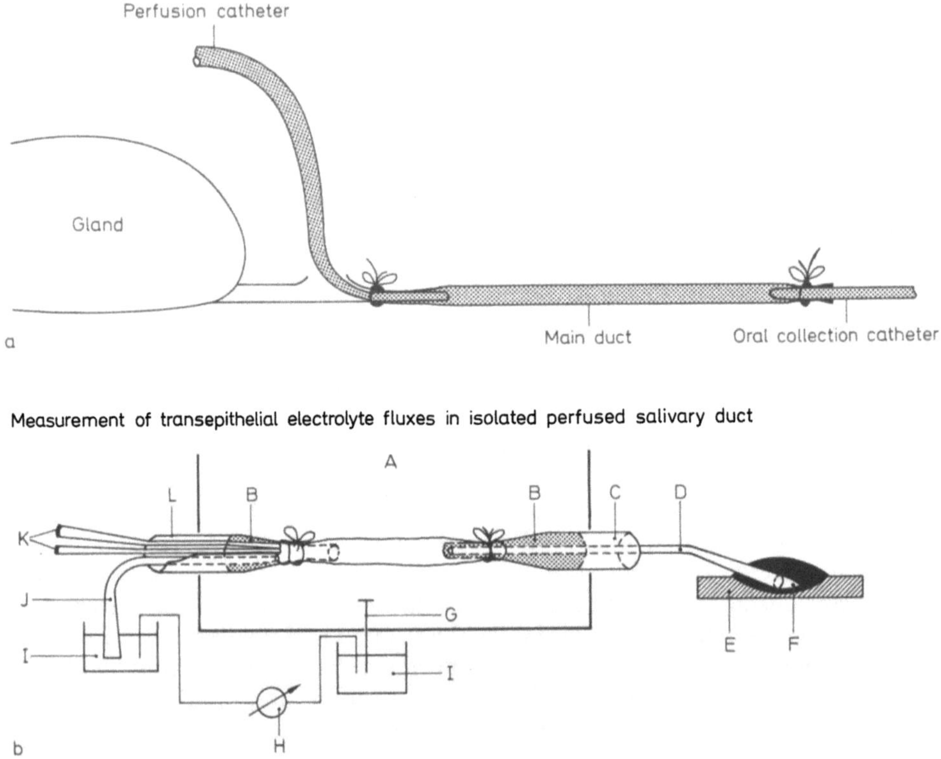

Fig. 8. Schematic diagrams illustrating techniques employed for microperfusion of salivary ducts *in vivo* (a) and *in vitro* (b). A = bathing medium; B = araldite seal; C = glass holding capillary (outflow); D = polyethylene collection catheter; E = glass cavity slide; F = sample collected under oil; G = indifferent electrode; H = electrometer; I = calomel half-cells; J = recording bridge electrode; K = perfusion catheters; L = glass holding capillary (inflow) (a: from YOUNG, 1973; b: after KNAUF, 1972a)

for the perfusion of the mandibular main duct of the rabbit *in vitro* (KNAUF, 1972a and b; MARTIN et al., 1973). The experimental arrangement is shown in Figure 8b. A 5- to 8-mm length of main duct (diameter 300 μm) is placed in plasma or another suitable bathing medium and is then mounted on two glass holding capillaries into which are sealed a variable number of polyethylene tubes. At the input end, we have one or more perfusion catheters and a bridge electrode (usually filled with agar dissolved in a volume of perfusion fluid), whilst at the output end there is usually a single collection catheter with dimensions designed to minimize dead-space. Transepithelial potential differences are monitored from the bridge electrode relative to an indifferent electrode in the plasma bath. The system can be modified so that an exploring electrode can be placed at various points along the lumen of the duct.

Using these techniques, the transport properties of the main ducts of rat and rabbit mandibular glands have been studied in some detail, both *in vivo* and *in vitro* (YOUNG et al., 1967; KNAUF, 1972a; MARTIN et al., 1973; SCHNEYER,

1967). In addition, good studies with a modified *in-vivo* procedure have also been published for the parotid and submandibular main ducts of human subjects (KNAUF and FRÖMTER, 1970b and c). In general terms, all the excretory ducts so far investigated have quite similar properties, although some striking differences amongst them have been noted.

Before examining what is at present known about these properties it seems desirable to summarize the distinctive anatomical features of the duct that may have a bearing on its transport properties. The ducts are lined by a simple layer of columnar cells with oval nuclei placed near the apex and with perpendicular striations in the basal zone. Small basal cells, having no direct connection with the lumen, lie scattered along the basement membrane at irregular intervals. The columnar cells can be subdivided into two types, light cells, which closely resemble the agranular cells of striated ducts, and dark cells, which are somewhat smaller than light cells and have a cytoplasm that is densely packed with microfilaments extending down from apically located microvilli. Both light and dark cells have extensively infolded baso lateral membranes enclosing short, but tortuous intercellular spaces. A special feature of the epithelium in the rat mandibular gland is the presence of deep, intercellular clefts, open to the lumen, that abut onto junctional complexes from the luminal side. For detailed descriptions of the histology and ultrastructure of these ducts the reader is referred to publications by TAMARIN and SREEBNY (1965), SHACKLEFORD and SCHNEYER (1971), YOUNG and VAN LENNEP (1978), TANDLER and POULSEN (1976b). Near the buccal end of the main duct, the epithelium often becomes stratified and, in the case of the rat mandibular gland, the duct expands into a salivary reservoir (SCHNEYER and FLATLAND, 1975) that can cause difficulties when duct perfusion *in vivo* is attempted.

a) Permeability Properties of Salivary Ducts

i) *Sodium.* In all salivary ducts studied so far, the transepithelial potential difference has shown a dependency on the luminal sodium concentration. In the rat mandibular duct (Fig. 9), the potential difference was –80 mV when the luminal concentration of sodium was 150 mmol l^{-1} and it fell to about –15 mV when the luminal sodium concentration was reduced to 2 mmol l^{-1} (YOUNG et al., 1967). The relation between the potential difference and the luminal sodium concentration is approximately logarithmic and, during perfusion with sodium chloride solutions, the potential difference diminishes by about 35 mV for each 10-fold fall in sodium concentration (35 mV/decade[Na^+]: YOUNG et al., 1967 SCHNEYER, 1968; FIELD and YOUNG, 1973). SCHNEYER (1968) reports little or no increase in the steepness of this slope during perfusion with solutions containing sulphate instead of chloride but, in experiments *in vitro*, we find an increase in the electrode slope almost up to the ideal value of 61 mV/decade[Na^+] (FIELD, 1972).

The human submandibular and parotid ducts, when perfused with sodium chloride solutions, also give slopes of about 35 mV/decade [Na^+] (KNAUF and FRÖMTER, 1970b). The rabbit mandibular duct behaves somewhat differently during perfusion with sodium chloride solutions, since the potential difference is

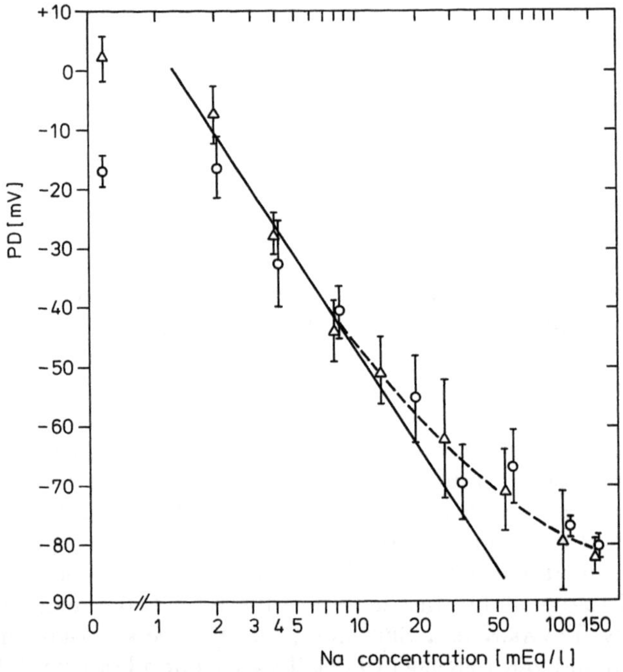

Fig. 9. Sodium-selectivity characteristics of the luminal face of epithelium of rat mandibular main duct. Transepithelial potential differences plotted as functions of luminal sodium concentrations from experiments where ducts were perfused *in vivo* with isotonic solutions containing sodium chloride and either potassium (circles) or choline (triangles) chloride. (From YOUNG et al., 1967)

low (15–25 mV) even when luminal sodium concentration is high (KNAUF, 1972a; MARTIN et al., 1973), and the drop in potential with reduction in luminal sodium concentration is only about 5 mV/decade[Na^+]. During perfusion with sulphate-containing solutions, however, the slope approaches 61 mV/decade-[Na^+] (KNAUF, 1972a; KNAUF and FRÖMTER, 1971; ENDRE, 1972; FRÖMTER et al., 1974; SLEGERS et al., 1975). From these studies one may conclude that the luminal membrane of the salivary duct epithelium behaves like a sodium-selective electrode, partly short-circuited by a chloride shunt in the case of man and rat, and much more strongly short-circuited in the case of rabbit.

Few studies are available on the sodium selectivity of the interstitial surface of the salivary duct epithelium, but it is generally agreed that this membrane is relatively impermeable to sodium. For example, KNAUF and FRÖMTER (1971) have shown that replacement of sodium by choline in the bathing medium of the isolated rabbit mandibular duct makes little difference to the transepithelial potential difference.

FIELD and YOUNG (1973) have attempted to estimate the transepithelial sodium permeability coefficient of the rat mandibular main duct, perfused *in vitro*. They report a value of about $5.27 \cdot 10^{-6}$ cm s^{-1}, which is considerably less than that reported for rat proximal tubule (ULLRICH, 1973). This value is likely to be somewhat higher than the permeability of the basal cell membrane itself, since the epithelial membrane permeability includes the shunting permeability of the intercellular junction. In view of the steep transepithelial concentration gradients that salivary ducts can establish for sodium (see below) it is not surprising that the overall epithelial permeability coefficient should be rather low.

ii) *Potassium.* The permeabilities of salivary ducts for potassium ions appear to be the reverse of those for sodium ions, *i. e.* the luminal surface of the epithelium is relatively impermeable and the basal surface quite permeable to potassium ions.

YOUNG et al. (1967) perfused the rat mandibular duct *in vivo* with isotonic mixtures of sodium chloride and either potassium chloride or choline chloride (Fig. 9). They found that the transepithelial potential difference was not greatly affected by the use of choline rather than potassium as a substitute for sodium ions, although at zero sodium concentration the potential difference during choline chloride perfusion was +3 mV, whereas during potassium chloride perfusion it was about −15 mV. They interpreted their results as indicating that the luminal membrane was extremely impermeable to both choline and potassium ions (relative to sodium ions), but that the permeability to potassium was greater than that to choline. This conclusion was also reached by SCHNEYER (1969), who studied the effects of changing luminal potassium concentration while maintaining a high, constant luminal sodium concentration, and by KNAUF, who assessed luminal membrane potassium conductance by measuring the effect on potential difference of changing the luminal perfusate from a 4-mmol l^{-1} to a 40-mmol l^{-1} solution of potassium sulphate (KNAUF and LÜBCKE, 1975; KNAUF et al., 1976b). Essentially similar findings have been reported for the rabbit mandibular duct perfused *in vitro* (KNAUF and FRÖMTER, 1971) and *in vivo* (ENDRE, 1972) and for the main ducts of both the parotid and submandibular glands in man (KNAUF and FRÖMTER, 1970b).

SCHNEYER (1969, 1974b) has also measured unidirectional potassium efflux (from lumen to interstitium) in the rat mandibular duct by perfusing it with solutions containing isotopic $^{42}K^+$. He found that unidirectional efflux, which occurs down an electrochemical gradient, was about one-twentieth the unidirectional influx, which is believed to be due to active transport (see below). Although his data did not permit him to arrive at a numerical estimate of the potassium permeability coefficient for the epithelium, it is apparent from them that the sodium permeability coefficient is greater by more than a factor of 50.

In contrast to the relative impermeability of the luminal face of the duct epithelium, the interstitial surface appears to the selectively permeable to potassium ions. KNAUF and FRÖMTER (1971) report that replacing sodium with potassium ions in the medium bathing the outer surface of the isolated rabbit mandibular duct results in a decrease in potential of 37 mV for a 10-fold increase in potassium concentration. This is best interpreted as indicating that there is a large potassium diffusion potential normally present across the basal cell membrane, which can be depolarized by increasing the potassium concentration in the bathing fluid. Similar results are obtained regardless of whether the principal anion in the bathing solution is sulphate or chloride, suggesting that either anion is capable of shunting the potassium diffusion potential to some extent

(SAMUELS and YOUNG, 1977). In the rat mandibular main duct perfused *in vitro,* FIELD (1972) has shown a similar dependence of the basal membrane potential on the external potassium concentration.

SAMUELS and YOUNG (1977) have attempted to study the passive movement of potassium across the epithelial cells of the isolated rabbit mandibular main duct. They perfused the ducts with solutions containing potassium in concentrations of 4 to 150 mmol l^{-1} and measured net fluxes into or out of the lumen. By keeping the sodium concentration constant at 2 mmol l^{-1}, the transepithelial potential difference was also held constant and, in consequence, net flux depended mainly on the sum of an active transport term and a passive diffusion term due to the transepithelial electrochemical gradient for potassium. Results of such experiments are shown in Figure 10, where it can be seen that net potassium efflux is related to luminal potassium concentration in an approximately linear manner. This suggests that passive flux across the epithelium is not carrier mediated but, rather, obeys Fick's Law of Diffusion. Under these circumstances, assuming negligible transepithelial water flux, it can be shown that

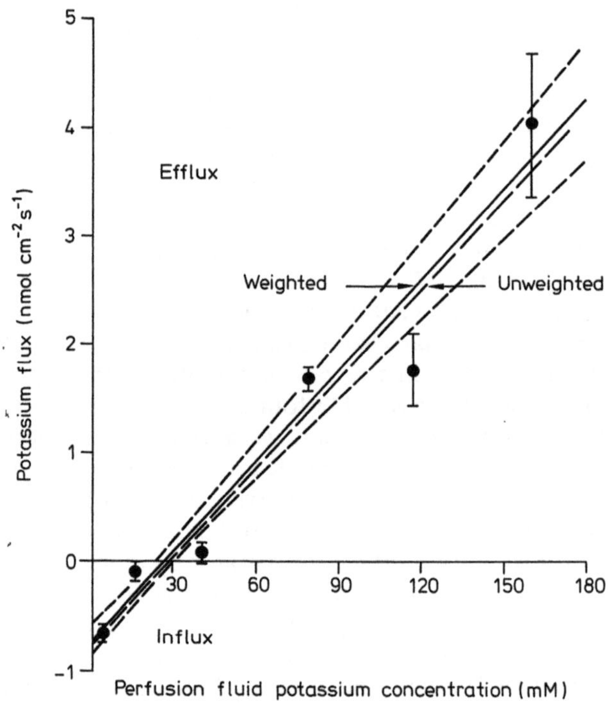

Fig. 10. Net potassium flux across the epithelium of isolated rabbit mandibular main duct plotted as a function of luminal potassium concentration. Isotonic perfusion solutions contained sodium chloride (2 mmol l^{-1}) and mixtures of potassium and choline chloride. In all experiments, the transepithelial potential difference was between 0 and −2 mV and there was a small constant efflux of sodium ($0.3 \cdot 10^{-9}$ mol cm^{-2} s^{-1}). Each point is mean (± S.E.M.) of results from 4–6 experiments. Solid line is least-squares regression calculated from data weighted inversely according to variance; broken line is the regression calculated from unweighted data. The 66% confidence limits of weighted regression also shown. (From SAMUELS and YOUNG, 1977)

the permeability coefficient of the duct wall to potassium is given by the following expression (based on the Goldman equation):

$$P = \frac{-V}{2\pi rL\,(1 - zF\Delta\Phi RT)} \ln\left(1 - \frac{2\pi rLS}{V}\right)$$

where P is the permeability coefficient (cm s^{-1})
r is the duct radius (cm)
L is the duct length (cm)
V is the perfusion rate (cm^3 s^{-1})
$\Delta\Phi$ is the potential difference between lumen and interstitium (mV)
S is the slope of the line relating nett flux to luminal concentration (cm s^{-1})
and R, T, z and F have their conventional meanings.

Using a value of –5.3 mV for the transepithelial potential difference (constant in these experiments), the authors estimated the transepithelial permeability coefficient for potassium to be 3.27 (± 0.48, S.E.M.) · 10^{-5} cm s^{-1}. This value is only slightly smaller than has been reported for kidney proximal tubule (ULLRICH, 1973). This is not necessarily incompatible with the conclusion, reached above, that the luminal face of the duct epithelium is impermeable to potassium, relative to sodium ions, but, if our estimate can be accepted as reasonably accurate, the large size of the coefficient implies that the duct permeability to other ions, such as sodium and chloride, should be even higher. Estimates of sodium permeability for the rabbit duct are not yet available, but FRÖMTER et al. (1974) have determined the partial conductance of the duct to chloride and report that it is about 85 · 10^{-3} Ω^{-1} cm^{-2}, a value that is very high when compared with those for other similar epithelia (*cf.* rat mandibular duct: SCHNEYER, 1968; YOUNG, 1968). Using their own estimate of the potassium permeability coefficient, SAMUELS and YOUNG (1977) have calculated that when the rabbit duct is perfused with a steady-state solution (see below), the potassium conductance is about 1.8 · 10^{-3} Ω^{-1} cm^{-2}, a value about 50 times smaller than the chloride conductance. It thus seems reasonable to conclude that the rabbit mandibular duct is, in general, more permeable to ions than the rat duct, but that it is far less permeable to potassium than to chloride.

iii) *Anions.* In contrast to sodium, potassium and bicarbonate ions, chloride ions seem to be distributed passively across the salivary duct epithelium (YOUNG, 1968). When the rat mandibular main duct is filled with an electrolyte solution, it is found that after a suitable contact time, the electrolyte concentrations in the luminal fluid reach steady-state values and the transepithelial potential difference stabilizes at about –11 mV, lumen-negative (see Section C. III. 2. b; Table 6). The steady-state chloride concentration is found to be about 78 mmol l^{-1}, so that the electrochemical gradient for chloride between lumen and interstitium is very close to zero (YOUNG, 1968). In the rabbit mandibular duct, under steady-state conditions, the potential difference is –47 mV and the luminal chloride concentration is 17 mmol l^{-1} (SAMUELS and YOUNG, 1977; GRANT et al., 1974). Hence, as in the case of the rat duct, the electrochemical gradient for chloride is near zero, and it can be concluded that the ions are distributed passively across the duct epithelium, according to the prevailing electrochemical gradient.

As mentioned above, YOUNG et al. (1967) and SCHNEYER (1968) reported that the luminal surface of the rat mandibular duct behaved like a sodium-selective electrode, partly short-circuited by chloride ions. In addition, SCHNEYER (1968) found that substitution of sulphate for chloride ions in the luminal perfusate made no difference to the electrode slope, from which he concluded that the luminal membrane was almost equally impermeable to chloride and sulphate. He calculated that the epithelium had a chloride conductance of about $1.87 \cdot 10^{-4}\ \Omega^{-1}$ per duct when perfused with chloride-Ringer and $1.99 \cdot 10^{-4}\ \Omega^{-1}$ per duct when perfused with sulphate-Ringer; assuming a duct surface area of 14 mm^2 (YOUNG et al., 1967), these conductances become $13.4 \cdot 10^{-4}$ and $14.2 \cdot 10^{-4}\ \Omega^{-1}\ cm^{-2}$, respectively. YOUNG (1968) perfused the ducts with isotonic potassium chloride and potassium bicarbonate solutions and reported values of $65 \cdot 10^{-4}$ and $27 \cdot 10^{-4}\ \Omega^{-1}\ cm^{-2}$, respectively. He attributed his higher values to the effect of potassium on the luminal membrane resistance (*cf.* frog skin; USSING et al., 1965). From his data one may calculate that the permeability coefficient for passive transepithelial chloride flux is about $1.5 \cdot 10^{-5}$ cm s^{-1}.

In contrast to the rat mandibular main duct, the rabbit duct appears to be rather permeable to chloride ions. Thus, while perfusion of the duct with sodium sulphate solutions reveals that the luminal cell membrane has near-ideal sodium electrode characteristics, perfusion with sodium chloride solutions indicates that chloride can shunt out the sodium diffusion potential and reduce the electrode slope from nearly 61 mV/decade $[Na^+]$ to less than 5 mV/decade $[Na^+]$ (KNAUF and FRÖMTER, 1971; KNAUF, 1972a; ENDRE, 1972; FRÖMTER et al., 1974; SLEGERS et al., 1975). Perfusion with isotonic solutions containing sodium chloride and sodium sulphate in various proportions, or with various concentrations of a simple sodium chloride solution, reveals that the luminal surface of the duct can behave like an ideal chloride electrode (Fig. 11; FRÖMTER et al., 1974; GRANT et al., 1974).[7]

Early attempts at measuring the electrical conductance of the duct wall were not entirely successful, since the duct exhibits a strong polarization effect when square-wave current pulses are passed across it (KNAUF, 1972a). However, FRÖMTER et al. (1974) were able to overcome this problem and obtained estimates of the so-called zero-current membrane resistance (FRÖMTER, 1972). They estimated the membrane conductance to be $104 \cdot 10^{-3}\ \Omega^{-1}\ cm^{-2}$ when the duct was perfused with NaCl-Ringer's solution, and when it was perfused with Na_2SO_4-Ringer, the value fell by 82 percent to $18.9 \cdot 10^{-3}\ \Omega^{-1}\ cm^{-2}$. Taking 0.82 as the transport number for chloride, they estimated that the partial chloride conductance of the epithelium was $85 \cdot 10^{-3}\ \Omega^{-1}\ cm^{-2}$, a value many times larger than that reported for the rat mandibular duct (see above).

7 GRANT et al. (1974) also studied the effect of replacing chloride with bicarbonate instead of sulphate ions (Fig. 11). Their results indicate that, at extremely low bicarbonate concentrations, the duct behaves like an ideal chloride electrode but, as bicarbonate concentration increases, the slope of the chloride electrode deviates markedly from the ideal curve. Nevertheless, at all concentrations studied, the luminal membrane seemed to show some chloride selectivity relative to bicarbonate.

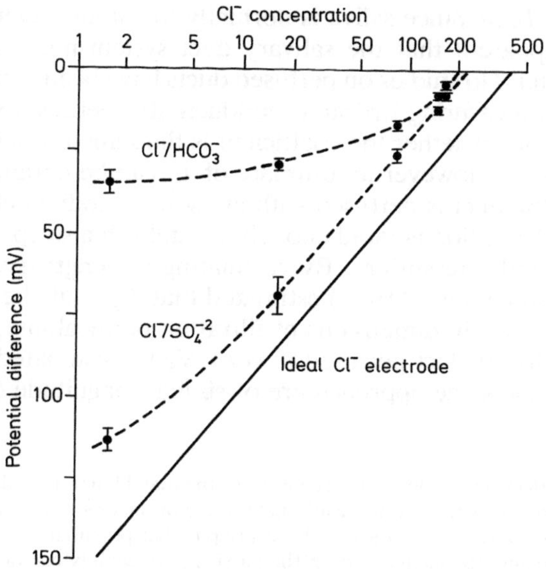

Fig. 11. Chloride-selectivity characteristics of the luminal face of epithelium of rabbit mandibular main duct, perfused *in vivo* with isotonic mixtures of sodium chloride and either sodium sulphate or sodium bicarbonate. Each point is the mean of observations from four ducts ± S.D. (GRANT et al., 1974)

Since the rabbit mandibular duct is extremely impermeable to water (see below) and transports hypertonically (MARTIN et al., 1973), it ought, like the rat duct, to belong to the group of so-called tight epithelia (FRÖMTER and DIAMOND, 1972), except that such epithelia are characterized by having a low electrical conductance. This led FRÖMTER et al. (1974) to investigate the cause of the high duct conductance in more detail. After poisoning active transport with ouabain, they showed that the duct developed symmetrical chloride diffusion potentials when anion gradients across the duct wall were suddenly reversed, a finding that pointed to the possibility that the chloride diffusion potentials were developed across a single symmetrical membrane (*i. e.* the intercellular junctional membrane). However, the diffusion barrier was found to have an anion selectivity sequence ($Br^- \geqq Cl^- > I^- > F^-$), which coincides almost exactly with that of frog muscle cell membrane, so FRÖMTER et al. (1974) were inclined, somewhat reluctantly, to reject the possibility of anion permeation through the tight junctions and to postulate that chloride crossed the epithelium *via* a transcellular route. SLEGERS et al. (1975) have performed similar experiments on the rabbit mandibular duct and confirm many of FRÖMTER's findings. However, they argue strongly for a paracellular pathway for chloride, because the unidirectional chloride flux was found to have a voltage-dependency that conformed to the Goldman-Hodgkin-Katz equation and because the permeability of the epithelium to mannitol (which is unlikely to be able to cross cell membranes) can be increased by duct perfusion with chloride-containing rather than sulphate-containing solutions. These authors calculated that the duct had a chloride permeability coefficient of about $5.5 \cdot 10^{-5}$ cm s^{-1}.

iv) *Water and Urea.* Since saliva is normally hypotonic, even at low rates of secretion, it is apparent that the salivary duct system must be relatively impermeable to water. No studies on perfused ducts have been carried out with the deliberate aim of measuring hydraulic conductivity coefficients (L_p) and it has not been determined whether the coefficient is the same for nett water flux into and out of the duct. However an estimate of L_p can be obtained from experiments in which the duct is perfused with an isotonic sodium chloride solution, since sodium reabsorption is rapid enough to establish a steep osmotic gradient between lumen and interstitium. By calculating a "length-averaged" osmotic gradient, FIELD and YOUNG (1973) estimated that L_p would be about $4.2 \cdot 10^{-6}$ ml cm^{-2} s^{-1} atm^{-1} (or, in dimensions of filtration permeability, $6 \cdot 10^{-3}$ cm s^{-1}) for the rat mandibular duct perfused *in vitro*. Values estimated for rabbit mandibular ducts by the same approach are of similar magnitude (Table 6).

Table 6. Water and electrolyte transport parameters of the mandibular main ducts of rat and rabbit perfused *in vivo* or *in vitro* with isotonic bicarbonate saline or under steady-state conditions of zero nett flux of ions and water. $\Delta\Phi$ indicates the transepithelial potential difference (mV), L_p the hydraulic conductivity and P the passive permeability coefficient

		Rat		*Rabbit*	
		in vivo [a,b,c,d,l]	*in vitro* [e]	*in vivo* [f,g]	*in vitro* [g,h,i,j,k]
Steady-state electrolyte concentrations (mmol l^{-1})					
Na	(mmol l^{-1})	2	2	1.8	2.0
K	"	135	140	87.7	62.5
Cl	"	78	–	11.7	17.0
HCO_3	"	55	–	79.7	47.5
$\Delta\Phi$	(mV)	−11	–	−33.5	−47.0
Nett electrolyte fluxes during perfusion with bicarbonate saline					
Na	(mol cm^{-2}s^{-1}10^{-9})	3.2	2.8	9.4	11.4
K	"	−2.5	−1.9	−0.5	−0.5
Cl	"	0.7	1.7	8.7	10.5
$\Delta\Phi$	(mV)	−71	−64	−13.5	−13.1 to −17.3
Permeability properties					
L_p	(ml cm^{-2}s^{-1}atm^{-1} 10^{-6})	1.55	4.20	2.91	14.0
P_{Na}	(cm s^{-1} 10^{-5})	–	0.53	–	–
P_K	"	2.91	–	–	3.3
P_{Cl}	"	1.50	–	–	5.5
Duct dimensions					
area	(cm^2)	0.130	0.032	0.140	0.051
radius	(cm)	0.01	0.01	0.015	0.015

Source:
[a] YOUNG et al. (1967)
[b] YOUNG (1968)
[c] YOUNG et al. (1970)
[d] MARTIN and YOUNG (1971b)
[e] FIELD and YOUNG (1973)
[f] GRANT et al. (1974)
[g] MARTIN et al. (1973)
[h] KNAUF (1972a)
[i] FRÖMTER et al. (1974)
[j] SAMUELS and YOUNG (1977)
[k] SLEGERS et al. (1975)
[l] RODRIGUEZ (1978)

These values of L_p are extremely low, and are similar to those reported for mammalian collecting duct in the absence of vasopressin and for the ascending limb of the loop of Henle (ULLRICH, 1973; YOUNG, 1973). Presumably the ducts also resemble loops of Henle in having no receptors for vasopressin, since low hydraulic conductivities were measured *in vivo* under conditions in which circulating vasopressin could be expected to be high. However, to date no direct study on the effect of vasopressin on L_p has been published.

It is worth noting that the salivary ducts seem also to be extremely impermeable to urea, a small water-soluble, non-polar molecule (see Section C. II. 1). From experiments in which the rat mandibular duct was perfused with a solution containing ^{14}C-labelled urea, the present authors estimate that the duct has a urea permeability coefficient of only $5 \cdot 10^{-7}$ cm s^{-1} (unpublished data). Such a low value accords with the expectations of an epithelium having an extremely low hydraulic conductivity.

b) *Active Transport and Carrier-Mediated Passive Transport by Salivary Ducts*

As mentioned above, when contact time is sufficient, salivary ducts are capable of lowering luminal sodium concentration and increasing luminal potassium and bicarbonate concentration so as to establish steep electrochemical potential gradients for these ions between the lumen and the interstitium. Table 6 includes estimates of the luminal ion concentrations and the transepithelial potential difference that are developed when the duct epithelium is permitted to establish steady-state conditions of zero nett flux of ions and water. From these values it can be calculated that while the electrochemical gradient for chloride is close to zero, there are steep gradients present for the other ions, *i. e.* sodium appears to have been reabsorbed and potassium and bicarbonate secreted by active transport mechanisms.

i) *Minimum Requirements for Maintenance of Active Transport.* The minimum requirements for maintenance of active transport by the duct perfused *in vitro* have been studied by KNAUF (1972b). Stability of function requires the presence of a macromolecular substance such as albumin or dextran in the periductal bathing medium, although removal of such a substance does not produce an immediate cessation of transport. Glucose alone is inadequate to support active transport, but good function can be maintained if the tricarboxylic cycle substrates acetate and pyruvate, or citrate alone, are used rather than glucose. Oxygen is needed to maintain transport and cyanide and dinitrophenol effectively block it (KNAUF und FRÖMTER, 1971; AUGUSTUS, 1976).

KNAUF (1972b) reported that both calcium and magnesium ions were *"ineffectual on Na^+ transport and on the permeability properties of the salivary duct when administered in physiological doses (2.5 and 1.5 mEq l^{-1}, respectively) both from the blood side and the luminal side of the epithelium".* However, this conclusion, which was not supported with published data, is at variance with results of other investigations. Figure 12 shows the results of two series of experiments in which isolated rabbit mandibular ducts were perfused with a potassium-rich, sodium-poor solution and the ability of the ducts to maintain

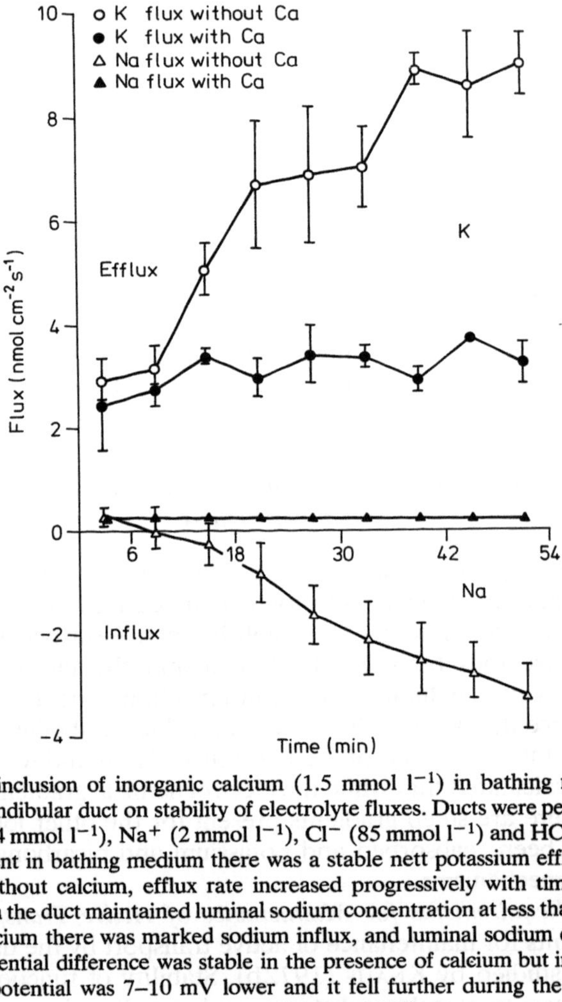

Fig. 12. Effect of inclusion of inorganic calcium (1.5 mmol l^{-1}) in bathing medium of isolated perfused rabbit mandibular duct on stability of electrolyte fluxes. Ducts were perfused with solution containing K^+ (144 mmol l^{-1}), Na^+ (2 mmol l^{-1}), Cl^- (85 mmol l^{-1}) and HCO_3^- (61 mmol l^{-1}). With calcium present in bathing medium there was a stable nett potassium efflux ($2.3 \cdot 10^{-9}$ mol cm^{-2} s^{-1}), but without calcium, efflux rate increased progressively with time. Similarly, in the presence of calcium the duct maintained luminal sodium concentration at less than 2 mmol l^{-1} but in the absence of calcium there was marked sodium influx, and luminal sodium concentrations rose. Transepithelial potential difference was stable in the presence of calcium but in the absence of the cation the initial potential was 7–10 mV lower and it fell further during the experiment. (From SAMUELS and YOUNG, 1977)

constant net ionic fluxes in the presence or absence of extracellular calcium ions was followed for about 1 hour (SAMUELS and YOUNG, 1977). It is clear that lack of calcium resulted in loss of the ability of the duct to maintain steady sodium and potassium concentration gradients; this effect could have been due to a decline in active transport rates, an increase in passive permeability coefficients, or a combination of these effects. SCHNEYER (1974a) has investigated the role of calcium ions applied from the luminal surface of the ducts in the regulation of electrolyte transport. He found that the removal of calcium from the lumen enhanced sodium reabsorption, and to a lesser extent, potassium secretion. His findings are compatible with the hypothesis that calcium regulates ionic conductances of the luminal membrane in a manner analogous to its postulated action on the basal membrane of endpiece cells (see Section C. II. 5. c. i). It is

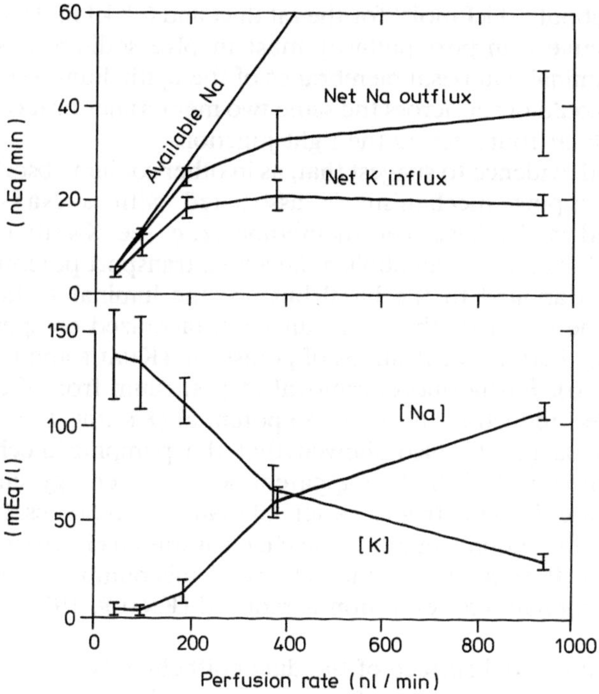

Fig. 13. Electrolyte concentrations in collected perfusate and calculated nett transepithelial electrolyte fluxes for the main duct of the rat mandibular gland perfused *in vivo* with physiological bicarbonate saline solution. As perfusion rate is reduced, contact time increases and luminal electrolyte concentrations approach steady-state values. As perfusion rate is increased, nett transport rates approach their maxima. (From YOUNG et al., 1967)

difficult to account for KNAUF's (1972b) original report but, whatever the explanation, it seems safe to conclude that calcium ions are required to regulate ductal electrolyte transport processes, at both the luminal and the basal membrane of the epithelium.

ii) *Sodium Reabsorption.* The net rates of reabsorption of sodium by rat and rabbit mandibular ducts perfused with isotonic bicarbonate saline are listed in Table 6 (see also Fig. 13). It is apparent that, in comparison with other mammalian epithelia, both ducts can reabsorb sodium powerfully, particularly the rabbit duct where the sodium transport rate ($11.4 \cdot 10^{-9}$ mol cm^{-2} s^{-1}) is one of the largest yet described for an epithelium. Equally remarkable is the fact that the ducts are capable of lowering the luminal sodium concentrations almost to zero and thereby of establishing transepithelial concentration gradients of around 140–150 mmol l^{-1}. As mentioned above, the mechanism responsible for this sodium transport is an active process capable of expending energy to develop an electrochemical potential gradient. When the active transport process is allowed to build up a transepithelial electrochemical gradient, a condition of steady state will be approached, in which active absorption from lumen to interstitium will be balanced by passive back flux down the electrochemical gradient established by the pump (about 10 kJ mol^{-1} for the rat duct and 6.5 kJ mol^{-1} for the rabbit). Clearly, the active transport pathway must involve sodium ions crossing two barriers, the luminal and basal membranes of the epithelium (see Fig. 15). Passive back flux could occur across the same two membranes or it could take place *via* an intercellular route across the tight junctions.

There is good evidence to suggest that, as in other sodium absorbing epithelia, the sodium-pumping mechanism is associated with a Na^+-,K^+-activated ATPase, bound in the basal cell membrane (*e. g.* see KNAUF et al., 1976a). Ouabain ($5 \cdot 10^{-6}$ mol l^{-1}) can abolish the active transport potential rapidly and reversibly when applied to the basal but not the luminal surface of the epithelium, and the effect of the drug can be antagonized competitively by the presence of high bath concentrations of potassium (KNAUF and FRÖMTER, 1971; AUGUSTUS, 1976). Furthermore, removal of potassium from the bathing fluid causes a marked reduction in the pump potential (KNAUF, 1972b; AUGUSTUS, 1976). AUGUSTUS (1976) also showed that the pumping mechanism, which could be inhibited at 0° C, had an apparent activation energy similar to that of Na^+, K^+-ATPases isolated from a variety of mammalian tissues. He reached the conclusion that the sodium-pumping mechanism was an electrogenic Na^+-, K^+-ATPase system. It appears that the activity of this pump is reduced when the periductal bicarbonate concentration is reduced (KNAUF, 1972b; KNAUF et al., 1975).

Although the luminal surface of the duct epithelium is sodium-selective (see above) it appears that passage of sodium across this membrane is largely by means of membrane carriers rather than by simple diffusion. Thus, perfusion of the duct with amiloride or triamterene in concentrations of 10^{-5} and 10^{-4} mol l^{-1} resulted in partial or complete blockage of transepithelial sodium transport, a sharp depolarization of the transepithelial potential difference and a decrease in luminal membrane sodium conductance (SCHNEYER, 1970; KNAUF and LÜBCKE, 1975; KNAUF et al., 1976). These effects were accompanied by a reduced secretion of potassium and an increased secretion of bicarbonate[8]. KNAUF proposes (see Fig. 15) that there is a carrier molecule located in the luminal membrane which couples sodium efflux with potassium or proton influx and, as in other epithelia such as frog skin or kidney tubule, amiloride and triamterene act by blocking the transport protein – such an action would reduce sodium reabsorption and potassium secretion directly, and increase bicarbonate secretion indirectly (by abolishing proton exchange for sodium). Since the intracellular electrolyte concentrations of duct cells and the potential difference across the luminal membrane have not been measured, it cannot be said with certainty whether sodium transport across the apical membrane is passive or active although, by analogy with other epithelia, it seems unnecessary to postulate active transport. Passage of sodium down its electrochemical gradient might well provide energy for the counter movement of potassium and protons.

8 MARTINEZ and MARTINEZ (1974) studied the effect *in vivo* of retrograde ductal injection of amiloride in the dog mandibular gland. They concluded that the drug blocked a component of ductal sodium reabsorption that was not linked to potassium secretion.

The kinetics of the transepithelial sodium transport process have been studied in the rat mandibular duct, perfused *in vitro*, by FIELD and YOUNG (1973). These authors measured the transepithelial electrochemical gradient for sodium under steady-state conditions at a number of different bath and luminal sodium concentrations. They found that ductal sodium transport exhibited saturation kinetics (Fig. 14) although they were unable fully to saturate the transport capacity of the duct even at a luminal sodium concentration of 86 mmol l^{-1} (corresponding to a bath concentration of 323 mmol l^{-1}). Using a value of 2.75 for the unidirectional flux ratio (SCHNEYER, 1969), FIELD and YOUNG (1973) estimated that the duct sodium permeability was about $5.3 \cdot 10^{-6}$ cm s^{-1} and the unidirectional active sodium transport rate was $4.4 \cdot 10^{-9}$ mol cm^{-2} s^{-1} when the luminal sodium concentration was about 133 mmol l^{-1}. If the normal range of luminal sodium concentration is taken as 2–140 mmol l^{-1}, the functional half-saturation constant of the transport system is equal to a luminal concentration of 29 mmol l^{-1}.

iii) *Potassium Secretion.* From the data on steady-state conditions (Table 6) and a knowledge of the periductal potassium concentrations it can be calculated that there is an electrochemical gradient for potassium from lumen to interstitium of about 7.5 kJ mol^{-1} in the case of the rat mandibular duct perfused *in vivo* and of about 2.5 kJ mol^{-1} in the case of the rabbit duct perfused *in vitro*. Clearly both ducts can secrete potassium actively, although, as is born out by the

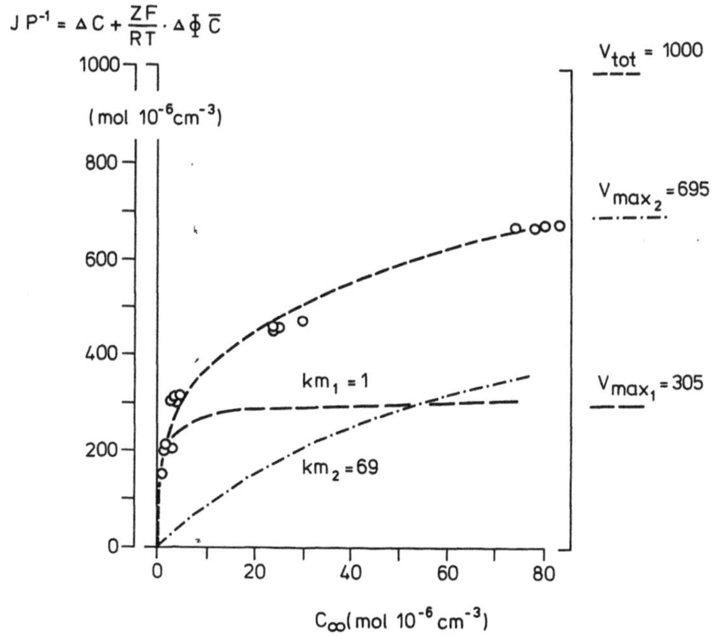

Fig. 14. Relation between active sodium transport rate (divided by sodium permeability coefficient), $J P^{-1}$, and luminal steady-state sodium concentration (C_∞) for rat mandibular main duct perfused *in vitro*. The curvilinear relation can be described in terms of two hyperbolic functions with K_m and V_{max} values as indicated. (From YOUNG, 1973)

measurement of net flux under non-steady-state conditions (Table 6), the rat duct is far more powerful in this respect (Fig. 13).

Since secreted potassium must cross both the basal and the luminal cell membrane in order to enter the duct lumen, the question arises as to whether the basally located Na^+-, K^+-ATPase referred to above will suffice to account for secretion, or whether an additional, active transport step must take place at the luminal cell membrane. There are insufficient data available to answer the questions but, as was the case for sodium, there is some evidence to indicate that potassium movement across the luminal membrane is at least carrier-mediated and coupled to the movement of other ions (Fig. 15). Thus, as mentioned earlier (Section C. III. 2. a. ii), the luminal surface of the duct epithelium in rat, rabbit and man has a relatively low potassium conductance. Furthermore, in the rat mandibular duct, SCHNEYER (1969) has measured unidirectional potassium fluxes and reports an influx/efflux ratio of 20.1 whereas the efflux/influx ratio for sodium was only 2.75. His data indicate that the epithelial sodium permeability is more than 50 times greater than the potassium permeability. Since the basal cell membrane has a relatively high potassium conductance (Section C. III. 2. a. ii), this makes it seem reasonable to conclude that the luminal membrane is highly impermeable to the ion and that potassium crosses by means of a carrier with high electrical resistance. Since ductal potassium secretion can be reduced to less than 50 percent of normal by removing sodium from the luminal perfusate (YOUNG et al. 1967; KNAUF, 1974) or by adding amiloride or triamterene to it (SCHNEYER, 1970; KNAUF and LÜBCKE, 1975; KNAUF et al., 1976b), it seems quite probable that the carrier protein couples potassium influx to sodium efflux at this membrane. In addition, however, Knauf has produced good evidence to indicate that a component of luminal potassium transport is coupled to hydrogen ion absorption, so that this fraction of secreted potassium is actually accompanied by secretion of bicarbonate ions rather than exchanged for sodium ions (KNAUF et al., 1975; KNAUF and LÜBCKE, 1975; KNAUF et al., 1976b). SCHNEYER (1974b) has shown that cytochalasin B inhibits a component of duct potassium secretion not coupled to sodium absorption.

From inspection of Table 6 it is apparent that although the rabbit mandibular duct secretes potassium, it does so at extremely low rates. Since both ducts possess Na^+-, K^+-ATPases (KNAUF et al., 1975), it is perhaps surprising that the rabbit duct, which absorbs sodium so much more powerfully than the rat duct, does not secrete potassium at a rate proportionate to its sodium reabsorptive rate. It is possible that the luminal cell membrane lacks the carrier system described above for the rat duct, but no direct evidence shedding light on this problem is available. Of course the electrical driving force favouring potassium secretion is much lower in the rabbit duct, and this alone might account for the difference, but only if the potassium conductances were extremely large, which seems not to be the case. It would be interesting to ascertain whether potassium secretion was increased during duct perfusion with sodium sulphate when the transepithelial potential difference exceeds –160 mV (cf. COOK and YOUNG, 1977).

SAMUELS and YOUNG (1977) have investigated the kinetics of potassium transport in the rabbit duct by determining the steady-state potassium gradient that the epithelium developed when the bathing fluid potassium concentration of the ion was 4, 8.3, or 13 mmol l^{-1}. They found that the ductal secretory rate for potassium was a saturable function of the bath potassium concentration with a half-saturation constant of approximately 7 mmol l^{-1} and a maximum transport capacity of about $2 \cdot 10^{-9}$ mol cm^{-2} s^{-1}.

The interrelations between potassium and bicarbonate secretion are discussed in the following section.

iv) *Transport of Bicarbonate or Protons*[9]. The steady-state bicarbonate concentrations developed by both rat and rabbit ducts (Table 6) indicate unmistakably that these ducts secrete bicarbonate ions actively. This fact is especially clear in the case of the rabbit duct studied *in vitro* (SAMUELS and YOUNG, 1977), where we calculate that the electrochemical gradient for bicarbonate (6.1 kJ mol^{-1}) is almost as large as for sodium. In contrast, however, the ducts appear to have secreted little or no bicarbonate when perfused under non-steady-state conditions with an isotonic bicarbonate-saline solution (MARTIN et al., 1973); indeed, in the rabbit duct perfused *in vivo* or *in vitro,* it is common to observe a slight but significant nett absorption of bicarbonate (MARTIN et al., 1973). In the rat duct undergoing perfusion with isotonic ($[Na^+]$ = 142 mmol l^{-1}) or hypotonic ($[Na^+]$ = 32 mmol l^{-1}) bicarbonate-saline, a small but significant degree of nett bicarbonate secretion (0.6 to 0.9 nmol min^{-1}) has been seen (YOUNG et al., 1970; MARTIN and YOUNG, 1971b) but during perfusion with isotonic mannitol (MARTIN and YOUNG, 1971b) or with a potassium chloride solution ($[Na^+]$ = 2 mmol l^{-1}; Field, 1972) the net bicarbonate secretory rates were much greater (1.4 to 2.6 nmol min^{-1}). A similar inverse relation between ductal bicarbonate secretory rate and luminal sodium concentration has been demonstrated in the rabbit duct perfused *in vivo* (ENDRE, 1972).

KNAUF (1974) and KNAUF et al. (1975) have studied the effect on secretion of potassium and bicarbonate by the rat duct *in vivo* of replacing luminal sodium with the impermeant choline ion. They report that net potassium secretion decreased and net bicarbonate secretion increased progressively as luminal sodium concentration was decreased. They attributed this effect to the change in transepithelial potential difference that occurs in the rat duct when luminal sodium concentration is lowered and concluded that, at zero potential difference, there was a small residual secretion of potassium and bicarbonate, which they thought was mediated by a tightly coupled K^+–H^+ exchange mechanism. Although their conclusion concerning a residual K^+–H^+ exchange mechanism can be accepted as plausible, the present authors do not accept the conclusion that the increased secretion of potassium and the reduced secretion of bicarbonate, observed when luminal sodium concentrations are high, were the result merely of passive fluxes driven by the transepithelial potential difference, but

9 The small volume of fluid contained in the mandibular duct makes it seem likely that CO_2 tension will be constant and equal to that in capillary blood; *e. g.* in human parotid saliva the pCO_2 ranges from 35 to 39 mm Hg (KREUSSER et al., 1972). Under these circumstances, reabsorption of H^+ will result in accumulation of luminal bicarbonate and it will not readily be possible to distinguish between H^+ absorption and HCO_3^- secretion. In the present paper bicarbonate secretion will be referred to often, although in fact the authors are of the opinion that this comes about through transport of protons.

are rather of the opinion that these fluxes are coupled in part to sodium efflux and are directly dependent on the luminal sodium concentration, since lowering the luminal sodium concentration in the rabbit duct has a similar effect on potassium and bicarbonate flux although potential difference is not reduced and may even be increased if chloride is also removed (ENDRE, 1972; FRÖMTER et al., 1974). Furthermore, from KNAUF's (1974) own data it can be calculated that, for the transepithelial potential difference to have driven the potassium flux at the observed rate, the epithelium would have had to have a potassium conductance in excess of $3.8 \cdot 10^{-3}\ \Omega^{-1}cm^{-2}$ and a potassium permeability coefficient greater than $21 \cdot 10^{-5}$ cm s^{-1}. This seems most improbable since the epithelium is remarkable for its low potassium conductance (Section C. III. 2. a. ii; see also Table 6). COOK and YOUNG (1977) have investigated this point further by studying potassium secretion in ducts perfused with isotonic solutions containing mixtures either of NaCl and sucrose (Na concentration 10 mmol l^{-1}) or of NaCl and Na_2SO_4 (Na concentration 246 mmol l^{-1}) chosen to give rise to the same transepithelial electrochemical gradient for potassium. They found that the availability of Na^+ ions in the luminal fluid for exchange with K^+ ions across the luminal membrane was more important for the determination of the rate of potassium secretion than was the magnitude of the potassium electrochemical gradient.

In other experiments, KNAUF has shown that partial inhibition of ductal sodium absorption by administration of an aldosterone antagonist, or total inhibition by luminal application of amiloride or triamterene, had effects on secretion of potassium and bicarbonate that are similar to the effects of lowering luminal sodium concentration (KNAUF, 1974; KNAUF and LÜBCKE, 1975; KNAUF et al., 1975, 1976b). These experiments, also, are best interpreted as indicating that some potassium and bicarbonate ions enter the lumen coupled to sodium efflux but that, when these fluxes are blocked, the presence of an additional mechanism for secretion of potassium and bicarbonate, probably a coupled K^+–H^+ exchange carrier, is revealed.

As might be expected, ductal secretion of electrolytes, particularly of bicarbonate, seems also to be influenced by body acid-base status. In metabolic alkalosis, the ductal secretion of both potassium and bicarbonate is increased while sodium transport remains constant; in metabolic acidosis potassium secretion is greatly reduced and bicarbonate secretion is abolished altogether (KNAUF, 1974). In ducts in which sodium transport has been completely blocked with amiloride the effect of acidosis is in some way to reverse the pump and cause the duct to secrete hydrogen ions (KNAUF et al., 1975). Acidosis seems also to cause a reduction in ductal sodium reabsorption both *in vivo* (KNAUF et al., 1975) and *in vitro* (KNAUF, 1972b).

c) A Transport Model for the Duct Epithelium

It is clear from the above that the interrelations between transport of sodium, potassium and bicarbonate are complex and no simple model can easily account for all observations. As a working hypothesis, KNAUF et al. (1976c) propose that there are two distinct transport mechanisms on the luminal cell membrane

(Fig. 15). One of these allows for coupling of sodium flux from lumen to cytoplasm, with potassium or hydrogen ion flux in the opposite direction. Amiloride and triamterene appear to act by blocking this carrier, although they may also have nonspecific actions on other enzyme systems (KNAUF et al., 1976a and b). Whether this transport mechanism is active or passive and whether the sodium diffusion potential generated across the luminal membrane is produced by passage of sodium through this site is unknown, but in view of the action of amiloride on potential difference, it is simplest to assume that the potential does arise at this locus. KNAUF postulates that there is a second luminal membrane receptor responsible for coupling of potassium flux from cytoplasm to lumen with hydrogen ion flux from lumen to cytoplasm (*i. e.* it secretes potassium bicarbonate). This mechanism seems to be blocked by cytochalasin B but not by amiloride (SCHNEYER, 1974b). It may be an active process associated with the presence of a bicarbonate-stimulated ATPase (WAIS und KNAUF, 1975; SIMON et al., 1972). It has been postulated that its pumping rate is controlled by the intracellular concentration of hydrogen and potassium ions. At the basal membrane there seems to be a sodium-potassium exchange pump. The activity of this pump is reduced by the application of ouabain, or by removal of potassium or bicarbonate ions from the periductal fluid. It appears to be an ATPase and operates electrogenically (see Section C. III. 2. b).

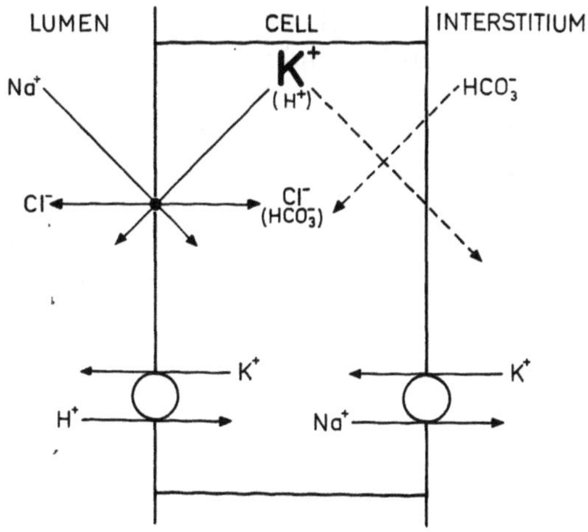

Fig. 15. Hypothetical model to account for electrolyte transport by excretory ducts of rat and rabbit mandibular gland. An Na^+-, K^+-ATPase is located on the basal plasma membrane, which extrudes Na^+ from cell and concentrates K^+ in it; pump is believed to operate electrogenically. Potassium back-diffusion from cell to interstitium and HCO_3^- diffusion from interstitium to cell are thought also to contribute to the basal membrane potential. On the luminal membrane there appears to be a pump mechanism, associated with an HCO_3^--ATPase, that exchanges K^+ for H^+ non-electrogenically; it may be inhibited by cytochalasin B. Also on the luminal membrane is a mechanism that allows for functional coupling of Na^+ influx to H^+ or K^+ efflux; Cl^- also may pass through this mechanism. Its operation seems to depend on the presence of a pre-existing electrochemical gradient (*i.e.* it is not an active pump); it can be blocked by luminal application of amiloride or triamterene. Sodium ions generate a diffusion potential when moving through this site. (Unpublished diagram from *Dr. H. Knauf* by kind permission; after KNAUF et al., 1976c)

During duct perfusion with a sodium-rich fluid, sodium normally crosses the luminal membrane down an electrochemical gradient *via* the postulated carrier, partly in exchange for potassium or hydrogen ions, and is then pumped out of the cell into the interstitium by the basal pump. At the same time, the second luminal carrier transfers potassium into the lumen down a gradient generated primarily by the basal Na^+–K^+ pump. Since hydrogen ion transport by the two luminal carriers is in the opposite direction, the nett effect is minimal and little bicarbonate secretion is observed. Sodium ions leaving the lumen are partly replaced by potassium and hydrogen ions and partly enter the interstitium accompanied by chloride. In the rabbit duct, where secretion of potassium and bicarbonate is minimal under these conditions, it is presumed that the initial efflux of sodium is accompanied largely by efflux of chloride.

During duct perfusion with a low sodium fluid (*e. g.* isotonic mannitol) there is no sodium influx across the luminal membrane, so the sodium carrier ceases to shift H^+ or K^+ into the lumen. The second potassium carrier still operates, however, and the counter movement of H^+ ions is now unmasked and is manifested as secretion of bicarbonate coupled to potassium. Similarly, amiloride or triamterene, by blocking the sodium carrier, will reduce K^+ and H^+ secretion and unmask the coupled potassium and bicarbonate secretory process.

When the duct is perfused with isotonic saline and perfusion is suddenly stopped, electrolyte transport will proceed until steady-state conditions develop. Initially only sodium reabsorption and potassium secretion will be observed but, as luminal sodium concentration falls and net sodium transport approaches zero, the bicarbonate secretory mechanism will be unmasked and luminal bicarbonate concentration will begin to rise. In the rabbit duct, where initially the nett bicarbonate secretion rate is zero, the transepithelial potential during development of steady-state conditions changes in a characteristic biphasic manner, first rising to a maximum when luminal sodium and chloride concentrations reach a minimum, and then falling again to a plateau when luminal bicarbonate concentration has risen and the bicarbonate diffusion potential has partly cancelled the chloride diffusion potential (Grant et al., 1974).

With the above model it is easy to interpret the effect of an aldosterone blocker, which reduces the sodium transport rate and thereby unmasks the bicarbonate secretory mechanism. Similarly, disturbances in acid-base status could be expected to operate by producing reciprocal changes in intracellular concentrations of hydrogen and potassium ions. One observation that cannot be explained, however, is the induction by acidosis of hydrogen-ion secretion by ducts in which the sodium carrier is blocked with amiloride. Knauf (1974) and Knauf et al. (1975) attempt to explain this by postulating that there are two different cell types in the duct, one of which secretes hydrogen ions while the other reabsorbs them. Since they recognize two different types of duct cell on the basis of the histochemical reactions, a dark cell and a light cell, and since the proportion of dark cells increases in acidosis, they were inclined to ascribe a hydrogen ion secretory role to the dark cell. This suggestion cannot be excluded with certainty but direct evidence in support of it is lacking.

Since salivary glands contain appreciable quantities of carbonic anhydrase

(YOSHIMURA et al., 1959; KARLER and WOODBURY, 1960; LEDER and TRITSCHLER, 1966; MAREN, 1967), it might be expected that the catalytic hydration of CO_2 would be involved in the process of salivary excretion of bicarbonate, either at the endpieces or in the ducts. In fact, however, administration of carbonic anhydrase inhibitors has no effect on bicarbonate excretion by the intact salivary gland (NIEDERMEYER et al., 1955; CHAUNCEY and WEISS, 1958; YOSHIMURA et al., 1959; YOUNG et al., 1970), although PETERSEN and POULSEN (1966) have observed that acetazolamide can reduce salivary flow rates and secretory potentials in the cat mandibular gland. YOUNG et al. (1970) report that acetazolamide has no effect on bicarbonate secretion by the rat mandibular duct, perfused *in vivo* with bicarbonate-saline, even when the secretory process had been stimulated with carbachol. Since the Knauf model for ductal bicarbonate secretion (Fig. 15) envisages an HCO_3^--ATPase on the luminal cell membrane that exchanges cytoplasmic K^+ for luminal H^+, it is perhaps not surprising that acetazolamide does not disturb its functioning, since carbonic anhydrase is not available in the lumen to catalyze formation of H^+ from CO_2. However, it is a little unexpected that the supply of H^+ to the luminal membrane sodium exchange carrier is not a rate-limiting step in the H^+ secretory process. More recently, SOMMER et al. (1975) report that luminal application of acetazolamide can block Na^+/H^+ exchange in rat parotid ducts.

3. Innervation and Autonomic Control of Ductal Transport

During the first half of the twentieth century, when most workers ascribed a secretory function to various parts of the salivary duct system (BABKIN, 1950), the notion that the duct cells must be under autonomic control seemed unexceptional. With the general acceptance of the view that duct cells were concerned mainly with the transport of electrolytes rather than organic compounds, interest in ductal innervation waned[10], despite the fact that LUNDBERG (1955) claimed to have obtained electrical recordings from duct cells and to have induced depolarizing responses in them by stimulating the autonomic nerves. As pointed out above (Section C. III. c), examination of salivary electrolyte excretory patterns suggests that ductal transport properties may be altered during autonomic stimulation, and for this reason, YOUNG et al. (1970) began studying the action of autonomic agonists on electrolyte transport by the perfused rat mandibular main duct. Before considering the outcome of these and subsequent, related experiments, it seems appropriate to comment on what is known about innervation of salivary ducts.

Histochemical studies provide abundant evidence to suggest that cholinergic nerves run close to striated and excretory ducts, although the pattern in different glands is quite variable. Thus cholinergic nerves are said to abound around excretory ducts in mandibular glands of the cat (GARRET, 1966a) but are rare in man (GARRETT, 1967). Similarly, adrenergic fibres are abundant around striated

10 See, however, BURGEN and TERROUX (1963), BLAIR-WEST et al. (1967), and YOUNG and SCHÖGEL (1966).

ducts in the salivary glands of the rabbit (EHINGER et al., 1967), man (GARRETT, 1967), hamster (ALM et al., 1973) and cat (GARRETT, 1974), but are said not to be near these ducts in the rat (NORBERG and OLSEN, 1965). GARRETT (1974) stresses that adrenergic fibres have never been found near the main duct cells of any salivary gland although, as will be discussed below, SCHNEYER (1976a) has shown convincingly that sympathetic nerve stimulation can alter potentials and electrolyte fluxes across the rat mandibular main duct epithelium. Ductal innervation is usually of the epilemmal type, although hypolemmal nerve endings have been described in association with both striated and intercalated ducts (see Section C. II. 1). It seems reasonable to conclude that, provided duct cells have appropriate receptors, the disposition of nerve terminals, particularly cholinergic terminals, is adequate to permit autonomic control of ductal function.

In 1970, YOUNG et al. studied the effects of parenteral administration of large doses of the autonomic agonist, carbamyl choline chloride (carbachol), on electrolyte transport by the rat mandibular main duct, perfused *in vivo* with an isotonic bicarbonate saline solution. They found that carbachol reduced the rate of reabsorption of sodium and the rate of secretion of potassium by about 40 percent and 30 percent, respectively; at the same time there was a five-fold increase in bicarbonate secretion. MARTIN and YOUNG (1971b) showed that these flux changes were associated with a depolarization of the duct potential. MARTIN and YOUNG (1971b) also showed that carbachol could stimulate potassium as well as bicarbonate secretion when the duct was perfused with sodium-poor solutions such as isotonic mannitol or a steady state (Table 6) solution. MARTIN et al. (1973) extended these studies to the rabbit mandibular duct, perfused *in vivo* or *in vitro*. In this duct, which secretes no bicarbonate and little potassium when perfused with bicarbonate saline (Table 6), carbachol depolarized the potential difference and reduced sodium reabsorption. From dose-response curves the authors concluded that a carbachol concentration as low as 10^{-8} mol l^{-1} inhibited sodium reabsorption but an effect on the duct potential was seen only at a concentration of 10^{-7} mol l^{-1}. FIELD and YOUNG (1973) confirmed that carbachol (10^{-5} mol l^{-1}) depolarized the potential difference and inhibited sodium transport in the rat duct perfused *in vitro* but, in contrast to MARTIN and YOUNG's (1971b) finding from studies *in vivo*, they observed no effect of carbachol on bicarbonate transport.

DENNISS and YOUNG (1975) set up dose-response curves for acetylcholine applied either to the luminal or the basal surface of the rabbit duct, perfused *in vitro*. Acetylcholine applied to the basal surface had the same effects as carbachol (Figs 16 an 17): it inhibited sodium reabsorption and depolarized the duct potential, but it was effective in concentrations as low as 10^{-12} mol l^{-1} and 10^{-10} mol l^{-1}, respectively. Its action was antagonized competitively by atropine (Fig. 17). When applied from the luminal surface, the drug was without effect except at much higher concentrations ($> 10^{-5}$ mol l^{-1}). KNAUF et al. (1972) have published preliminary data to indicate that the effect of cholinergic agonists is to diminish the sodium conductance of the luminal face of the epithelium. Recently SCHNEYER (1976b) has shown that parasympathetic nerve stimulation alters duct potential difference.

MARTIN and YOUNG (1971b) also studied the action of the β-adrenergic ag-

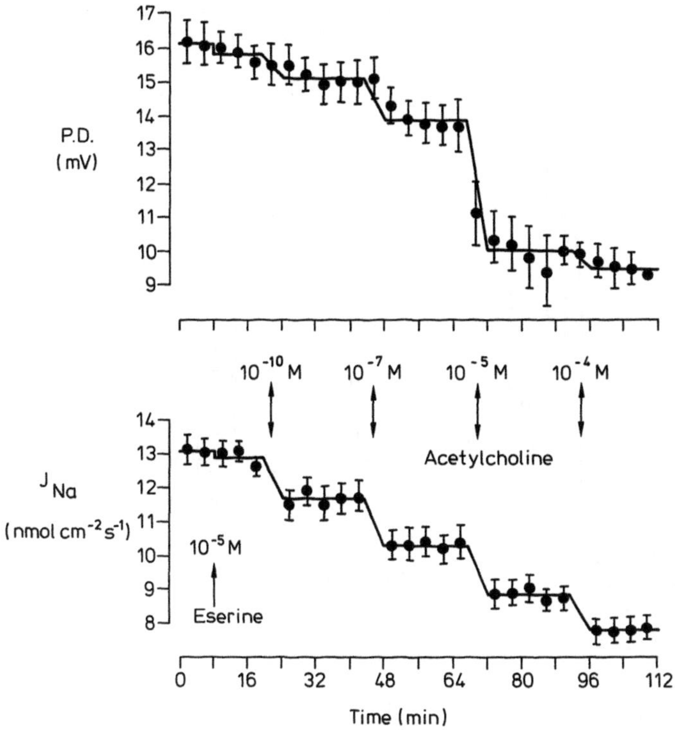

Fig. 16. Effect of acetylcholine, administered in the bathing fluid in concentrations indicated, on transepithelial potential difference and nett reabsorption of sodium in the isolated rabbit mandibular main duct perfused *in vitro* with bicarbonate saline. (From DENNISS and YOUNG, 1975)

onist, isopropyl noradrenalin (isoproterenol) on the rat mandibular duct perfused *in vivo*. Qualitatively the effects of the drug were similar to those of carbachol *i. e.* inhibition of sodium reabsorption and potassium secretion and stimulation of bicarbonate secretion. Subsequently, SCHNEYER and THAVORNTHON (1973) performed similar studies, except that they administered much lower doses than used by MARTIN and YOUNG (1971b). Under these circumstances, sodium reabsorption, instead of being inhibited, was enhanced by 40 percent, although potassium secretion was still inhibited (bicarbonate secretion was not studied). Recent studies (DENNISS et al., 1978) on the rat duct perfused *in vitro* confirm that isoproterenol in low concentrations (10^{-9} mol l^{-1}) stimulates sodium reabsorption without altering the potential difference significantly, whereas in high concentrations (10^{-4} mol l^{-1}), it depolarizes the duct and reduces sodium reabsorption. SCHNEYER (1976a) has studied the effects of sympathetic nerve stimulation on electrolyte transport by the rat duct perfused *in vivo*. He finds that supramaximal nerve stimulation mimics the effect of large doses of isoproterenol (*i. e.* inhibits Na^+ reabsorption and K^+ secretion and depolarizes the epithelium) but the responses are blocked by administration of α-adrenergic, rather than β-adrenergic antagonists. DENNISS and YOUNG (1975) have also studied the effects of isoproterenol and noradrenalin on the rabbit

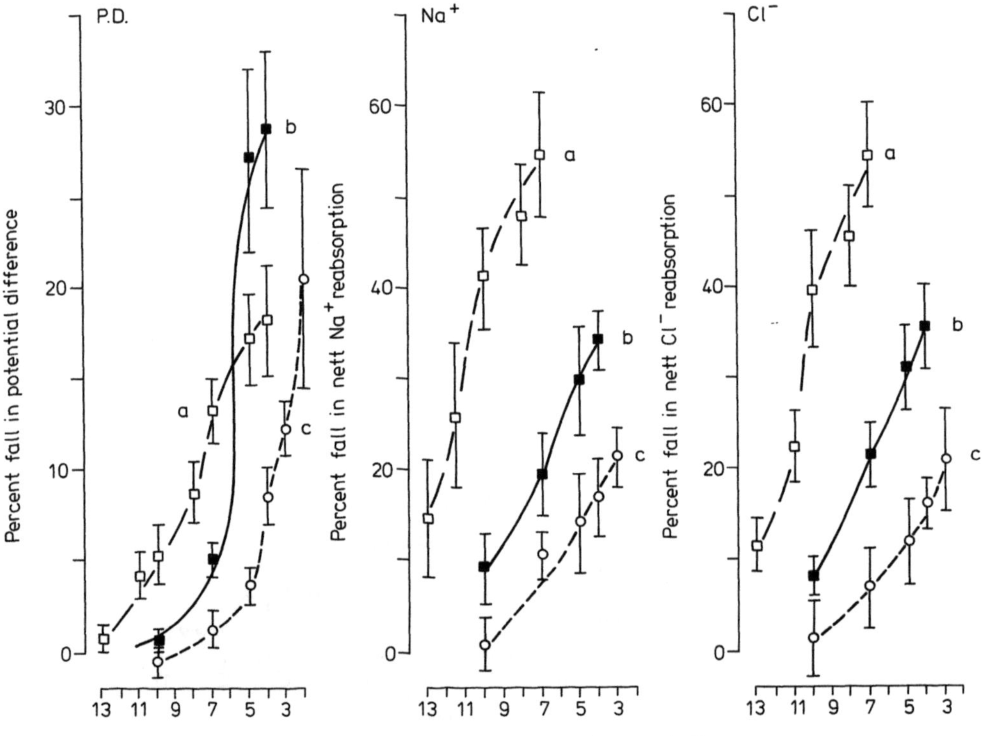

Fig. 17. Dose-response curves showing effects of acetylcholine on rabbit mandibular main duct perfused *in vitro* with bicarbonate saline solution. Each point is mean of results from 4 or 5 experiments ± S.E.M. *a* indicates pretreatment of the ducts with theophylline (10^{-5} mol l^{-1}); *b* indicates control ducts; *c* indicates pretreatment with atropine ($5 \cdot 10^{-5}$ mol l^{-1}). (From DENNISS and YOUNG, 1975)

mandibular duct perfused *in vitro*. Both compounds depolarized the duct and reduced sodium reabsorption, but isoproterenol was only effective at concentrations above 10^{-3} mol l^{-1}, whereas noradrenalin acted at concentrations down to 10^{-7} mol l^{-1}. It appears that this duct possesses adrenergic receptors even though it lacks a sympathetic secretomotor innervation.[11]

From the above it seems clear (a) that salivary ducts are innervated and (b) that activation of either division of the autonomic nervous system or administration of autonomic agonists can alter ductal electrolyte transport properties. Stimulation seems to have little effect on ductal water permeability (MARTIN and YOUNG, 1971b; SCHNEYER, 1976a). The simplest explanation of the action of

11 Stimulation of the sympathetic nerves supplying the dog mandibular gland also causes no secretion, but if they are stimulated simultaneously with the parasympathetic nerves the effect is to cause an increase in salivary sodium concentration above that seen when the parasympathetic nerves are stimulated alone (IMAI et al., 1970). This probably indicates that, as in the rabbit mandibular gland, ductal sodium reabsorption is inhibited by sympathetic nerve activity.

acetylcholine (and carbachol) and large doses of isoproterenol is that, in some way, they block the luminal membrane sodium carrier (Fig. 15). This would account for the observed depolarization and the decrease in luminal sodium conductance. Similarly it would account for the effects on Na^+, K^+ and HCO_3^- flux. The action of carbachol on ducts perfused with a sodium-free solution or a steady-state solution (stimulated secretion of both K^+ and HCO_3^-) suggests that, in addition to inhibiting the luminal sodium carrier, the drugs may also stimulate the potassium carrier; such an effect would not be obvious during duct perfusion with sodium-rich fluids when the effect on the sodium carrier would mask an action on the potassium carrier.

One important observation is not easily explained by such a model: in all experiments *in vitro,* on both rat and rabbit ducts, the inhibitory effect of the agonists on sodium transport was manifested at lower drug concentrations than the effect on transepithelial potential difference (MARTIN et al., 1973; DENNISS and YOUNG, 1975; MARTIN et al., 1974; DENNISS et al., 1978). If the only action of the drugs on sodium transport were mediated by blocking the luminal sodium carrier one might expect voltage and flux changes to occur simultaneously although, since the current voltage relationship of the carrier has not been characterized, this objection may prove to be illusory. It may be that the agonists have a direct inhibitory effect on the basal membrane Na^+–K^+ pump (see DENNISS and YOUNG, 1975, for discussion of this point).

Since the available experimental evidence indicates, as one would expect, that acetylcholine and noradrenalin act only at specific receptors on the basal membrane of the epithelium, one might anticipate that the drug-receptor interaction would give rise to changes in the Na^+ and K^+ conductances of the basal membrane as they do in endpiece cells (Section C. III. 5. b) and at synaptic junctions. Such changes may well occur and could lead to reduced sodium absorption and duct depolarization, but they could not directly account for a decrease in the epithelial sodium conductance, nor could they explain the action of acetylcholine on potassium and bicarbonate secretion. Hence it becomes necessary to postulate the existence of a system of intracellular messengers that ultimately induce changes in the properties of the luminal membrane.

DENNISS and YOUNG (1975) have studied the effects of two putative messengers, cyclic AMP and cyclic GMP, and of the phosphodiesterase inhibitor, theophylline, on salivary duct electrolyte transport. Cyclic AMP in a concentration of $4 \cdot 10^{-5}$ mol l^{-1} (and probably much lower) caused depolarization and inhibition of sodium transport when applied from the luminal but not from the basal surface of the duct. Theophylline applied to both surfaces in a concentration of 10^{-5} mol l^{-1} also depolarized the duct and reduced sodium transport. In addition, this drug caused a marked potentiation of the acetylcholine response (Fig. 17), so that the agonist now became effective in concentrations of 10^{-13} mol l^{-1}, perhaps even as low as 10^{-15} mol l^{-1}. In high concentrations (10^{-2} mol l^{-1}), theophylline imitated the effect of carbachol but blocked further action of the agonist (MARTIN et al., 1974). Cyclic GMP had effects similar to those of cyclic AMP but only at concentrations of $4 \cdot 10^{-5}$ mol l^{-1}. DENNISS and YOUNG (1975) interpret these results as indicating that cyclic AMP (rather than cyclic GMP) is a mediator of the action of acetylcholine and leads ultimately to a change in the conductance properties of the apical membrane, mediated in turn perhaps *via* a protein kinase. However, their studies have been confined to the rabbit duct which is not very sensitive to adrenergic agonists; it is important that these studies be extended to the rat duct, which is unequivocally sensitive both to cholinergic and adrenergic agonists, in order to disentangle the possible messenger roles of the two cyclic nucleotides. To date, no studies are available on the possible role of calcium ions in the mediation of the duct responses to acetylcholine.

4. Endocrine Control of Ductal Transport[12]

A large literature exists on the effects of various endocrine disturbances on salivary gland morphology, but few studies are available which shed light on the possible roles that hormones may play in the regulation of the transport functions of salivary glands (RAYNAUD, 1960; SREEBNY and MEYER, 1964). An exception to this statement is that a large literature exists on the role of salt balance and adrenal corticosteroids in determining the electrolyte composition of the salivary secretions of man and sheep (SREEBNY and MEYER, 1964; BLAIR-WEST et al., 1967). Since the publication of BLAIR-WEST's excellent review of the topic of steroid control of salivary glands (BLAIR-WEST et al., 1967), only a few studies have been published in which the newer investigative techniques of micropuncture and duct microperfusion have been employed with the aim of elucidating salivary hormonal control mechanisms. In the present article attention is confined mainly to these recent studies.

a) Mineralocorticoids

In man it is well established that the sodium and potassium content of saliva is influenced by adrenocortical hormones, particularly aldosterone: administration of ACTH or a mineralocorticoid usually reduces Na^+ concentration and increases K^+ concentration. Similar changes in electrolyte concentration are seen in salt deprivation, in primary and secondary hyperaldosteronism, and in pregnancy: in Addison's disease changes in the reverse direction are seen. For a time, salivary sodium-potassium concentration ratios were used diagnostically in patients with suspected disorders of aldosterone secretion (LAULER et al., 1962),

12 No studies have been published in which the effect of vasopressin (ADH) on the isolated salivary duct has been reported. Several studies have been published in which ADH (as well as cyclic AMP, theophylline and imidazole) were injected *in vivo* into the arteries of the mandibular glands of dogs and rabbits (MARTINEZ and MARTINEZ, 1972a, b; SMAJE, 1974). Such *in vivo* studies, however, are open to the criticism that any effect produced may not have arisen from the direct action of the drugs on glandular cells but, rather, from some systemically mediated effect such as a fall in blood pressure. MARTINEZ and MARTINEZ (1972a, b) interpreted their results as indicating that ADH affects both sodium and water transport in the ducts of the dog mandibular gland and that two separate pathways, only one of which is mediated via cyclic AMP, are involved. SMAJE (1974) observed a decrease in the rate of formation of spontaneous secretion by the rabbit mandibular gland which he attributed to an increase in ductal reabsorption of fluid.

but the overlap between normal and abnormal ratios has proven too great for them to be considered reliable for diagnostic purposes and their determination is now seldom undertaken (BLAIR-WEST et al., 1967).

With general acceptance of a two-stage mechanism for production of saliva, it is clearly of interest to determine whether mineralocorticoids induce changes in salivary sodium and potassium content at the site of formation of the primary secretion or at the sites of ductal modification of this precursor saliva. From micropunture studies on the parotid glands of adrenalectomized rats, MANGOS and McSHERRY (1969) concluded that aldosterone did not affect the rate of production or the electrolyte composition of primary fluid, although a small, indirect effect resulted from changes in serum sodium and potassium concentrations (Table 1). On the other hand, their studies provided convincing evidence to suggest that lack of the hormone resulted in decreased sodium absorption and potassium secretion by the striated ducts.

Micropuncture studies on the effects of aldosterone on the composition of the primary fluid of the rat mandibular gland have not been published, but a number of studies on the effects of the hormone on the perfused main duct are available (GRUBER et al., 1973; KNAUF, 1974; KNAUF et al., 1975, 1976c). These studies make it clear that adrenalectomy or administration of a spironolactone reduces sodium reabsorption and potassium secretion and results in an increased rate of accumulation of bicarbonate in the luminal fluid. KNAUF et al (1976c) interpret the results as indicating that aldosterone has two actions: it increases the conductance of the luminal sodium carrier (which exchanges Na^+ for K^+ and H^+) and increases the transport rate of the basal membrane sodium pump. It is not clear whether these two actions are wholly independent or whether the stimulated basal transport of sodium results merely from an increased intracellular sodium concentration consequent on increased conductance of the luminal membrane to sodium. The hormone does not seem to have any direct action on the luminal potassium/hydrogen carrier – its effects on transport of these ions seem merely to be a consequence of its effect on the sodium carrier.

The rabbit mandibular gland also alters its electrolyte excretory patterns in response to salt deprivation or steroid treatment (COMPTON, 1976; COMPTON and YOUNG, 1977; COMPTON et al., 1975). Figure 18 shows the sodium excretory curves for groups of rabbits deprived of salt or treated with DOCA or dexamethasone for periods of 50–180 days. The control group show a typical sodium excretory curve (Pattern B in Fig. 3) with low concentrations of about 20 mmol l^{-1} at low salivary flow rates, rising progressively to about 90 mmol l^{-1} as flow rate approaches a maximum. Sodium deprivation (Fig. 18 D) caused a marked reduction in salivary sodium concentration, which ranged between 20 and 40 mmol l^{-1} and showed no definite flow-dependency. Treatment with DOCA (0.5 mg, every other day) had a similar effect (Fig. 18C). Unexpectedly, treatment with the glucocorticoid, dexamethasone (0.2 mg, every other day) had an even more dramatic effect (Fig. 18B) and gave rise to salivary sodium concentrations below 20 mmol l^{-1} at all salivary secretory rates. None of the treatments had an effect on the salivary excretion curves for potassium. Micropuncture studies (Table 1) make it clear that the changes in sodium excretion

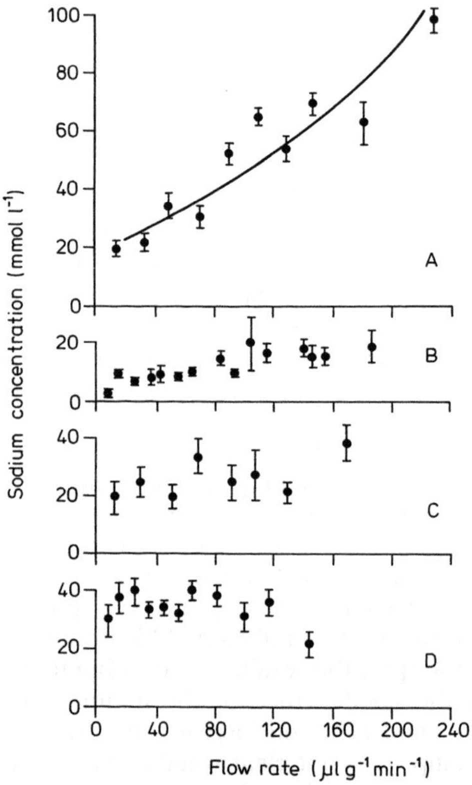

Fig. 18. Sodium excretory curves for the mandibular gland of the rabbit stimulated to secrete by i. p. injection of pilocarpine and carbachol. A. indicates control group, fed diet of oats (naturally poor in sodium) and given salt supplements for 118–219 days; B. indicates animals fed oats plus salt but given injections of 0.2 mg dexamethasone every other day for 88–108 days; C. indicates animals fed oats plus salt but given injections of 0.5 mg desoxy-corticosterone acetate every other day for 94–114 days; and D. indicates animals given oats without sodium supplementation. (COMPTON and YOUNG, 1977; COMPTON, 1976)

cannot be explained on the basis of changes induced in the composition of the primary secretion. Hence it seems reasonable to conclude that sodium deprivation promotes increased ductal sodium reabsorption, probably mediated *via* the action of aldosterone. Why dexamethasone should also cause an increased ductal sodium reabsorption is obscure. The simplest explanation would be that the steroid receptors in the rabbit mandibular gland lack specificity.

In wild rabbits feeding on salt-poor vegetation, the proportions of ducts to endpieces in the mandibular gland changes, so that ducts become extremely conspicuous in histological sections (BLAIR-WEST et al., 1968); a similar change is observed when laboratory rabbits are deprived of salt (COMPTON et al., 1975; COMPTON and YOUNG, 1977; COMPTON, 1976). It is tempting to postulate that this increased prominence of ductal elements is associated with the increased ability of the ducts to reabsorb sodium in salt-deprived rabbits. However, prolonged treatment of rabbits with DOCA, which promotes ductal sodium reabsorption, does not cause duct hypertrophy although chronic treatment with dexamethasone does (COMPTON, 1976). The present authors are of the opinion that stress factors leading to an increased secretion of glucocorticoids are responsible for the apparent hypertrophy of the ducts, and that this hypertrophy is unrelated to mineralocorticoid-induced increases in ductal sodium transport.

It has long been known that a change in sodium balance or the administration of mineralocorticoids causes especially dramatic changes in the sodium and potassium concentrations of parotid saliva in ruminants (BLAIR-WEST et al., 1967; THAYSEN and TARDING, 1974). In sheep fed a sodium-rich diet, the parotid saliva is isotonic with high concentrations of sodium (~ 160 mmol l^{-1}) and relatively low concentrations of potassium (< 20 mmol l^{-1}). When the animal is made sodium-deficient, for example by the surgical establishment of a unilateral parotid fistula, the salivary composition is changed dramatically with a time course of two to seven days (Fig. 19), so that the final saliva, while remaining isotonic, becomes sodium-poor and potassium-rich (Fig. 20). In extreme cases, sodium deprivation may cause salivary potassium concentrations to increase to more than 120 mmol l^{-1} (BLAIR-WEST et al., 1967).

In view of the fact that aldosterone has been demonstrated to control ductal Na^+ and K^+ fluxes in rat and rabbit salivary glands, it seems highly likely that in sodium-deprived sheep, aldosterone (released in increased amounts) acts on salivary ducts in a similar manner, although BLAIR-WEST et al., (1969) have argued that the principal effect of the hormone is on the composition of the primary fluid. Recent micropuncture experiments (COMPTON and YOUNG, 1976) show that sodium deprivation does indeed result in a fall in primary fluid sodium concentration, but not in a rise in potassium concentration (Table 1). Compari-

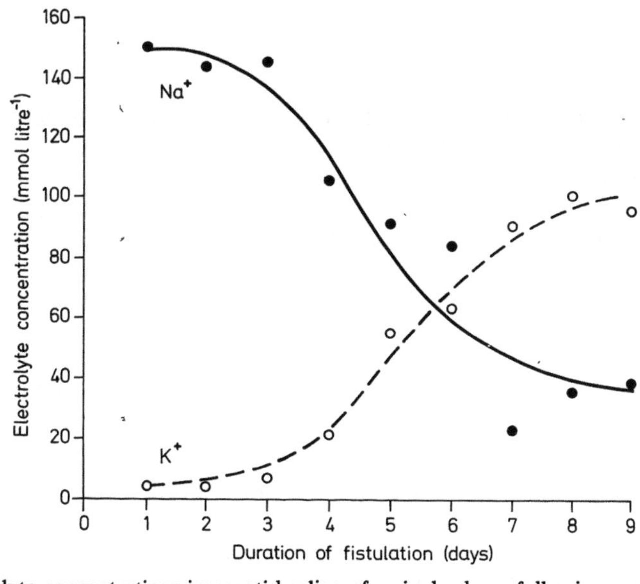

Fig. 19. Electrolyte concentrations in parotid saliva of a single sheep following creation of a unilateral parotid fistula. (From COMPTON, 1976)

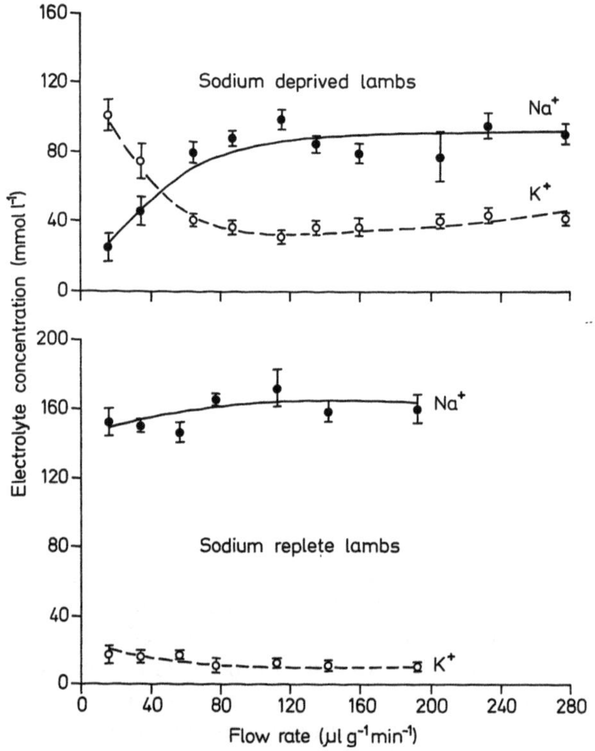

Fig. 20. Electrolyte excretory patterns in sodium replete sheep and in sheep rendered sodium-deficient by creation of unilateral parotid fistulae. (From COMPTON, 1976)

son of primary fluid composition with that of final saliva, collected at rest and during stimulation, suggests that in the sodium-replete sheep the ducts neither reabsorb sodium nor secrete potassium appreciably (although they do secrete bicarbonate and reabsorb phosphate). In salt-deprived sheep however, the ducts develop the ability to reabsorb sodium and secrete potassium so that final saliva differs markedly in composition from the primary secretion. Thus it is clear that sodium deprivation brings about changes in electrolyte transport, both by the endpieces and by the ducts of the sheep parotid. Since sodium deprivation has no effect at all on the composition of parotid saliva in adrenalectomized sheep (BLAIR-WEST et al., 1967), it can be concluded that adrenal steroids are involved in the regulation of electrolyte transport by both endpieces and ducts. However, since administration of supramaximal doses of aldosterone to sodium-replete sheep only causes salivary sodium concentration to fall to around 120 mmol l^{-1} and potassium to rise to 30–40 mmol l^{-1} (BLAIR-WEST et al., 1967; THAYSEN and TARDING, 1974), it is apparent that some additional factor must be involved. What this could be is unknown, but THAYSEN and TARDING (1974) speculate that it might operate by regulating salivary gland blood flow.

b) Angiotensin

Salivary glands, at least in some species, have very high tissue concentrations of renin (WERLE et al., 1957; BING et al., 1967; BING and FÅRUP, 1965; BHOOLA and HEAP, 1970; COHEN et al., 1972). Although there is no evidence to suggest that this renin plays any role in the regulation of the concentration of circulating angiotensin (BING and POULSEN, 1975), it would occasion no surprise if locally produced angiotensin were involved in the regulation of salivary gland function. Experiments in which angiotensin has been infused into the salivary gland *via* the arterial blood indicate that this substance acts on salivary endpieces only by provoking the release of acetylcholine from nerve endings (KREUSSER et al., 1974; KREUSSER, 1974). It does not alter the electrolyte composition of parasympathomimetically evoked saliva when infused simultaneously with a parasympathomimetic drug (CORONEO, 1976), nor does it modulate the action of simultaneously administered aldosterone (BLAIR-WEST et al., 1965). Recently, however, it has been shown that, in concentrations similar to those in normal rabbit plasma (McDONALD et al., 1970), angiotensin can reduce sodium reabsorption and depolarize the potential difference across the epithelium of the isolated rabbit mandibular duct perfused *in vitro* (HEALY et al., 1976). These experiments show quite clearly that angiotensin could play a role in the regulation of salivary electrolyte excretion, but it remains to be shown whether such regulation actually occurs under physiological conditions.

c) Gastrointestinal Polypeptide Hormones and Related Substances

It has been known for over 10 years that certain naturally occurring polypeptides such as eledoisin and physalaemin, as well as some structurally related synthetic peptides, are powerful sialogogues when injected into the circulation in extremely low concentrations (BERTACCINI and DE CARO, 1965; BERTACCINI et al., 1965a and b; ERSPAMER and ERSPAMER, 1962; SCHNEYER and HALL, 1968a; EMMELIN et al., 1973). Similarly powerful sialogogic polypeptides (*e. g.* sialogen and substance P) can be extracted from mammalian tissues such as the rat hypothalamus (LEEMAN and HAMMERSCHLAG, 1967; LEMBECK et al., 1968). To date, however, no one has described a naturally circulating polypeptide that stimulates salivary secretion.

DENNISS and YOUNG (1978) have studied the action of physalaemin on the main ducts of both rat and rabbit mandibular gland, perfused *in vitro*. In a bath concentration of 10^{-10} mol l^{-1} the peptide was without effect, but at a concentration of $4 \cdot 10^{-8}$ mol l^{-1} it caused a sharp increase in ductal sodium transport. Applied from the luminal surface in a similar concentration, the drug had no action. The effects of several gastrointestinal polypeptide hormones on the isolated rabbit duct have also been investigated by these authors (DENNISS and YOUNG, 1978; MARTIN et al., 1974). Both VIP (vasoactive intestinal peptide) and GIP (gastric inhibitory polypeptide) inhibit ductal sodium transport when applied from the bath surface of the duct in concentrations of 10^{-11} mol l^{-1}; gastrin and secretin also have such an action but at considerably higher concentrations (2–$5 \cdot 10^{-7}$ mol l^{-1}). Caerulein is without effect at bath concentrations of 10^{-6} and 10^{-8} mol l^{-1}.

These results certainly do not prove that any gastrointestinal hormone plays a role in regulating salivary electrolyte transport, but the low concentrations in which VIP and GIP are effective should definitely encourage further investigation.

Similarly, the remarkable stimulatory effect of physalaemin on salivary duct sodium transport warrants further investigation.

5. Relative Roles of Granular, Striated, and Excretory Ducts in Saliva Formation

From the foregoing it is apparent that most of the available information on salivary duct function has come from microperfusion studies on the main excretory ducts of mandibular glands in rat, rabbit, and man. Micropuncture and microcatheterization studies make it appear likely that striated and excretory ducts function similarly but differences of either a qualitative or a quantitative type have not been excluded. Even if the transport properties of both duct segments were identical, functional differences could arise from differences in the type of innervation or vascularization that each might have.

It is now clear that the excretory ducts play a major role in rendering saliva hypotonic, but the relative importance to be ascribed to the striated ducts in the fulfilment of this task is difficult to determine. Despite frequent statements to the contrary in the literature, a correlation between the degree of morphological development of the striated ducts in a particular gland and the extent of the hypotonicity that develops in the saliva is lacking. For example, both the rat mandibular gland (in which striated ducts in the male are almost completely transformed into granular ducts) and the rat sublingual gland (in which striated ducts are sparse, and granular ducts are lacking) produce profoundly hypotonic final saliva (MARTIN and YOUNG, 1971a; YOUNG and SCHÖGEL, 1966; YOUNG and MARTIN, 1971); similarly the human sublingual gland lacks striated ducts (LEESON, 1967) but forms a hypotonic secretion (KNAUF and FRÖMTER, 1970a). In contrast, although the isotonic-secreting sublingual glands of carnivores are admittedly largely devoid of striated ducts, the parotids of ruminants, which are also isotonic-secreting, have these structures in abundance (BLAIR-WEST et al., 1969; COATS and WRIGHT, 1957; YOUNG and VAN LENNEP, 1978). Attempts to demonstrate an inverse correlation between salivary osmolality (or sodium content) and the number or mass of striated ducts (*e. g.* JUNQUEIRA and FAVA-DE-MORAES, 1965) in various glands have led to false conclusions, because a sufficient number of species was not considered. These arguments are not intended to suggest that striated ducts do not render saliva hypotonic by a process of sodium reabsorption – the present authors are of the opinion that they do – they are merely to stress that the excretory ducts are of equal if not greater importance in this respect.

It also seems relevant to stress that in those animals where striated ducts are converted into granular ducts there is little evidence to suggest that the ducts lose their normal ability to transport electrolytes. TAMARIN and SREEBNY (1965) point out that *"the relative complexity of the granular cell basal plasma mem-*

branes and their association with great numbers of mitochondria leads one to suspect that active transport" (of ions) "*may also be an important function of these cells. In this regard they are similar to striated ducts*". The only hint to suggest that they do not absorb sodium stems from the failure of micropuncture workers to obtain hypotonic samples when puncturing from the surface of the rat mandibular gland (YOUNG and MARTIN, 1971).

The reader's attention is also drawn to the fact that saliva, of necessity, is handled sequentially by the various segments of the salivary duct tree. Since transepithelial potential difference is dependent on the luminal sodium concentration and since ductal sodium reabsorption is a saturable function of luminal sodium concentration, it is clear that the actual handling of sodium by each duct segment would differ even if the basic properties of the epithelial cells were uniform. A similar argument would apply to other ions. In addition, it is known that the ductal response to neurotransmitter agonists depends on the luminal electrolyte concentration, so that, for example, acetylcholine will retard ductal secretion of potassium when the luminal sodium concentration is high (*e. g.* at the beginning of the striated ducts), but will stimulate it when the luminal sodium concentration is very low (*e. g.* in the excretory ducts). Finally, since the blood supply to these ducts is thought to run in a countercurrent direction (SCHNEYER, 1968; YOUNG and VAN LENNEP, 1978), a detailed analysis of integrated ductal function would be incomplete if it did not allow for the possibility of countercurrent diffusion exchange between blood and saliva.

D. Transport of Proteins

Disc gel-electrophoresis and immunoelectrophoresis studies on saliva from various mammalian glands have shown that these fluids contain quite a large number of protein species (KÖSTLIN and RAUCH, 1957; PATTON and PIGMAN, 1959; D'SILVA and FERGUSON, 1962; ELLISON et al., 1960; MASSON et al., 1965; ELLISON, 1967; CALDWELL and PIGMAN, 1965; MAYO and CARLSON, 1974; MENAKER et al., 1974; CASTLE et al., 1975). For this discussion the following three groups must be distinguished:

1. Serum proteins passing through the salivary gland epithelium into the saliva.
2. Proteins synthesized mainly by stromal cells (plasma cells) and partly by salivary gland epithelial cells; only one protein is known to belong to this group (immunoglobulin A).
3. Proteins wholly synthesized in salivary gland epithelium. These comprise by far the largest group. They include the glycoproteins, important as the lubricating element of saliva, and a wide range of digestive enzymes. Many animals, although the sheep is an obvious exception, have relatively high concentrations of amylase in their parotid saliva. The amylase content of mandibular saliva is very low and may be almost absent in some species of bat and in the laboratory white rat. In the parotid and mandibular glands of carnivores, also, there is relatively little amylase. Apart from amylase, saliva

contains proteases, including kallikrein, and several other enzymes (YOUNG and VAN LENNEP, 1978).

I. Transport of Albumin

Albumin has been found in large quantities in human whole (mixed) saliva, but studies by OPPENHEIM (1970) have shown that most of this albumin originates as an exudate of gingival mucosa. Pure human parotid and submandibular salivas contain only small amounts of albumin. Its immunological identity with serum albumin was demonstrated by McKEAN and BEELEY (1974) and by BRANDTZAEG and HENSTEN-PETTERSEN (1976). In a much earlier paper, JUNQUEIRA et al. (1965) showed that Evans blue, bound to serum albumin, passes through the mandibular gland epithelium of the rat. A limited permeability of the salivary gland epithelium to serum albumin seems therefore to be well established. This is probably due to a passive leaking through the junctional complexes of the epithelium rather than to transcellular transport. Studies with extracellular tracers such as lanthanum hydroxide and horse radish peroxidase have shown that the *zonulae occludentes* in rabbit striated duct epithelium (GARRETT and PARSONS, 1974a and b) and in rat parotid acini (YOUNG and VAN LENNEP, 1978) may be rather leaky for these tracers. Stimulation of the rat mandibular gland is stated by JUNQUEIRA et al. (1965b) to cause an increase in serum albumin leakage into the saliva, particularly after adrenalin stimulation. The data are difficult to evaluate, however, since flow rates were not given in this paper. MANDEL and KHURANA (1969) showed that stimulation of human parotid secretion (by oral application of citric acid) caused a slight increase in the rate of serum albumin discharge by the gland; this discharge by no means paralleled the discharge of secretory proteins. An increased leakiness of the glandular epithelium is in line with the observation by MARTIN and BURGEN (1962) of an increase in permeability of salivary gland epithelium after sympathetic stimulation.

II. Transport of Immunoglobulin A (IgA)

Salivary IgA differs from serum IgA in possessing a so-called secretory component. IgA monomeres and J-chains are synthesized and secreted in the interstitium of salivary glands by plasma cells, which are quite numerous in the glandular stroma, particularly around the striated ducts (BRANDTZAEG, 1974; BRANDTZAEG et al., 1970; HÜRLIMANN and ZUBER, 1968; KRAEHENBUHL et al., 1971; ROSSEN et al., 1968). Immunohistochemical studies have shown the presence of free and IgA-bound secretory component (SC) in the striated or granular ducts of the mouse mandibular gland (COMOGLIO and GUGLIELMONE, 1973). Immunohistochemical investigations on human salivary glands were done by KRAUS and MESTECKI (1971) and by BRANDTZAEG (1974). In the study by KRAUS and MESTECKI (1971) on the human parotid, the authors found IgA synthesis only in periductal plasma cells and secretory component in the basolateral regions of the striated duct cells. BRANDTZAEG (1974), on the other hand, observed a faint reaction for secretory component only in the seromucous cells of the human submandibular gland. In any case, there seems little doubt that the secretory component is the only part of the salivary IgA that is synthesized by salivary gland parenchymal cells. It seems likely that SC forms the vehicle for the passage of IgA from the interstitium across the epithelium into the lumen. The rate of this transepithelial passage of IgA is slightly increased by stimulation of the gland but, as with serum albumin, is not equivalent to the rate of discharge of salivary secretory proteins proper. BRANDTZAEG (1974) suggests that SC synthesized in the rough endoplasmic reticulum and Golgi apparatus of the epithelial cell becomes incorporated in the basolateral plasma membrane of the epithelial cell. In this position it binds the IgA dimer (with J-chain) and the complete secretory IgA then moves laterally, *i. e.* in the plane of the plasma membrane, towards the apical surface, whence it is subsequently released into the lumen. (See YOUNG and VAN LENNEP, 1978, for a detailed treatment of this topic.)

III. Transport of Secretory Proteins

In salivary glands, secretory proteins are produced mainly by the secretory endpieces, but in many mammalian species, *e. g.* various small rodents, parts of the striated duct epithelium contain secretory granules as well. The secretory proteins are temporarily stored in the apical cytoplasm of the glandular cell in the form of discrete, membrane-bound secretion granules or droplets. Each secretion granule, probably without exception, contains a number of species of protein, which in many cases will be inhomogeneously distributed and at least partly aggregated in the granule, judging from the morphological heterogeneity of many secretion granules. The morphological appearance of secretion granules reflects to some extent the predominant types of protein present and particularly the amount of carbohydrate present in the form of glycoproteins. Thus, as a general rule, granules with a large proportion of carbohydrate in the form of acid glycoproteins (*i. e.* sialomucins and sulphomucins) are large, pale-staining with ordinary dyes in light microscopy, and appear electron-lucent under the electron microscope. Small, homogeneously electron-dense granules, on the other hand, usually contain relatively little carbohydrate. The former type of granule is usually referred to as a mucous granule and the latter as a serous granule. Granules with a heterogeneous appearance mostly contain a moderate amount of glycoprotein; they are referred to as seromucous.

The now classical view of the cellular processes involved in protein secretion has recently been reviewed by one of the pioneers in this field (PALADE, 1975). One of the first steps in the process is the synthesis of polypeptides on the ribosomes of the rough endoplasmic reticulum (RER). In a recent paper, BLOBEL and DOBBERSTEIN (1975) suggest that after free cytoplasmic ribosomes have become attached to the initiation codon of messenger ribonucleic acid (mRNA), a following short series of "signal codons" of this mRNA results in the formation of a signal chain of amino acids, which in turn causes the ribosome to become attached to "attachment proteins" of the endoplasmic reticulum membrane. The ribosome and the attachment proteins form a tunnel through which the signal chain, followed by the secretory polypeptide itself, enters the lumen of the endoplasmic reticulum cisterna. The ribosome detaches itself from the endoplasmic reticulum once the final codon of the mRNA has passed the ribosome. The signal chain becomes split off from the secretory polypeptide proper, presumably by the action of enzymes present in the cisternal lumen.

The polypeptides are then transferred by means of so-called transitional vesicles, either directly to condensing vacuoles or first to the Golgi complex and from the latter to condensing vacuoles. In the early papers by JAMIESON and PALADE (1967a and b) on guinea-pig pancreas, only the direct pathway from the RER *via* peripheral Golgi elements to condensing vacuoles was considered. This is probably the main pathway by which simple proteins are transported in salivary gland cells. LEBLOND and his associates have demonstrated the importance of the Golgi apparatus for the synthesis of glycoproteins (BENNETT et al., 1974; NEUTRA and LEBLOND, 1966a and b), and the latter form an important constituent of many salivary gland secretions. Thus, in salivary gland cells the main pathway is probably through the Golgi cisternae. This was indeed confirmed by CASTLE et al. (1972) for the rat parotid. Studies on the incorporation of ^{35}S in parotid secretory proteins (BERG and AUSTIN, 1976) showed that sulphation occurs in the Golgi cisternae.

As their name implies, the main function of condensing vacuoles is the concentration of the secretion product with a loss of water. In a typical serous glandular cell this results in the formation of a discrete homogeneous electron-dense granule. Very little condensation occurs in typical mucous glandular cells and a distinction between a condensing vacuole and a mature granule cannot very well be made. The secretion granules in the apical cytoplasm of a salivary gland cell are usually stored for a limited time and either discharged into the lumen, or, if no stimulation occurs, reabsorbed by lysosomal action (HAND, 1972a). Discharge of serous and seromucous granules is by exocytosis, *i. e.* the membrane of the granule fuses with the apical plasma membrane, bordering the central lumen, or an intercellular secretion canaliculus; fusion of the two membranes is followed by perforation, allowing the granule contents to escape into the lumen without disruption of the integrity of the plasma membrane. Chain exocytosis is a common feature in salivary gland cells; this involves fusion of the membranes of a whole row of granules, each granule acting as an extension of the lumen, or in fact, as an intracellular secretion canaliculus for the next granule (AMSTERDAM et al., 1969; see also ICHIKAWA, 1965).

The secretion of proteins by glandular cells involves transepithelial transport with concomitant synthesis and packaging of the secretion product. This transport process can be divided into three phases:

1. Uptake of amino acids and carbohydrate units through the basal plasma membrane.
2. Intracellular transport of the synthesized polypeptides from their place of

synthesis in the RER in the basal cytoplasm to the apex of the cell where they are stored as secretion granules.
3. Transport by exocytosis of the secretory proteins through the apical plasma membrane.

1. Uptake of Amino Acids into Secretory Cells

The first step in the transport of organic substances leading ultimately to the discharge of proteins into the gland lumen is the uptake of amino acids and monosaccharides. Although amino acid uptake has been studied extensively in many micro-organisms and in tissues such as liver, brain, intestinal epithelium, and kidney, the present authors are unaware of any systematic study of this transport in salivary gland cells. A number of transport systems have been described for the various amino acids, and it can be presumed that similar systems also occur in salivary glands. Systematic studies on amino acid uptake by salivary glandular cells under various conditions in which the effects (if any) are observed of stimulation by secretagogues, cyclic AMP and cations, on the uptake of amino acids, are badly needed. VAN VENROOIJ et al. (1973) studied the uptake and incorporation of isotope-labelled (^{3}H and ^{14}C) leucine in unstimulated rat mandibular glands. These workers found that leucine is taken up quite rapidly into the cells during the first 10 minutes of incubation and that the concentration of labelled soluble leucine in the cells increased at a slower rate during the remaining time of the experiment (30 minutes). Their results suggest that the source of amino acids to be used in protein synthesis is the extracellular rather than the intracellular pool of amino acids. BARKA (1971) found that a single injection of isoproterenol stimulated the uptake of the non-metabolizable amino acid, α-aminoisobutyric acid, into rat mandibular and sub-lingual gland cells.

2. Intracellular Transport

Polypeptides are synthesized and sequestered in the cisternae of the rough endoplasmic reticulum, which is mostly located in the basal region of the cyto-plasm. These polypeptides end up in secretion granules in the apical cytoplasm near the luminal plasma membrane. In their passage from the RER to secretion granule the secretion products are continuously segregated from the cytoplasm (for an alternative view, see ROTHMAN, 1975) by the membranes, respectively, of the RER, transitional vesicles, Golgi cisternae, Golgi vesicles, condensing vacuole, and finally granule membrane. If all parts of this system formed a continuous lumen, *i. e.* if RER cisternae, Golgi cisternae and condensing vac-uoles were in open communication with one another, one would have to assume a more or less constant flow of polypeptide molecules through this system of cisternae and tubules.

There is little doubt that the condensing vacuoles are not permanently in open communication with either Golgi or RER cisternae. It would otherwise be difficult to imagine how they could concentrate the secretion product with loss of water. The question whether RER cisternae and Golgi cisternae communicate directly with one another is rather controversial. An old view of cell morphology had all cisternal systems, including the nuclear envelope, in open communication with one another and ultimately with the cell exterior. This must be very rare, at least in animal cells, since entry of extracellular tracers into Golgi and RER cisternae of intact cells has, to the present authors' knowledge, never been recorded (the tubular systems in chloride cells of fish gills, sometimes referred to as "endoplasmic reticulum", which take up lanthanum, should be considered as tubular invaginations of the lateral plasma membrane; they have no open communcation with the sparse RER in such cells).

The classical theory of the secretion process, as proposed by PALADE and his collaborators, considered the RER and the peripheral cisternae of the Golgi apparatus as separate compartments with transitional vesicles ferrying polypeptides from one compartment to the next. More recently, studies with the high-voltage (1000 kV) electron microscope (RAMBOURG et al., 1973, 1974), which allows three-dimensional viewing of thick (1–3μm) sections of cells in which the Golgi cisternae are stained with osmium tetroxide, have suggested that the so-called transitional vesicles in fact form a network of narrow tubules (the "primary network") connecting the RER cisternae with the outermost Golgi cisternae; the latter also communicate with one another. If these observations are confirmed, it will mean that the rough endoplasmic reticulum, including the nuclear envelope, and the whole of the Golgi apparatus form a single compartment, allowing continuous flow with perhaps a restricted passage through the primary network. The flow of polypeptides through this system must be essentially unidirectional and one can only speculate about its mechanism.

All stages of the intracellular transport involved in the secretion process require ATP (JAMIESON and PALADE, 1968, 1971). This includes the transport from RER to Golgi apparatus, a fact that is difficult to explain when continuity between the two systems is accepted. The process of pinching off the vesicles from a membrane is known to require energy, as is the vectorial movement of vesicles and particles in the cytoplasm. It is assumed that contractile proteins play a part in both processes, and microtubules have been implicated in directional transport and cellular polarity (ALLISON, 1973; PORTER, 1973). Although microtubules are often found in the Golgi region of salivary gland cells, there is no evidence to suggest that they are directly involved in the movement of either transitional vesicles or Golgi vesicles, the latter transferring secretion products from the Golgi complexes to the condensing vacuoles. Experiments with colchicine or vinblastine, drugs known to block the action of microtubules, have proven inconclusive in this respect, because their effects are difficult to distinguish from those on the process of discharge. This point will be discussed later. The fact that condensing vacuoles are mostly found in the neighbourhood of the Golgi apparatus and that mature secretion granules are found preferentially near the apical plasma membrane in exocrine cells suggests a vectorial transport of the immature granules. A transport directed towards the plasma membrane must certainly occur in endocrine cells where secretion granules are either randomly scattered throughout the cytoplasm as in the B-cells of the islets of

Langerhans, where the role of microtubules has been established (LACY et al., 1968), or are only found in the outer periphery of the cytoplasm, *e. g.* the pituitary corticotrophs of the rat. In exocrine cells such as the pancreatic and parotid acinar cells, there exists a strong degree of polarity – itself probably controlled by microtubules. The basal cytoplasm is fully occupied by the rough endoplasmic reticulum, and immature granules have nowhere else to go but to the apical cytoplasm. This is even more obvious in typical mucous cells such as the goblet cell, where there is an almost continuous Golgi apparatus separating the RER in the basal cytoplasm from the immature and mature granules in the apical cytoplasm. Secretion granules newly formed in the uppermost dilated cisternae of the Golgi complex would simply push the previously formed granules up towards the apex.

3. Discharge of Secretion Granules

The process of exocytosis requires the granule to approach the plasma membrane so closely that fusion of the granule membrane with the plasma membrane becomes possible. In parotid glands of starved rats, the apical cytoplasm is packed with secretion granules and many granules can be seen to lie in very close proximity to the apical plasma membrane without actually touching it and without apparent fusion taking place. Presumably a second condition must be met before such fusion can occur. It is not clear what this condition is, but it probably involves a molecular rearrangement within either or both membranes, possibly of the kind described for the mucocyst of *Tetrahymena* (SATIR et al., 1973). Fusion of granule membrane with plasma membrane is probably a fairly fast process, leading rapidly to perforation, considering the comparative rarity of electron micrographs showing the fused membranes. TANDLER and POULSEN (1976a) observed both pentalaminar and trilaminar membranes at the region of fusion of mucous granules with the apical plasma membrane in the mucous cells of the cat mandibular gland.

The molecular rearrangements occurring during the fusion process may lead to changes in the permeability to water and electrolytes of the membranes (*cf.* ALLISON and DAVIES, 1974). This would result in different conditions with regard to ion activities within the granule. The present authors have observed in a number of salivary glands that granules that were apparently about to be discharged showed a considerable amount of swelling and concomitant loss of electron density, although admittedly they could not definitely establish whether in these cases perforation had or had not already occurred. Nevertheless, the whole granule membrane almost certainly undergoes a change during this process, since the phenomenon of chain exocytosis shows that other granules, in close contact with the discharging one, react with the membrane of the discharging granule as if it were part of the apical plasma membrane. Freeze-fracture studies on the rat parotid, on the other hand, have shown marked differences in distribution of membrane particles between the granule membrane and the apical plasma membrane (CAMILLI et al., 1976). These differences remain even whilst the granule membrane is temporarily incorporated in the apical plasma membrane.

Exocytosis of secretion granules has sometimes been called "reverse pinocytosis" on the assumption that the same processes occur in exocytosis as in pinocytosis but in reverse order. This is probably correct if pinocytosis is used to mean macropinocytosis, and not the much more common micropinocytosis seen in endothelial cells. Macropinocytosis as it occurs in the process of reuptake of colloid into the thyroid follicular cell is inhibited by cytochalasin B (ALLISON et al., 1971; WILLIAMS and WOLFF, 1971), but micropinocytosis is not inhibited by this drug (ALLISON et al., 1971; WILLS et al., 1972). Cytochalasin B has been shown to cause approximately 50-percent inhibition of stimulated discharge of rat parotid amylase *in vivo* (BUTCHER and GOLDMAN, 1972) and *in vitro* (personal observations). It is thought by many authors, but by no means all (*cf.* DOUGLAS, 1974), that cytochalasin B inhibits the activity of microfilaments of the terminal web and/or membrane-associated contractile proteins. Although the effect of cytochalasin B on single cells in tissue culture seems to be primarily on membrane phenomena, it is difficult to dissociate these phenomena completely from what happens to the underlying terminal web (GOLDMAN et al., 1973). The difference between the sensitivity of macropinocytosis and micropinocytosis to cytochalasin B might be explained by the hypothesis that the terminal web is involved in the former but not in the latter process.

Where granules lie at some distance from the apical plasma membrane, they must first be moved to that membrane. Microfilaments and microtubules may be involved in this process. The fact that cytochalasin B results in no more than 50-percent inhibition, even at high dosage, and that the results of several attempts to block discharge with colchicine (BUTCHER and GOLDMAN, 1972; TEMPLE et al., 1972) and with vinblastine (unpublished) have been inconsistent, may possibly be due to the fact that in the fasted rat parotid a large proportion of the granules already lie in close proximity either to the apical plasma membrane, including the membrane lining the secretion canaliculi, or to other granules that are in that position, and that therefore only isolated granules need the help of microfilaments and microtubules to bring them to the apical plasma membrane.

IV. Control of Protein Secretion

1. Control of Amylase Release

Most work on salivary gland protein secretion has been carried out on serous or seromucous glands, particularly on fragments of rat parotid gland agitated and incubated *in vitro.* The available information on control of parotid amylase secretion has been summarized in the section of the review dealing with stimulus-secretion coupling in salivary endpiece cells (Section C. II. 5. i) and need not be repeated here. However, since there are obvious differences in the secretory processes of serous and mucous glands, what is known about the control of mucous secretion will be discussed at this point. In addition, secretion by the granular ducts of rodent mandibular glands will be discussed, since the topic has generated considerable interest.

2. Control of Mucoprotein Release

Very little is known concerning the control and mechanism of discharge of secretion droplets from mucous glandular cells. The exact mode of discharge of mucous droplets has long been in doubt. Since, in most electron micrographs, mucous granules or droplets appear to fill the apical cytoplasm almost completely, leaving only small bits of cytoplasm between the often partly fused droplets, a typical exocytosis as seen in the pancreas or parotid acinar cell is seldom, if ever, observed. Most workers therefore believe that a number of droplets are discharged *en masse,* taking bits of cytoplasm with them. Thus, TANDLER et al. (1969) observed an explosive rupture of "ripe" mucous cells in the labial gland, which they considered might be due to a contraction of myoepithelial cells.

KIM et al. (1972) described the coalescence of mucous droplets in unstimulated rat sublingual glands. Stimulation with isoproterenol and pilocarpine together resulted in total fusion of all droplets; the resulting large mucous droplet was then discharged through a large gap in the apical plasma membrane, leaving a wide empty space in the cell, open to the lumen. Such a discharge mechanism would be rather unique, even when considered as an apocrine secretion. However, TANDLER and POULSEN (1976a) observed, after careful fixation of the cat mandibular gland, that the mucous droplets of the mucous-secreting cells show a normal fusion of the granule membrane and the apical plasma membrane. After good fixation they also failed to find any fusion of individual droplets. We are left wondering whether the differences between the observations of TANDLER and POULSEN (1976a) and those of KIM et al. (1972) and, *inter alia,* of TANDLER and his co-workers in their earlier work, really were the result of different fixation procedures or the use of different glandular cells for study.

Control of glycoprotein secretion by the mucous cells of rat sublingual glands has been investigated in this laboratory by KENNERSON (unpubl.). In most experiments the secretion rate was assessed by expressing the amount of hexose released into the medium as a percentage of total hexose in the gland and incubation medium after 60 min incubation of small tissue slices. Unstimulated glands release approximately 10 percent of their hexose content in 60 minutes. Carbachol (20 μM) raised the discharge rate by 150 to 200 percent, whereas isoproterenol and noradrenalin alone had no effect on the rate of discharge. Carbachol-stimulated discharge was completely blocked by atropine. Incubation in a calcium-free medium without prior EDTA treatment did not affect carbachol-stimulated discharge, but with previous EDTA treatment discharge was inhibited. No stimulation was observed with either cyclic AMP or theophylline alone. It appears therefore that glycoprotein secretion by the rat sublingual gland depends on cholinergic stimulation and the presence of intracellular calcium, but that it does not involve cyclic AMP.

3. Protein Secretion by Rodent Granular Duct Cells

The cells of the granular ducts in the mandibular glands of the rat and the mouse contain conspicuous granules.[13] Many authors have doubted the secretory nature of these granules because they have failed to see any signs of extrusion, nor did they find the prominence of the rough endoplasmic reticulum normally associated with protein-secreting cells. However, the present author's experience has shown that morphological signs of exocytosis are not as uncommon as has been suggested in the earlier papers and a number of more recent publications have confirmed the secretory nature of the granular duct cells. MATTHEWS (1974b) studied the uptake of tritiated tryptophan by the granular duct cells of the rat mandibular gland. He found that the peak of grain density over the base of the cells occurred one hour after injection of the isotope-labelled amino acid, whereas the apical region, which contains the granules, became maximally labelled after 4 to 12 hours, indicating a very slow turnover of secretory protein. Stimulation of the gland with pilocarpine has little effect on tritiated leucine uptake by the granular duct cells (KUIJPER-LENSTRA et al., 1975). Secretion may be stimulated with pilocarpine (MATERAZZI and VITAIOLI, 1969; CUTLER and CHAUDHRY, 1973), but MATTHEWS (1974c) found that stimulation of parasympathetic nerves had no effect on secretion; stimulation of sympathetic nerves caused release of protease and degranulation. The use of α- and β-receptor blockers showed that α-receptors are involved. CUTLER and CHAUDHRY (1973) described the secretory cycle of rat granular duct cells after stimulation with pilocarpine. They found that degranulation is followed by a rapid build-up of the RER, followed by reconstitution of the secretion granules, the whole process taking approximately eight hours. Their findings are hard to reconcile with those of KUIJPER-LENSTRA et al. (1975), who found no increase in tritiated leucine uptake after stimulation.

The possible functions of these ducts and the details of the occurrence, histology, histochemistry and ultrastructure have been reviewed in detail elsewhere (YOUNG and VAN LENNEP, 1978).

Note added in proof

Addendum to Table 5

Since Table 5 was composed, additional excretory curves have been obtained for rat salivary glands stimulated with acetylcholine and physalaemin (CORONEO, DENNISS and YOUNG, in preparation 1979) and with phenylephrine (YOUNG, THOMPSON and JONES, in preparation 1979).

Mandibular Gland:	*acetylcholine*	Na 40 (B), K 40 (A) –
	physalaemin	Na 25 (E), K 30 (A), HCO_3 40 (B)
	phenylephrine	Na 40 (E), K 40 (A), HCO_3 60 (B)
Sublingual Gland:	*acetylcholine*	Na 80 (A), K 45 (A) –
	physalaemin	Na 10 (D), K 40 (A) –
Parotid Gland:	*phenylephrine*	Na 80 (A), K 25 (A), HCO_3 20 (E)

13 The secretion granules of the granular ducts might be defined as serous granules since they fulfil the usual criteria of being homogeneously electron-dense and of containing little glycoprotein. They possess some properties, however, that are not normally encountered in other serous granules. They are quite polymorphic, suggesting a fluid, rather than a solid consistency, and the limiting membrane of two granules may fuse over a considerable distance, thus forming a pentalaminar membrane. On a number of occasions the present authors have also noted an apparent fusion of the granule membrane with the apical plasma membrane without any further sign of exocytosis.

References

Part I

ALBANO, J., BHOOLA, K. D., CROCKER, M., HEAP, P. F., LEMON, M. J. C.: J. Physiol. (Lond.) **254**, 10 P (1976).
ALLISON, A. C.: In: Locomotion of Tissue Cells (R. Porter and D. W. Fitzsimons, Eds). Amsterdam: Associated Scientific Publishers 1973, p. 109.
ALLISON, A. C., DAVIES, P., de PETRIS, S.: Nature New Biol. **232**, 153 (1971).
ALLISON, A. C., DAVIES, P.: In: Advances in Cytopharmacology, Vol. 2 (B. Ceccarelli, F. Clementi, and J. Meldolesi, Eds). New York: Raven 1974, p. 237.
ALM, P., BLOOM, G. D., CARLSÖÖ, B.: Z. Zellforsch. **138**, 407 (1973).
AMSTERDAM, A., OHAD, I., SCHRAMM, M.: J. Cell Biol. **41**, 753 (1969).
AUGUSTUS, J.: Biochim. biophys. Acta **419**, 63 (1976).
BABAD, H., BEN-ZVI, R., BDOLAH, A., SCHRAMM, M.: Europ. J. Biochem. **1**, 96 (1967).
BABKIN, B. P.: Secretory Mechanism of the Digestive Glands, 2nd ed. New York: Hoeber 1950.
BARKA, T.: Exp. Cell Res. **64**, 371 (1971).
BATZRI, S., SELINGER, Z.: J. biol. Chem. **248**, 356 (1973).
BATZRI, S., AMSTERDAM, A., SELINGER, Z., OHAD, I., SCHRAMM, M.: Proc. nat. Acad. Sci. (Wash.) **68**, 121 (1971).
BATZRI, S., SELINGER, Z., SCHRAMM, M., ROBINOVITCH, M. R.: J. biol. Chem. **248**, 361 (1973).
BDOLAH, A., SCHRAMM, M.: Biochem. biophys. Res. Commun. **18**, 452 (1965).
BENNETT, H., HODGSON, C., SPEIRS, R. L.: J. Physiol. (Lond.) **242**, 145P (1974).
BERG, N. B., AUSTIN, B. P.: Cell Tiss. Res. **165**, 215 (1976).
BERRIDGE, M. J.: In: Advances in Cyclic Nucleotide Research, Vol. 6 (P. Greengard and G. A. Robison, Eds). New York: Raven 1975, p. 1.
BERRIDGE, M. J., OSCHMAN, J. L.: Transporting Epithelia. New York, London: Academic 1972.
BERTACCINI, G., DE CARO, G.: J. Physiol. (Lond.) **181**, 68 (1965).
BERTACCINI, G., CEI, J. M., ERSPAMER, V.: Brit. J. Pharmacol. **25**, 363 (1965a).
BERTACCINI, G., CEI, J. M., ERSPAMER, V.: Brit. J. Pharmacol. **25**, 380 (1965b).
BHOOLA, K. D., HEAP, P. F.: J. Physiol. (Lond.) **210**, 421 (1970).
BHOOLA, K. D., DOREY, G., JONES, C. W.: J. Physiol. (Lond.) **235**, 503 (1973).
BING, J., FÅRUP, P.: Acta path. microbiol. scand. **64**, 203 (1965).
BING, J., POULSEN, K.: Acta path. microbiol. scand. Sect. A. **83**, 733 (1975).
BING, J., ESKILDSEN, P. C., FAARUP, P., FREDERIKSEN, O.: Circulat. Res. (Suppl. II) **20–21**, 3 (1967).
BLAIR-WEST, J. R., COGHLAN, J. P., DENTON, D. A., GODING, J. R., WRIGHT, R. D.: Aust. J. exp. Biol. med. Sci. **43**, 489 (1965).
BLAIR-WEST, J. R., COGHLAN, J. P., DENTON, D. A., WRIGHT, R. D.: In: Handbook of Physiology, Section 6, Vol. 2 (C. F. Code, Ed.). Baltimore, Md: Williams and Wilkins 1967, p. 633.
BLAIR-WEST, J. R., COGHLAN, J. P., DENTON, D. A., NELSON, J. F., ORCHARD, E., SCOGGINS, B. A., WRIGHT, R. D., MYERS, K., JUNQUEIRA, C. L.: Nature (Lond.) **217**, 922 (1968).
BLAIR-WEST, J. R., COGHLAN, J. P., DENTON, D. A., NELSON, J., WRIGHT, R. D., YAMAUCHI, A.: J. Physiol. (Lond.) **205**, 563 (1969).
BLOBEL, G., DOBBERSTEIN, B.: J. Cell Biol. **67**, 835 (1975).
BLOMFIELD, J., WARTON, K. L., BROWN, J. M.: Arch. Dis. Childh. **48**, 267 (1973).
BLOOD, L. A., KENNERSON, A. R., VAN LENNEP, E. W.: Anat. Rec. **188**, 489, (1977).
BOGART, B. I., de LEMOS, C. L.: Anat. Rec. **177**, 219 (1973).
BOLANDE, R. P., TOWLER, W. F.: Arch. Path. **95**, 172 (1973).
BRANDTZAEG, P.: J. Immunol. **112**, 1553 (1974).
BRANDTZAEG, P., FJELLANGER, I., GJERULDSEN, S. T.: Scand. J. Haemat. Suppl. 12, 3 (1970).
BRANDTZAEG, P., HENSTEN-PETTERSEN, A.: Arch. oral Biol. **21**, 69 (1976).
BRETTEL, C.: Eckhard's Beitr. Anat. Physiol. (Giessen) **4**, 89 (1869).
BRUCKMAN, J. E., MANGOS, J. A.: Proc. Soc. exp. Biol. **134**, 606 (1970).

Brusilow, S. W., Cooke, R. E.: Amer. J. Physiol. **196**, 831 (1959).
Bundgaard, M., Møller, M., Poulsen, J. H.: J. Physiol. (Lond.) **273**, 339 (1977).
Burgen, A. S. V.: The secretion of potassium in saliva. J. Physiol. (Lond.) **132**, 20 (1956).
Burgen, A. S. V.: In: Salivary Glands and Their Secretions (L. M. Sreebny and J. Meyer, Eds). Oxford: Pergamon 1964, p. 197.
Burgen, A. S. V.: In: Handbook of Physiology, Section 6, Vol. 2 (C. F. Code, Ed.). Baltimore, Md: Williams and Wilkins 1967, p. 561.
Burgen, A. S. V., Emmelin, N. G.: Physiology of the Salivary Glands. London: Arnold 1961.
Burgen, A. S. V., Seeman, P.: Canad. J. Biochem. Physiol. **36**, 119 (1958).
Burgen, A. S. V., Terroux, K. G.: J. Physiol. (Lond.) **161**, 399 (1962).
Burgen, A. S. V., Terroux, K. G.: J. Physiol. (Lond.) **169**, 663 (1963).
Butcher, F. R.: Metabolism **24**, 409 (1975).
Butcher, F. R., Goldman, R. H.: Biochem. biophys. Res. Commun. **48**, 23 (1972).
Butcher, F. R., Goldman, J. A., Nemerovski, M.: Biochim. biophys. Acta **392**, 82 (1975).
Caldwell, R. C., Pigman, W.: Arch. Biochem. Biophys. **110**, 91 (1965).
Camilli, P. de, Peluchetti, D., Meldolesi, J.: J. Cell Biol. **70**, 59 (1976).
Caramia, F.: J. Ultrastruct. Res. **16**, 505 (1966a).
Caramia, F.: J. Ultrastruct. Res. **16**, 524 (1966b).
Castle, J. D., Jamieson, J. D., Palade, G. E.: J. Cell Biol. **53**, 290 (1972).
Castle, J. D., Jamieson, J. D., Palade, G. E.: J. Cell Biol. **64**, 182 (1975).
Chauncey, H. H., Quintarelli, G.: Amer. J. Anat. **108**, 263 (1961).
Chauncey, H. H., Weiss, P. A.: Arch. int. Pharmacodyn. **113**, 377 (1958).
Chrétien, M.: C. R. Acad. Sci. (Paris) **274**, 3428 (1972a).
Chrétien, M.: J. Microscop. (Paris) **14**, 35 (1972b).
Chrétien, M.: J. Microscop. (Paris) **14**, 55 (1972c).
Chrétien, M., Zajdela, F.: C. R. Acad. Sci. (Paris) **260**, 4263 (1965).
Coats, D. A., Wright, R. D.: J. Physiol. (Lond.) **135**, 611 (1957).
Cohen, S.: J. biol. Chem. **237**, 1555 (1962).
Cohen, S., Elliott, G. A.: J. invest. Derm. **40**, 1 (1963).
Cohen, S., Taylor, J. M., Murakami, K., Michelakis, A. M., Inagami, T.: Biochemistry **11**, 4286 (1972).
Comoglio, P. M., Guglielmone, R.: Immunology **25**, 71 (1973).
Compton, J. S.: Hons. Thesis, University of Sydney, 1976.
Compton, J. S., Young, J. A.: Proc. Aust. physiol. pharmacol. Soc. **7**, 143P (1976).
Compton, J. S., Young, J. A.: Clin. exp. Pharmacol. Physiol. **4**, 479 (1977).
Compton, J. S., Denniss, A. R., Hung, P., Kennerson, A. R., van Lennep, E. W., Young, J. A.: Proc. Aust. physiol. pharmacol. Soc. **6**, 132 (1975).
Compton, J., Nelson, J., Wright, R. D., Young, J. A.: J. Physiol. (Lond.), in press, 1979.
Cook, D. I., Young, J. A.: Proc. Aust. physiol. pharmacol. Soc. **8**, 3P (1977).
Coroneo, M. T.: Hons. Thesis, University of Sydney, 1976.
Cowley, L. H., Shackleford, J. M.: J. Morph. **132**, 117 (1970).
Creed, K. E.: In: Secretory Mechanisms of Exocrine Glands (N. A. Thorn and O. H. Petersen, Eds). Copenhagen: Munksgaard 1974, p. 254.
Creed, K. E., McDonald, I. R.: Comp. gen. Pharmacol. **1**, 285 (1970).
Creed, K. E., McDonald, I. R.: J. Physiol. (Lond.) **231**, 327 (1973).
Creed, K. E., Wilson, J. A. F.: Aust. J. exp. Biol. med. Sci. **47**, 135 (1969).
Cutler, L. S., Chaudhry, A. P.: Anat. Rec. **176**, 405 (1973).
Dawes, C.: Arch. oral Biol. **14**, 277 (1969).
Dawes, C.: Arch. oral. Biol. **19**, 887 (1974).
Denniss, A. R., Young, J. A.: Pflügers Arch. **357**, 77 (1975).
Denniss, A. R., Young, J. A.: Pflügers Arch. **376**, 73 (1978).
Denniss, A. R., Schneyer, L. H., Sucanthapree, C., Young, J. A.: Amer. J. Physiol. **235**, F 548 (1978).
Diamond, J. M., Bossert, W. H.: J. gen. Physiol. **50**, 2061 (1967).
Dische, Z., Kahn, N., Rothschild, C., Danilchenko, A., Licking, J., Wang, S. C.: J. Neurochem. **17**, 649 (1970).
Dorey, G., Bhoola, K. D.: II. Z. Zellforsch. **126**, 335 (1972).

DOUGLAS, W. W.: In: Secretory Mechanisms of Exocrine Glands (N. A. Thorn and O. H. Petersen, Eds). Copenhagen: Munksgaard 1974, p. 116.
DOUGLAS, W. W., POISNER, A. M.: J. Physiol. (Lond.) **165**, 528 (1963).
DREISBACH, R. H.: Amer. J. Physiol. **196**, 645 (1959).
DREISBACH, R. H.: J. dent. Res. **39**, 1133 (1960).
DROZ, B., PISAM, M., CRÉTIEN, M.: Acta endocr. (Kbh.), Suppl. 180, 13 (1973).
D'SILVA, J. L., FERGUSON, D. B.: Arch. oral Biol. **7**, 563 (1962).
DURHAM, J. P., BUTCHER, F. R.: FEBS Letters **47**, 218 (1974).
EHINGER, B., GARRETT, J. R., OHLIN, P.: Experientia (Basel) **23**, 924 (1967).
ELLISON, S. A.: In: Handbook of Physiology, Section 6, Vol. 2, (C. F. Code, Ed.). Baltimore, Md: Williams and Wilkins 1967, p. 531.
ELLISON, S. A., MASHIMO, P. A., MANDEL, I. D.: J. dent. Res. **39**, 892 (1960).
EMMELIN, N.: Acta physiol. scand. **34**, 11 (1955).
EMMELIN, N.: In: Handbook of Physiology, Section 6, Vol. 2 (C. F. Code, Ed.). Baltimore, Md: Williams and Wilkins 1967, p. 665.
EMMELIN, N., ENGSTRÖM, J.: J. Physiol. (Lond.) **153**, 1 (1960).
EMMELIN, N., GJÖRSTRUP, P.: J. Physiol. (Lond.) **230**, 185 (1973).
EMMELIN, N., GJÖRSTRUP, P.: In: Secretory Mechanisms of Exocrine Glands (N. A. Thorn and O. H. Petersen, Eds). Copenhagen: Munksgaard 1974, p. 29.
EMMELIN, N., STRÖMBLAD, R.: Acta physiol. scand. **31**, Suppl 114, 12 (1954).
EMMELIN, N., MUREN, A., STRÖMBLAD, R.: Acta physiol. scand. **32**, 325 (1954).
EMMELIN, N., HOLMBERG, J., OHLIN, P.: Brit. J. Pharmacol. **25**, 134 (1965).
EMMELIN, N., GARRETT, J. R., OHLIN, P.: J. Physiol. (Lond.) **200**, 539 (1969a).
EMMELIN, N., OHLIN, P., THULIN, A.: Brit. J. Pharmacol. **37**, 666 (1969b).
EMMELIN, N., SCHNEYER, C. A., SCHNEYER, L. H.: In: Pharmacology of Gastrointestinal Motility and Secretion. International Encyclopedia of Pharmacology and Therapeutics, Section 39A, Vol. 1 (P. Holton, Ed.). Oxford: Pergamon 1973, p 1.
EMMELIN, N., GARRETT, J. R., OHLIN, P.: Arch. oral. Biol. **19**, 275 (1974).
ENDRE, Z. H.: Hons. Thesis, University of Sydney, 1972.
ENEROTH, C.-M., HÖKFELT, T., NORBERG, K.-A.: Acta oto-laryng. (Stockh.) **68**, 369 (1969).
ERSPAMER, V., ERSPAMER, G. F.: Brit. J. Pharmacol. **19**, 337 (1962).
FAHRENHOLZ, C.: In: Handbuch der vergleichenden Anatomie der Wirbeltiere, Vol. 3 (new priting) (L. Bolk, E. Göppert, E. Kallius, and N. Lubosch, Eds). Amsterdam: Asher 1967, p. 115.
FEINSTEIN, H., SCHRAMM, M.: Europ. J. Biochem. **13**, 158 (1970).
FIELD, M. J.: Hons. Thesis, University of Sydney 1972.
FIELD, M. J., YOUNG, J. A.: Pflügers Arch. **345**, 207 (1973).
FRASER, P. J.: Brit. J. Pharmacol. **12**, 47 (1957).
FRIEDMANN, I., HODGES, G. M.: Acta oto-laryng. (Stockh.) **79**, 197 (1975).
FRITZ, M. E.: Proc. Soc. exp. Biol. **137**, 776 (1971).
FRITZ, M. E.: In: Secretory Mechanisms of Exocrine Glands (N. A. Thorn and O. H. Petersen, Eds). Copenhagen: Munksgaard 1974, p. 230.
FRITZ, M. E., BOTELHO, S. Y.: Amer. J. Physiol. **216**, 1180 (1969a).
FRITZ, M. E., BOTELHO, S. Y.: Amer. J. Physiol. **216**, 1392 (1969b).
FRÖMTER, E.: J. Membrane Biol. **8**, 259 (1972).
FRÖMTER, E., DIAMOND, J.: Nature New Biol. **235**, 9 (1972).
FRÖMTER, E., MÜLLER, C. W., WICK, T.: In: Electrophysiology of Epithelial Cells (G. Giebisch, Ed.). Stuttgart: Schattauer 1971, p. 119.
FRÖMTER, E., GEBLER, B., SCHOPOW, K., POCKRANDT-HEMSTEDT, H.: In: Secretory Mechanisms of Exocrine Glands (N. A. Thorn and O. H. Petersen, Eds). Copenhagen: Munksgaard 1974, p. 496.
GARRETT, J. R.: J. roy. micr. Soc. **85**, 135 (1966a).
GARRETT, J. R.: J. roy. micr. Soc. **85**, 149 (1966b).
GARRETT, J. R.: Arch. oral Biol. **12**, 1417 (1967).
GARRETT, J. R.: In: Oral Physiology (N. Emmelin and Y. Zotterman, Eds). Oxford: Pergamon 1972, p. 83.
GARRETT, J. R.: In: Secretory Mechanisms of Exocrine Glands (N. A. Thorn and O. H. Petersen, Eds). Copenhagen: Munksgaard 1974, p. 17.

GARRETT, J. R.: Brit. med. Bull. **31**, 152 (1975).
GARRETT, J. R.: In: Scientific Foundations of Dentistry, Section X. (B. Cohen and I. Kramer, Eds). London: Heinemann 1976, p. 499.
GARRETT, J. R., PARSONS, P. A.: J. Physiol. (Lond.) **237**, 3P (1974).
GARRETT, J. R., PARSONS, P. A.: In: Secretory Mechanisms of Exocrine Glands (N. A. Thorn and O. H. Petersen, Eds). Copenhagen: Munksgaard 1974, p. 487.
GARRETT, J. R., THULIN, A.: Cell Tiss. Res. **159**, 179 (1975).
GEORGE, W. J., WILKERSON, R. D., KODOWITZ, P. J.: J. Pharmacol. exp. Ther. **184**, 228 (1973).
GOLDMAN, R. D., BERG, G., BUSHNELL, A., CHENG-MING, C., DICKERMAN, L., HOPKINS, N., MILLER, M. L., POLLACK, R., WANG, E.: In: Locomotion of Tissue Cells (R. Porter and D. W. Fitzsimons, Eds). Amsterdam: Associated Scientific Publishers 1973, p. 83.
GRANT, S. B. J., ENDRE, Z. H., YOUNG, J. A.: Proc. Aust. physiol. pharmacol. Soc. **5**, 206 (1974).
GREGERSEN, M. I., INGALLS, E. N.: Amer. J. Physiol. **98**, 441 (1931).
GRUBER, W. D., KNAUF, H., FRÖMTER, E.: Pflügers Arch. **344**, 33 (1973).
GUIDOTTI, A., WEISS, B., COSTA, E.: Molec. Pharmacol. **8**, 521 (1972).
HAND, A. R.: J. Cell Biol. **47**, 540 (1970).
HAND, A. R.: Anat. Rec. **173**, 131 (1972a).
HAND, A. R.: Amer. J. Anat. **135**, 71 (1972b).
HARFIELD, B., TENENHOUSE, A.: Canad. J. Physiol. Pharmacol. **51**, 997 (1973).
HARRISON, J. D., GARRETT, J. R.: J. oral Path. **4**, 297 (1975).
HARRISON, J. D., GARRETT, J. R.: J. Path. **118**, 245 (1976a).
HARRISON, J. D., GARRETT, J. R.: Histochem. J. **8**, 35 (1976b).
HEALY, J. K., FRASER, P. A., YOUNG, J. A.: Pflügers Arch. **363**, 69 (1976).
HEBB, C. O., STAVRAKY, G. W.: Quart. J. exp. Physiol. **26**, 141 (1936).
HEGEL, U., FRÖMTER, E., WICK, T.: Pflügers Arch. **294**, 274 (1967).
HEIDENHAIN, R.: Pflügers Arch. **17**, 1 (1878).
HENRIKSEN, A.: Z. Säugetierkunde **37**, 185 (1972).
HENRIQUES, B. L., SPERLING, A. L.: J. appl. Physiol. **21**, 1247 (1966).
HILDES, J. A., FERGUSON, M. H.: Canad. J. Biochem. Physiol. **33**, 217 (1955).
HODGKIN, A. L., HOROWICZ, P.: J. Physiol. (Lond.) **145**, 405 (1959).
HOLZGREVE, H., MARTINEZ, J. R., VOGEL, A.: Pflügers Arch. **290**, 134 (1966).
HURLIMANN, J., ZUBER, C.: Immunology **14**, 809 (1968).
ICHIKAWA, A.: J. Cell Biol. **24**, 369 (1965).
ICHIOKA, M., SHIMIZU, H., SHIMIZU, T.: Z. Biol. **115**, 1 (1965).
IGNARRO, L. J.: J. Immunol. **112**, 210 (1974).
IMAI, Y.: J. physiol. Soc. Jap. **27**, 304 (1965a).
IMAI, Y.: J. physiol. Soc. Jap. **27**, 313 (1965b).
IMAI, Y.: In: Secretory Mechanisms of Exocrine Glands (N. A. Thorn and O. H. Petersen, Eds). Copenhagen: Munksgaard 1974, p. 199.
IMAI, Y., SUEKI, M., YOSHIMURA, H.: J. physiol. Soc. Jap. **32**, 35 (1970).
ITO, Y.: Ann. N.Y. Acad. Sci. **85**, 228 (1960).
JACOBY, F., LEESON, C. R.: J. Anat. (Lond.) **93**, 201 (1959).
JAMIESON, J. D., PALADE, G. E.: J. Cell Biol. **34**, 577 (1967a).
JAMIESON, J. D., PALADE, G. E.: J. Cell Biol. **34**, 597 (1967b).
JAMIESON, J. D., PALADE, G. E.: J. Cell Biol. **39**, 589 (1968).
JAMIESON, J. D., PALADE, G. E.: J. Cell Biol. **48**, 503 (1971).
JUNQUEIRA, L. C. U., FAVA-DE-MORAES, F.: In: Funktionelle und morphologische Organization der Zelle, Vol. 2 (K. E. Wohlfarth-Bottermann, Ed.). Berlin-Heidelberg-New York: Springer 1965, p. 36.
JUNQUEIRA, L. C. U., TOLEDO, A. M. S., FERRI, R. G.: Arch. oral Biol. **10**, 863 (1965).
KAGAYAMA, M., NISHIYAMA, A.: Tohoku J. exp. Med. **108**, 179 (1972).
KAGAYAMA, M., NISHIYAMA, A.: J. Physiol. (Lond.) **242**, 157 (1974).
KAHN, N., MANDEL, I., LICKING, J., WASSERMANN, A., MOREA, D.: Proc. Soc. exp. Biol. **130**, 314 (1969).
KAISER, D., PIVETTA, O., RENNERT, O. M.: Life Sci. **15**, 803 (1974a).
KAISER, D., SCHÖNI, M., DRACK, E.: Helv. paediat. Acta **29**, 145 (1974b).
KALADELFOS, G.: Hons. Thesis, University of Sydney, 1971.

KALADELFOS, G., YOUNG, J. A.: Pflügers Arch. **341**, 143 (1973).
KALADELFOS, G., YOUNG, J. A.: Aust. J. exp. Biol. med. Sci. **52**, 67 (1974).
KARLER, R., WOODBURY, D. M.: Biochem. J. **75**, 538 (1960).
KATZ, B., MILEDI, R.: Proc. roy. Soc. B. **161**, 483 (1965).
KATZ, R. L., MANDEL, I. D.: Proc. Soc. exp. Biol. **128**, 1140 (1968).
KAY, R. N. B.: J. Physiol. (Lond.) **150**, 515 (1960).
KIM, S. K.: J. supramolec. Struct. **4**, 185 (1976).
KIM, S. K., HAN, S. S.: Amer. J. Anat. **144**, 467 (1975).
KIM, S. K., NASJLETI, C. E., HAN, S. S.: J. Ultrastruct. Res. **38**, 371 (1972).
KNAUF, H.: Pflügers Arch. **333**, 82 (1972a).
KNAUF, H.: Pflügers Arch. **333**, 326 (1972b).
KNAUF, H.: In: Secretory Mechanisms of Exocrine Glands (N. A. Thorn and O. H. Petersen, Eds). Copenhagen: Munksgaard 1974, p. 535.
KNAUF, H., FRÖMTER, E.: Pflügers Arch. **316**, 213 (1970a).
KNAUF, H., FRÖMTER, E.: Pflügers Arch. **316**, 238 (1970b).
KNAUF, H., FRÖMTER, E.: Pflügers Arch. **316**, 259 (1970c).
KNAUF, H., FRÖMTER, E.: In: Electrophysiology of Epithelial Cells (G. Giebisch, Ed.). Stuttgart: Schattauer 1971, p. 187.
KNAUF, H., LÜBCKE, R.: Pflügers Arch. **361**, 55 (1975).
KNAUF, H., GEBLER, B., MARTIN, C. J., YOUNG, J. A.: Pflügers Arch. **335**, R60 (1972).
KNAUF, H., RÖTTGER, P., WAIS, U., BAUMANN, K.: Fortschr. Zool. **23**, 307 (1975).
KNAUF, H., SIMON, B., WAIS, U.: Naunyn-Schmiedebergs Arch. exp. Path. Pharmak. **292**, 189 (1976a).
KNAUF, H., WAIS, U., LÜBCKE, R., ALBIEZ, G.: Europ. J. clin. Invest. **6**, 43 (1976b).
KNAUF, H., PANDER, K., FRANCK, S.: Europ. J. clin. Invest. **6**, 17 (1976c).
KÖSTLIN, A., RAUCH, S.: Helv. med. Acta **24**, 600 (1957).
KRAEHENBUHL, J. P., de GRANDI, P. B., CAMPICHE, M. A.: J. Cell Biol. **50**, 432 (1971).
KRAUS, F. W., MESTECKY, J.: Arch. oral Biol. **16**, 781 (1971).
KREUSSER, W.: Inaugural Dissertation, University of Würzburg, 1974.
KREUSSER, W., HEIDLAND, A., HENNEMANN, H., WIGAND, M. E., KNAUF, H.: Europ. J. clin. Invest. **2**, 398 (1972).
KREUSSER, W., PINNOW, E., HENNEMANN, H., HEIDLAND, A.: Verh. dtsch. Ges. inn. Med. **80**, 204 (1974).
KUIJPER-LENSTRA, A. H., KRAMER, M. F.: Cell Tissue Res. **164**, 457 (1975).
KUIJPER-LENSTRA, A. H., KRAMER, M. F., VAN VENROOIJ, W. J.: Cell Tissue Res. **164**, 447 (1975).
KURTZ, S. M. In: Electron Microscopic Anatomy (S. M. Kurtz, Ed.). New York: Academic 1964, p. 97.
LACASSAGNE, A.: C. R. Soc. Biol. (Paris) **133**, 180 (1940a).
LACASSAGNE, A.: C. R. Soc. Biol. (Paris) **133**, 539 (1940b).
LACY, P. E., HOWELL, S. L., YOUNG, D. A., FINK, C. J.: Nature (Lond.) **219**, 1177 (1968).
LANGLEY, J. N.: Untersuch. Physiol. Inst. Heidelberg **1**, 476 (1878).
LANGSTROTH, G. O., MCRAE, D. R., STAVRAKY, G. W.: Arch. int. Pharmacodyn. **58**, 61 (1938).
LAUGESEN, L. P., NIELSEN, J. O. D., POULSEN, J. H.: Pflügers Arch. **364**, 167 (1976).
LAULER, D. P., HICKLER, R. B., THORN, G. W.: New Engl. J. Med. **267**, 1136 (1962).
LEDER, O., TRITSCHLER, P.: Pflügers Arch. **292**, 229 (1966).
LEE, T.-P., KUO, J. F., GREENGARD, P.: Proc. nat. Acad. Sci. (Wash.) **69**, 3287 (1972).
LEEMAN, S. E., HAMMERSCHLAG, R.: Endocrinology **81**, 803 (1967).
LEESON, C. R.: In: Handbook of Physiology, Section 6, Vol. 2 (C. F. Code, Ed.). Baltimore, Md: Williams and Wilkins 1967, p. 463.
LEESON, C. R., JACOBY, F.: J. Anat. (Lond.) **93**, 287 (1959).
LEMBECK, F., OBERDORF, A., STARKE, K., HETTICH, R.: Naunyn-Schmiedebergs Arch. exp. Path. Pharmak. **261**, 338 (1968).
LENNEP, E. W. VAN, KENNERSON, A. R., COMPTON, J. S.: Cell Tiss. Res. **179**, 377 (1977).
LEVI-MONTALCINI, R., ANGELETTI, P. U.: Quart. Rev. Biol. **36**, 99 (1961).
LEVI-MONTALCINI, R., ANGELETTI, P. U.: In: Salivary Glands and Their Secretions (L. M. Sreebny and J. Meyer, Eds). Oxford: Pergamon 1964, p. 129.
LEVI-MONTALCINI, R., COHEN, S.: Ann. N.Y. Acad. Sci. **85**, 324 (1960).

LING, G., GERARD, R. W.: J. cell. comp. Physiol. **34**, 383 (1949).
LUDWIG, C.: Mittheilung Nr. 50 der Zürch. Naturf. Ges. (1851).
LUNDBERG, A.: Acta physiol. scand. **35**, 1 (1955).
LUNDBERG, A.: Acta physiol. scand. **40**, 21 (1957a).
LUNDBERG, A.: Acta physiol. scand. **40**, 35 (1957b).
LUNDBERG, A.: Acta physiol. scand. **40**, 101 (1957c).
LUNDBERG, A.: Physiol. Rev. **38**, 21 (1958).
MALAMUD, D.: Biochim. biophys. Acta **279**, 373 (1972).
MAMELAK, M., WISSIG, S. L., BOGOROCH, R., EDELMAN, I. S.: J. Membrane Biol. **1**, 144 (1969).
MANDEL, I. D., KHURANDA, H. S.: Arch. oral Biol. **14**, 1433 (1969).
MANGOS, J. A., BRAUN, G.: Pflügers Arch. **290**, 184 (1966).
MANGOS, J. A., MCSHERRY, N. R.: Pediat. Res. **2**, 378 (1968).
MANGOS, J. A., MCSHERRY, N. R.: Proc. Soc. exp. Biol. **132**, 797 (1969).
MANGOS, J. A., MCSHERRY, N. R.: Amer. J. Physiol. **218**, 1329 (1970).
MANGOS, J. A., MCSHERRY, N. R.: Amer. J. Physiol. **221**, 496 (1971).
MANGOS, J. A., BRAUN, G., HAMANN, K. F.: Pflügers Arch. **291**, 99 (1966).
MANGOS, J. A., BENKE, P. J., MCSHERRY, N.: Pediat. Res. **3**, 562 (1969a).
MANGOS, J. A., MCSHERRY, N. R., BENKE, P. J., SPOCK, A.: Proceedings of the Fifth International Cystic Fibrosis Conference (LAWSON, D., Ed.). Cystic Fibrosis Research Trust 1969b, p. 25.
MANGOS, J. A., MARAGOS, N., MCSHERRY, N. R.: Amer. J. Physiol. **224**, 1260 (1973a).
MANGOS, J. A., MCSHERRY, N. R., NOUSIA-ARVANITAKIS, S., IRWIN, K.: Amer. J. Physiol. **225**, 18 (1973b).
MANGOS, J. A., MCSHERRY, N. R., IRWIN, K., HONG, R.: Amer. J. Physiol. **225**, 450 (1973c).
MANGOS, J. A., MCSHERRY, N. R., ARVANITAKIS, S. N.: Amer. J. Physiol. **225**, 683 (1973d).
MAREN, T. H.: Physiol. Rev. **47**, 595 (1967).
MARTIN, C. J., YOUNG, J. A.: Pflügers Arch. **324**, 344 (1971a).
MARTIN, C. J., YOUNG, J. A.: Pflügers Arch. **327**, 303 (1971b).
MARTIN, C. J., FRÖMTER, E., GEBLER, B., KNAUF, H., YOUNG, J. A.: Pflügers Arch. **341**, 131 (1973).
MARTIN, C. J., DENNISS, A. R., ENDRE, Z. H., YOUNG, J. A.: In: Secretory Mechanisms of Exocrine Glands (N. A. Thorn and O. H. Petersen, Eds). Copenhagen: Munksgaard 1974, p. 549.
MARTIN, K., BURGEN, A. S. V.: J. gen. Physiol. **46**, 225 (1962).
MARTINEZ, J, R.: J. Pharmacol. exp. Ther. **178**, 616 (1971).
MARTINEZ, J. R., MARTINEZ, A. M.: Europ. J. Pharmacol. **18**, 375 (1972a).
MARTINEZ, J. R., MARTINEZ, A. M.: Europ. J. Pharmacol. **18**, 386 (1972b).
MARTINEZ, J. R., MARTINEZ, A. M.: Arch. oral Biol. **19**, 57 (1974).
MARTINEZ, J. R., QUISSELL, D. O.: J. Pharmacol. exp. Ther. **199**, 518 (1976).
MARTINEZ, J. R., HOLZGREVE, H., FRICK, A.: Pflügers Arch. **290**, 124 (1966).
MARTINEZ, J. R., QUISSELL, D. O., WOOD, D. L., GILES, M.: J. Pharmacol. exp. Ther. **194**, 384 (1975a).
MARTINEZ, J. R., ADSHEAD, P. C., QUISSELL, D. O., BARBEDO, G. J.: Pediat. Res. **9**, 470 (1975b).
MARTINEZ, J. R., QUISSELL, D. O., GILES, M.: J. Pharmacol. exp. Ther. **198**, 385 (1976).
MASSON, P. L., CARBONARA, A. O., HEREMANS, J. F.: Biochim. biophys. Acta **107**, 485 (1965).
MATERAZZI, G., VITAIOLI, L.: J. Anat. (Lond.) **105**, 163 (1969).
MATTHEWS, E. K.: In: Secretory Mechanisms of Exocrine Glands (N. A. Thorn and O. H. Petersen, Eds). Copenhagen: Munksgaard 1974, p. 185.
MATTHEWS, R. W.: Arch. oral Biol. **19**, 985 (1974a).
MATTHEWS, R. W.: Arch. oral Biol. **19**, 989 (1974b).
MAYO, J. W., CARLSON, D. M.: Arch. Biochem. Biophys. **161**, 134 (1974).
MCDONALD, G. J., LOUIS, W. J., RENZINI, V., BOYD, G. W., PEART, W. S.: Circulat. Res. **27**, 197 (1970).
MCKEAN, F. J., BEELEY, J. A.: Arch. oral Biol. **19**, 103 (1974).
MENAKER, L., SHEETZ, J. H., COBB, C. M., NAVIA, J. M.: Lab. Invest. **30**, 341 (1974).
NAUGHTON, M. A., KOCH, J., HOFFMAN, H., BENDER, V., HAGOPIAN, H., HAMILTON, E.: Exp. Cell Res. **57**, 95 (1969).
NEUTRA, M., LEBLOND, C. P.: J. Cell Biol. **30**, 119 (1966a).
NEUTRA, M., LEBLOND, C. P.: J. Cell Biol. **30**, 137 (1966b).

NIEDERMEIER, W., STONE, R. E., DREIZEN, S., SPIES, T. D.: Proc. Soc. exp. Biol. **88**, 273 (1955).
NIELSEN, S. P., PETERSEN, O. H.: Pflügers Arch. **318**, 63 (1970).
NIELSEN, S. P., PETERSEN, O. H.: J. Physiol. (Lond.) **223**, 685 (1972).
NISHIYAMA, A., PETERSEN, O. H.: J. Physiol. (Lond.) **242**, 173 (1974).
NISHIYAMA, A., PETERSEN, O. H.: J. Physiol. (Lond.) **244**, 431 (1975).
NORBERG, K.-A., OLSON, L.: Z. Zellforsch. **68**, 183 (1965).
NORDENFELT, I., OHLIN, P.: Acta physiol. scand. **41**, 12 (1957).
OHLIN, P.: Acta Univ. Lund. **2**, 7: 1 (1966).
OHLIN, P., PEREC, C.: Experientia (Basel) **21**, 408 (1965).
OPPENHEIM, F. G. Helv. odont. Acta **14**, 10 (1970).
OSCHMAN, J. L., BERRIDGE, M. J.: Tissue and Cell **2**, 281 (1970).
PALADE, G.: Science **189**, 347 (1975).
PATTERSON, J., TITCHEN, D. A.: J. Physiol. (Lond.) **252**, 58P (1975).
PATTON, J. R., PIGMAN, W.: J. Amer. chem. Soc. **81**, 3035 (1959).
PEARSON, O. P.: J. Mammal. **23**, 159 (1942).
PEARSON, O. P.: Anat. Rec. **107**, 161 (1950).
PEDERSEN, G. L., PETERSEN, O. H.: J. Physiol. (Lond.) **234**, 217 (1973).
PETERSEN, O. H.: J. Physiol. (Lond.) **208**, 431 (1970a).
PETERSEN, O. H.: J. Physiol. (Lond.) **210**, 205 (1970b).
PETERSEN, O. H.: Acta physiol. scand. **80**, 117 (1970c).
PETERSEN, O. H.: Pflügers Arch. **323**, 91 (1971a).
PETERSEN, O. H.: J. Physiol. (Lond.) **216**, 129 (1971b).
PETERSEN, O. H.: Phil. Trans. B **262**, 307 (1971c).
PETERSEN, O. H.: In: Electrophysiology of Epithelial Cells (G. Giebisch, Ed.). Stuttgart: Schattauer 1971d, p. 207.
PETERSEN, O. H.: Acta physiol. scand. Suppl. **381**, 1 (1972).
PETERSEN, O. H.: In: Transport Mechanisms in Epithelia (H. H. Ussing and N. A. Thorn, Eds). Copenhagen: Munksgaard 1973, p. 478.
PETERSEN, O. H.: Experientia (Basel) **30**, 130 (1974a).
PETERSEN, O. H.: Experientia (Basel) **30**, 1105 (1974b).
PETERSEN, O. H.: Proc. Aust. physiol. pharmacol. Soc. **6**, 1 (1975).
PETERSEN, O. H.: In: Stimulus-Secretion Coupling in the Gastrointestinal Tract (R. M. Case and H. Goebell, Eds). Lancaster: MTP Press 1976a, p. 281.
PETERSEN, O. H.: Experientia (Basel) **32**, 471 (1976b).
PETERSEN, O. H.: Physiol. Rev. **56**, 535 (1976c).
PETERSEN, O. H., PEDERSEN, G. L.: J. Membrane Biol. **16**, 353 (1974).
PETERSEN, O. H., POULSEN, J. H.: Experientia (Basel) **22**, 821 (1966).
PETERSEN, O. H., POULSEN, J. H.: Acta. physiol. scand. **70**, 158 (1967a).
PETERSEN, O. H., POULSEN, J. H.: Acta. physiol. scand. **70**, 293 (1967b).
PETERSEN, O. H.: POULSEN, J. H.: Acta physiol. scand. **73**, 93 (1968).
PETERSEN, O. H., POULSEN J. H.: In: Exocrine Glands (S. Y. Botelho, F. P. Brooks, and W. B. Shelley, Eds). Philadelphia: University of Pennsylvania Press 1969, p. 3.
PETERSEN, O. H., UEDA, N.: J. Physiol. (Lond.) **254**, 38P (1976a).
PETERSEN, O. H., UEDA, N.: J. Physiol. (Lond.) **254**, 583 (1976b).
PORTER, K. R.: In: Locomotion of Tissue Cells (R. Porter and D. W. Fitzsimons, Eds). Amsterdam: Associated Scientific Publishers 1973, p. 149.
POULSEN, J. H.: Pflügers Arch. **338**, 201 (1973a).
POULSEN, J. H.: Acta physiol. scand. **87**, 51A (1973b).
POULSEN, J. H.: Pflügers Arch. **349**, 215 (1974).
PRINCE, W. T., RASMUSSEN, H., BERRIDGE, M. J.: Biochim. biophys. Acta **329**, 98 (1973).
RAMBOURG, A., MARRAUD, A., CHRÉTIEN, M.: J. Micr. (Oxf.) **97**, 49 (1973).
RAMBOURG, A., CLERMONT, Y., MARRAUD, A.: Amer. J. Anat. **140**, 27 (1974).
RASMUSSEN, H.: Science **170**, 404 (1970).
RASMUSSEN, H., TENENHOUSE, A.: Proc. nat. Acad. Sci. (Wash.) **59**, 1364 (1968).
RAYNAUD, J.: Bull. biol. Franc. Belg. **94**, 400 (1960).
ROBBERECHT, P.: In: Stimulus-Secretion Coupling in the Gastrointestinal Tract (R. M. Case and H. Goebell, Eds). Lancaster: MTP Press 1976, p. 203.

ROBISON, G. A., BUTCHER, R. W., SUTHERLAND, E. W.: Cyclic AMP. New York, London: Academic 1971.
RODRIGUEZ, M. L.: Hons. Thesis, University of Sydney 1977.
ROSSEN, R. D., MORGAN, C., HSU, K. C., BUTLER, W. T., ROSE, H. M.: J. Immunol. **100**, 706 (1968).
ROTHMAN, S. S.: Science **190**, 747 (1975).
RUBIN, R. P.: Pharmacol. Rev. **22**, 389 (1970).
SALOMON, Y., SCHRAMM, M.: Biochem. biophys. Res. Commun. **38**, 106 (1970).
SAMUELS, P. J. A., YOUNG, J. A.: (in preparation).
SATIR, B., SCHOOLEY, C., SATIR, P.: J. Cell Biol. **56**, 153 (1973).
SCHAFFER, J.: In: Handbuch der mikroskopischen Anatomie des Menschen, Vol. 2, Pt 1 W. (von Möllendorff, Ed.). Berlin: Springer 1927, p. 1.
SCHNEYER, C. A.: Amer. J. Physiol. **203**, 232 (1962).
SCHNEYER, C. A.: In: Regulation of Organ and Tissue Growth (R. Goss, Ed.). New York, London: Academic 1972, p. 211.
SCHNEYER, C. A., HALL, H. D.: Amer. J. Physiol. **209**, 484 (1965).
SCHNEYER, C. A., HALL, H. D.: Proc. Soc. exp. Biol. **127**, 1245 (1968a).
SCHNEYER, C. A., HALL, H. D.: Amer. J. Physiol. **214**, 808 (1968b).
SCHNEYER, C. A., SCHNEYER, L. H.: Amer. J. Physiol. **201**, 939 (1961).
SCHNEYER, L. H.: In: Secretory Mechanisms of Salivary Glands (L. H. Schneyer and C. A. Schneyer, Eds). New York, London: Academic 1967, p. 32.
SCHNEYER, L. H.: Amer. J. Physiol. **215**, 664 (1968).
SCHNEYER, L. H.: Amer. J. Physiol. **217**, 1324 (1969).
SCHNEYER, L. H.: Amer. J. Physiol. **219**, 1050 (1970).
SCHNEYER, L. H.: Amer. J. Physiol. **226**, 821 (1974a).
SCHNEYER, L. H.: Amer. J. Physiol. **227**, 606 (1974b).
SCHNEYER, L. H.: Amer. J. Physiol. **229**, 1056 (1975).
SCHNEYER, L. H.: Amer. J. Physiol. **230**, 341 (1976a).
SCHNEYER, L. H.: Fed. Proc. **35**, 764 (1976b).
SCHNEYER, L. H., FLATLAND, R. F.: J. appl. Physiol. **39**, 519 (1975).
SCHNEYER, L. H., SCHNEYER, C. A.: Amer. J. Physiol. **199**, 649 (1960).
SCHNEYER, L. H., SCHNEYER, C. A.: Proc. Soc. exp. Biol. **116**, 813 (1964).
SCHNEYER, L. H., SCHNEYER, C. A.: Amer. J. Physiol. **209**, 1304 (1965a).
SCHNEYER, L. H., SCHNEYER, C. A.: Amer. J. Physiol. **209**, 111 (1965b).
SCHNEYER, L. H., SCHNEYER, C. A.: In: Handbook of Physiology, Section 6, Vol. 2. (C. F. CODE, Ed.). Baltimore, Md: Williams and Wilkins 1967, p. 497.
SCHNEYER, L. H., THAVORNTHON, T.: Amer. J. Physiol. **224**, 136 (1973).
SCHNEYER, L. H., YOSHIDA, Y.: Proc. Soc. exp. Biol. **130**, 192 (1969).
SCHNEYER, L. H., SCHNEYER, C. A., YOSHIDA, Y.: Amer. J. Physiol. **215**, 1146 (1968).
SCHNEYER, L. H., YOUNG, J. A., SCHNEYER, C. A.: Physiol. Rev. **52**, 720 (1972).
SCHRAMM, M., NAIM, E.: J. biol. Chem. **245**, 3225 (1970).
SCHRAMM, M., SELINGER, Z.: In: Advances in Cytopharmacology, Vol. 2, (B. Ceccarelli, F. Clementi and J. Meldolesi, Eds). New York: Raven 1974, p. 29.
SCHRAMM, M., SELINGER, Z.: In: Stimulus-Secretion Coupling in the Gastrointestinal Tract (R. M. Case and H. Goebell, Eds). Lancaster: MTP Press 1976, p. 49.
SCHRAMM, M., SELINGER, Z., SALOMON, Y., EYTAN, E., BATZRI, S.: Nature New Biol. **240**, 203 (1972).
SCHULTZ, G., HARDMAN, J. G., SCHULTZ, K., BAIRD, C. E., SUTHERLAND, E. W.: Proc. nat. Acad. Sci (Wash.) **70**, 3889 (1973).
SCHULTZ, S. G.: In: Transport Mechanisms in Epithelia (H. H. Ussing and N. A. Thorn, Eds). Copenhagen: Munksgaard 1973, p. 281.
SCHWARTZ, A., MOORE, C. A.: Amer. J. Physiol. **214**, 1163 (1968).
SCHWARTZ, A., LASETER, A. H., KRAINTZ, L.: J. cell. comp. Physiol. **62**, 193 (1963).
SELINGER, Z., NAIM, E.: Biochim. biophys. Acta **203**, 335 (1970).
SELINGER, Z., NAIM, E., LASSER, M.: Biochim. biophys. Acta **203**, 326 (1970).
SELINGER, Z., BATZRI, S., EIMERL, S., SCHRAMM, M.: J. biol. Chem. **248**, 369 (1973).
SELINGER, Z., EIMERL, S., SCHRAMM, M.: Proc. nat. Acad. Sci. (Wash.) **71**, 128 (1974).
SEWELL, W., YOUNG, J. A.: Proc. Aust. physiol. pharmacol. Soc. **4**, 80 (1973).
SEWELL, W. A., YOUNG, J. A.: J. Physiol. (Lond.) **252**, 379 (1975).

SHACKLEFORD, J. M., SCHNEYER, L. H.: Anat. Rec. **169**, 679 (1971).
SIEGEL, I. A.: Arch. oral Biol. **17**, 1385 (1972).
SIEGEL, I. A.: Amer. J. Physiol. **231**, 974 (1976).
SIMON, B., KINNE, R., KNAUF, H.: Pflügers Arch. **337**, 177 (1972).
SLEGERS, J. F. G., MOONS, W. M., IDZERDA, P. P., STADHOUDERS, A. M.: J. Membrane Biol. **25**, 213 (1975).
SMAJE, L. H.: J. Physiol. (Lond.) **231**, 179 (1973).
SMAJE, L. H.: In: Secretory Mechanisms of Exocrine Glands (N. A. Thorn and O. H. Petersen Eds). Copenhagen: Munksgaard 1974, p. 608.
SMITH, R. J., FROMMER, J.: Arch. oral Biol. **17**, 1375 (1972a).
SMITH, R. J., FROMMER, J.: Arch. oral Biol. **17**, 1561 (1972b).
SMITH, R. J., FROMMER, J., SCHIFF, R.: J. Histochem. Cytochem. **19**, 310 (1971).
SOMMER, H. M., KAISER, D., DRACK, E.: Pflügers Arch. **355**, 353 (1975).
SREEBNY, L. M., MEYER, J.: Salivary Glands and Their Secretions. Oxford: Pergamon 1964.
TAMARIN, A., SREEBNY, L. M.: J. Morph. **117**, 295 (1965).
TANDLER, B.: In: Proceedings of Symposium on Salivary Glands and Their Secretions (N. H. Rowe, Ed.). Ann Arbor: Univ. Michigan 1972, p. 8.
TANDLER, B., MACCALLUM, D. K.: J. Anat. (Lond.) **117**, 117 (1974).
TANDLER, B., POULSEN, J. H.: J. Cell Biol. **68**, 775 (1976a).
TANDLER, B., POULSEN, J. H.: J. Morph. **149**, 183 (1976b).
TANDLER, B., DENNING, C. F., MANDEL, I. D., KUTSCHER, A. H.: J. Morph. **127**, 383 (1969).
TEMPLE, R., WILLIAMS, J. A., WILBUR, J. F., WOLFF, J.: Biochem. biophys. Res. Commun. **46**, 1454 (1972).
THAYSEN, J. H.: In: The Alkali Metal Ions in Biology. (Handbuch exp. Pharmakol. Bd. **13**) (H. H. Ussing, P. Kruhöffer, J. H. Thaysen, and N. A. Thorn, Eds). Berlin-Göttingen-Heidelberg: Springer 1960, p. 424.
THAYSEN, J. H., TARDING, F.: In: Secretory Mechanisms of Exocrine Glands (N. A. Thorn and O. H. Petersen, Eds). Copenhagen: Munksgaard 1974, p. 464.
THAYSEN, J. H., THORN, N. A., SCHWARTZ, I. L.: Amer. J. Physiol. **178**, 155 (1954).
ULLRICH, K. J.: In: Handbook of Physiology, Section 8. (J. ORLOFF and R. W. BERLINER, Eds). Baltimore, Md: Williams and Wilkins 1973, p. 377.
USSING, H. H., BIBER, T. U. L., BRICKER, N. S.: J. gen. Physiol. **48**, 425 (1965).
VENROOIJ, W. J. VAN, KUIJPER-LENSTRA, A. H., KRAMER, M. F.: Biochim. biophys. Acta **312**, 392 (1973).
VOGEL, J. J., NAUJOKS, R., BRUDEVOLD, F.: Arch. oral Biol. **10**, 523 (1965).
WAIS, U., KNAUF, H.: Pflügers Arch. **361**, 61 (1975).
WALLACH, D., SCHRAMM, M.: Europ. J. Biochem. **21**, 433 (1971).
WASER, P. G.: Experientia (Basel) **14**, 356 (1958).
WATSON, E. L., IZUTSU, K. T., SIEGEL, I. A.: Arch. oral Biol. **19**, 13 (1974).
WEIMAR, V. L., HARAGUCHI, K. H.: Physiol. Chem. Phys. **7**, 7 (1975).
WERLE, E., VOGEL, R., GÖLDEL, L. F.: Naunyn-Schmiedelbergs Arch. exp. Path. Pharmak. **230**, 236 (1957).
WIESMANN, U. N., BOAT, T. F., DI SANT'AGNESE, P. A.: Lancet **172** 2, 510 (1972).
WILLIAMS, J.A.: J. theoret. Biol. **28**, 287 (1970).
WILLIAMS, J. A., WOLFF, J.: Biochem. biophys. Res. Commun. **44**, 422 (1971).
WILLS, E. J., DAVIES, P., ALLISON, A. C., HASWELL, A. D.: Nature New Biol. **240**, 58 (1972).
WOODBURY, D. M., WOODBURY, J. W.: J. Physiol. (Lond.) **169**, 553 (1963).
YAMASHITA, K., FIELD, J. B.: J. biol. Chem. **247**, 7062 (1972).
YOHRO, T.: J. Anat. (Lond.) **108**, 409 (1971).
YOSHIDA, Y., SPRECHER, R. L., SCHNEYER, C. A., SCHNEYER, L. H.: Proc. soc. exp. Biol. **126**, 912 (1967).
YOSHIMURA, H., IMAI, Y.: Jap. J. Physiol. **17**, 280 (1967).
YOSHIMURA, H., IWASAKI, H., NISHIKAWA, T., MATSUMOTO, S.: Jap. J. Physiol. **9**, 106 (1959).
YOUNG, J. A.: Pflügers Arch. **303**, 366 (1968).
YOUNG, J. A.: Electrolyte transport by salivary epithelia. Proc. Aust. physiol. pharmacol. Soc. **4**, 101–121 (1973).
YOUNG, J. A., VAN LENNEP, E. W.: In: MTP International Review of Science (Gastrointestinal

Physiology), Series II, Vol. 12 (R. K. Crane, Ed.). Baltimore, Md.: University Park Press 1977, p. 105.
YOUNG, J. A., VAN LENNEP, E. W.: The Morphology of Salivary Glands. London: Academic 1978.
YOUNG, J. A., MARTIN, C. J.: Pflügers Arch. **327**, 285 (1971).
YOUNG, J. A., SCHÖGEL, E.: Pflügers Arch. **291**, 85 (1966).
YOUNG, J. A., FRÖMTER, E., SCHÖGEL, E., HAMANN, K. F.: Pflügers Arch. **295**, 157 (1967).
YOUNG, J. A., MARTIN, C. J., ASZ, M., WEBER, F. D.: Pflügers Arch. **319**, 185 (1970).
ZIMMERMANN, K. W.: In: Handbuch der mikroskopischen Anatomie des Menschen, Vol. **5**, Pt 1 (W. von Möllendorff, Ed.). Berlin: Springer 1927, p. 61.

Part II: Salt Glands

E.W. VAN LENNEP and J.A. YOUNG

A. Introduction

The need for marine vertebrates to possess an organ for extrarenal excretion of salt was first recognized in the nineteen-thirties. Thus, SMITH (1930) came to the conclusion that marine teleosts needed an organ that could either absorb fresh-water from the sea, or excrete chloride into it. Experiments on the "heart-gill" preparation demonstrated conclusively that, in teleosts, the gills fulfilled this role by excreting sodium chloride (KEYS, 1931). In a subsequent paper, KEYS and WILLMER (1932) described large cells in the fish gill epithelium which they called "chloride cells". Both in appearance and in histochemical properties these cells seemed similar to the parietal cells of the mammalian stomach.

It was not until much later, when SCHMIDT-NIELSEN and his associates investigated the ability of marine birds such as cormorants to maintain the osmolality of their blood at a level similar to that of terrestrial and freshwater birds, that the significance of the supraorbital nasal gland as a salt excreting organ was appreciated (SCHMIDT-NIELSEN et al., 1957, 1958). Glands with a similar function were subsequently discovered in marine turtles by SCHMIDT-NIELSEN and FÄNGE (1958). The term *"salt gland"* was coined by SCHMIDT-NIELSEN (1960) to describe *"any gland in the head region of marine birds and reptiles which, irrespective of anatomic origin, has an osmoregulatory function and secretes highly hypertonic sodium chloride solutions".*

Since then, glands with a similar function and ultrastructure have been found in elasmobranchs (BÜRGER and HESS, 1960), marine catfish (van LENNEP and LANZING, 1967), terrestrial birds (SCHMIDT-NIELSEN et al., 1963), and reptiles (SCHMIDT-NIELSEN et al., 1963; TEMPLETON, 1963), although in elasmobranchs and catfish the glands were not in the head region. Consequently, the term salt gland is now used for all extrarenal glandular organs when their main function is the excretion of a hypertonic salt solution, irrespective of the anatomical position of the gland, the habitat of the animal, or the nature of the electrolytes excreted (Table 7). Chloride cells are not included in this definition of salt glands, although they clearly have a related function in marine teleosts. However, the dendritic organs of the marine catfish, which are intermediate in structure and function between "true" salt glands and teleost chloride cells, will be included.

Several comprehensive reviews on the structure and function of avian and reptilian salt glands have appeared recently (PEAKER and LINZELL, 1975; DUNSON, 1976, 1977). Since there seems to be little point in repeating all that has already been said, this review will present only a survey of the most important findings relevant to a discussion of transport in salt glands. We will present the information from a slightly different viewpoint arising particularly from our morphological experience with salt glands in nonavian vertebrates.

Table 7. Occurrence of salt-excreting glands in vertebrates, their anatomical position, habitat of animal, and electrolyte concentrations in secreted fluid (mmol l^{-1}, mean ± S. D., or range)

	Habitat	Cytology		Na^+	K^+	Source
Elasmobranchs (rectal gland)						
Squalus acanthias (dogfish)	M	L(B)F	(a)	410±10	10±1.1	1
Raja ocellata (skate)	M	L(B)F	(b)	490± 2	11.2±0.3	2
Teleosts						
Plotosidae (marine catfish) dendritic organ	M-E	BT	(c)			
Reptiles						
Marine turtles (lacrimal gland)		LF				
Caretta caretta	M		(d)	732–878	18–31	3
Lepidochelys olivacea	M		(b)	713±3.2	18.8±3.3	4
Terrapins						
Malaclemys terrapin	B		(d)	616–784		3
Sea snakes (posterior sublingual gland)	M	L(B)F				
Aipysurus laevis			(d)	798±49	28±6	5
Lapemis hardwickii			(d)	676±40	23±2	5
Hydrophis elegans			(d)	509±20	20±0.5	5
Lizards (lateral nasal gland)		L(B)F				
Amblyrhynchus cristatus (marine iguana)	M–T		(e)	1434	235	4
Ctenosaura pectinata	T		(f)	72–124	380–706	6
			(g)	274–627	228–272	6
Iguana iguana	T		(h)	340–894	316–862	7
			(i)	728	290	7
Dipsosaurus dorsalis	T		(h)	494	1387	7
			(i)	1032	640	7
Uromastix acanthinurus	T		(h)	639	1398	7
Birds (supraorbital nasal gland)		BF				
Larus argentatus (herring gull)	M		(j)	718	24	8
Phalacrocorax auritus (double-crested cormorant)	M-E		(j)	500–600		8
Pelecanus occidentalis (brown pelican)			(j)	600–750		8
Spheniscus humboldti (Humboldt's penguin)	M		(j)	725–900		8
Diomeda nigripes (black-footed albatross)	M		(j)	800–900		8
Oceanodroma leucorrhea (Leach's petrel)	M		(j)	900–1100		8
Anas platyrhynchos (domestic duck)	F		(d)	450–587		9
			(k)	471±666	14.4±2.1	10
			(l)	523±37	13.2±2.0	10

Explanation of symbols
Habitat: B brackish water, F fresh water, M marine, M-E marine and estuarine, T terrestrial.
Cytology: BF basal folds, BT basal tubular invaginations, LF lateral folds only, L(B)F mainly lateral folds with some broad basal folds.

Electrolyte concentrations:
a) gland perfused with standard shark saline solution containing NaCl 280 mmol · l^{-1}
b) unstimulated gland (no extra salt loading)
c) no ducts present; electrolytes secreted directly into seawater
d) after salt loading
e) values probably too high because of evaporation
f) animals recently caught; salt-loaded with NaCl
g) animals kept in laboratory for 3 months, salt-loaded with NaCl
h) samples collected 2–4 hours after heavy NaCl load
i) samples collected 2 days after heavy NaCl load
j) stimulation of gland not reported
k) samples collected one hour after salt loading
l) samples from animals submitted to 4 days salt loading

Source:
1 SIEGEL et al. (1976)
2 HOLT and IDLER (1975)
3 SCHMIDT-NIELSEN and FÄNGE (1958)
4 DUNSON (1969)
5 DUNSON and DUNSON (1974)
6 TEMPLETON (1967)
7 SCHMIDT-NIELSEN et al. (1963)
8 SCHMIDT-NIELSEN (1960)
9 SMITH (1972)
10 WOOD and BALLANTYNE (1968)

B. Structure of Salt Glands

I. Microanatomy

With the exception of the catfish dendritic organ, whose status as a salt gland is not quite clear, salt glands, irrespective of embryonic origin or anatomical position, have essentially the same microanatomical structure. They are compound tubular glands, with the main duct branching into secondary ducts and these into tertiary ducts. These tertiary ducts form the central canal of a cylindrical lobule[1], the substance of which is formed by the numerous, radially orientated, secretory tubules. The rectal glands of some elasmobranchs, such as *Squalus acanthias, Galeorhinus galeus, Scylliorhinus canicula,* and *Raja batis,* have a single central canal that does not branch, so that the whole gland is composed of a single lobule (BULGER, 1963; KOMNICK and WOHLFARTH-BOTTERMANN, 1966). Other elasmobranchs have more compound glands (FÄNGE and FUGELLI, 1963; BONTING, 1966). The primitive elasmobranch *Chimaera monstrosa* does not possess a proper rectal gland but, instead, has a number of small glands in the wall of the rectum, probably with only a very limited capacity for salt secretion (FÄNGE and FUGELLI, 1963).

The glands are usually surrounded by dense connective tissue capsules (partly by cartilage in lizards), from which connective tissue septa enter the parenchyma, carrying the major blood vessels with them. The blood supply in the

1 FÄNGE et al. (1958a) called this structure a "lobe", but according to generally accepted histological usage it is more properly described as a lobule.

avian gland has been investigated by FÄNGE et al. (1958a). Arteries from the interlobular connective tissue pass through the wall of the lobule and give off arterioles in the connective tissue surrounding the central canal. These arterioles break up into capillaries that run along the secretory tubules towards the periphery of the lobules, whence they are drained by veins. Blood flow through the capillaries in avian glands is therefore in the opposite direction to the flow of fluid in the secretory tubules. In the dogfish *Squalus acanthias*, on the other hand, blood flow is in the same direction as the flow of secretory fluid (HAYSLETT et al., 1974).

II. Ultrastructure

Several ultrastructural studies on vertebrate salt glands have been published since 1960, for example on herring gulls (KOMNICK, 1963a and b; KOMNICK and KNIPRATH, 1970); marine turtles and terrapins (ELLIS and ABEL, 1964; ABEL and ELLIS, 1966; COWAN, 1971); lizards (*Uromastyx acanthinurus* [VAN LENNEP and KOMNICK, 1970] and *Dipsosaurus dorsalis* [ELLIS and GOERTEMILLER, 1974]); sea snakes (DUNSON and DUNSON, 1974); elasmobranchs (DOYLE, 1962; BULGER 1963, 1965; KOMNICK and WOHLFARTH-BOTTERMANN, 1966; STOCKEM et al., 1968); and the marine catfish (VAN LENNEP and LANZING, 1967). In all these glands the fully differentiated secretory cells show a striking similarity in ultrastructure. A spheroidal nucleus lies in the centre or slightly above the centre of the cell, the basolateral plasma membranes are extensively folded, and junctional complexes are present at the apical side of the lateral plasma membranes. The apical plasma membrane is relatively smooth with short microvilli present only in some cases. Mitochondria are abundant. There are a few cisternae of rough endoplasmic reticulum in the perinuclear cytoplasm and the apical cytoplasm contains variable numbers of smooth-membrane vesicles.

A number of points are worth noting. First, although all salt gland secretory cells show an extensive folding of basolateral plasma membranes and an abundance of mitochondria, there are important differences in the pattern of folding and in the distribution of mitochondria (Table 7). In the rectal glands of elasmobranchs and the lacrimal glands of marine turtles, the basal membrane is relatively smooth but the lateral membranes are extensively plicated. The folds from adjacent cells interdigitate with one another, and the short ovoid mitochondria are closely packed in the perinuclear cytoplasm. The other pattern is shown by avian salt glands[2] and the principal cells of the catfish gland. In these cells the basal and the lower part of the lateral membrane is deeply invaginated to form narrow folds (birds) or thin tubules (catfish). The mitochondria are elongated and sited within the basal folds or along the tubules. The basal regions of these cells thus resemble the basal regions of the cells of mammalian striated

2 Although the basal folds are typical for salt gland cells of marine birds and salt-adapted geese and ducks, this is not necessarily true for salt gland cells of terrestrial birds. Thus, the nasal glands of the road runner, *Geococcyx californianus*, and the quail, *Coturnix coturnix*, show typical lateral folding (DUNSON et al., 1976).

ducts or kidney distal tubules. In the proximal part of the secretory tubules in the nasal glands of the lizards *Uromastyx* and *Dipsosaurus,* two types of cells can be distinguished: light cells resembling the secretory cells of elasmobranchs and marine turtles, and dark cells in which the basal cytoplasm is divided into a small number of basal processes packed with mitochondria.

A second point worth noting about the morphology of salt glands is the total absence of secretion canaliculi. In this respect also, the secretory tubules of salt glands resemble electrolyte transporting epithelia such as striated ducts and kidney tubules. In point of fact, secretion canaliculi are also absent from secretory tubules of lateral nasal glands in the lizard, *Egernia cunninghami,* in which the gland does not function as a salt gland (unpublished observations). Although tight junctions (*zonulae occludentes*)[3] have been described in some salt glands (KOMNICK and WOHLFARTH-BOTTERMANN, 1966; ERNST and ELLIS, 1969; VAN LENNEP and KOMNICK, 1970), in other species they seem to be lacking or at least to be permeable for extracellular tracers (VAN LENNEP, 1968; MARTIN and PHILPOTT, 1973; DUNSON and DUNSON, 1974). It is possible in these latter cases that the short "lips" formed by the latero-apical edges of adjacent cells above the junctional complex can be pressed together to act as a kind of valve. On physiological grounds one would expect very tight *zonulae occludentes* to be present, so as to allow the high electrochemical gradients that exist in the avian salt glands to be maintained (PEAKER and LINZELL, 1975). Since the existing descriptions of *zonulae occludentes* in salt glands have been based on observations made on routine transmission electron micrographs, and since in such preparations *zonulae occludentes* can easily be confused with gap junctions, which do not form a barrier, the whole matter of the existence of *zonulae occludentes* in salt glands needs to be re-examined both by use of extracellular tracers and by application of freeze-fracture methods. A recent freeze-fracture study would seem to indicate that the *zonulae occludentes* in avian salt glands are exceptionally leaky (ELLIS et al., 1977); see also Section F.

The cells of the epithelium of the branched secretory tubules in the avian salt gland show a progressive change in appearance, starting from the distal (blind) end at the periphery of the lobule (ELLIS, 1965): the distal[4] cells are relatively

3 A *zonula occludens* ("tight junction") is a belt-like region of contact between the lateral plasma membranes of adjacent epithelial cells, which appears in the transmission electron microscope as a partial fusion of the plasma membranes and in freeze-fracture replicas as a meshwork of intersecting ridges (and corresponding grooves on the opposite membrane). A measure of the tightness of a *zonula occludens* is the transepithelial electrical resistance. There is generally a good correlation between the depth of the *zonula* and the number of ridges, on the one hand, and the transepithelial resistance, on the other hand (CLAUDE and GOODENOUGH, 1973). Another means of testing the tightness or leakiness of a tight junction is by the use of extracellular tracers, *e. g.* horse radish peroxidase and lanthanum hydroxide. If the large molecules of such tracers can pass through the *zonula occludens* it may be assumed that relatively small ions may also do so. It should be pointed out that horse-radish peroxidase is introduced into the animal before fixation and that its presence in tight junctions is therefore unlikely to be induced during the process of fixation.

4 ERNST and ELLIS (1969) use the term "peripheral cells", but this term might lead to some confusion when used in connection with the very irregular arrangement of lobules in other animals *e. g.* lizards. The present authors prefer to designate the cells at the blind end of the tubule "distal", and those closest to the central canal "proximal".

small and undifferentiated with few mitochondria and virtually no basal infoldings while the more proximal cells become larger and progressively more differentiated in appearance (ERNST and ELLIS, 1969). The distal portion of the secretory tubules in the nasal salt glands of *Uromastyx* and *Dipsosaurus* also contains a number of "mucoid" cells, which lack the characteristics of salt-secreting cells but possess secretion granules and a well-developed rough endoplasmic reticulum. They also contain other cells, rich in mitochondria, each of which is provided with an apical tuft of microvilli (VAN LENNEP and KOMNICK, 1970; ELLIS and GOERTEMILLER, 1974). Mucoid cells are also present in the sublingual salt glands of sea-snakes.

C. Adaptation to Salt Loading

Many birds such as domestic ducks, although reared in a fresh water habitat, nevertheless possess a salt gland that will respond immediately to salt loading (orally or intravenously) with a fairly copious secretion. Keeping such animals on a saltwater diet leads to an increase in the weight of the glands (ELLIS et al., 1963; ERNST et al., 1967; FLETCHER et al., 1967; HOLMES and STEWART, 1968), resulting from an increase in the length of the secretory tubules and differentiation of the tubule cells (ERNST and ELLIS, 1969). The differentiation of the secretory tubule consists mainly of an increase in basolateral plasma membrane and in the number of mitochondria. The former arises from the Golgi complex (LEVINE et al., 1972). There is a concomitant rise in total ATPase activity (ERNST et al., 1967; FLETCHER et al., 1967; STEWART et al., 1976), in sulphatides (KARLSSON et al., 1974) and in sialic acid (MARTIN and PHILPOTT, 1974).

Less is known concerning adaptive changes in reptilian and ichthyan salt glands. Most data come from the work by TEMPLETON (TEMPLETON, 1964, 1967, 1972; TEMPLETON et al., 1968, 1972) on four species of terrestrial lizard. The main cation excreted by the salt glands of these animals is potassium (Table 7). The short-term effect of an intraperitoneal injection of either potassium or sodium chloride is an increased rate of potassium excretion. The long-term effect of sodium chloride loading differs in various species. For example, in *Ctenosaurus pectinata*, sodium excretion increases gradually whereas potassium excretion declines. BRAYSHER (1971) observed a similar increase in sodium excretion after sodium chloride loading in the Australian skink, *Trachydosaurus rugosus*. SHOEMAKER et al. (1972) confirmed TEMPLETON's findings in *Dipsosaurus dorsalis;* these authors also found that chloride is only excreted if the animal is loaded with chloride, but that bicarbonate is the anion excreted if secretion is stimulated by loading with potassium bicarbonate, potassium acetate, or potassium succinate. COWAN (1974b) could find no differences in either K^+-dependent or K^+-independent p-nitrophenyl-phosphatase (NPPase) in salt glands of the euryhaline terrapin, *Malaclemys terrapin,* kept for 40 days in either freshwater or seawater. DUNSON and DUNSON (1975), on the other hand, observed values for Na^+-, K^+-dependent ATPase that were three times higher in terrapins kept in seawater and salt-loaded than in terrapins kept in freshwater. In contrast to birds, where salt water adaptation leads to an increase in the size of the salt gland, in terrapins the lacrimal (salt) glands show no increase in mass with salt water adaptation (COWAN, 1974a; DUNSON and DUNSON, 1975).

There are only a few reports of adaptive morphological changes in nonavian salt glands. ELLIS and GOERTEMILLER (1974) compared the cytology of the nasal glands in *Dipsosaurus dorsalis* under control conditions and after salt loading with both sodium and potassium for four to eight days. They found that the difference between dark and light cells became more marked and the lateral intercellular spaces appeared wider after salt loading. A similar increase in the width of the intercellular space had earlier been observed by KOMNICK and KNIPRATH (1970) in the herring gull salt gland after salt water adaptation. OGURI (1964) compared the histological structure of rectal glands of bull sharks (*Carcharhinus leucas*) caught in the freshwater of Lake Nicaragua and Rio San Juan with those from animals caught in seawater. Rectal glands from freshwater animals were much less in weight and the secretory tubules were atrophied and partly replaced by connective tissue. The author suggests that bull sharks may migrate periodically from the sea into the river, up to the lake. Complete dedifferentiation of the principal cells in the dendritic gland of the marine catfish *Cnidoglanis macrocephala* occurred when this animal was gradually adapted to fresh water (R. HOLST, pers. comm.).

D. Neural and Hormonal Control of Salt Gland Secretion

The secretory response of avian salt glands to salt loading is mediated by cholinergic postganglionic fibres. The response is initiated by osmoreceptors, probably located in the heart or great vessels (see PEAKER and LINZELL, 1975). Although adrenal cortical hormones were originally thought to be directly involved in the initiation of secretion (HOLMES et al., 1961, 1963; WRIGHT et al., 1966), more recent studies have shown that the effects of these hormones are probably indirect (*e. g.* through stimulation of differentiation of the secretory tubule epithelium). Prolactin, which is known to be involved in osmoregulation in fish, may also play a part in the control of secretion in the avian salt gland (PHILLIPS and ENSOR, 1972, cited by PEAKER and LINZELL, 1975).

Where investigated, reptile salt secretion seems also to have been under cholinergic control (SCHMIDT-NIELSEN and FÄNGE, 1958; DUNSON, 1970; NORRIS and DAWSON, 1964; TEMPLETON, 1964; LEMIRE et al., 1972). Adrenal glucocorticoids appear to be necessary for the continued secretory activity of the lacrimal glands of green turtles (HOLMES and MCBEAN, 1964). On the other hand, however, SHOEMAKER *et al.* (1972) found that cortisone reduced the ability of the nasal gland of the terrestrial lizard *Dipsosaurus dorsalis* to excrete sodium after loading with sodium chloride, and that aldosterone completely abolished sodium excretion but that neither hormone affected potassium excretion (see also TEMPLETON et al., 1968, 1972).

Although in marine birds an increase in plasma osmotic activity, irrespective of the cause, seems to be the initial stimulus that triggers the excretion of sodium and chloride, the situation is probably different in terrestrial birds such as the ostrich (which normally excrete more potassium: SCHMIDT-NIELSEN et al., 1963) and in terrestrial reptiles. It is not known what acts as the trigger in these cases. The fact that the ratio of sodium to potassium excretion in long-term experiments (*i. e.* over several days) depends on the cation with which the animals is loaded suggests that there are separate receptors for sodium and potassium; the studies by SHOEMAKER et al. (1972) on *Dipsosaurus* point also to a separate receptor for chloride. On the other hand, the work by these same authors on the dependence of sodium excretion on the level of corticosteroids and the independence of potassium excretion of these hormones suggests two different pathways of stimulation for the two cations.

The rectal gland in elasmobranchs does not appear to be under nervous control (BURGER, 1962; CHAN and PHILLIPS, 1967). Although periodic fluctuations in flow rate (probably induced by extraneous influences since they appear to occur simultaneously in all fishes kept in a laboratory) make controlled experiments somewhat hazardous, HOLT and IDLER (1973) managed to demonstrate a significant decrease in the volume of fluid secreted and in the sodium and chloride concentrations of the secretion after removal of the interrenal gland in the skate *Raja ocellata*. Sham-operated animals did not show this effect. The elasmobranch corticosteroid, 1α-hydroxycorticosterone, which has a strong mineralocorticoid effect in rat and mouse (IDLER et al., 1967), increased the rectal gland secretory rate and sodium and chloride concentrations in the secretion obtained from the interrenalectomized skate. Interrenalectomy has no effect on either total ATPase or Na^+-, K^+-dependent ATPase activity (IDLER and KANE, 1976).

E. Flow Rates and Electrolyte Concentrations

Very few data have been published showing the relation between the electrolyte concentration in and the rate of formation of the secretions of salt glands. This is probably because it has proven extremely difficult to cannulate the main ducts of salt glands, at least in most animals. Of course, even when such studies are performed, the interpretation of the data is much more difficult than in the case of salivary glands since results of micropuncture experiments are not available. Furthermore, since the stimulus for secretion is usually an increase in the osmotic activity of the plasma, or a rise in the plasma concentration of a particular electrolyte, it is difficult to disentangle the effects of stimulation *per se* on the secretion and the effects of a changed composition of the interstitial fluid.

HANWELL et al. (1971) studied the electrolyte excretion patterns in the secretion of the salt gland of the domestic goose. When results from all geese studied were considered together, the authors observed an increase in sodium chloride concentration in the secretion as secretory rates increased. In contrast, however, when results from geese were evaluated individually, it was found that the salt concentration decreased as the flow rate increased. The difference arises, of course, from the fact that the salt glands in different geese exhibit different degrees of salt-adaptation so that glands secreting at high rates (salt-adapted) have high salt concentrations whereas nonadapted glands secrete at low rates and have low electrolyte concentrations. The excretory pattern that is of interest, however, is that seen in individual geese, since this pattern suggests a likely secretory model.

As mentioned above, the cells near the blind end of the secretory tubules are relatively undifferentiated, ultrastructurally, when compared with the secretory cells situated nearer the central canal of the lobule. This fact has prompted PEAKER and LINZELL (1975) to postulate that the less well-differentiated cells are more water permeable than the differentiated cells so that stimulation causes them to produce a primary secretion that is only slightly hypertonic. They postulate that this secretion is then made more hypertonic by secretion into it of salt, accompanied by little or no water, as the fluid passes along the part of the tubule lined by differentiated cells. Stimulation could then cause a decrease in osmotic activity of the final secretion if the salt-secreting response of the undifferentiated cells, where water secretion accompanies salt transport, were greater than the response of the differentiated cells. Their hypothesis is plausible although there is no direct evidence to support it.

Apart from the above-mentioned studies on birds, other studies of this kind have been performed on the salt glands of sea snakes where samples of secreted fluid were collected by cannulation of the tongue sheath (DUNSON and DUNSON, 1974). The results revealed a positive relation between flow rate and chloride concentration of the excreted fluid in the snakes, *Aipysurus duboisii* and *Hydrophis elegans,* following salt loading.

The flow rate of the fluid secreted from the rectal gland of the dogfish, *Squalus acanthias,* may vary a great deal in control animals (BURGER and HESS, 1960), but sodium chloride concentrations remain constant. This was confirmed in perfusion experiments carried out *in vitro* (SIEGEL et al., 1976), in which the gland was stimulated by perfusion with solutions in which sodium or chloride concentrations were varied. High sodium or chloride concentrations in the perfusion fluid cause a high flow rate but do not affect the actual concentration of the salts in the secreted fluid.

F. The Mechanism of Salt Secretion by the Tubular Endpiece

Any discussion of the secretory mechanisms operating in the tubular endpiece cells of salt glands must be considered merely as speculation, since there are insufficient data available on which to base a model with any confidence. There are a number of reasons to account for this lack of data. The excretory ducts of most salt glands are rather inaccessible and they have only been successfully cannulated in a few instances, so that the electrolyte excretory patterns for most glands are unknown. This inaccessibility has also defeated attempts to employ micropuncture techniques, and only one report exists in which microelectrode recordings of secretory cell potentials are described (in the salt gland of the herring gull: THESLEFF and SCHMIDT-NIELSEN, 1962). It should be remembered that the secretory tubules are not composed of a homogeneous cell population but rather, at least in birds and lizards, contain a population of cells ranging from the very undifferentiated to the highly differentiated (see Section B. II.). Hence, in the absence of micropuncture data, we cannot know whether the electrolyte concentrations along a secretory tubule are constant or exhibit a profile and, without such information it is impossible to build up a detailed secretory hypothesis analogous to that developed for salivary glands. Finally, it needs to be stressed that there are obvious differences among the salt glands of various animals that may make it mandatory to postulate different secretory models for each gland type. Thus the salt glands of marine vertebrates secrete virtually a pure sodium chloride solution, whereas the glands of some terrestrial birds and reptiles normally secrete a potassium bicarbonate solution with little sodium chloride and they can alter the concentration ratios for both K^+/Na^+ and HCO_3^-/Cl^- in the secretion, according to the dietary salt intake.

The high salt concentration found in the secretions of most salt glands allows us to exclude on thermodynamic grounds the possibility that they function as filtering organs. This conclusion has been reinforced by the demonstration that intravascularly administered inulin and ^{14}C-labelled sucrose do not enter the avian salt gland secretion (SCHMIDT-NIELSEN, 1960; PEAKER and HANWELL, 1974). It has always been argued that the gland epithelium must constitute a reasonably impermeable barrier between interstitium and lumen, so that the high salt concentrations that develop in the secretion are not dissipated.

A recent freeze-fracture demonstration (ELLIS et al., 1977) that *zonulae occludentes* between the principal cells lining the secretory tubules of avian salt glands are of an exceptionally 'leaky' type (see Section C.) has necessitated a reappraisal of this view. ELLIS et al. (1977) postulate that the glands secrete by using a two-stage process, reminiscent of that believed to operate in salivary glands: in the first stage, an isotonic primary fluid is produced by the undifferentiated peripheral cells constituting the blind terminal section of the secretory tubule (see Section C.) whilst, in the second stage, water is reabsorbed across the *zonulae occludentes* between the more differentiated principal cells that make up the remainder of the secretory tubule. They postulate that standing salt gradients are developed in the lateral inter-cellular spaces between the principal cells by the activity of a conventional, membrane-bound, Na^+-, K^+-ATPase and that these gradients provide the osmotic force needed to remove water from the lumen, across the *zonulae occludentes*. For the model to work, it is essential that the luminal membrane of the principal cells, in contrast to the *zonulae occludentes* separating them, be relatively impermeable to salt and water so that salt entering the intercellular spaces from the cytoplasm will be derived, ultimately, from the interstitium, not from the lumen.

THESLEFF and SCHMIDT-NIELSEN (1962) measured the potential difference between the blood (or the gland surface) and the interior of the excretory duct of the herring gull salt gland during stimulation of the secretomotor nerve or following administration of the cholinergic drug methacholine. They found that stimulation caused the recorded potential to shift by 40–60 mV (lumen-positive) after a delay of 15–30 seconds. Although these results are often quoted in the literature it is difficult to interpret them meaningfully. It is quite likely that the potential difference change they recorded reflected a change in the ionic composition of the luminal fluid bathing the tip of the recording electrode (*i. e.* it reflected a change in the electrode liquid junction potential) and had little to do with a change in the transepithelial potential difference. Even if the potential change did reflect a change in the epithelial potential difference, it must be emphasized that the electrical space constant of a small duct filled with an electrolyte solution would be so short that the potential recorded would have been derived only from the epithelial cells in the immediate vicinity of the electrode tip and would not reflect potential changes arising across the cells of the tubular secretory endpieces (see Pt. I, Section C. II. 4). These objections also apply to more recent experiments on the rectal gland of the dogfish, where duct potentials of 10–18 mV (lumen-negative) have been reported (SIEGEL et al., 1976) and misinterpreted as secretory potentials.

THESLEFF and SCHMIDT-NIELSEN (1962) also attempted to record potentials across the basal plasma membranes of secretory cells by inserting electrodes into the surface layer of the salt gland of the herring gull. Insertion of the electrode was accompanied by a sudden potential shift of 40–80 mV (recording electrode negative to interstitium), presumably as a result of impalement of a secretory cell. This initial negativity dropped within a few seconds to –20 mV. Such recordings of the membrane potential can hardly be said to constitute successful impalements according to the criteria set by Petersen (see Pt I, Section C. II. 4). This is perhaps not surprising when the morphology of an avian salt gland cell with its deep basal infoldings is considered. THESLEFF and SCHMIDT-NIELSEN apparently recognized this difficulty and did not attach too much significance to their recordings, but their values for "membrane potentials" have nevertheless frequently been quoted.

These same authors found that retrograde injection of ouabain up the ducts abolished both secretion and the "secretory" potential. It was therefore suggested that a ouabain-senitive, Na^+-, K^+-dependent ATPase system was responsible for the transport of sodium across the apical (luminal) membrane of the cell (*cf.* BONTING, 1970). These results cannot, however, be taken as evidence for the existence of an apical pump system. Retrograde injection of ouabain does not necessarily mean that the drug acts only on the luminal membrane; the argument put forward by PEAKER and LINZELL (1975) that morphological demonstrations of the leakiness of tight junctions are the result of fixation artifacts holds even more strongly for retrograde injections. There can be little doubt that some of the ouabain would leak through the junctional complexes into the lateral intercellular spaces[5]. The present authors' own attempts at retrograde injection of ouabain into the central canal of the shark

5 See footnote 1 in Part I of this article (p. 570).

rectal gland resulted in the loss of the ultrastructural integrity of the secretory epithelium (unpubl.). Of the many attempts undertaken to demonstrate ATPase histochemically in salt gland epithelium, the results of only two suggest that a reaction product can be found on the apical plasma membrane (ABEL and ELLIS, 1966; BALLANTYNE and WOOD, 1970). In both attempts, the Wachstein-Meissel method, now considered unspecific for Na^+-, K^+-ATPase at the light-microscope level, was employed. All cytochemical studies (*i. e.* studies performed at the electron-microscope level) show reaction product on the basolateral membranes but not on the apical membrane, regardless of whether the Wachstein-Meissel method (VAN LENNEP, 1968; ABEL, 1969) or the more specific p-nitrophenylphosphate method (ERNST, 1972; ELLIS and GOERTEMILLER, 1974) is used.

The intracellular electrolyte concentrations in avian salt glands have been calculated by a number of authors (HOKIN, 1967; PEAKER, 1971a; PEAKER and STOCKLEY, 1973; SCHMIDT-NIELSEN, 1976). Except for HOKIN (1967), who found very high levels of intracellular sodium, the sodium concentrations calculated are similar to those reported for other tissues. Thus SCHMIDT-NIELSEN (1976), in a careful study, found the following intracellular concentrations in the secreting salt gland of the herring gull (calculated from plasma concentrations as representative of intercellular concentrations): sodium, 40 mmol $l^{-1}_{H_2O}$; potassium, 125 mmol $l^{-1}_{H_2O}$; and chloride, 63 mmol $l^{-1}_{H_2O}$. No significant differences in the concentrations of these ions were observed between secreting and inactive glands. Values of intracellular sodium and potassium content estimated for sea snake salt glands (DUNSON and DUNSON, 1974) and for the rectal gland of the dogfish (HAYSLETT et al., 1974) are comparable with the figures quoted above for salt gland of the herring gull.

Attempts to demonstrate sodium concentrations histochemically (by precipitation as sodium antimonate) in the salt gland of the herring gull (KOMNICK and KOMNICK, 1963), in the rectal glands of elasmobranchs, and in the dendritic gland of the marine catfish (VAN LENNEP, 1968) suggested high intracellular concentrations of sodium. In the light of later studies, it now seems likely that the large amounts of antimonate precipitate seen in the salt gland cells may have been caused by relatively high concentrations of divalent cations (calcium and magnesium).

The answer to the question of whether the passage of sodium ions through the basolateral plasma membrane takes place by passive diffusion or by active transport obviously hinges to some extent on the intracellular sodium concentration. Assuming that SCHMIDT-NIELSEN's figures are correct, sodium may diffuse into the cell and acetylcholine could stimulate salt transport by increasing the permeability of the basolateral membrane for sodium (VAN ROSSUM, 1966). PEAKER (1971b) proposes that sodium and chloride are transported across the basal membrane *via* exchange pumps by which sodium is exchanged for hydrogen, and chloride for bicarbonate ions, as has been postulated for the chloride cells of fish gills (MAETZ, 1971). There do appear to be some, albeit slight grounds for postulating the existence of the exchange pumps. Salt gland cells have high carbonic anhydrase activity (NECHAY et al., 1960; BONTING et al., 1964), and intravenous injection of acetazolamide or methazolamide blocks the secretory response normally seen after salt loading (FÄNGE et al., 1958a and b; NECHAY et al., 1960). On the other hand, THESLEFF and SCHMIDT-NIELSEN (1962) found no effect of acetazolamide on the secretory response to nerve stimulation or cholinergic drug administration, and consequently, suggested that acetazolamide's action in blocking the response to salt loading was due to an action somewhere in the reflex pathway rather than in the gland tissue.

In addition to the proposed basal Na^+-H^+ exchange pump, PEAKER and STOCKLEY (1974) argue for the existence of an apically located Na^+-extrusion mechanism, whose transport rate can be increased by the application of cholinergic drugs. Their evidence comes from experiments with incubated gland slices; Li^+ ions in a concentration of 10 mmol l^{-1} were incorporated in the bathing medium with the aim of stimulating CO_2 production and thereby increasing the turnover of the postulated basal Na^+/H^+ pump. The data they obtained can be said to be in accord with their model, but it must be stressed that there is no direct evidence to establish the existence of either of these pumps unequivocally. If the apical sodium pump really does exist it might be expected to prove to be sensitive to ethacrynic acid, since SMITH (1972) has shown that this drug can inhibit salt gland secretion in the duck; it is believed by some that an ethacrynic acid-sensitive, apical Na^+ pump is responsible for secretion in salivary glands (Section C. II. 6 of Pt. I.).

Two other pumps have been postulated for some salt glands, a potassium pump and a chloride pump. HUGHES (1970) reports potassium concentrations as high as 68 mmol l^{-1} in the salt gland secretions of salt-loaded gulls (*Larus glaucescens*), and even higher concentrations have been found in the salt secretions of certain terrestrial lizards (Table 7)[6]. Whether such a potassium pump, if it actually exists, has to be located apically or basally will depend on what the cell membrane potentials prove to be and what the intracellular potassium concentration is.

SIEGEL et al. (1976) have argued for the existence of an active chloride pump in the rectal gland of the dogfish. Their argument was based on calculations of the electrochemical gradient against which they believed chloride was secreted. However, their measurements of the transepithelial electrical potential difference were made *via* their collection catheter, placed in the central canal of the gland. Even in so simple a gland as the one they studied, the cable properties of a tubular system filled with an electrolyte solution make it extremely unlikely that the potential difference across the secretory tubule cells could be recorded from an electrode placed in the central canal (see Section F). If this criticism is valid then one can no longer claim to know what the secretory cell potential is or even what its polarity might be, and hence one cannot calculate the electrochemical potential gradients against which Na^+ and Cl^- were secreted.

It is interesting to note that the ouabain-sensitive, Na^+-, K^+-dependent ATPase, originally considered by most authors to be the only or at least the main enzyme system involved in sodium transport across the secretory cell, has

6 For a discussion of the possible role of the duct epithelium in the secretion of potassium in lizard salt glands see Section G.

been relegated by some workers merely to the role of maintaining the intracellular electrolyte concentrations under resting conditions (*e. g.* see SCHMIDT-NIELSEN, 1976). This function appears to have little to do directly with the transepithelial transport of sodium; on the contrary, at first sight it would seem to oppose the effect of the sodium exchange pumps in the basal membrane. However, in the model proposed by ELLIS et al. (1977) the enzyme plays a vital role (see above), similar to that seen in other epithelia.

FLETCHER et al. (1967) found a strong correlation between the maximum rate of salt secretion and the ATPase activity of salt glands of ducks adapted either to fresh water or to salt water; the number of moles of sodium tranported per mole of ATP hydrolyzed by the ouabain-sensitive enzyme system was virtually constant (approximately 4.5). PEAKER and LINZELL (1975), appreciating the inconsistency of calculating Na^{+}/ATP ratios from the activity of an enzyme system that did not appear in their model to be directly involved in the sodium transport, calculated instead the work required to pump sodium across the apical plasma membrane (using data from the literature) and arrived at approximately the same figure as FLETCHER et al. (1967). Some of the data used by PEAKER and LINZELL (1975) in their calculations seem likely to prove rather unreliable *e. g.* the assumption that the potential difference across the apical membrane is +90 mV (based on THESLEFF and SCHMIDT-NIELSEN, 1962).

The main function of both the ouabain-sensitive Na^{+}-, K^{+}-ATPase and the ouabain-insensitive Mg^{2+}-ATPase in salt glands remains a vexed question. A parallel increase in the quantity of basolateral plasma membrane present, in the activities of both ATPases and in the secretory capacity of the gland, is seen during salt water adaptation, and this seems to point to an important function of the ATPases in salt secretion. An alternative explanation would be that these ATPases, as normal constituents of basolateral plasma membranes, simply increase in total activity *pari passu* with the increase in the quantity of plasma membrane and that their function, at least that of the ouabain-sensitive system, is to cope with the increase in total passive fluxes of sodium and potassium that result from the increased membrane suface area. However, STEWART et al. (1976) showed that during salt water adaptation the membrane fraction had a higher ATPase activity than in controls, which they attributed to a higher content of catalytic protein. Apart from the ATPase discussed above, none of the enzymes involved in the pumps postulated by PEAKER (1971 b) has been demonstrated unequivocally, either biochemically or histochemically.

The avian salt gland can concentrate anions other than chloride in its secretions *e. g.* iodide (CAREY and SCHMIDT-NIELSEN, 1962) and thiocyanate (DOUGLAS, 1966). Although chloride seems to be preferred to iodide, the avian salt gland does not seem to discriminate between chloride and thiocyanate (DOUGLAS, 1966). Since thiocyanate is known to be an inhibitor of chloride pumps in the chloride cells of fish gills (EPSTEIN et al., 1973), the fact that the anion does not inhibit chloride excretion in the penguin salt gland (DOUGLAS, 1966) suggests that a chloride pump is probably not present.

The effect of acetylcholine and of cholinergic drugs on the avian, and probably also reptilian, salt gland may be due partly to an increase in the permeability of the basal plasma membrane to sodium (VAN ROSSUM, 1966); however the work by PEAKER and STOCKLEY (1974) with methacholine suggests that the cholinergic drugs stimulate an apical sodium pump. It is attractive to postulate a mechanism similar to that suggested for fluid and electrolyte secretion in salivary glands, with cyclic GMP and calcium as second messengers (see Pt I, Section C. II. 5. c), but neither the necessity of calcium for secretory activity nor the presence of cyclic GMP has so far been studied in salt glands. The relatively long delay (15–30 seconds) between stimulation and onset of secretion is noteworthy.

In summary, it may be said that, despite a rather large literature, relatively little is known with certainty about salt gland secretory mechanisms. The major difficulty is the lack of reliable estimates of the potentials across the apical and basal membranes of the secretory cell. This difficulty is compounded by the total lack of information on the electrolyte content of the secretion actually formed in the secretory tubules. Flux studies on incubated slices will not provide direct proof of the existence of active or passive transepithelial transport mechanisms; data from such experiments are unlikely to prove more fruitful than corresponding data obtained from kidney slices studies and, ultimately, as in the kidney only micropuncture experiments will provide the answers. It is also worth noting that studies on salt glands seem to be plagued by the same problem that plagues studies on salivary glands *viz.* a plethora of different species, each with glands having highly specialized features not seen in the glands of other species.

G. The Role of the Duct System

The part played by the duct system in determining the composition of the final secretion is unknown, but most authors assume that it is unlikely that appreciable electrolyte transport takes place in them. The ducts in the avian salt gland have a stratified cuboidal epithelium (KOMNICK, 1964), which has been described as transitional and has been compared to that of the mammalian ureter (PEAKER and LINZELL, 1975). Ultrastructurally such a comparison is not valid; furthermore it is not permissible to conclude from the fact that the epithelium is stratified, or even cornified, and contains few mitochondria, that it is inert. There are many examples of such epithelia that are known to absorb sodium and chloride ions powerfully *e. g.* frog skin and rumen epithelium.

PEAKER and LINZELL (1975) believe that the secretory tubules of the nasal glands in lizards secrete a concentrated sodium chloride solution similar to that formed in other salt glands and that the sodium can be replaced with potassium during passage of the primary secretion along the duct system. They base this belief entirely on the observation that duct cells in these glands have a pseudostratified epithelium consisting of cells with some lateral folding and a fairly large number of mitochondria, which they consider indicates some degree of transport activity. It must be pointed out, however, that the duct system in these nasal glands is very short (VAN LENNEP and KOMNICK, 1970), and it is difficult to believe that these very short duct segments could exchange nearly all the sodium for potassium, and the chloride for bicarbonate, even at the rather low flow rates observed in lizard salt glands. However, in the absence of quantitative data the question must be considered an open one — the morphology alone cannot justify postulation of specific electrolyte transport processes in the epithelium.

The technical difficulties in the way of performing micropuncture on avian or reptilian salt glands leave the question of ductal electrolyte transport unanswered for the time being. However, it should be possible to perform such experiments on elasmobranch rectal glands, which have a single central canal lined with a stratified epithelium although, in some species, the cells of the epithelium are of several distinct types (BULGER, 1965; STOCKEM et al., 1968). Of course one could not automatically extrapolate from elasmobranchs to other species, but it would be interesting if a transport function were ascribed to these ducts and would discourage their being dismissed as mere passive conduits.

Acknowledgements

We are indebted to Miss *K. E. Wilson* for the meticuluous care with which she prepared and checked the bibliographies. The authors' experimental work reported in the present article has been supported by the National Health and Medical Research Council of Australia, The Max Planck Gesellschaft, The Deutsche Forschungsgemeinschaft, The Cystic Fibrosis Association of N. S. W. and by research funds voted by the University of Sydney.

References

Part II

ABEL, J. H.: J. Histochem. Cytochem. **17**, 570 (1969).
ABEL, J. H., ELLIS, R. A.: Amer. J. Anat. **118**, 337 (1966).
BALLANTYNE, B., WOOD, W. G.: Cytobios **2**, 41 (1970).
BONTING, S. L.: Comp. Biochem. Physiol. **17**, 953 (1966).
BONTING, S. L.: In: Membranes and Ion Transport, Vol. 1 (E. E. Bittar, Ed.). London: Wiley-Interscience 1970, p. 257.
BONTING, S. L., CARAVAGGIO, L. L., CANADY, M. R., HAWKINS, N. M.: Arch. Biochem. Biophys. **106**, 49 (1964).
BRAYSHER, M.: Physiol. Zool. **44**, 129 (1971).
BULGER, R. E.: Anat. Rec. **147**, 95 (1963).
BULGER, R. E.: Anat. Rec. **151**, 589 (1965).
BURGER, J. W.: Physiol. Zool. **35**, 205 (1962).
BURGER, J. W.: Physiol. Zool. **38**, 191 (1965).
BURGER, J. W., HESS, W. N.: Science **131**, 670 (1960).
CAREY, F. G., SCHMIDT-NIELSEN, K.: Science **137**, 866 (1962).
CHAN, D. K. O., PHILLIPS, J. G.: J. Anat. (Lond.) **101**, 137 (1967).
CLAUDE, P., GOODENOUGH, D. A.: J. Cell Biol. **58**, 390 (1973).

COWAN, F. B. M.: Canad. J. Zool. **49**, 691 (1971).
COWAN, F. B. M.: Comp. Biochem. Physiol. **48A**, 489 (1974a).
COWAN, F. B. M.: Comp. Biochem. Physiol. **49A**, 775 (1974b).
DOUGLAS, D. S.: Nature (Lond.) **209**, 1150 (1966).
DOYLE, W. L.: Amer. J. Anat. **111**, 223 (1962).
DUNSON, M. K., DUNSON, W. A.: J. comp. Physiol. **101**, 89 (1975).
DUNSON, W. A.: Amer. J. Physiol. **216**, 995 (1969).
DUNSON, W. A.: Comp. Biochem. Physiol. **32**, 161 (1970).
DUNSON, W. A.: In: Biology of the Reptilia. Physiol. A, Vol. 5 (W. R. Dawson and C. Gans, Eds). New York: Academic 1976.
DUNSON, W. A.: In: Mechanisms of Osmoregulation in Animals (R. Gilles, Ed.). New York: Wiley-Interscience 1977.
DUNSON, W. A., DUNSON, M. K.: Amer. J. Physiol. **227**, 430 (1974).
DUNSON, W. A., DUNSON, M. K., OHMART, R. D.: J. exp. Zool. **198**, 209 (1976).
ELLIS, R. A.: Amer. Zool. **5**, 648 (1965).
ELLIS, R. A., ABEL, J. H.: Science **144**, 1340 (1964).
ELLIS, R. A., GOERTEMILLER, C. C.: Anat. Rec. **180**, 285 (1974).
ELLIS, R. A., GOERTEMILLER, C. C., DELELLIS, R. A., KABLOTSKY, Y. H.: Develop. Biol. **8**, 286 (1963).
ELLIS, R. A., GOERTEMILLER, C. C., STETSON, D. L.: Nature (Lond.) **268**, 555 (1977).
EPSTEIN, F. H., MAETZ, J., de RENZIS, G.: Amer. J. Physiol. **224**, 1295 (1973).
ERNST, S. A.: J. Histochem. Cytochem. **20**, 23 (1972).
ERNST, S. A., ELLIS, R. A.: J. Cell Biol. **40**, 305 (1969).
ERNST, S. A., GOERTEMILLER, C. C., ELLIS, R. A.: Biochim. biophys. Acta **135**, 682 (1967).
FÄNGE, R., FUGELLI, K.: Sarsia **10**, 27 (1963).
FÄNGE, R., SCHMIDT-NIELSEN, K., OSAKI, H.: Biol. Bull. (Woods Hole, Mass.) **115**, 162 (1958a).
FÄNGE, R., SCHMIDT-NIELSEN, K., ROBINSON, M.: Amer. J. Physiol. **195**, 321 (1958b).
FLETCHER, G. L., STAINER, I. M., HOLMES, W. N.: J. exp. Biol. **47**, 375 (1967).
HANWELL, A., LINZELL, J. L., PEAKER, M.: J. Physiol. (Lond.) **213**, 373 (1971).
HAYSLETT, J. P., SCHON, D. A., EPSTEIN, M., HOGBEN, C. A. M.: Amer. J. Physiol. **226**, 1188 (1974).
HOKIN, M. R.: J. gen. Physiol. **50**, 2197 (1967).
HOLMES, W. N., MCBEAN, R. L.: J. exp. Biol. **41**, 81 (1964).
HOLMES, W. N., STEWART, D. J.: J. exp. Biol. **48**, 509 (1968).
HOLMES, W. N.: PHILLIPS, J. G., BUTLER, D. G.: Endocrinology **69**, 483 (1961).
HOLMES, W. N., PHILLIPS, J. G., CHESTER JONES, I.: Rec. Progr. Horm. Res. **19**, 619 (1963).
HOLT, W. F., IDLER, D. R.: Comp. Biochem. Physiol. **50C**, 111 (1975).
HUGHES, M. R.: Comp. Biochem. Physiol. **32**, 807 (1970).
IDLER, D. R., KANE, K. M.: Gen. comp. Endocr. **28**, 100 (1976).
IDLER, D. R., FREEMAN, H. C., TRUSCOTT, B.: Gen. comp. Endocr. **9**, 207 (1967).
KARLSSON, K.-A., SAMUELSSON, B. E., STEEN, G. O.: Europ. J. Biochem. **46**, 243 (1974).
KEYS, A. B.: Z. vergl. Physiol. **15**, 364 (1931).
KEYS, A., WILLMER, E. N.: J. Physiol. (Lond.) **76**, 368 (1932).
KOMNICK, H.: Protoplasma (Wien) **56**, 274 (1963a).
KOMNICK, H.: Protoplasma (Wien) **56**, 605 (1963b).
KOMNICK, H.: Protoplasma (Wien) **58**, 96 (1964).
KOMNICK, H., KNIPRATH, E.: Cytobiologie **1**, 228 (1970).
KOMNICK, H., KOMNICK, U.: Z. Zellforsch. **60**, 163 (1963).
KOMNICK, H., WOHLFARTH-BOTTERMANN, K. E.: Z. Zellforsch. **74**, 123 (1966).
LEMIRE, M., DELOINCE, R., GRENOT, C.: C. R. Acad. Sci. (Paris) **274**, 3413 (1972).
LENNEP, E. W. VAN: J. Ultrastruct. Res. **25**, 94 (1968).
LENNEP, E. W. VAN, KOMNICK, H.: Cytobiologie **2**, 47 (1970).
LENNEP, E. W. VAN, LANZING, W. J. R.: J. Ultrastruct. Res. **18**, 333 (1967).
LEVINE, A. M., HIGGINS, J. A., BARRNETT, R. J.: J. Cell Sci. **11**, 855 (1972).
MAETZ, J.: Phil. Trans. B **262**, 209 (1971).
MAETZ, J., BORNANCIN, M.: Fortschr. Zool. **23**, 322 (1975).
MARTIN, B. J., PHILPOTT, C. W.: J. exp. Zool. **186**, 111 (1973).
MARTIN, B. J., PHILPOTT, C. W.: Cell Tissue Res. **150**, 193 (1974).

NECHAY, B. R., LARIMER, J. L., MAREN, T. H.: J. Pharmacol. exp. Ther. **130**, 401 (1960).
NORRIS, K. S., DAWSON, W. R.: Copeia 638 (1964).
OGURI, M.: Science **144**, 1151 (1964).
PEAKER, M.: J. Physiol. (Lond.) **213**, 399 (1971a).
PEAKER, M.: Phil. Trans. B. **262**, 289 (1971b).
PEAKER, M., HANWELL, A.: Pflügers Arch. **352**, 363 (1974).
PEAKER, M., LINZELL, J. L.: Salt Glands in Birds and Reptiles. Cambridge: Cambridge University Press 1975.
PEAKER, M., STOCKLEY, S. J.: Nature (Lond.) **243**, 297 (1973).
PEAKER, M., STOCKLEY, S. J.: Experientia (Basel) **30**, 158 (1974).
ROSSUM, G. D. V. VAN: Biochim. biophys. Acta **126**, 338 (1966).
SCHMIDT-NIELSEN, B.: Amer. J. Physiol. **230**, 514 (1976).
SCHMIDT-NIELSEN, K.: Circulation **21**, 955 (1960).
SCHMIDT-NIELSEN, K., FÄNGE, R.: Nature (Lond.) **182**, 783 (1958).
SCHMIDT-NIELSEN, K., JÖRGENSEN, C. B., OSAKI, H.: Fed. Proc. **16**, 113 (1957).
SCHMIDT-NIELSEN, K., JÖRGENSEN, C. B., OSAKI, H.: Amer. J. Physiol. **193**, 101 (1958).
SCHMIDT-NIELSEN, K., BORUT, A., LEE, P., CRAWFORD, E.: Science **142**, 1300 (1963).
SHOEMAKER, V. H., NAGY, K. A., BRADSHAW, S. D.: Comp. Biochem. Physiol. **42A**, 749 (1972).
SIEGEL, N. J., SCHON, D. A., HAYSLETT, J. P.: Amer. J. Physiol. **230**, 1250 (1976).
SMITH, D. P.: Cytobios **5**, 217 (1972).
SMITH, H. W.: Amer. J. Physiol. **93**, 480 (1930).
STEWART, D. J., SEMPLE, E. W., SWART, G. T., SEN, A. K.: Biochim. biophys. Acta **419**, 150 (1976).
STOCKEM, W., KOMNICK, H., WOHLFARTH-BOTTERMANN, K. E.: Helgol. wiss. Meeresunters. **18**, 424 (1968).
TEMPLETON, J. R.: Amer. Zool. **3**, 530 (1963).
TEMPLETON, J. R.: Texas Rep. Biol. Med. **22**, 206 (1964).
TEMPLETON, J. R.: Copeia 136 (1967).
TEMPLETON, J. R.: Symp. zool. Soc. (Lond.) **31**, 61 (1972).
TEMPLETON, J. R., MURRISH, D., RANDALL, E., MUGAAS, J.: Amer. Zool. **8**, 818 (1968).
TEMPLETON, J. R., MURRISH, D. E., RANDALL, E. M., MUGAAS, J. N.: Z. vergl. Physiol. **76**, 255 (1972).
THESLEFF, S., SCHMIDT-NIELSEN, K.: Amer. J. Physiol. **202**, 597 (1962).
WOOD, W. G., BALLANTYNE, B.: J. Anat. (Lond.) **103**, 277 (1968).
WRIGHT, A., PHILLIPS, J. G., HUANG, D. P.: J. Endocr. **36**, 249 (1966).

Chapter 13

Gastric Secretion

T.E. MACHEN and J.G. FORTE

A. Introduction

The stomach is a specialized region of the alimentary tract lying between the esophagus and small intestine and is found in all vertebrates with the exception of a few fish (ANDREW, 1959). Like most exocrine glands, the secretory function of the stomach is carefully regulated; in response to appropriate neural or hormonal influences, the stomach secretes relatively large volumes of isotonic hydrochloric acid (HCl). The acidity of the gastric juice can reach 160 mN, *i. e.*, pH 0,8, in mammals, thus representing an enormous chemical gradient which must be provided with a ready supply of energy from cellular metabolism. In addition to its active and well-adapted energy metabolism, when stimulated to secrete HCl, very dramatic ultrastructural changes occur in the apical membranes of the oxyntic (acid-secreting) cells. The stomach is therefore an interesting transporting organ from a number of different perspectives: those of neurohumoral control (both extracellular and intracellular) of secretion, the energetic basis of transport, and, finally, the "barrier" function of the stomach, since the healthy gastric epithelium will not be destroyed or digested by its own secretory product. The stomach also secretes digestive enzymes (pepsinogen) and mucus, but we will not be concerned with these organic secretions here. In this chapter, we will place particular emphasis on the molecular mechanisms by which H^+, Cl^-, Na^+, and water are transported across the gastric epithelium. Many different techniques have been used to study these problems, but we will be discussing a rather limited number of these, primarily the *in-vitro* biophysical, electrophysiological, and biochemical approaches. We will discuss work which has approached the gastric mucosa as a "black box", *e. g.,* transepithelial electrical and ion-flux data analyzed as though all the cells of the tissue are identical, though the mucosal and serosal membranes are assumed to be quite distinct. We will also concentrate on more recent experimental approaches, which have attempted to look within the black box. Such experiments include making intracellular recordings of potential differences (PD's) and resistances with microelectrodes so as to develop a more comprehensive picture as to the location and nature of the principal transport functions. Finally, we will review those biochemical approaches that have been used to probe the energetic requirements and transport mechanisms at the subcellular level.

B. Morphological Features of Gastric Mucosa

I. Histology

As seen in Figure 1, the whole single-cell layer of the gastric epithelium is thrown into elaborate folds (rugae) and can be roughly divided into regions with histological and functional distinctions. Secretion of acid occurs in the fundus and main body (corpus) of the stomach, which also secretes mucus and digestive enzymes into the lumen. The antrum does not secrete acid nor enzymes, but is very important for the hormonal control of gastric secretion (see BROOKS, 1967).

Figure 2 shows schematically the most important features of the gastric mucosa itself. It is composed of a single layer of epithelial cells, attached to each

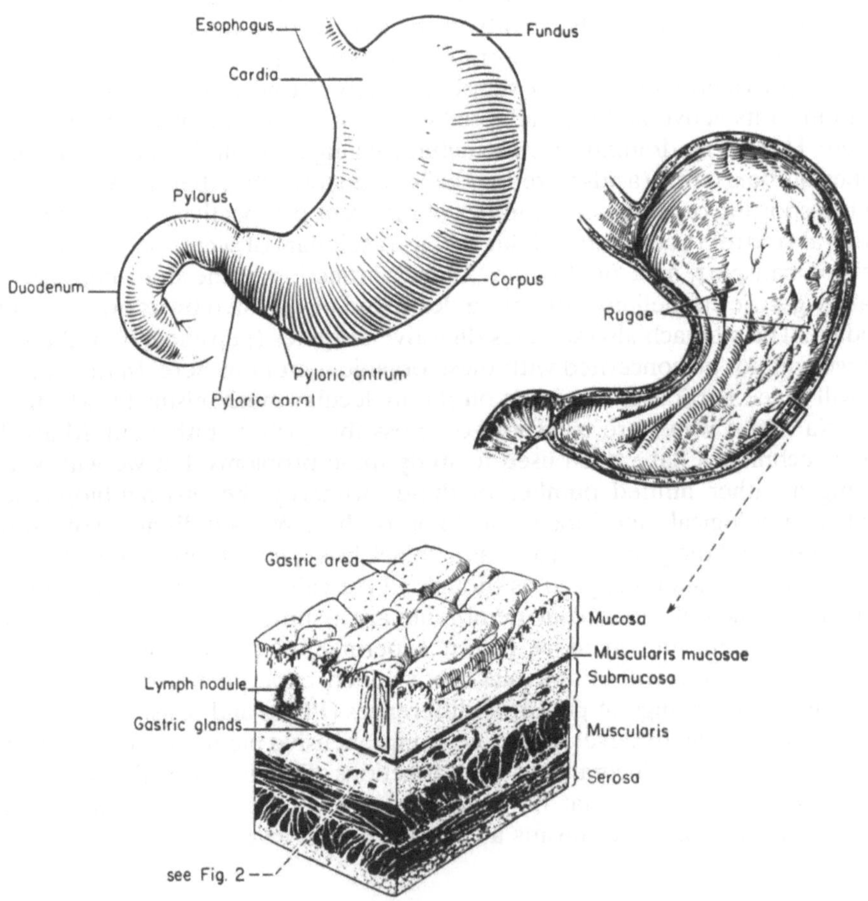

Fig. 1. Diagram of surface and interior of a simple glandular stomach. Lower: slightly enlarged view of stomach wall and its four layers. (Taken from ITO, 1967, with permission)

other by typical junctional complexes (FARQUHAR and PALADE, 1963; SEDAR and FORTE, 1964). The epithelium is folded into many inpocketings or glands, which extend below the central gastric lumen. Mucus-secreting, surface epithelial cells cover the luminal aspect and project down into the gastric pits, while the subadjacent gastric glands—usually four or five—lead into the gastric pits. In the

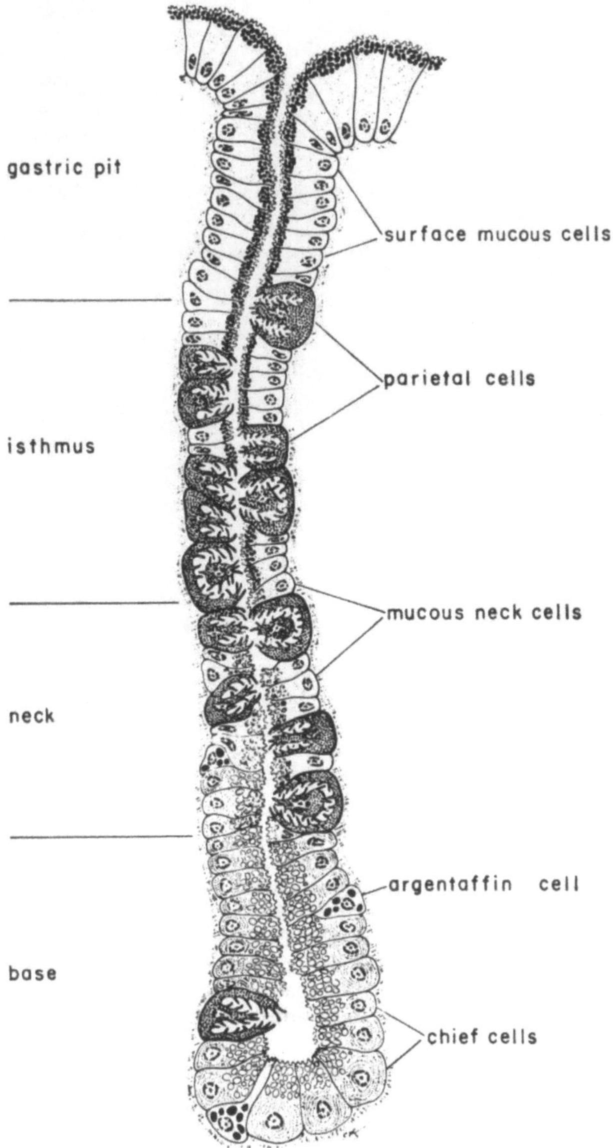

Fig. 2. Diagram of a tubular oxyntic gland from corpus or fundus of bat stomach. Structural arrangement is very similar to this in all mammalian species. Frog and lizard gastric mucosae do not have separate parietal (oxyntic) and chief cells. Their function is combined into one cell type in these species. (From ITO, 1967, with permission)

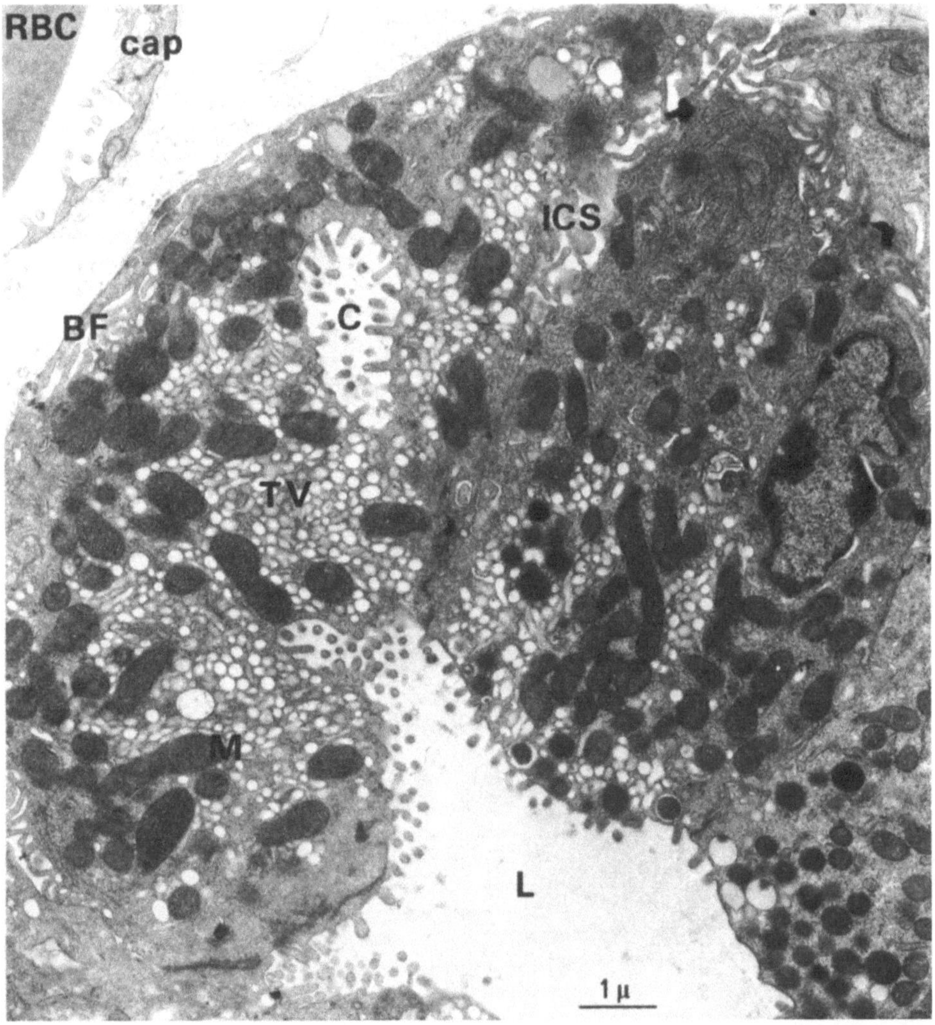

Fig. 3. Electron micrograph of oxyntic cells from nonsecreting piglet gastric mucosa. Tubulovesicles (TV) and large mitochondria (M) are prominent within the cells. Another specialized feature of mammalian oxyntic cells are the intracellular canaliculi (C) extending from luminal (L) aspect to regions deep within cell. Canaliculus indicated by C makes contact with luminal surface in another plane of section. Luminal and canalicular surfaces are studded with short, stubby microvilli (contrast with Fig. 4). Also shown are lateral intercellular spaces (ICS), basal infoldings (BF) and a red blood cell (RBC) in a basally adjacent capillary (cap)

mammal, the gastric gland is lined with enzyme secreting (chief) cells and acid-secreting (oxyntic) cells. In amphibians and reptiles, gastric glands are constituted of a single cell type that secretes both acid and digestive enzymes. Other cell types (*e. g.*, argentaffin cells, see ITO, 1967) are also present, but very scarce. These latter cells may play some role in control of secretion (RANGACHARI, 1975b), but they probably do not contribute directly to the electrophysiological properties of the tissue.

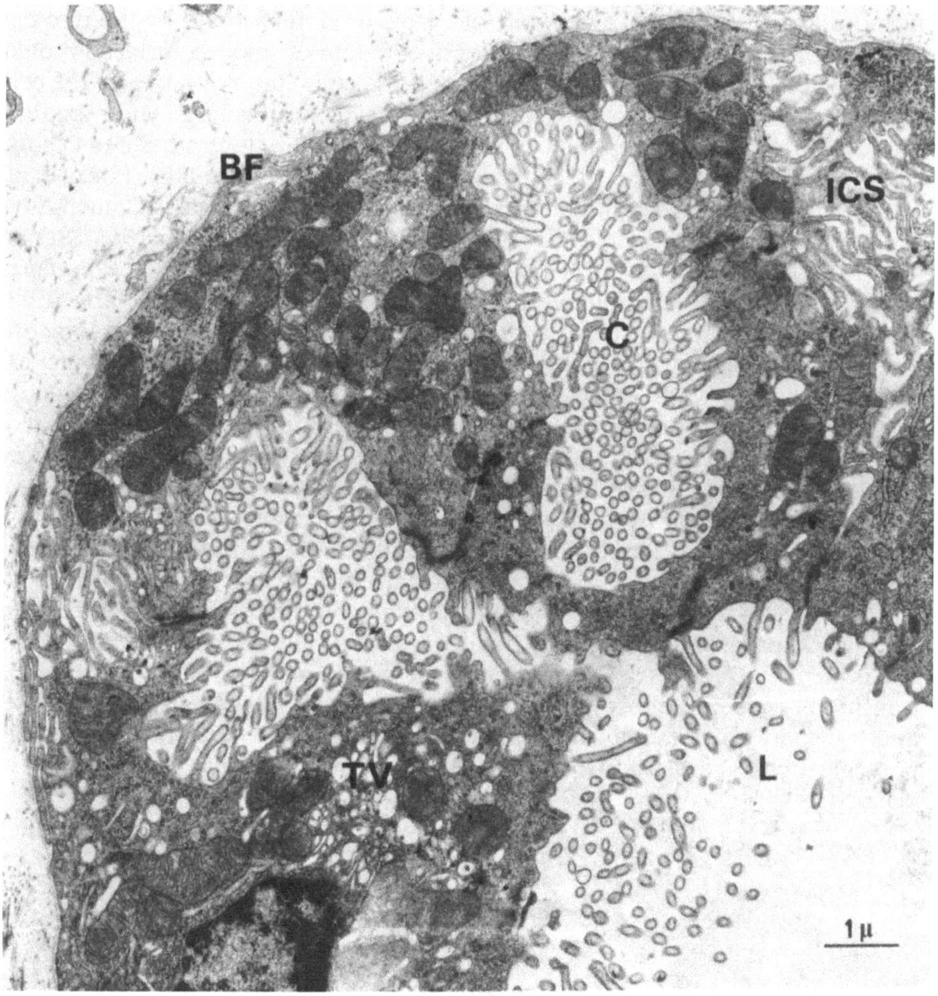

Fig. 4. Oxyntic cells from piglet gastric mucosa stimulated by histamine. This preparation was the paired half to that in Fig. 3 and was secreting at rate of 7,5 μEq HCl $cm^{-2} \cdot h^{-1}$. Note numerous long cytoplasmic extensions at canalicular (C) and luminal (L) aspects of cell; these greatly increase apical surface area. Tubulovesicles (TV) within cell are drastically reduced compared to the nonsecreting state (cf. Fig. 3)

II. Ultrastructure of the Oxyntic Cell

The oxyntic cells provide another level of morphological complexity since they undergo marked transformations when stimulated to secrete acid (VIAL and ORREGO, 1960; ITO, 1961; SEDAR and FORTE, 1964; HELANDER and HIRSCHOWITZ, 1972; FORTE et al., 1975). As shown in Figure 3, the resting or nonsecreting oxyntic cell has a rather elaborate membrane ultrastructure, which includes apical microvilli, a network of canals or clefts formed by the infolded apical surface (canaliculi), and an extensive system of smooth-surfaced tubules and vesicles located in the apical portion of the cells (tubulovesicles). The cells also contain a well-developed Golgi apparatus, as well as large and numerous mitochondria. The development of the tubulovesicular system has been shown to be directly correlated to the stomach's capacity to produce HCl (FORTE et al., 1969), and its arrangement changes dramatically when the gastric mucosa has been stimulated to secrete HCl (VIAL and ORREGO, 1960; ITO, 1961; SEDAR, 1961a and b; FORTE et al., 1975). Specifically, as shown for the stimulated oxyntic cell in Fig. 4, tubulovesicles decrease in area while the apical and intracellular canalicular surfaces become very much engrossed with long and extensive membranous projections. It is generally accepted that the tubulovesicles communicate with—and contribute directly to—the increased surface area of the apical plasma membrane during HCl secretion. Since HELANDER and HIRSCHOWITZ (1972) have shown that there is conservation of total membrane surface area, such ultrastructural transformations between tubulovesicles and apical surface form an important basis for a membrane recycling hypothesis underlying gastric HCl secretion. Future definition of the mechanisms of membrane fusion and recycling promises important insight into secretory function.

Thus, the gastric mucosa is a geometrically complex tissue due to the extensive glandular infoldings, the presence of multiple cell types in parallel, and the elaborate membrane transformations that occur in oxyntic cells during the transition between the resting and actively secreting states. Analysis of electrical and transport phenomena in this tissue must ultimately correlate to these morphological characteristics before a coherent picture of gastric secretion can be claimed.

C. Ion Transport

I. General

The gastric mucosa of most vertebrate species maintains an electrical potential difference with the mucosal side negative to the serosal, or blood, side by -10 to -60 mV. Adult rabbit gastric mucosa maintained *in vitro* has a very small PD, and in the dogfish the PD is actually reversed, with the mucosal side being $+2$ to $+5$ mV with respect to the serosa. These latter cases may be exceptional, though, for other elasmobranch (HOGBEN, 1959a) and teleost (REHM, 1962) fish, and also the newborn rabbit (MACHEN et al., 1974), develop significant PDs with the typical polarity. The gastric PD is generated by the active transport current of monovalent ions traversing a transepithelial resistance with a range of about 100–600 Ω-cm^2. Active Cl^- secretion from the serosal to the mucosal solution is a principal component for the generation of the PD (HOGBEN, 1955a). Indeed, for the well-studied frog mucosa, the short-circuit current (I_{sc}) is equivalent to that Cl^- that is secreted in *excess* of the H^+ produced (Table 1).

Table 1. Chloride transport by short-circuited bullfrog gastric mucosa

Chloride transfer		Net charge transfer	
$J^{ns}_{Cl^-}$	10.65	Short-circuit current	3.05
$J^{sn}_{Cl^-}$	6.38	H^+ secretion	1.20
Net Cl^- flux	4.27	Sum	4.25

Mean conductivity
Start: 3.21 mmhos cm^{-2}
End: 2.28 mmhos cm^{-2}

Values for $J^{ns}_{Cl^-}$ and $J^{sn}_{Cl^-}$ are the unidirectional fluxes of Cl^- from nutrient to secretory (ns) and secretory to nutrient (sn), respectively. All values of flux and current are given in $\mu Eq\ cm^{-2} \cdot h^{-1}$. (Data from HOGBEN, 1959b).

Active H^+ secretion (HEINZ and DURBIN, 1959) and Na^+ absorption (KITAHARA et al., 1969; FORTE and MACHEN, 1975) may also contribute to the PD, the exact percentage depending on the particular circumstances of bathing solutions employed, species being studied, and the secretory state of the tissue. For example, in Cl^--free solutions, the frog gastric mucosa exhibits a mucosa-positive PD, and I_{sc} is proportional to the H^+-secretion rate (HEINZ and DURBIN, 1959; FORTE, 1971). The H^+ mechanism thus seems to have the ability to generate a net current (*i. e.*, it is electrogenic), and the EMF of this system is normally masked by Cl^- secretion. Further, Na^+ absorption contributes significantly to the observed PD in mammalian gastric mucosa (KITAHARA et al., 1969; FORTE and MACHEN, 1975), especially during resting conditions. The magnitude of the PD is also directly dependent on the secretory activity, as well as the transepithelial resistance of the tissue.

When a resting preparation of gastric mucosa is stimulated to secrete HCl, there is a decrease in resistance. In some species transmucosal PD is observed to decrease after stimulation of HCl secretion; DURBIN (1967) has noted that the drop in PD can be almost entirely attributed to the decrement in resistance with no appreciable change in current generated by the tissue. In a few species PD and short-circuit current are both significantly increased when HCl secretion is stimulated. The decrease in resistance can be attributed, at least partially, to increases in the apical membrane surface area of the oxyntic cells (FORTE et al., 1975), but other qualitative changes in the membranes, such as in carriers or pumps, must also be considered (see Section E. I. 1.).

II. H^+ Secretion

1. Locus of H^+ Secretion

H^+ secretion by the gastric mucosa occurs down an electrical gradient, but against a huge concentration gradient from blood to gastric juice. From an analysis of these factors, it is clear that H^+ is actively transported against a gradient of electrochemical potential (see REHM, 1950). It is also generally accepted that the locus of this H^+ secretion is at, or near, the apical membranes of the oxyntic cells. If H^+ secretion were localized to the serosal membrane, then upon stimulation one would expect an intracellular milieu of extremely low pH. This is not consistent with what is known about intracellular pH in general, nor is it supported by experiments which have attempted to probe this question. For instance, HERSEY (1971) and VILLEGAS (1975), using separate techniques of pH indicator dyes and distribution of DMO, respectively, both noted that the pH of the cellular mass actually increased when the tissue was stimulated to secrete H^+. The logical conclusion is that secretion of free H^+ is localized to the mucosal membrane.

2. Acid-Base Balance of Oxyntic Cells

Concomitant with the secretion of H^+, an equivalent amount of base is necessarily formed. The asymmetric liberation of equivalent amounts of H^+ and OH^- (or HCO_3^-) by gastric mucosa has been observed by many workers (*e. g.*, DAVIES, 1948; TEORRELL, 1951). Accepting that the H^+ secretion mechanism is located at the mucosal membrane of the oxyntic cells, the OH^- is produced within the cell, and mechanisms must be operating to buffer or neutralize this OH^-. DAVIES (1948) has shown that the OH^- is neutralized by the CO_2/HCO_3^- system, possibly *via* a carbonic anhydrase-catalyzed reaction. It is known that the oxyntic cells have a high content of carbonic anhydrase (DAVENPORT, 1939; MAREN, 1967). The HCO_3^- generated in the reaction of CO_2 with OH^- is liberated at the serosal side of the cell (Fig. 5a) and is the phenomenon that accounts for what is known as the "alkaline tide of a meal."

A few interesting features regarding this whole scheme are pertinent at this point. First, several types of experiments have emphasized the importance of exogenous CO_2 for maintenance of H^+ secretion. The rate of HCl secretion is sensitive to the availability of CO_2 (DELRUE, 1931; GRAY et al., 1940). For the isolated frog gastric mucosa in the absence of external CO_2 supplies, H^+ secretion occurs in an apparently normal fashion so long as the rate of H^+ secretion does not exceed the rate of CO_2 production from metabolism (DAVIES, 1948; DAVIES and LONGMUIR, 1948). However, during maximal stimulation, when the rate of H^+ secretion can be stoichiometrically two to six times in excess of endogenous CO_2 production, the gastric mucosa will ulcerate, possibly due to the intracellular accumulation of alkali and the resulting altered cell pH (DAVIES and LONGMUIR, 1948). Both IMAMURA (1967) and KIDDER (1974) have re-emphasized the importance of adequate supplies of CO_2 for optimal maintenance of H^+ secretion.

Second, it has been found recently that in the absence of exogenous CO_2, normal secretion rates could be maintained without ulceration in frog gastric mucosa if the serosal solution was changed to approximately pH 4 (SANDERS et al., 1973; SILEN et al., 1975). This would seem to violate the earlier premise of a direct role for CO_2. However, an interesting mechanism has been proposed for these specialized circumstances (SANDERS et al., 1973), whereby the low serosal

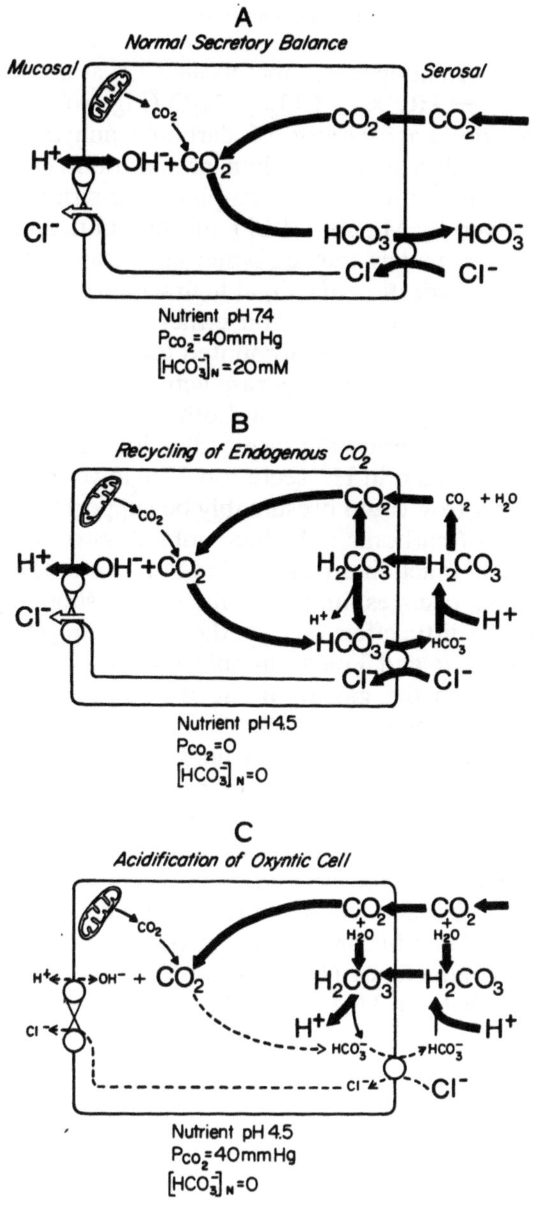

Fig. 5. Proposed schema for role of CO_2 under normal and unusual circumstances. (A) Under normal conditions, exogenous CO_2 neutralizes base formed in cell by production of H^+. HCO_3^- thus formed exits from cell in exchange for Cl^-. (B) In absence of exogenous CO_2, endogenous CO_2 leaves cell as HCO_3^- in usual exchange for Cl^-. Limited endogenous CO_2 is recycled when nutrient solution is acidic. HCO_3^- is converted to H_2CO_3 close to nutrient membrane. Either H_2CO_3 or CO_2, both readily diffusible, reenters cell to neutralize base formed when H^+ is secreted; thus intracellular acid-base balance is maintained. (C) When exogenous CO_2 is introduced while nutrient solutions acidic, intracellular pH tends to fall because recycling of endogenous CO_2 occurs *pari passu* with acidification of cell by readily diffusible exogenous CO_2. If H^+ secretion and consequent formation of base decreases, amount of acid neutralized intracellularly is less and pH inside cell is lowered. Amount of exogenous acid (CO_2, H_2CO_3, or H^+ in nutrient solution) tolerated will depend on rate of H^+ secretion. (From SILEN et al. 1975, with permission)

pH allows a recycling of endogenous, metabolic CO_2 at the serosal membrane by the reaction of $H^+ + HCO_3^- \rightarrow CO_2 + H_2O$ (Fig. 5b).

Third, acetazolamide, an inhibitor of carbonic anhydrase, causes severe reduction of HCl secretion in mammalian preparations (MAREN, 1967), but variable effects have been observed with amphibian gastric mucosa. These findings have been reviewed by MAREN (1967) and by FORTE and SOLBERG (1973), and fit in with the conclusions that carbonic anhydrase activity is absolutely required to catalyze the hydration of CO_2 at high secretory rates. Acetazolamide does not completely inhibit H^+ secretion either *in vivo* or *in vitro*. Residual secretory rates can most probably be accounted for by uncatalyzed CO_2 utilization, or even by residual carbonic anhydrase activity (DAVIES, 1951; FORTE and SOLBERG, 1973). For frog gastric mucosa, both HOGBEN (1955b) and DURBIN and HEINZ (1958) have observed that acetazolamide causes a much more pronounced reduction in I_{sc} than in H^+ secretion. The lower rates of H^+ secretion commonly observed in frog could presumably be supported by the uncatalyzed reaction. The effects of carbonic anhydrase inhibitors on active Cl^- transport may be mediated by their action on a specific Cl^--receptor (HOGBEN, 1967), but an indirect effect *via* changes in intracellular pH should not be ignored. Acetazolamide has very little effect on I_{sc} in the resting frog (DURBIN and HEINZ, 1958) or piglet (MACHEN and FORTE, unpubl.) gastric mucosa, or in actively secreting tissues in which the gassing phase is 100 percent O_2 instead of the usual 95 percent O_2 −5 percent CO_2 (SILEN et al., 1975).

Finally, it is generally accepted that a neutral nonconductive exchange of HCO_3^- for Cl^- occurs across the serosal membrane of the oxyntic cells and serves to maintain cellular acid-base balance. REHM (1967) was the first to offer convincing evidence for neutral HCO_3^-/Cl^- exchange across the serosal surface of the oxyntic cell, which more recently has been acknowledged as an important exchange mechanism for other cell types (*e. g.*, red cells; see Vol. II, Chapter 2). It appears that the total conductance of the gastric serosal surface is equal to the sum of K^+ and Cl^- conductance (HARRIS and EDELMAN, 1964; SPANGLER and REHM, 1968; and see below). HCO_3^- (or OH^-) has been eliminated as a significant conductive species since rapid changes in $[HCO_3^-]$ or pH within the serosal solution have no effect on the transepithelial PD (SANDERS et al., 1973). These experiments and others involving analysis of the effects of Ba^{2+} on tissue conductance (PAÇIFICO et al., 1969), indicate that the simplest explanation is that there is a nonconductive carrier exchange mechanism for Cl^- and HCO_3^- between the cell and the serosal medium.

3. Anion Dependence of H^+ Secretion

Many authors have reported the Cl^- requirement for H^+ secretion in frog (HOGBEN, 1955a; DURBIN, 1964), lizard (HANSEN et al., 1975) and piglet (FORTE and MACHEN, 1975) gastric mucosa. This requirement is almost exclusively from the serosal side; complete replacement of Cl^- in the serosal solution with impermeant anions (SO_4^{--}, glucuronate, or isethionate) causes H^+ secretion to drop by 75 percent or more (HEINZ and DURBIN, 1959; DURBIN, 1964; KANEKO-MOHAM-

MED and HOGBEN, 1964; FORTE and MACHEN, 1975; HANSEN et al., 1975). Replacement of Cl^- with the less permeable anions is accompanied by a large decrease in mucosal conductance and the transepithelial PD often reverses in sign, the mucosal side becoming positive with respect to the serosal side.

The dependency of H^+ secretion on the concentration of Cl^- in the serosal solution, $[Cl^-]_s$, has been shown to occur in a manner resembling the familiar MICHAELIS-MENTEN kinetics. DURBIN (1964) plotted $1/H^+$ rate vs. $1/[Cl_s]$ and showed linear correlations except at extremely low $[Cl^-]_s$; analysis of the data gave $V_{max} = 3$ μEq $cm^{-2} \cdot h^{-1}$ with an apparent K_a for $[Cl^-]_s$ of about 6 mM. The data of FORTE, ADAMS and DAVIES (1963) yield an even lower $K_a \approx 3$ mM. Since we know that Cl^- concentration within the cells, $[Cl^-]_c$, is likely to be much less than $[Cl^-]_s$ (DAVENPORT and ALZAMORA, 1962), these data suggest that the apparent activation of the H^+ secretory mechanism by Cl^- at the mucosal membrane is probably much lower even than 1 mM. This mechanism obviously has a high affinity for Cl^-.

However, the impression should not be given that the H^+ secretion mechanism is necessarily coupled in a one-for-one fashion with Cl^- secretion. In fact, H^+ secretion can operate in the presence of anions other than Cl^-, the selectivity of the H^+ mechanism being in the order $Br^- \geqslant Cl^- > I^-$ (DURBIN, 1964). It thus appears that the H^+ transport mechanism requires at least the presence of a permeable anion to achieve maximum secretory capacity, but that the selectivity for Cl^- is not very strict. We stress that the anion selectivity may be only apparently that of the H^+ mechanism. Even when relatively non-permeating anions such as SO_4^{2-} or isethionate completely replace Cl^-, H^+ secretory rates are 20 to 30 percent of those observed in normal Cl^- solutions. In order to balance charge requirements of the H^+ transport mechanism, some anion must move in the same direction, or a cation must move from mucosal to serosal solution as an exchange ion. The interpretation of these data from anion substitution experiments has been usefully applied to the theories of HCl transport detailed below.

4. Cation Dependence of H^+ Secretion

Although the gastric mucosa can secrete HCl in the absence of K^+, Na^+, and Ca^{2+} in the mucosal solution, the tissue requires all three of these cations in the serosal solution for normal function. For example, removal of K^+ from the serosal solution causes H^+ secretion to decrease to zero (DELRUE, 1930; GRAY and ADKISON, 1941; DAVENPORT, 1963). Concomitantly, the transepithelial PD transiently increases and then decreases toward zero while the resistance increases to very high levels (DAVIS et al., 1965). Because of the large decrements in H^+ secretion and PD, and the rise in resistance, the current-generating mechanisms (primarily Cl^- secretion) must also decrease to very low levels during K^+-free conditions. The secretory and electrical effects of K^+ removal are reversed by restoring K^+ to the serosal solution and ameliorated by addition of high K^+ (104 mM) to the mucosal bath.

Although DAVIS et al. (1965) could find no significant difference in K^+ and water contents of the gastric epithelium upon changing to K^+-free solutions, other workers have found consistent decreases in tissue K^+ (SEDAR and WIEBELHAUS, 1972; TAKEGUCHI et al., 1976). These data can best be explained on the basis of the analysis by TAKEGUCHI et al. (1976), who propose compartments of free and bound K^+ within the epithelial cells. Their experiments show that in certain regions of the cell $[K^+]$ is normally relatively low and that it is reduced to essentially zero by removal of external K^+; furthermore, the shutoff of H^+ secretion results from a depletion of K^+ from its critical exchangeable pool. Thus, most of the intracellular K^+ appears to be bound in an osmotically inactive form—either within intracellular organelles or to some fixed anionic charges.

The presence of Na^+ and Ca^{2+} in the serosal solution is also critical for gastric function. Na^+-free solutions (replaced with choline) cause a decrease (to 20–40%) in H^+ secretion after two hours. PD and short-circuit current also decrease (SACHS et al., 1966; HANSEN et al., 1975). Approximately 20 percent of tissue Na^+ remains after two hours. Ca^{2+}-free solutions have even more drastic effects on gastric secretory activity. At concentrations below 10^{-5} M (FORTE, unpubl.), PD, resistance, and H^+ secretion all decrease nearly to zero. It has been concluded from both physiological (FORTE and NAUSS, 1963) and ultrastructural (SEDAR and FORTE, 1964) studies that Ca^{2+} is required for maintenance of structural and functional integrity of the junctional complex that binds adjacent cells together. That is, Ca^{2+} is required to maintain the shunt conductance at low levels. When $[Ca^{2+}] \approx 10^{-5}$ to 10^{-4}M, H^+ secretion and PD are both decreased below control levels, while resistance increases (JACOBSON et al., 1965). Thus, in addition to maintaining the integrity of intercellular junctions, Ca^{2+} is required in the serosal solution for functional secretory capacity. There may also be some role for Ca^{2+} in secretagogue-receptor interaction and/or control of cellular cAMP levels, as suggested by KASBEKAR (1974). Finally, the effects of Ca^{2+}-free media, like those of K^+ and Na^+-free media, may be only indirect—elimination of one specific cation causing subsequent changes in the intracellular ionic milieu, which in turn leads to the observed secretory inhibition.

III. Cl^- Transport

1. Active Transport

In most of the early theories of the mechanism of gastric acid secretion, it was assumed that the chloride ions followed the secreted H^+ ions passively so as to maintain electroneutrality (DAVIES, 1951; HEINZ and ÖBRINK, 1954). It has now been established that Cl^- movement through the gastric epithelium occurs by three distinct processes: active transport, exchange diffusion, and simple ionic diffusion. (For a discussion of these pathways and some of the influencing conditions in frog mucosa, see FORTE, 1969). Proof for active transport of Cl^- has been derived both *in situ*, from an analysis of the electrical p. d. and the respective Cl^- concentrations in the blood and gastric juice (REHM, 1950), and by flux analysis *in vitro*, with the short-circuit current technique (HOGBEN, 1951). HOGBEN demonstrated that with identical bathing solutions on each side of the tissue and no electrical PD (short-circuit), there was a net secretion of Cl^- from the serosal to the mucosal solution. Table 1 gives his results, among which he recorded unidirectional fluxes of Cl^-. As shown, the short-circuit current (I_{sc}) was equal to net transport of Cl^- ($J_{Cl^-}^{net}$), in excess of that secreted as HCl, *i. e.*:

$$I_{sc} = J_{Cl^-}^{net} - \text{HCl production} = J_{Cl^-}^{net} - J_{H^+}^{net} \quad (1)$$

A similar relationship also holds for the mammalian gastric mucosa, with the exception that there is also a net Na^+ absorption ($J_{Na^+}^{net}$), which must be algebraically added into the summation of ionic flows across the tissue. Thus, as shown from the data of Fig. 6 for the mammalian preparation during both resting and actively secreting conditions:

$$I_{sc} = J_{Cl^-}^{net} - J_{H^+}^{net} + J_{Na^+}^{net} \quad (2)$$

The selectivity of this active anion transport mechanism for the halides, as measured by the net transport of anions during the short-circuited state in frog gastric mucosa, was in the order $Cl^- \geqslant Br > I^-$ (HOGBEN and GREEN, 1958). There is no evidence for active transport of either SO_4^{--} (HOGBEN, 1955) or isethionate (FORTE, unpubl.).

Looking again at Table 1 and Fig. 6, we find that the data are consistent with the idea that a component of active Cl^- secretion is coupled to H^+ production (acidic Cl^-) while a certain amount of the Cl^- secretion seems to be H^+-independent (nonacidic Cl^-) (DAVENPORT, 1940; DURBIN and HEINZ, 1958; FORTE and MACHEN, 1975). Dividing active Cl^- transport into acidic and nonacidic portions is admittedly simplistic and perhaps a little arbitrary, but a number of experimental observations indicate that these terms may, in fact, be realistic descrip-

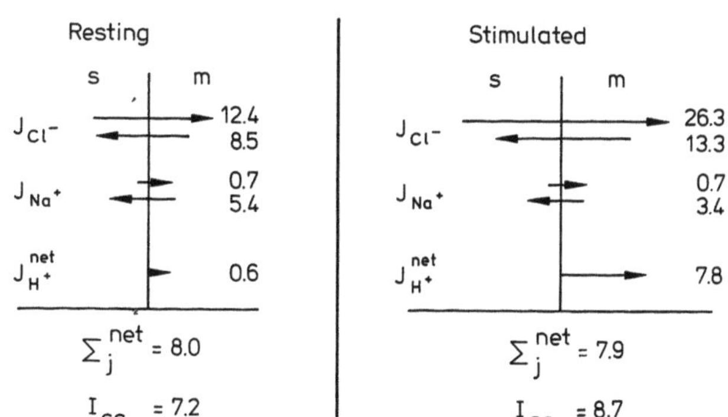

Fig. 6. Unidirectional ionic fluxes across short-circuited piglet gastric mucosa during resting and actively secreting conditions. In both cases, sum of net ionic fluxes ($\Sigma^{net}_{H^+, Cl^-, Na^+}$) is approximately equal to short-circuit current (I_{sc}). (Data from FORTE and MACHEN, 1975)

tions. For example, active Cl^- secretion occurs in the gastric mucosa of developing animals in which oxyntic cells have not yet differentiated (tadpole: FORTE et al., 1969a; fetal rabbit: WRIGHT, 1962). In addition, during resting conditions in adult animals, net Cl^- secretion is far in excess of H^+ production (Fig. 6). After histamine stimulation, there are increases in H^+ production, unidirectional fluxes of Cl^-, and net Cl^- secretion, with the increase in H^+ secretion being matched rather closely by the increase in net Cl^- flux (Fig. 6).

There are also interesting effects of two agents, acetazolamide and Ba^{2+}, which can be interpreted in terms of these two types of Cl^- secretion, acidic and nonacidic. In the frog gastric mucosa, 10 mM acetazolamide in the serosal solution inhibits I_{sc}, which is due exclusively to Cl^- transport, without affecting H^+ production (HOGBEN, 1967). A similar effect is also noted when 1 mM Ba^{2+} is added to the serosal solution. In the original experiments it was reported that Ba^{2+} caused only an increase in resistance, presumably due to competition with K^+ for conductance sites in the serosal membrane; there were no changes in H^+ secretion or PD (PACIFICO et al., 1969). In later experiments on the resting frog gastric mucosa, though, RANGACHARI (1975a) showed that Ba^{2+} caused a decrease in PD as well as an increase in resistance. There was a net decrease in I_{sc}, which could be accounted for mainly by a decrease in $J^{net}_{Cl^-}$. MACHEN and FORTE (1976) have found that Ba^{2+} also inhibits net Cl^- secretion in histamine-stimulated piglet gastric mucosa, with no effect on the rate of H^+ secretion. The specific molecular mechanism by which Ba^{2+} inhibits Cl^- secretion has not yet been specified.

The results from all these experiments imply that there are indeed two separate mechanisms for Cl^- secretion by gastric mucosa. In this regard, it was mentioned above that the Cl^- secretion associated with H^+ must have $K_a < 1$ mM. From the data of FORTE et al (1963) it can be calculated that the K_a of the current-generating (*i. e.*, nonacidic Cl^-) mechanism must be at least a factor of 10 higher than this. Thus, Ba^{2+} and acetazolamide may act by limiting Cl^- entry into gastric epithelial cells, thereby inducing competition for the available cellular Cl^- between the acidic (low-K_a) and nonacidic (high-K_a) Cl^- secretion mechanisms. Another possibility is that acidic and nonacidic Cl^- secretions are mediated by different cell types, and that Ba^{2+} and acetazolamide act on the nonoxyntic cell type.

2. Exchange Diffusion

A significant component of the isotopically measured flux of Cl^- across gastric mucosa has been shown to occur by the phenomenon of exchange diffusion (see Vol. I, Chapter 5). This was first pointed out by HOGBEN (1955a) through his analysis of data similar to those shown in Table 1. The partial ionic conductance due to Cl^- (G_i) can be calculated from the equation developed by USSING (1949) and HODGKIN (1951):

$$G_i = \frac{z^2 F^2}{RT} \cdot J_i \tag{3}$$

where J_i is the unidirectional flux of the ion (in this case Cl^-); z, F, R and T have their usual meanings. HOGBEN found that the partial ionic conductance of Cl^-, calculated from Eq. (3) from the passive mucosal-to-serosal Cl^- flux, was greater than the measured total transepithelial conductance. Thus, he concluded that a significant fraction of Cl^- movement across the gastric epithelium must occur *via* an uncharged, nonconductive combination of Cl^- with some positively charged membrane carrier.

An independent demonstration of this proposed exchange diffusion for Cl^- was provided by HEINZ and DURBIN (1957), who showed that the unidirectional flux of isotopic Cl^- (from serosa to mucosa) was dependent on the presence of Cl^- in the trans-solution. Thus, during short-circuit conditions, replacement of Cl^- in the mucosal solution with a variety of anions and nonelectrolytes caused serosal-to-mucosal flux of Cl^- to decrease. HEINZ and DURBIN modeled and characterized the system as an ion exchange operating through some mobile membrane carrier. The exchange process exhibited the familiar selectivity for the halides, namely $Cl^- \geqslant Br^- > I^-$ (DURBIN et al., 1964).

The components of Cl^- flux across bullfrog gastric mucosa were further studied by FORTE (1969), who characterized the fractional flux occurring by active transport, passive diffusion and exchange diffusion. Since these studies with the frog and later work with piglet gastric mucosa (FORTE and MACHEN, 1975) showed some correlation between the magnitude of the HCl component of Cl^- transport and Cl^- exchange diffusion, it was suggested that these processes may be occurring at the same site, even possibly using the same carrier.

Several experiments have been designed to specify the locus, or at least the membrane interface, at which Cl^- exchange occurs. For example, COTLOVE and HOGBEN (1956) measured the rate of uptake of ^{36}Cl across the gastric epithelial surfaces and showed that the relative flux across the mucosal border was 16-fold greater than that across the serosal border. Under normal conditions, equilibration of ^{36}Cl with tissue Cl^- was rapid when the isotope was added to the mucosal solution. After the addition of 2, 4-dinitrophenol, Cl^- exchange across the mucosal surface was severely depressed, while movement of Cl^- across the serosal surface was hardly affected (COTLOVE et al., 1959). Inhibition of the active component of Cl^- transport by the metabolic inhibitor is to be expected, but it is of interest that Cl^- exchange diffusion localized at the mucosal interface was also reduced, or abolished. VILLEGAS (1965) has come to an opposite conclusion in proposing that the serosal surface is the principal site for Cl^- exchange; however, his analysis was based upon a questionable treatment of efflux kinetics (see FORTE, 1971). It must be pointed out that the data of both COTLOVE and HOGBEN (1956) and VILLEGAS (1965) have the technical shortcomings of not completely accounting for the heterogeneity of cell types within the gastric mucosa.

In actual fact, the locus and molecular mechanisms of active transport and exchange diffusion by gastric mucosa still remain far from clear. By way of summary, we can say that these two transport processes show a number of correlative characteristics (*e. g.*, enhancement by histamine, inhibition by anoxia and dinitrophenol, and the same selectivity for halides) which support some commonalities of mechanism and/or location. With regard to the locus, one body of evidence is consistent with the notion that a major nonconductive exchange pathway for Cl^-/HCO_3^- is present at the serosal surface (see REHM, 1967), while the above-mentioned experiments of HOGBEN and his colleagues support the occurrence of a principal Cl^- exchange diffusion pathway at the mucosal surface. One would expect that the cellular isotopic exchange studies should be able to identify or distinguish these possibilities, and, in fact, there may be separate mechanisms operating at both cell surfaces. However, the final analysis of these problems will require careful considerations of the ionic fluxes and electrochemical differences between the principal compartments: the serosal solution, the cell interior and the mucosal solution.

IV. Na^+ Absorption

The predominant active transport processes occurring in the gastric mucosa are those of H^+ and Cl^- secretion. A residual active Na^+ absorption process (from the mucosa to serosa) has also been demonstrated in frog gastric mucosa (HOGBEN, 1965). This Na^+ transport seems to be much more resistant to anoxia than either H^+ or Cl^- transport (FLEMSTRÖM, 1971). The isolated stomach or gastric mucosa from both mammals (KITAHARA et al., 1969) and reptiles (HANSEN et al., 1975) absorb Na^+ and, in contrast to the situation in the amphibian preparations, this active transport contributes significantly to the electrical properties of these tissues. It has been suggested that the Na^+ transport observed for the *in-vitro* mammalian preparations is an artifactual result of these mammalian tissues being hypoxic (FLEMSTRÖM, 1971). In keeping with this idea, in the early experiments with mammalian gastric mucosae (CUMMINS and VAUGHAN, 1965; KITAHARA et al., 1969), acid secretion rates were very low and could not be stimulated with secretagogues. However, Na^+ transport has been observed *in situ* (BORNSTEIN et al., 1959), and recently FORTE and MACHEN (1975) have demonstrated active Na^+ transport in the piglet gastric mucosa, a preparation that can be stimulated by histamine to secrete H^+ at very high rates (up to 19 $\mu Eq\ cm^{-2} \cdot h^{-1}$). It has also been observed that replacement of Na^+ with choline (or treatment with amiloride; see below) in the mucosal solution has approximately the same effects on the PD, *in vivo* and *in vitro* (MACHEN et al., 1978). Active Na^+ transport thus seems to make a significant contribution to the current-generating capabilities of the mammalian stomach, especially during resting conditions (FORTE and MACHEN, 1975). The exact percentage that Na^+ contributes to the I_{sc} depends on the particular species (KITAHARA et al., 1969).

The cell type that effects net Na^+ transport has not been specified, but it is surely not the oxyntic cell. Fetal rabbit stomach, that has not yet developed oxyntic cells (these develop relatively late during gestation) can still actively absorb Na^+ (WRIGHT, 1962). In addition, the lizard gastric mucosa, which, like the frog, has only two basic cell types – the surface epithelial cells and the oxyntic cells – exhibits a Na^+ transport system very similar to that of the mammal. The implication is that Na^+ absorption is accomplished by the surface cells and not by the oxyntic or chief cells.

This active Na^+ transport system has much in common with those in other "tight" epithelia (FRÖMTER and DIAMOND, 1972) such as frog skin (ERLIJ, 1976) and rabbit urinary bladder (LEWIS and DIAMOND, 1975). For example, this mechanism is sensitive to amiloride, a K^+-sparing diuretic (BAER et al., 1967; NAGEL and DÖRGE, 1970). Addition of amiloride (10^{-5} M) to the mucosal solution of the piglet preparation produced a prompt inhibition of net Na^+ transport, which was closely matched by the decrement in I_{sc} associated with this system; H^+ and Cl^- transport were unaffected (FORTE and MACHEN, 1975; MACHEN et al., 1978). Amiloride produced its action by specifically, and reversibly, blocking the entry of Na^+ from the mucosal solution into the gastric cells. The Na^+ entry step was also found to be pH-sensitive such that at pH 2.0 Na^+ transport was blocked (KITAHARA et al., 1969; MACHEN, SILEN and FORTE, 1975, 1978). From analysis of the tissue-generated current vs. pH of the mucosal solution, an apparent "pK" ≈ 4.0 was found for the groups governing the channel for Na^+ entry into the cells (Fig. 7), which would be consistent with either carboxyl and/or phosphoryl groupings as the predominant functional moieties controlling Na^+ entry.

The transport functions of gastric mucosa are also sensitive to cardiac glycosides, but the effects of these well-known inhibitors of Na^+-K^+-ATPase are slow ($t_{1/2} \approx$ 10–15 min) and are manifest on H^+ and Cl^- transport, as well as Na^+ transport. The inhibitory effects of ouabain on Na^+ transport in the mammalian preparation (FORTE and MACHEN, 1975) are to be expected, but the observed

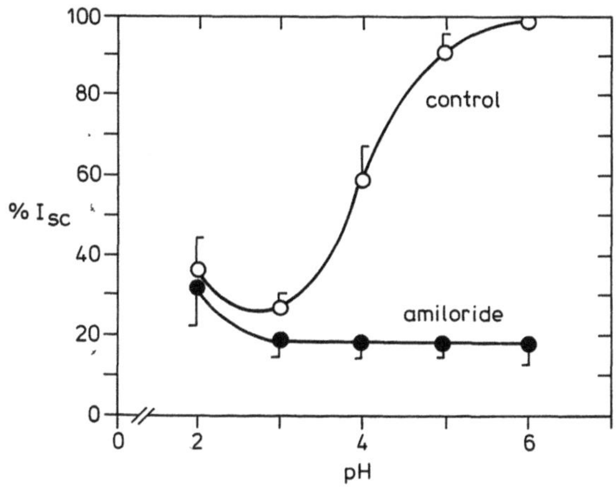

Fig. 7. Effect of pH within mucosal solution on Na^+ current delivered by rabbit gastric mucosa. In this case mucosa was in Cl^--free solutions (isethionate substitution) and nonsecreting, so as to eliminate current due to Cl^- and H^+ transport. I_{sc} is very nearly equal to current of Na^+ transport and is expressed as percentage of that observed at pH 6, which was arbitrarily set equal to 100 percent. The pH of mucosal solution was lowered by addition of H_2SO_4 during control conditions (○) and then repeated after blocking Na^+ transport by adding 10^{-5} M amiloride to mucosal solution (●). During control conditions, I_{sc} decreases below pH 6 with a pK ≈ 4. Increase in I_{sc} below pH 3 may be due to a mucosa-to-serosa H^+ flux. Results obtained in Cl^- solutions are essentially similar except that Cl^- transport current is not affected by amiloride. (Results of MACHEN, SILEN and FORTE, 1975, 1978)

decrements of H^+ and Cl^- transport for both amphibian and mammalian gastric mucosa (COOPERSTEIN, 1959; DAVENPORT, 1962; FORTE and MACHEN, 1975) require some further comment. It has been proposed that these latter effects of ouabain may be not a direct effect on the H^+ and Cl^- pumps, but rather an "indirect" effect due to the disruption of the intracellular milieu (DAVENPORT, 1962; FORTE and SOLBERG, 1973). That is, inhibition of the serosally directed Na^+ pump would lead to elevation of Na^+ and decrease of K^+ within the cell, which has been observed for frog by DAVENPORT (1962). Since reduction of cellular K^+ does lead to reduced H^+ and Cl^- transport (TAKEGUCHI et al., 1976; HARRIS et al., 1958), the proposed "indirect" effects of ouabain appear quite reasonable.

Thus, Na^+ transport in gastric mucosa can be summarized as follows: as for most cells, a Na^+ pump, directed from cell interior to blood, is located on the basolateral membrane of gastric epithelial cells and operates to maintain the intracellular ionic milieu. In the case of frog stomach, where there is very low permeability of the apical membrane to Na^+, there would also be very low levels of net Na^+ transport and the pump would primarily be a leak/pump system at the serosal surface. For those species where a significant transepithelial Na^+ transport occurs, Na^+ enters some cells (most likely the surface epithelial cells) from the mucosal solution *via* an amiloride- and H^+-sensitive channel whose predominant functional groupings are carboxyl and/or phosphoryl. Net trans-mucosal transport of Na^+ is subsequently tied to the activity of the Na^+ pump located at the basolateral surface. The function of this pump for cellular ionic balance is readily apparent; the physiological role of the transepithelial component of Na^+ transport has not been determined.

D. Water Transport

I. Introduction

The dog gastric mucosa can secrete approximately 2000 μl $cm^{-2} \cdot h^{-1}$ of nearly isotonic HCl (REHM et al., 1970), the highest rate of fluid transport recorded for any epithelium concerned with secretion or absorption (see HOUSE, 1974)[1]. The most abundant component of this gastric juice is obviously water. The specific mechanisms that give rise to these large water flows are not known in any detail, but it is generally assumed that the water transport across the pithelium occurs passively in response to osmotic gradients created within the epithelium by active transport of H^+ and Cl^- ions. Measurements of passive movements of water are therefore of some interest.

1 It should be noted, though, that these rates were recorded in the intact dog stomach, where there is extensive folding of the epithelium. If correction is made for this folding, rates will be decreased by approximately a factor of 5. (See ALTAMIRANO, 1969).

II. Diffusional Permeability to Water

The diffusional permeability (P_d) of the gastric mucosa has been measured with the aid of tritiated water, and the values recorded range from 1.9×10^{-5} cm s^{-1} (ÖBRINK, 1956) to 4.9×10^{-5} cm s^{-1} (DURBIN et al., 1956). These values are somewhat lower than those obtained in several other transporting epithelia (*e. g.*, frog skin: 15×10^{-5} cm s^{-1}, DAINTY and HOUSE, 1966; toad bladder treated with antidiuretic hormone: 110×10^{-5} cm s^{-1}, HAYS and FRANKI, 1970). The reason for this discrepancy may lie in the fact that unstirred layers adjacent to the cell membranes of the epithelium can cause dramatic underestimates in permeability measurements of very permeant species like water. This can be appreciated by reference to the following equation, which takes into account the effects of unstirred layers on permeability measurements (DAINTY, 1963):

$$\frac{1}{P_{\text{measured}}} = \frac{1}{P_{\text{true}}} + \frac{\delta_m}{D} + \frac{\delta_s}{D}. \tag{4}$$

P_{measured} represents the measured permeability of the permeating species, P_{true} is the actual permeability of the epithelium to the species, δ is the thickness of the unstirred layer next to the serosal (δ_s) and mucosal (δ_m) membranes, and D is the free-solution diffusion coefficient of the species. The equation shows that for substances that have large permeabilities, like water, the P_{measured} can become limited by the thickness of the unstirred layers. The P_{true} will then be drastically underestimated if corrections are not made for these unstirred layer effects.

This effect of unstirred layers on measurements of P_d has been very clearly shown in the gallbladder epithelium, where original measurements for fish gallbladder were $P_d = 8.3 \times 10^{-5}$ cm s^{-1} (DIAMOND, 1962), but in later measurements on rabbit gallbladder, where unstirred layers were accounted for, $P_d = 184 \times 10^{-5}$ cm s^{-1}(VAN OS and SLEGERS, 1973). Because the unstirred layers in the folded gastric epithelium are even larger than in the gallbladder, we conclude that P_d measurements summarized in Table 2 are underestimated by at least an order of magnitude.

III. Hydraulic Conductivity

It has been found that the measured hydraulic conductivity (L_p) depends on whether the tissue is resting or actively secreting and also on whether osmotic or hydrostatic methods are used for the measurement. For example, L_p has been determined from osmotic experiments by means of the equation:

$$J_v = -L_p RT \Delta C_s \tag{5}$$

where J_v is the net volume flow and ΔC_s is the experimentally imposed osmotic gradient of solute(s) across the epithelium. MOODY and DURBIN (1969) meas-

Table 2. Water transport parameters for gastric mucosa

Species	Rate of transport (μl cm^2 · h)	P_d (cm s^{-1} · 10^5)	Lp (cm s^{-1} · 10^3)
Frog, *in vitro*			
resting			1.21[a*]
secreting	8.6[a], 12[b]	4.9[b]	2.04[a*], 1.1[b*]
Mouse, *in vitro*			
resting		1.9[c]	
Dog, *in situ*			
resting		3.8[f]	4.87[f*], 134[d+]
secreting	75[dx], 2000[e]		2.85[d*], 306[d+]

x calculated from secretory rates on the basis that the actual area of the mucosa is 4.7 times the apparent area (see ALTAMIRANO, 1969).
+ measured by application of hydrostatic pressures to mucosal surface (see text).
* measured by increasing osmolality of mucosal solution with an impermeant solute (see text).

[a] VILLEGAS (1963)
[b] DURBIN et al (1956)
[c] ÖBRINK (1956)
[d] MOODY and DURBIN (1969)
[e] THULL and REHM (1956)
[f] ALTAMIRANO and MARTINOYA (1966)

ured the volume flow in the intact, actively secreting dog gastric mucosa in response to increased tonicity of the mucosal solution and found values for L_p that were about half those observed by ALTAMIRANO and MARTINOYA (1966) in the nonsecreting dog stomach (Table 2). These results demonstrate the "sweeping away" effect described by REHM et al. (1962), who pointed out that volume flow of HCl tends to carry exogenously added solute out from the gastric pits, thus decreasing the concentration of these solutes at the base of the pits; and hence the ΔC_s of Eq. (5). The concentration, C, of the substance at the base of the pits is given by:

$$C = C_o^{-vx/D} \tag{6}$$

where C_o is the solute concentration in the bulk phase, x is the pit depth, v the secretion velocity in the pits and D the diffusion coefficient of the solute. MOODY and DURBIN (1969) calculated that the effective concentration of solute tending to elicit water flow at the oxyntic cell level could be reduced by 75–98 percent by the sweeping-out effect of the secreted HCl.

It should also be pointed out that changes in solute concentrations in unstirred layers next to the serosal membranes can also cause underestimates in measurements of the true L_p of the epithelium. For example, when exogenous solutes are added to the mucosal solution to increase its tonicity, the resulting serosal-to-mucosal osmotic flux will tend to alter solute concentration on both the serosal and the mucosal side due to a sweeping-away effect (DAINTY, 1963; WRIGHT et al., 1972) similar to that mentioned above for volume flow of HCl out of the gastric pits. The volume flow reduces the osmotic concentration at the mucosal membrane and enhances it at the serosal membrane by amounts given by Equation (5) if we let x be the unstirred layer thicknesses at the mucosal and serosal interfaces instead of the pit depth. These effects obviously cause a large decrease in the osmotic driving force (ΔC_s), and consequently an underestimation of the calculated L_p. It should be further mentioned that if the permeability of the surface epithelial cells to water is low (REHM et al., 1970), then most of the water flow due to osmotic gradients will be through the oxyntic cell region alone, and because of the large flow velocities that would occur in the gastric pits and lateral spaces between oxyntic cells, artifacts due to changes in solute concentration in unstirred layers will be even further exacerbated. Because no one has yet made correction for these unstirred layer effects in gastric mucosa, the values measured for by L_p by osmotic methods (Table 2) are most probably underestimated by at least a factor of 10, and probably more (see WRIGHT et al., 1972).

That this may be the case can be inferred from the measurements of L_p by means of hydrostatic pressure by MOODY and DURBIN (1969). They applied hydrostatic pressures (25–75 cm H_2O) to the mucosal surface of intact dog stomachs through a closed air column from a water manometer and found, quite surprisingly, that the L_p measured in this way was more than 100-fold greater than those measured with osmotic methods (Table 2). They further found that the L_p of the actively secreting preparation was about two times larger than that of the resting tissue.

MOODY and DURBIN (1969) have suggested that the difference between the osmotic and hydrostatic L_p measurements might be explained in terms of a small number of large pores in parallel with a relatively large number of small pores, and, indeed, from studies on nonelectrolyte permeation in gastric mucosa, it has been proposed that there may exist just such a distribution of large and small pores (*e. g.*, 7% 60 Å, 93% 2.5 Å: DURBIN et al., 1956). However, some of the discrepancy between the two may also arise from the fact that the hydrostatic methods are more immune from the artifacts induced by the instirred layers than are the osmotic methods.

IV. Water Flow During Secretion

1. Hydrostatic Pressure and Ultrafiltration

Early theories suggested that the energy to move water during acid secretion was derived from blood pressure in the mucosal capillaries. However, it is now generally accepted that this factor is probably of minimal importance for the HCl-related secretion. For example, DAVIES and TERNER (1949) showed that net water flow could be observed in the isolated, secreting frog gastric mucosa with no blood supply. This water transport could also occur against small hydrostatic pressure gradients (5 cm H_2O). Hydrostatic pressure may, however, be important for water secretion in the intact stomach, especially in the case of the so-called alkaline component of gastric secretion. In this latter regard, ALTAMIRANO (1963, 1969) has shown that the secretion produced by cells other than the oxyntic cells of the intact gastric mucosa is most probably due to ultrafiltration.

The concentrations of certain solutes in this alkaline secretion (*e. g.*, Na^+, Ca^{2+}, Cl^-, HCO_3^-, and urea) were typical of those found in plasma. ALTAMIRANO found that this alkaline secretion could be increased by intraarterial injection of acetylcholine (1963) and reduced by mucosal application of hydrostatic pressure at the mucosa surface (1969). He also concluded that the hydrostatic pressure in the capillaries and arterioles of the gastric mucosa attained the rather high level of 90 mm Hg. A luminal pressure of about 90 cm H_2O is sufficient to halt both H^+ and water secretion in the intact canine gastric mucosa (MOODY and DURBIN, 1969), but at present it is not known whether this inhibition is due to a direct effect on water and H^+ secretion or is mediated *via* restriction of local blood supply.

2. Endogenous HCl Gradients and Water Secretion

There are at least two basic reasons for assuming that an osmotic gradient of HCl created somewhere *within* the gastric epithelium is the driving force for water transport during secretion: (1) The isolated frog gastric mucosa can transport water against an imposed transepithelial osmotic gradient of 100 mOsm kg^{-1} (DURBIN et al., 1956). (2) The measured L_p values are about 50-fold too small to allow the observed water fluxes in the face of only 10 percent hypertonicity of the secreted fluid (e. g., see REHM et al., 1970). During normal rates of secretion the HCl is hypertonic by 2–10 percent (THULL and REHM, 1956; MOODY and DURBIN, 1965)[2]. Several schemes, including the asymmetric double-membrane model (CURRAN, 1960; DURBIN, 1960; PATLAK et al., 1963) and the standing local osmotic gradient model (DIAMOND and BOSSERT, 1967), have been proposed to account for these observations (see Fig. 8). Both of the above-mentioned formulations require the generation of a local osmotic (*i. e.*, endogenous) gradient by active transport of some solute into a restricted space. In the gastric mucosa, this restricted space may correspond to the spaces of the intracellular canaliculi of oxyntic cells (Fig. 8e) and/or the lumina of the glands[3].

Although there have been rather few experiments designed to test either of these hypotheses, MOODY and DURBIN (1965) have invoked the asymmetric double-membrane model to explain (1) the large volume flows in the face of only small HCl gradients, (2) the differences between hydrostatic and osmotic L_p measurements, and (3) the effects of instilling isosmotic glycine solutions on the mucosal surface. These glycine instillates caused an increase of up to 100

2 At low rates of the secretion, gastric juice is hypotonic by about 3 percent. The water fluxes under these circumstances can be explained by a model in which ionic concentration differences develop due to Na^+-H^+ exchange in the gastric pits, and this in turn induces a local osmosis resulting in the water flux. Any HCO_3^- secretion by ultrafiltration would tend to induce a hypotonic secretion due to the neutralization of HCO_3^- by H^+ and formation of CO_2.

3 It is noted in this regard that WINSHIP and CAFLISCH (1973) have measured pH at various depths within the lumina of the glands with microglass electrodes. The lowest pH observed was pH 0.77, and this was found throughout the depth of the gland. Thus, if there is a gradient of HCl within the glands, it is certainly rather small.

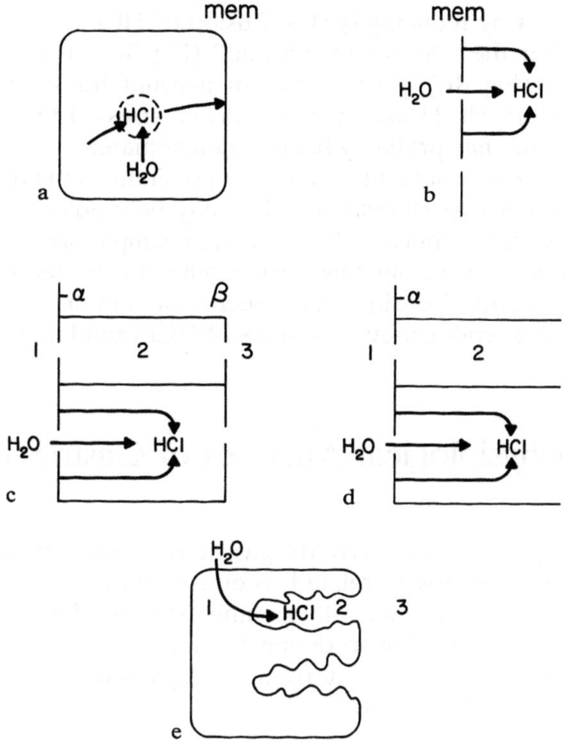

Fig. 8a–e. Proposed schemes to account for H_2O flux accompanying HCl secretion by gastric mucosa. (a) Pinocytotic mechanism whereby HCl is secreted into membrane-bound vesicles in cytoplasm, drawing water into vesicular space. Isotonic HCl would then be released by exocytosis-type mechanism. (b) Simple osmosis where osmotic force of secreted HCl drives water transport. (c) Asymmetric double-membrane model, which assumes different L_p's, σ's, and permeabilities for membranes α and β, and also assumes that compartment 2 is unstirred. (d) Local osmosis scheme in which membrane β of the double membrane model is removed. Standing osmotic gradient within compartment 2 has been proposed (DIAMOND and BOSSERT, 1967). (e) A drawing of the oxyntic cell to show possible ultrastructural counterparts of the compartments and membranes of the local osmosis scheme. (Redrawn from DURBIN and MOODY, 1965)

percent in [H^+] in gastric juice (TEORELL, 1940; MOODY and DURBIN, 1965). The increase in [H^+] was due to an increased production of protons and a decreased volume flow, the latter effect being the predominant one. Isosmotic solutions of NaCl, sucrose and mannitol induced similar, but quantitatively smaller effects. These results are important because they virtually exclude the possibility that a pinocytotic mechanism (Fig. 8a) can account for H^+ production. They do, however, support the proposal that endogenous osmotic gradients of H^+ and Cl^- provide the driving force for net water flow. Thus, glycine may reduce the endogenous gradient of H^+ by neutralizing H^+ ions and also by "diluting" the secreted HCl; whereas isosmotic solutions of mannose, sucrose, and NaCl lead to a similar effect by dilution alone. This dilution of the HCl reduces a net volume flow due to reduction of the osmotic gradient as well as by increasing effective HCl output by reducing back-diffusion of HCl.

At present, either the "double membrane" (Fig. 8c) or the "standing local osmotic gradient" (Fig. 8d) mechanism can account for solute-linked water secretion by the stomach. However, it should be stressed that since the measured L_p of this tissue has probably been underestimated by at least a factor of 10, and perhaps by even more than that, the possibility should not be ignored that the 2–10 percent hypertonicity of HCl may be gradient enough to drive water secretion by gastric mucosa. In this case, a simple one membrane model (Fig. 8b) with very high L_p but low permeability to the secreted HCl could account for water transport during gastric secretion, and the need for proposing the formation of large endogenous gradients of HCl would then be eliminated.

E. Electrophysiological Analyses of Gastric Transport

From the preceding considerations of the gastric H^+ and Cl^- transport, it is clear that these functions are closely related to electrical manifestations within the epithelial cells. In this section, we will examine the basis of the electrical activity in gastric mucosa in an attempt to define the nature and site of origin of the transepithelial PD as well as the interrelationships between the secretory and electrical activity within the tissue.

I. Electrogenic vs. Electroneutral H^+ and Cl^- Pumps

A variety of mechanisms have been proposed to relate the electrophysiological data to H^+ and Cl^- secretion, and these can be categorized as falling into one of two theoretical generalizations: an electrogenic mechanism in which the H^+ and Cl^- pumps have the separate and independent capacity to generate an EMF by transport of the respective species; and an electrically neutral mechanism in which H^+ and Cl^- are coupled. The electrogenic and neutral mechanisms are schematically compared in Figure 9.

1. The Electrogenic Hypothesis

As originally formulated and further developed by REHM (1945, 1950), the electrogenic hypothesis holds that the gastric H^+ and Cl^- pumps can be represented by an equivalent circuit including the respective EMFs (E_H, E_{Cl}) and resistances (R_H, R_{Cl}) of the H^+ and Cl^- limbs (Fig. 9A). In his expanded conceptual schema, REHM proposes that the H^+ and Cl^- transport mechanisms would operate at separate sites within the apical cell membrane and with distinct sets of energy providing reactions, that is to say, the pump systems are not directly

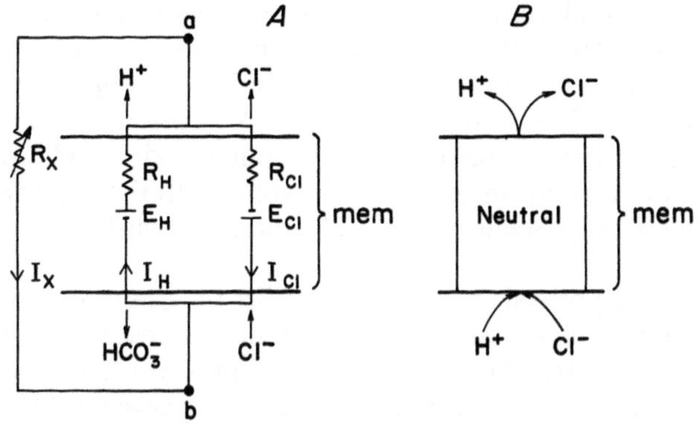

Fig. 9. (A) Electrogenic theory of HCl secretion represented by equivalent circuit within conceptual single-membrane model. Separate limbs, or pathways, through membrane (mem) are indicated for H^+ and Cl^-. I's, E's and R's refer to currents, emfs and resistances, respectively. R_x represents external resistance for system between points a and b; current flow in external circuit is indicated by I_x. (B) Neutral mechanism for HCl secretion. (Redrawn from REHM and SANDERS, 1975)

coupled through a biochemical reaction (*e. g.*, see REHM and SANDERS, 1975). During open-circuit conditions, these pumps at the mucosal membrane would be electrically coupled, in that, to maintain electroneutrality, the H^+ current (I_H) must equal the Cl^- current (I_{Cl}) since these species are the principal ions being transported. Thus, according to the law of circuits the open-circuit PD (*i. e.*, $I_x = 0$) across the system (V_{ab}) represented in Fig. 9A is given by:

$$V_{ab} = E_{Cl} - R_{Cl} I_{Cl} \tag{7}$$

and

$$V_{ab} = E_H - R_H I_H \tag{8}$$

Since in this open-circuit condition $I_{Cl} = I_H = I$, then by substitution, combining (7) and (8):

$$V_{ab} = \frac{R_{Cl} E_H - R_H E_{Cl}}{R_{Cl} - R_H} \tag{9}$$

The electroneutrality requirement is relaxed when external current is passed ($I_x \neq 0$) or the system is short-circuited ($R_x = 0$), and net current will move through the membrane so that:

$$I_x = I_{Cl} - I_H \tag{10}$$

and combining (7), (8) and (10), we have:

$$I_x = [E_{Cl} - V_{ab})/R_{Cl}] - [(E_H - V_{ab})/R_H] \tag{11}$$

One requirement of the electrogenic hypothesis is that current passage through the system will change the outputs of H^+ and Cl^-, and this point was fundamental to Rehm in his formulation of the electrogenic hypothesis, since for both the dog stomach flap (Rehm, 1945) and the frog gastric mucosa (Crane et al., 1948), the rate of H^+ secretion could be changed by passing an external current through the tissue. Current sent from mucosa to serosa decreased the H^+ rate, while current passed from serosa to mucosa enhanced the H^+ rate. It is also noteworthy that the formalism of Eq. (10) agrees with the experimental observations of Hogben (see Section C. III.1) when the gastric mucosa is short-circuited.

The electrogenic hypothesis would also predict the observed inverse relationship between the rate of H^+ secretion and mucosal resistance (Rehm, 1967). That is, a change in the Rs of the electrogenic pumps would provide an effective basis for modulating secretory output. Thus, with the onset of H^+ secretion, the tissue resistance has been observed to decrease drastically (Rehm, 1953; Forte et al., 1975). Conversely, when H^+ secretion is inhibited with anoxia, dinitrophenol or CN^-, the tissue resistance increased (LeFevre et al., 1964; Rehm, 1967; Forte, 1969; Forte et al., 1975). However, it should be recalled that major changes in membrane surface ultrastructure are associated with the state of HCl secretion (Section B. II), and it is reasonable that transmucosal resistance would reflect the changes in surface area (apical oxyntic cell surface or the basolateral cell surface) that have been correlated with stimulation or inhibition of secretion (Forte et al., 1975). The degree to which these quantitative changes in total epithelial membrane surface area can be distinguished from any "qualitative" change in the nature of the pumps (*i. e.*, changes in the Rs or Es of Fig. 9) for the secretory-dependent parameter of resistance remains to be established.

Some of the most convincing evidence in favor of the electrogenic hypothesis comes from studies in which the Cl^- in the bathing solutions was replaced by impermeant anions (SO_4^{--}, isethionate etc.). Under these conditions and according to the circuit represented in Figure 9a, the mucosal surface should become less negative and the tissue resistance should increase. Heinz and Durbin (1959) and others (Rehm et al., 1963; Forte, 1969) have found this to be the case. In fact, the PD reversed and became positive. The measured I_{sc} was proportional to H^+ secretion (Heinz and Durbin, 1959). In addition, when positive current is passed from the serosal to the mucosal sides in SO_4^{--} solutions, the rate of H^+ secretion is enhanced exactly as predicted. However, current passage could not increase the rate of H^+ secretion to the level reached in Cl^--containing solutions. Following the analysis of an equivalent circuit for the mucosa in Cl^--free solutions, Rehm and LeFevre (1965) showed that the addition of the inhibitor dinitrophenol produced a direct and linear relationship between the transmucosal PD and the rate of H^+ secretion. Their results, which have been confirmed with several other inhibitors (Sachs et al., 1965; Schwartz and MacKrell, 1969), are shown in Figure 10 and represent strong support for the electrogenic nature of gastric H^+ secretion.

Several objections to the electrogenic hypothesis have been effectively dealt with by Rehm (Rehm, 1975; Rehm and Sanders, 1975) and can be explained when one considers more realistic, but more complicated, circuit schemes for the gastric epithelium. However, there are a few experimental challenges for

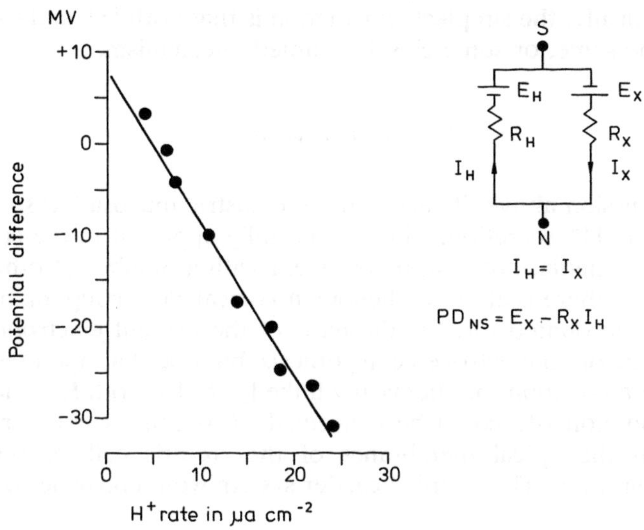

Fig. 10. PD vs H^+ secretory rate in Cl^--free media; −PD means nutrient side negative. Farthest point to right gives PD and H^+ rate before addition of dinitrophenol to nutrient fluid; with addition of aliquots of dinitrophenol, PD decreases linearly with decrease in H^+ rate. Equivalent circuit on right used for analysis of linear relationship. (Reprinted from REHM and SANDERS, 1975, with permission)

which a separate site-electrogenic hypothesis does not appear to be wholly satisfactory (RANGACHARI, 1975a; SOLBERG and FORTE, 1971), and we will consider an alternative mechanism below.

2. The Electroneutral Hypothesis

An electroneutral mechanism involves either a forced one-for-one exchange of H^+ with another cation between mucosal solution and cytoplasm, or a simultaneous transfer of an anion by a single coupled carrier (*e. g.*, Fig. 9B). In its simplest formulation, the transepithelial PD would not be generated directly by the transport mechanism, but would be the result of diffusion PDs at the mucosal and serosal membranes (HARRIS and EDELMAN, 1964). These PDs may be due to differential permeabilities of the mucosal and serosal membranes to K^+ and Cl^- and to the concentration gradients of these ions at these interfaces, which are maintained by the various ion pumps.

Several experimental observations are generally consistent with the electroneutral hypothesis, although they do not exclude the operation of an electrogenic process. Thus, during open-circuit conditions, net H^+ secretion is identical to Cl^- secretion. In normal Cl^- bathing media, this equality holds when one compares ion transport changes associated with stimulation or inhibition of secretion (FORTE, 1968; FORTE and MACHEN, 1975). Since both H^+ and Cl^- are actively transported across the gastric mucosa, and since neither Na^+ nor K^+ is required in the mucosal fluid, and since Cl^--free solutions cause H^+ secretion to decrease drastically, the simplest explanation is that both H^+ and Cl^- are being pumped by the same, or some closely coupled, mechanism.

3. Conclusions

From the discussion above, it seems that the gastric mucosa has some capacity for electrogenic H^+ secretion. This is especially apparent in the conditions of Cl^--free bathing media, but is also consistent with a number of other observations. However, there is also the phenomenological close coupling between H^+ and Cl^- to be accounted for. On the basis of the present electrophysiological evidence, we favor an electrogenic hypothesis, but one that operates within the framework of a common, or chemically linked, site for both H^+ and Cl^- transport. This common site could be conceived of as some carrier translocation mechanism in the apical membranes of the oxyntic cell, as schematically depicted in Fig. 11A. The coupled carrier system would provide some physical

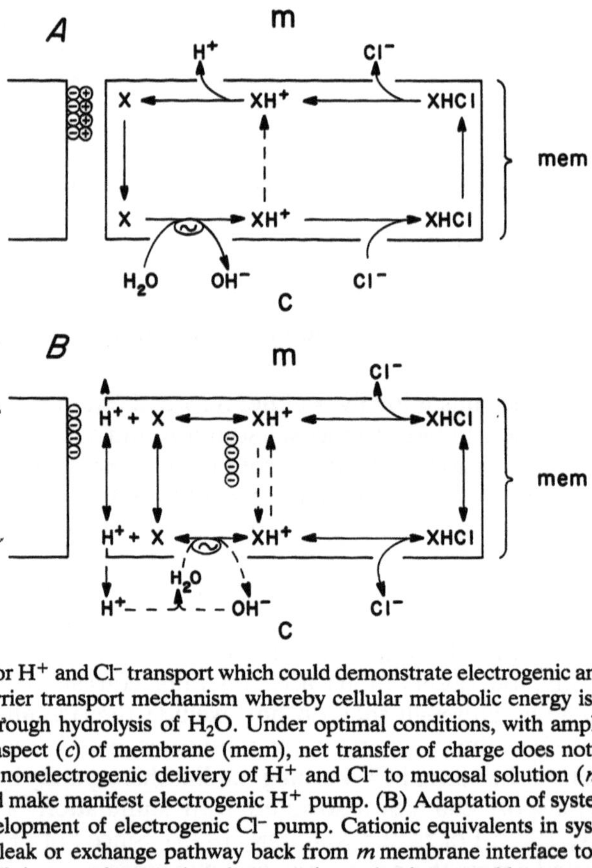

Fig. 11. Models for H^+ and Cl^- transport which could demonstrate electrogenic and nonelectrogenic function. (A) Carrier transport mechanism whereby cellular metabolic energy is put into protonation of carrier through hydrolysis of H_2O. Under optimal conditions, with ample supply of Cl^- at internal cellular aspect (*c*) of membrane (mem), net transfer of charge does not occur and system would appear as nonelectrogenic delivery of H^+ and Cl^- to mucosal solution (*m*). Deprivation of cellular Cl^- would make manifest electrogenic H^+ pump. (B) Adaptation of system shown in (A) to demonstrate development of electrogenic Cl^- pump. Cationic equivalents in system (XH^+ and/or H^+) would have leak or exchange pathway back from *m* membrane interface to *c* side. Net movements of H^+ or Cl^- in case of respective electrogenic mechanisms would occur only if shunt leakage (or exchange) were present, or if current were passed in appropriate direction

analogy compatible with REHM's electrogenic model whereby equal and oppositely oriented flows of current through the pump site deposit HCl with no net charge transfer. Furthermore, the basic carrier mechanism is adaptable to allow "slippage" under specialized circumstances and thus forms the basis for experimentally observed electrogenic H^+, and even Cl^-, transport. The protonated carrier is generated at one membrane interface *via* endergonic reactions and in this form will act in any one of the following modes:

(i) as an electrically neutral pump for H^+ and Cl^- when the system is saturated with its component substrates and operating in a steady state for HCl secretion;

(ii) as a vehicle for electrogenic H^+ transport when the availability of Cl^- to the pump is limited, *i. e.*,low-Cl^- solutions;

(iii) as a locus for isotopic exchange diffusion of Cl^- (if the mobility, or translocation time, of the XHCl form within the membrane is higher than the other carrier forms).

An adaptation of the scheme has been suggested (Fig. 11B), whereby a specific and localized H^+ leak pathway within the membrane could dissipate the local proton gradient, thus providing an apparent electrogenic Cl^- pump (FORTE, 1971). It is clear that the stomach can support a component of Cl^- transport that is independent of net H^+ secretion. We have referred to this as the nonacidic component, and it could be driven by a Cl^- pump entirely separate from H^+ secretion. Present evidence is incomplete, so that we can specify neither the epithelial cell type nor the cellular interface for this Cl^- transport.

II. Intracellular PDs

Measurement of the PD across a specific cellular membrane of gastric mucosa has been complicated by the number of cell types and the geometrical configuration of the gastric tubules, which does not afford ready visual acess to a particular cell. VILLEGAS (1962) has provided some provocative data from his measurements of intracellular PDs with microelectrodes followed by histological localization of an injected spot of dye. VILLEGAS concluded that in Cl^--containing RINGER's solutions, the transpithelial PD of −30 mV (serosal solution as reference) was the result of two different PD profiles, depending upon the cell type. In oxyntic cells, there was a two-step PD profile at the mucosal and serosal membranes. That is, as the electrode penetrated from the lumen into the cell, the first jump was from −30 to −18 mV. The remaining 12-mV drop occurred at the serosal membrane. In surface epithelial cells the profile was entirely different, the initial PD jump being from −30 to −49 mV. The PD rose to zero mV when the electrode penetrated to the *lamina propria.*

Workers in the Alabama groups (*e. g.*, CANOSA and REHM, 1968; SPENNEY et al., 1975) have confirmed the existence of the PD profile characteristic of the surface epithelial cells, but have disputed the results obtained from oxyntic cells. These latter groups suggest that the PD profile of the surface cells and oxyntic cells is the same, basing this suggestion on their analysis, which assumes that all cells of the epithelium are electrically coupled (LOEWENSTEIN and KANNO, 1964). Recently, a wide range of PD profiles has been observed in the surface epithelial cells (O'CALLAGHAN et al., 1974) as well as deeper in the gastric epithelium (WINSHIP and CAFLISCH, 1973).

The obvious conclusion from these studies is that the PD profile across the oxyntic cells is in doubt at this time. Specification of this PD is of great consequence, especially when one is interpreting transport functions. For example, if there is a negative potential drop from cell interior to mucosal solution, then Cl^- transport at this boundary must be active since it would be occurring against both an electrical and chemical gradient. On the other hand, if this step is positive, then Cl^- movement at the mucosal border could occur by passive means, which might implicate active transport at the serosal membrane. Obviously, such conclusions would depend on a careful analysis of the electrochemical profile within the gastric epithelium. Repetition of the microelectrode experiments is required.

III. Permeability and Conductance Pathways

1. Transcellular and Paracellular Conductances

Definition of conductance pathways is important for a complete characterization of the function of the gastric mucosa, but the anatomical complexity of this tissue has made a comprehensive analysis very difficult. As will be shown below, the different cell types may have markedly different conductances, and this fact will have important consequences for the interpretation of the electrical activity of the tissue. In addition, the magnitude of the paracellular "shunt" pathway (FRÖMTER and DIAMOND, 1972) may be of importance for the function of the epithelium. We will discuss both the problem of the conductance of this heterogeneous tissue with several different cell types and also the analysis of the paracellular shunt conductance.

REHM and his colleagues (O'CALLAGHAN et al., 1974) have made an interesting analysis of the K^+ conductances of the two major cell types by exploiting the fundamental observation that Ba^{2+} competes with K^+ for permeation sites at the serosal side of the epithelium. They made intracellular recordings in surface cells and measured the resulting PD changes across the mucosal (ΔV_{mc}) and serosal (ΔV_{cs}) cell borders when a constant pulse of current was passed through the epithelium. The principal observation was that Ba^{2+}, added to the serosal solution, caused the ΔV_{mc} to increase in response to the current pulse. They interpreted this as showing that Ba^{2+} was principally acting on the serosal membrane of the *oxyntic* cell; thus, for the division of the current pulse flowing through the epithelium as shown in diagram form in Fig. 12, a greater proportion of current would flow through the surface cells, increasing the resulting ΔV_{mc} and ΔV_{cs}. Their quantitative assessment of the current and voltage changes led to the conclusion that the surface cell population was contributing less than 25 percent to the total epithelial conductance and the oxyntic cells must therefore contribute at least 75 percent of the conductance pathway. This is perhaps not too surprising, given the large surface area the gastric glands provide (BLUM et al., 1971). However, for those who have followed the above argument through the circuit diagram of Figure 12, two very critical assumptions of REHM and his colleagues should be pointed out. First, the assumption was made that the paracellular shunt resistance was very high compared to that through the cells (*i. e.*, $R_5 \gg R_1 + R_2$ or $R_3 + R_4$). Experiments cited below show this to be a reasonable assumption. Secondly, REHM's group assumed that Ba^{2+} exerted its effects exclusively on the serosal-facing membranes. On this latter point, recent evidence has shown that Ba^{2+} also inhibits active Cl^- secretion, and this action seems to occur indirectly through competition with K^+ sites at the serosal side of the epithelium (RANGACHARI, 1975a; MACHEN and FORTE, 1976). Thus, it is possible that Ba^{2+} could have "indirect" effects on the mucosal-facing membranes through its action on $[Cl^-]$ in the cells and/or the Cl^- conductive pathways through the apical membranes of either cell type.

This latter analysis does not necessarily preclude the hypothesis of O'CALLAGHAN et al. (1974) regarding the relative conductances of the surface and tubular cell regions of the epithelium. Indeed, much of the available experimental data fits in with this concept. However, we wish to point out that the complete picture is not yet settled; further innovative analyses will be required for a satisfactory solution of this question (see Discussion, O'CALLAGHAN et al., 1974).

As mentioned above, it has been assumed by most workers that the conductance of the gastric mucosa is determined by cellular, rather than paracellular, or shunt, pathways. Indeed, with respect to the ability of the stomach to maintain a large concentration gradient of H^+ ions, this appears to be true: the loss of H^+

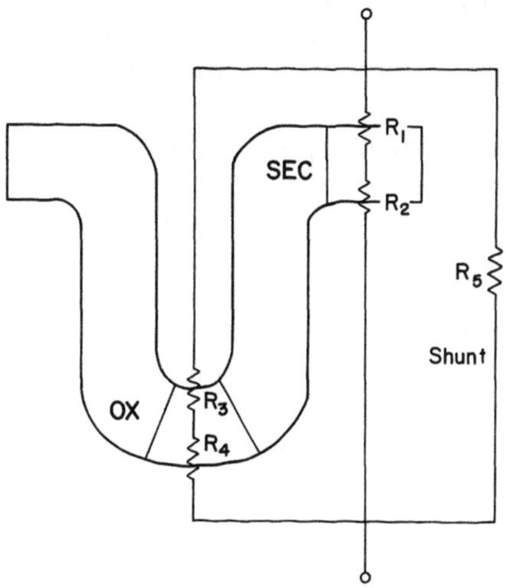

Fig. 12. Equivalent electrical circuit of gastric mucosa. R_1 and R_2 are resistances of apical and basal membranes of surface epithelial cells; R_3 and R_4 are resistances of apical and basal membranes of oxyntic cells. Resistance of parallel shunt pathwaxs (R_5) is assumed to be very large compared to other resistors. (Redrawn from O'CALLAGHAN et al., 1974)

ions from the gastric fundus has been found to be 15 times slower than from the antrum, 45 times slower than from the duodenum, and 60 times slower than from the rest of the small intestine (DYCK et al., 1969; HIMAL et al., 1969).

However, the predominance of the paracellular shunt (through the tight junctions) as the determinant of transepithelial conductance in many transporting epithelia (*e. g.*, gallbladder: WRIGHT et al., 1971; BARRY et al., 1971; FRÖMTER, 1972; small intestine: CLARKSON, 1967; FRIZZELL and SCHULTZ, 1972; proximal tubule: GIEBISCH et al., 1971) has attracted attention to this problem in the gastric mucosa. Those epithelia which have high shunt conductance have been termed "leaky" (FRÖMTER and DIAMOND, 1972) and are characterized by having low transepithelial PDs and resistance, linear current-voltage (*I–V*) relationships and high rates of isotonic fluid transport. Since the gastric mucosa has at least two of these characteristics (relatively low resistance and high rate of fluid transport) in common with leaky epithelia, the magnitude of the shunt is of some importance.

A unique approach to a concrete definition of the magnitude of the shunt in gastric mucosa was to use an enzymatic digestion technique to isolate individual cells and gastric glands from *Necturus*, and to compare the resistances of the individual cell membranes (calculated from the *I–V* curves of individual cells), to the total transepithelial resistance (BLUM et al., 1971). If, after correction for area, the resistance measured on individual cells was much greater than that of the intact mucosa, the implication would be that a significant shunt conductance was present. These authors were forced to make a number of estimates and assumptions, *e. g.*, of the folding of the glands and the percentage of surface epithelial cells and oxyntic cells in the epithelium (as well as their surface area), as judged from measurements using phase-contrast and electron microscopy, but when they finally calculated the transcellular resistance of the intact epithelial cell layer, it was approximately 2000 $\Omega \cdot cm^2$. Since transepithelial resistance measured directly was 710 $\Omega \cdot cm^2$, the implication was that the shunt conductance was of approximately the same magnitude as the transcellular conductance.

More quantitative experiments have recently been reported by SPENNEY et al. (1975). They used two different techniques, which involved the measurement of intracellular PDs with respect to the mucosal and serosal solutions and analysis of an equivalent electrical circuit: (1) The change in PD across these faces in response to constant product KCl changes ($[K^+]_s \times [Cl^-]_s = 400$ mM) was analyzed assuming coupling resistance between all the cells to be very small compared to the resistance of the cell membranes. They found that the resistance of the shunt was 3.7 times higher than the resistance of the transcellular pathway. (2) Measurement of the space constant for current spread within the epithelium, solving the differential equation with a Bessel function (see FRÖMTER, 1972), and correcting for the increase in area due to infolding of the gastric pits gave a cell: shunt conductance ratio of 3.8 : 1. The implication from both of these studies was that the shunt contributes no more than 25 percent of the total transepithelial conductance in *Necturus* gastric mucosa.

As mentioned, the underlying assumption in these experiments was that coupling resistance between all the cells (in the LOEWENSTEIN sense) was very low. SPENNEY et al. (1975) have clearly shown cell-to-cell coupling among surface epithelial cells, and BLUM et al. (1971) showed coupling among oxyntic cells within gastric glands, but no one has clearly shown that the critical coupling exists between oxyntic and surface cells. Indeed, if the results of VILLEGAS (1962) are correct (see Section E. II), then these two different cell types are undoubtedly *not* coupled, or the space constant for electrical coupling is small compared to the distances encountered between the surface cells and the base of the gastric pits.

Thus, an independent confirmation of a low shunt conductance in mammalian gastric mucosa is of some interest. MACHEN et al. (1978) have found the backflux of Na^+ (serosa-to-mucosa) to be a rather accurate measure of passive Na^+ flux through gastric tissue. Neither amiloride nor ouabain, agents that are known to affect Na^+ fluxes across cell membranes, had any appreciable effect on this backflux in the short-circuited state (MACHEN et al., 1975). In addition, from measurements of diffusion PDs in Na^+-isethionate solutions, it was concluded that this backflux was primarily through a free-solution pathway (MACHEN et al., 1975). When this backflux of Na^+ was used as a measure of the shunt conductance and all ions in the bathing solutions were assumed to be moving at a rate proportional to their free-solution mobility, it was calculated that the shunt contributed only 25 percent to the total tissue conductance of newborn rabbit or pig stomachs during the resting state. The contribution of the shunt was only 12 percent in the actively secreting piglet tissue (MACHEN et al., 1978). It is noted, though, that this calculation assumes that both Na^+ and also Cl^- move with free-solution mobility through the shunt. The possibility exists that Na^+ movement is through free solution but that Cl^- has a conductance much greater than that in free solution. A similar situation is found in the Th^{4+}-treated gallbladder (MACHEN and DIAMOND, 1972). In this case, the above calculation would underestimate the contribution of the paracellular shunt to the total tissue conductance. However, at present the evidence points toward the fact that the stomach is indeed a "tight" epithelium, *i. e.*, that the predominant (> 75%) route of ion permeation is transcellular and not paracellular. This is teleologically sensible for tissue that has as one of its primary duties the maintenance of a million-fold concentration gradient of H^+ ions.

2. Cell Membrane Permeability Characteristics

Since the gastric mucosa seems to have a very low component of transepithelial conductance through shunt pathways, changing ionic concentrations in the bathing solutions causes changes in transepithelial PD and resistance, which are attributable to changes at the mucosal and/or serosal interfaces of the cells. Consequently, changing concentrations of the constituents in the bathing solutions can yield information regarding the relative ionic permeability of the mucosal and serosal membranes. However, as will be shown below, analysis of the permeability properties of these membranes is not always straight forward. For example, the measurements often become complicated by the presence of parallel pathways (*e. g.*, surface cells and oxyntic cells) with different conductances. In addition, the permeability of the mucosal and serosal membranes may be dependent on the secretory state of the tissue.

In Cl^- solutions, variations in $[K^+]_m$, $[HCO_3^-]_m$, or $[H^+]_m$[4] (FORTE et al., 1963; HARRIS and EDELMAN, 1964; DENNIS and REHM, 1958) are without significant effect on the PD or resistance. Decreasing $[Na^+]_m$ has no effect on the PD or resistance in amphibian gastric mucosa, but causes a decrease in PD and increase in resistance in mammalian and lizard gastric mucosae. On the other hand, replacement of Cl^- with the impermeant anion isethionate (Ise^-) in the mucosal solution causes an increase in resistance and PD in all preparations studied (FORTE et al., 1963; FORTE and MACHEN, 1975). At the serosal surface, HARRIS and EDELMAN (1964) have found that varying $[Cl^-]_s$ and $[K^+]_s$ at a constant product produced slopes of 50–55 mV for a 10-fold concentration change over the range where the H^+ secretion rate was very little affected (20–90 mM Cl^-). Changing $[Na^+]_s$ between 10 and 120 mM has very little effect on PD or resistance of frog gastric mucosa (SACHS et al., 1966). All these observations taken together are consistent with the hypothesis that the mucosal membrane of the gastric mucosa has a significant permeability to Cl^- (and Na^+ in the lizard and mammal) and only slight permeability to K^+, H^+, HCO_3^-, and Na^+; the serosal membrane is predominantly permeable to K^+ and Cl^- with the ratio $P_K/P_{Cl} \approx 2$–6 (SPANGLER and REHM, 1968; HARRIS and EDELMAN, 1964; TAKEGUCHI et al., 1976).

Experiments involving Cl^--free solutions and the use of various inhibitors indicate that the apparent permeability properties of the membranes for K^+ and Cl^- are dependent on the secretory state of the tissue. For example, for frog mucosa bathed in isethionate solutions, FORTE et al. (1963) found that increasing the $[Cl^-]_s$ from 1 to 10 mM produced a 17-mV change in PD and was accompanied by a three-fold stimulation of H^+ output, whereas further increases from 10 to 100 mM produced a 50 mV change in PD with rather little additional enhancement of H^+ secretion. For these same conditions variations in $[Cl^-]_m$ were essentially without effect on the gastric PD (FORTE et al., 1963). In addition, when H^+ secretion is inhibited by substances that interfere with energy production and H^+ secretion by the tissue *(e. g.*, anoxia, dinitrophenol or amytal), the transepithelial PD response to changes in $[Cl^-]$ or $[K^+]$ in the mucosal and/or serosal solutions is much reduced compared to control conditions (REHM, 1968; FORTE, 1971; SACHS et al., 1967).

The straightforward interpretation of all these results is that these experimental maniuplations cause an actual physical change in the membranes such that their permeabilities have been altered. For example, to explain the results obtained in Cl^--free solutions, FORTE (1971) has proposed that the diffusion potential for Cl^- at the mucosal interface may be the result of two boundary potentials (DONNAN equilibrium) between Cl^- and a carrier at the inner and outer surfaces of the mucosal membrane. In essence, the proposal is that the permeability to Cl^- may be dependent on the presence of Cl^- on both sides of

4 Decreasing the pH of the mucosal solution to pH 2 causes an increase in resistance and decrease in PD in the mammalian gastric mucosa, and these changes may be attributed to inhibition of Na^+ transport (see Section C. IV).

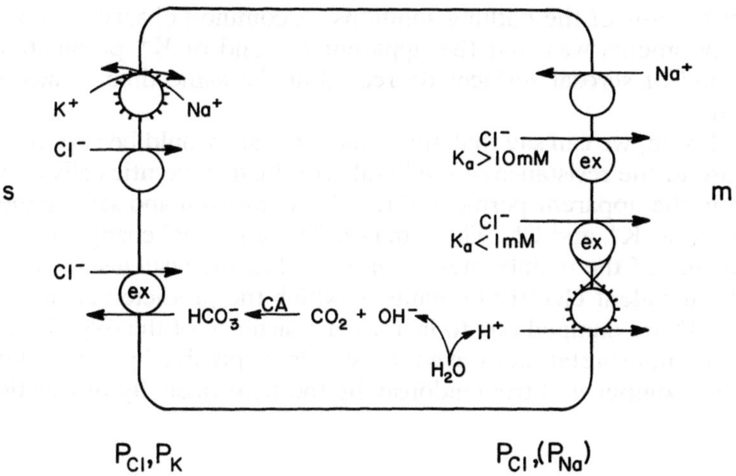

Fig. 13. Schematic drawing of ion transport processes in gastric mucosa. The apical membrane has significant permeability (*P*) only to Cl^- (and to Na^+ in the mammal and lizard) while the serosal membrane has predominantly Cl^- and K^+ permeability. Cogged wheels represent the fact that H^+ secretion and Na^+ extrusion out of the cell are both energy-dependent and are most probably coupled in some way to ATP breakdown. Smooth circles represent membrane-mediated events, specific for ions involved, but which have not been specified as to their requirement for metabolic energy. "ex" within smooth circles indicates that these processes may have exchange-diffusion characteristics (Cl^--Cl^- and/or Cl^--HCO_3^-)

this apical membrane. If the membrane concentration of effective Cl^- carrier were dependent on the access of Cl^- *via* the serosal membrane, then in Cl^--free solutions the amount of carrier at the mucosal interface would be severely limited, and variations in $[Cl^-]_m$ might not establish an interfacial potential. Further, with regard to the effect of inhibitors it was suggested that intracellular concentrations of K^+ and Cl^- may be dependent on energy stores (FORTE, 1971). Thus, because during anoxia there may be a relatively rapid equilibration of extracellular Cl^- and/or K^+ with intracellular contents, changes in $[K^+]_s$ and/or $[Cl^-]_s$ may have only transient and reduced effects on the PD.

Alternative explanations for the effect of inhibitors on membrane permeability involving equivalent electrical circuits of the gastric mucosa have also been proposed by two groups from Alabama. REHM (1968) has argued that during metabolic inhibition of H^+ secretion, the resistance of the mucosal membrane would be expected to increase and may approach or even exceed the value of the shunt resistance. Following an equivalent circuit for this system, this increased apical resistance might in turn alter the PD response of the *serosal* membrane to changes in $[K^+]_s$. Therefore, during metabolic inhibition of H^+ secretion, small PD changes in response to changes in $[K^+]_s$ do not necessarily reflect real changes in passive permeability of the gastric mucosa. SACHS et al. (1969) have used a slightly more complex circuit with two pathways for K^+ movement across the epithelium, the surface cells and the oxyntic cells, to come to the same conclusion, *i. e.*, that changes in resistance of the oxyntic cell will cause alterations in the PD response of the whole gastric mucosa to changes of ionic constituents of the bathing solutions. A common observation of all of the above experiments was that the apparent Cl^- and/or K^+ permeability at the mucosal and/or serosal surfaces decreased at the same time as acid secretion decreased.

In conclusion, we can say that any maneuver that would be expected to cause an increase in the resistance of the highly conductive oxyntic cells also causes a decrease in the apparent permeabilities of the mucosal and serosal membranes of the tissue to K^+ and Cl^-. These may reflect an actual change in the physical characteristics of the membranes themselves, but they can also be explained in terms of equivalent electrical circuits in which the predicted changes in transepithelial PD are damped due to increased resistance of the oxyntic cells. It can be seen that interpretation of even the simplest "passive" permeability characteristics are complicated tremendously by the heterogeneity of this tissue.

IV. Black Box Model of Gastric Mucosa

Model systems are always highly simplified and idealized. The main reason for constructing a model like that in Fig. 13 is to provide a conceptual framework with which to examine data and propose new experiments. At this stage, it also allows us the opportunity to summarize many of the above sections in a coherent manner.

The gastric mucosa has a large permeability to Cl^- at the mucosal interface. In mammalian and lizard gastric mucosae, there is also a significant Na^+ permeability at this boundary. The serosal membrane is predominantly permeable to K^+ and Cl^-. There are problems in accurate calculation of $[Cl^-]_c$ and $[K^+]_c$, but the available data for the ionic composition of gastric tissue are generally consistent with the interpretation that a Cl^- gradient (from serosal solution to cell interior) and a K^+ gradient (cell interior to serosal solution) contribute to the PD at the serosal interface; this serosal membrane potential also contributes at least a portion of the total transepithelial PD.

Since the H^+ secretory mechanism seems to have an electrogenic capacity, we have shown this pump as being able to operate independently from the so-called acidic Cl^- pump. However, to account for the phenomenological coupling between H^+ and Cl^- secretion which occurs under normal physiological conditions, our scheme also shows these processes to be very closely associated. At this stage, it is not possible to decide whether this close association between active transport of H^+ and acidic Cl^- by oxyntic cells is due to a tight electrical or a biochemical coupling of the two processes. However, since the H^+ pump appears to have some ability to operate electrogenically, one could propose that in the absence of Cl^-, independent transport of H^+ at the mucosal interface of oxyntic cells generates an EMF by separation of charge; the transpeithelial PD is then proportional to H^+ secretion. In Cl^- solutions an obligatory Cl^- transfer to maintain electroneutrality would result in a stoichiometric coupling between H^+ and Cl^-. Whatever the mechanism of HCl secretion, acidic Cl^- transport has $K_a < 1$ mM and is very closely related to the Cl^- exchange diffusion mechanism.

Neutralization of OH^- produced within the cell is accomplished by a carbonic anhydrase-catalyzed reaction with subsequent removal of the HCO_3^- from the cell *via* a Cl^--HCO_3^- exchanger located at the serosal membrane.

There also exists a fraction of Cl^- secretion that is independent of H^+ secretion, the so-called nonacidic Cl^- transport. Formulating a model for nonacidic Cl^- transport has not proven to be straightforward. What we can say with some degree of certainty is that this mechanism has a higher K_a than the acidic Cl^- mechanism and most probably possesses exchange diffusion characteristics as well. However, the precise localization of the mechanism remains an open question. If nonacidic Cl^- secretion is provided by the oxyntic cells, and if the intracellular PD profile in these cells is oriented as VILLEGAS says it is (*i. e.*, -18 mV with respect to the serosal solution but $+12$ mV with respect to the mucosal solution), then the simplest, and most commonly presented, model would have Cl^- enter the cells passively, and active secretion would occur at the mucosal membrane. However, if the REHM and SACHS groups are correct and the intracellular PD of all cells is -80 mV with respect to the serosal solution and -50 mV with respect to the mucosal solution (or if nonacidic Cl^- secretion occurs through the surface epithelial cells), then it is conceivable that this Cl^- secretion involves an active step at the serosal border and passive movement across the mucosal border. Thus, in Fig. 13 we have shown that there may be mediated steps for nonacidic Cl^- at either mucosal or serosal (or both) membranes. A certain fraction of nonacidic Cl^- secretion undoubtedly originates in the surface epithelial cells.

At the serosal membrane, in addition to the $Cl^- - HCO_3^-$ exchanger and possible mediated Cl^- transport, there also exists a Na^+-K^+ pump that aids in the maintenance of constant intracellular ionic milieu. This pump may also provide a means of effecting net transepithelial movement of Na^+ in the mammalian and lizard gastric mucosae. In these latter cases, entry of Na^+ into the cells is through an amiloride- and H^+ ion-sensitive "channel" that has pK $\approx$ 4. Net Na^+ absorption is most probably a contribution of the surface epithelial cells.

F. Biochemical Basis of Gastric HCl Secretion

I. Metabolic Requirements

The interrelationship of gastric H^+ and Cl^- transport with energy metabolism has been a vigorously debated problem. The area of research is of special interest since the energetic requirements for the transport functions are so high and the fundamental mechanisms for ion translocation are obviously tightly coupled to the metabolic processes. Under physiological conditions, the stomach is capable of transporting H^+ against a gradient of more than a millionfold, and under appropriately maintained stimulation this secretory function occurs at rather voluminous rates. We can evaluate the minimum energy requirement of gastric HCl transport by calculating the work associated with the moving of a given number of equivalents of $H^+(n_H)$ and $Cl^-(n_{Cl})$ through the known electrochemical gradients. For the secreting stomach, we would have to consider the serosal solution, the mucosal solution (gastric juice), and the electrical potential difference across the gastric epithelium. Thus, the work required for H^+ and Cl^- transport (W_{H^+,Cl^-}) is given as:

$$W_{H^+,Cl^-} = n_H \left[RT \ln \frac{[H^+]_{lu}}{[H^+]_{bl}} + z_H F(\Psi_{lu} - \Psi_{bl})\right] + n_{Cl} \left[RT \ln \frac{[Cl^-]_{lu}}{[Cl^-]_{bl}} + z_{Cl} F(\Psi_{lu} - \Psi_{bl})\right] \quad (12)$$

where the concentrations of H^+ and Cl^- are given for the stomach lumen (lu) and the blood (bl); z is the charge on the ion; $\Psi_{lu} - \Psi_{bl}$ is the transepithelial PD; and R, T and F have their usual meanings. Thus, for the secretion of 0.15 N HCl, with the blood at 37° C containing H^+ at pH 7.4 ($10^{-7.4}$ N) and $[Cl^-]_{bl} =$ 0.12 N, and a PD $= -$ 0.04 V, lumen negative with respect to blood, we have:

$$W_{H^+,Cl^-} = n_H \left(RT \ln \frac{0.15}{10^{-7.4}} - 0.04\,F\right) + n_{Cl} \left(RT \ln \frac{0.15}{0.12} + 0.04\,F\right) \quad (13)$$

$$W_{H^+,Cl^-} = n_H (8417) + n_{Cl} (1060) \quad (14)$$

Therefore, we can conclude that for gastric secretion under the above stated conditions 8417 cal per equivalent of H^+ and 1060 cal per equivalent of Cl^- represent the minimum energy requirement of the transport processes. You will note that the electrical gradient $F(\Psi_{2u} - \Psi_{62})$ $(\Psi_{lu} - \Psi_{bl})$ favors H^+ transport and accordingly diminishes the calculated caloric requirement of H^+ transport, whereas Cl^- is transported against the electrical gradient.

The energy requirements for transport clearly must come from metabolic activity of the tissue, and most, if not all, of the energy is derived from oxidative metabolism. In fact, it is this close coupling between metabolism and transport which has been the source of development of much of the theoretical background and experimental approaches to the elucidation of the gastric secretory processes. At the present time, two general theoretical bases are available to account for the input of metablic energy into the gastric H^+ transport process: the redox hypothesis and the ATP utilization hypothesis. In addition, there are a number of proposals that suggest complementary involvement of both oxidation-reduction and high-energy phosphate mechanisms.

II. Redox Hypotheses

Several different forms of the redox hypothesis have been proposed. The simplest form was suggested simultaneously by three different groups of workers (ROBERTSON and WILKINS, 1948; CONWAY and BRADY, 1948; CRANE and DAVIES, 1948) and essentially proposes that substrate H atoms are oxidized to H^+, the electrons being passed to molecular osygen *via* a suitable *membrane* redox system, *e. g.*, cytochrome chain.

$$\begin{matrix} 2SH_2 & & Fe^{3+} & & 4OH^- \\ & \times & & \times & \\ 4H^+ + 2S & & Fe^{2+} & & O_2 + 2H_2O \end{matrix}$$

Two things are immediately apparent from such a system. First, there is a maximum stoichiometric limit of four equivalents of H^+ produced per mol O_2 consumed. Secondly, if the reducing power of the system, *e. g.*, through NADH, is channeled into the production of the proton gradient for secretion, there would not be an equivalent production of ATP by the electron transport system.

KIDDER (1970) and HERSEY (1974) have independently suggested the possibility that some system of "variable coupling" might exist for mitochondria of the oxyntic cell whereby (i) during resting conditions reducing power is funneled into ATP production (for cell maintenance and synthetic processes), and (ii) during stimulated conditions the system would divert its energetic potential into proton production through some redox mechanism.

Finally, REHM (1972) has proposed a somewhat more complex, but more flexible, scheme. He has suggested some system of shuttle of NADH to a plasma membrane site for the liberation of H^+ with the concomitant shuttling back through a cytoplasmic electron carrier (*e. g.*, cytochrome *c*) to the subsequent steps (cyto *c*, cyto *a*, cyto a_3, O_2) in the mitochondria. A major feature of REHM's scheme is that ATP produced from this phase of mitochondrial electron transport, or for that matter from other sources such as glycolysis, could be used to drive reversed electron transport through the cytochrome chain and generate additional equivalents of reducing power for the H^+ transport potential. From his balance sheet of H atoms, electron transport and ATP participation, a maximum of 34 H^+ could be produced per six O_2 consumed, *i. e.*, $H^+/O_2 \cong 5.7$.

While all of these schemes are theoretically possible, we should examine the evidence that might stand in support of, or contradiction to, the various models. For a simple redox system, we could examine the stoichiometry between the number of ions transported and the moles of oxygen consumed by functioning tissue. In measuring the ratio of H^+ production to the total tissue oxygen consumption, a number of different investigators have shown that the stoichiometric efficiency is less than $4\,H^+/O_2$ (range 1–3; see FORTE, 1971). Since the secreted H^+ is by definition from the acid-secreting cells, while O_2 uptake includes all the cells and metabolic processes of the entire gastric preparation in addition to those responsible for H^+ secretion, the measured values of H^+/O_2 are clearly minimal ones. Attempts to measure the changes in H^+ secretion and O_2 consumption associated with stimulation or inhibition of secretion as a more appropriate index of stoichiometric efficiency has produced variable results, some treatments producing ratios of less than 4.0 and others greater than 4.0. However, it has been pointed out that if both H^+ and Cl^- transport are included in the stoichiometric calculations (*e. g.*, as a $(H^+ + Cl^-)/O_2$ ratio), then values above 4.0 most certainly occur even when the O_2 consumption is for the entire tissue (FORTE and DAVIES, 1963). Thus, for a simple redox scheme to operate, it would have to be exclusive for H^+ transport; Cl^- transport would be driven by some alternative mechanism such as an ATP utilizing system (see below), and this would require a certain amount of the oxidative metabolism for its sustained operation.

Modifications of the simple redox scheme that would allow a secretory-dependent alternative of electron flow directed toward either proton production or mitochondrial ATP synthesis (HERSEY, 1974; KIDDER, 1970) also have problems, which represent tentative objections to a casual acceptance of the idea. First, mitochondria isolated from the stomach do not appear to be radically different from those of other tissue: for example, rabbit gastric mitochondria utilize substrates, phosphorylate ADP and show respiratory control, as classically demonstrated for preparations from other sources (FORTE et al., 1967a). In addition, 2, 4-dinitrophenol, which effectively uncouples oxidative phosphorylation in gastric mitochondria (FORTE et al., 1967a), has the same dose-dependent effects in inhibiting gastric H^+ secretion and reducing cellular levels of ATP (DURBIN and KASBEKAR, 1965). Thus, within the framework of classic mitochondrial metabolism the gastric mitochondria seem to follow the typical functions of providing a source of ATP from oxidative phosphorylation. Furthermore, no one has identified a specific pathway or system to move free H^+ from its proposed mitochondrial source of production to the tubular system or apical plasma membrane of the acid-secreting cell.

REHM's (1972) proposed scheme can, in principle, account for all of the above-mentioned criticisms, since it incorporates the fundamental basis of normal mitochondrial function, as well as a plasma membrane redox system; however, it depends on suitable systems within (a) the plasma membrane and (b) cytoplasm for the proposed shuttling processes. Cytochemical experiments with frog oxyntic cells have shown that cytochrome oxidase is localized only within mitochondria, thereby negating this cytochrome as a significant extra-mitochondrial electron shuttle component (HERSEY et al., 1975). Although there have been reports of some extramitochondrial redox pathways in mammalina stomach (FORTE et al., 1967b; 1976), attempts to demonstrate significant quantities of NADH oxidizing systems, such as cyto b_5 and cyto P450, in the tubular membrane system of bullfrog oxyntic cells have been unsuccessful (FORTE, unpubl.). Thus, until more definitive evidence is provided for the existence and nature of specific plasma membrane-mitochondrial electron shuttle components, REHM's scheme must be held as strictly tentative.

III. ATP Utilization Hypotheses

Recognizing possible stoichiometric inadequacies of the simple redox hypothesis, DAVIES and OGSTON (1950) published the details of a mechanism whereby phosphate bond potential could be coupled into an electron-transport scheme to produce H^+ and which would be limited only by the overall thermodynamic constraints of the system. Thus, ATP, CrP, *etc.*, would be used to drive an electron-transport system in liberating protons whose ultimate source was water. You might recognize this system as analogous to that proposed by MITCHELL (1961) as the chemiosmotic hypothesis for oxidative phosphorylation. MITCHELL has refined the principle to the following set of generalities. Electron transfer through the mitochondrial electron-transport chain results in the separation of H^+ and OH^- in adjacent submitochondrial compartments. The electrochemical gradient of H^+ activity between two compartments, 1 and 2 ($[\bar{a}_H]_1$ and $[\bar{a}_H]_2$), would be the force for production of ATP from ADP and Pi. The latter would operate through an asymmetric ATPase (or ATP synthetase) in accordance with the thermodynamic conditions of equilibrium. The asymmetry of the ATPase is defined as a membrane-bound system with side-dependent differential accessibilities to H^+ and OH^-. The phosphorylating system at equilibrium would be represented as:

$$\frac{[\mathrm{ATP}]}{[\mathrm{ADP}]} = \frac{[\mathrm{Pi}]}{K[\mathrm{H_2O}]} \cdot \frac{[\bar{a}_H]_1}{[\bar{a}_H]_2} \tag{15}$$

where the brackets indicate the electrochemical activity of the species and K is the equilibrium constant. It is obvious that for such an asymmetrically oriented system the equilibrium ratio of [ATP]/[ADP] will depend upon the electrochemical potential gradient of H^+ across the membrane. Conversely, for application to the problem of gastric H^+ secretion, the H^+ gradient across the membrane would depend on the [ATP]/[ADP] ratio, and by simple rearrangement of Eq. (15) we have:

$$\frac{[\bar{a}_H]_1}{[\bar{a}_H]_2} = K \cdot \frac{[\mathrm{ATP}][\mathrm{H_2O}]}{[\mathrm{ADP}][\mathrm{Pi}]} . \tag{16}$$

Substituting known conditions of ATP, ADP and Pi for whole gastric tissue (FORTE et al., 1965; DURBIN et al., 1974) and designating the hydrolysis constant, $K[H_2O]$, under physiological conditions as 10^5 (BENZINGER et al., 1959), we have:

$$\frac{[\bar{a}_H]_1}{[\bar{a}_H]_2} = 10^5 \cdot \frac{[1.1 \cdot 10^{-3}]}{[0.2 \cdot 10^{-3}][1.6 \cdot 10^{-3}]} \tag{17}$$

$$\frac{[\bar{a}_H]_1}{[\bar{a}_H]_2} = 3.4 \cdot 10^8 . \tag{18}$$

That is to say, the gross concentrations of the constituents would be thermodynamically consistent with supporting an electrochemical gradient of H^+ up to $3.4 \cdot 10^8$-fold between the two sides of the membrane-bound ATPase system. The electrochemical potential gradient term includes both chemical activities and membrane potential such that:

$$\log \frac{[\bar{a}_H]_1}{[\bar{a}_H]_2} = (pH_2 - pH_1) + \frac{F}{2.3\,RT}(\Psi_1 - \Psi_2), \tag{19}$$

where $[H^+]$ is the concentration of H^+, pH = -log $[H^+]$ and Ψ the electrical potential, with the subscripts 1 and 2 referring to the individual sides of the membrane. We do not know what value to insert for the electrical potential difference across the system ($\Psi_1 - \Psi_2$), but even if this were as large as +100 mV, the $[H^+]$ gradient would still be greater than 10^6, which is what would be required for H^+ secretion. Thus, we could conclude that for the observed conditions, there is sufficient energy available to provide gastric H^+ transport by an ATP-driven reaction with a stoichiometry of at least one equivalent H^+/mol ATP hydrolyzed. If H^+ and Cl^- transport show a degree of coupling similar to that discussed in Section E. I. 3. (Fig. 11a), then the Cl^- transport directly associated with H^+ transport could be accomplished with relatively little additional energy cost. Since the energetic relationships are such that they provide at least one H^+ per ATP hydrolyzed, and we assume a typical ratio for oxidative phosphorylation of $ATP/O_2 = 6$, the stoichiometry between acid secretion and oxygen consumption would also be six. Possibilities for variable, and even higher, stoichiometry have also been suggested (DAVIES, 1957).

IV. The Search for Gastric ATPases

1. HCO_3^--Stimulated ATPase

From the above discussion, it is reasonable to conclude that gastric H^+ and Cl^- transport might be coupled to an ATP-utilizing system, as has been demonstrated for other ion transport processes. Thus, it remains to demonstrate the presence and location of such a system within the gastric epithelial cells. KASBEKAR and DURBIN (1965) were the first to report the existence of such a gastric ATPase in the "microsomal" fraction from frog stomach. Interest was especially aroused since this ATPase had the properties of being stimulated by HCO_3^- and inhibited by SCN^-. A number of investigators pursued the characterization of this enzyme system (SACHS et al., 1972; DE PONT et al., 1972; SPENNEY et al., 1973). However, it has been demonstrated that the so-called gastric HCO_3^--stimulated ATPase was very probably a mitochondrial contamination within the microsomal fraction (GANSER and FORTE, 1973a; FORTE et al., 1975; SOUMARMON et al., 1974). Because this enzyme is typically expressed in ruptured mitochondria (RACKER, 1962; FANESTIL et al., 1963; MITCHELL and MOYLE,

1971), it was probably released by vigorous homogenization of the gastric tissue and harvested along with the microsomes.

2. ($Na^+ + K^+$)-ATPase

A number of investigators have examined gastric membrane fractions to localize the well-known enzyme associated with Na^+ transport. ($Na^+ + K^+$)-ATPase has been demonstrated in amphibian (LIMLOMWONGSE and FORTE, 1970), reptilian (HANSEN et al., 1972) and mammalian (MÓZSIK and OYE, 1969; SACHS et al., 1971; FORTE et al., 1975) stomach tissue; however, specific activities and total amounts of enzyme are not high compared with tissue whose primary function is Na^+ transport. From the characteristics of membrane fractions by enzyme marker studies or sedimentation (SACHS et al., 1972; FORTE et al., 1975), it would appear that the gastric ($Na^+ + K^+$)-ATPase is associated with a plasma membrane fraction, most probably the basolateral membranes. At the serosal surface, the gastric ($Na^+ + K^+$)-ATPase may be operating to maintain the intracellular ionic milieu. MÓZSIK and his collaborators have found interesting stimulatory effects of cholinergic agents on ($Na^+ + K^+$)-ATPase (MÓZSIK et al., 1974), and inhibition by epinephrine and cyclic AMP (MÓZSIK, 1969, 1970), and they have suggested the possibility that this membrane ATPase system may have some regulatory role for activation of the stomach by secretagogues.

3. K^+-Stimulated ATPase

When gastric microsomes are purified on a sucrose density gradient, a low-density band of membranes is separated, which is free of mitochondrial contamination (Ganser and FORTE, 1973a; FORTE et al., 1975). This purified microsomal fraction contains ATPase activity, and, interestingly, it is stimulated by

Table 3. K^+-stimulated ATPase activity of gastric microsomal preparations from frog oxyntic cells and pig fundic mucosa

	ATPase activity (μmol · mg prot^{-1} · h^{-1})	
	Frog	Pig
Basal	19.6	15.1
K^+	36.0	105.3
Na^+	20.4	21.2
$Na^+ + K^+$	35.2	92.8
$Na^+ + K^+$ + ouabain	34.1	91.6
K^+ + gramicidin	55.9	187.5
$Na^+ + K^+$ + gramicidin	–	174.2

Basal activity refers to ATPase in the presence of Mg^{2+} alone. Where indicated, other additions were 10^{-4} M ouabain, 10^{-6} M gramicidin; for frog microsomes, [K^+] was 25 mM, [Na^+] was 60 mM; for pig microsomes, [K^+] was 5 mM, [Na^+] was 10 mM.

K^+ but not by Na^+ (Table 3). Some of the principal features of the K^+-stimulated ATPase are summarized in Table 4. The enzyme reaction can be simplified according to the following two-step reaction:

$$ATP + E \underset{}{\overset{Mg^{2+}}{\rightleftharpoons}} E \sim P + ADP \tag{20}$$

$$E \sim P \underset{}{\overset{K^+, Mg^{2+}}{\rightleftharpoons}} E + Pi \tag{21}$$

The first step involves the Mg^{2+}-catalyzed kinase reaction of enzyme (E) with ATP to form a phosphorylated intermediate (E ~ P). The subsequent hydrolysis of E ~ P to free enzyme and Pi is stimulated by K^+. It is of interest that the gastric K^+-stimulated ATPase has several features in common with the well-known ($Na^+ + K^+$)-ATPase: a K^+-stimulated phosphatase step; a phosphorylated intermediate with an acyl phosphate linkage; a 90000–100000 dalton phosphorylated protein subunit. However, very distinctive differences, such as the absence of Na^+ stimulation and the absence of inhibition by ouabain, clearly show that this gastric enzyme is not to be confused with the ($Na^+ + K^+$)-ATPase (FORTE et al., 1974).

Several findings suggest that the rather unique K^+-stimulated ATPase in oxyntic cells from frog and mammalian fundic mucosa has some specific role in gastric secretory function. a) It has long been known that K^+ is required for gastric secretion: in K^+-free media, HCl secretion is abolished (GRAY and ADKISON, 1941; HARRIS et al., 1958; DAVIS et al., 1965). b) The enzyme appears to be associated with apical plasma membranes and/or tubulovesicles of oxyntic cells (GANSER and FORTE, 1973a; FORTE et al., 1975). c) Moreover, the ontogenetic appearance of the enzyme (K^+-phosphatase) in tadpole stomach is well correlated with the development of the oxyntic cell and the appearance of functional HCl secretion (FORTE et al., 1969b; LIMLOMWONGSE et al., 1970). A number of inhibitors have been identified for the enzyme system (*e. g.*, N-ethylmaleimide, F^-, Zn^{2+}); however, these agents have general inhibitory function and until some specific inhibitor is found, such as ouabain for the Na^+ pump, the above points of correlation between the enzyme and acid secretion simply remain suggestive.

Table 4. General properties of gastric microsomal K^+-stimulated ATPase

1. Stimulated by K^+; requires Mg^{2+}.
2. Further activated by K^+ ionophores.
3. Not affected by Na^+ or ouabain.
4. Has an associated K^+-stimulated phosphatase.
5. Found in fundic mucosa of all species examined; not present in other tissues examined (*i. e.* brain, liver, kidney, antral mucosa).
6. Localized to oxyntic cells (bullfrog stomach); most probably associated with apical plasma membrane and/or tubulovesicular membranes.
7. Enzyme function segregated into at least two steps: a phosphokinase (not requiring K^+); and a K^--dependent phosphoprotein phosphatase.

The gastric microsomal ATPase has the very interesting feature of being stimulated by ionophores in the presence of K^+ (GANSER and FORTE, 1973b). Interpretation of this ionophoretic effect and its abolition by gentle disruption techniques, such as freeze-drying and treatment with detergents, have led to the conclusion that the isolated gastric membranes are largely vesicular in nature. In fact, the transport properties of these vesicles may provide some useful clues for their involvement in gastric secretion.

V. Transport Properties of Gastric Microsomal Vesicles

Using a preparation of microsomes from dog fundic mucosa, LEE et al. (1974) demonstrated that, in the presence of K^+ and Mg^{2+}, the pH of the bulk medium increased after the addition of ATP (Fig. 14). The logical conclusion was that the observed alkalinization was matched by an equivalent accumulation of H^+ within the vesicles and they suggested an ATP-generated transport of H^+ into the vesicles *via* a H^+/K^+ exchange mechanism. This interpretation was supported by the fact that the H^+ gradient could be prevented or abolished either by the addition of ionophores for K^+ and H^+ (*e. g.*, valinomycin plus FCCP) or by the disruption of vesicular integrity by the detergent, Triton X-100 (Fig. 14).

The observations with gastric microsomal vesicles have interesting similarities

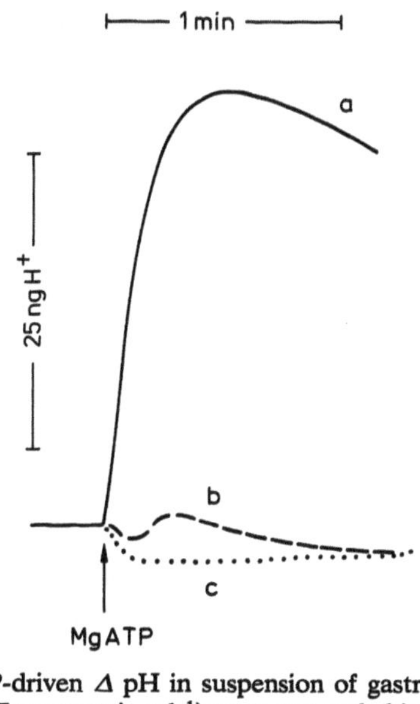

Fig. 14. Time course of ATP-driven Δ pH in suspension of gastric microsomal vesicles. Vesicles from dog gastric mucosa (0.67 mg protein ml^{-1}) were suspended in medium of 150 mM KCl, 5 mM glycyl glycine, pH 6.1 at 25° C. Reaction was initiated with 100 nmol of MgATP where indicated. Trace *a* shows alkalinization of external medium under control conditions. The pH change is markedly damped or prevented in presence of valinomycin and FCCP (trace *b*) or Triton X 100 (trace *c*). (Redrawn from LEE et al., 1974)

to studies with other subcellular transport systems. For instance, vesicles prepared from bacteria (HAROLD, 1972), mitochondria (MITCHELL and MOYLE, 1965; CHANCE and MELA, 1967) and chloroplasts (JAGENDORF and NEUMANN, 1965) demonstrate a H^+-transport function, while sarcoplasmic reticulum vesicles (INESI, 1972) and liposomes containing ($Na^+ + K^+$)-ATPase (HILDEN and HOKIN, 1976) show ATP-generated gradients of the associated transported cations.

The work from SACHS' laboratory has been productive in extending the transport studies of gastric microsomal vesicles. They have used purified (mitochondrial-free) hog microsomal preparations to confirm the earlier work of LEE et al. (1974), and have concentrated studies on several aspects of the system: (i) characterization and quantitation of the ATP-driven transport of vesicular ion transport (SACHS et al., 1975); (ii) evaluation of the vesicular permeability characteristics (SCHACKMANN et al., 1976); (iii) measurements of transport-associated membrane potential (LEWIN and SACHS, 1975); and (iv) evaluation of transport by reconstitution of vesicles into artificial thin lipid membranes (GOODALL, 1976). SACHS and his colleagues have shown that ATP-driven proton transport into gastric vesicles operates best in a medium of high KCl. Na^+ is a very poor substitute for K^+, and SO_4^{--} substitution reduces the H^+ accumulation by about 50 percent. As LEE et al. (1974) found, complete dissipation of the H^+ gradient required both a protonophore (*e. g.*, FCCP) and a K^+ inophore (*e. g.*, valinomycin). It was argued that because valinomycin produced a greater dissipation of the proton gradient than FCCP, the "nonenergized" H^+ conductance pathway was greater than the K^+ conductance (SACHS et al., 1975). With a measured intravesicular space of 2.2 μl mg^{-1} vesicle protein, it was calculated that ATP-generated proton transport could theoretically bring the intravesicular [H^+] to pH 1.7 (SACHS et al., 1976). Buffering capacity within the vesicles had not been accounted for, so this value is just a rough estimate. They have also made estimates of the stoichiometry between H^+ transport/ATP hydrolyzed and found this to be 1.0 or greater.

Measurements of vesicular ion permeability from uptake studies with $^{86}Rb^+$ (as a radiolabeled substitute for K^+) or $^{36}Cl^-$ showed approximately similar time courses for Rb^+ and Cl^-, with the half-time ($t_{1/2}$) being about 30–35 min at room temperature (SCHACKMANN et al., 1976). Rb^+ uptake was enhanced about twofold in the presence of valionomycin and decreased by 50 percent when ATP was added. When ATP was added to membranes previously equilibrated with Rb^+, there was a fairly rapid (5–10 min) efflux of Rb^+ followed by a slow leak ($t_{1/2} \approx 1$ h) back into the vesicles and a return to the equilibrium concentration state. From a correlation of these results with the H^+ uptake and ATP utilization data it was concluded that the gastric vesicular ATPase energized a one-for-one, electroneutral K^+/H^+ exchange. Attempted measurements of vesicular membrane potential with ANS as a fluorophore produced results consistent with this interpretation (LEWIN and SACHS, 1975), but further work is required before we can firmly establish the significance of the electrical data at the vesicular level. GOODALL (1976) has recently studied the reincorporation of microsomal membranes into artificial systems, but this work is presently in its germinative stages, and it will not be reviewed here, except to say that such an approach may provide some useful conclusions, as it has for the $(Na^+ + K^+)$-ATPase system (HILDEN and HOKIN, 1976).

Recently, our laboratory has undertaken a different methodological approach to study H^+ transport by gastric microsomal vesicles (LEE et al., 1976; LEE and FORTE, 1978). The approach is to follow either spectral changes for certain metachromatic dyes (*e. g.* acridine orange or neutral red) or fluorescence quenching of a number of probes (*e. g.* acridine orange, 9-amino acridine or quinacrine). These dyes are all weak bases, and the membrane vesicles are permeable to the dyes in the neutral form but impermeable in the charged, protonated form. Thus, the optical properties of dye-membrane mixtures can be used as an index of H^+ transport into the vesicles as the dyes simply accumulate or are bound in their protonated form (LEE and FORTE, 1978).

This technique is advantageous because it avoids the problem of having to perform experiments at a less-than-optimal pH of 6.1, the $[H^+]$ at which ATP hydrolysis is "isoionic" (*i. e.*, no net H^+ production except that due to membrane translocation processes [MITCHELL and MOYLE, 1967; ALBERTY, 1968]). In addition, because the dye-binding technique affords a high degree of sensitivity, a much lower concentration of membrane protein can be used.

Addition of ATP to gastric microsomal membranes produces a large shift in the dye spectra (acridine organe and neutral red) toward the blue region. The ATP-induced spectral shift is absolutely dependent upon K^+ and Mg^{2+}, and from additional studies with a number of inhibitors we can conclude that the shift is dependent on a functional K^+-ATPase. It has been shown that the spectral shift is not simply due to a pH change in the bulk medium, but is due to accumulation of the dyes within the low intravesicular pH environment and to an orderly stacking to negative sites within the membrane vesicles (LEE et al., 1976); DELL'ANTONE et al., 1975). For at least one of the dyes (9-amino acridine) there was no measurable binding, so that the kinetics of accumulation of this fluorophore simply followed the intravesicular pH and can be used as a quantitative probe for the pH gradient (LEE and FORTE, 1978).

A comparison of the ATP-induced effects on dye binding and ΔpH is presented in Fig. 15. It can also be seen that upon addition of a protonophore and K^+-ionophore, the dye binding and pH are restored to their respective initial values. Thus, the dye binding has been interpreted in terms of an ATP-induced energization of the membranes concomitant with active H^+ accumulation. This is supported by an alternative set of experiments in which dye binding is induced by an artificial proton gradient. When vesicles are preincubated at pH 4.0 in succinate buffer, then rapidly transferred to pH 8.0, a Mg^{2+}-dependent dye binding is observed; addition of protonophores, such as CCCP, collapses the H^+ gradient and the dye is released (LEE et al., 1976). The dye-binding studies are thus consistent with pH-electrode studies showing that the K^+-ATPase can energize the membranes to support H^+ transport, very likely through a K^+/H^+ exchange system.

Using the dye accumulation assay, several properties of the vesicular H^+ pump, the K^+-ATPase, and their interrelationships have been described and will briefly be listed here. The K^+ requirement for the pump is at an intramembrane or intravesicular site, and the pump is of a general H^+/K^+ exchange

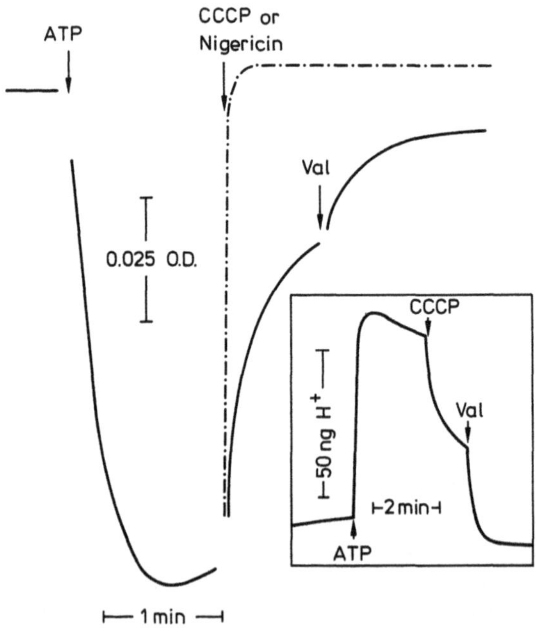

Fig. 15. Time course of ATP-induced binding of acridine orange to purified microsomal vesicles prepared from hog gastric mucosa. Membrane vesicles (0.1 mg protein ml^{-1}) were preincubated for 2 h at room temperature in medium containing 150 mM KCl, 2 mM $MgCl_2$, 10 mM Pipes buffer (pH 6.1) and 0.2 mM EDTA. 10 μM acridine orange was added and O. D. recorded by dual-wavelength spectroscopy, 493 nm being used as sample wavelength and 550 nm as reference. At time indicated by arrow, 0.2 mM ATP was added and decrease in recorded O. D. indicates dye binding associated with energization of membrane vesicles. Energization was dissipated rapidly, by (a) sequential addition of CCCP and valinomycin (solid line) or (b) addition of nigericin (dotted line), which is an ionophore for both H^+ and K^+, or (c) dissipated slowly as ATP was used up (not shown). Insert illustrates analogous experiment in which pH was recorded (cf. Fig. 14). Conditions were as above except that membrane protein concentration was 1.0 mg ml^{-1} and no buffer was included. Addition of 0.8 mM ATP caused alkalinization of medium, which returned to near baseline levels after successive additions of CCCP and valinomycin. Neither the dye binding nor the pH changes occurred if ATP, K^+, or Mg^{2+} was excluded from medium or if known inhibitors of the K^+-ATPase were used. (From LEE et al., 1976, with permission)

(exact stoichiometry uncertain). Anions participate passively in the process; their ability to stimulate K^+-ATPase and promote vesicular acidification is directly related to their permeability with the observed sequence, $Br^- > Cl^- > I^- >$ acetate ≈ isethionate. Interestingly, this is an anion sequence similar to that described by DURBIN (1964) for promoting acid secretion by *in vitro* frog gastric mucosa. With a permeant anion (*e. g.* Cl^-), the net effect of the K^+-ATPase pump is the accumulation of HCl within the vesicles. Distribution studies with pH gradient probes (*e. g.* 9-amino acridine) shows that intravesicular pH can be lowered to at least pH 2.5 by the pump. This observed pH is still more than an order of magnitude from that required to account for gastric HCl secretion, but there may be an increased leakiness to protons in the isolated vesicles preparation as compared to membrane function *in vivo*.

The work on H^+ transport by isolated microsomal vesicles forms a very

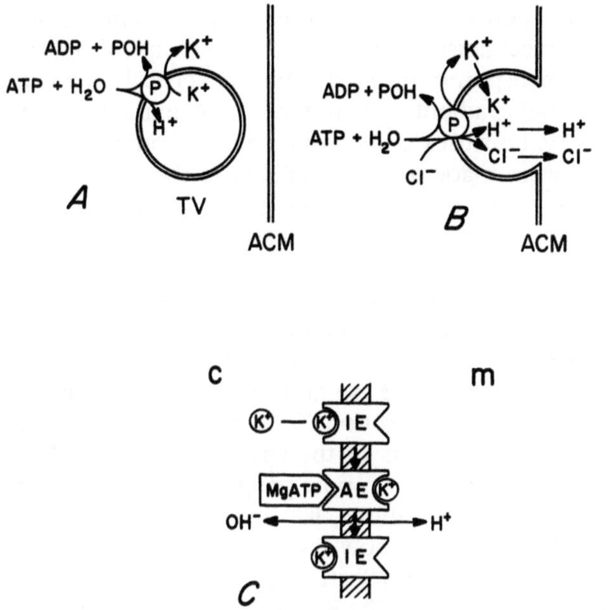

Fig. 16. Hypothetical scheme showing how gastric microsomal K^+-ATPase could be involved in HCl secretion. (A) Diagram showing K^+-ATPase as pump enzyme (P) in tubulovesicle membrane (TV) in unstimulated oxyntic cell. Cellular conditions of ATP, ADP, K^+, etc. would operate through enzyme to poise system at electrochemical potential of transport reaction, but because system is not in contact with apical cell membrane (ACM), net transport across epithelium does not occur. (B) After appropriate stimulation of oxyntic cell, tubulovesicles fuse with apical cell membrane and H^+ and Cl^- can diffuse away from their immediate site of production. System would require some mechanism of K^+ conductance (or recycling as in C) from cell to outer surface to activate proton pump. Cl^- is shown as having some association with H^+ transport step, but precise mechanisms cannot be specified. Details of interaction within translocation step will determine whether or not pump can operate electrogenically or as a neutral system. (C) Possible recycle mechanism for K^+ involvement in the ATPase and proton transport. K^+ on mucosal surface (m) would serve to activate proton transport step in presence of MgATP. K^+ might be required on cytoplasmic aspect (c) of membrane to dephosphorylate phosphoprotein intermediate and re-establish cycle. IE, inactive enzyme; AE, activated enzyme

promising basis for pursuing questions of gastric acid secretion. The requirement of the ATPase for K^+ correlates with the well-known K^+ requirement for gastric secretion (Section C. II.4). The purified microsomal fraction is derived from the tubulovesicle membrane system of the oxyntic cell (GANSER and FORTE, 1973a), and, as we have pointed out, these membranes are involved in fundamental ultrastructural transformations associated with HCl secretion (Section B. II). We present here a hypothetical scheme that conforms to many of the experimental findings and represents a plausible model for H^+ transport (Fig. 16). When the oxyntic cell is resting, the tubulovesicular membranes are not in communication with the apical cell surface; the K^+-ATPase is primed to function and cellular ATP establishes an energized state such that the electrochemical potential gradient across the tubulovesicular membranes and the chemical driving force are at equilibrium, with no net flow of energy or materials through the system. Stimulation of the oxyntic cell by appropriate secretagogue and cellular activation mechanisms is known to involve apical cell membrane proliferation by fusion incorporation of the tubulovesicles. This could relax the electrochemical potential gradient, which would now exist between cytoplasm and gastric lumen, as HCl diffuses away from the membrane surface. If an appropriate supply of K^+ were made locally available to the ATPase, net H^+ secretion would occur at the expense of ATP driving the reaction. Oxidative metabolism would, of course, be responsible for the resynthesis of ATP. Cessation of secretion would be accomplished by reincorporation of the apical surface into tubulovesicles (*e. g.,* secretagogue withdrawal), interference with metabolism (*e. g.,* anoxia), or some specific pump enzyme inhibitor.

The above proposal is a highly simplified attempt to incorporate the K^+/H^+ exchange transport, the K^+-ATPase and oxyntic cell membrane turnover into an hypothesis of gastric HCl secretion. There are a number of serious questions to be raised so that further tests of the specific or general hypotheses can be developed. First, virtually all transport properties of the purified gastric microsomal vesicles concern the cations, K^+ and H^+; active Cl^- transport has not been directly involved, yet we know the stomach actively transports Cl^-. However, as was suggested earlier (Section C. III.1), there may be a separate, nonacidic Cl^- pump mechanism; the K^+-ATPase system may simply account for the Cl^- directly associated with H^+ transport. Another problem is that, as proposed, K^+ would have to be supplied to the pump site at the extracellular membrane surface, and unless we were to suppose a local, and fairly tight, recycling of K^+, this would represent a serious challenge. We know that K^+ is required for acid secretion, but it is more effective on the serosal side than on the mucosal side. There is a well-established relationship between increased K^+ output into gastric juice and HCl secretion, which might be the result of certain leakiness or inefficiency of the K^+/H^+ exchange site; however, this problem will bear careful scrutiny in future analyses. Full interpretation of the gastric ATPase system also suffers from not having a specific and effective inhibitor available, such as ouabain for the $(Na^+ + K^+)$-ATPase. SCN^- is a well-known, and fairly specific, inhibitor of gastric HCl secretion. However, this anion has very little effect on the microsomal K^+-ATPase activity (GANSER and FORTE, 1973a and b), and it has been reported to have no effect on the ATP-induced proton gradient of microsomal vesicles (SACHS et al., 1976). This latter refractoriness is curious, since SCN^- is a fairly lipophilic anion and one might expect it to collapse the vesicular proton gradient somewhat; but, within the framework of available data, we cannot specifiy the action of this inhibitor. There are also a number of additional questions persisting, such as those concerning the correlation of electrophysiological data from intact mucosa with vesicular ion transport properties. Perhaps future studies with the isolated gastric vesicles that would closely monitor (a) the kinetics of H^+ and K^+ transport, (b) the electrochemical limits of the K^+-ATPase as a driving force, (c) the specific role of transport processes in the generation of vesicular membrane potentials, and (d) the detailed biochemical events which lead to activation and/or fusion of the oxyntic cell tubulovesicular membrane system will provide a more complete basis on which to make correlations between *in vitro* and *in vivo* systems.

Acknowledgements

We gratefully acknowledge support from NIH grants AM19520 and AM10141.

References

ALBERTY, R. A.: J. biol. Chem. **243**, 1337 (1968).
ALTAMIRANO, M.: J. Physiol. **168**, 787 (1963).
ALTAMIRANO, M.: Amer. J. Physiol. **216**, 33–40 (1969).
ALTAMIRANO, M., MARTINOYA, C.: J. Physiol. **184**, 771 (1966).
ANDREW, W.: In: Textbook of Comparative Histology. New York: Oxford 1959, p. 138.
BAER, J. E., JONES, C. B., SPITZER, A. S., RUSSO, H. F.: J. Pharmacol. exp. Ther. **157**, 472 (1967).
BARRY, P. H., DIAMOND, J. M., WRIGHT, E. M.: J. Membrane Biol. **4**, 358 (1971).
BENZINGER, T., KITZINER, C., HEMS, R., BURTON, K.: Biochem. J. **71**, 400 (1959).
BLUM, A. L., HIRSCHOWITZ, B. I., HELANDER, H. F., SACHS, G.: Biochim. Biophys. Acta **241**, 261 (1971).
BORNSTEIN, A. M., DENNIS, W. H., REHM, W. S.: Amer. J. Physiol. **197**, 332 (1959).
BROOKS, F. P.: In: Handbook of Physiology, Section 6: Alimentary Canal, Vol. 2: Secretion (C. F. Code, Ed.). Washington, D. C.: Amer. Physiol. Soc. 1967, p. 805.
CANOSA, C. A., REHM, W. S.: Biophys. J. **8**, 415 (1968).
CHANCE, B., MELA, L.: J. biol. Chem. **242**, 830 (1967).
CLARKSON, T. W.: J. gen. Physiol. **50**, 695 (1967).
CONWAY, E. J., BRADY, T. G.: Nature (Lond.) **162**, 456 (1948).
COOPERSTEIN, I. L.: J. gen. Physiol. **42**, 1233 (1959).
COTLOVE, E., HOGBEN, C. A. M.: Fed. Proc. **15**, 41 (1956).
COTLOVE, E., GREEN, N. D., HOGBEN, C. A. M.: Fed. Proc. **18**, 31 (1959).
CRANE, E. E., DAVIES, R. E.: Biochem. J. **43**, xliii (1948).
CRANE, E. E., DAVIES, R. E., LONGMUIR, N. M.: Biochem. J. **43**, 321 (1948).
CUMMINS, B. E., VAUGHAN, J. T.: Biochim. biophys. Acta **94**, 280 (1965).
CURRAN, P. F.: J. gen. Physiol. **43**, 1137 (1960).
DAINTY, J.: Advanc. bot. Res. **1**, 279 (1963).
DAINTY, J., HOUSE, C. R.: J. Physiol. **183**, 66 (1966).
DAVENPORT, H. W.: J. Physiol. **97**, 32 (1939).
DAVENPORT, H. W.: Amer. J. Physiol. **129**, 505 (1940).
DAVENPORT, H. W.: Proc. Soc. exp. Biol. **110**, 613 (1962).
DAVENPORT, H. W.: Amer. J. Physiol. **204**, 213 (1963).
DAVENPORT, H. W., ALZAMÓRA, F.: Amer. J. Physiol. **202**, 711 (1962).
DAVIES, R. E.: Biochem J. **42**, 609 (1948).
DAVIES, R. E.: Biol. Rev. **26**, 87 (1951).
DAVIES, R. E.: In: Metabolic Aspects of Transport across Cell Membranes. Madison: University of Wisconsin Press 1957, p. 240.
DAVIES, R. E., LONGMUIR, N. M.: Biochem. J. **42**, 621 (1948).
DAVIES, R. E., OGSTON, A. G.: Biochem. J. **46**, 324 (1950).
DAVIES, R. E., TERNER, C.: Biochem. J. **44**, 379 (1949).
DAVIES, R. E., CRANE, E. E., LONGMUIR, N. M.: Biochem. J. **43**, 321 (1948).
DAVIS, T. L., RUTLEDGE, J. R., KEESEE, D. C., BAJANDES, F. J., REHM, W. S.: Amer. J. Physiol. **209**, 146 (1965).
DELL'ANTONE, P., COLONNA, R., AZZONE, G. F.: Biochim. biophys. Acta **234**, 541 (1971).
DELRUE, G.: Arch. internat. Physiol. **33**, 196 (1930).
DENNIS, W. H., REHM, W. S.: Amer. J. Physiol. **193**, 15 (1958).
DE PONT, J. J. H. H. M., HANSEN, T., BONTING, S. L.: Biochim. biophys. Acta **274**, 189 (1972).

DIAMOND, J. M.: J. Physiol. **161**, 503 (1962).
DIAMOND, J. M.: J. gen. Physiol. **48**, 15 (1964).
DIAMOND, J. M., BOSSERT, W. H.: J. gen. Physiol. **50**, 2061 (1967).
DURBIN, R. P.: J. gen. Physiol. **44**, 315 (1960).
DURBIN, R. P.: J. gen. Physiol. **4**, 735 (1964).
DURBIN, R. P.: In: Handbook of Physiology, Section 6; Alimentary Canal, Vol. 2: Secretion. (C. F. Code, Ed.). Washington, D. C.: Amer. Physiol. Soc. 1967, p. 879.
DURBIN, R. P., HEINZ, E.: J. gen. Physiol. **41**, 1035 (1958).
DURBIN, R. P., KASBEKAR, D. K.: Fed. Proc. **24**, 1377 (1965).
DURBIN, R. P., MOODY, F. G.: Symp. Soc. exp. Biol. **19**, 299 (1965).
DURBIN, R. P., FRANK, H., SOLOMAN, A. K.: J. gen. Physiol. **39**, 535 (1956).
DURBIN, R. P., KITAHARA, S., STAHLMANN, K., HEINZ, E.: Amer. J. Physiol. **207**, 1177 (1964).
DURBIN, R. P., MICHELANGELI, F., NICKEL, A.: Biochim. biophys. Acta **367**, 177 (1974).
DYCK, W. P., WERTHER, J. L., RUDICK, J.: Gastroenterology **56**, 488 (1969).
ERLIJ, D.: Pflügers Arch. **364**, 91 (1976).
FANESTIL, D. D., HASTINGS, A. B., MAHOWALD, T. A.: J. biol. Chem. **238**, 836 (1963).
FARQUHAR, M. G., PALADE, G. E.: J. Cell Biol. **17**, 375 (1963).
FLEMSTRÖM, G.: Biochim. biophys. Acta **225**, 35 (1971).
FORTE, G. M., LIMLOMWONGSE, L., FORTE, J. G.: J. Cell Sci. **4**, 709 (1969).
FORTE, J. G.: Biochim. biophys. Acta **150**, 136 (1968).
FORTE, J. G.: Amer. J. Physiol. **216**, 167 (1969).
FORTE, J. G.: In: Membranes and Ion Transport, Vol. 3 (E. Bittar, Ed.). New York: Wiley 1971, p. 111.
FORTE, ADAMS and DAVIES. Nature **197**, 874 (1963).
FORTE, J. G., DAVIES, R. E.: Amer. J. Physiol. **204**, 812 (1963).
FORTE, J. G., NAUSS, A. H.: Amer. J. Physiol. **205**, 631 (1963).
FORTE, J. G., SOLBERG, L. A.: In: International Encyclopedia of Pharmacology and Experimental Therapeutics, Section 29 A, Vol. 1. New York: Pergamon 1973, p. 195.
FORTE, J. G., ADAMS, P. H., DAVIES, R. E.: Biochim. biophys. Acta **104**, 25 (1965).
FORTE, J. G., FORTE, G. M., GEE, R., SALTMAN, P.: Biochem. biophys. Res. Commun. **28**, 215 (1967a).
FORTE, J. G., FORTE, G. M., SALTMAN, P.: J. Cell Physiol. **69**, 293 (1967b).
FORTE, J. G., LIMLOMWONGSE, L., KASBEKAR, D. K.: J. gen. Physiol. **54**, 76 (1969a).
FORTE, J. G., LIMLOMWONGSE, L., KASBEKAR, D. K.: J. gen. Physiol. **59**, 79 (1969b).
FORTE, J. G., GANŞER, A. L., TANISAW, A. S.: Ann. N. Y. Acad. Sci. **242**, 255 (1974).
FORTE, J. G., GANSER, A., BEESLEY, R., FORTE, T. M.: Gastroenterology **69**, 175 (1975).
FORTE, J. G., MACHEN, T. E.: J. Physiol. (Lond.) **224**, 33 (1975).
FORTE, T. M., MACHEN, T. E., FORTE, J. G.: Gastroenterology **69**, 1208 (1975).
FRIZZELL, R. A., SCHULTZ, S. G.: J. gen. Physiol. **59**, 318 (1972).
FRÖMTER, E.: J. Membrane Biol. **8**, 259 (1972).
FRÖMTER, E., DIAMOND, J. M.: Nature New Biol. **235**, 9 (1972).
GANSER, A. L., FORTE, J. G.: Biochim. biophys. Acta **307**, 169 (1973a).
GANSER, A. L., FORTE, J. G.: Biochem. biophys. Res. Commun. **54**, 690 (1973b).
GIEBISCH, G., BOULPAEP, E. L., WHITTEMBURY, G.: Phil. Trans. roy. Soc. B. **262**, 175 (1971).
GOODALL, M.: Biophys. J. **16**, 83a (1976).
GRAY, J. S., ADKISON, J. L.: Amer. J. Physiol. **134**, 27 (1941).
GRAY, J. S., ADKISON, J. L., ZELLE, K.: Amer. J. Physiol. **130**, 327 (1940).
HANSEN, T., BONTING, S. L., SLEGERS, J. F. G., De PONT, J. J. H. H. M.: Pflügers Arch. **334**, 141 (1972).
HANSEN, T., SLEGERS, J. F. G., BONTING, S. L.: Biochim. biophys. Acta **382**, 590 (1975).
HAROLD, F. M.: Bact. Rev. **36**, 172 (1972).
HARRIS, J. B., EDELMAN, I. S.: Amer. J. Physiol. **190**, 205 (1960).
HARRIS, J. B., EDELMAN, I. S.: Amer. J. Physiol. **206**, 769 (1964).
HARRIS, J. B., FRANK, H., EDELMAN, I. S.: Amer. J. Physiol. **195**, 499 (1958).
HAYS, R. M., FRANKI, J.: J. Membrane Biol. **2**, 263 (1970).
HEINZ, E., DURBIN, R. P.: J. gen. Physiol. **41**, 101 (1957).
HEINZ, E., DURBIN, R.: Biochim. biophys. Acta **31**, 246 (1959).

HEINZ, E., ÖBRINK, K. J.: Physiol. Rev. **34**, 643 (1954).
HELANDER, H. F., HIRSCHOWITZ, B. I.: Gastroenterology **63**, 951 (1972).
HERSEY, S. J.: Phil. Trans. roy. Soc. B. **262**, 261 (1971).
HERSEY, S. J.: Biochim. Biophys. Acta **344**, 157 (1974).
HERSEY, S. J., SIMON, T. W., GASTE, C.: J. Histochem. Cytochem. **23**, 271 (1975).
HILDEN, S., HOKIN, L.: Biochem. biophys. Res. Commun. **69**, 521 (1976).
HIMAL, H. S., YOUNG, S. B., RUDICK, J.: Gastroenterology **58**, 959 (abstract) (1969).
HODGKIN, A. L.: Biol. Rev. **26**, 339 (1951).
HOGBEN, C. A. M.: Proc. nat. Acad. Sci. (Wash.) **37**, 393 (1951).
HOGBEN, C. A. M.: Amer. J. Physiol. **180**, 641 (1955a).
HOGBEN, C. A. M.: In: Electrolytes in Biological Systems (A. M. Shanes, Ed.). Washington, D. C.: Amer. Physiol. Soc. 1955b, p. 576.
HOGBEN, C. A. M.: Science **129**, 1224 (1959a).
HOGBEN, C. A. M.: Amer. J. digest. Dis. **4**, 184 (1959b).
HOGBEN, C. A. M.: Fed. Proc. **24**, 1353 (1965).
HOGBEN, C. A. M.: Molec. Pharmacol. **3**, 318 (1967).
HOGBEN, C. A. M.: Amer. J. digest. Dis. **14**, 221 (1969).
HOUSE, C. R.: Water Transport in Cells and Tissues. London: Arnold 1974.
IMAMURA, A.: Biochim. biophys. Acta **135**, 155 (1967).
INESI, G.: Amer. Rev. Biophys. Bioeng. **1**, 191 (1972).
ITO, S.: J. Biophys. Biochem. Cytol. **11**, 333 (1961).
ITO, S.: In: Handbook of Physiology, Section 6: Alimentary Canal, Vol. 2 (C. F. Code, Ed.). Washington, D. C.: Amer. Physiol. Soc. 1967, p. 705.
JACOBSON, A., SCHWARTZ, M., REHM, W. S.: Amer. J. Physiol. **209**, 134 (1965).
JAGENDORF, A. J., NEUMANN, J.: J. biol. chem. **240**, 3210 (1965).
KANEKO-MOHAMMAD, S., HOGBEN, C. A. M.: Amer. J. Physiol. **207**, 117 (1964).
KASBEKAR, D. K.: Proc. Soc. exp. Biol. **145**, 234 (1974).
KASBEKAR, D. K., DURBIN, R. P.: Biochim. biophys. Acta **105**, 472 (1965).
KIDDER, G. W.: Amer. J. Physiol. **219**, 641 (1970).
KIDDER, G. W.: Amer. J. Physiol. **227**, 300 (1974).
KIDDER, G. W.: Amer. J. Physiol. **231**, 1240 (1976).
KITAHARA, S., FOX, K. R., HOGBEN, C. A. M.: Amer. J. digest. Dis. **14**, 221 (1969).
LEE, J., SIMPSON, G., SCHOLES, P.: Biochem. biophys. Res. Commun. **60**, 824 (1974).
LEE, H. C. and FORTE, J. G.: Biochim. Biophys. Acta, in press (1979).
LEE, H. C., QUINTANILHA, A. T., FORTE, J. G.: Biochem. biophys. Res. Commun. **72**, 1179 (1976).
LEFEVRE, M. E., GOHMANN, E. J., REHM, W. S.: Amer. J. Physiol. **207**, 613 (1964).
LEWIN, M., SACHS, G.: The Physiologist **18**, 290 (1975).
LEWIS, S., DIAMOND, J. M.: Nature **253**, 747 (1975).
LIMLOMWONGSE, L., FORTE, J. G.: Amer. J. Physiol. **219**, 1717 (1970).
LOEWENSTEIN, W. R., KANNO, Y.: J. Cell Biol. **22**, 565 (1964).
MACHEN, T. E., FORTE, J. G.: Biophys. J. **16**, 129a (1976).
MACHEN, T. E., ERLIJ, D., WOODING, F. B. P.: J. Cell Biol. **54**, 302 (1972).
MACHEN, T. E., SILEN, W., FORTE, J. G.: In: International Congress of Physiological Science, New Delhi 1974, Abs. xxvi.
MACHEN, T. E., SILEN, W., FORTE, J. G.: 5th International Biophysics Congress, Copenhagen, 1975, Abstract S-2.
MACHEN, T. E., SILEN, W., FORTE, J. G.: Amer. J. Physiol. **234**, E 228 (1978).
MAREN, T.: Physiol. Rev. **47**, 595 (1967).
MITCHELL, P.: Nature (Lond.) **191**, 144 (1961).
MITCHELL, P., MOYLE, J.: Nature (Lond.) **208**, 1205 (1965).
MITCHELL, P., MOYLE, J.: Biochem. J. **104**, 588 (1967).
MITCHELL, P., MOYLE, J.: Bioenergetics **2**, 1 (1971).
MOODY, F. G., DURBIN, R. P.: Amer. J. Physiol. **209**, 122 (1965).
MOODY, F. G., DURBIN, R. P.: Amer. J. Physiol. **217**, 255 (1969).
MÓZSIK, G.: Europ. J. Pharmacol. **7**, 319 (1969).
MÓZSIK, G.: Europ. J. Pharmacol. **9**, 207 (1970).
MÓZSIK, G. and I. OYE, Biochim. Biophys. Acta. **183**, 640 (1969).

Mózsik, G., Nagy, L., Kutas, J., Tárnok, F.: Scand. J. Gastroenterol. **9**, 741 (1974).

Mózsik, G., Nagy, L., Kutas, J., Tárnok, F.: Scand. J. Gastroenterol. **9**, 741 (1974).

Nagel, W., Dörge, A.: Pflügers Arch. **317**, 84 (1970).
Öbrink, K. J.: Acta physiol. scand. **36**, 229 (1956).
O'Callaghan, J., Sanders, S. S., Shoemaker, R. L., Rehm, W. S.: Amer. J. Physiol. **227**, 273 (1974).
Pacifico, A. D., Schwartz, M., MacKrell, T. N., Spangler, S. G., Rehm, W. S.: Amer. J. Physiol. **216**, 536 (1969).
Patlak, C. S., Goldstein, D. A., Hoffman, J. F.: J. theoret. Biol. **5**, 426 (1963).
Racker, E.: Fed. Proc. **21**, 54 (1962).
Rangachari, P. K.: Amer. J. Physiol. **220**, 551 (1975a).
Rangachari, P. K.: Amer. J. Physiol. **229**, 627 (1975b).
Rehm, W. S.: Amer. J. Physiol. **144**, 115 (1945).
Rehm, W. S.: Gastroenterology **14**, 401 (1950).
Rehm, W. S.: Amer. J. Physiol. **172**, 689 (1953).
Rehm, W. S.: Amer. J. Physiol. **203**, 63 (1962).
Rehm, W. S.: Amer. J. Physiol. **203**, 1091 (1962).
Rehm, W. S.: Fed. Proc. **26**, 1303 (1967).
Rehm, W. S.: J. gen. Physiol. **51**, 250s (1968).
Rehm, Butler, Spangler and Sanders: J. Theor. Biol. **27**, 433 (1970).
Rehm, W. S.: In: Metabolic Transport, Vol. 6 (L. E. Hokin, Ed.). New York: Academic 1972, p. 187.
Rehm, W. S.: In: Current Topics in Membranes and Transport, Vol. 7. New York: Academic 1975, p. 217.
Rehm, W. S., LeFevre, M. E.: Amer. J. Physiol. **208**, 922 (1965).
Rehm, W. S., Sanders, S. S.: Ann. N. Y. Acad. Sci. **264**, 442 (1975).
Rehm, W. S., Davis, T. L., Chandler, C., Gohmann, E., Bashirelahi, A.: Amer. J. Physiol. **204**, 233 (1963).
Rehm, W. S., Schlesinger, H., Dennis, W. H.: Amer. J. Physiol. **175**, 473 (1963).
Robertson, R. N., Wilkins, M. J.: Aust. J. Sci. Res. **1**, 17 (1948).
Sachs, G., Shoemaker, R. L., Hirschowitz, B. I.: Amer. J. Physiol. **209**, 461 (1965).
Sachs, G., Shoemaker, R. L., Hirschowitz, B. I.: Proc. Soc. exp. Biol. **123**, 47 (1966).
Sachs, G., Shoemaker, R. L., Hirschowitz, B. I.: Biochim. biophys. Acta **143**, 522 (1967).
Sachs, G., Collier, R. H., Pacifico, A., Shoemaker, R. L., Zweig, R. A., Hirschowitz, B. I.: Biochim. biophys. Acta **173**, 509 (1969).
Sachs, G., Shah, G., Strych, A., Cline, G., Hirschowitz, B. I.: Biochim. Biophys. Acta **266**, 625 (1972).
Sachs, G., Rabon, E., Saccomani, G., Sarau, H. M.: Ann. N. Y. Acad. Sci. **264**, 456 (1975).
Sachs, G., Chang, Rabon, E., Schackmann, R., Lewin, M., Saccomani, G.: Biol. Chem. **251**, 7690 (1976).
Sanders, S. S., Haynes, V. B., Rehm, W. S.: Amer. J. Physiol. **225**, 1311 (1973).
Sanders, S. S., O'Callaghan, J., Butler, C. F., Rehm, W. S.: Amer. J. Physiol. **222**, 1348 (1972).
Schackmann, R., Schwartz, A., Sachs, G.: J. Membrane Biol. **32**, 361 (1977).
Schwartz, M., MacKrell, T. N.: Proc. Soc. exp. Biol. **130**, 1048 (1969).
Sedar, A. W.: J. Biophys. Biochem. Cytol. **9**, 1 (1961a).
Sedar, A. W.: J. Biophys. Biochem. Cytol. **10**, 47 (1961b).
Sedar, A. W., Forte, J. G.: J. Cell Biol. **22**, 173 (1964).
Sedar, A. W., Wiebelhaus, V. D.: Amer. J. Physiol. **223**, 1088 (1972).
Silen, W., Machen, T. E., Forte, J. G.: Amer. J. Physiol. **299**, 721 (1975).
Solberg, L. A., Forte, J. G.: Amer. J. Physiol. **220**, 1404 (1971).
Soumarmon, A., Lewin, M., Cheret, A., Bonfils, S.: Biochim. biophys. Acta **339**, 403 (1974).
Spangler, S. G., Rehm, W. S.: Biophys. J. **8**, 1211 (1968).
Spenney, J. G., Strych, A., Price, A. H., Helander, H. F., Sachs, G.: Biochim. biophys. Acta **311**, 545 (1973).
Spenney, J. G., Shoemaker, R. L., Sachs, G.: J. Membrane Biol. **19**, 109 (1975).
Takeguchi, N., Horikoshi, L., Hatton, M.: Amer. J. Physiol. **232** (3): E 294 (1977).
Teorell, T.: J. Physiol. **97**, 308 (1940).

TEORELL, T.: J. Physiol. **114**, 267 (1951).
THULL, N. B., REHM, W. S.: Amer. J. Physiol. **185**, 317 (1956).
USSING, H. H.: Acta physiol. scand. **19**, 43 (1949).
VAN OS, C. H., SLEGERS, J. F. G.: Biochim. biophys. Acta **29**, 197 (1973).
VIAL, J. D., ORREGO, H. J.: Biochem. Biophys. Cytol. **7**, 367 (1960).
VILLEGAS, L.: Biochim. biophys. Acta **64**, 359 (1962).
VILLEGAS, L.: Biochim. biophys. Acta **75**, 131 (1963a).
VILLEGAS, L.: Biochim. biophys. Acta **75**, 377 (1963b).
VILLEGAS, L.: Amer. J. Physiol. **208**, 380 (1965).
VILLEGAS, L.: Biophys. J. **15**, 227a (1975).
WINSHIP, D. H., CAFLISCH, C. R.: Biochim. biophys. Acta **291**, 280 (1973).
WRIGHT, E. M., BARRY, P. H., DIAMOND, J. M.: J. Membrane Biol. **4**, 331 (1971).
WRIGHT, E. M., SMULDERS, A. P., TORMEY, J. M.: J. Membrane Biol. **7**, 190 (1972).
WRIGHT, G. H.: J. Physiol. **163**, 201 (1962).

Chapter 14

Transport Across Small Intestine

S.G. SCHULTZ

A. Introduction

The mechanisms responsible for ion, nonelectrolyte and water transport by the small intestine have been the subjects of extensive investigations for several decades. This chapter will not exhaustively review the numerous transport functions of the small intestine, many of which are as yet poorly understood; rather than this, its purpose is to summarize briefly our current understanding of several central topics that have been the focus of considerable attention in recent years. Specifically, this chapter will be concerned with (a) the characteristics and physiologic consequences of low-resistance paracellular ("shunt") pathways that traverse the small intestine; (b) current views regarding the mechanisms responsible for transcellular Na transport; and, (c) recent developments dealing with the interactions between Na transport and the transport of other solutes. The emphasis will be on data derived from studies on *in-vitro* preparations of small intestine, particularly rabbit ileum (for a recent discussion of transport by small intestine *in situ* the reader is referred to SCHULTZ et al. [1974]). The aim is to bring unresolved problems into sharper focus and to suggest important areas for future research. It is, therefore, inevitable that discussions of controversial issues will be colored by the author's personal views; he hopes the readers will indulge him this license.

More comprehensive treatments of the transport of ions and water (SCHULTZ and CURRAN, 1974; SCHULTZ, 1976; ARMSTRONG, 1975; HOUSE, 1974), sugars (CRANE, 1960; SCHULTZ and CURRAN, 1970; CSAKY, 1975), and aminoacids (SCHULTZ and FRIZZELL, 1975; MATTHEWS, 1975) can be found in the references cited. Finally, we will not consider the newly "rediscovered" and important phenomenon of intestinal secretion, which has been the subject of several recent reviews (HENDRIX and BAYLESS, 1970; FIELD, 1974; SCHULTZ et al., 1974).

B. The Paracellular Pathway

It is now generally accepted that the small intestine falls within the category of "leaky epithelia", which includes renal proximal tubule and gallbladder (FRÖMTER and DIAMOND, 1972; DIAMOND, 1974). All of these tissues are characterized by low spontaneous transepithelial electrical potential differences (PD), low transepithelial electrical resistances (R_e), and high hydraulic conductivities (L_p). All are capable of rapidly absorbing relatively large amounts of solute and solvent in near-isotonic proportions and, with the exception of the gallbladder, are not normally faced with large transepithelial concentration differences or electrochemical potential gradients (*i. e.* physiologically they operate close to the condition referred to as "level flow" [KEDEM and CAPLAN, 1965; ESSIG and CAPLAN, 1968]). Although these characteristics have been recognized for many years, it has only recently become clear that the low transepithelial PDs and resistances are not due to unusually leaky cell membranes or areas of cell denudation but, instead, are direct consequences of the presence of low resistance, transepithelial, paracellular pathways. Electrophysiologic (FRÖMTER, 1972) and histochemical studies (MACHEN et al., 1972; WHITTEMBURY and RAWLINS, 1971) have established beyond reasonable doubt that the anatomic counterparts of these paracellular or "shunt" pathways are the lateral intercellular spaces and the *zonulae occludens*, which bind the epithelial cells together at their apical borders. These so-called "tight junctions" are impermeable to macromolecules (FARQUHAR and PALADE, 1963) but are readily permeated by lanthanum (MACHEN et al., 1972; WHITTEMBURY and RAWLINS, 1971; TISHER and YARGER, 1973), whose crystal radius is similar to those of the alkali metal ions and halide anions. Freeze-fracture electronmicroscopic studies (CLAUDE and GOODENOUGH, 1973) appear to have found ultrastructural differences between tight junctions of "leaky epithelia," such as renal proximal tubule and small intestine, and those of "tight epithelia," such as amphibian urinary bladder, but the molecular basis for the differences among the permeabilities of "tight" junctions is unknown. STAEHELIN (1974) has recently reviewed the structure of intercellular junctional complexes.

I. Permeability to Ions

The transepithelial electrical characteristics, the estimated ratios of the conductance of passive transepithelial pathways[1] to total tissue conductances (G_L/G_e), and the hydraulic conductivities (L_p) of several preparations of small intestine are given in Table 1; for comparative purposes, data are also tabulated for some other leaky and some considerably "tighter" epithelia. Clearly, the defining characteristic of a leaky epithelium is G_L/G_e. In all of the leaky epithelia, the conductance of the shunt pathway accounts for more than 75 percent of the total tissue conductance, and for the highly leaky epithelia G_L/G_e exceeds 0.95. In

1 The term "passive conductance pathway(s)" refers to routes that permit transepithelial ion diffusion as defined by conformity with flux equations derived from the Nernst-Planck formulation. These routes may be transcellular or extracellular; the techniques employed to characterize these pathways cannot distinguish between different anatomic locations and thus yield "lumped" properties. However, it appears quite clear that for leaky epithelia most if not all transepithelial ionic diffusion is paracellular.

Table 1. Some properties of small intestine and other epithelia

Tissue	ψ_{ms} (mV)	R_e (Ω cm^2)	L_p (cm s^{-1} Osm)	G_L/G_e –
Rabbit ileum	2–3 (a)	30 (a)	–	>0.90 (b)
Rat jejunum	1–2 (c)	30–50 (c)	$2 \cdot 10^{-4}$ (d)	>0.85 (e)
Human jejunum	<2 (f, g)	50 (f)	$9 \cdot 10^{-5}$ (h)	–
Human ileum	5 (g, i, j)	50–70 (i)	$4 \cdot 0^{-5}$ (h)	–
Necturus proximal tubule	15 (k)	70 (k)	1–10 $\cdot 10^{-4}$ (l, m)	>0.95 (k)
Rat proximal tubule	1 (n)	5 (n)	$4 \cdot 10^{-3}$ (o)	>0.95 (n)
Rabbit gallbladder	0 (p, g)	30 (q, r)	1–10 $\cdot 10^{-4}$ (s, t)	0.95 (q)
Necturus gallbladder	3 (u, v)	300–500 (u, v)		0.96 (u, v)
Frog skin	>100 (w)	>1–4 $\cdot 10^3$ (w)	$1 \cdot 10^{-5}$ (x)	0.1 (w)
Toad urinary bladder	60 (y)	>3 $\cdot 10^3$ (y)	$7 \cdot 10^{-6}$ (z)	0.25 (y)

ψ_{ms} is the transpeithelial electrical potential difference with respect to the mucosal or outer solution; R_e is the transepithelial electrical resistance; L_p is the transepithelial hydraulic conductivity; and, G_L/G_e is the ratio of the shunt conductance (G_L) to the total tissue conductance (G_e). All values have been rounded off. In many instances there is considerable variation in the reported values for the transepithelial electrical parameters and hydraulic conductivities depending upon species, season, techniques etc; the selected values given should be considered approximate and representative.

(a) ROSE and SCHULTZ, 1971; (b) FRIZZELL and SCHULTZ, 1972; (c) MUNCK, 1972; (d) SMYTH and WRIGHT, 1966; (e) MUNCK and SCHULTZ, 1974; (f) ROHDE and ANDERSON, 1973; (g) FORDTRAN et al., 1968; (h) SOERGEL et al., 1968; (i) AL-AWQATI et al., 1973; (j) GRADY et al., 1967; (k) BOULPAEP, 1971; (l) BENTZEL et al., 1969; (m) GRANDCHAMP and BOULPAEP, 1974; (n) FRÖMTER et al., 1971; (o) ULLRICH, 1973; (p) DIAMOND, 1964; (q) FRIZZELL et al., 1975; (r) WRIGHT et al., 1971; (s) WRIGHT et al., 1972; (t) van OS, 1974; (u) FRÖMTER, 1972; (v) REUSS and FINN, 1975; (w) USSING and WINDHAGER, 1964; (x) DAINTY and HOUSE, 1966; (y) REUSS and FINN, 1974; (z) HAYS and LEAF, 1962.

contrast, in very tight epithelia, such as frog skin and toad urinary bladder, ionic diffusion through passive conductance pathways generally accounts for less than 25 percent of the total tissue conductance[2]. Tissues such as rabbit colon (FRIZZELL et al., 1976) and *Necturus* gastric antrum (SPENNEY et al., 1975) display intermediate values of G_L/G_e in the range of 0.50 to 0.75. The results of the few studies on leaky and tight epithelia in which the transepithelial resistances and the resistances of the two limiting membranes have been determined (Table 2) indicate that the resistances of the two limiting plasma membranes do not differ markedly, so that the pathways responsible for the low transepithelial resistances of the leaky epithelia must be paracellular.

The ion-selective properties of the shunt pathways across small intestine and

2 A precise evaluation of G_L/G_e for tight epithelia is complicated by the almost inevitable presence of some "edge damage" in most *in-vitro* studies (HELMAN and MILLER, 1973), which inserts an "artificial" low-resistance pathway in parallel with the tissue. This complication is less likely to distort, significantly, electrical or ion flux measurements on leaky epithelia, where the transepithelial resistance approaches that of an equivalent layer of Ringer's solution.

Table 2. Transcellular and paracellular resistances of leaky and tight epithelia

	R_m	R_s	$(R_m + R_s)$	R_L
Leaky epithelia				
Necturus gallbladder (a)	4500	2900	7400	320
Necturus gallbladder (b)	3350	2750	6100	480
Necturus renal proximal tubule (c, d)	6000	2000	7900	70
Tight epithelia				
Frog skin (e)	–	–	3800	24000,
Toad urinary bladder (f)	3800	3600	7400	12320

R_m is the apparent resistance of the mucosal membrane; R_s is the apparent resistance of the serosal menbrane, and R_L is the apparent resistance of the paracellular pathway. All values are in Ω cm^2.

(a) FRÖMTER, 1972; (b) REUSS and FINN, 1975; (c) WINDHAGER et al., 1966; (d) BOULPAEP, 1971; (e) USSING and WINDHAGER, 1964; (f) REUSS and FINN, 1974

several other leaky epithelia have been explored either by examining the effect of an imposed PD on net or unidirectional ion movements or by analyzing transepithelial diffusion potentials with the aid of the Goldman-Hodgkin-Katz "constant-field" equation (GOLDMAN, 1943; HODGKIN and KATZ, 1949); as shown by FRIZZELL and SCHULTZ (1972) for rabbit ileum, both approaches appear to be valid. The relative permselective properties of the shunt pathways across several preparations of small intestine are given in Table 3; data are also

Table 3. Relative ionic permeabilities of the shunt pathway across several epithelia

	P_{Cs}	P_{Rb}	P_K	P_{Na}	P_{Li}	P_{TEA}	P_{Cl}
Rabbit ileum (a)*	1.43	1.43	1.14	1.00	0.57	0.02	0.55
Rat jejunum (b)	–	1.4	1.6	1.00	–	~0	0.2
Rat jejunum (c)	–	–	1.2	1.00	–	–	0.1
Rat ileum (d)	–	–	–	1.00	–	–	0.8
Human ileum (e, f)	–	–	1.2	1.00	–	–	0.3
Rabbit gallbladder** (g, h)	0.96	1.4	1.7	1.00	0.86	~0	0.33
Dog proximal tubule (i)	–	–	1.1	1.00	–	–	0.72
Rat proximal tubule (j)	–	–	1.1	1.00	–	–	0.65
Relative free-solution mobility at 25° C (k)	1.54	1.55	1.47	1.00	0.77	0.65	1.52

* and unpublished observations
** Data given for rabbit gallbladder were obtained 120 min after dissection; immediately after dissection the gallbladder appears to be much less permeable to anions and $P_{Cl}/P_{Na} < 0.1$

(a) FRIZZELL and SCHULTZ, 1972; (b) MUNCK and SCHULTZ, 1974; (c) WRIGHT, 1966; (d) CLARKSON, 1967; (e) TURNBERG et al., 1970; BARRY et al., 1971; (f) TURNBERG, 1971; (g) BARRY et al., 1971; (h) WRIGHT et al., 1971; (i) BOULPAEP and SEELEY, 1971; (j) FRÖMTER et al., 1971; (k) ROBINSON and STOKES, 1969

tabulated for some other leaky epithelia for comparative purposes. One striking feature that emerges from these data is that the shunt pathways across a variety of leaky epithelia appear to have very similar properties. In all instances the shunts are cation-selective inasmuch as P_K/P_{Cl} significantly exceeds the value of unity which would be observed in free solution. Further, unlike the situation for cell membranes, the permeabilities of the shunt pathways to Cs, Rb, K and Na do not differ by as much as a factor of two, and the relative permeabilities of these cations closely parallel their relative mobilities in free solution. These observations strongly suggest that the shunt pathways offer a watery environment for the transepithelial diffusion of ions in their hydrated forms (EISENMAN, 1961). The finding that the shunts across rabbit and rat small intestine are essentially impermeable to tetraethylammonium ion (TEA), a spherical structure with a hydrated radius of 4 Å, suggests that the pathway has an "equivalent diameter" of approximately 10–15 Å. Similar dimensions have been estimated by MORENO and DIAMOND (1975a and b) for rabbit and bullfrog gallbladder, and are suggested by the results of a number of studies on canine and human small intestine (FORDTRAN et al., 1965; LEVITT et al., 1969; SOERGEL et al., 1968).

Finally, there is considerable evidence that the cation selectivity of the shunt pathway is not a consequence of the presence of a high density of *net* fixed or mobile charged groups. Instead, the behavior of the pathway resembles that of a neutral pore; that is, electroneutrality within the pathway appears to be principally maintained by fixed membrane components of opposite charge and is not dependent upon the presence of excess mobile counterions. For example, the channel may be lined with fixed dipoles (BARRY et al., 1971) or with dissociated oppositely charged groups in approximately equal numbers, forming a complex of dipolar ions or zwitterions (FRIZZELL and SCHULTZ, 1972). Cation selectivity could result from the alignment of the electronegative groups so that they restrict anion entry into and/or mobility within the channel. These anionic groups appear to have a pK_a of 3–4 (SMYTH and WRIGHT, 1966; WRIGHT and DIAMOND, 1968; MORENO and DIAMOND, 1975a), interact strongly with lanthanum (WRIGHT and DIAMOND, 1968), and behave like strong proton-acceptor sites (*e.g.* the permeability of the shunt pathways across rabbit and bullfrog gallbladder to organic cations increases with increasing proton-doner or hydrogen-binding activity of the probe molecule [MORENO and DIAMOND, 1975b]). Reasonable candidates for these strongly acidic membrane components are carboxyl, phosphoryl, and/or sulfonyl groups.

MORENO (1974, 1975a and b) has recently demonstrated that triaminopyrimidinium (TAP), which in its mono-protonated form has strong hydrogen-bonding activity, reversibly reduces the Na conductance of the shunt pathways across rabbit, bullfrog and *Necturus* gallbladder, rabbit and bullfrog small intestine, and bullfrog choroid plexus. In each instance TAP increases the transepithelial resistance and decreases P_{Na} without affecting P_{Cl}. These findings are consistent with the notion that TAP blocks a highly cation-selective channel through the formation of strong hydrogen-bonded complexes with anionic membrane ligands. These observations also support the notion that, in the tissues studied, Na (and other alkali metal ions) and Cl traverse separate and different paracellular pathways (BARRY et al., 1971; MORENO and DIAMOND,

1975 a)[3]. Moreno's findings open a potentially exciting and rewarding avenue for a detailed analysis of the physico-chemical properties of shunt pathways and the contribution of these routes to transepithelial transport.

II. Permeability to Nonelectrolytes

The extent to which paracellular pathways across small intestine contribute to the transepithelial movements of nonelectrolytes, and in particular water, is as yet an unsettled issue. In human and canine small intestine the movements of small hydrophilic molecules such as urea, xylose and arabinose appear to be significantly influenced by solvent-drag or convective flow, and the "apparent" reflection coefficients[4] of these molecules are significantly less than unity (LEVITT et al., 1969; LIFSON and HAKIM, 1966; HAKIM and LIFSON, 1964; FORDTRAN et al., 1968). The reflection coefficients are consistent with the notion that these solutes are entrained in a fluid stream that traverses pathways with equivalent pore radii of 7–15 Å. These findings certainly suggest that the paracellular pathway is a significant route for the transepithelial movements of water and small nonelectrolytes; it seems very improbable that pathways of this size traverse *both* limiting cell membranes. In addition, there is a striking, direct correlation between L_p and G_e in a wide variety of epithelia; that is, tissues with high G_e also, generally, have high L_p and *vice versa* (FRÖMTER and DIAMOND, 1972; SCHULTZ, 1977). In view of the data in Table 2, which indicate that the limiting cell membranes of tight and leaky epithelia have similar electrical resistances, it seems unlikely that these membranes are responsible for differences in L_p that may exceed two orders of magnitude. Instead, it seems more reasonable to infer that low-resistance paracellular pathways are responsible for *both* the high G_e and the high L_p of leaky epithelia (SCHULTZ, 1977).

On the other hand, WRIGHT et al. (1972) and van Os (1974) have argued that the contribution of paracellular pathways to osmotic water flow across rabbit gallbladder is minimal. Furthermore, MORENO (1975b) has shown that the reduction of P_{Na} of the shunt pathway across bullfrog gallbladder brought about by TAP is associated with a decrease in the permeability of the tissue to urea and glycerol but that neither the osmotic nor the diffusional permeability to

3 DIAMOND and his collaborators have provided fairly compelling evidence that there are two paracellular pathways across *in-vitro* rabbit and bullfrog gallbladder (cf. MORENO and DIAMOND, 1975). The first is highly cation-selective and presumably characterizes the "native" or *in-situ* tissue. The second behaves like a "free-solution shunt" whose permeabilities to cations and anions parallel free-solůtion mobilities and which develops with time following dissection. This interpretation is stronly supported by the findings of MORENO (1975) and, if correct, suggests that the properties of the low-resistance shunt across gallbladder differ from those of some other leaky epithelia that have been studied *in vivo* (*e. g.* proximal tubule, human ileum).

4 As discussed previoulsy elsewhere (SCHULTZ et al., 1974), the methods employed to determine the "reflection" or "sieving" coefficients of *in-vivo* mammalian small intestine for small, hydrophilic solutes are not strictly valid, so that the reported values should be considered "apparent" or "approximate".

water is significantly affected. These findings suggest that, whereas urea and glycerol can traverse the cation-selective shunt pathway across gallbladder, this route does not contribute significantly to transepithelial water movements (*i. e.* water movement takes place predominantly through the transcellular pathway).

Thus, it seems reasonable to conclude at this time that the shunt pathways across small intestine and other leaky epithelia are permeable to small nonelectrolytes such as urea and glycerol and perhaps, sparingly, to somewhat larger molecules such as pentoses and hexoses (LEVITT et al., 1969; ROSS et al., 1972; LOEHRY et al., 1970). They are certainly permeable to water but their relative contribution to the overall hydraulic conductivity of the tissue is unknown. As discussed recently (SCHULTZ, 1977; SACKIN and BOULPAEP, 1975; HILL, 1975) the resolution of this problem will have an important bearing on the mechanism of coupling between water and solute movements during isotonic fluid absorption.

III. Electrophysiologic Implications

It is intuitively obvious that the low spontaneous transepithelial PDs characteristic of *in-vitro* and *in-vivo* preparations of small intestine (Table 1) can be attributed, at least in part, to the presence of low-resistance paracellular pathways that shunt the electrical effects of transcellular ion (particularly Na) transport. There are, however, some additional consequences of low-resistance paracellular pathways that affect the electrophysiology of leaky epithelia that can be readily appreciated from an analysis of equivalent electrical circuit models.

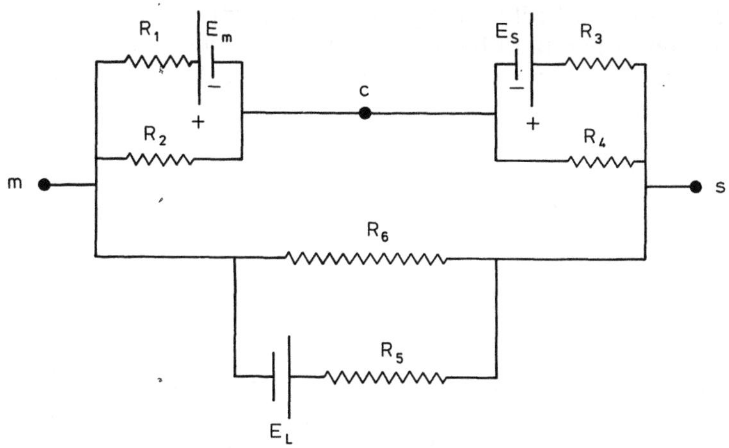

Fig. 1. Equivalent electrical circuit model generally applicable to all epithelia comprising a single layer of transporting cells. Provisions not made for cell-to-cell electrical coupling (LOWENSTEIN et al., 1965); if such coupling exists one must assume that neighboring cells have identical electrical properties. m, c and s designate mucosal, intracellular and serosal compartments respectively; other symbols defined in text. (From ROSE and SCHULTZ, 1971)

An equivalent electrical circuit model that is generally applicable to all epithelia characterized by a single layer of transporting cells is illustrated in Fig. 1 (ROSE and SCHULTZ, 1971). E_m is the electromotive force (emf) across the mucosal membrane, R_1 is the internal resistance of this battery, and R_2 represents a shunt resistance across the mucosal membrane; E_s, R_3 and R_4 and E_L, R_5 and R_6 are the corresponding electrical analogues for the serosal (or basolateral) membrane and the permselective paracellular pathway. The solutions of this circuit for the transepithelial electrical potential difference, ψ_{ms}, and the PD across the mucosal membrane, ψ_{mc} are:

$$\psi_{ms} = [R_5 R_L(E_s R_s - E_m R_m) + E_L R_L(R_1 R_m + R_3 R_s)]/R_t \quad (1)$$

and

$$\psi_{mc} = -[E_m R_m(R_3 R_s + R_5 R_L) + (E_s R_s - E_L R_L) R_1 R_m]/R_t \quad (2)$$

where

$$R_m = R_2/(R_1 + R_2) \quad , R_s = R_4/(R_3 + R_4) \quad , R_L = R_6/(R_5 + R_6)$$

and

$$R_t = R_1 R_m + R_3 R_s + R_5 R_L.$$

All electrical potential differences are with respect to the mucosal solution (m) and the orientations of the emfs as illustrated in Fig. 1 are included in the solution. Clearly, $R_5 R_L$ is the resistance of the paracellular pathway and $(R_1 R_m + R_3 R_s)$ is the total resistance of the transcellular pathway.

Several electrophysiologic consequences of the presence of a low-resistance transepithelial shunt pathway are evident from an examination of Equations 1 and 2.

First, for a tight epithelium where $(R_1 R_m + R_3 R_s) << R_5 R_L$, ψ_{ms} is a reasonable measure of $(E_s R_s - E_m R_m)$ even in the presence of transepithelial ionic asymmetries. However, in a leaky epithelium where $R_5 R_L << (R_1 R_m + R_3 R_s)$, $(E_s R_s - E_m R_m)$ is markedly attenuated and ψ_{ms} approaches $E_L R_L$. Thus, in a leaky epithelium, when both bathing solutions have identical compositions, (so that $E_L R_L$ is zero or very small), ψ_{ms} represents only a small fraction of the difference between the electromotive forces across the two limiting membranes. On the other hand, in the presence of transepithelial ionic asymmetries, ψ_{ms} will be dominated by diffusion potentials established across the paracellular pathway $(E_L R_L)$ and provides no information regarding the values of emf across the cell membranes. For rabbit ileum, where $R_5 R_L < 0.1\ R_t$, Eq. (1) reduces to

$$\psi_{ms} \cong 0.1\ (E_s R_s - E_m R_m) + 0.9\ E_L R_L.$$

Thus, the ψ_{ms} of approximately 3 mV (serosa-positive) that is observed when the tissue is bathed by identical glucose-free solutions means that the difference between $E_s R_s$ and $E_m R_m$ is greater than 30 mV and may be as large as 100 mV (see footnote 2). Furthermore, when the ionic compositions of the two bathing solutions differ, ψ_{ms} does not differ markedly from transepithelial diffusion potentials calculated from the Goldman-Hodgkin-Katz constant field equation and the relative permeabilities of the shunt pathway given in Table 3 (FRIZZELL and SCHULTZ, 1972).

Second, according to Eq. 2, when $R_5R_L >> (R_1R_m + R_3R_s)$, ψ_{mc} approaches $-E_mR_m$ so that the intracellular potential with respect to the mucosal solution is determined predominantly by the emf and resistance across the mucosal membrane. However, when $R_5R_L << (R_1R_m + R_3R_s)$, ψ_{mc} is influenced by E_mR_m, E_LR_L and E_sR_s; that is there is coupling among the electromotive forces across the two limiting membranes and the shunt by virtue of current flow through the shunt pathway. Consequently, changes in ψ_{mc} need not reflect changes in any of the properties of the mucosal membrane, but could result from changes in the properties of the shunt and/or the basolateral membranes. These interactions obviously complicate the interpretation of changes in the intracellular electrical potential. However, some of these difficulties may be circumvented by simultaneous studies of ψ_{ms} and ψ_{mc}. It can readily be shown (ROSE and SCHULTZ, 1971) that if a change in ψ_{mc} is due *solely* to a change in E_mR_m

$$\Delta\psi_{ms}/\Delta\psi_{mc} = 1/[1+(R_3R_s/R_5R_L)].$$

Thus, if $R_5R_L >> R_3R_s$, $\Delta\psi_{ms} \cong \Delta\psi_{mc}$. However, if $R_5R_L << R_3R_s$ the change in ψ_{ms} will be less than the change in ψ_{mc} but in the same direction. If a change in ψ_{mc} is due *solely* to a change in E_sR_s then

$$\Delta\psi_{ms}/\Delta\psi_{mc} = -(R_5R_L/R_1R_m).$$

Thus, for example, an increase in E_sR_s would lead to an increase in ψ_{ms} (more positive) and a decrease in ψ_{mc} (more negative). If $R_5R_L << R_1R_m$ an increase in E_sR_s may simply lead to a marked hyperpolarization of the cell interior with little change in the transepithelial PD. Finally, for a change in E_LR_L *alone*

$$\Delta\psi_{ms}/\Delta\psi_{mc} = 1 + (R_3R_s/R_1R_m),$$

so that if E_LR_L increases, both the transepithelial PD and the intracellular PD will become more positive but the change in the former will exceed that of the latter.

As will be illustrated below, these relations provide an approach to interpreting changes in ψ_{ms} and ψ_{mc} and, in addition, permit estimates of the ratios of the three (lumped) resistances. However, in order to employ these relations one must be certain that a change in ψ_{mc} or ψ_{ms} is attributable to a change in only *one* of the emfs; if two or more emfs change simultaneously the interpretation of $\Delta\psi_{ms}$ and $\Delta\psi_{mc}$ becomes extremely complex.

Finally, electrical coupling between the two limiting cell membranes *via* the shunt pathway seriously complicates the interpretation of the electrical potential profile across small intestine and other leaky epithelia. In some tight epithelia, such as frog skin and toad urinary bladder, the intracellular electrical potential is positive with respect to the outer or mucosal solution and is a measure of E_m; in leaky epithelia such as small intestine, gallbladder, and renal proximal tubule the intracellular potential is negative with respect to the mucosal or luminal solution (SCHULTZ, 1972). However, these observations do not *necessarily* imply that the emfs across the mucosal membranes of leaky epithelia are oriented in a direction opposite to those in tight epithelia; E_m may be oriented with the positive pole facing the intracellular compartment (*i. e.* opposite to that illustrated in Fig. 1) but ψ_{mc} may still be negative due to the influence of E_s (SCHULTZ, 1972). In short, the findings that the intracellular Pds in rabbit ileum (ROSE and SCHULTZ, 1971), rat jejunum (BARRY and EGGENTON, 1972; LYON and SHEERIN, 1971) and ileum (LYON and SHEERIN, 1971), hamster jejunum (WRIGHT, 1966), and bullfrog small intestine (WHITE and ARMSTRONG, 1971) are negative with respect to the mucosal solution do not *a priori* imply that the properties of the mucosal membranes of these tissues differ markedly from those of tight epithelia. The resistances of the two limiting membranes and the shunt pathway must be known in order to qualitatively evaluate the extent to which coupling between E_s and E_m may affect the electrical potential profile across a leaky epithelium. Because small intestine is a villous structure comprised of at least five different cell types (TRIER and RUBIN, 1965), and because of the strong likelihood of low-resistance electrical coupling between cells (LOWENSTEIN et al., 1965), R_1R_m and R_sR_s cannot be directly determined with existing methods and theory. Data reported for *Necturus* gallbladder (FRÖMTER, 1972; REUSS and FINN, 1975) indicate that the low-resistance shunt pathway *cannot by itself* account for the finding that ψ_{mc} is negative, and that the emf across the mucosal membrane *must* be oriented as illustrated in Fig. 1, in a direction opposite to those of frog skin and toad urinary bladder. Although this issue is not resolved for small intestine[5] it is evident that the presence of a low-resistance shunt pathway must, in general, markedly influence the electrical potential profile across the epithelial cell and, as will be discussed below, may thereby significantly affect transcellular transport.

C. Active Sodium Transport

It is clear from the above considerations that two routes are available for Na transport across small intestine, a paracellular route that circumvents the limiting cell membranes and a transcellular route, which presumably involves movements of Na across at least two membranes arranged in series. Clearly, the contribution of the shunt pathway to the bidirectional transepithelial (diffusional and convective) movements of Na must be defined quantitatively before the mechanisms responsible for (active) transcellular Na transport can be analyzed.

5 ROSE and SCHULTZ (1971), using indirect techniques, estimated that for rabbit ileal mucosa stripped of the underlying muscle layers $R_1R_m = 300\ \Omega\ cm^2$, $R_3R_s = 600\ \Omega\ cm^2$ and $R_5R_L = 28\ \Omega\ cm^2$. Using their values for ψ_{ms} (3 mV) and ψ_{mc} (−36 mV) and Eqs. 1 and 2 (assuming that $E_L = 0$), the calculated values for E_mR_m and E_sR_s are approximately 4 and 104 mV respectively. Thus, in the absence of a shunt pathway, ψ_{mc} would be only −4 mV and ψ_{ms} would be +100 mV. The low effective resistances of the mocusal and serosal membranes, compared to the data reported in Table 2, may be due to the fact that in the intestinal cell the area of the mucosal membrane is markedly increased by microvilli and the area of the basolateral membrane is markedly increased by infoldings.

I. Paracellular Na Transport

The diffusional movement of Na through the neutral, shunt pathway conforms to the Goldman constant-field flux equation (GOLDMAN, 1943; FRIZZELL and SCHULTZ, 1972). For small intestine and other leaky epithelia where $\psi_{ms} < 10$ mV, this equation can be approximated by (SCHULTZ, 1974)

$$_{d}J^{Na} = P_{Na}\{[Na]_m \exp(F\,\psi_{ms}/2RT) - [Na]_s \exp(-F\,\psi_{ms}/2RT)\} \quad (3)$$

where [Na] is the concentration of Na in the mucosal (m) or serosal (s) solutions, P_{Na} is the Na permeability of the shunt pathway, and $_{d}J^{Na}$ is the net diffusional flux of Na from mucosa to serosa *via* the shunt pathway. When $[Na]_m = [Na]_s$ Eq. 3 reduces to:

$$_{d}J^{Na} = -P_{Na}[Na]F\,\psi_{ms}/RT. \quad (4)$$

For rabbit ileum, $P_{Na} = 0.035$ cm h^{-1} and $\psi_{ms} = 3$ mV when $[Na]_m = [Na]_s =$ 140 mM; thus, $_{d}J^{Na} = 0.6\ \mu$ Eq cm^{-2}h. Under these conditions the rate of net transepithelial Na absorption is 2–3 μ eq cm^{-2} h, so that even in the presence of a small transepithelial PD, backflux through the shunt pathway reduces the rate of Na absorption by 20–25 percent. In addition, we can describe $_{d}J^{Na}$ in terms of the difference between the two, transepithelial, unidirectional diffusional fluxes as follows:

$$_{d}J^{Na} = {_{d}J^{Na}_{ms}} - {_{d}J^{Na}_{sm}}$$

where $_{d}J^{Na}_{sm}$ is the unidirectional diffusional flux from mucosa to serosa and $_{d}J^{Na}_{sm}$ is the diffusional flux in the opposite direction. From Eq. 3 it follows that when $[Na]_m = [Na]_s$

$$_{d}J^{Na}_{ms} = P_{Na}[Na]\exp(F\,\psi_{ms}/2RT) \quad (5)$$

and

$$_{d}J^{Na}_{sm} = P_{Na}[Na]\exp(F\,\psi_{ms}/2RT) \quad (6)$$

We now define $_{n}J^{Na}_{sm}$ as the unidirectional flux of Na from serosa to mucosa that is not influenced by small imposed transepithelial PDs. Then, from Eq. 6, the *total* unidirectional Na flux from serosa to mucosa is given by

$$J^{Na}_{sm} = {_{n}J^{Na}_{sm}} - P_{Na}[Na]\exp(F\,\psi_{ms}/2RT). \quad (7)$$

SCHULTZ and ZALUSKY (1964a) and NELLANS et al. (1974) have shown that Eq. 7 describes J^{Na}_{sm} across rabbit ileum and that $_{n}J^{Na}_{sm} = 0$; that is, they found a linear relation between J^{Na}_{sm} and $\exp(F\,\psi_{ms}/2RT)$ that extrapolates to the origin. Their findings suggest that the entire unidirectional flux of Na from serosa to mucosa is diffusional and is restricted to the shunt pathway. However, DESJEUX et al. (1974) have reported a linear relation between J^{Na}_{sm} and $\exp(F\,\psi_{ms}/2RT)$ for rabbit ileum that does not extrapolate to the origin; according to these investigators, approximately 20 percent of the J^{Na}_{sm} across rabbit ileum under short-circuit conditions cannot be attributed to simple ionic diffusion. The discrepancy between the findings of SCHULTZ and ZALUSKY (1964) and NELLANS et al. (1974) and those of DESJEUX et al. (1974) is unresolved, and possible explanations have been discussed previously (NELLANS et al., 1974; DESJEUX et al., 1974; SCHULTZ and CURRAN, 1974)[6]. The conclusion of NELLANS et al. (1974) and FRIZZELL and SCHULTZ (1972), that most if not all of the unidrectional Na flux from serosa to mucosa is attributable to simple ionic diffusion through the shunt, is in keeping with emerging views on other epithelia such as rabbit gallbladder (FRIZZELL et al., 1975), rabbit descending colon (FRIZZELL et al., 1976), isolated frog skin (MANDEL and CURRAN, 1972), and toad urinary bladder (CIVAN et al., 1966; CIVAN, 1970; SAITO et al., 1974). If correct, these findings suggest that under physiologic conditions transcellular Na transport across rabbit ileum is rectified in the mucosa-to-serosa direction. CLARKSON (1967) has suggested a similar model for active Na transport by rat ileum.

II. Transcellular Na Transport

Transcellular Na transport is generally considered to be the result of at least two distinct steps, entry into the cell across the mucosal membrane followed by extrusion from the cell across the basolateral membrane. In the presence of a normal Ringer solution the volume-average or "overall" intracellular Na concentration in rabbit ileum is approximately 50 mM (SCHULTZ et al., 1966; KOOPMAN and SCHULTZ, 1969) and that in bullfrog small intestine (ARMSTRONG et al., 1970) is approximately 30 mM. NELLANS and SCHULTZ (unpubl.) have shown that only 50 percent of the total intracellular Na exchanges with Na in the mucosal solution after one hour, and LEE and ARMSTRONG (1972), using cation-selective microelectrodes, have demonstrated that the intracellular Na *activity* in bullfrog small intestine is approximately 15 mM, one-half of the overall chemical concentration. Furthermore, ψ_{mc} in rabbit ileum (ROSE and SCHULTZ, 1971) and bullfrog small intestine (WHITE and ARMSTRONG, 1971; ARMSTRONG, 1975) is approximately −35 mV. Although somewhat smaller absolute values have been reported for hamster (WRIGHT, 1966) and rat (BARRY and EGGENTON, 1972; LYON and SHEERIN, 1971) small intestine, in all instances the cell interior is negative with respect to the mucosal solution. Hence, Na entry apparently takes place down an electrochemical potential gradient, which for bullfrog small intestine is estimated to be approximately 6000 joules/Eq. (ARMSTRONG et al., 1973; ARMSTRONG, 1975)[7]. Nevertheless, Na entry into rabbit ileal epithelium,

6 The resolution of this issue is experimentally difficult because J_{sm}^{Na} varies considerably from tissue to tissue. Thus, it is difficult to establish the statistical significance of a non-zero intercept of a plot of J_{sm}^{Na} vs. exp ($F\ \psi_{ms}/2RT$).

7 These considerations assume that the Na that enters the cell destined for transepithelial transport mixes with the bulk intracellular Na pool. As discussed below, this does not appear to be the case for rabbit ileum; the "active Na transport pool" in this tissue, as in some other epithelia (USSING et al., 1974; MACKNIGHT et al., 1975) appears to be a small fraction of the total intracellular Na and the thermodynamic activity of Na in this pool is entirely unknown.

where it has been most extensively studied, is not entirely diffusional but is a complex process involving at least three, apparently parallel, mechanisms. First, it is well established that Na may enter the cell coupled to the entry of some sugars and L-aminoacids (SCHULTZ and CURRAN, 1970). Second, more recently, it has been demonstrated that there is a coupled NaCl transport mechanism in the mucosal membranes of rabbit ileum that mediates a one-for-one entry of NaCl in the form of a neutral complex (NELLANS et al., 1973); this mechanism is inhibited by intracellular cyclic 3'-5' adenosinemonophosphate (cAMP) (NELLANS et al., 1973) and acetazolamide (NELLANS et al., 1975). A similar mechanism has been identified in rabbit gallbladder (FRIZZELL et al., 1975). Finally there appears to be a route for Na entry apparently uncoupled to the *entry* of other solutes; it is not known whether this process, albeit "downhill", is diffusional, carrier-mediated and/or part of an exchange for an intracellular cation (*e. g.* H; see SCHULTZ, 1974). Recent studies by MURER et al. (1976) on brush-border membrane vesicles isolated from rat small intestine suggest that Na entry may be, in part, the result of an electroneutral, carrier-mediated exchange with cell H. These investigators also found a small component of Na entry that is apparently uncoupled to the counter-transport of H or co-transport of an anion. These intriguing findings support the notions of TURNBERG et al. (1970) (see also SCHULTZ et al., 1974), but the extent to which the properties of brush-border vesicles precisely parallel the behavior of the brush borders of intact cells is still unclear (*e. g.* the Na entry uncoupled to H efflux may be due to incomplete sealing of the vesicle and not physiologically significant). Finally, it is unclear whether these parallel entry processes reside in a single cell or are properties of the mucosal membranes of different cell types. The mature villus cells are almost certainly involved in Na-entry coupled to sugars and aminoacids (KINTER and WILSON, 1965), but the site(s) of the other Na entry processes have not been identified.

The neutrality of the coupled NaCl entry process in rabbit ileum has been established by both kinetic analysis of the stoichiometry (NELLANS et al., 1973) and the fact that inhibition of this process (either by replacement of Cl with the nontransported anion sulfate, or exposure of the tissue to acetazolamide) has no effect on ψ_{mc} (ROSE and SCHULTZ, 1971). In contrast, there is strong evidence that Na entry coupled to the entry of sugars and L-aminoacids is *electrogenic* or, preferably, *rheogenic* (ROSE and SCHULTZ, 1971). As shown in Fig. 2, the addition of alanine to the mucosal solution bathing rabbit ileum results in a prompt depolarization of ψ_{mc}; similar results are obtained with glucose, galactose, 3-0-methylglucose, and other L-aminoacids, but no effect on ψ_{mc} is observed when fructose, or D-valine is added to the mucosal solution or when glucose is added to a mucosal solution containing phlorizin. Thus, the depolarizing effect on ψ_{mc} appears to be limited to those sugars and aminoacids whose entry into the enterocyte is dependent upon Na and enhances transepithelial Na transport. Furthermore, as shown in Fig. 3, the addition of alanine to a Na-free mucosal solution has no effect on ψ_{mc}; identical results have been obtained with glucose and 3-0-methylglucose. Thus it seems reasonable to conclude that the depolarization of ψ_{mc}, shown in Fig. 2, is due to Na-entry coupled to the entry of sugars or aminoacids and that this entry process does not involve a *one-for-one*

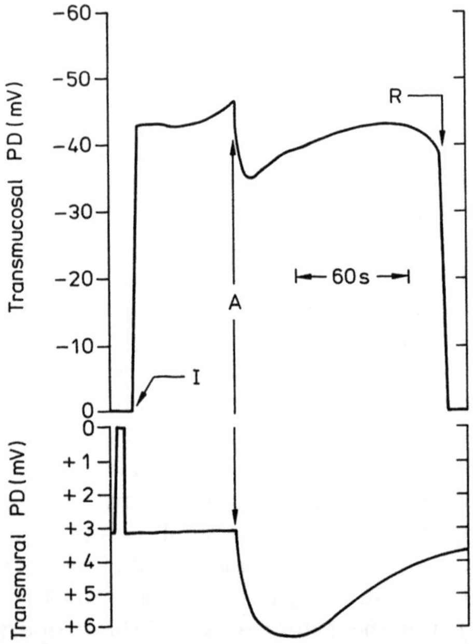

Fig. 2. Effect of alanine on transmucosal PD (ψ_{mc}) and transmural PD (ψ_{ms}) of rabbit ileum. (I) designates time of cell impalement and (R) retraction of microelectrode from cell. At (A), alanine added directly to mucosal chamber, resulting in a prompt depolarization of ψ_{mc} and hyperpolarization ψ_{ms}; slow return to baseline values is due to washout of alanine from the mucosal solution by continously flowing alanine-free solution. (From Rose and Schultz, 1971)

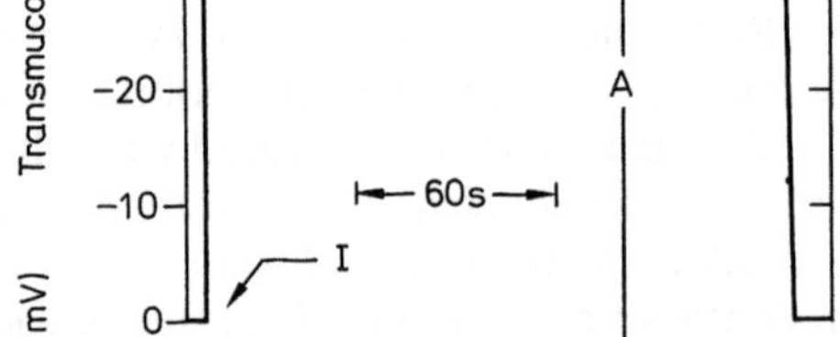

Fig. 3. Effect of alanine on transmucosal PD (ψ_{mc}) and transmural PD (ψ_{ms}) when Na bathing media is completely replaced with choline. (From Rose and Schultz, 1971)

exchange of Na with an intracellular cation. Similar results have been reported for bullfrog small intestine (White and Armstrong, 1971) and the proximal renal tubule of the newt (Maruyama and Hoshi, 1972) and rat (Frömter and Luer, 1973; Samarzija and Frömter, 1975). On the other hand, studies on tortoise (Wright, 1966), hamster (Wright, 1966), and rat (Barry and Eggenton, 1972; Lyon and Sheerin, 1971) small intestine failed to demonstrate a change in ψ_{mc} following the addition of sugars to the mucosal solution. However, these results are (in the opinion of this author) not compelling, inasmuch as the values reported for ψ_{mc} were very low and unstable, prolonged impalements were not achieved, and the effects of the added sugar could not be studied *during* a single, successful impalement. As discussed by Rose and Schultz (1971) and Armstrong (1975), it seems likely that these "negative findings" are due to the fact that the mucosal membranes of the cells were damaged during impalement. The possibility that species differences are responsible for these discrepant findings is rendered somewhat unlikely, at least for the case of rat small intestine, by the finding that Na-coupled glucose (Murer and Hopfer, 1974) and alanine (Sigrist-Nelson et al., 1975) uptake by brush-border vesicles prepared from rat jejunum appears to be rheogenic.

In summary, it appears that Na entry into small intestinal cells involves at least three parallel processes: (a) an electroneutral co-transport of Na and Cl; (b) an electroneutral counter-transport of Na and H; and, (c) a rheogenic co-transport of Na coupled to sugars, aminoacids and perhaps other nonelectrolytes. The extent to which the Na gradient is involved in energizing some of these movements will be discussed below.

Sodium exit from the cell is apparently directed against an electrochemical potential difference and is dependent upon coupling to a supply of metabolic energy. Further, active transepithelial Na transport is abolished by ouabain only when this glycoside is present in the serosal solution; the presence of ouabain in the mucosal solution alone is ineffective (Schultz and Zalusky, 1964a). The results of autoradiographic studies (Stirling, 1972) and of enzymic analyses of fractionated intestinal cells (Quigley and Gotterer, 1969; Fujita et al., 1972; Murer et al., 1974; Douglas et al., 1972) have localized ouabain-binding sites and Na-K ATPase activity to the basolateral membranes; little or no activity is found in the brush border (Fig. 4). All of these findings *would appear to suggest* that active Na extrusion from the cell across the basolateral membranes is mediated by the ubiquitous, Na-K-dependent ATPase and is coupled to K uptake, and that transcellular Na transport and homocellular regulation of cell Na and K composition are accomplished by the same mechanism. These notions were originally suggested by Koefoed-Johnsen and Ussing (1958) for Na transport by isolated frog skin, and have been widely adopted for other epithelia since then. However, studies on rabbit ileum indicate that the situation may be somewhat more complex.

First, the results of electrophysiological studies (Rose and Schultz, 1971) strongly suggest that the Na-extrusion mechanism at the basolateral membrane is not neutral and thus cannot simply mediate a one-for-one exchange between cell Na and K in the serosal solution. The evidence and reasoning behind this conclusion are as follows. As already seen in Fig. 2, the addition of alanine to

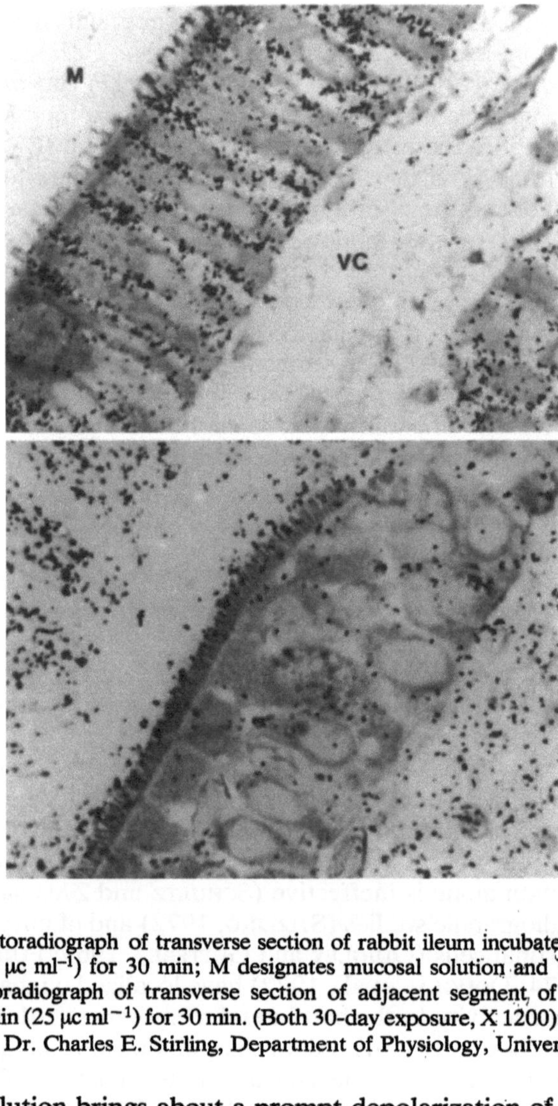

Fig. 4. Above: Autoradiograph of transverse section of rabbit ileum incubated in presence of ^{3}H-ouabain (1 μM, 25 μc ml^{-1}) for 30 min; M designates mucosal solution and VC designates villus core. Below: Autoradiograph of transverse section of adjacent segment of ileum incubated in presence of ^{3}H-inulin (25 μc ml^{-1}) for 30 min. (Both 30-day exposure, X 1200). (Photomicrographs kindly provided by Dr. Charles E. Stirling, Department of Physiology, University of Washington)

the mucosal solution brings about a prompt depolarization of the intracellular PD and a simultaneous hyperpolarization of the transepithelial PD, and is associated with an increased rate of active Na absorption. At first sight one might argue that the increase in ψ_{ms} is secondary to the increase in ψ_{mc} and due to a change in $E_m R_m$ alone. As discussed above, if the change in ψ_{ms} is due solely to a change in $E_m R_m$

$$\Delta\psi_{ms}/\Delta\psi_{mc} = 1/[1+(R_3 R_s/R_5 R_L)],$$

Rose and Schultz (1971) found that, on average, in response to the addition of sugars or aminoacids to the mucosal solution, $\Delta\psi_{ms} = +3$mV and $\Delta\psi_{mc} = +10$ mV. If these changes were due to an increase in $E_m R_m$ *alone,* secondary to rheogenic, coupled Na-nonelectrolyte entry, the effective resistance of the basolateral membranes (R_3R_s) would only be twice that of the low-resistance shunt pathway (R_5R_L); this conclusion seemed unreasonable. The problem was clarified, however, by the results of studies on the effect of sugars and aminoacids on ψ_{mc} and ψ_{ms} in tissues poisoned with ouabain and/or cyanide. As shown in Fig. 5, in poisoned tissues ψ_{ms} initially does not differ significantly from zero. The addition of alanine to the mucosal solution brings about a prompt depolarization of ψ_{mc} and the magnitude of this depolarization does not differ significantly from that observed in non-poisoned tissue. However, this depolarization of ψ_{mc} is associated with only a minimal increase in ψ_{ms}. The average value of $\Delta\psi_{ms}/\Delta\psi_{mc}$ in poisoned tissue was only 0.06; this is consistent with the notion that R_3R_s is approximately twenty times greater than R_5R_L, in closer accord with the data tabulated in Table 3. These findings suggest that (a) although Na entry coupled to the entry of sugars and aminoacids is rheogenic, the depolarization of ψ_{mc} has a minimal effect on ψ_{ms} due to attenuation by the low-resistance shunt; and (b) the normal increase in ψ_{ms} is almost entirely the result of an increase in E_sR_s associated with an increased rate of Na extrusion from the cell across the basolateral membranes. Thus, these results stronly suggest that the Na extrusion mechanism cannot be coupled one-for-one to K uptake but that it is rheogenic. However, this conclusion, *per se*, does not exclude the possibility that Na extrusion is mediated by a Na-K exchange mechanism, since it is well established for erythrocytes (GLYNN et al., 1971; SKOU, 1975) and nerve and muscle (THOMAS, 1972) that, in general, the stoichiometry of the Na-K exchange pump is not one-for-one but that the Na:K ratio exceeds unity.

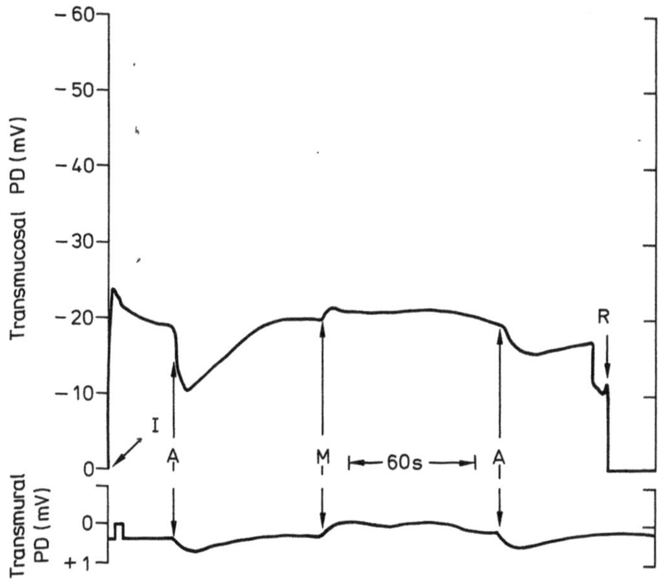

Fig. 5. Effect of alanine on transmucosal PD (ψ_{mc}) and transmural PD (ψ_{ms}) in tissue exposed to metabolic inhibitors and ouabain. (From ROSE and SCHULTZ, 1971)

To explore this problem further, NELLANS and SCHULTZ (1976) directly examined the relation between transepithelial Na transport and K uptake across the basolateral membranes of a preparation of rabbit ileum stripped of the underlying musculature and connective tissues. As shown in Table 4, under control conditions, the unidirectional influx of K from the serosal solution into the epithelium, J^K_{sc}, averages close to 2 μmol cm^{-2}h; under identical conditions the rate of active transepithelial Na transport is approximately 3 μmol cm^{-2}h. The presence of ouabain in serosal solution abolishes active Na transport, significantly inhibits J^K_{sc}, and results in a decrease in cell K and an equivalent increase in cell Na. In addition, depletion of cell Na by exposure of the tissue to Na-free media results in a similar inhibition of J^K_{sc}. These results are consistent with the presence of a ouabain-sensitive Na-K exchange mechanism at the basolateral membranes that is responsible for the maintenance of normal intracellular Na and K concentrations. *However,* J^K_{sc} is not affected when active Na absorption is *abolished* by rendering only the mucosal solution Na-free. Moreover, when active Na absorption is stimulated by the presence of glucose or alanine in the mucosal solution, J^K_{sc} is not affected. Thus, inhibition (abolition) or a near two fold stimulation of active Na absorption does not affect J^K_{sc}.

Finally, as shown in Table 4, when the tissue is bathed with a Na-free mucosal solution for 45 minutes $[K]_c$ and (not shown) cell K and H_2O contents do not differ significantly from control, in spite of the fact that Na absorption is abolished. The rate of active Na absorption by this preparation is approximately 3 μEqcm^{-2} h. If Na efflux across the basolateral membranes is coupled to K influx and the Na:K exchange ratio is approximately 3:2, K influx driven by active transepithelial Na transport would be 2 μEqcm^{-2}h. The K content of the epithelium is approximately 3 μEq/cm^{-2}. Thus, 45 minutes would see the turnover of approximately 50 percent of the total cell K as a result of transepithelial Na transport, and inhibition of Na absorption should result in a readily detectable decline in cell K content.

Thus, neither inhibition nor stimulation of Na transport affects cell K content or the rate of K exchange across the basolateral membranes; similar findings

Table 4. Unidirectional influxes of potassium across the basolateral membranes

Condition	$[Na]_c$ (mM)	$[K]_c$ (mM)	J^K_{sc} (μEq cm^{-2}h)
Ringer (Control)	61±1	125±2	1.8±0.1
Ringer + 10^{-3}M ouabain	122±3	48±3	0.6±0.0
Na-free (choline) media	7±0	126±5	0.6±0.1
$[Na]_m$ = 0; $[Na]_s$ = 140	24±2	134±7	1.5±0.1
Ringer + 20 mM alanine	65±2	125±2	1.9±0.2
Ringer + 10 mM glucose	62±4	124±6	1.7±0.1

$[Na]_c$ and $[K]_c$ are intracellular Na and K concentrations in mucosal strips of rabbit ileum determined by using inulin as a marker for the extracellular space. $[Na]_m$ and $[Na]_s$ designate the Na concentrations in the mucosal and serosal solutions respectively. The Na concentration in the Ringer solution is 140 mM. From NELLANS and SCHULTZ (1967)

have been reported by MACKNIGHT et al. (1975) and ROBINSON and MACKNIGHT (1976) for toad urinary bladder. However, the interpretation of these negative findings is complicated by at least two problems. First, the small intestinal epithelium is comprised of several cell types, some of which are not likely to be involved in transepithelial Na transport or responsible for the stimulation of Na absorption associated with the absorption of sugars or aminoacids. If the population of cells involved in transepithelial Na transport is only a small fraction of the total number of cells present, changes in the rate of K exchange or cell K content in response to changes in the rate of Na absorption could be masked. However, the autoradiographic studies of KINTER and WILSON (1965) and STIRLING and KINTER (1967) indicate quite clearly that virtually all of the cells lining the villus are involved in absorption of sugars and aminoacids; most of the cells in the stripped preparation of rabbit ileum employed by NELLANS and SCHULTZ (1976) are of this type, so that it seems unlikely that only a small fraction of the total population of cells is involved in sugar- and aminoacid-enhanced Na absorption.

The second problem stems from the possibility that Na contained within cells that are involved in transepithelial Na transport is compartmentalized into an Na transport pool and an intracellular pool that is not involved in transepithelial Na transport. If the Na transport pool is small, Na extrusion from this pool into the serosal solution could be coupled to K uptake into an equally small pool. The size and turnover time of this K pool could be so small as to be undetectable with currently available techniques.

There is some evidence suggesting that the Na transport pool in small intestine comprises only a small fraction of the total Na contained within the stripped preparation and, indeed, the total Na contained by the villus cells that are probably responsible for sugar- and aminoacid-stimulated Na absorption. Thus:

(a) Exposure of mucosal strips of small intestine to sufficiently high concentrations of sugars or aminoacids to double or triple the rate of transepithelial Na transport has no effect on intracellular Na or K contents; the accumulation of these osmotically active nonelectrolytes is accompanied by an increase in cell water content and a secondary decline in cell Na and K concentrations (SCHULTZ et al., 1966; KOOPMAN and SCHULTZ, 1969; CSAKY and ESPOSITO, 1969; ARMSTRONG et al., 1970).

(b) LEE and ARMSTRONG (1972) have shown that the activities of Na and K in villus cells of bullfrog small intestine, determined using ion-selective microelectrodes, *decrease* in the presence of sugars that bring about a marked *increase* in transepithelial Na transport. The decrease in cell Na and K activities paralleled the decrease in the intracellular Na and K concentrations of the stripped epithelial preparation, which, in turn, paralleled the increase in epithelial cell water content. These findings, on single villus cells, suggest that the negative findings on epithelial strips are not seriously distorted by technical problems or cellular inhomogeneity.

(c) The addition of sugars or aminoacids to the solution bathing the mucosal surface of rabbit ileum brings about increases in the transepithelial PD and I_{sc}, which are complete within 10 seconds (SCHULTZ and ZALUSKY, 1964b and 1965;

ROSE and SCHULTZ, 1971; NELLANS and SCHULTZ, unpubl.). As discussed above, the increase in ψ_{ms}, even during the rapid transient period, appears to be due to an energy-dependent, ouabain-inhibitable increase in $E_s R_s$, and thus is likely to be due to an almost immediate increase in the rate of active Na extrusion across the basolateral membranes. If so, it seems reasonable to infer that the rate of active Na transport can be *doubled* within 10 seconds following exposure of the mucosal surface of the tissue to the appropriate concentrations of sugars or aminoacids. The data obtained by CURRAN et al. (1967) and GOLDNER et al. (1969) indicate that the increment in Na influx into the epithelium within 10 seconds after the addition of 10 mM alanine or glucose to the mucosal solution is only $10 \cdot 10^{-9} \mathrm{Eq\,cm^{-2}}$. This represents only 1 percent of the total cell Na contained within the epithelial strip and is certainly less than 5 percent of the Na contained within the villus cells. Thus, if this line of reasoning is correct, it seems that a very small increase in the intracellular Na content of the villus cells is sufficient to double the rate of transepithelial Na transport. These inferences are obviously very indirect and subject to considerable error. Nonetheless, even if the estimates were in error by an order of magnitude, they would still be consistent with the notion that the Na transport pool is only a small fraction of the Na contained within the epithelial cells and the mature villus cell. These considerations are entirely consistent with the inability to find any relation between the rate of transepithelial Na transport by small intestine and either the Na content of the epithelial sheet or the Na activity of mature villus cells.

Thus it seems entirely possible, if not likely, that the Na transport pool in rabbit ileal epithelial cells represents only a small fraction of the total Na contained within those cells involved in transepithelial Na transport. If so, the failure to find a relation between transepithelial Na transport and the K content of the cells or the rate of exchange of K across the basolateral membranes need not exclude the possibility that a Na-K exchange process is responsible for Na extrusion across the basolateral membranes; if Na extrusion from a small intracellular pool were coupled to K uptake into an equally small pool, changes in K content or exchange by the epithelial strip in response to changes in transepithelial Na transport could go undetected. However, if this is the case, a double-membrane model of the epithelial cell which explicitly relates transcellular Na transport with the homocellular regulation of Na and K composition is untenable.

In summary, although the mechanisms responsible for Na entry into the intracellular Na pool appear to be well defined, the size of this pool and the mechanism(s) responsible for Na extrusion from this pool are poorly understood. It seems that: (a) Na extrusion from the cell across the basolateral membrane is rheogenic and is principally, if not entirely, responsible for the transepithelial PD. Although Cl is actively absorbed and HCO_3 is actively secreted by rabbit ileum, these processes appear to be coupled to or dependent upon Na movements, and they are electrically silent (SCHULTZ et al., 1974; FIELD, 1971; NELLANS et al., 1974; SHEERIN and FIELD, 1975); in the absence of Na the transepithelial PD is abolished. (b) The intracellular Na pool represents only a small fraction of the Na contained within those cells responsible for transepithelial Na transport. (c) Whereas there is little reason to doubt that an ouabain-sensitive Na-K exchange pump at the basolateral membrane is responsible for maintaining the high intracellular K and low intracellular Na that are characteristic of the epithelial cells, it is not certain whether this mechanism is also directly involved in transcellular Na transport.

The data available at present cannot be readily reconciled with any model that invokes a close relation between transepithelial Na transport and the homocellular regulation of Na and K composition or with the notion that the epithelial cell can be adequately represented by a double-membrane model.

Finally, if is of interest to consider the possible influence of the low-resistance *paracellular* pathway on *transcellular* Na transport. As discussed above, the presence of a low-resistance shunt *must* affect the transepithelial electrical potential profile, *i. e.* ψ_{mc} and ψ_{cs}. Figure 6a illustrates the electrical potential profile across *Necturus* gallbladder reported by Frömter (1972) and confirmed by Reuss and Finn (1975). As noted above, this is a very leaky epithelium; the transepithelial PD is only 3 mV and the paracellular pathway accounts for more than 95 percent of the total tissue conductance. Figure 6b illustrates the electrical potential profile that *would* be observed *if* there were no paracellular shunt, calculated from the data provided by these investigators and Eqs. 1 and 2. Clearly, the effect of the shunt is to make ψ_{mc} significantly more negative and ψ_{cs} significantly less positive. The same *must* be true for small intestine although

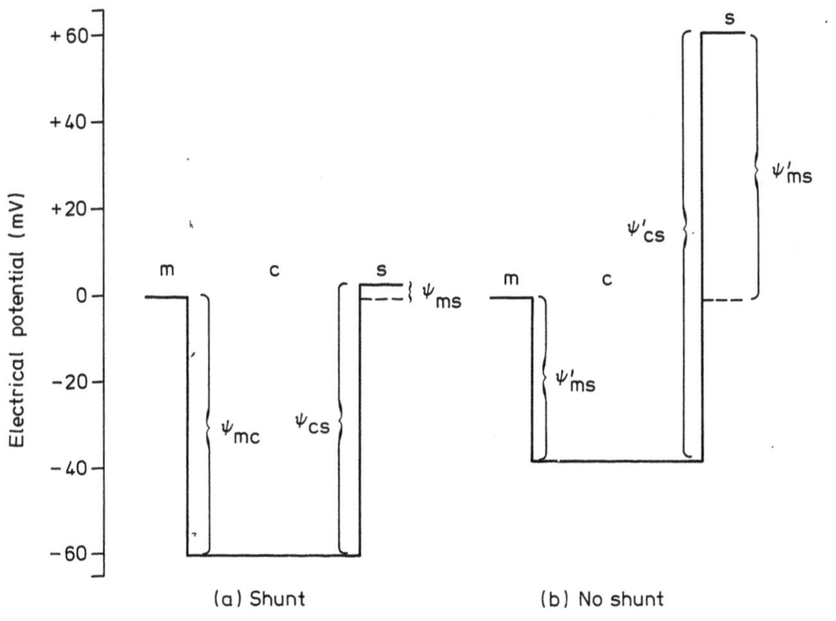

Fig. 6. Effect of low-resistance passive conductance ("shunt") pathway on electrical potential profile across *Necturus* gallbladder. ψ_{cs} is electrical potential difference across serosal (or basolateral) membrane with reference to cell interior. (a) Electrical potential profile reported by Frömter (1972) and Reuss and Finn (1975). (b) Electrical potential profile that would be observed in absence of low-resistance shunt pathway calculated from Eqs. 1 and 2

direct data for a quantitative analysis are not available[8]. Since Na entry coupled to the entry of sugars and aminoacids is rheogenic, an increase in intracellular negativity due to the shunt increases the electrochemical potential difference for Na across the mucosal membrane and may increase the rate of net Na entry due to the coupled entry process. Similarly, since Na extrusion from the cell is rheogenic, at least for rabbit ileum, a decrease in the electrical potential difference across the basolateral membrane (a positive step) decreases the "work" of the Na pump and may accelerate Na extrusion[9]. Thus, although the low-resistance paracellular pathway permits a backflux of Na (from serosa to mucosa) and thereby decreases the overall efficiency of Na absorption by small intestine, the influence of these shunts on the electrical potential profile in all likelihood has a positive effect on transcellular Na transport and, as will be discussed below, Na-coupled nonelectrolyte transport processes.

8 A definitive interpretation of these data is complicated by the presence of several cell types in rabbit ileum, some of which are not likely to be involved in transepithelial Na transport. The studies of KINTER and WILSON (1965) and of STIRLING and KINTER (1967) strongly suggest that almost all of the enterocytes that line the villus epithelium accumulate sugars and aminoacids, and thus probably participate in transepithelial Na transport stimulated by these nonelectrolytes. These cells comprise the large majority of cells that form the epithelium, so that it seems unlikely that the results of NELLANS and SCHULTZ (1976) (Table 4) and the estimates of the Na transport pool size are critically distorted by the heterogeneity of cells that line the epithelium.

9 Since the entry of Na into the cell coupled to the entry of sugars and aminoacids, and the extrusion of Na from the cell into the serosal solution are rheogenic, nondiffusional movements, the effect of ψ_{mc} on the rate of coupled Na entry and the effect of ψ_{cs} on the rate of active Na extrusion cannot be predicted with certainty. That is, although in both instances Na traverses the limiting membrane as part of a charged complex, these movements are coupled to other flows, and the extent to which they are influenced by conjugate driving forces (*i. e.* the transmembrane electrochemical gradient for Na) is unpredictable over a given operating range. The effect of the degree of coupling between conjugate and non-conjugate forces on flow processes is discussed by KEDEM and CAPLAN (1965) and ESSIG and CAPLAN (1968).
MURER and HOPFER (1974) and SIGRIST-NELSON et al. (1975) have presented evidence suggesting that the rate of Na-coupled sugar and aminoacid entry across the brush border of rat small intestine is markedly affected by the transmembrane PD. Thus, under steady-state conditions, an increase in intracellular negativity would be expected to increase the rate of Na entry coupled to the entry of sugars and aminoacids and (unless the serosal pump is saturated) the rate of transepithelial Na transport.
On the other hand, there is no evidence that the rate of rheogenic Na extrusion from cells mediated by the Na-K exchange pump is affected by the transmembrane PD. MARMOR (1971) has demonstrated that the rate of rheogenic Na extrusion by a molluscan neurone is not affected by the transmembrane PD over the range of −30 to −100 mV, and BRINLEY and MULLINS (1974) have shown that active, rheogenic Na efflux from perfused squid axon is not affected by the transmembrane PD over the range −10 to −90 mV (cell interior negative). Thus the Na-K exchange pump, albeit rheogenic, may function as a constant current generator over a relatively wide range of transmembrane PDs.

D. Relations Between Sodium Transport and the Transport of Other Solutes

There is now no doubt that the transport of a variety of water-soluble solutes is influenced by, or dependent upon, active transepithelial Na transport. The interactions between Na absorption and the transport of other solutes fall into two categories. The first includes direct interactions in which the movement of Na is coupled to the movements of other solutes by what appear to be carrier-mediated mechanisms located at the mucosal membranes. Such direct interactions have been demonstrated for sugar and aminoacid absorption by jejunum and ileum (SCHULTZ and CURRAN, 1970; FRIZZELL et al., 1973b), and for Cl absorption by rabbit ileum (NELLANS et al., 1973). In addition there is suggestive evidence that Na may be directly involved in the absorption of dipeptides (RUBINO et al., 1971; MATTHEWS et al., 1974; KOHN et al., 1968; HELLER et al., 1973) and biotin (BERGNER et al., 1972) by mammalian small intestine, and HCO_3 secretion by rabbit ileum (SHEERIN and FIELD, 1975) though compelling evidence for carrier-mediated coupling is lacking.

The second category includes transport processes which appear to be indirectly influenced by Na transport through its role in promoting water absorption. There is compelling evidence that an increase in the rate of water absorption by small intestine (LOESCHKE et al., 1970; 1971; LEVITT et al., 1969; DIBONA et al., 1974; HAKIM and LIFSON, 1969; HUMPHREYS and EARLEY, 1971), and other leaky epithelia (SMULDERS et al., 1972; WRIGHT et al., 1972; BERRY and BOULPAEP, 1975; BENTZEL et al., 1969; BENTZEL, 1972; BOULPAEP, 1972) is associated with a widening of the lateral intercellular spaces, an increase in L_p and, in all likelihood, an increase in the permeability of the tight junctions to ions and small nonelectrolytes. Conversely, a decrease in water absorption (or osmotically induced water secretion) is associated with a narrowing (or collapse) of the interspaces, a decrease in L_p, and a decrease in the permeability of the epithelium to small solutes. Thus, water absorption, osmotically linked to Na absorption (plus the absorption of the accompanying anions as well as other solutes directly coupled to Na transport) will enhance the diffusional and convective (*i. e.* solvent-drag) absorption of other small solutes. In the absence of Na, water absorption is essentially abolished so that (a) the absorptive movements of other solutes resulting from solvent-drag are essentially abolished and (b) the permeability of the epithelium to small solutes decreases and diffusional movements are inhibited. As noted above, there is good evidence that the movements of small nonelectrolytes, such as urea, xylose and arabinose (HAKIM and LIFSON, 1964; LIFSON and HAKIM, 1966; FORDTRAN et al., 1968; LEVITT et al., 1969), as well as ions (FORDTRAN et al., 1968; TURNBURG, 1971; MODIGLIANI and BERNIER, 1972) across small intestine *in vivo* are significantly influenced by bulk flow and hence, indirectly, by active Na transport.

The indirect influences of Na on the absorption of other solutes appear to be relatively straightforward, so the remainder of this section will be addressed to the problem of direct Na-coupled solute transport.

I. Na-Coupled Sugar and Aminoacid Transport

It is well established that intracellular accumulation and transcellular transport of some D-sugars and L-aminoacids by *in-vitro* small intestine are dependent upon the presence of Na in the mucosal solution and that, as discussed above, the entry of these nonelectrolytes into the epithelium from the mucosal solution is coupled to Na entry; kinetic descriptions of these Na-coupled influx processes have been reviewed in detail (SCHULTZ and CURRAN, 1970; SEMENZA, 1967; HEINZ et al., 1972). Furthermore, it appears that the movements of sugars and aminoacids from the absorptive cells across the basolateral membranes into the serosal solution are mediated by Na-independent "carrier" systems (BIHLER and CYBULSKY, 1973; NAFTALIN and CURRAN, 1974; CURRAN, 1975; MURER et al., 1974). Thus, "uphill"[10] intestinal absorption of sugars and aminoacids appears to be the result of (a) Na-coupled, carrier-mediated transport into the absorptive cells which leads to (b) intracellular concentrations of these solutes that exceed those in the serosal solution, and (c) "downhill" exit from the cells by means of Na-independent, carrier-mediated processes. The central question, which is still unresolved, is whether the energy that is necessary for the transport of these solutes *into* the cell against concentration differences is derived *solely* from ion-coupled movements across the mucosal membrane *or* whether there must be an *additional direct* coupling to a source of metabolic energy. There is compelling evidence that "ion-gradients" do *contribute* energy to the "uphill" entry step *but* the pivotal question is whether this contribution is enough.

The definitive resolution of this question faces at least two formidable unknowns. First, the specific "ion-gradients" that contribute energy to the net, uphill movements of sugars and aminoacids from the mucosal solution into the transporting cells have not been clearly identified. The unidirectional influxes of sugars and aminoacids into these cells across their mucosal membranes appear to be predominantly influenced by the Na concentration in the mucosal solution. However, the influence of intracellular ions other than Na (*e. g.* K, H) on the unidirectional effluxes of these solutes from the cell into the mucosal solution is uncertain (FRIZZELL and SCHULTZ, 1970). Thus, it is not known whether the maximum energy available for these transport processes *from ion gradients alone* is derived solely from the electrochemical potential difference for Na across the mucosal membrane or whether other ion gradients are also involved. The finding that Na-coupled entry processes are rheogenic (ROSE and SCHULTZ, 1971; WHITE and ARMSTRONG, 1971) indicates that Na entry does not involve an obligatory one-for-one exchange with an intracellular cation and is consistent

10 As discussed previously (SCHULTZ and CURRAN, 1970) the term "active" should be reserved for transport processes that are directly coupled to an exergonic biochemical reaction regardless of the direction of flow. To date there is no compelling evidence that the accumulation of sugars or aminoacids by small intestine is directly coupled to a biochemical reaction, and the notion that accumulation can be energized by coupling to ionic gradients cannot be dismissed. Thus, it seems more appropriate to describe the movements of sugars and aminoacids into the enterocyte and across the epithelium as "uphill" rather than "active" until this issue is resolved.

with the notion that the Na-gradient *alone* contributes energy to these processes. However, the possibility that these processes involve a *partial* exchange of Na for intracellular K (or H) cannot be excluded; if this is the case the maximal energy available from ion gradients would be derived from some combination of Na, K, H (?etc.) gradients. Studies employing brush-border vesicles that reversibly accumulate sugars and aminoacids and that can be preloaded with a desired ion composition (HOPFER et al. 1973; SIGRIST-NELSON et al., 1975) could assist in resolving this important question. The second obstacle stems from the fact that it is extremely difficult to determine *ionic activities in the cells that are responsible for ion-coupled sugar and/or aminoacid transport* even if the specific ions involved are clearly identified.

ARMSTRONG et al. (1973) have recently attempted to evaluate the adequacy of ion-gradients for the case of galactose accumulation by bullfrog small intestine using cation-sensitive microelectrodes to determine intracellular Na and K activities. Their data indicate that if the coupling ratio between galactose entry and Na entry is unity, as suggested for rabbit ileum by GOLDNER et al. (1969), the electrochemical potential difference of Na across the mucosal membrane could provide sufficient energy but the process would require an efficiency of approximately 60 percent[11]. If the combined electrochemical potential differences of both Na and K contribute to the energy required for galactose accumulation, the efficiency of energy transfer would only have to be approximately 30 percent. The data available on rabbit ileum suggest that if the coupling ratio for Na-solute entry is unity, the electrochemical potential difference of Na across the mucosal membrane can energize a ten-fold accumulation of an uncharged solute providing that the efficiency of energy transfer is 75 percent; thus, this gradient alone would be sufficient to energize alanine transport by this preparation (SCHULTZ et al., 1966; ROSE and SCHULTZ, 1971; NELLANS and SCHULTZ, unpubl.). On the basis of these considerations the Na (or ion)-gradient hypothesis for intestinal sugar and aminoacid transport by small intestine cannot be rejected on strictly thermodynamic grounds. HEINZ and GECK (1974) and REID et al. (1974) have arrived at similar conclusions for α-aminoiosobutyrate and methionine accumulation by Ehrlich ascites cells.

Finally, the finding that Na-coupled sugar and aminoacid influx across the mucosal membrane is rheogenic may provide the explanation for observations that have challenged investigators for almost a decade. It is well established that actively transported sugars inhibit the intracellular accumulation and transepithelial transport of aminoacids by small intestine *in vitro* and *vice versa* (c. f. KIMMICH, 1973). Three explanations have been proposed to account for these mutual, inhibitory interactions; namely, that (a) sugars and aminoacids compete

11 KEDEM and CAPLAN (1965) have analyzed the efficiency of coupled flow processes, using the formalism of irreversible thermodynamics. They demonstrated that the maximum efficiency is very stronly dependent upon the degree of coupling between the two flows. For the efficiency of energy transfer to exceed 50% the two flows must be very tightly coupled. For the present purposes this implies that the bidirectional movements of the nonelectrolyte across the mucosal membrane must be almost entirely restricted to the Na or ion-coupled mechanism, *i. e.* there can be little or no ion-independent back-leak of the nonelectrolyte across the mucosal membrane.

for entry into the cells on an Na-dependent "polyfunctional" carrier (ALVARADO, 1966); (b) these solutes compete for a limited, common-energy supply (NEWEY and SMYTH, 1964); and (c) Na entry coupled to the entry of a sugar or aminoacid reduces the Na *concentration* gradient across the mucosal membrane and thereby inhibits any other Na-coupled entry process (CHEZ et al., 1966; READ, 1967; FRIZZELL and SCHULTZ, 1971; SEMENZA, 1971 and 1973). Clearly, if the Na-gradient hypothesis is correct (b) and (c) are synonymous.

The demonstration of mutual inhibitory interactions between sugar and aminoacid uptake by ATP-depleted brush-border vesicles from rat small intestine (MURER and HOPFER, 1973) would seem to exclude the possibility that these interactions are due to competition for an intracellular supply of metabolic energy.

Since Na entry, coupled to the entry of sugars and neutral amino acids is rheogenic, the intracellular electrical potential must influence both the rate of entry (see footnote 9) and the steady-state level of accumulation of the transported sugar and/or aminoacid. Indeed, MURER and HOPFER (1974) have demonstrated that valinomycin markedly increases Na-dependent glucose accumulation by isolated brush-border vesicles from rat intestine if the vesicles are preloaded with K, and similar findings have been reported by SIGRIST-NELSON et al. (1975) for L-alanine. Similarly, REID et al. (1974) have shown that treatment of ATP-depleted ascites tumor cells with valinomycin markedly increases the rate of accumulation and the final steady-state level of methionine. These effects appear to be attributable to the fact that the ionophore essentially converts the membrane into a K-electrode and increases the transmembrane PD (interior negative). Since Na-coupled sugar and/or aminoacid entry normally depolarizes ψ_{mc}, *any Na-coupled entry process will reduce the electrochemical potential difference for Na across the mucosal membrane and, in turn, reduce the rate of entry and the steady-state level of accumulation of any other uncharged solute that enters the cell via a rheogenic Na-coupled mechanism.* Thus the mutual inhibitory interactions between sugars and aminoacids may be attributable to the effect of these transport processes on ψ_{mc}. This explanation is clearly similar to (c), above, but instead of postulating a decrease in the Na concentration difference across the mucosal membrane, which has not been detected, it is based on the observed decrease in ψ_{mc}. Needless to say, although the effect on ψ_{mc} probably contributes to the mutual inhibitory interactions between Na-coupled sugar and aminoacid transport and is the simplest explanation for these phenomena, it remains to be seen whether this effect is sufficient to *quantitatively* account for these interactions or whether additional mechanisms must be invoked.

For more detailed, recent comprehensive reviews of Na-coupled sugar and aminoacid transport by small intestine and other cells the reader is referred to KIMMICH (1973), CHRISTENSEN et al., 1973; HEINZ, 1972).

II. Na-Coupled Cl Transport

It is now well established that *in-vivo* and *in-vitro* preparations of mammalian ileum and *in-vitro* bullfrog small intestine (SCHULTZ et al., 1974; QUAY and ARMSTRONG, 1969) actively transport Cl from mucosa to serosa. Studies on *in-vitro* rabbit ileum (NELLANS et al., 1974; SHEERIN and FIELD, 1975) bullfrog small intestine (QUAY and ARMSTRONG, 1969) and human ileum (TURNBURG et al., 1970) have demonstrated that Cl transport is abolished in the absence of Na and that Na absorption is reduced in the absence of Cl. In rabbit ileum, these observations appear to be due to the fact that Cl entry into the cell is mediated entirely by a neutral, one-for-one NaCl influx process. This process in rabbit ileum is inhibited by acetazolamide (NELLANS et al., 1975) and by elevated intracellular levels of cAMP (NELLANS et al, 1973). Recently, FRIZZELL et al. (1975) have demonstrated what appears to be an identical coupled influx process in the mucosal membranes of rabbit gallbladder which is responsible for neutral NaCl absorption by this tissue (DIAMOND, 1964).

The coupled NaCl entry processes in rabbit ileum and gallbladder appear to closely parallel the behavior of Na-coupled sugar transport by rabbit ileum except that entry is not rheogenic. Thus: (a) The coupled entry process appears to be capable of transporting Cl into the cell against an "apparent" electrochemical potential difference; the energy available from the Na-gradient appears to be more than sufficient to account for the uphill movement of Cl (NELLANS et al., 1973; FRIZZELL et al., 1973; 1975).

Some support for this notion is derived from (a) the finding of de JESUS et al. (1974), using Cl-selective liquid ion exchanger microelectrodes, that under some conditions the intracellular Cl activity in rabbit terminal ileum is greater than that predicted by a simple Nernst distribution.[12]

(b) In the absence of Na, transepithelial Cl transport is abolished (NELLANS et al., 1974) and the intracellular Cl concentration declines. In rabbit gallbladder, intracellular [Cl] approaches the level expected for an equilibrium distribution given by the Nernst equation (Fig. 7) (FRIZZELL et al., 1975).[12]

(c) The coupled NaCl entry process conforms to a kinetic model (NELLANS et al., 1973) that is identical with that suggested for the Na-sugar entry process (GOLDNER et al., 1969).

Is is uncertain at present whether Cl exit from the cell down an apparent electrochemical potential gradient into the serosal solution is attributable to simple diffusion or whether a carrier-mediated process must be invoked; information regarding the intracellular Cl activity, the Cl conductance of the basolateral membranes, and the kinetics of exit is needed to resolve this issue. In any

12 More recent measurements of intracellular Cl activities in Rabbit gallbladder and flounder small intestinal cells incicate that in the presence of Na the intracellular Cl activity is approximately three times that predicted by the Nernst equation for a passively distributed monovalent anion. When these tissues are bathed by a Na-free Ringer solution, the cell Cl activity declines to be expected equilibrium value. These findings strongly support the inferences drawn from studies on cell Cl concentration (Fig. 7). DUFFEY et al., J. Membrane Biol.m 1978, in press).

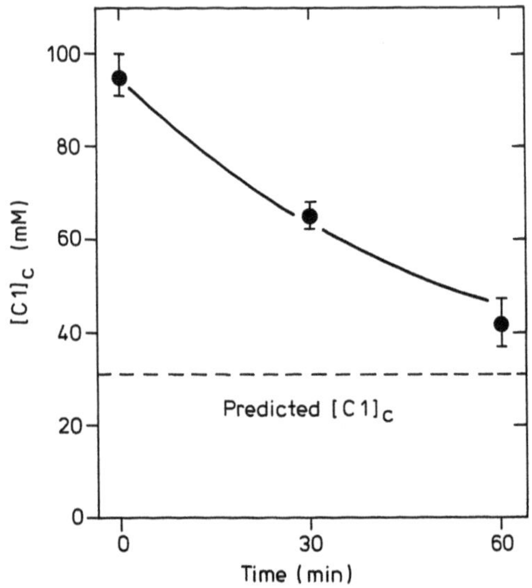

Fig. 7. Effect of incubation in Na-free medium on intracellular Cl ($[Cl]_c$) concentration in rabbit gallbladder. Dashed line represents value of $[Cl]_c$ predicted for passive distribution using Nernst equation. (From FRIZZELL et al., 1975)

event it seems clear that Cl is one more solute whose transport by ileum is directly coupled to Na transport and can be energized by the Na-gradient across the mucosal membrane. It will be of interest to see how many other solutes (e. g. dipeptides, biotin) fall in this category. The energetic advantages of such coupled processes for an organ whose physiologic function is the absorption of large quantities of diverse solutes against minimal transepithelial electrochemical potential differences are obvious.

E. Summary

As pointed out above, the purpose of this chapter is not to paraphrase recent reviews dealing with intestinal transport or to attempt to capsulize our current understanding of the many absorptive and secretory functions of the small intestine within a very limited space. Instead, I have tried to focus on several central issues that have been explored in considerable detail within the past two decades. As a result of these explorations some purely descriptive features appear to have been explained, and the remaining questions deal with underlying mechanisms. Attempts to answer these basic questions are faced by formidable conceptual and technical barriers.

References

AL-AWQATI, Q., CAMERON, J. L., GREENOUGH, W. B., III: Amer. J. Physiol. **224**, 818 (1973).
ALVARADO, F.: Science **151**, 1010 (1966).
ARMSTRONG, W. McD.: In: Intestinal Absorption and Malabsorption, (T. Z. Csaky, Ed.). New York: Raven 1975, p. 45.
ARMSTRONG, W. McD., MUSSELMAN, D. L., REITZUG, H. C.: Amer. J. Physiol. **219**, 1023 (1970).
ARMSTRONG, W. McD., BYRD, B. J., HAMANG, P. M.: Biochim. biophys. Acta **330**, 237 (1973).
BARRY, R. J. C., EGGENTON, J.: J. Physiol. (Lond.) **227**, 201 (1972).
BARRY, P. H., DIAMOND, J. M., WRIGHT, E. M.: J. Membrane Biol. **4**, 358 (1971).
BENTZEL, C. J.: Kidney International **2**, 335 (1972).
BENTZEL, C. J., PARSA, B., HARE, D. K.: Amer. J. Physiol. **217**, 570 (1969).
BERGER, E., LONG, E., SEMENZA, G.: Biochim. biophys. Acta **255**, 873 (1972).
BERRY, C. A., BOULPAEP, E. L.: Amer. J. Physiol. **228**, 581 (1975).
BIHLER, I., CYBULSKY, R.: Biochim. biophys. Acta **298**, 429 (1973).
BOULPAEP, E. L.: In: Electrophysiology of Epithelial Cells. (G. Giebisch, Ed.) Stuttgart: Schattauer 1971, p. 91.
BOULPAEP, E. L.: Amer. J. Physiol. **222**, 517 (1972).
BOULPAEP, E. L., SEELY, J. F.: Amer. J. Physiol. **221**, 1084 (1971).
BRINLEY, F. J., Jr, MULLINS, L. J.: Ann. N. Y. Acad. Sci. **242**, 406 (1974).
CHEZ, R. A., SCHULTZ, S. G., CURRAN, P. F.: Science **153**, 1012 (1966).
CHRISTENSEN, H. N., DEGESPEDES, C., HANDLOGTEN, M. E., RONQUIST, G.: Biochim. biophys. Acta **300**, 487 (1973).
CIVAN, M. M.: Amer. J. Physiol. **219**, 234 (1970).
CIVAN, M. M., KEDEM, O., LEAF, A.: Amer. J. Physiol. **211**, 569 (1966).
CLARKSON, T. W.: J. gen. Physiol. **50**, 695 (1967).
CLAUDE, P., GOODENOUGH, D. A.: J. Cell Biol. **58**, 390 (1973).
CRANE, R. K.: Physiol. Rev. **40**, 789 (1960).
CSAKY, T. Z.: Intestinal Absorption and Malabsorption. New York: Raven 1975.
CSAKY, T. Z., ESPOSITO, G.: Amer. J. Physiol. **217**, 753 (1969).
CURRAN, P. F.: In: Intestinal Absorption and Malabsorption (T. Z. Csaky, Ed.). New York: Raven 1975, p. 113.
CURRAN, P. F., SCHULTZ, S. G., CHEZ, R. A., FUISZ, R. E.: J. gen. Physiol. **50**, 1261 (1967).
DAINTY, J., HOUSE, C. R.: J. Physiol. **185**, 172 (1966).
DE JESUS, C. H., ELLORY, J. C., SMITH, M. W.: J. Physiol. (Lond.) **244**, 31P (1974).
DESJEUX, J.-F., TAI, Y. H., CURRAN, P. F.: J. gen. Physiol. **64**, 274 (1974).
DIAMOND, J. M.: J. gen. Physiol. **48**, 1 (1964).
DIAMOND, J. M.: Fed. Proc. **33**, 2220 (1974).
DIBONA, D. R., CHEN, L. C., SHARP, G. W. G.: J. clin. Invest **53**, 1300 (1974).
DOUGLAS, A., KERLEY, R., ISSELBACHER, K. J.: Biochem. J. **128**, 1329 (1972).
EISENMAN, G.: In: Membrane Transport and Metabolism (A. Kleinzeller and A. Kotyk, Eds.). New York: Academic 1961, p. 163.
ESSIG, A., CAPLAN, S. R.: J. gen. Physiol. **8**, 1434 (1968).
FARQUHAR, M. G., PALADE, G. E.: J. Cell Biol. **17**, 375 (1963).
FIELD, M.: New Engl. J. Med. **284**, 1137 (1971).
FIELD, M.: Gastroenterology **66**, 1063 (1974).
FORDTRAN, J. S., RECTOR, F. C., Jr, EWTON, M. F., SOTER, N., KINNEY, J.: J. clin. Invest. **44**, 1935 (1965).
FORDTRAN, J. S., RECTOR, F. C., Jr, CARTER, N. W.: J. clin. Invest. **47**, 884 (1968).
FRIZZELL, R. A., SCHULTZ, S. G.: J. gen. Physiol. **56**, 462 (1970).
FRIZZELL, R. A., SCHULTZ, S. G.: Biochim. biophys. Acta **233**, 485 (1971).
FRIZZELL, R. A., SCHULTZ, S. G.: J. gen. Physiol. **59**, 318 (1972).
FRIZZELL, R. A., NELLANS, H. N., ROSE, R. C., MARKSCHEID-KASPI, L., SCHULTZ, S. G.: Amer. J. Physiol. **224**, 328 (1973a).
FRIZZELL, R. A., NELLANS, H. N., SCHULTZ, S. G.: J. clin. Invest. **52**, 215 1973b).
FRIZZELL, R. A., DUGAS, M. C., SCHULTZ, S. G.: J. gen. Physiol. **65**, 769 (1975).

FRIZZELL, R. A., KOCH, M. J., SCHULTZ, S. G.: J. Membrane Biol. **27**, 297 (1976).
FRÖMTER, E.: J. Membrane Biol. **8**, 259 (1972).
FRÖMTER, E., DIAMOND, J.: Nature New Biol. **235**, 9 (1972).
FRÖMTER, E., LUER, K.: Pflüger's Arch. **343**, R47 (1973).
FRÖMTER, E., MÜLLER, C. W., WICK, T.: In: Electrophysiology of Epithelial Cells (G. Giebisch, Ed.). Stuttgart: Schattauer 1971, p. 119.
FUJITA, M., OHTA, H., KAWAI, K., MATSUI, H., NAKAO, M.: Biochim. biophys. Acta **274**, 336 (1972).
GLYNN, I. M., HOFFMAN, J. F., LEW, V. L.: Phil. Trans. roy. Soc. **262**, 91 (1971).
GOLDMAN, D. E.: J. gen. Physiol. **27**, 37 (1943).
GOLDNER, A. M., SCHULTZE, S. G., CURRAN, P. F.: J. gen. Physiol. **53**, 362 (1969).
GRADY, G. F., MADOFF, M. A., DUHAMEL, R. C., MOORE, E. W., CHALMERS, T. C.: Gastroent. **53**, 737 (1967).
GRANDCHAMP, A., BOULPAEP, E. L.: J. clin. Invest. **54**, 69 (1974).
HAKIM, A. A., LIFSON, N.: Amer. J. Physiol. **206**, 1315 (1964).
HAKIM, A. A., LIFSON, N.: Amer. J. Physiol. **216**, 276 (1969).
HAYS, R. M., LEAF, A.: J. gen. Physiol **45**, 905 (1962).
HEINZ, E.: Na-Linked Transport of Organic Solutes. Berlin: Springer 1972.
HEINZ, E., GECK, P.: Biochim. biophys. Acta **339**, 426 (1974).
HEINZ, E., GECK, P., WILDBRANDT, W.: Biochim. biophys Acta **255**, 442 (1972).
HELLER, M. D., THIRUMALI, C., HOLDSWORTH, C. D.: Gut **14**, 41 (1973).
HELMAN, S. L., MILLER, D. A.: Amer J. Physiol. **225**, 972 (1973).
HENDRIX, T. R., BAYLESS, T. M.: Ann. Rev. Physiol. **32**, 139 (1970).
HILL, A. E.: Proc. roy. Soc. **190**, 115 (1975).
HODGKIN, A. L., KATZ, B.: J. Physiol. **108**, 37 (1949).
HOPFER, U., NELSON, K., PERROTTO, J., ISSELBACHER, K. J.: J. biol. Chem **248**, 25 (1973).
HOUSE, C. R.: Water Transport in Cells and Tissues. London: Arnold 1974.
HUMPHREYS, M. H., EARLEY, L. E.: J. clin. Invest. **50**, 2355 (1971).
KEDEM, O., CAPLAN, S. R.: Trans Faraday Soc. **61**, 1897 (1965).
KIMMICH, G. A.: Biochim. biophys. Acta **300**, 31 (1973).
KINTER, W. B., WILSON, T. H.: J. Cell Biol. **25**, 19 (1965).
KOEFOED-JOHNSEN, V., USSING, H. H.: Acta physiol. scand. **42**, 298 (1958).
KOHN, P. G., SMYTH, D. H., WRIGHT, E. M.: J. Physiol. (Lond.) **196**, 723 (1968).
KOOPMAN, W., SCHULTZ, S. G.: Biochim. biophys. Acta **173**, 338 (1969).
LEE, C. O., ARMSTRONG, W. MCD.: Science **175**, 1261 (1972).
LEVITT, D. G., HAKIM, A. A., LIFSON, N.: Amer. J. Physiol. **217**, 777 (1969).
LIFSON, N., HAKIM, A. A.: Amer. J. Physiol. **211**, 1137 (1966).
LOEHRY, C. A., AXON, A. T. R., HILTON, P. J., HIDER, R. C., CREAMER, B.: Gut **11**, 466 (1970).
LOESCHKE, K., BENTZEL, C. J., CSAKY, T. Z.: Amer. J. Physiol. **218**, 1723 (1970).
LOESCHKE, K., HARE, D., CSAKY, T. Z.: Pflügers Arch. **328**, 1 (1971).
LOWENSTEIN, W. R., SOCOLAR, S. J., HIGASHINO, S., KANNO, Y., DAVIDSON, N.: Science **149**, 295 (1965).
LYON, I., SHEERIN, H. E.: Biochim. biophys. Acta **249**, 1 (1971).
MACHEN, T. E., ERLIJ, D., WOODING, F. B. P.: J. Cell Biol. **54**, 302 (1972).
MACKNIGHT, A. D. C., CIVAN, M., LEAF, A.: J. Membrane Biol. **20**, 365 (1975a).
MACKNIGHT, A. D. C., CIVAN, M. M., LEAF, A.: J. Membrane Biol. **20**, 387 (1975b).
MANDEL, L. J., CURRAN, P. F.: J. gen. Physiol. **59**, 503 (1972).
MARMOR, M. F.: J. Physiol. (Lond.) **218**, 599 (1971).
MURUYAMA, T., HOSHI, T.: Biochim. biophys. Acta **282**, 214 (1972).
MATTHEWS, D. M.: In: Intestinal Absorption and Malabsorption (T. Z. Csaky, Ed.). New York: Raven 1975, p. 95.
MATTHEWS, D. M., ADDISON, J. M., BURSTON, D.: Clin. Sci. molec. Med. **46**, 693 (1974).
MODIGLIANI, R., BERNIER, J. J.: Biologie et Gastro-Entérologie **5**, 165 (1972).
MORENO, J. H.: Nature **251**, 150 (1974).
MORENO, J. H.: J. gen. Physiol. **66**, 97 (1975a).
MORENO, J. H.: J. gen. Physiol. **66**, 117 (1975b).
MORENO, J. H., DIAMOND, J. M.: In: Membranes-A Series of Advances. (G. Eisenman, Ed.). New York: Dekker 1975a, p. 383.

Moreno, J. H., Diamond, J. M.: J. Membrane Biol. **21**, 197 (1975b).
Munck, B. G.: J. Physiol. (Lond.) **223**, 699 (1972).
Munck, B. G., Schultz, S. G.: J. Membrane Biol. **16**, 163 (1974).
Murer, H., Hopfer, U.: In: Biochemical and Clinical Aspects of Peptide and Amino Acid Absorption. (K. Rommel and H. Goebell, Eds.). Stuttgart: Schattauer 1973, p. 461.
Murer, H., Hopfer, U.: Proc. nat. Acad. Sci. (Wash.) **71**, 484 (1974).
Murer, H., Hopfer, U., Kinne-Saffran, E., Kinne, E.: Biochim. biophys. Acta **345**, 170 (1974).
Murer, H., Hopfer, U., Kinne, R.: Biochem. J. **154**, 597 (1976).
Naftalin, R., Curran, P. F.: J. Membrane Biol. **16**, 257 (1974).
Nellans, H. N., Schultz, S. G.: J. gen. Physiol. **68**, 441 (1976).
Nellans, H. N., Frizzell, R. A., Schultz, S. G.: Amer. J. Physiol. **225**, 467 (1973).
Nellans, H. N., Frizzell, R. A., Schultz, S. G.: Amer. J. Physiol. **226**, 1131 (1974).
Nellans, H. N., Frizzell, R. A., Schultz, S. G.: Amer. J. Physiol. **228**, 1808 (1975).
Newey, H., Smyth, D. H.: Nature (Lond.) **202**, 400 (1964).
Quay, J. F., Armstrong, W. McD.: Amer. J. Physiol. **217**, 694 (1969).
Quigley, J. P., Gotterer, G. S.: Biochim. biophys. Acta **173**, 456 (1969).
Read, C. P.: Biol. Bull. **133**, 630 (1967).
Reid, M., Gibb, L. E., Eddy, A. A.: Biochem. J. **140**, 383 (1974).
Reuss, L., Finn, A. L.: J. Gen. Physiol. **64**, 1 (1974).
Reuss, L., Finn, A. L.: J. Membrane Biol. **25**, 115 (1975).
Robinson, B. A., Macknight, A. D. C.: J. Membrane Biol. **26**, 269 (1976).
Robinson, R. A., Stokes, R. H.: Electrolyte Solutions, 2nd ed. New York: Academic 1959.
Rohde, J. E., Andersen, B.: J. appl. Physiol. **35**, 557 (1973).
Rose, R. C., Schultz, S. G.: J. gen. Physiol. **57**, 639 (1971).
Ross, A., Rubin, A. W., Deren, J. J.: J. clin. Invest. **51**, 2414 (1972).
Rubino, A., Field, M., Shwachman, H.: J. biol. Chem. **246**, 3542 (1971).
Sackin, H., Boulpaep, E. L.: J. gen. Physiol. **66**, 671 (1975).
Saito, T., Lief, P. D., Essig, A.: Amer. J. Physiol. **226**, 1265 (1974).
Samarzija, I., Frömter, E.: Pflüger's Arch. **359**, R119 (1975).
Schultz, S. G.: J. gen. Physiol. **59**, 794 (1972).
Schultz, S. G.: In: Gastrointestinal Physiology (E. D. Jacobson and L. L. Shanbour, Ed.). MTP International Review of Science London: Butterworths, Baltimore: University Park Press, 1974, p. 69.
Schultz, S. G.: Yale J. Biol. Med. **50**, 99 (1977).
Schultz, S. G., Curran, P. F.: Physiol. Rev. **50**, 637 (1970).
Schultz, S. G., Curran, P. F.: In: Current Topics in Membrane Transport (F. Bronner and A. Kleinzeller, Eds). New York: Academic 1974, p. 225.
Schultz, S. G., Frizzell, R. A.: In: Intestinal Absorption and Malabsorption (T. Z. Csaky, Ed.). New York: Raven 1975, p. 77.
Schultz, S. G., Zalusky, R.: J. gen. Physiol. **47**, 567 (1964a).
Schultz, S. G., Zalusky, R.: J. gen. Physiol. **47**, 1043 (1964b).
Schultz, S. G., Zalusky, R.: Nature (Lond.) **204**, 292 (1965).
Schultz, S. G., Fuisz, R. E., Curran, P. F.: J. gen. Physiol. **49**, 848 (1966).
Schultz, S. G., Frizzell, R. A., Nellans, H. N.: Ann. Rev. Physiol. **36**, 51 (1974).
Semenza, G.: J. theoret. Biol. **15**, 145 (1967).
Semenza, G.: Biochim. biophys. Acta **241**, 637 (1971).
Semenza, G.: In: Biochemical and Clinical Aspects of Peptide and Amino Acid Absorption (K. Rommel and H. Goebell, Eds). Stuttgart: Schattauer 1973, p. 45.
Sheerin, H. E., Field, M.: Amer. J. Physiol. **228**, 1065 (1975).
Sigrist-Nelson, K., Murer, H., Hopfer, U.: J. biol. Chem. **250**, 5674 (1975).
Skou, J. C.: Quart. Rev. Biophys. **3**, 401 (1975).
Smulders, A. P., Tormey, J. McD., Wright, E. M.: J. Membrane Biol. **7**, 164 (1972).
Smyth, D. H., Wright, E. M.: J. Physiol. (Lond.) **182**, 591 (1966).
Soergel, K. H., Whalen, G. E., Harris, J. A.: J. appl. Physiol. **24**, 40 (1968).
Speeney, J. G., Flemstrom, G., Shoemaker, R. L., Sachs, G.: J. gen. Physiol. **65**, 645 (1975).
Staehelin, L. A.: Internat. Rev. Cytol. **39**, 191 (1974).
Stirling, C. E.: J. Cell Biol. **53**, 704 (1972).

STIRLING, C. E., KINTER, W. B.: J. Cell Biol. **35**, 585 (1967).
THOMAS, R. C.: Physiol. Rev. **52**, 563 (1972).
TISHER, C. C., YARGER, W. E.: Kidney International **3**, 238 (1973).
TRIER, J. S., RUBIN, C. E.: Gastroenterology **49**, 574 (1965).
TURNBERG, L. A.: Gut **12**, 811 (1971).
TURNBERG, L. A., BIEBERDORF, F. A., MORAWSKI, S. G., FORDTRAN, J. S.: J. clin. Invest. **49**, 557 (1970).
ULLRICH, K. J.: In: Handbook of Physiology, Section 8: Renal Physiology (J. Orloff and R. W. Berliner, Eds). Baltimore, Md: Williams American Physological and Wilkins 1973, p. 337.
USSING, H. H., ERLIJ, D., LASSEN, U.: Ann. Rev. Physiol. **36**, 17 (1974).
USSING, H. H., WINDHAGER, E.: Acta Physiol. Scand. **61**, 484 (1964).
VAN OS, C. H.: Doctoral Thesis. University of Nijmegen. Nijmegen, The Netherlands, 1974.
WHITE, J. F., ARMSTRONG, W. McD.: Amer. J. Physiol. **221**, 194 (1971).
WHITTEMBURY, G., RAWLINS, F. A.: Pflügers Arch. **330**, 302 (1971).
WINDHAGER, E. E., BOULPAEP, E. L., GIEBISCH, G.: Proc. Intern. Congr. Nephrol., 3rd, Washington, D. C. **1**, 35 (1966).
WRIGHT, E. M.: Nature (Lond.) **212**, 189 (1966a).
WRIGHT, E. M.: J. Physiol. (Lond.) **185**, 486 (1966b).
WRIGHT, E. M., DIAMOND, J. M.: Biochim. biophys. Acta **163**, 57 (1968).
WRIGHT, E. M., BARRY, P. H., DIAMOND, J. M.: J. Membrane Biol. **4**, 331 (1971).
WRIGHT, E. M., SMULDERS, A. P., TORMEY, J. McD.: J. Membrane Biol. **7**, 198 (1972).

Chapter 15

Transport in Large Intestine

D.W. POWELL

A. Introduction

Until recently, there has been relatively less understanding of fundamental mechanisms of electrolyte transport in the colonic epithelium than in the small intestine. The reason for this lag in research effort is not entirely clear. It may have been for the practical reason that "almost any study of colonic function involves work on colonic contents" (WRONG, 1971). Alternatively, it may have been for the more philosophical reasons cited – "The small intestine, after all, provides the main reason for the existence of the gut; all the rest is prologue or epilogue, for man can live without his stomach or his esophagus, and may thrive without his colon . . ." (SPIRO, 1975). In the present author's opinion, the reason was probably scientific – the colonic epithelium was thought not to do as many interesting things as the small intestine. Thus, the colonic epithelium is incapable of active monosaccharide transport (PARSONS and PATERSON, 1960, 1965; LONG et al. 1967; HEATON, 1972), or active amino acid transport[1] (BAILLIEN and SCHOFFENIELS, 1961; BINDER, 1970); and there is neither coupled sodium-glucose nor sodium-amino acid transport (POWELL and MALAWER, 1968; BILLICH and LEVITAN, 1969; GRADY et al. 1970). However, unlike the small intestine, the colonic epithelium can transport sodium against a high concentration gradient and responds to aldosterone by increasing sodium transport. In addition, its internal milieu is such that nitrogen transport and organic anion transport may be among its important functions. These are the reasons why this epithelium is now beginning to receive attention. Many of these aspects are discussed in several excellent reviews of the general absorptive and secretory functions of the colon (WRONG, 1971; HAGIHARA and GRIFFEN, 1972; SHIELDS, 1972; CUMMINGS, 1975).

Although it has been suggested that calcium is actively secreted by the mammalian colon (PHILLIPS and GILLER, 1973), there has been little investigation of divalent cation transport by the large intestine (CASSIDY and TIDBALL, 1969; HARRISON and HARRISON, 1969; BEHAR, 1974; PITITH and SCHEDL, 1976). Until very recently, there has also been little information about the colonic absorption of other important nutrients, such as oxalate (BINDER, 1974; DOBBINS and BIN-

1 SMITH and JAMES (1976) have demonstrated active methionine transport, which is maintained for at least 24 hours after birth, in the heliocoidal colon of the new-born pig.

DER, 1976) or vitamin K (HOLLANDER and TRUSCOTT, 1974; HOLLANDER et al., 1976). Therefore, this review will be addressed toward our current understanding of the mechanisms of transport of water and monovalent electrolytes, as well as the colonic transport of the weak electrolytes – ammonia and the volatile fatty acids.

B. Water and Electrolyte Transport

The colon aids in the conservation of body fluid and electrolytes and does so with over 90 percent efficiency. In the human 500–1500 ml of electrolyte-rich fluid enters the colon each day (PHILLIPS and GILLER, 1973), while in larger animals, such as the pony, over 20 liters are presented to the colon (ARGENZIO et al., 1974a). An equal amount may be secreted by the colon itself and thus require resorption (ARGENZIO et al., 1974a). No less impressive is the bird, whose large intestine must also absorb water from urine which is moved in a retrograde fashion into the colon from the cloaca where it is stored (SKADHAUGE, 1967, 1968). The reader is well aware, from personal experience, of the consequence of failure to absorb this fluid (PHILLIPS, 1972). This section of this review will be concerned with the physiological properties of the colonic epithelium and the mechanisms whereby the colon transports this fluid and electrolyte.

I. Epithelial Properties

The colonic epithelium does not have a uniform transport capacity; there appears to be greater absorption from the proximal than from the distal colon (LEVITAN et al., 1962; EDMONDS, 1967b; DEVROEDE and PHILLIPS, 1970; DEVROEDE et al., 1971; RASK-MADSEN, 1973; DOLMAN and EDMONDS, 1975; YAU and MAKHLOUF, 1975). However, when compared to the small intestine, the entire colon can be described as a "tight" epithelium. This gives rise to several of its electrical properties, which have been the subject of previous reviews (BARRY and EGGENTON, 1974; RASK-MADSEN, 1974; EDMONDS, 1975). These properties are summarized in Table 1. The electrical resistance of this epithelium ranges between 80 and 1400 Ω cm^2, at a minimum some three times that of the small intestine. The electrical potential difference (PD, Ψ_{ms}) across this epithelium is oriented with the blood or serosal solution positive with respect to the lumen or mucosal solution and is also generally 3–10 times that in the small intestine. Under normal circumstances, the short-circuit current (I_{sc}) of the colonic epithelium is of similar magnitude to the small intestine I_{sc}, and appears to be primarily due to the active transport of sodium from the lumen to blood (see below). The PD and I_{sc} are dependent on luminal sodium, possibly serosal potassium, and on oxygen, and are inhibited by metabolic inhibitors such as dinitrophenol (DNP) and ATPase inhibitors such as ouabain (BAILLIEN and

Table 1. Electrical parameters in colonic epithelium, *in vivo* and *in vitro*

Species and References	Ψ_{ms} In Vivo	Ψ_{ms} In Vitro$_{mV}$	Ψ_{mc}	Ψ_{cs}	I_{sc} µAcm^{-2}	R Ω cm^2
Toad (Bufo *Marinus*)[1]		4–40			20–90	300–500
Toad (Bufo *Arenarum*)[2]		15–60			75	467
Frog (Rana *Catesbiana*)[3, 4]		40–100			100	450
Greek tortoise[5, 6]		20–50	−6	29	15	1438
Guinea-pig cecum[7, 8]	4–10	8–11			60	172
Rabbit[9]		10–50	−39	54	75–150	190–550
Rat[10, 11, 12]	10–30	10–15	−45	55	140	108
Rat[13, 14]		2–4			35	80
Pig, infant[15]		8–20			60	160
Dog[16]	10–40					
Human[17, 18]	25–50					
Human[19, 20]		10–20			85, 185	175

[1] Cofré and Crabbé, 1967 (unstripped colon); Dawson and Curran, 1976 (stripped colon); [2] Lew, 1970; [3] Cooperstein and Hogben, 1959; [4] Perheentupa et al., 1972; [5] Baillien and Schoffeniels, 1961; [6] Gilles-Baillien and Schoffeniels, 1967a and b; [7] Powell et al., 1968; [8] Ussing and Andersen, 1956; [9] Frizzell et al., 1976; Yorio and Bentley, 1977; Schultz et al., 1977, [10] Edmonds and Marriott, 1968; [11] Edmonds and Nielsen, 1968; [12] Edmonds and Marriott, 1970; [13] Binder and Rawlins, 1973a; [14] Stoebel and Goldner, 1975; [15] Bentley and Smith, 1975; [16] Cooperstein and Brockman, 1959; [17] Edmonds, 1975; [18] Rask-Madsen, 1974; [19] Grady et al., 1970; [20] Archampong et al., 1972.

Schoffeniels, 1961b; Bentley, 1962; Gilles-Baillien and Schoffeniels, 1967a; Edmonds and Marriott, 1968b; Dalmark, 1970; Binder and Rawlins, 1973a; Frizzell et al., 1976; Frizzell and Jennings, 1977).

The transepithelial PD, which ranges from 3 to 100 mV in the various species studied, is the sum of potential steps across the mucosal and serosal cell borders, the cell interior being negative with respect to both the mucosal and serosal solutions. Thus, there is a "well-type" potential with a negative potential step across the mucosal membrane (Ψ_{mc}) of 10–45 mV and a positive step across the serosal membrane (Ψ_{cs}) that is sufficient to account for the magnitude of Ψ_{ms}. (See Schultz et al., 1977 for a discussion of an equivalent electrical circuit for this epithelium). Early studies of the effects of various ions on these separate membrane potentials (Gilles-Baillien and Schoffeniels, 1967a) are difficult to interpret because of failure to consider the influence of the shunt path on the potentials generated at either cell border (Schultz, 1972). However, these studies do suggest that sodium entry across the mucosal membrane plays a role in Ψ_{mc}. This has been confirmed in recent studies with amiloride, which blocks unidirectional sodium entry into the colonic cells. Addition of this drug to the mucosal solution of the *in-vitro* rabbit colon abolishes Ψ_{ms} and yet hyperpolarizes Ψ_{mc}, indicating that sodium entry across the mucosal cell membrane is a conductive process (Frizzell et al., 1976; Schultz et al., 1977; Yorio and Bentley, 1977).

The "tightness" of the colonic epithelium can be domonstrated in other ways than by the electrical resistance. For example, the osmotic permeability of the colon (BILLICH and LEVITAN, 1969) is much less than that of the small intestine (FORDTRAN et al., 1965), with a calculated reflection coefficient for urea of approximately 1.0 in the colon. In addition to accounting, in large part, for the magnitude of the PD, this "tightness" also explains the ability of this epithelium to transport against an osmolar gradient of greater than 50 mOsmol (GOLDSCHMIDT and DAYTON, 1919a and b; BILLICH and LEVITAN, 1969), and its ability to transport sodium from luminal concentrations as low as 15 mM (LEVITAN et al., 1962; SHIELDS and MILES, 1965; DEVROEDE and PHILLIPS, 1969). Finally, an interesting, and probably teleologically useful, consequence of these functional properties is that the colon transports a hypertonic absorbate. That is, during absorption from isotonic solutions, the ratio of transported solute to solvent is hypertonic (PARSONS, 1956; SKANDHAUGE, 1967; POWELL and MALAWER, 1968; BILLICH and LEVITAN, 1969; PHILLIPS and SCHMALZ, 1970; BINDSLEV and SKADHAUGE, 1971a und b; BOWN et al., 1972; PERHEENTURA, 1972). Low values of water permeability, among other factors, could account for hypertonic transport *via* either the serial membrane hypothesis or the standing gradient osmotic model of water transport (PATLAK et al., 1963; DIAMOND and BOSSERT, 1967).

Estimates of the passive permeability of this epithelium to electrolytes have yielded rat colonic permeabilities of chloride greater than sodium (CURRAN and SCHWARTZ, 1960), and of potassium greater than both sodium and chloride (EDMONDS, 1967c; EDMONDS and MARRIOTT, 1968a). *In-vitro* studies in rabbit colon have confirmed these impressions, with calculated relative permeabilities of $P_K : P_{Na} : P_{Cl} = 1.0 : 0.07 : 0.11$ (FRIZZELL et al., 1976). The similar permeabilities of sodium and chloride lend an explanation for the finding that streaming potentials cannot be developed in the rat colon in response to osmotic water flows (EDMONDS and MARRIOTT, 1968a). These studies of FRIZZELL et al. further indicate that the ratio of 0.64 for $P_{Na} : P_{Cl}$ is the same as the ratio of their free-solution mobilities ($\lambda_{Na} : \lambda_{Cl} = 0.7$). Thus, rather than only restricting diffusional flows of electrolytes, the passive conductance pathway in the colon may, in effect, accelerate the diffusion of potassium. This finding has obvious implications with regard to the passive entry of potassium into the colonic lumen in response to electrical driving forces (see below). Such a pathway would also prevent the passive backflux of sodium and chloride, thus allowing transport against a high concentration gradient, while simultaneously allowing excretion of potassium.

Following the paradigm recently proposed for epithelia – cells attached near their apical surfaces by tight-junction intercellular complexes with relatively higher, selective passive permeability (DIAMOND, 1974) – a unique picture emerges for the colonic epithelium. The studies cited above suggest that it is a relatively tight epithelium consisting of cells with sodium, but little potassium, permeability, which are short-circuited by shunt paths with relatively low permeability to sodium and chloride, but high potassium permeabilities. The accuracy of this synopsis remains to be determined.

II. Sodium, Cloride, and Bicarbonate Transport

Although not necessarily true in the amphibian, when studied either *in vivo* or *in vitro* the mammalian colon absorbs sodium and chloride and secretes potassium and bicarbonate[2] (D'AGOSTINO et al., 1953; PARSONS, 1956; COOPERSTEIN and BROCKMAN, 1959; LEVITAN et al., 1962; see also the excellent reviews by SHIELDS and MILES, 1965; LEVITAN, 1968, 1969: PHILLIPS, 1969; TURNBERG, 1970; EDMONDS and PILCHER, 1972; SHIELDS, 1972). The mechanisms whereby electrolyte transport occurs in this epithelium remain to be clearly defined. The *in-vitro* data summarized in Table 2, when combined with information from *in-vivo* studies, does allow a tentative definition of electrolyte transport based on various combinations of the mechanisms depicted in Fig. 1.

With reference to Table 2, transport by the colon can be used to separate the species studied into three groups: first, species in which the investigators felt that net sodium transport (J^{Na}_{net}) was clearly less than the simultaneously measured I_{sc} (numbers 1 and 2); secondly, species in which J^{Na}_{net} was equal to I_{sc} (3 and 4, probably 5, possibly 6); and thirdly, species in which J^{Na}_{net} was clearly greater than the I_{sc} (possibly 6, 7–11).

Table 2. Electrolyte fluxes in colonic epithelium, *in vitro*

Species and References[a]	J^{Na} ms	J^{Na} sm	J^{Na} net (µEq h·cm⁻²)	I_{sc}	$100\,J^{Na}_{net}/I_{sc}$ %	J^{Cl} ms	J^{Cl} sm	J^{Cl} net (µEq h·cm⁻²)	$J^{R\,d}$ net
1. Toad (B. *Arenarum*)[2]	2.2	0.2	2.0	2.9	69	1.0	1.0	0.0	0.9
2. Frog (R. *Catesbiana*)[3]	4.3	0.6	3.7	4.0	93	1.5	1.4	0.1	0.4
3. Toad (B. *Marinas*)[1]	3.5	1.3	2.2	2.3	96	–	–	–	–
4. Rabbit[b9]	4.4	1.4	3.0	2.8	107	5.6	4.0	1.4	1.6
5. Guinea pig cecum[8]	7.7	5.0	2.7	2.2	122	–	–	–	–
6. Greek tortoise[6]	0.9	0.2	0.7	0.5	140	2.6	1.4	1.2	1.0
7. Rat[c13]	11.4	8.8	2.5	1.2	208	14.6	12.7	1.9	0.6
8. Rat[14]	10.0	6.3	3.6	0.7	514	16.1	12.6	3.6	0.7
9. Pig, infant[15]	12.0	3.8	8.2	2.8	293	4.7	3.4	1.3	−4.1
10. Human[19]	7.1	3.1	4.0	3.2	124	–	–	–	–
11. Human[20]	21.0	10.9	10.1	7.0	145	11.0	8.4	2.4	−0.7

[a] See Table 1 for references (page 783)

[b] Fluxes in rabbit colon reported by YORIO and BENTLEY (1977) were performed in colon with the muscle layers intact and are not shown here. Na fluxes in that unstripped preparation were almost identical to those shown here, but there was no net transport of Cl

[c] Fluxes in rat colon by CURRAN and LINDEMANN, unpublished observations (quoted by SCHULTZ and CURRAN, 1968) have not been included since they may have been performed in HCO_3-free Ringer. Fluxes in rat colon by EDMONDS and MARRIOTT (1970) have not been included since they were performed *in vivo*

[d] $J^{R}_{net} = I_{sc} - (J^{Na}_{net} - J^{Cl}_{net})$. In rabbit ileum there is compelling evidence that a positive $J^{R}_{net} = HCO_3^-$ secretion (DIETZ and FIELD, 1973)

2 The term bicarbonate secretion or transport, in this review, does not necessarily mean transport of the HCO_3^- species itself. Secretion of the hydroxyl ion or absorption of hydrogen ion can give the same end result (BRODSKY and SCHILB, 1967; HUBEL, 1968, 1974).

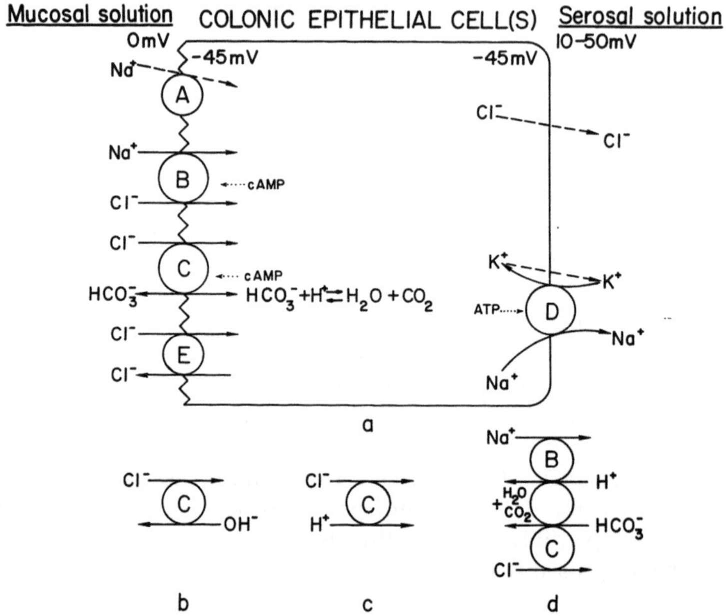

Fig. 1. General, theoretical model to account for Na, Cl, and HCO_3 transport by colonic epithelium. (a) *A* represents either diffusional or carried-mediated electrogenic Na entry process; *B* is neutral, coupled NaCl transport mechanism; *C* is $Cl^-:HCO_3^-$ exchange; *D* is basolaterally located, ATP-dependent rheogenic Na-K pump; and *E* is $Cl^-:Cl^-$ exchange mechanism. It should be obvious that these various mechanisms might not all reside in the same cells in the colonic epithelium; and that "pumps" *B, C,* and *E* could be located in the basolateral cell membrane rather than in the brush border. Finally, it should be noted that the HCO_3^- secretion depicted here could just as well be (b) OH^- secretion, or (c) H^+ absorption, and that neutral, coupled NaCl transport cannot be easily distinguished from a double recriprocal exhange mechanism, (d), involving $Na^+:H^+$ exchange associated with either $Cl^-:HCO_3^-$ (OH^-) exchange or coupled $Cl^-:H^+$ absorption. Cyclic AMP-mediated changes in colonic electrolyte transport can be explained by cAMP inhibition of *B* and/or cAMP alterations of *C* (Modified from SCHULTZ and CURRAN, 1974; based on the concepts of FRIZZELL et al., 1976 and SCHULTZ et al., 1977, and the data in the text)

The amphibians absorb sodium but transport little chloride, and may either secrete bicarbonate (COOPERSTEIN and HOGBEN, 1959; LEW and CARLISKY, 1967; CARLISKY and LEW, 1970; LEW, 1970) or absorb bicarbonate (CHALFIN et al., 1958; PERHEENTUPA et al., 1972). In the case of the frog and toad, *B. arenarum,* (Group 1), the difference in J^{Na}_{net} and I_{sc} appears to be made up by active bicarbonate transport from s to m. This discrepancy between J^{Na}_{net} and I_s in the frog disappears when fluxes are measured in bicarbonate-free solution (COOPERSTEIN and HOGBEN, 1959), but does not disappear in bicarbonate-free studies in the toad (LEW and CARLISKY, 1969; CARLISKY and LEW, 1970; LEW, 1970). Thus, LEW et al. have postulated that in the toad the bicarbonate is produced in the cell and preferentially transported into the mucosal solution by electrogenic mechanisms. Although the mechanism of sodium entry across the luminal cell border has been investigated only in the rabbit (FRIZZELL et al., 1976; SCHULTZ et al., 1977), it is probable that a similar entry mechanism exists in Group 1 and in the other species. Likewise, there have been no formal studies of the mechanism(s) of sodium exit at the serosal border of the colonic cell, other than noting the susceptibility of the transepithelial PD to cardiac glycosides and a possible dependence of the PD on the serosal solution potassium concentration (see above). However, it seems reasonable to assume that this serosally located sodium pump in the colonic cell is a Na-K-ATPase-mediated process, as in other sodium transporting epithelia. Finally, in both frog and toad there is evidence for chloride: chloride exchange (COOPERSTEIN and HOGBEN, 1959; LEW, 1970). Thus, with reference to Figure 1, transport in Group 1 could be explained by the operation of mechanisms *A, D, E,* and the bicarbonate-secreting half of *C.*

When *B. arenarum* are maintained without food for 20–30 days, flux measurements show that J^{Na}_{net} equals I_{sc} (LEW, 1970), as it does in *B. marinus* (COFRÉ and CRABBÉ, 1967; DAWSON and CURRAN, 1976), in the rabbit (FRIZZELL et al., 1976; YORIO and BENTLEY, 1977), and possibly in the guinea-pig cecum (USSING and ANDERSEN, 1956). *In-vitro* studies in rabbit colon, stripped of its muscle layers, also reveal apparent active chloride absorption and a residual net flux, J^{R}_{net}, that is not accounted for by the sum of J^{Na}_{net} and J^{Cl}_{net} (FRIZZELL et al., 1976). This J^{R}_{net} was equal in magnitude to J^{Cl}_{net}. By inference from studies in rabbit ileum (DIETZ and FIELD, 1973), J^{R}_{net} was probably s to m (or c to m) bicarbonate transport (see footnote 2). Net chloride absorption and equal bicarbonate secretion (J^{R}_{net}) was also present in sodium-free solutions, strengthening the idea that this was chloride-bicarbonate exchange. The finding of J^{R}_{net} in bicarbonate-free solutions has led these investigators to propose that the bicarbonate in this species also comes from endogenous production within the cell. Finally, FRIZZELL et al. (1976) found evidence for chloride: chloride exchange in the rabbit. Thus, transport in Group 2 could be explained by a combination of mechanisms *A, C, D* and *E.*

The reptile (turtle and tortoise) may belong in Group 2 or Group 3, since there is conflicting data as to whether J^{Na}_{net} is equal to (DAWSON and BRANDES, 1976) or greater than (GILLES-BAILLIEN and SCHOFFENIELS, 1967a and b) the I_{sc}.

The nonherbivorous mammals more clearly make up the third transport group, where J^{Na}_{net} is significantly greater than the I_{sc}. Sodium entry into the cell may be electrogenic (GILLES-BAILLIEN and SCHOFFENIELS, 1967a; SCHULTZ et al. 1977), and the I_{sc} appears to be due to rheogenic sodium transport (see the preceding section), indicating the presence of mechanisms *A* and *D.* In addition to finding a residual flux *in vitro,* several *in-vivo* studies indicate the presence of chloride: bicarbonate exchange. For example, *in-vivo* studies in human (DEVROEDE and PHILLIPS, 1969; BOWN et al., 1972) and rat colon (PHILLIPS and SCHMALZ, 1970) have shown that bicarbonate secretion is dependent on the presence of luminal chloride. This is also true in *in-vitro* rat studies (BINDER and RAWLINS, 1973a; STOEBEL and GOLDNER, 1975). Chloride absorption in rat[3] and man also continues, or is increased, at low colonic pH, a condition that would

3 CURRAN and SCHWARTZ (1960) reported that chloride transport in rat colon was entirely passive. However, those experiments were performed *in vivo* with bicarbonate-free solutions, a condition that might significantly alter chloride movement in the colonic epithelium.

favor bicarbonate secretion (ROUSSEAU and SLADEN, 1971; BOWN et al., 1972). Finally, nature has provided experimental evidence for this mechanism in man in the inherited form of diarrhea called congenital chloridorrhea. Stools in this condition have high chloride content and low pH. Evidence has been presented that this can be explained by a defect in the ileal and colonic chloride: bicarbonate exchange mechanism (TURNBERG, 1971; BIEBERDORF et al., 1972; HOLMBERG et al., 1975). Although chloride: chloride exchange diffusion has been reported in the dog colon (COOPERSTEIN and BROCKMAN, 1959), its presence has not been adequately sought in either rat or man. Thus, mechanisms *A, C,* and *D,* and possibly *E,* are present in these species, as in the rabbit. What differentiates transport in these species from that in the rabbit is evidence for *B,* a coupled, sodium chloride absorption mechanism. This mechanism is necessary to explain a J^{Na}_{net} greater than I_{sc}, and a J^{Cl}_{net} greater than J^{R}_{net}, in the short-circuited rat colon (BINDER and RAWLINS, 1973a; STOEBEL and GOLDNER, 1975; BINDER, 1977). A precedent for the existence of such a transport mechanism has been well documented in other intestinal segments and animal species such as the rabbit ileum (NELLANS et al., 1973, 1974). It is difficult to be entirely sure that mechanism *B* is present in human colon, since J^{Na}_{net} is only 20–40 percent greater than I_{sc} (Table 2) and, considering the difficulties in obtaining human colon for *in-vitro* studies, this discrepancy could be due to experimental error. In favor of the existence of a coupled NaCl absorptive mechanism in man is the finding that *in-vivo* sodium (and water) absorption are reduced in the absence of luminal chloride, and chloride absorption is reduced in the absence of luminal sodium (BILLICH and LEVITAN, 1969; HOLMBERG et al., 1975).

Although it may be useful to categorize colonic electrolyte transport according to the relationship of net sodium transport to the simultaneously measured short-circuit current, this is probably somewhat of an oversimplification. *In-vitro* studies in the pony large intestine, which is long and multicompartmental, indicate that J^{Na}_{net} may be less than, equal to, or greater than the I_{sc}, depending on the source of the mucosa studied (ARGENZIO et al., 1976). The more proximal part of this large intestine had a $J^{Na}_{net} < I_{sc}$, the middle segments had $J^{Na}_{net} \simeq I_{sc}$, and the distal, small colon showed $J^{Na}_{net} > I_{sc}$. In addition, changes in solution pH or the presence of small chain fatty acids would further alter these relationships. It is interesting to speculate whether fatty acid transport might in some way be responsible for the species differences noted above.

It should be obvious that it is difficult to distinguish coupled sodium-chloride absorption (mechanism *B* in Fig. 1a) from a process (Fig. 1d) where sodium is exchanged for hydrogen and this is linked to chloride: bicarbonate exchange. Such a model was first proposed to explain chloride absorption and bicarbonate secretion in the human ileum (TURNBERG et al., 1970) and it has been invoked to explain rat colonic electrolyte transport as well (BINDER and RAWLINS, 1973a). If bicarbonate comes from endogenous sources, the hydrogen ion thus formed must exit *via* the mucosal membrane of the cell, or the chloride: bicarbonate exchange would not be an electrically neutral process. Operationally these processes also appear to be a model similar to that of TURNBERG et al. Since this model depends on the movement of hydrogen into the lumen in exchange for sodium, when the lumen is perfused with chloride-free solutions, the process of sodium absorption should be accompanied by luminal acidification and an increase in pCO_2 (as bicarbonate is neutralized by the secreted hydrogen ion). This has been demonstrated in studies in three humans (HOLMBERG et al., 1975). However, one would think that if both hydrogen and bicarbonate ions were being secreted simultaneously into the lumen, it should also increase the pCO_2 of a Ringer solution. When measured, the pCO_2 of Ringer solution in the colonic lumen is less than plasma (PARSONS, 1956; SWALLOW and CODE, 1967) which is a point against H^+-Na^+ exchange. These issues remain to be settled.

Additional aspects that need further investigation are the mechanisms of action of acetazolamide on electrolyte transport, a possible sodium-sodium exchange mechanism in this epithelium, and the interrelationship between organic anions and electrolyte transport. Acetazolamide inhibits bicarbonate secretion and sodium absorption in the rat colon (PARSONS, 1956; PHILLIPS and SCHMALZ, 1970; BINDER and RAWLINS, 1973a). Since carbonic anhydrase, which is present in high concentrations in colonic mucosa (CARTER and PARSONS, 1971), might be involved in endogenous bicarbonate production, this drug may work through the action traditionally considered to be its primary one. However, inhibition of carbonic anhydrase would not necessarily account for the reduction in sodium transport noted in those studies unless chloride: bicarbonate exchange is operationally linked to a sodium: hydrogen exchange (Fig. 1d). This could also be explained if one postulates that acetazolamide inhibits the coupled sodium-chloride transport mechanism in rat colon (mechanism *B*) as it does in rabbit ileum (NELLANS et al., 1975).

Several investigators have shown in *in-vivo* studies that the s to m flux of sodium was dependent on the mucosal sodium concentration (transconcentration effect) (VISSCHER et al., 1944; CURRAN and SCHWARTZ, 1960; EDMONDS, 1967b; EDMONDS and MARRIOTT, 1970). CURRAN and SCHWARTZ postulated a sodium-sodium exchange to account for this. Whether this represents a true exchange-diffusional mechanism for sodium remains to be determined.

Finally, an area in critical need of further investigation is the relationship between organic anion (volatile fatty acids) and colonic electrolyte transport. These anions are present in high concentration in the mammalian colon and have been shown to both increase (ARGENZIO et al., 1975) and decrease (GIDDINGS et al., 1974; ARGENZIO et al., 1976) colonic Na transport. Existing largely as anions at colonic pH, these small fatty acids could also interact with other anionic (Cl or HCO_3) carriers. It is equally likely (see below) that the processes of H and/or HCO_3 transport by this epithelium could create the necessary conditions for efficient colonic fatty acid absorption (ARGENZIO et al., 1976).

III. Potassium Transport

When perfused with potassium concentrations less than 15–25 mM, the mammalian colon secretes potassium into the lumen (PHILLIPS and CODE, 1966; DEVROEDE and PHILLIPS, 1969); and in the human this is the cation of highest concentration (75 mM) in fecal water (WRONG et al., 1965). Colonic potassium secretion can be increased under three circumstances: (1) aldosterone administration or release of endogenous aldosterone through sodium depletion or potassium loading (SHIELDS et al., 1966; ELMSLIE et al., 1966; EDMONDS, 1967c; EDMONDS and MARRIOTT, 1967; DOLMAN and EDMONDS, 1975); (2) by maneuvers that increase colonic cell Na-K-ATPase activity, such as glucocorticoid administration (CHARNEY et al., 1975; BINDER, 1976) or in renal insufficiency (BASTL et al., 1977), where the mechanism apparently involves more than endogenous aldosterone release; or (3) by maneuvers that might increase the permeability of the luminal cell border, such as lowering luminal pH (BOWN et al., 1972), the action of amphotericin B (FRIZZELL and TURNHEIM, 1978), perhaps the action of aldosterone itself (EDMONDS, 1969; BARNABY and EDMONDS, 1969), and perhaps the effects of inflammatory bowel disease (ARCHAMPONG et al., 1972; HARRIS and SHIELDS, 1970; EDMONDS and PILCHER, 1973).

The mechanism of this potassium secretion by the colon remains uncertain. The argument revolves around whether potassium secretion is due entirely to passive movement down an electrical gradient, or whether it is an active process. Evidence in favor of an active mechanism comes largely from the investigations of EDMONDS, who has demonstrated under a number of circumstances that the steady-state luminal potassium concentration was greater than would be expected from the Nernst equation (EDMONDS, 1967a–c, 1969; EDMONDS and MARRIOTT, 1967; EDMONDS and NIELSEN, 1968; BARNABY and EDMONDS, 1969; SALAS-COLL et al., 1976). This finding has also been reported by other investigators (PHILLIPS and CODE, 1966; COOPERSTEIN and BROCKMAN, 1969; PERHEENTUPA et al., 1972). Several investigators have also demonstrated net K transport (secretion) in the *in-vitro,* short-circuited colon when the transmural PD has been abolished (ARCHAMPONG et al., 1972; BENTLEY and SMITH, 1975; YORIO and BENTLEY 1977).

In contrast, others have found that potassium movement was explicable by the large colonic PD. Indeed, the potassium secretion stimulated by aldosterone, glucocorticoids, or potassium adaptation are all accompanied by an increase in the colonic PD. In addition, transepithelial flux measurements in isolated stripped rabbit colon (FRIZZELL et al., 1976; FRIZZELL and SCHULTZ, 1978) have failed to show net potassium transport in the absence of an electrochemical gradient. Apparent active potassium secretion has been found, however, when amphotericin B was present in the mucosal solution *in vitro* (FRIZZELL and TURNHEIM, 1978). This is perhaps attributable to amphotericin-induced alterations in the luminal membrane potassium permeability, allowing potassium pumped into the cell by the activity of the basolateral cell membrane Na-K-ATPase to exit passively across the luminal cell border.

It is obvious that both active and passive potassium secretion could be occurring simultaneously and to varying degrees depending on the circumstances. On the basis of measurements of steady-state potassium fluxes, isotopic potassium efflux and influx, and potassium pool size, EDMONDS has postulated that potassium crosses the epithelium *via* two routes (EDMONDS, 1969; BARNABY and EDMONDS, 1969): (1) a transmucosal shunt through which potassium movement could be altered by the PD, and (2) a transcellular route, which is made up to two compartments, a large one with slow turnover ($T^{1}/_{2} = 30$ minutes), and a small pool (5–10% of total mucosal K^+) with rapid turnover ($T^{1}/_{2} = 6$ minutes). EDMONDS believes this small pool to be responsible for the active component of potassium secretion. Such a compromise view, where both active and passive mechanisms are considered to play a role, would be in keeping with our understanding of distal renal tubular potassium secretion, a secretion process that is very similar to that in the colon (see GIEBISCH, 1975 and chapter 5 of this volume, and the comparison of potassium secretion by these two epithelia in FISHER et al., 1976).

Finally, to further complicate the issue, it should be pointed out that under certain conditions potassium "secretion" might come from goblet cells rather than colonic absorptive cells. These goblet cells make up a considerable proportion of the colonic epithelium, and colonic mucus has a high potassium content – up to 150 mM (CRANE, 1965; GILLER and PHILLIPS, 1972). Potassium may be secreted from the goblet cells during mucus extrusion, a process that appears to be under cholinergic and β-adrenergic control (SMITH and BUTLER, 1974).

IV. Intestinal Secretion

Like the epithelium of the small intestine, the colonic epithelium is capable of net plasma-to-lumen electrolyte and fluid transport (secretion) in addition to the movements of ions associated with absorption, as noted above (FLOREY et al., 1941; HENDRIX and BAYLESS, 1970; FIELD, 1974; KIMBERG, 1974). Various reported stimuli of colonic secretion, many identical to those affecting the small intestine, are summarized in Table 3. The mechanism of the secretory response to each of these stimuli remains to be determined, particularly *in vivo*, where changes in passive permeability and hydrostatic pressure may play an important role. However, in instances where these stimuli have been studied *in vitro*, although changes in passive permeability may be present, it appears that active electrolyte transport is a major cause of the secretory phenomenon (ARCHAMPONG et al., 1972; BRIGHT-ASARE and BINDER, 1973). In at least four of the stimuli listed in Table 3, bile acids, dioctyl sodium sulfosuccinate (DSS), cholera toxin, and vasoactive intestinal polypeptide, the process seems to be mediated by cyclic AMP (BINDER and RAWLINS, 1973b; YAU and MAKHLOUF, 1974; BINDER et al., 1975; DONOWITZ and BINDER, 1975; DONOWITZ and BINDER, 1976; RACUSEN and BINDER, 1977). These agents increase colonic cyclic AMP levels and all affect *in-vitro* rat colonic electrolyte transport in a fashion similar to that seen with theophylline and dibutyryl cyclic AMP. Bile acids and DSS differ from cyclic AMP in that they increase the transepithelial conductance and thus increase the unidirectional fluxes of all ions. Inhibition of bile acid-stimulated colonic adenylate cyclase and secretion by β-adrenergic blocking agents has been reported, but the dose of blocking agent required and inability to demonstrate a role for catecholamines in the secretory process suggests that this is a direct, membrane effect of the drug, rather than blockade of adrenergic receptors (COYNE et al., 1976; CONLEY et al., 1976; TAUB et al., 1977).

Cyclic AMP affects electrolyte transport differently in rabbit and rat colon. In

Table 3. Stimuli of colonic water and electrolyte secretion

Stimulus and Species	*References*
1. Spontaneous secretion (dog)	1,2
2. Cholinergic nerve stimulation and cholinergic drugs (dog, cat)	3
3. Saline adaptation (toad)	4
4. Hormones	
Cyclic AMP and theophylline (rat, rabbit)	15, 17, 29
Glucagon (dog)	5
? Secretin (man)	6
Cholera toxin (rat)	30
Vasoactive intestinal polypeptide (rat)	31
5. Laxatives	
Bisacodyl (dog, man)	7, 8, 9
Oxyphenisatin (dog, rat)	7, 10
Long-chain fatty acids and hydroxy fatty acids, *e. g.* ricinoleic acid castor oil (rat, man)	11, 12
Dihydroxy bile acids (rat, man)	13, 14, 15, 16, 17, 18
Dioctyl sodium sulfosuccinate (rat)	19
6. Colonic tumors such as villous adenomas (man)	20, 21, 22
7. Inflammatory bowel disease such as ulcerative colitis or regional enteritis (man)	23, 24, 25, 26, 27
8. Increased intraluminal pressure (rat)	28

[1] COOPERSTEIN and BROCKMAN, 1959; [2] CLARKE et al., 1967; [3] FLOREY et al., 1941; [4] FERRIERA and SMITH, 1968; [5] BARBEZAT, 1973; [6] JOHANSEN et al., 1968; [7] FORTH 1966; [8] EWE, 1972; [9] EWE and HÖLKER, 1974; [10] NELL et al., 1973; [11] BRIGHT-ASARE and BINDER, 1973; [12] AMMON and PHILLIPS, 1973a; [13] MEKHJIAN and PHILLIPS, 1970; [14] MEKHJIAN et al., 1971; [15] BINDER et al., 1975; [16] WINGATE et al., 1973; [17] BINDER and RAWLINS, 1973b; [18] SCHWITER et al., 1975; [19] DONOWITZ and BINDER, 1975; [20] DUTHIE and ATWELL, 1963; [21] SHIELDS, 1966; [22] DACRUZ et al., 1968; [23] HEAD et al., 1969; [24] HARRIS and SHIELDS, 1970; [25] ARCHAMPONG et al., 1972; [26] EDMONDS and PILCHER, 1973; [27] RASK-MADSEN and JENSEN, 1973; [28] BURY et al., 1974; [29] FRIZZELL and SCHULTZ, 1977; [30] DONOWITZ and BINDER, 1976; [31] RACUSEN and BINDER, 1977.

rabbit, cyclic AMP stimulates net chloride secretion, abolishes the residual ion flux (bicarbonate secretion), and increases the I_{sc} (FRIZZELL et al., 1976). Under these conditions, the I_{sc} can be accounted for by net sodium absorption plus chloride secretion. With reference to Fig. 1, the effects of cyclic AMP on transport in the rabbit colon are explicable on the basis of cyclic AMP operating on pump *C* in such a fashion as to interrupt the chloride : bicarbonate exchange and cause the electrogenic secretion of chloride either by converting pump *C* from a neutral anion exchange process to an electrogenic anion secretory mechanism, or by stimulating a new chloride secretory pump.

In contrast, in the rat colon, theophylline and dibutyryl cyclic AMP reduce net sodium transport to zero in addition to stimulating chloride secretion, abolishing the residual flux, and increasing the I_{sc} (BINDER and RAWLINS, 1973b; BINDER et al., 1975). Thus, cyclic AMP in rat appears to inhibit neutral sodium chloride transport in addition to converting chloride : bicarbonate exchange into electrogenic chloride secretion. In Fig. 1, cyclic AMP could be considered as exerting its effect on *B* as well as on *C*. There is a precedent for such an action on coupled NaCl transport – the rat ileum, where cyclic AMP inhibits coupled sodium chloride influx into the cell (NELLANS et al., 1974). The localization of adenylate cyclase in the brush border of the colonic epithelium, as well as along the lateral cell membranes (CORRIVEAU and ROJO-ORTEGA, 1975), would also be consistent with this hypothetical scheme.

Recently, FRIZZELL (1977) has demonstrated that the calcium ionophore A23187 mimics the effect of cyclic AMP on electrolyte transport in rabbit colon. The ionophore had no effect on cyclic AMP levels. Further, addition of cyclic AMP to colonic mucosa preloaded with ^{45}Ca caused an increased efflux of Ca from the tissue. FRIZZELL has thus suggested that the secretory mechanisms in the colonic cell are stimulated by increases in intracellular calcium, and that cyclic AMP may release calcium from intracellular stores for this purpose.

V. Hormones and Electrolyte Transport

Investigation of the effects of hormones on colonic epithelial transport appears to be complicated by erratic responses, which may be due to the pre-study hormonal status of the animal and perhaps also to species differences. There is almost uniform agreement that aldosterone alters colonic transport, but the responses to antidiuretic hormone and angiotensin are less clear (see below). Even less is known about the effects of other hormones. For example, glucagon stimulates colonic secretion in dog (BARBEZAT, 1973), but not in rat (YAU and MAKHLOUF, 1974); there is morphologic evidence of colonic secretion in response to secretin in man (JOHANSEN et al., 1968), but no physiological evidence of altered transport in dog (BARBEZAT, 1973); and finally, attempts to demonstrate a significant effect of gastrin or prostaglandins on colonic transport have been unrewarding (YAU and MAKHLOUF, 1974; MILTON-THOMPSON et al., 1975). Recent work suggests that the colon secretes in response to vasoactive intestinal polypeptide (RACUSEN and BINDER, 1977).

The colons of several, but not all, species have been shown to respond to aldosterone, or other salt-retaining adrenal hormones, with an increase in water, sodium, and chloride absorption, usually an increase in potassium secretion, and a significant increase in the transepithelial PD (BERGER et al., 1960; BENTLEY, 1962; COFRÉ and CRABBÉ, 1965, 1967; LEVITAN and INGELFINGER, 1965; ELMSLIE et al., 1966; SHIELDS et al., 1966; LEVITAN, 1967; EDMONDS, 1967a–c, 1969; EDMONDS and MARRIOTT, 1967, 1968a and b, 1970; FERREIRA and SMITH, 1968; BARNABY and EDMONDS, 1969; DOLMAN and EDMONDS, 1975). The colon apparently does not "escape" from the effect of continued aldosterone stimulation (EDMONDS and RICHARDS, 1970), and this organ has been shown to contribute to sodium and water conservation *via* the aldosterone effect (RUBENS and LAMBERT, 1972).

When administered *in vitro*, a 30–90 minute lag period follows addition of the hormone to the serosal solution before there is any effect. Aldosterone then stimulates net sodium transport, with an equivalent increase in the I_{sc} (COFRÉ and CRABBÉ, 1965, 1967; FERREIRA and SMITH, 1968; DAWSON and CURRAN, 1976; FRIZZELL and SCHULTZ, 1978). *In vivo* (open-circuit conditions) this increase in sodium transport is accompanied by increased chloride transport, with little effect on bicarbonate, and an increase in water absorption (LEVITAN and INGELFINGER, 1965; EDMONDS and MARRIOTT, 1967, 1968a and b; DOLMAN and EDMONDS, 1975). The changes in potassium secretion have been discussed above. Whether the stimulation of sodium absorption occurs *via* a change in luminal cell membrane permeability with increased sodium entry into the colonic absorptive cell, or whether aldosterone stimulates the basolaterally located sodium pump remains to be determined in the colon, as in other epithelia. Evidence has been presented that the "sodium transport pool" in toad colonic epithelium increases with aldosterone (COFRÉ and CRABBÉ, 1967), suggesting that the site of action is the luminal cell membrane. As with other studies of this type, the crucial question is whether one has truly defined a transport pool. More direct evidence for a luminal cell membrane site of action has been obtained by FRIZZELL and SCHULTZ (1978). They found in the *in-vitro* rabbit colon that the unidirectional influx of Na from mucosal secretion across the luminal membrane into the cell was increased by aldosterone, and that this stimulated influx accounted for the increased net Na transport and the increase in I_{sc}. The accessibility of colonic tissue does offer another useful epithelium in which to explore the mechanism of aldosterone's action.

The effect of antidiuretic hormone (ADH) is less clear. USSING and co-workers demonstrated a transient increase in short-circuit current on addition of ADH to the bathing solution of the *in-vitro* guinea-pig cecum (USSING and ANDERSEN, 1955). An increase in the PD and an increase in osmotic water flow in toad colon has also been demonstrated by COFRÉ and CRABBÉ (1967), but not by GREEN and MATTY (1966). Rat colon (AULSEBROOK, 1961) and mouse colon (GREEN and MATTY, 1966) apparently increase sodium absorption in response to ADH, although this response has not been uniform (GREEN and MATTY, 1966). Tortoise colon does not respond to ADH (BENTLEY, 1962). *In-vivo* experiments in man have shown the opposite effect – a decrease in sodium, chloride, and water absorption in the colon upon intravenous administration of ADH (LEVITAN and MAUER, 1968), a finding also reported in the small intestine (SOERGEL et al., 1968). Interpretation of these *in-vivo* experiments in man is difficult because of the depressant effect of ADH on mesenteric blood flow.

Even more controversial is the response of colonic epithelium to angiotensin. This hormone has been studied in the rat, where its effects can best be shown in adrenalectomized and nephrectomized animals (PARSONS and MUNDAY, 1972). Angiotensin seems to have a biphasic dose-response, with high concentrations inhibiting and lower concentrations stimulating sodium, chloride, and water absorption *via* an electrically neutral, or nonrheogenic, mechanism by non-cyclic AMP-dependent pathways (DAVIES et al., 1970, 1972; PARSONS and MUNDAY, 1972). Recent studies complicate the story even further, suggesting that angiotensin has different effects on the ascending than on the descending colon of normal rats (HORNYCH et al., 1973), or no significant effect at all *in vivo* (DOLMAN and EDMONDS, 1975).

C. Weak Electrolyte Transport
(Ammonia and Volatile Fatty Acids)

Bacterial hydrolysis of urea and carbohydrates within the colonic lumen creates significant quantities of ammonia[4] and fecal organic anions (volatile fatty acids, VFA)[5], which are available for transport and metabolism by the animal (AOYAGI et al., 1966; EVANS et al., 1966; METCALFE-GIBSON et al., 1967; HOUPT and HOUPT, 1968; RUBINSTEIN et al., 1969; WOLPERT et al., 1970, 1971; WRONG, 1971; CUMMINGS, 1975). The nitrogen from ammonia absorption may re-enter the urea cycle and be used for albumin or body-protein synthesis. In ruminants, absorption of ammonia from the rumen allows the use of nonprotein nitrogen as feed, but in monogastric mammals the usefulness of colonically absorbed ammonia for protein synthesis is less clear (CHALUPA, 1972, 1973; SALTER, 1973). In man, bacterial hydrolysis of urea (O'GRADY, 1966; VINCE et al., 1973) occurs at the rate of 6–9 g urea per day (20% of daily urea synthesis), yielding some 3–4 g potentially usable nitrogen daily (WRONG, 1971; CUMMINGS, 1975). Thus, urea could be an important source of nitrogen in people on low-protein or parenteral diets (RICHARDS et al., 1967; Leading Article, Lancet, 1975), as it appears to be for other monogastric animals (ROGOECZI et al., 1965; WOOTTON and ARGENZIO, 1975). Even if it is of marginal importance nutritionally, colonic ammonia absorption in man is significant with regard to the pathogenesis of hepatic coma (CHALMERS, 1960; SCHENKER et al., 1974).

Similarly, the VFAs play a major role as an energy source in ruminants, where their metabolism may account for 70–80 percent of total energy expenditures (BERGMAN et al., 1965; DOBSON and PHILLIPSON, 1968; STEVENS, 1973). In monogastric animals the role of VFAs in energy metabolism is less clear, but colonic absorption of these substances may provide 10–30 percent of the basal body energy output of herbivorous animals such as rabbit or porcupine, and as much as 5 percent in mice and rats (MCBEE, 1970; YANG et al., 1970; CUMMINGS, 1975). In man the conversion of soluble carbohydrate to VFAs by colonic bacteria and the absorption of these products may be an important mechanism for reducing the osmotic load in the colon and for salvaging the calories of carbohydrate not absorbed by the small intestine (BOND and LEVITT, 1976). If not utilized by the organism as a whole, it is possible that VFAs might be an important metabolic fuel for the colonic mucosa itself. VFAs are metabolized by the rumen (STEVENS, 1973) and the small intestine (BARRY and SMYTH 1960), and their metabolism by the large intestine of both pony and pig has been demonstrated (ARGENZIO and SOUTHWORTH, 1974; ARGENZIO et al.,

4 In this review the term ammonia or T_{NH_3} will be used to indicate the total ammonium ion-free ammonia system ($NH_4^+ - NH_3$). Ammonium or ammonium ion will refer to NH_4^+, and free ammonia or gaseous ammonia to NH_3.

5 Fecal organic anion is defined as the titratable acidity of stool between the pH levels 2.7 and 8.0, except that due to bicarbonate and phosphate. Fifty percent of this organic anion in man is composed of short-chain fatty acids. These steam-volatile fatty acids (VFAs) are usually defined as those aliphatic, monocarboxylic acids that are moderately polar and water-soluble, and have chain lengths less than six carbon atoms.

1974b). As with the rumen (FERREIRA et al., 1966), VFAs will support water (and solute) transport by *in-vitro* rat colon, but not by small intestine (PARSONS and PATERSON, 1965; BINDER, 1977).

In the monogastric mammal the concentration of VFAs range from 50–250 mM per liter of colon contents. The predominant acid is acetate, constituting 50–70 molar percent of the total, with propionate (20–35%) and butyrate (5–15%) making up the majority of the remainder (WRONG et al., 1965; RUBINSTEIN et al., 1969; ARGENZIO et al., 1974b; ARGENZIO and SOUTHWORTH, 1974). In man an additional 3 percent or so of the short-chain acids are lactate, fumarate, formate, succinate and valerate (RUBINSTEIN et al., 1969).

I. Nonionic Diffusion

An understanding of colonic ammonia and VFA transport requires familiarity with nonionic diffusion as it relates to the epithelial transport of weak electrolytes. There are several excellent general discussions of weak acid and weak base transport (JACOBS, 1940; SHORE et al., 1957; SCHANKER et al., 1958; MILNE et al., 1958; HOGBEN et al., 1959; JACKSON et al., 1974; JACKSON and MORGAN, 1975), and more specifically of ammonia transport (JACOBS, 1940; ORLOFF and BERLINER, 1956; MANNING, 1964), and VFA transport (JACKSON, 1974).

On the basis of concepts developed by OSTERHOUT (1925), TEORELL (1933), and INGRAHAM and VISSCHER (1935), JACOBS (1940) formalized the following equations relating to weak electrolyte transport:

$$\text{(a)}\quad NH_4^+ \rightleftharpoons NH_3 + H^+ \qquad \text{(b)}\quad HAc \rightleftharpoons H^+ + Ac^-; \qquad (1)$$

$$\text{(a)}\quad \frac{[NH_3]\,[H^+]}{[NH_4^+]} = Ka \qquad \text{(b)}\quad \frac{[H^+]\,[Ac^-]}{[HAc]} = Ka; \qquad (2)$$

$$\text{(a)}\quad pH = pKa + \log\frac{NH_3}{NH_4^+} \qquad \text{(b)}\quad pH = pKa + \log\frac{Ac^-}{HAc}\,; \qquad (3)$$

$$\text{(a)}\quad NH_4^+ = NH_3 \cdot 10^{(pKa-pH)} \qquad \text{(b)}\quad Ac^- = HAc \cdot 10^{(pH-pKa)}. \qquad (4)$$

Equation (1) describes the two forms in which the weak base (NH_3) or the weak acid (HAc) exist at the physiological pH, and Eq. (2) indicates that a dissociation constant can be calculated for this relationship. These systems can be further described by the Henderson-Hasselbalch equation (3), where the pKa for the ammonia system is approximately 9.0 (BANK and SCHWARTZ, 1960; BROMBERG et al., 1960) and the pKa for C_{2-6} acids approximately 4.8 (WEAST, 1971). Equation (4) is simply another form of (3), which points out that at pH 8 only one out of every 10 molecules of the weak base exists as gaseous ammonia (NH_3) and one out of every 1000 molecules of the weak acid exists as undissociated acid (HAc). At pH 6, the proportion of NH_3 decreases to one out of every 1000 molecules, and that of HAc increases to one out of 10.

Theoretically, the transepithelial flux of a weak electrolyte can be described as (JACKSON et al., 1974):

$$J_{ms} = J_{ms}^{ni} + J_{ms}^{i} \text{ and } J_{sm} = J_{sm}^{ni} + J_{sm}^{i}, \tag{5}$$

where J_{ms} and J_{sm} are the unidirectional fluxes from the mucosal to serosal or from the serosal to mucosal sides of the epithelium, and J^{ni} and J^{i} refer to the fluxes of the nonionized and ionized forms of the weak electrolytes respectively.

Since

$$J_{ms}^{ni} = P_{ni}[NI_m] \quad \text{and} \quad J_{sm}^{ni} = P_{ni}[NI_s] \tag{6a}$$

and

$$J_{ms}^{i} = \frac{P_i[I_m]\xi}{[1 - e^{-\xi}]} \quad \text{and} \quad J_{sm}^{i} = \frac{P_i[I_s]\xi e^{-\xi}}{[1 - e^{-\xi}]}, \tag{6b}$$

where P_{ni} and P_i are the transepithelial permeabilities of the nonionized and ionized forms; respectively: $[NI_m]$, $[NI_s]$, $[I_m]$, and $[I_s]$ are the concentrations of the nonionized and ionized forms in the m and s compartments; and $\xi = zF(\Psi_m - \Psi_s)/RT$, in which z is the valence of the ionized form, F is the Faraday, $\Psi_m - \Psi_s$ is the transepithelial potential difference and R and T have their usual meaning, it can be shown (JACKSON et al., 1974) that the flux ratio equation becomes:

$$J_{ms}/J_{sm} = \frac{1 + \left[\frac{P_i}{P_{ni}}\right]\left[\frac{\xi 10^{\alpha}}{1 - e^{-\xi}}\right]}{1 + \left[\frac{P_i}{P_{ni}}\right]\left[\frac{\xi e^{-\xi 10\alpha}}{1 - e^{-\xi}}\right]} \tag{7}$$

where $\alpha = (\mathrm{pH} - \mathrm{pKa})$ for weak acids and $(\mathrm{pKa} - \mathrm{pH})$ for weak bases. If $P_{ni} >>> P_i$, then Eq. (7) reduces to (JACOBS, 1940; SHORE et al., 1957; JACKSON et al., 1974):

$$J_{ms}/J_{sm} = \frac{1 + 10^{\alpha s}}{1 + 10^{\alpha m}}. \tag{8}$$

Thus, at the heart of this question is whether the epithelial permeability to the free base (NH_3) or the free acid (HAc), which are noncharged and lipid-soluble, is indeed far greater than the charged, hydrated, water-soluble NH_4^+ or Ac^- form of the electrolyte, as is generally assumed for all lipid membranes and observed whenever measured. In the colon the $P_{NH_3}/P_{NH_4^+}$ is at least 5:1 (CASTELL and MOORE, 1971; BOWN et al., 1975); and although not determined in the colon, the P_{HAc}/P_{Ac^-} in the small intestine is also large (SALLEE and DIETSCHY,

1973; JACKSON et al., 1974). Therefore, transport of these weak electrolytes, if by passive mechanisms, would presumably occur *via* nonionic diffusion and be pH-dependent. For example, the distribution of T_{NH_3} across the epithelium would depend on the pH on each side of the epithelium. NH_3 would pass from the compartment of higher pH (alkaline) to the side of lower pH (acid) even if T_{NH_3} was equal on the two sides of the epithelium at the start. At equilibration, the nonionized NH_3 would be the same on both sides of the epithelium, but the T_{NH_3} on each side would be determined by the pH (see Fig. 2a). The same considerations apply to weak acid systems, except the direction of net transport is the opposite, from acid to alkaline.

II. Colonic Transport

What then is the evidence that ammonia and VFA transport occurs by nonionic diffusion? In spite of the potential importance of VFA transport by mammalian colon, there are few reports concerned with VFA absorption by the large intestine. VFAs appear to be rapidly absorbed from the large intestine of several animal species (ARGENZIO et al., 1974; ARGENZIO and SOUTHWORTH, 1974; ARGENZIO et al., 1975; BOND and LEVITT, 1976). VFAs are also absorbed by the human colon and the rates are directly proportional to chain length and chloroform : water partition coefficient (DAWSON et al., 1966). This direct relationship between fatty acid chain length and its passive permeability coefficient has been confirmed in the rat jejunum (SALLEE and DIETSCHY, 1973) and human small intestine (SCHMITT et al., 1976). Additional studies of colonic VFA transport are crucial.

With regard to the colonic transport of ammonia, available evidence indicates that it is due primarily to nonionic diffusion. First, ammonia transport appear to be pH-dependent. A comparison of colonic ammonia absorption at two different pH levels in human (CASTELL and MOORE, 1971; BOWN et al., 1975) and in rat (SWALES et al., 1970; WRONG, 1971) demonstrated greater transport from solutions at pH 7–9 than from solutions at pH 5–6. Secondly, from considerations similar to Equation 8, a relationship with a slope of -1 can be predicted between the log $[T_{NH_3}]$ and pH for the urinary excretion of ammonia (ORLOFF and BERLINER, 1956). The necessary conditions for this are that the epithelium be impermeable to the ionized form, NH_4^+, and that complete equilibrium of NH_3 be achieved. Applying this to colonic ammonia transport, DOWN et al. (1972) described a negative relationship between the log $[T_{NH_3}]$ and pH of fecal dialyzate, and attributed the fact that the slope was not unity to a failure to meet the mandatory conditions mentioned above.

There is a special type of transport, which might be termed "paired" or "coupled" nonionic diffusion (ROSENFELD et al., 1963), which seems to apply to ammonia transport by the colon (Fig. 2). This system works because the dissociation products of NH_4^+ and HCO_3^- (H^+ and OH^-) neutralize each other and thus the process of weak electrolyte dissociation does not alter the mucosal solution pH. As a consequence, the dissociation reactions can proceed to the right,

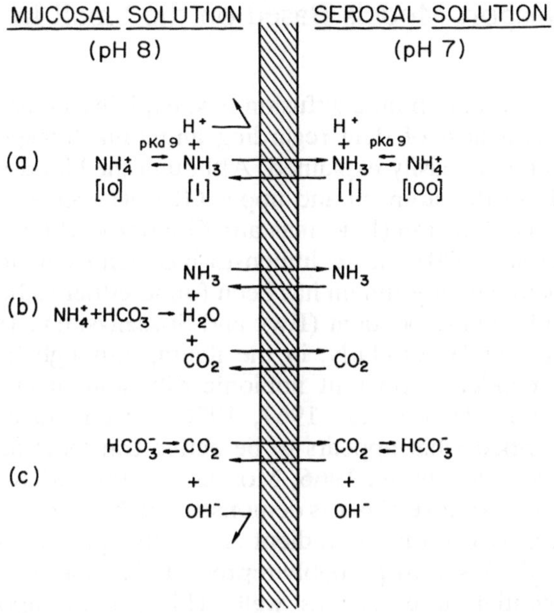

Fig. 2. (a) Ammonia transport by nonionic diffusion is somewhat self-limiting because the trapped hydrogen ion decreases the pH in mucosal solution, thus inhibiting the dissociation of ammonium. However, with a mucosal solution-serosal solution pH difference of one unit, if one assumes an epithelium that is impermeable to hydrogen ion and freely diffusable to NH_3, at equilibrium there will be a 10-fold accumulation of total ammonia on the serosal side of the membrane. (c) CO_2 created from dissociation of bicarbonate ion will freely diffuse across epithelium, but this diffusion will also be limited because of resulting pH change. The hydroxyl ion left behind in the mucosal solution will increase mucosal pH, thus limiting CO_2 formation from bicarbonate. (b) When ammonium ion and bicarbonate ion are present together in same solution, and particularly when available together in the unstirred layer next to colonic epithelium, NH_3 and CO_2 formed will freely diffuse across epithelium and no self-limiting pH change in microclimate will result. (Adapted from SWALES et al., 1970; PRICE et al., 1970; WRONG, 1971. Based on the concepts presented by ORLOFF and BERLINER, 1956 and ROSENFELD et al., 1963)

creating more NH_3 and CO_2 which can diffuse across the epithelium. Thus, the colon, which is one of the few locations where high concentrations of ammonia and bicarbonate co-exist, is ideally suited for such transport. In this scheme, bicarbonate secretion by the colon would have a dual function in increasing ammonia absorption: (1) creating an alkaline mucosal solution and (2) providing a co-substance for "paired diffusion". Indirect evidence that paired diffusion occurs in the gastrointestinal (GI) tract is suggested by the finding that ammonia transport in hamster ileum *in vitro* is dependent on mucosal bicarbonate (MOSSBERG, 1967), and the presence of luminal ammonia prevents the detection of bicarbonate secretion, perhaps by efficiently promoting bicarbonate reabsorption, in the vascularly perfused rat colon (PARSONS and POWIS, 1971). More directly, it has been shown in both dog urinary bladder (ROSENFELD et al., 1963) and mammalian colon (SWALES et al., 1970; WRONG, 1971) that the presence of bicarbonate will increase ammonia absorption, and the presence of ammonia will increase bicarbonate absorption (or decrease secretion) from these organs.

III. Transport Models Based on the Small Intestine

Although it appears that nonionic diffusion best explains ammonia transport by the colon, there is a host of data regarding ammonia transport in the small intestine which is not so easily explained. Although the blood-to-lumen movement of ammonia in the stomach and upper GI tract seems to be due to pH-dependent, nonionic diffusion (FLESHLER and GABUZDA, 1965; SUMMERSKILL et al., 1966; PRICE et al., 1970), the net lumen-to-blood movement (absorption) of ammonia from mammalian jejunum has been found either to be (SWALES et al., 1970) and not to be pH-dependent (EWE and SUMMERSKILL, 1965; KETTERING and SUMMERSKILL, 1967). Similarly, in the ileum, although there are several reports of pH-dependent, apparent nonionic diffusion of ammonium in the human and hamster (PRICE et al., 1967, 1970; SWALES et al., 1973), other reports have described what appears to be active transport of ammonium by hamster ileum (MOSSBERG et al., 1966; MOSSBERG, 1967; MOSSBERG and ROSS, 1967). The last group of investigators demonstrated the serosal accumulation of ammonia against a concentration gradient and, although they did not measure the intraluminal PD, this most probably represented ammonia transport against an electrical potential difference as well. This process appeared to obey Michaelis-Menten kinetics, and was inhibited by anaerobiosis and by metabolic inhibitors such as CN and DNP. Even with the creation of adverse pH gradients in the serosal solution, active accumulation of ammonia could be demonstrated. However, as pointed out below, the interpretation of these experiments in gut sacs could be complicated by the presence of unstirred layers.

In addition to evidence supporting "active" ammonia transport by the hamster ileum, there is evidence for active VFA transport in rat jejunum. Although the peculiar phenomenon of more rapid absorption of even-chain than odd-chain fatty acids (DEUELL et al., 1941) has not been confirmed, accumulation of C_{2-4} fatty acids against a concentration gradient has been demonstrated (BARRY and SMYTH, 1960). This accumulation was inhibited by anaerobiosis and by DNP (SMYTH and TAYLOR, 1958; BARRY and SMYTH, 1960). In addition, it has been shown that VFA transport appears to be sodium-dependent (although it does not promote electrogenic sodium transport (JACKSON and SMYTH, 1968)), obeys Michaelis-Menten kinetics, and is inhibited by other fatty acids (BARRY and SMYTH, 1960). VFA transport seems to be species-specific in that propionate is transported by the rat jejunum, but not by the jejunum of hamster or guinea-pig (JACKSON, 1974), and intestinal segment-specific in that it is transported by the rat jejunum but not by rat ileum (CLARKSON et al., 1961). Furthermore, BARRY et al. (1966) have shown that transport of VFAs cannot be accounted for by solvent drag and appears to take place against an electrochemical potential difference, a finding that has been confirmed by applying USSING's flux ratio equation (JACKSON, 1974) to the data of others (CLARKSON et al., 1961). BARRY et al. (1966) went a step further in showing transfer of VFAs against an adverse pH gradient. These findings, many of which have been reproduced for C_8 fatty acids (BLOCH et al., 1972), all suggest active transport, and indicate that VFA transport across intestine does not occur by simple passive diffusion.

Recently, investigation of monocarboxylic acid transport in isolated rat jejunal cells showed competitive inhibition, countertransport effects, and a high temperature coefficient, which indicated carrier-mediated exchange diffusion of VFAs in these cells (LAMERS, 1975). However, this anionic exchange diffusion process could just as well be located on the basolateral cell membrane of these isolated cells and this carrier might have little to do with transepithelial transport or VFA. When the experiments are designed to study transport across only the luminal cell membrane, the unidirectional, cellular uptake of fatty acid is linear rather than saturating; there is no competition for uptake by other fatty acids; and there is no effect of temperature on uptake (SALLEE and DIETSCHY, 1973; JACKSON, 1974). In addition, JACKSON (1974) has shown some dependence of transport on the pH of the mucosal solution, a point in favor of nonionic diffusion. Finally, hyperpolarization or depolarization of the epithelium had no effect on the flux ratio of weak acids or weak bases, which is inconsistent with significant permeability of the ionic species (JACKSON et al., 1974).

In a recent review, JACKSON (1974) has indicated a way to account for the seemingly active transport of ammonia in the ileum and weak organic acids in the jejunum. All of these phenomena can be explained by a model of weak electrolyte transport that depends on the presence of three compartments, with the middle compartment existing at a pH different from the others and separated by barriers with dissimilar ionic permeabilities (Fig. 3). Here, the pH gradient provides the driving force for solute movement, and the barrier permeabilities dictate the direction of transport. This model, which was first devised by HOGBEN (1955), has been used to explain weak electrolyte transport in the intestine, primarily as it relates to drugs (SHORE et al., 1957; SCHANKER et al., 1958; HOGBEN et al., 1959), and has been used to explain fatty acid transport in the rumen (ASH and DOBSON, 1963; STEVENS et al., 1969; STEVENS, 1973) and DMO transport by small intestine (DIETSCHY and CARTER, 1965; BUTLER, 1966). As proposed originally, the middle compartment has a pH lower than that of the bulk phase of either the mucosal or serosal compartments, and the barrier facing the mucosa is equally permeable to both the ionized and the unionized species of the weak electrolyte in question. However, the recent model proposed by JACKSON (JACKSON, 1974; JACKSON et al., 1974; JACKSON and MORGAN, 1975) indicates that an intermediate compartment of higher pH, in which the serosally facing membrane is permeable to both species, would also account for weak electrolyte transport. JACKSON also indicates the possible relationship of these models to the "acid-base metabolism" of the intestine. Accordingly, the jejunum of rat and man (but not hamster or guinea-pig) "absorbs" bicarbonate by virtue of hydrogen ion secretion into the luminal solution, while the ileum "secretes" bicarbonate either by bicarbonate or hydroxyl ion secretion or by hydrogen ion absorption (WILSON and KAZYAK, 1957; PARSONS, 1956; TURNBERG et al., 1970; POWELL et al., 1971; HUBEL, 1973, 1974). These directions of transport, in which acid goes to one side of the epithelium and base to the other, could create, in unstirred layers or regions, intermediate compartments of differing pH from the bulk solutions on either side of the epithelium. Unstirred layers can certainly account for weak electrolyte transport, as elegantly demonstrated in lipid-membrane systems (GUTKNECHT and TOSTESON, 1973). This mid-

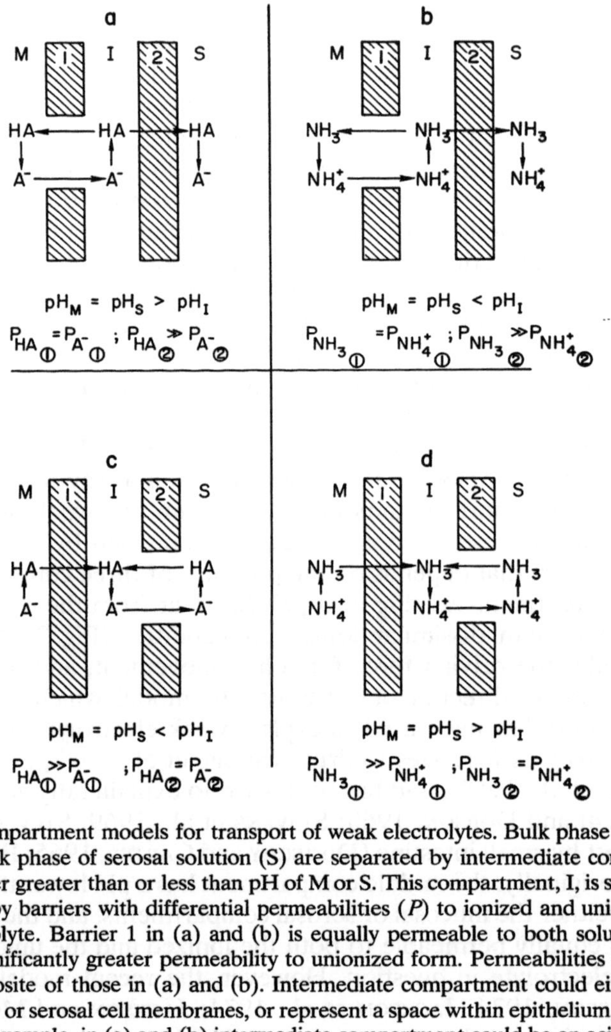

Fig. 3. Three-compartment models for transport of weak electrolytes. Bulk phase of mucosal solution (M) and bulk phase of serosal solution (S) are separated by intermediate compartment (I) in which pH is either greater than or less than pH of M or S. This compartment, I, is separated from M and S solutions by barriers with differential permeabilities (*P*) to ionized and unionized species of the weak electrolyte. Barrier 1 in (a) and (b) is equally permeable to both solute species, while barrier 2 has significantly greater permeability to unionized form. Permeabilities of barriers in (c) and (d) are opposite of those in (a) and (b). Intermediate compartment could either be unstirred layers at mucosal or serosal cell membranes, or represent a space within epithelium, such as intercellular space. For example, in (a) and (b) intermediate compartment could be an acidic or an alkaline unstirred layer adjacent to mucosal cell membrane as result of mucosal acidification in jejunum (a) or mucosal alkalinization in colon (b). Conversely, intermediate compartment in (c) and (d) could represent an alkaline intercellular compartment in jejunum (c) and an acidic intercellular compartment in colon (d), as result of different "acid-base" metabolism of these segments of intestine. (Adapted from JACKSON, 1974; JACKSON et al., 1974; JACKSON and MORGAN 1975)

dle compartment was thought initially to be the unstirred layer at the mucosal surface (HOGBEN, 1955). Indeed, LUCAS et al. (1975) have reported some direct evidence, from pH-microelectrode studies, for such a low-pH microclimate in rat jejunum. However, the intermediate compartment need not necessarily be a luminal or serosal unstirred layer, but could be the intercellular compartment (JACKSON, 1974; JACKSON et al., 1974; JACKSON and MORGAN, 1975). And, as noted above, JACKSON has presented data indicating that the intermediate compartment is of high pH rather than low pH. In any case, either of the models presented in Figure 3 could account for the apparent active transport of VFA and ammonia by the small intestine, and these concepts could apply to the colon as well.

It has also been shown that in the ileum weak acids are preferentially transported from s to m, and weak bases are preferentially transported from m to s (JACKSON and MORGAN, 1975). Since the colon is a bicarbonate-secreting organ like the ileum, one might expect this to occur in colon as well. Thus, one would predict efficient and what might appear to be active transport of weak bases (ammonia), but not weak acids (VFAs) in mammalian colon at the usual colonic pH of 6–8. This would not preclude net absorption of VFA by the colon if the intraluminal concentrations were high enough. In addition, if the mechanism of HCO_3 secretion by the colon were one of CO_2 hydration in the luminal unstirred layer with subsequent dissocation and H^+ absorption, the H^+ thus formed could generate the lipid-soluble, undissociated VFA, which would be rapidly absorbed (ARGENZIO et al., 1976). Thus, as is the case with NH_3 absorption, colonic VFA absorption could be intimately related to H^+–HCO_3^- transport by this epithelium. Finally, these models point out also that experimental artifacts could arise if there is inadequate solution stirring. Poor stirring could create unstirred-layer compartments of differing pH, whose effects might be superimposed on any truly physiologically important third compartment.

D. Conclusion

The present author has attempted to synthesize current knowledge of colonic water and electrolyte transport in order to present, where possible, models that explain these facts within the framework of our understanding of epithelial transport processes. These models are often based on concepts developed from studies of small intestine, and are presented more as ideas that can be tested with current techniques rather than as proven entities. Although many of the questions raised in this review are fundamental to our understanding of epithelial transport in general, many questions raised regarding colinic transport are unique to this epithelium. For instance, an additional intriguing question is the relationship between the high intraluminal bacterial population, with all of its metabolic machinery, and the transport mechanisms of this organ. Is the bicarbonate-rich, alkaline milieu of this tight epithelium promoted by other regulatory factors in order to encourage bacterial growth, or do the bacteria, with their cyclic nucleotide-stimulating toxins and other metabolic activities, in some way exert control over water and electrolyte transport in this epithelium? Differences in electrolyte transport and in the electrical parameters of the colon of germ-free animals favor the latter hypothesis (ASANO, 1969; NAKAMURA and GORDON, 1973; SIMONETTA et al., 1975). This is but one of many unanswered issues regarding this interesting epithelium.

Acknowledgements

The author would like to thank *Kathy Dodson* for her invaluable help in preparing this manuscript, and Drs. *Michael Jackson, Raymond Frizzell,* and *Henry Binder* for useful discussions.

References

AMMON, H. V., PHILLIPS, S. F.: Gastroenterology **65**, 744 (1973).
AOYAGI, T., ENGSTROM, G. W., EVANS, W. B., SUMMERSKILL, W. H. J.: Gut **7**, 631 (1966).
ARCHAMPONG, E. Q., HARRIS, J., CLARK, C. G.: Gut **13**, 880 (1972).
ARGENZIO, R. A., SOUTHWORTH, M.: Amer. J. Physiol. **228**, 454 (1974).
ARGENZIO, R. A., LOWE, J. E., PICKARD, D. W., STEVENS, C. E.: Amer. J. Physiol. **226**, 1035 (1974a).
ARGENZIO, R. A., SOUTHWORTH, M., STEVENS, C. E.: Amer. J. Physiol. **226**, 1043 (1974b).
ARGENZIO, R. A., MILLER, N., von ENGELHARDT, W.: Amer. J. Physiol. **229**, 997 (1975).
ARGENZIO, R. A., SOUTHWORTH, M., LOWE, J. E., STEVENS, C. E.: Amer. J. Physiol. **233**, E469 (1977).
ASANO, T.: Proc. Soc. exp. Biol. **131**, 1201 (1969).
ASH, R. W., DOBSON, A.: J. Physiol. **169**, 39 (1963).
AULSEBROOK, K. A.: Endocrinology **68**, 1063 (1961).
BAILLIEN, M., SCHOFFENIELS, E.: Biochim. biophys. Acta **53**, 521 (1961a).
BAILLIEN, M., SCHOFFENIELS, E.: Biochim. biophys. Acta **53**, 537 (1961b).
BANK, N., SCHWARTZ, W. B.: J. appl. Physiol. **15**, 125 (1960).
BARBEZAT, G. O.: Scand. J. Gastroent. **8**, 1 (Suppl. 22) (1973).
BARNABY, C. F., EDMONDS, C. J.: J. Physiol. **205**, 647 (1969).
BARRY, R. J. C., EGGENTON, J.: Biomembranes **4B**, 917 (1974).
BARRY, R. J. C., SMYTH, D. H.: J. Physiol. **152**, 48 (1960).
BARRY, R. J. C., JACKSON, M. J., SMYTH, D. H.: J. Physiol. **182**, 150 (1966).
BASTL, C., HAYSLETT, J. P., BINDER, H. J.: Kidney International **12**, 916 (1977).
BEHAR, J.: Amer. J. Physiol. **227**, 334 (1974).
BENTLEY, P. J.: Gen. comp. Endocrinol. **2**, 323 (1962).
BENTLEY, P. J., SMITH, M. W.: J. Physiol. **249**, 103 (1975).
BERGER, E. Y., KANZAKI, G., STEEL, J. M.: J. Physiol. **151**, 352 (1960).
BERGMAN, E. N., REID, R. S., MURRAY, M. G., BROCKWAY, J. M., WHITELAW, F. G.: Biochem. J. **97**, 53 (1965).
BIEBERDORF, F. A., GORDEN, P., FORDTRAN, J. S.: J. clin. Invest. **51**, 1958 (1972).
BILLICH, C. O., LEVITAN, R.: J. clin. Invest. **48**, 1336 (1969).
BINDER, H. J.: Biochim. biophys. Acta **219**, 503 (1970).
BINDER, H. J.: Gastroenterology **67**, 441 (1974).
BINDER, H. J.: Gastroenterology **70**, A6/864 (1976).
BINDER, H. J.: In: Coupled Transport Phenomena in Cells and Tissues. (I. F. Hoffman, Ed.). New York: Raven Press 1977, p. 309.
BINDER, H. J., RAWLINS, C. L.: Amer. J. Physiol. **225**, 1232 (1973a).
BINDER, H. J., RAWLINS, C. L.: J. clin. Invest. **52**, 1460 (1973b).
BINDER, H. J., FILBURN, C., VOLPE, B. T.: Gastroenterology **68**, 503 (1975).
BINDSLEV, N., SKADHAUGE, E.: J. Physiol. **216**, 735 (1971a).
BINDSLEV, N., SKADHAUGE, E.: J. Physiol. **216**, 753 (1971b).
BLOCH, R., HABERICH, F. J., LORENZ-MEYER, H.: Pflügers Arch. **335**, 198 (1972).
BOND, J. H., Jr, LEVITT, M. O.: J. clin. Invest. **57**, 1158 (1976).

BOWN, R. L., SLADEN, G. E., ROSASSEAU, B., GIBSON, J. A., CLARK, M. L., DAWSON, A. M.: Clin. Sci. **43**, 891 (1972).
BOWN, R. L., GIBSON, J. A., FENTON, J. C. B., SNEDDEN, W., CLARK, M. L., SLADEN, G. E.: Clin. Sci. molec. Med. **48**, 279 (1975).
BRIGHT-ASARE, P., BINDER, H. J.: Gastroenterology **64**, 81 (1973).
BRODSKY, W. A., SCHILB, J. P.: Fed. Proc. **26**, 1314 (1967).
BROMBERG, P. A., ROBBIN, E. D., FORKNER, C. E., Jr: J. clin. Invest. **39**, 332 (1960).
BURY, K. D., MCCLURE, R. L., WRIGHT, H. K.: Arch. Surg. **108**, 854 (1974).
BUTLER, T. C.: Science **151**, 1564 (1966).
CARLISKY, N. J., LEW, V. L.: J. Physiol. **206**, 529 (1970).
CARTER, M. J., PARSONS, D. S.: J. Physiol. **215**, 71 (1971).
CASSIDY, M. M., TIDBALL, C. S.: Amer. J. Physiol. **217**, 674 (1969).
CASTELL, D. O., MOORE, E. W.: Gastroenterology **60**, 33 (1971).
CHALFIN, D., COOPERSTEIN, I. L., HOGBEN, C. A. M.: Proc. Soc. exp. Biol. **99**, 746 (1958).
CHALUPA, W.: Fed. Proc. **31**, 1152 (1972).
CHALUPA, W.: Proc. Nutr. Soc. **32**, 99 (1973).
CHALMERS, T. C.: New Engl. J. Med. **263**, 23 and 77 (1960).
CHARNEY, A. N., KINSEY, M. D., MYERS, L., GIANNELLA, R. A., GOTS, R. E.: J. clin. Invest. **56**, 653 (1975).
CLARKE, A. M., MILLER, M., SHIELDS, R.: Gastroenterology **52**, 846 (1967).
CLARKSON, T. W., ROTHSTEIN, A., CROSS, A.: Amer. J. Physiol. **200**, 781 (1961).
COFRÉ, G., CRABBÉ, J.: Nature **207**, 1299 (1965).
COFRÉ, G., CRABBÉ, J.: J. Physiol. **188**, 177 (1967).
CONLEY, D., COYNE, M., CHUNG, A., BONORRIO, G., SCHOENFIELD, L.: Gastroenterology **71**, 72 (1976).
COOPERSTEIN, I. L., BROCKMAN, S. K.: J. clin. Invest. **38**, 435 (1959).
COOPERSTEIN, I. L., HOGBEN, C. A. M.: J. gen. Physiol. **42**, 461 (1959).
CORRIVEAU, M., ROJO-ORTEGA, J. M.: Virchows Arch. B. Cell. Path. **18**, 129 (1975).
COYNE, M. J., BONORRIO, G. G., CHUNG, A., CONLEY, D. R., CROKE, J., SCHOENFIELD, L. J.: Gastroenterology **71**, 68 (1976).
CRANE, C. W.: Gut **6**, 439 (1965).
CUMMINGS, J. H.: Gut **16**, 323 (1975).
CURRAN, P. F., SCHWARTZ, G. F.: J. gen. Physiol. **43**, 555 (1960).
DaCRUZ, G. M. G., GARDNER, J. D., PESKIN, G. W.: Amer. J. Surg. **115**, 203 (1968).
D'AGOSTINO, A., LEADBETTER, W. F., SCHWARTZ, W. B.: J. clin. Invest. **32**, 444 (1953).
DALMARK, M.: Scand. J. Gastroent. **5**, 421 (1970).
DAVIES, N. T., MUNDAY, K. A., PARSONS, B. J.: J. Endocrinol. **48**, 39 (1970).
DAVIES, N. T., MUNDAY, K. A., PARSONS, B. J.: J. Endocrinol. **54**, 483 (1972).
DAWSON, A. M., HOLDSWORTH, C. D., WEBB, J.: Proc. Soc. exp. Biol. **117**, 97 (1966).
DAWSON, D. C., BRANDES M. M.: Fed. Proc. **35**, 618 (1976).
DAWSON, D. C., CURRAN, P. F.: J. Membrane Biol. **28**, 295 (1976).
DEUEL, H. J., Jr, HOLLMAN, L., REIFMAN, A.: J. Nutr. **21**, 373 (1941).
DEVROEDE, G. J., PHILLIPS, S. F.: Gastroenterology **56**, 101 (1969).
DEVROEDE, G. J., PHILLIPS, S. F.: Gut **11**, 438 (1970).
DEVROEDE, G. J., PHILLIPS, S. F., CODE, C. F., LIND, J. F.: Canad. J. Physiol. Pharm. **49**, 1023 (1971).
DIAMOND, J. M.: Fed. Proc. **33**, 2220 (1974).
DIAMOND, J. M., BOSSERT, W. H.: J. gen. Physiol. **50**, 2061 (1967).
DIETSCHY, J. M., CARTER, N. W.: Science **150**, 1294 (1965).
DIETZ, J., FIELD, M.: Amer. J. Physiol. **225**, 858 (1973).
DOBBINS, J. W., BINDER, H. J.: Gastroenterology **70**, 1096 (1976).
DOBSON, A., PHILLIPSON, A. J.: In: Handbook of Physiology, Section 6: Alimentary Canal, Vol. 5: Bile; Digestion; Ruminal Physiology (C. F. Code, Ed.). Baltimore, Md: Williams and Wilkins 1968, p. 2761.
DOLMAN, D., EDMONDS, C. T.: J. Physiol. **250**, 597 (1975).
DONOWITZ M., BINDER, H. J.: Gastroenterology **69**, 941 (1975).
DONOWITZ, M., BINDER, H. J.: J. infect. Dis. **134**, 135 (1976).

Down, P. F., Agostini, L., Murison, J., Wrong, D. M.: Clin. Sci. **43**, 101 (1972).
Duthie, H. L., Atwell, J. D.: Gut **4**, 373 (1963).
Edmonds, C. J.: J. Physiol. **193**, 571 (1967a).
Edmonds, C. J.: J. Physiol. **193**, 589 (1967b).
Edmonds, C. J.: J. Physiol. **193**, 603 (1967c).
Edmonds, C. J.: J. Physiol. **203**, 533 (1969).
Edmonds, C. J.: Gut **16**, 315 (1975).
Edmonds, C. J., Marriott, J. C.: J. Endocrinol. **39**, 517 (1967).
Edmonds, C. J., Marriott, J.: J. Physiol. **194**, 457 (1968a).
Edmonds, C. J., Marriott, J.: J. Physiol. **194**, 479 (1968b).
Edmonds, C. J., Marriott, J.: J. Physiol. **210**, 1021 (1970).
Edmonds, C. J., Nielsen, O. E.: Acta physiol. scand. **72**, 338 (1968).
Edmonds, C. J., Pilcher, D.: In: Transport Across the Intestine (A Glaxo Symposium) (W. L. Burland and P. D. Samuel, Eds). Baltimore, Md: Williams and Wilkins 1972, p. 43.
Edmonds, C. J., Pilcher, D.: Gut **14**, 784 (1973).
Edmonds, C. J., Richards, P.: Lancet **1970 2**, 624.
Elmslie, R. G., Mulholland, A. T., Shields, R.: Gut **7**, 697 (1966).
Evans, W. B., Aoyagi, T., Summerskill, W. H. J.: Gut **7**, 635 (1966).
Ewe, K.: Europ. J. clin. Invest. **2**, 283 (1972).
Ewe, K., Hölker, B.: Klin. Wschr. **52**, 827 (1974).
Ewe, K., Summerskill, W. H. J.: J. Lab. clin. Med. **65**, 839 (1965).
Ferreira, H. G., Smith, M. W.: J. Physiol. **198**, 329 (1968).
Ferreira, H. G., Harrison, F. A., Keynes, R. D.: J. Physiol. **187**, 631 (1966).
Field, M.: Gastroenterology **66**, 1063 (1974).
Fisher, K. A., Binder, H. J., Hayslett, J. P.: Amer. J. Physiol. **231**, 987 (1976).
Fleshler B., Gabuzda, G. J.: Gut **6**, 349 (1965).
Florey, H. W., Wright, R. D., Jennings, M. A.: Physiol. Rev. **21**, 36 (1941).
Fordtran, J. S., Rector, F. C., Jr, Newton, M. F., Soter, N., Kinney, J.: J. clin. Invest. **44**, 1935 (1965).
Forth, W.: Ann. Univ. Sarav. Med. **13**, 163 (1966).
Frizzell, R. A.: J. Membrane Biol. **35**, 175 (1977).
Frizzell, R. A., Jennings, B. L.: Fed. Proc. **36**, 300 (1977).
Frizzell, R. A., Schultz, S. G.: J. Membrane Biol. (in press) (1978).
Frizzell, R. A., Turnheim, K.: J. Membrane Biol. **401**, 193 (1978).
Frizzell, R. A., Koch, M. J., Schultz, S. G.: J. Membrane Biol. **27**, 297 (1976).
Giddings, R. F., Argenzio, R. A., Stevens, C. E.: Amer. J. vet. Res. **35**, 1511 (1974).
Giebisch, G.: Yale J. Biol. Med. **48**, 315 (1975).
Giller, J., Phillips, S. F.: Amer. J. dig. Dis. **17**, 1003 (1972).
Gilles-Baillien, M., Schoffeniels, E.: Comp. biochem. Physiol. **23**, 95 (1967a).
Gilles-Baillien, M., Schoffeniels, E.: Arch. Int. Physiol. Biochim. **75**, 754 (1967b).
Goldschmidt, S., Dayton, A. B.: Amer. J. Physiol. **48**, 419 (1919a).
Goldschmidt, S., Dayton, A. B.: Amer. J. Physiol. **48**, 459 (1919b).
Grady, G. F., Duhamel, R. C., Moore, E. W.: Gastroenterology **59**, 583 (1970).
Green, K., Matty, A. J.: Life Sci. **5**, 205 (1966).
Gutknecht, J., Tosteson, D. C.: Science **182**, 1258 (1973).
Hagihara, P. F., Griffen, W. O., Jr: Surg. Clin. N. America **52**, 797 (1972).
Harris, J., Shields, R.: Gut **11**, 27 (1970).
Harrison, H. C., Harrison, H. E.: Amer. J. Physiol. **217**, 121 (1969).
Head, L. H., Heaton, J. W., Jr, Kivel, R. M.: Gastroenterology **56**, 571 (1969).
Heaton, J. W., Jr: Amer. J. dig. Dis. **17**, 7 (1972).
Hendrix, T. R., Bayless, T. M.: Ann. Rev. Physiol. **32**, 139 (1970).
Hogben, C. A. M.: In: Electrolytes in Biological Systems (A. M. Shanes, Ed.). Washington, D. C.: American Physiological Society 1955, p. 176.
Hogben, C. A. M., Tocco, D. J., Brodie, B. B., Schanker, L. S.: J. Pharmacol. exp. Ther. **125**, 275 (1959).
Hollander, D., Truscott, T. C.: J. Lab. clin. Med. **83**, 648 (1974).
Hollander, D., Muralidhara, K. S., Rim, E.: Amer. J. Physiol. **230**, 251 (1976).

HOLMBERG, C., PERHEENTUPA, J., LAUNIALA, K.: J. clin. Invest. **56**, 302 (1975).
HORNYCH, A., MEYER, P., MILLIEZ, P.: Amer. J. Physiol. **224**, 1223 (1973).
HOUPT, T. R., HOUPT, K. A.: Amer. J. Physiol. **214**, 1296 (1968).
HUBEL, K. A.: Gastroenterology **54**, 647 (1968).
HUBEL, K. A.: J. clin. Invest. **52**, 3172 (1973).
HUBEL, K. A.: J. clin. Invest. **53**, 964 (1974).
INGRAHAM, R. C., VISSCHER, M. B.: J. gen. Physiol. **18**, 695 (1935).
JACKSON, M.: Biomembranes **4B**, 673 (1974).
JACKSON, M. J., MORGAN, B. N.: Amer. J. Physiol. **228**, 482 (1975).
JACKSON, M. J., SMYTH, D. H.: Nature **219**, 388 (1968).
JACKSON, M. J., SHIAU, Y. F., BANE, S., FOX, M.: J. gen. Physiol. **63**, 187 (1974).
JACOBS, M. H.: Cold Spr. Harbor Symp. quant. Biol. **8**, 30 (1940).
JOHANSEN, P. G., HADORN, B., ANDERSON, C. M.: Nature **217**, 468 (1968).
KETTERING, R. F., SUMMERSKILL, W. H. J.: Medicine **46**, 91 (1967).
KIMBERG, D. V.: Gastroenterology **67**, 1023 (1974).
LAMERS, J. M.: Biochim. biophys. Acta **413**, 265 (1975).
Leading Article. Lancet **1975 2**, 214.
LEVITAN, R.: J. Lab. clin. Med. **69**, 558 (1967).
LEVITAN, R.: Israel J. Med. Sci. **4**, 140 (1968).
LEVITAN, R.: Amer. J. clin. Nutr. **22**, 315 (1969).
LEVITAN, R., INGELFINGER, F. J.: J. clin. Invest. **44**, 801 (1965).
LEVITAN, R., MAUER, I.: J. Lab. clin. Med. **72**, 739 (1968).
LEVITAN, R., FORDTRAN, J. S., BURROWS, B. A., INGELFINGER, F. J.: J. clin. Invest. **41**, 1754 (1962).
LEW, V. L.: J. Physiol. **206**, 509 (1970).
LEW, V. L., CARLISKY, N. J.: Biochim. biophys. Acta **135**, 793 (1967).
LONG, C. L., GEIGER, J. W., KINNEY, J. M.: Metabolism **16**, 413 (1967).
LUCAS, M. L., SCHNEIDER, W., HABERISH, F. J., BLAIR, J. A.: Proc. roy. Soc. Med. (Lond.) **192**, 39 (1975).
MCBEE, R. H.: Amer. J. clin. Nutr. **23**, 1514 (1970).
MANNING, R. T.: Biochem. Clin. **3**, 225 (1964).
MEKHJIAN, H. S., PHILLIPS, S. F.: Gastroenterology **59**, 120 (1970).
MEKHJIAN, H. S., PHILLIPS, S. F., HOFMANN, A. F.: J. clin. Invest. **50**, 1569 (1971).
METCALFE-GIBSON, A., ING, T. S., KUIPER, J. J., RICHARDS, P., WARD, E. E., WRONG, O. M.: Clin. Sci. **33**, 89 (1967).
MILNE, M. D., SCHRIBNER, B. H., CRAWFORD, M. A.: Amer. J. Med. **24**, 709 (1958).
MILTON-THOMPSON, G. J., CUMMINGS, J. H., NEWMAN, A., BILLINGS, J. A., MISIEWICZ, J. J.: Gut **16**, 42 (1975).
MOSSBERG, S. M.: Amer. J. Physiol. **213**, 1327 (1967).
MOSSBERG, S. M., ROSS, G.: J. clin. Invest. **46**, 490 (1967).
MOSSBERG, S. M., ROSS, G., WEINGARTEN, B.: Nature **212**, 1588 (1966).
NAKAMURA, S., GORDON, H. A.: Proc. Soc. exp. Biol. **142**, 1336 (1973).
NELL, G., OVERHOFF, H., KULENKAMPFF, H., SPECHT, W., RUMMEL, W.: Naunyn-Schmiedebergs Arch. Pharm. **277**, 53 (1973).
NELLANS, H. N., FRIZZELL, R. A., SCHULTZ, S. G.: Amer. J. Physiol. **225**, 467 (1973).
NELLANS, H. N., FRIZZELL, R. A., SCHULTZ, S. G.: Amer. J. Physiol. **226**, 1131 (1974).
NELLANS, H. N., FRIZZELL, R. A., SCHULTZ, S. G.: Amer. J. Physiol. **228**, 1808 (1975).
O'GRADY, F.: Proc. roy. Soc. Med. **59**, 1246 (1966).
ORLOFF, J., BERLINER, R. W.: J. clin. Invest. **135**, 223 (1956).
OSTERHOUT, W. J. V.: J. gen. Physiol. **8**, 131 (1925).
PARSONS, B. J., MUNDAY, K. A.: In: Transport Across the Intestines (A Glaxo Symposium) (W. L. Burland and P. D. Samuel, Eds). Baltimore, Md: Williams and Wilkins 1972, p. 59.
PARSONS, D. S.: Quart. J. exp. Physiol. **41**, 410 (1956).
PARSONS, D. S., PATERSON, C. R.: Biochim. biophys. Acta **41**, 173 (1960).
PARSONS, D. S., PATERSON, C. R.: Quart. J. exp. Physiol. **50**, 220 (1965).
PARSONS, D. S., POWIS, G.: J. Physiol. **217**, 641 (1971).
PATLAK, C. S., GOLDSTEIN, D. A., HOFFMAN, J. A.: J. theoret. Biol. **5**, 426 (1963).
PERHEENTUPA, J., HARRISON, H. C., HARRISON, H. E.: Biochim. biop phys. Acta **255**, 858 (1972).

PHILLIPS, S. F.: Gastroenterology **56**, 966 (1969).
PHILLIPS, S. F.: Gastroenterology **63**, 495 (1972).
PHILLIPS, S. F., CODE, C. F.: Amer. J. Physiol. **211**, 607 (1966).
PHILLIPS, S. F., GILLER, J.: J. Lab. clin. Med. **81**, 733 (1973).
PHILLIPS, S. F., SCHMALZ, P. F.: Proc. Soc. exp. Biol. **135**, 116 (1970).
PITITH, M. M., SCHEDL, H. P.: Gastroenterology **71**, 1039 (1976).
POWELL, D. W., MALAWER, S. J.: Amer. J. Physiol. **215**, 49 (1968).
POWELL, D. W., MALAWER, S. J., PLOTKIN, G. R.: Amer. J. Physiol. **215**, 1226 (1968).
POWELL, D. W., SOLBERG, L. I., PLOTKIN, G. R., CATLIN, D. H., MAENZA, R. M., FORMAL, S. B.: Gastroenterology **60**, 1076 (1971).
PRICE, J. B., SCHWARTZ, G. F., MOLAVI, A., BRITTON, R. C., VORHEES, A. B. Jr: Surg. Forum **18**, 331 (1967).
PRICE, J. B., SAWDA, M., VOORHEES, A. B., Jr: Amer. J. Surg. **119**, 595 (1970).
RACUSEN, L., BINDER, H. J.: Gastroenterology **73**, 790 (1977).
RASK-MADSEN, J.: Scand. J. Gastroent. **8**, 327 (1973).
RASK-MADSEN, J.: Scand. J. Gastroent. **9**, 223 (1974).
RASK-MADSEN, J., JENSEN, B.: Scand. J. Gastroent. **8**, 169 (1973).
RICHARDS, P., METCALFE-GIBSON, A., WORD, E. E., WRONG, O., HOUGHTON, B. J.: Lancet **1967 2**, 845.
ROGOECZI, E., IRONS, L., KOJ, A., MCFARLAND, A. S.: Biochem. J. **95**, 521 (1965).
ROSENFELD, J. B., ABOULAFIA, E. D., SCHWARTZ, W. B.: Amer. J. Physiol. **204**, 568 (1963).
ROUSSEAU, B., SLADEN, G. E.: Biochim. biophys. Acta **233**, 591 (1971).
RUBENS, R. D., LAMBERT, H. P.: Gut **13**, 915 (1972).
RUBINSTEIN, R., HOWARD, A. V., WRONG, O. M.: Clin. Sci. **37**, 549 (1969).
SALAS-COLL, C., KERMODE, J. C., EDMONDS, C. J.: Clin. Sci. molec. Med. **51**, 287 (1976).
SALLEE, V. L., DIETSCHY, J. M.: J. Lipid Res. **14**, 475 (1973).
SALTER, D. N.: Proc. Nutr. Soc. **32**, 65 (1973).
SCHANKER, L. S., TOCCO, D. J., BRODIE, B. B., HOGBEN, C. A. M.: J. Pharmacol. exp. Ther. **123**, 81 (1958).
SCHENKER, S., BREEN, K. J., HOYUMPA, A. M., Jr: Gastroenterology **66**, 121 (1974).
SCHMITT, M. G., SOERGEL, K. H., WOOD, C. M.: Gastroenterology **70**, 211 (1976).
SCHULTZ, S. G.: J. gen. Physiol. **59**, 794 (1972).
SCHULTZ, S. G., CURRAN, P. F.: In: Handbook of Physiology, Section 6: Alimentary Canal, Vol. 2: Intestinal Absorption (C. F. Code, Ed.). Baltimore, Md: Williams and Wilkins 1968, p. 1245.
SCHULTZ, S. G., CURRAN, P. F.: Current Topics in Membranes and Transport **5**, 225 (1974).
SCHULTZ, S. G., FRIZZELL, R. A., NELLANS, H. N.: J. Membrane Biol. **33**, 351 (1977).
SCHWITER, E. J., HEPNER, G. W., ROSE, R. C.: Gut **16**, 477 (1975).
SHIELDS, R.: Brit. J. Surg. **53**, 893 (1966).
SHIELDS, R.: In: Recent Advances in Gastroenterology (J. Badenoch and B. N. Brooke, Eds). Baltimore, Md: Williams and Wilkins 1972, p. 215.
SHIELDS, R., MILES, J. B.: Postgrad. med. J. **41**, 435 (1965).
SHIELDS, R., MULHOLLAND, A. T., ELMSLIE, R. G.: Gut **7**, 686 (1966).
SHORE, P. A., BRODIE, J. B., HOGBEN, C. A. M.: J. Pharmacol. exp. Ther. **119**, 361 (1957).
SILVA, P., CHARNEY, A. N., EPSTEIN, F. H.: Amer. J. Physiol. **229**, 1576 (1975).
SIMONETTA, M., FAELLI, A., CREMASCHI, D., GORDON, H. A.: Proc. Soc. exp. Biol. **150**, 541 (1975).
SKADHAUGE, E.: Comp. Biochem. Physiol. **23**, 483 (1967).
SKADHAUGE, E.: Comp. Biochem. Physiol. **24**, 7 (1968).
SMITH, B., BUTLER, M.: Brit. J. exp. Path. **55**, 615 (1974).
SMITH, M. W., JAMES, P. S.: Biochim. biophys. Acta **419**, 391 (1976).
SMYTH, D. H., TAYLOR, E. B.: J. Physiol. **141**, 73 (1958).
SOERGEL, K. H., WHALEN, G. E., HARRIS, H. A., GEENEN, J. E.: J. clin. Invest. **47**, 1071 (1968).
SPIRO, H. M.: New Engl. J. Med. **293**, 83 (1975).
STEVENS, C. E.: In: Transport Mechanisms In Epithelia (H. H. Ussing and N. A. Thorn, Eds). Copenhagen: Munksgaard, New York: Academic 1973, p. 404.
STEVENS, C. E., DOBSON, A., MAMMANO, J. H.: Amer. J. Physiol. **216**, 983 (1969).
STOEBEL, D. P., GOLDNER, A. M.: The Physiologist **18**, 410 (1975).
SUMMERSKILL, W. H. J., AOYAGI, T., EVANS, W. B.: Gut **7**, 497 (1966).

SUMMERSKILL, W. H. J., WOLPERT, E.: Amer. J. clin. Nutr. **23**, 633 (1970).
SWALES, J. D., PAPADIMITRIOU, M., WRONG, O. M.: Gut **14**, 697 (1973).
SWALES, J. D., TANGE, J. D., WRONG, O. M.: Clin. Sci. **39**, 769 (1970).
SWALLOW, J. H., CODE, C. F.: Amer. J. Physiol. **212**, 717 (1967).
TAUB, M., BONORRIS, G., CHUNG, A., COYNE, M. J., SCHOENFIELD, L. S.: Gastroenterology **72**, 101 (1977).
TEORELL, T.: Skand. Arch. Physiol. **66**, 225 (1933).
TURNBERG, L. A.: Gut **11**, 1049 (1970).
TURNBERG, L. A.: Gut **12**, 544 (1971).
TURNBERG, L. A., BIEBERDORF, F. A., MORAWSKI, S. G., FORDTRAN, J. G.: J. clin. Invest. **49**, 557 (1970).
USSING, H. H., ANDERSEN, B.: In: Proceedings of the 3rd International Congress of Biochemistry, Brussels, 1955. New York: Academic 1956, p. 434.
VINCE, A., DAWSON, A. M., PARK, N., O'GRADY, F.: Gut **14**, 171 (1973).
VISSCHER, M. B., VARCO, R. H., CARR, C. W., DEAN, R. B., ERICHSON, D.: Amer. J. Physiol. **141**, 488 (1944).
WEAST, R. C. (Ed.): CRC Handbook of Chemistry and Physics. Cleveland, Ohio: The Chemical Rubber Company 1971.
WILSON, D. R., ING, T. S., METCALFE-GIBSON, A., WRONG, O. M.: Clin. Sci. **34**, 211 (1968).
WILSON, T. H., KAZYAK, L.: Biochim. biophys. Acta **24**, 124 (1957).
WILSON, T. H.: Biochem. J. **56**, 521 (1954).
WINGATE, D. L., KRAG, E., MEKHJIAN, H. S., PHILLIPS, S. F.: Clin. Sci. molec. Med. **45**, 593 (1973).
WOLPERT, E., PHILLIPS, S. F., SUMMERSKILL, W. H. J.: New Engl. J. Med. **283**, 159 (1970).
WOLPERT, E., PHILLIPS, S. F., SUMMERSKILL, W. H. J.: Lancet **1971 2**, 1387.
WOOTON, J. F., ARGENZIO, R. A.: Amer. J. Physiol. **229**, 1062 (1975).
WRONG, O. M.: In: The Scientific Basis of Medicine, Annual Reviews (I. Gilliland and J. Francis, Eds). London: Athlone 1971, p. 192.
WRONG, O., METCALFE-GIBSON, A., MORRISON, R. B. I., NG, S. T., HOWARD, A. V.: Clin. Sci. **28**, 357 (1965).
YANG, M. G., MANOHARAN, K., MICKELSEN, O.: J. Nutr. **100**, 545 (1970).
YAU, W. M., MAKHLOUF, G. M.: Gastroenterology **67**, 662 (1974).
YAU, W. M., MAKHLOUF, G. M.: Amer. J. Physiol. **228**, 191 (1975).
YORIO, T., BENTLEY, P. J.: Amer. J. Physiol. **232**, F5 (1977).

Chapter 16

Transport Processes in the Exocrine Pancreas

I. SCHULZ and K.J. ULLRICH

A. Introduction

This review is concerned with the transport of small solutes in the exocrine pancreas and the processes triggering electrolyte and enzyme secretion. No attempt will be made to cover the literature on synthesis, storage, and intracellular transport of enzymes, which has been extensively studied by SIEKEVITZ and PALADE (1960), PALADE et al. (1962), CARO and PALADE (1964), REDMAN et al. (1966), JAMIESON and PALADE (1967a and b) and reviewed by JAMIESON (1972 and 1973) and HOLTZMAN et al. (1977).

The transport of small solutes within the exocrine pancreas serves two principal functions: 1. Secretion of electrolytes and water, which occurs mainly in the acini and small excretory ducts; and 2. Triggering of enzyme secretion in the acini. Furthermore, the contraluminal uptake of amino acids is necessary for enzyme synthesis and therefore related to the latter category.

The luminal diameters of the acini and the small intercalated ducts are only about 3 μm. Since in addition the smallest ducts are covered by acini, it is very difficult to perform micropuncture and electrophysiological experiments on those duct segments that show the highest electrolyte and water transport. Extensive electrophysiological studies on acinar cells, however, have been performed in the intact organ (PETERSEN, 1976), and separation of viable acinar cells allowed biochemical studies on hormone binding and hormone action (CHRISTOPHE et al., 1976a; GARDNER et al., 1975; ROBBERECHT et al., 1976 and 1977; KONDO and SCHULZ, 1976a and b). Furthermore, separation of isolated duct cells made it possible to characterize the hormone specificity of their adenylate cyclase which is different from that of acinar cells (MILUTINOVIĆ et al., 1977a). It has also been possible to prepare plasma membrane vesicles for studies on amino acid uptake and electrolyte permeability (TYRAKOWSKI et al., 1978; SCHULZ and HEIL, 1978). Thus, recent progress in the understanding of pancreatic transport processes has been remarkable, although on some points the experimental data are still too incomplete to allow safe conclusions.

B. Secretion of Electrolytes and Water

I. Stimulatory Processes

Although the pancreas of some herbivores such as horse, calf, sheep, and rabbit, secretes fluid and electrolytes at rest, especially under *in-vitro* conditions (COMBINE et al., 1969; TERNOUTH and BUTTLE, 1973; CAPLE and HEATH, 1972; BAXTER, 1931; ROTHMAN, 1964; RIDDERSTAP, 1969a), in most mammalians fluid secretion is evoked by the proteohormone secretin, the main physiological stimulant for electrolyte and water secretion. Secretin was discovered and first extracted from the dog jejunum by BAYLISS and STARLING (1902). Pure secretin was prepared and its amino acid sequence established by JORPES and MUTT (1961) and MUTT et al. (1965, 1970); it contains 27 amino acids and was synthesized by BODANSZKY et al. (1967) and WÜNSCH (1972). It was found that secretin stimulates a membrane-bound adenylate cyclase (RUTTEN et al., 1972; SCHULZ et al., 1974; BONTING and DE PONT, 1974; BONTING et al., 1974; MILUTINOVIĆ et al., 1976) and that the same adenylate cyclase is also stimulated by pancreozymin, the peptide hormone that elicits enzyme release (RUTTEN et al., 1972; KEMPEN et al., 1974; MILUTINOVIĆ et al., 1976). Since acinar cells make up approximately 90 percent of the total pancreas, it seemed reasonable to assume that this adenylate cyclase is present in acinar cell membranes. Separation of duct cells from acinar cells made a distinction between two different adenylate cyclases possible. While membranes isolated from acinar cells possess a secretin-stimulated adenylate cyclase which is also stimulated by pancreozymin, adenylate cyclase in duct cell membranes cannot be stimulated by pancreozymin, but shows a higher sensitivity for secretin than that in acinar cell membranes (Fig. 1; MILUTINOVIĆ et al., 1977a). The finding of a secretin-stimulated adenylate cyclase in both cell types parallels the observation that adenosine 3′5′-cyclic monophosphate (cAMP) production is stimulated by secretin in a purified preparation of acinar cells (ROBBERECHT et al., 1976) as well as in duct cells obtained by selective degradation of acinar cells *in vivo* and removal of fat cells which had replaced acinar cells in the course of degradation (FÖLSCH and CREUTZFELDT, 1975).

Furthermore, the effect of secretin on fluid and bicarbonate secretion can be mimicked by dibutyryl adenosine 3′5′-cyclic monophosphate (dbcAMP) (CASE and SCRATCHERD, 1972; WIZEMANN et al., 1973). Alloxan, an inhibitor of adenylate cyclase, also inhibits electrolyte and fluid secretion (SCRATCHERD, 1974) while theophylline, an inhibitor of phosphodiesterase augments secretion (CASE and SCRATCHERD, 1972; SCRATCHERD et al., 1975). Cholera toxin, a specific activator of the adenylate cyclase in many cells also stimulates adenylate cyclase activity and increases cAMP levels in the pancreas. Moreover, it stimulates fluid and electrolyte secretion in a secretin-like fashion, whereas enzyme secretion is not affected. These observations indicate that the cAMP-adenylate cyclase system is a rate-limiting factor in the regulation of fluid secretion (KEMPEN et al., 1975; SMITH and CASE, 1975). A parallel influence on fluid secretion and adenylate cyclase activity was also observed with the SH-group reagent p-chloromer-

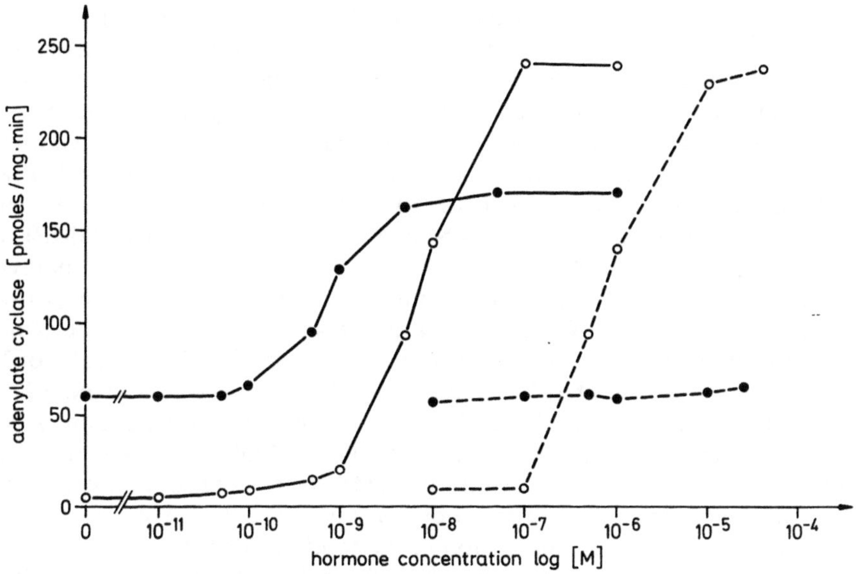

Fig. 1. Localization of hormone receptors as revealed by adenylate cyclase stimulation of isolated acinar cells and duct cells from rats. (○) acinar cells, (●) duct cells, —— secretin, --- pancreozymin. (MILUTINOVIĆ et al., 1977a)

curibenzoate (PCMB), which stimulates both events at concentrations lower than 10^{-9} M and inhibits them above 10^{-7} M. Secretin binding, too, was influenced by PCMB in the same concentration-dependent manner (SCHULZ and MILUTINOVIĆ, 1977).

The characteristics of secretin binding to a crude preparation of isolated cat pancreatic membranes were studied (MILUTINOVIĆ et al., 1976) and the dissociation constant of secretin binding of $4 \cdot 10^{-9}$ M was found to be in the same concentration range as the apparent K_m value for adenylate cyclase stimulation in the same preparation ($8 \cdot 10^{-9}$ M). From data on binding of vasoactive intestinal polypeptide (VIP, a hormone similar to secretin) to isolated acinar cells, a half-maximal secretin concentration for VIP displacement was determined (CHRISTOPHE et al., 1976a), which compares very well with the secretin binding data in isolated membranes (MILUTINOVIĆ et al., 1976). The former value also corresponds to the secretin concentration for half-maximal stimulation of cAMP production in acinar cells (ROBBERECHT et al., 1976). The secretin concentration for stimulation of half-maximal fluid secretion in the perfused cat pancreas ($3 \cdot 10^{-10}$ M), however, is lower (WÜNSCH et al., 1977) and might represent the higher sensitivity of the bicarbonate-secreting duct cells (MILUTINOVIĆ et al., 1977a). The involvement of acinar cAMP in fluid secretion is not yet clear. Although it is possible that acinar cells participate in electrolyte and fluid secretion on stimulation with secretin, direct proof is still lacking. The implication of a pancreozymin-stimulated adenylate cyclase in acinar cells also remains undecided. The observation that the secretin effect on fluid secretion can be potentiated by pancreozymin and that the pancreozymin effect on enzyme secretion is potentiated by secretin led to the assumption of a close mutual interaction of both hormones in the same cell (Fölsch and Wormsley, 1973). Recently a study on amylase release from dispersed pancreatic acinar cells of guinea-pig has shown that the responses to cholecystokinin-octapeptide (CCK-OP) or carbamylcholine were potentiated by secretin, vasoactive intestinal peptide (VIP) or dibutyryl cyclic AMP (dbcAMP). Secretin or VIP also caused amylase release. This effect was further potentiated by CCK-OP, carbamylcholine or dibutyryl guanosine 3′5′-cyclic monophosphate (dbcGMP). The authors interpreted their observations as due to the presence of two pathways for the regulation of amylase release from pancreatic acinar cells: one pathway can be stimulated by CCK-OP or cholinergic agonists and this response is mediated by cGMP, whereas the other pathway can be stimulated by secretin or VIP and this response is mediated by cyclic AMP (Gardner and Jackson, 1977).

Although the mode whereby cAMP induces electrolyte secretion is completely unknown, one could consider possible analogies to other electrolyte transport systems that are regulated by cAMP. For instance, one can assume that, as in phosphate reabsorption in the proximal tubule, (Agus et al., 1971; Evers et al., 1978) or Na^+ absorption in the toad bladder (Leaf and Dempsey, 1960), the entrance step of the solute into the cell is affected by cAMP. Thus, the increased mucosal Na^+ entry in the toad bladder was observed to coincide with dephosphorylation of membrane proteins (Walton et al., 1975), while the decreased luminal phosphate entry in the proximal tubule of the kidney seems to be correlated with phosphorylation of membrane proteins (Kinne et al., 1975). The suggestion that cAMP influences the electrolyte permeability of pancreatic cells is supported by the finding that after stimulation with secretin a potassium transient occurs in the perfusate of the isolated perfused cat pancreas (Case et al., 1969). This means that potassium ions leave the cells at the contraluminal cell side. The same phenomenon was also observed in the acetylcholine-stimulated salivary gland, where electrical measurements had demonstrated that acetylcholine increased the K^+ permeability of the contraluminal cell side (Nishiyama and Petersen, 1974a, 1975a).

Other hormones that elicit electrolyte and fluid secretion are dopamine (Hashimoto et al., 1971; Iwatsuki et al., 1973) and vasoactive intestinal polypeptide (VIP) (Scratcherd et al., 1975; Wünsch et al., 1977). An inhibitory effect on the secretin-evoked fluid and electrolyte secretion was ascribed to motilin (Konturek et al., 1976) and somatostatin (Creutzfeldt et al., 1975; Konturek et al., 1976; Boden et al., 1976; Regal et al., 1977). However, the latter two studies did not exclude an indirect effect of somatostatin due to the suppression of secretin release from the intestine. Other authors, however, have not found an inhibitory effect of somatostatin on pancreatic fluid and bicarbonate secretion (Bloom et al., 1975). Binding studies with ^{125}I-labeled VIP (Christophe et al., 1976a; Gardner et al., 1976) and with ^{125}I-labeled secretin (Milutinović et al., 1976) revealed that secretin and VIP bind to distinct receptors to which small cross-affinities for both hormones exist. Analogous experiments with dopamine, somatostatin, and motilin are still lacking.

It should be mentioned, however, that large species differences exist concerning responsiveness to hormones. The cat pancreas, for instance, secretes fluid only in response to secretin and not to pancreozymin, caerulein, or the octapeptide of pancreozymin. The last three substances elicit enzyme secretion, which in the cat can only be assessed when superimposed on a secretin-evoked fluid secretion (BROWN et al., 1967). In the dog, however, these three substances evoke, beside enzyme secretion, a vigorous fluid secretion (STENING and GROSSMAN, 1969; DEBAS and GROSSMAN, 1973) and in the rat the fluid secretion evoked by the enzyme secretagogues is even more copious than that evoked by secretin (SEWELL and YOUNG, 1975). As in rat, maximal fluid secretion can also be evoked by pilocarpine in the mouse (MANGOS et al., 1973, 1974). Consequently, although secretin-stimulated electrolyte and fluid secretion have been ascribed to the cells of the small excretory ducts (SCHULZ et al., 1969; SWANSON and SOLOMON, 1973), the above-mentioned data suggest that the acinar cells, which are the target for the enzyme secretagogues, are also involved in electrolyte and fluid secretion.

II. Fluid Secretion and Ionic Requirement

Fluid secretion and ionic requirement can best be studied in the artificially perfused cat pancreas (CASE et al., 1968b) or in the rabbit pancreas kept *in vitro* (ROTHMAN and BROOKS 1965a). The presence of CO_2-bicarbonate or a related buffer (CASE et al., 1970; SCHULZ, 1971), Ca^{++} ions (ARGENT et al., 1973); and glucose (WIZEMANN and SCHULZ, 1973) in a well-oxygenated saline perfusate (CASE et al., 1968b) is satisfactory for the maintenance of pancreatic fluid secretion. Even under anaerobic conditions glucose is able to support a small but finite fluid secretion, lasting several hours (WIZEMANN and SCHULZ, 1973). If the NaCl concentration in the perfusate is varied, the secretory fluid rate varies inversely, the secreted fluid being isosmotic with the perfusate all the time (CASE et al., 1968b). Thus, the secretory rate of sodium remains fairly constant despite variation of the perfusate Na^+ concentration between 70 and 250 mEq l^{-1} and even higher; on the other hand the water permeability of the secretory epithelial cell layer must be quite high, so that the concentration difference of the anions between secreted and interstitial fluid at different reflection coefficients ($\sigma_{HCO_3} > \sigma_{Cl}$) provides enough driving force for the secretion of water in isotonic proportions (ULLRICH, 1974). In this regard, the exocrine pancreas seems to behave like other leaky epithelia, for instance the rabbit gallbladder (DIAMOND, 1965) or the renal proximal tubule (BALDAMUS et al., 1969). Indeed, recent determinations of reflection coefficients of non-electrolytes relative to sucrose showed that the secretin-stimulated exocrine cat pancreas is more permeable to these solutes than the gall-bladder of rabbit, the proximal tubule of rat kidney, and probably similar to that of small intestine of guinea pig and man (DEWHURST et al., 1978). If the secretory net water movement is stopped by the addition of sucrose to the pancreatic perfusate a sodium concentration difference of 110 mEq l^{-1} is built up, $[Na^+]_{secretion} > [Na^+]_{perfusae}$ (CASE et al., 1968b). A lumen-negative electrical potential difference of 25 mV could account for this. However, the electrical potential difference that has been measured under free-flow conditions in the main pancreatic duct of the cat (WAY and DIAMOND, 1970) and along the duct system of the rabbit pancreas (SCHULZ et al., 1969; SWANSON and SOLOMON, 1973) is only around −5 to −9 mV after stimulation with secretin. From these data, unfortunately, it is not possible to decide whether or not there is a finite electrochemical potential difference for sodium at zero net flux of solutes and water, and therefore an active transport of sodium. The electrical potential difference at the site of maximal fluid and electrolyte secretion may deviate considerably from the values published in the papers mentioned. The question of which ion species could be actively transported will be discussed later.

If the sodium in the perfusate is replaced by lithium, fluid secretion is reduced (CASE et al., 1968a), although a small secretion rate can be maintained for long periods. Under these conditions the concentration of lithium in the pancreatic juice is slightly higher than that in the perfusate (CASE and SCRATCHERD, 1974). The omission of potassium from the perfusate reduces secretion by 65 percent (CASE and SCRATCHERD, 1974). Secretion could, however, be completely restored by substituting rubidium for potassium, but only partially when potassium was replaced by caesium. When the potassium concentration in the perfusate is varied at the expense of sodium, at 10 mEq l^{-1} K^+ in the perfusate maximal sodium secretion, and at 50 mEq l^{-1} maximal potassium secretion is observed. At higher K^+ concentrations sodium secretion is diminished (Fig. 2), paralleling the effect on secretory flow rate (CASE and SCRATCHERD, 1974). The concentration of potassium in the pancreatic juice is identical with that of the perfusate over the whole concentration range. The fact that the concentrations of Na^+ and Li^+ in the secreted fluid are always a little higher and of K^+ the same as the respective concentrations in the pancreatic perfusate is in keeping with the hypothesis that Na^+ and K^+ are only secondarily involved in pancreatic

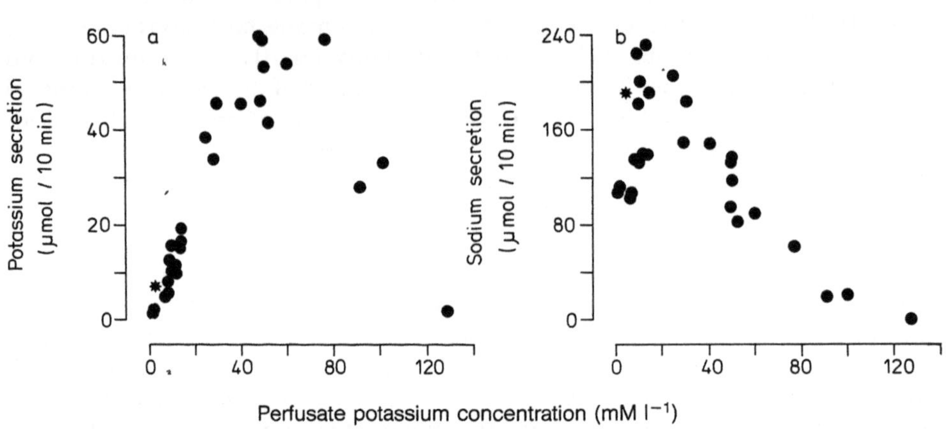

Fig. 2. Relationship between perfusate potassium concentration and secretion of (a) potassium and (b) sodium by isolated pancreas under isosmolal conditions, achieved by replacing perfusate sodium chloride by potassium chloride. Asterisks (*) indicate mean secretion in these experiments during perfusion with fluid of normal composition (mean $[K]_p$ = 4,5 mM l^{-1}). (CASE and SCRATCHERD, 1974)

electrolyte and fluid secretion. If one considers, however, that the fluid secretion parallels the Na^+ secretion, one is inclined to attribute an active role to Na^+ transport. If, on the other hand, the Cl^- concentration in the perfusate is replaced by bromide or iodide, only a slight increase or decrease in secretory rate was observed (CASE et al., 1968a). Even when chloride is replaced by sulfate or methylsulfate, secretion is still maintained, though after some time there is a marked reduction in rate. In this situation higher concentrations of bicarbonate appear at a given flow rate in the secretory fluid, accompanied by only a little sulfate. The behavior of these tested anions indicates that they, too, play only a secondary role in electrolyte and fluid secretion and that the active role could most probably be ascribed to the transport of HCO_3^- or to one of the HCO_3^-/CO_2 buffer components (*i.e.* HCO_3^-, OH^-, or H^+ ions). Finally, it should be mentioned that the omission of phosphate from the perfusate does not influence the rate of fluid secretion (CASE et al., 1968a).

III. Flow-Dependent Concentration Pattern in the Secreted Fluid

The most striking changes in the ionic composition of the secreted pancreatic juice are observed with HCO_3^- and Cl^-. At low secretory rates the HCO_3^- concentration has been found to be similar to that in plasma. But after stimulation with secretin it rises with increasing flow rates asymptotically to a maximum, which is different for different species. In cat and dog it reaches 130 mEq l^{-1} (HART and THOMAS 1945; CASE et al., 1969) and in rabbit and rat 80 mEq l^{-1} (ROTHMAN and BROOKS, 1965b; SEWELL and YOUNG, 1975). The Cl^- concentration pattern is a mirror image of that of HCO_3^-, because the sum of both anions is approximately the same at all flow rates. As pointed out above, in rat it is possible to stimulate fluid secretion with the enzyme secretagogues pancreozymin and caerulein even more than with secretin (SEWELL and YOUNG, 1975), and in rat and mouse maximal fluid secretion can be evoked by pilocarpine (MANGOS et al., 1973, 1974). But in these cases the HCO_3^- concentration in the secreted fluid stays constant, similar to that in plasma or somewhat higher. The acetylcholine- and pancreozymin- or caerulein-evoked fluid secretion, in contrast to that evoked by secretin, is independent of the presence of extracellular bicarbonate (PETERSEN and UEDA, 1977) or partially affected by the deprivation of bicarbonate (KANNO and YAMAMOTO, 1977; UEDA and PETERSEN, 1977) but much more sensitive to extracellular Ca^{++} deprivation (PETERSEN and UEDA, 1976a; KANNO and YAMAMOTO, 1977). This is in accordance with the concept that the triggering mechanism of fluid and enzyme secretion by the acinar cells is the same (see below). If a combination of secretin and pancreozymin is given to rats, a mixed pattern is obtained with a maximum HCO_3^- concentration at intermediate flow rates (SEWELL and YOUNG, 1975). The same pattern is observed if an impure secretin preparation is applied that is not completely free of pancreozymin, as is the case with the widely used Boots secretin. Such a mixed response apparently exists not only for rat (MANGOS and MCSHERRY, 1971), but also for man (WORMSLEY, 1968). SEWELL and YOUNG (1975) conclude from their data that the two hormones secretin and pancreozymin have totally independent actions on two different receptor systems, one assumed to be present in the duct cells stimulating fluid secretion high in HCO_3^- and low in Cl^-, and the other in the acinar cells, where a low HCO_3^- and high Cl^- secretion is evoked.

The Na^+ concentration in the secreted fluid is independent of the rate of fluid secretion, remaining constant at a level about 10 mEq l^{-1} higher than in the perfusate (WAY and DIAMOND, 1970; CASE and SCRATCHERD, 1974). In some species (cat, dog, rabbit) the K^+ concentration in the secretion stays constant at the plasma level or somewhat higher (BRO-RASMUSSEN et al., 1956; HEIDBREDER et al., 1972; CASE and SCRATCHERD, 1974), while in others (rat) it increases from a plasma level to twice that value when stimulated with secretin, but stays constant when stimulated with pancreozymin (SEWELL and YOUNG, 1975). The reason for this peculiar behavior in rat is unknown, but may be related to the very small fluid-secreting response elicited by secretin in that animal.

For the calcium excretory pattern, two components could be evaluated, one linked to protein secretion and amounting to 20–50 nmol per mg of protein, and one linked to the rate of fluid secretion and dependent directly on the concentration of calcium in the medium (GOEBELL et al., 1972; CECCARELLI et al., 1975; SCHREURS et al., 1976b). The latter component is considered to be a passive calcium movement through the tight junctions between the ductular cells. Thus, neglecting protein bound calcium, the ratio of total calcium concentration in the perfusate to total calcium concentration in the secreted fluid remains approximately constant if the perfusate calcium concentration is raised (ARGENT et al., 1973). The concentration of total calcium in the secreted fluid is smaller than in plasma and the concentration of *ionised* calcium might be as small as 0.075 mmol l^{-1} (ZIMMERMANN et al., 1971). It is noteworthy that with the same experimental procedure in some species (rat), the calcium concentration in the secreted fluid is found to be flow-independent while in others (rabbit) it falls with increasing flow rate (HEIDBREDER et al., 1972).

The behavior of magnesium secretion is similar to that of calcium, one component being linked to protein secretion and the other related to the magnesium concentration in the medium (SCHREURS et al., 1976b). In rabbits the concentration of magnesium in the secreted fluid (0.5 mEq l^{-1} is only one-third of the ultrafiltrable magnesium concentration in the plasma (1.6 mEq l^{-1}). With the increasing secretory rate induced by Boots secretin there is a tendency for the magnesium concentration in the secreted fluid to fall (HEIDBREDER et al., 1972).

The concentration of inorganic phosphate in the secreted fluid is only one-tenth that in the serum and has, like that of calcium and magnesium, a tendency to fall with increasing secretory flow rate (HEIDBREDER et al., 1972). It is likely that the flow-dependent excretion of calcium, magnesium and phosphate is a simple expression of the sieving function of the epithelial cell layer. Corresponding values for nonelectrolytes suggest that the pancreas is a leaky epithelium (DEWHURST et al., 1978) and that stimulants of enzyme secretion increase the permeability of a paracellular pathway (DE PONT et al., 1978).

IV. Local Transport Events as Revealed by Micropuncture and Microperfusion Techniques

1. Electrolytes and Water

Due to the heterogeneity of the pancreas and the difficulty of access to the duct system, the exact localization of HCO_3^- transport was one of the main unsolved problems in the understanding of the mechanisms responsible for pancreatic electrolyte secretion. Micropuncture techniques applied to the pancreas similar as are to kidney tubules (GIEBISCH, 1972) have helped to clarify some aspects of secretin action. Thus the site of HCO_3^- and fluid secretion was localized in duct cells (SCHULZ et al., 1969). However, it should be considered that with this technique it is still difficult to assign secretin-stimulated secretion to one cell type and to assess the possible contribution of acinar cells in electrolyte and fluid secretion when puncturing intralobular small ductlets near to acini. The finding of a secretin-stimulated adenylate cyclase in isolated acinar cells (BONTING and de PONT, 1974; KEMPEN et al., 1977b; ROBBERECHT et al., 1976; MILUTINOVIĆ et al., 1977a) make an involvement of acinar cells in electrolyte and fluid secretion likely. Another problem in the comparison of data obtained by micropuncture studies is the use of different preparations of secretin with varying degrees of purity. A crude secretin preparation could cause fluid secretion in rat, due to a response of acinar cells to pancreozymin with a secretion high in Cl^- and low in HCO_3^- concentration in this species (SEWELL and YOUNG, 1975). Discrepancies in data from different laboratories could also be due to species differences of the animals used.

So far four types of method have been used: free-flow micropuncture of the acini and the ductal system (REBER and WOLF, 1968; SCHULZ et al., 1969; MANGOS and MCSHERRY, 1971; SWANSON and SOLOMON, 1973; MANGOS et al., 1973, 1974), stop-flow microperfusion of the small pancreatic ducts (SCHULZ et al., 1969; MANGOS et al., 1974; LIGHTWOOD and REBER, 1977), microcanulation of small ducts (REBER and LIGHTWOOD, 1976) and perfusion of the main excretory duct (CASE et al., 1969; WAY and DIAMOND, 1970; CASE and SCRATCHERD, 1970; MOQTADERI et al., 1972). In the rabbit pancreas *in vivo*, the mean sodium concentration of the acinar fluid has been found to be about 157 mEq l^{-1} which is significantly higher than in plasma (141 mEq l^{-1}), while the potassium concentration is 3.1 mEq l^{-1} which is slightly lower. Both are flowrate-independent (MANGOS and MCSHERRY, 1971). In the experiments of SCHULZ et al. (1969), the Cl^- concentrations found in the acinar fluid of the unstimulated gland varied from animal to animal between 53 and 109 mEq l^{-1} but in the same animals, they were not changed after stimulation with secretin. MANGOS and MCSHERRY (1971) found Cl^- values between 38 and 47 mEq l^{-1} in rabbit; these are also independent of the secretory rate induced by Boots secretin. In the unstimulated glands the Cl^- concentration along the duct system increases between the acini (and intralobular duct) and the interlobar duct from 66 to 88 mEq l^{-1} in the experiments of SCHULZ et al. (1969) and from about 40 to 87 mEq l^{-1} in the experiments of MANGOS and MCSHERRY (1971). But between the interlobar duct and final secretion it increases only by a few mEq l^{-1}. If, however, the gland is stimulated by secretin (SCHULZ et al., 1969) the Cl^- concentrations decrease along the duct system, in one series of animals from 61 to 54 (summer experiments), and in another set of animals from 92 to 62 (winter experiments). The data from the summer experiment of SCHULZ et al. (1969) agree with those of REBER and WOLF (1968) in that there is no change in Cl^- and HCO_3^- concentration between the extralobular collecting system and final secretion. The decrease in Cl^- concentration along the duct system by secretin stimulation was also observed in a micropuncture study of the cat pancreas (LIGHTWOOD and REBER, 1977). The Cl^- concentration in the intralobular small ducts, however, is higher (112 mEq l^{-1}) in cat than in rabbit as measured by SCHULZ et al. (1969) and by MANGOS and MCSHERRY (1977). In an *in-vitro* preparation of rabbit pancreas, SWANSON and SOLOMON (1973) found in the unstimulated gland no change of the Cl^- concentration ($\sim$ 74 mEq l^{-1}) along the smaller and medium-sized duct segments. In the secretin-stimulated gland, however, the Cl^- concentration decreases in these duct segments by about 15 mEq l^{-1}. In both situations, unstimulated and stimulated glands, the Cl^- concentration in the main duct increases by about 10 mEq l^{-1}. By measuring the protein concentration profile, SWANSON and SOLOMON (1973) found that in the secretin stimulated gland the small extralobar ducts are the major source of fluid secretion. The same segments were shown in stop-flow microperfusion experiments of SCHULZ et al. (1969) to secrete a fluid containing 52 mEq l^{-1} Cl^- after secretin stimulation, while SWANSON and SOLOMON (1973) calculated from their experiments a value of 58 mEq l^{-1}.

Taken together, the micropuncture data from the rabbit pancreas give the following picture: the primary secretion produced in the acini or the small ductlet near the acini already has a low Cl^- and a high HCO_3^- concentration, and the composition is not changed by secretin. Unfortunately it is not known for certain whether the amount of fluid secreted by the acini is increased by that hormone. But it is certain that after secretin stimulation, fluid is secreted especially by the small extralobular ducts, and that this fluid has a lower chloride concentration and consequently a higher HCO_3^- concentration than the fluid secreted by the acini. Therefore the Cl^- concentration decreases and the HCO_3^- concentration increases in the extralobular duct segments of the secretin-stimulated gland. If, however, the gland is not stimulated by secretin, the relatively high HCO_3^- and low Cl^- concentration of the fluid secreted within or near the acini re-equilibrates towards the plasma concentrations. This re-equilibration, which is also seen in other salivary glands at low fluid secretion rates (YOUNG and SCHÖGEL, 1966) takes place in the *in-vivo* preparation mainly in the small extralobular ducts, while in the *in-vitro* preparation it was seen in the main collecting duct. That this re-equilibration, which is evident as an Cl^--HCO_3^- exchange, alters the flow-dependent concentration pattern was convincingly shown by SWANSON and SOLOMON's (1973) micropuncture data and also by perfusion of the cat's main excretory duct (CASE et al., 1969; CASE and SCRATCHERD, 1970; MOQTADERI et al., 1972).

In the *in-vivo* preparation of rat and mouse pancreas (MANGOS and MCSHERRY, 1971; MANGOS et al., 1973, 1974) the Na^+ concentration of the acinar fluid is somewhat higher than in blood plasma. In mouse the K^+ values in the acinar fluid are the same as in blood plasma; in rat they are somewhat higher. In the acinar fluid of both species Cl^- concentrations below 100 mEq l^{-1} have never been measured, and, like the Na^+ and K^+ concentrations, they never change significantly when the glands are stimulated with secretin, pancreozymin or pilocarpine. Stop-flow microperfusion experiments of the interlobar ducts of rats reveal that HCO_3^- ions are secreted in that duct segment after the application of secretin, but not after pancreozymin and pilocarpine or when secretin is added to the luminal microperfusion fluid (MANGOS et al., 1974). The increase in the HCO_3^- concentration, which is reflected by a decrease in the Cl^- concentration, is not however, accompanied by a change in the concentration of the volume marker, inulin, which had been added to the microperfusion fluid. This led to the suggestion that secretin stimulates the secretion of a low-HCO_3^- high-chloride fluid within the acini, which is subsequently changed within the ducts toward a high-HCO_3^- final secretion by secretion of HCO_3^- in exchange for Cl^-. Since the micropuncture experimenters were only able to measure the composition of the acinar fluid before and after application of secretagogues and not the fluid volume secreted by the acini, a generalizing hypothesis can be put forward only with reservations (Fig. 3):

1. If the secretagogues of enzyme secretion, pancreozymin, acetylcholine, or pilocarpine, are applied, the acini of some species (rat, mouse) react with a copious fluid secretion, but those of others (cat) react only to a small extent or

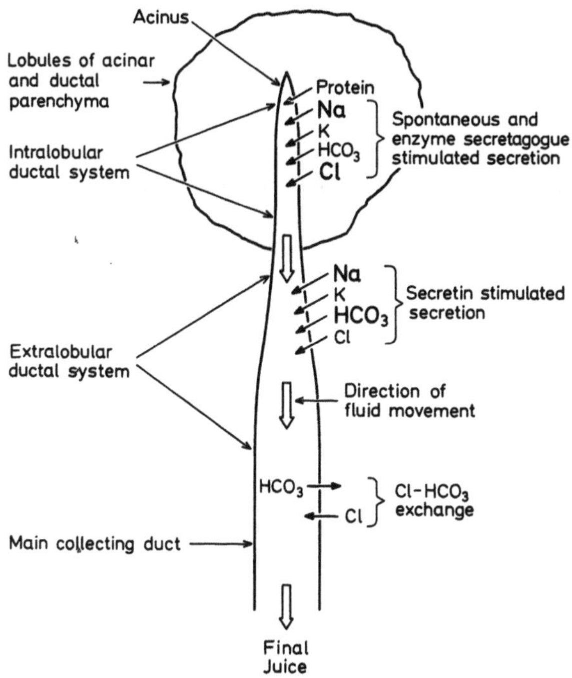

Fig. 3. Schematic representation of pancreatic electrolyte secretion. Size of the letters indicates that in acini mainly NaCl and in the ducts $NaHCO_3$ is secreted. (Modified from SWANSON and SOLOMON, 1973)

not at all. The composition of the acinar fluid shows a HCO_3^- concentration that in some species (rat, mouse, cat) is only a few mEq l^{-1} higher than that of plasma and remains stable when the fluid secretion increases. In other species (rabbit) the acinar fluid HCO_3^- concentration is higher than in plasma. In rabbit the HCO_3^- concentration of the acinar fluid increases when pancreozymin is given (SCHULZ et al., 1969). The enzyme secretagogues so far tested do not affect the function of the ducts.

2. On the other hand, the fluid and electrolyte secretagogue, secretin, induces HCO_3^- secretion in the small ducts. It is an open question, however, whether it induces fluid secretion only within the ducts or whether it induces fluid and electrolyte secretion in the acini also. In a micropuncture study in the cat pancreas, the experimenters felt that fluid was secreted by the acini, since there was no fluid in the acinar lumen before stimulation (LIGHTWOOD and REBER, 1977). The composition of the acinar fluid is not changed by secretin in rabbit, rat, or mouse pancreas. In the cat pancreas this point could not be decided because of lack of fluid secretion without secretin. But with secretin stimulation the Cl^- concentration is not much different from that of the extracellular fluid (LIGHTWOOD and REBER, 1977).

3. In more distal duct segments a re-equilibration towards the composition of blood takes place, *i. e.* a back-diffusion of HCO_3^- in exchange for Cl^-, which enters the ductal fluid. This re-equilibration process is responsible at least in part for the flow-dependent secretory pattern, *i. e.* its influence will be most apparent at low secretory rates.

2. Electrical Potential Differences

The transductal electrical potential differences (PD) were evaluated by SCHULZ et al. (1969), WAY and DIAMOND (1970), and SWANSON and SOLOMON (1973). In the first two reports, respectively, saline-filled microelectrodes and bridges were used and the liquid junction potential was considered carefully. The corrected values for the intralobar duct-hilus of the rabbit *in vivo* are -5 mV (lumen-negative), for the interlobar duct -2.4 mV and for the main duct -5.3 mV (SCHULZ et al., 1969). Unfortunately after secretin only the PD change in the main duct had been measured and not that in the small ducts. It increases slightly by about 1 mV. In the rabbit pancreas *in vitro*, using KCl-filled microelectrodes, SWANSON and SOLOMON (1973) recorded values of -9.1 mV for the intralobar duct and -7.4 mV for the extralobar (interlobar) ducts. They observed no change after secretin application, although such an effect may have been obscured, since no correction for junction potential was made. In the main excretory duct of the cat under free-flow conditions, WAY and DIAMOND (1970) measured $+2.0$ mV at rest and -4.9 mV after secretin stimulation. If, however, the duct was perfused with saline low-Na^+, high-HCO_3^- or high-K^+ solutions, the resting potentials differed from each other, but after secretin the PD shifted by between 2.3 and 3.0 mV, becoming more lumen-negative. A similar potential change of the perfused duct after application of secretin was also observed by MOQTADERI et al. (1972). From the PD changes, WAY and DIAMOND (1970) calculated the relative permeabilities for the main pancreatic duct as $P_{Cl}:P_K:P_{HCO_3}:P_{Na} = 1.00:0.90:0.59:0.52$, which are almost identical with the free solute mobilities. Since secretin causes no change in the relative permeabilities and produces a PD in the absence of any change in ion concentration gradients, the secretin-induced PD must be due directly to an active ion transport. Calculations based on the Nernst equation showed that the ion influxes caused by secretin are against the electrochemical gradient for HCO_3^- but down tha gradient for Cl^-, Na^+ and K^+; consequently one can say that secretin causes an active HCO_3^- transport from the interstitium into the duct lumen.

V. Mechanism of Ion Secretion

1. Buffer Secretion

In order to decide what ion species are secreted actively, the electrochemical potential differences of the different ions between duct lumen and interstitium have been determined (SCHULZ et al., 1969; WAY and DIAMOND, 1970; SWANSON and SOLOMON, 1973). All three groups of authors agree that, of the ion fluxes stimulated by secretin, that of HCO_3^- (or the other buffer components H^+ or OH^-) takes place against its electrochemical gradient, while those of Na^+, K^+ and Cl^- take place down their gradients. There are, however, other arguments, which favor an active secretory mechanism for Na^+ (RIDDERSTAP and BONTING, 1969a; SWANSON and SOLOMON 1972, 1973, 1975); these will be discussed later.

a) What Buffer Components in the Perfusate Determine Buffer and Fluid Secretion

Since it became clear that bicarbonate secreted by the pancreas is derived from the blood and only to a small extent (< 6%) from the CO_2 produced by the pancreatic cells (BALL et al., 1941; ROTHMAN and BROOKS, 1965b; CASE et al., 1970; SCHULZ et al., 1971; SWANSON and SOLOMON, 1973; HADI et al., 1976), the main interest has been focussed on the question of what buffer component is transported actively. In order to distinguish between the possible primary events, namely direct HCO_3^- ion secretion or H^+ ion reabsorption (OH^- ion secretion), the relative CO_2 pressures in the final secretion and in the perfusate were considered. HUBEL (1967) found in the rabbit pancreas kept *in vitro* that the pCO_2 in the secreted fluid was higher than in the bathing fluid, and attributed this to a higher pCO_2 of the pancreatic tissue or of lactic acid, which could have been released into the ductal fluid. RAWLS et al. (1963) studied the secretin-stimulated dog pancreas and found that the pCO_2 of the pancreatic juice was more than 10 to 18 mm Hg less than in plasma, except when the secretory rate was strongly inhibited by metabolic acidosis and simultaneous application of a carbonic anhydrase inhibitor. This finding, as well as that of CASE et al. (1970) that the pCO_2 of the effluent perfusate increases by 5 mm Hg when the gland is stimulated, indicates that H^+ ions are transported from the glandular lumen toward the perfusate, creating diffusion gradients of CO_2 in the secretory direction. In the small intestine, where after application of choleratoxin HCO_3^- appears in the lumen and the luminal pCO_2 falls, the same mechanism was proposed (HUBEL, 1974). The strongest evidence, however, that H^+/OH^- is transported primarily comes from the finding that weak organic acids such as sulfonamides, sulfamerazine or glycodiazine (SCHULZ et al., 1971; SCHULZ, 1971) or the organic acid anions formate, propionate, and butyrate (SWANSON, 1971) and especially acetate (SWANSON and SOLOMON, 1975; HADI et al., 1976) can substitute for bicarbonate in promoting pancreatic buffer and fluid secretion. Simultaneously the effect of the buffer components on fluid secretion and the concentration of buffer anions in the secreted fluid were studied by CASE et al. (1970), SCHULZ et al. (1971), SCHULZ (1971), and SWANSON and SOLOMON (1975). If the concentration of bicarbonate in the perfusate is augmented, the rate of fluid secretion and the concentration of bicarbonate in the secreted fluid increase proportionally (Table 1 and Fig. 4). The same, although to a much smaller degree, is observed when the buffer anion concentration (A^-) of glycodiazine in the perfusate is raised (SCHULZ, 1971). Recently it was shown that this relation also holds true for acetate (HADI et al., 1976). The effect of pCO_2 changes in the perfusate on secretory flow rate and buffer anion concentration, however, are rather small, and furthermore conflicting. This seems to be caused by the fact that at constant bicarbonate concentration, which is required because of the large HCO_3^- effects, a change in pCO_2 causes a change in pH as well. Thus SWANSON and SOLOMON (1975) attributed the pCO_2 effects they observed to pH changes, and claimed that an increase in perfusate pH increases the secretion rate of fluid but decreases the buffer concentration in the secreted fluid. The rate of buffer secretion, however, increases because the stimulatory effect on

Table 1. Effect of increased buffer components (A^-, pH, AH) in the perfusate on fluid secretion, or A^- concentration in the secreted fluid as indicated by the signs

Author	Animal	Buffer	Effect on secretion of	Rise in perfusate A^- (AH = const.)	Rise in perfusate pH (A = const.)	Rise in perfusate AH (A = const.)
SCHULZ, 1971	Cat	HCO_3^-/CO_2	Fluid	↑		↑
			$[A^-]$	↑		↑
SWANSON and SOLOMON 1975	Rabbit	HCO_3^-/CO_2	Fluid	↑	↑	
			$[A^-]$	↑	↓	
SCHULZ et al., 1971	Cat	Sulfamerazine	Fluid	~		~
			$[A^-]$	~		↑
SCHULZ, 1971	Cat	Glycodiazine	Fluid	↑	↑	
			$[A^-]$	↑	←→	
SWANSON and SOLOMON, 1975	Rabbit	Acetate	Fluid	↑		
HADI et al., 1976	Cat	Acetate and others	Fluid			↑

↑ fluid and/or A^- secretion increased; ↓ decreased; ~ were variable; ←→ were approximately equal; index = says what buffer component was held constant

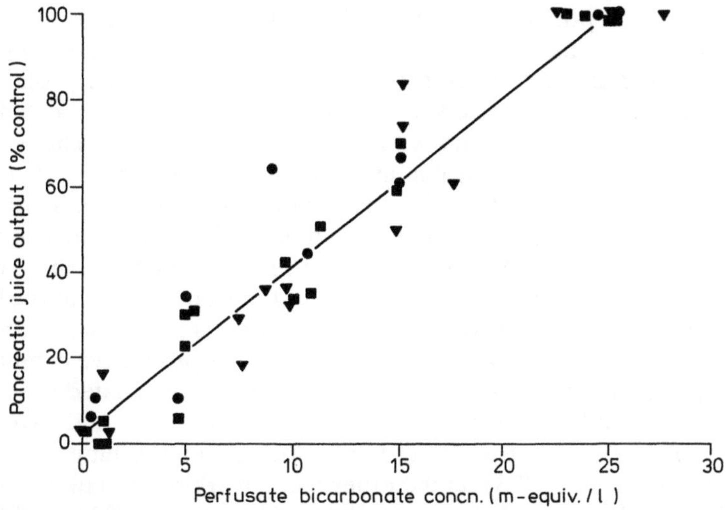

Fig. 4. Effect of perfusate bicarbonate concentration on rate of fluid secretion in perfused cat pancreas. ● Krebs-Ringer as perfusate, ■ Krebs-Ringer with 6% Dextran, ▼ Krebs-Ringer with 2% hemoglobin. Rate of secretion is expressed as percentage of rate at 25 mmolar HCO_3^-. (CASE et al., 1970)

fluid secretion is much larger than the inhibitory effect on the buffer concentration. At constant buffer anion concentrations a similar pH effect was seen with acetate (SWANSON and SOLOMON, 1975; HADI et al., 1976) and with glycodiazine (SCHULZ, 1971). On the other hand, SCHULZ (1971) observed in some rare cases that a rise in pCO_2 increased the bicarbonate concentration in the secreted fluid and to a larger extent the fluid secretion. This is in agreement with the sulfamerazine experiments, in which the buffer-anion concentration of the secretory fluid is determined by the concentration of the undissociated buffer component (AH) in the perfusate (SCHULZ et al., 1971).

Thus one can consider different components of the perfusate that stimulate buffer secretion: in the case of sulfamerazine it is an increased AH only; with acetate and glycodiazine it is an increase of A^- or an increase of pH. In the case of CO_2 it is a mixture of both: HCO_3^- and pH both stimulate, especially at high HCO_3^- concentrations; at low HCO_3^-, CO_2 can be stimulatory. Here it should be mentioned that the data that RAWLS et al. (1963) obtained in the dog pancreas after changing blood HCO_3^- or pH agree in general with those obtained *in vitro* in the rabbit pancreas (SWANSON and SOLOMON, 1975).

b) Sidedness of the Buffer Transport

As the actively transported buffer must take the transcellular route and undergo at least one active transport step, the question arises at which cell side this active step is located and how the buffer crosses the other cell side. Furthermore, it seemed desirable to establish what transport step is triggered by the secretagogue secretin. To answer this question several attempts to determine the transcellular electrochemical potential profile for H^+ (HCO_3^-) have been made, but so far none is really satisfactory. SCHULZ et al. (1969, 1971) and SWANSON and SOLOMON (1973) measured the transcellular electrical potential profile of the small ducts, where the main electrolyte secretion occurs: they reported values for the lumen and cell of −5 mV and −50 mV and of −9.1 and −40 mV, respectively. Knowing the intraluminal HCO_3^- concentration of maximal 100 mEq l^{-1} at a contraluminal HCO_3^- concentration of 28 mEq l^{-1}, SCHULZ et al. (1974) argued that the active transport step must be located at the contraluminal cell side if the intracellular HCO_3^- concentration is higher than 18 mEq l^{-1} *i. e.* if the intracellular pH is higher than 7.27, but must be located at the luminal cell side if the intracellular bicarbonate concentration is lower than 4.2 mEq l^{-1}, *i. e.* if the intracellular pH is lower than 6.65. From the subjective estimation of the cell color in experiments with bromthymol blue they estimated the intracellular pH of the small duct cells to be far lower than 7.0, and favored the hypothesis that the active transport step is located at the luminal cell side. SWANSON and SOLOMON (1973) on the other hand, working with pieces of rabbit pancreatic tissue, determined the intracellular pH with ^{14}C-DMO and found a value of 7.25 (HCO_3^- concentration = 18 mEq l^{-1}) at the pCO_2 they had applied. They assumed that the intracellular pH determined on whole tissue reflects that of the buffer-secreting ductular cells, and claimed that the active transport step must be located at the contraluminal cell side and overcome an electrochemical potential difference of 30 mV. They calculated in addition a maximum electrochemical potential difference of 15 mV against which H^+ (HCO_3^-, or OH^-) must be transported at the luminal cell membrane, and suggested an active transport step at this cell side as well. As far as the effect of secretin on the electrochemical potential profile for H^+ (HCO_3^-, or OH^-) of the buffer-secreting cells is concerned, no reliable data are available. There are, however, electrical potential measurements recorded in mouse pancreatic cells that respond to secretin with hyperpolarization (GREENWELL, 1975). Membrane potential frequency histograms led to the suggestion that this hyperpolarization is as large as 18 mV, *i. e.* from −23 to −41 mV. Furthermore, it is assumed that this electrical response to secretin occurs in the buffer-secreting duct cells. At least three different events could be suggested as possible causes of this hyperpolarization: 1. An increase of the K^+ permeability of the contraluminal cell side. This would be in keeping with the finding of CASE et al. (1969) that after secretin application a transient increase of K^+ concentration appears in the perfusate of the gland. 2. An increase in the intracellular HCO_3^- concentration at a high peritubular HCO_3^- permeability. 3. An increased electrogenic pump potential, caused either by increased active electrogenic Na^+ transport from the cell into the lumen or by H^+ transport out of the cell at the contraluminal cell side.

2. Secretion of Na^+ Ions

As pointed out above, Na^+ is secreted from the interstitium into the lumen along its electrochemical potential gradient. Since, however, Na^+ secretion stops when net water flux is prevented by adding sucrose to the peritubular perfusion fluid at a transepithelial concentration difference for Na^+ of 110 mEq l^{-1} (perfusate 55, secreted fluid 165 mEq l^{-1}) (CASE et al., 1968b) a transtubular electrical PD of up to 29 mV (lumen-negative) would be expected if Na^+ were to be transported passively. Although in these Na^+-replacement studies (CASE et al., 1968b; SWANSON and SOLOMON, 1975) a rise in the ratio of the K^+ concentration in the secreted fluid to that in the perfusate (S/P) by up to 3 could be taken as indicative of a PD increase to such a value, the published S/P ratios for Cl^-, which remain near 1.0, do not support this suggestion. Since the magnitude of the PD at the site of electrolyte secretion under the various secretory conditions is unknown, it is not possible to decide whether Na^+ is transported exclusively passively and could use the paracellular route, or whether it is transported actively and has therefore to pass through the cell. There are, however, other arguments, some of which support passive and others active Na^+ transport. The fact that Na^+ in the perfusate of the pancreas could be to a large extent substituted by K^+ (CASE and SCRATCHERD, 1974) and that both ions are always in the same or a somewhat higher concentration in the secreted fluid than in the perfusate, suggests a passive behavior of both ions. The fact, however, that under these circumstances the fluid secretory rate parallels that of Na^+ but not that of K^+ strongly supports a special role of Na^+. Furthermore, a strong argument that Na^+ is involved in some active route comes from the finding of RIDDERSTAP and BONTING (1969a) that the dose-response curve of ouabain on pancreatic Na^+-K^+-ATPase coincides with that on resting fluid secretion of the isolated rabbit pancreas. Both are inhibited by 50 percent with $5 \cdot 10^{-6}$ M ouabain, but stimulated by 15–20 percent at 10^{-9} M. SWANSON and SOLOMON (1975) observed the same strong inhibition of fluid secretion, but in contrast to RIDDERSTAP and BONTING (1969c) they also saw a marked decrease in the HCO_3^- concentration of final and ductal fluid, which they felt to be greater than would be expected on the basis of the secretion rate changes alone. Thus they suggested that ouabain inhibits the secretion of HCO_3^- more strongly than that of Na^+. Despite this strong inhibitory effect of ouabain on the rate of HCO_3^- secretion, no effect on intracellular pH (measured in whole pancreatic tissue) or on the intracellular PD was observed (SWANSON, 1971; SWANSON and SOLOMON, 1972). Since the transtubular electrochemical potential profile for H^+ was therefore not altered by ouabain, SWANSON and SOLOMON (1975) argue that ouabain must inhibit H^+ (HCO_3^-, or OH^-) transport across both cell membranes, the luminal and the contraluminal (Fig. 5). For the contraluminal inhibition they assume that the coupled Na^+-influx/H^+-outflux from the cell is decreased. Because of inhibition of the Na^+-K^+ pump, the intracellular Na^+ concentration rises after the administration of ouabain, thereby diminishing the driving force for the contraluminal Na^+ influx. At the luminal cell membrane they assume an inhibitition of the mucosal Na^+-H^+ pump due to the changed intracellular ionic composition.

In parallel experiments with low Na^+ in the bathing medium, where the HCO_3^- secretion is inhibited to the same extent as by ouabain, it was observed that the bath/cell Na^+ concentration ratio is reduced less (from 2.8 to 1.5) than with ouabain (to 1.1), but that the cell pH is raised (from 7.28 to 7.48) (SWANSON and SOLOMON, 1972). The cell PD with low Na^+ in the bathing medium is

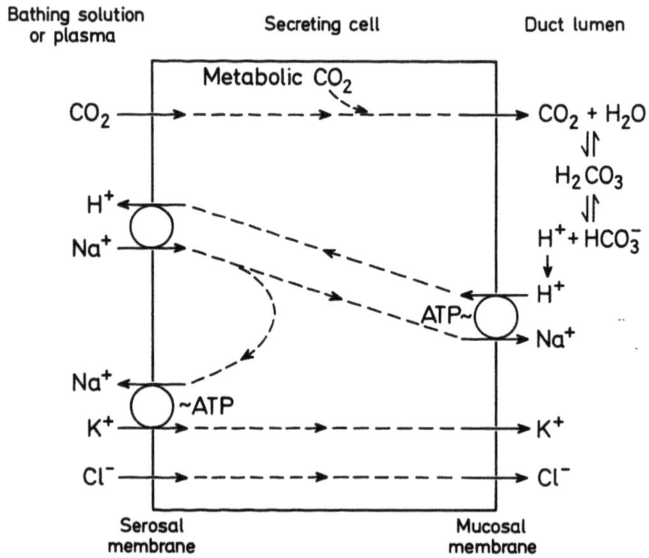

Fig. 5. Schematic representation of proposed cellular mechanism for pancreatic electrolyte secretion. Serosal Na^+-H^+ exchange is linked indirectly to metabolism through Na^+-K^+ pump (= Na^+ driven secondary active H^+ transport). Mucosal Na^+-H^+ pump requires direct metabolic coupling *i. e.* primary active H^+ transport. (SWANSON and SOLOMON, 1975)

unchanged (SWANSON, 1971). From these findings the authors concluded that at low Na^+ the net transport of H^+ ions is inhibited at the mucosal cell membrane. They propose an ATP-driven Na^+/H^+ pump at this cell site, which pumps Na^+ out of the cell into the duct lumen in exchange for H^+ ions, and operates at a reduced rate when the Na^+ gradient at this membrane is raised, a situation prevailing with low Na^+ in the perfusate. Finally another finding should be mentioned which supports a transcellular route of the secreted Na^+ ions, namely that the Na^+ concentration in the gland increases after secretin stimulation, while the K^+ content decreases (CLARK et al., 1967). This indicates that after secretin stimulation Na^+ enters the cells while K^+ leaves them. Simultaneously a change in tissue conductance occurs, which could be interpreted as cell swelling.

The problem of whether the paracellular pathway at the site of buffer secretion is tight or not, plays a big role in the argument as to whether Na^+ ions are secreted actively or passively. Thus SWANSON and SOLOMON (1973) mention that the composition of the relatively high-HCO_3^-, low-Cl^- juice secreted spontaneously in the interlobular ducts is not changed at rest when it flows through the small extralobular ducts — which respond to secretin with increased buffer secretion. The maintenance of the large electrochemical potential difference between the bathing solution and the ductal juice suggests that the intercellular pathway is rather impermeable to anions. On the other hand, sucrose, which is added to low-Na^+ media, appears in considerable amounts within the secreted fluid (CASE et al., 1968b; SWANSON and SOLOMON, 1975) and mannitol is found in the secreted fluid even in concentrations half that in the incubation medium (SCHREURS et al., 1975). Reflection coefficient measurements with nonelectrolytes showed that the small ducts are indeed amongst the leakiest epithelia (DEWHURST et al., 1978).

3. Possible Mechanism of Buffer Secretion

Since reliable data for buffer secretion in the pancreas are lacking, it may be useful to consider possible sidedness arrangements and molecular events of

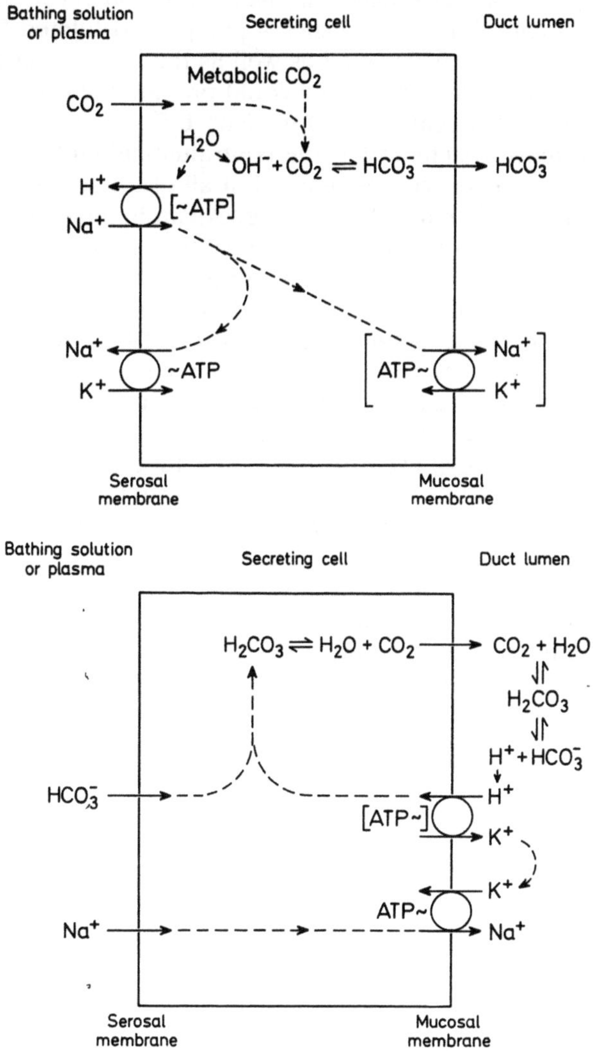

Fig. 6. Schematic representation of possible cellular mechanism for pancreatic $NaHCO_3$ secretion. A. Na^+ gradient or ATP-driven H^+ pump at contraluminal or serosal cell side, Na^+-K^+-ATPase at both cell sides creating Na^+ gradient. Exit of HCO_3^- at luminal or mucosal cell side is passive. B. ATP-driven H^+-K^+ and Na^+-K^+ pump at luminal cell side; entrance of HCO_3^- and Na^+ at contraluminal cell side is passive

transpithelial buffer transport. For the secretion of buffers in the pancreas we must postulate at least one active transport step through one cell side, combined with a passive step through the other. If we consider the strong Na^+ dependence and ouabain sensitivity of the buffer transport, beside the model depicted in Fig. 5, two others are possible (Fig. 6A and B).

Three combinations are possible:

1. Active transport step (H^+ pump) at the contraluminal cell side combined with a passive one (HCO_3^- exit) at the luminal cell side (Fig. 6A). In this situation we could have Na^+-H^+ countertransport at the contraluminal cell side. In analogy to the proximal renal tubule, where such a pump is postulated for the luminal cell side (ULLRICH et al., 1977), ATP as well as the Na^+ ion gradient directed from the interstitium to the cytoplasma could provide the driving force for H^+ transport, the electrochemical gradient for Na^+ being established by the Na^+-K^+-ATPase transport system within the contraluminal and luminal cell membrane. The latter, however, can be omitted, if all of the sodium is presumed to move passively through a paracellular route. Again in analogy to the proximal renal tubule, where a Na^+-H^+-exchange system and a HCO_3^- stimulated ATPase are located in the same cell membrane (KINNE-SAFFRAN and KINNE, 1974; LIANG and SACKTOR, 1976) and could be identical, HCO_3^--ATPase was found in pancreatic tissue (SIMON and THOMAS, 1972; SIMON et al., 1972; WIZEMANN and SCHULZ, 1973), and in higher concentrations in the small secreting ducts than in the acini and the large duct system (WIZEMANN et al., 1974). The passive buffer transfer at the luminal cell side could be similar to that at the basolateral cell membrane of the kidney proximal tubule (FRÖMTER et al., 1976).
2. An ATP driven H^+-K^+-transport system similar to that found in the stomach could be located at the luminal cell side. (GANSER and FORTE, 1973; SACHS et al., 1976) The combination of a K^+-H^+ exchange with a Na^+-K^+ ATPase could achieve a net Na^+ secretion and H^+ reabsorption (Fig. 6B). Theoretically, however, ATP would not be needed for a luminal K^+-H^+ exchange where the potassium gradient could provide the necessary driving force. The passive entry of HCO_3^- ions at the contraluminal cell site could then proceed by a similar system as described under 1. A K^+-stimulated, ouabain-insensitive ATPase has not yet been found in the pancreas (SCHULZ, I. unpublished observations).

Since the bicarbonate transfer step in the proximal renal tubule and in the turtle bladder is sensitive to SITS (ULLRICH et al., 1977; EHRENSPECK and BRODSKY, 1976; COHEN et al., 1978), a similar inhibitory effect was also expected from the pancreas. Indeed a strong inhibition of bicarbonate secretion in the cat pancreas after application of SITS was observed (SCHULZ I., unpubl.). The sidedness of this inhibition, however, has not yet been established.

C. Secretion of Enzymes

I. Transport Processes in the Acini Involved in Stimulus-Secretion Coupling

1. Secretagogues of Enzyme Secretion

The physiological stimulants of enzyme secretion from the acini are the neurotransmitter acetylcholine and the oligopeptide pancreozymin (cholecystokinin) (JORPES and MUTT, 1973). Gastrin, a peptide consisting of 17 amino acids and similar to pancreozymin in its C-terminal amino acid sequence also stimulates enzyme secretion (GREGORY and TRACY, 1964; ANDERSON et al., 1964). Pancreozymin, which in some species also stimulates electrolyte and fluid secretion (SEWELL and YOUNG 1975), was purified by JORPES and MUTT (1962) and JORPES et al. (1964) from a crude preparation that not only caused enzyme secretion in the pancreas but also gallbladder contraction. The latter effect was referred to another hormone called cholecystokinin. Further purification and analysis of the amino acid sequence revealed that the activities attributed to two hormones are exerted by one and the same substance, a peptide of 33 amino acids. Furthermore, enzyme secretion is stimulated by the decapeptide caerulein, which was isolated from the skin of Australian amphibia *Hyla caerulea* (ANASTASI et al., 1967). Caerulein is structurally similar to the C-terminal octapeptide of pancreozymin, except that it contains threonin instead of methionin (MUTT, 1970). Caerulein and the synthetic C-terminal octapeptide of pancreozymin (ONDETTI et. al., 1970) both exert all actions known for both gastrin and pancreozymin with a potency 8 to 16 times higher than that of pancreozymin. Recently another family of peptides, including bombesin, has been isolated from amphibian skin; these differ totally from the pancreozymin family, including caerulein (ERSPARMER and MELCHIORRI, 1975). Although the pancreatic response to these peptides had been attributed to their ability to release pancreozymin from the duodenal mucosa (ERSPARMER et al., 1968, 1974; ERSPARMER and MELCHIORRI, 1975), a direct effect of bombesin and litorin on amylase release and on Ca^{++} efflux from pancreatic fragments has been shown (DESCHODT-LANCKMAN et al., 1976). All these substances act on receptors on the outer surface of the acinar cells. They trigger a chain of reactions that culminate in fusion of the zymogen granule membrane with the luminal plasma membrane. Binding studies with ^{3}H-labeled octapeptide of pancreozymin and with [^{3}H]-caerulein to acinar cells and plasma membranes have been performed (MILUTINOVIĆ et al., 1977a; ROBBERECHT et al., 1977). The apparent K_D value for ^{3}H-octapeptide binding is 30 nM and that for the adenylate cyclase stimulation by the octapeptide 20 nM. To obtain the same half-maximal adenylate cyclase stimulation with pancreozymin as with octapeptide a concentration 30 times higher of the former hormone is necessary (MILUTINOVIĆ et al., 1977a). The K_D for caerulein binding is 100 times higher than the concentration required for amylase secretion (ROBBERECHT et al., 1977).

2. Electrophysiological Measurements

The effects of the enzyme secretagogues on the electrical parameters of the acinar cells have been investigated extensively (KANNO, 1972; PETERSEN and MATTHEWS, 1972; MATTHEWS and PETERSEN, 1973; MATTHEWS et al., 1973; NISHIYAMA and PETERSEN 1974a and b, 1975a and b; PETERSEN and UEDA, 1975a and b; PETERSEN, 1976; IWATSUKI and PETERSEN, 1976). Brief exposure of the superfused pancreas to acetylcholine, pancreozymin, gastrin, caerulein, or bombesin results in a short-term depolarization of the acinar cells, which have a resting potential of about −40 mV (MATTHEWS and PETERSEN, 1973; NISHIYAMA and PETERSEN, 1974a and b, 1975a and b; PETERSEN and UEDA, 1975b; IWATSUKI and PETERSEN, 1977a and b, 1978). Atropine blocks the depolarizing action of acetylcholine but not that of pancreozymin. Secretin and cyclic AMP have no effect on the membrane potential. A detailed analysis of the electrical events reveals that the acetylcholine-induced depolarization coincides with a reduction of the input resistance caused by an increased conduction of the plasma membrane to Na^+ and K^+, and possibly also to other ions (Fig. 7) (NISHIYAMA and PETERSEN, 1975a; IWATSUKI and PETERSEN, 1976). A subsequent analysis of the ionic dependence of the acetylcholine zero potential showed indeed that the acetylcholine-opened pathways in the acinar plasma membrane are mainly permeable to Cl^- and Na^+, i. e. stimulation causes uptake of Na^+ and Cl^- into the acinar cells (IWATSUKI and PETERSEN, 1977a). The electrical measurements, unfortunately, were not sensitive enough to detect electrical signals that could be attributed to a change in Ca^{++} conductance (NISHIYAMA and PETERSEN, 1975a). Thus the authors originally suggested that

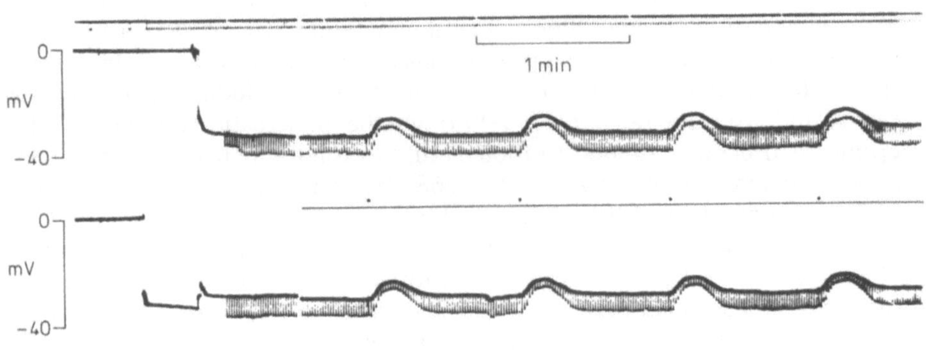

Fig. 7. Effect of short pulses of iontophoretic acetylcholine application on membrane potential and resistance of mouse pancreactic acinar cells. Two microelectrodes – upper and lower trace – are impaled in neighboring acinar cells. Through electrode 1 – upper trace – square wave current pulses were injected. Simultaneously, electrical potential was measured by this electrode. Short-duration current injections gave rise to short-lasting membrane hyperpolarizations both in the cell into which current was injected (upper trace) and in the neighboring cell (electrode 2, lower trace). Coupling ratio as evaluated by current pulse-induced potential change in cell of current injection divided by electrotonic potential change in neighboring cell is 1, *i. e.* there exists perfect coupling. After acetylcholine pulses, same degree of depolarization is seen in both cells. Furthermore, current pulse-induced potential changes decrease, again in both cells to the same degree, which indicates an increase in membrane conductivity. (IWATSUKI and PETERSEN, 1976)

Ca^{++} influx does not play any role in the acetylcholine-evoked depolarization, but they state explicitly that their results do not exclude the presence of a very small acetylcholine-induced conductance change for Ca^{++}.

In contrast to other granule-secreting tissues like the adrenal medulla and the neurohypophysis (DOUGLAS, 1968) in which depolarization by K^+ causes secretion, in the pancreas there is no direct relationship between amylase release and membrane potential. Irrespective of whether the resting acinar membrane potential is lowered artificially below the level normally produced by physiological stimulants (−25 to −30 mV) by an increase in extracellular K^+ concentration or by removal of extracellular Ca^{++} (ARGENT et al., 1971; PETERSEN and UEDA, 1976b; POULSEN and WILLIAMS, 1977a and b), the rate of amylase release does not exceed the previous basal level. Nor is amylase release directly related to the magnitude of stimulation-induced depolarization, since an almost normal amylase release can be elicited by secretagogues in the presence of high extracellular K^+ concentrations, at which the depolarizing effect of secretagogues was abolished (POULSEN and WILLIAMS, 1977b). Here it should be mentioned that the acetylcholine-evoked depolarization changes to a biphasic response (depolarization-hyperpolarization) if the resting potential of the acinar cell is lower than −40 mV (NISHIYAMA and PETERSEN, 1975b). This secondary hyperpolarization, which had apparently already been observed by KANNO (1972), seems to be caused by membrane permeability changes rather than by an electrogenic sodium pump, since ouabain has no effect on it (NISHIYAMA and PETERSEN, 1975b).

3. Role of Calcium Ions

As long ago as 1966 it was observed by HOKIN that acetylcholine-induced enzyme secretion does not occur if the incubation medium is free of Ca^{++} ions (HOKIN, 1966). This observation was later confirmed by ROBBERECHT and CHRISTOPHE (1971), HEISLER et al. (1972), KANNO (1972), BENZ et al. (1972), CASE and CLAUSEN (1973), and ARGENT et al. (1973). The key question was whether the enzyme secretagogues increased the intracellular Ca^{++} by stimulating Ca^{++} influx into the cells or by liberating Ca^{++} from the inside of the plasma membrane or other intracellular stores. In the latter case, release of calcium by direct action of the neurotransmitter or hormone or by a second messenger, such as cyclic adenosine 3'-5'-monophosphate (cAMP) or inflowing Na^+ ions was considered (MATTHEWS et al., 1973; CASE and CLAUSEN, 1973; WILLIAMS and CHANDLER, 1975; SCHREURS et al., 1976a). The stimulatory effect of extracellular Ca^{++} on enzyme secretion was clearly demonstrated by sudden increase of the extracellular Ca^{++} concentration (SCHULZ, 1975a and b; PETERSEN and UEDA, 1976b) and by use of the calcium ionophore A 23187 (EIMERL et. al., 1974; WILLIAMS and LEE, 1974; SCHREURS et al., 1976a). Although the ionophore can penetrate into the cell and deplete an intracellular pool of Ca^{++} (WILLIAMS et al., 1976a), its effect on enzyme secretion could be attributed to an increased Ca^{++} permeability of the cell surface membrane since its action was completely dependent on extracellular Ca^{++} (SCHREURS et al., 1976a; WILLIAMS et al., 1976b; POULSEN and WILLIAMS, 1977a). All authors who measured tracer $^{45}Ca^{++}$ fluxes[1] found that the enzyme secretagogues increased the $^{45}Ca^{++}$ efflux from the cells (HEISLER et al., 1972; CASE and CLAUSEN, 1973; MATTHEWS et al., 1973; SCHREURS et al., 1975, 1976a and b; KONDO and SCHULZ, 1976b; GARDNER and HAHNE 1977), but only HEISLER et al. (1972) and KONDO and SCHULZ (1976a) – the latter authors working with isolated cells and thereby able to remove interstitial $^{45}Ca^{++}$ – observed an increased $^{45}Ca^{++}$ influx as well as an increased $^{45}Ca^{++}$ pool size in the presence of secretagogues in the incubation medium. This indicates that the secretagogues increase the Ca^{++} permeability, supposedly of the outer cell membrane. Recently POULSEN and WILLIAMS (1977c) and dePONT et al. (personal communication) have confirmed the augmentation of $^{45}Ca^{++}$ uptake into isolated acini and acinar cells as stimulated by secretagogues of enzyme secretion. The authors, however, interpret their findings as due to an increased turnover rate of intracellular Ca^{++} rather than to a permeability change of the cell membrane for Ca^{++}. By comparing the amount of extra Ca^{++} flowing into a pancreas cell after acetylcholine stimulation (KONDO and SCHULZ, 1976a) with the electrical resistance of the acinar cell membrane (PETERSEN, 1976) it could be calculated that this Ca^{++} influx would diminish the membrane potential by 0.01 mV only. Such a small difference in potential difference cannot be detected by electrical measurements. On the other hand, PETERSEN and UEDA (1976b) observed that the electrical coupling between neighboring acinar cells was diminished after supermaximal stimulation, which is an infallible indication of elevated cytoplasmic Ca^{++} concentration (LOEWENSTEIN et al., 1967; DÉLÈZE and LOEWENSTEIN, 1976; ROSE and LOEWENSTEIN, 1976). In attemps at elucidating whether Ca^{++} liberated from intracellular stores participates in stimulus-secretion coupling, experiments with the Ca^{++}-complexing agent EGTA are of special interest. Acute application of EGTA *per se* induces a small increase in amylase secretion (SCHULZ, 1975b; SCHREURS et al., 1976a). If EGTA is given 1 min before the application of acetylcholine, it reduces the acetylcholine effect by 30 percent, while the effect of pancreozymin is either unchanged or even augmented

1 There is some confusion in the literature concerning ^{45}Ca flux measurement since it is often tacitly assumed that unidirectional influx or efflux values indicate net fluxes (GARDNER and HAHNE, 1977). The rate of net flux, however, equals the tracer flux only if the specific activity is the same in all compartments (SAUER, 1978). Otherwise the net flux is clearly the difference of the unidirectional fluxes. The unidirectional flux *per see* depends on possible coupling with many simultaneously occurring net fluxes and with the metabolism. Thus working at the condition of zero net flux creates a simpler situation but influence by tracer coupling and active transport is not eliminated. When pancreatic acinar cells are for the first time stimulated by secretagogues, burst secretion of enzymes with large amounts of bound Ca^{++} – and thus of ^{45}Ca, if preloaded with this isotope – occurs. This is a clear case of flux coupling which runs on the exocytosis pathway and must be regarded as an additional route parallel to that through the plasma membranes. Another problem is the correction for $^{45}Ca^{++}$ adhering extracellularly to cells. Any kind of washing with a fluid deviating in composition from that present during incubation creates artifacts. Thus the best, but because of the complexity of the system not unequivocally interpretable data, are those of $^{45}Ca^{++}$ efflux gained in a quasi steady state with zero net flux (KONDO and SCHULZ, 1976b). Here the outer face of the cell is never in contact with any other solution as the incubation solution and burst secretion is long over when the flux measurements start.

(SCHULZ, 1975b). When the gland was superfused with a Ca^{++} free fluid containing 10^{-4} M EGTA, the effect of subsequent acetylcholine stimulations declined more rapidly (PETERSEN and UEDA, 1976b). Nevertheless, after superfusion for more than 90 min the stimulatory effect of acetylcholine pulses is still present, although only half as large as in a subsequent control after readmission of Ca^{++}. With permanent superfusion with acetylcholine, PETERSEN and UEDA (1976b) made the important observation that the acetylcholine effect declines rapidly in a Ca^{++}-free solution containing EGTA, but recovers and remains stable when Ca^{++} is readmitted. This indicates that the secretagogues first act by liberating membrane-bound Ca^{++}, but that for a permanent action, Ca^{++} coming from outside into the cell is needed. This resembles the situation observed during cholinergic stimulation of smooth muscle (CHANG and TRIGGLE, 1973). The suggestion that secretagogues first liberate membrane-bound Ca^{++} is supported by the finding of SCHREURS et al. (1976a) that carbachol leads to ^{45}Ca release from preloaded cells while the Ca^{++} ionophore A 23187 does not. A similar lack of an effect of the ionophore on $^{45}Ca^{++}$ efflux was observed using Ca^{++}-deficient media. Nontheless, under the same Ca^{++}-poor conditions the ionophore did appear to stimulate cellular Ca^{++} loss as demonstrated indirectly by its ability to inhibit the secretory responses to carbachol, an effect which may have been due to an outward transport of cholinergic mobilized intracellular Ca^{++} by the ionophore (HEISLER, 1976). In this context it should be mentioned that experiments on isolated acetylcholine receptors of *Torpedo* are in agreement with the hypothesis that acetylcholine liberates a considerable amount of calcium from the receptor protein (RÜBSAMEN et al., 1976; MARTINEZ-CARRION et al., 1976). Concerning the small direct effect of EGTA in inducing enzyme secretion, it is possible that EGTA, by removing Ca^{++} from the extracellular side, leads to release of a small amount of membrane-bound Ca^{++} from the intracellular side, which increases the intracellular Ca^{++} concentration.

PETERSEN and UEDA (1976b) showed that the conductivity of the acinar cell membrane is governed by the Ca^{++} concentration of the incubation medium and that Mg^{++} and especially Mn^{++} can substitute for Ca^{++} in this effect. Furthermore, when shot superfusions with Mg^{++} and Mn^{++} are made on a gland in Ca^{++}-free EGTA solutions, enzyme secretion similar to that obtained with shot superfusions with Ca^{++} is provoked (PETERSEN and UEDA, 1976b). In explaining these results, the authors argue that considering the effect of Mg^{++} and Mn^{++} in other tissues (MILEDI, 1973; KANNO et al., 1973), it is more likely that both ions liberate Ca^{++} from the membrane than that they act directly. Although Mn^{++} alone, when flushed over the gland, stimulates enzyme secretion, Mn^{++} has an inhibitory action on the pancreozymin-induced release of enzyme in media containing Ca^{++} (KANNO and NISHIMURA, 1976). A kinetic scheme proposed by these authors is in agreement with the assumption that Mn^{++} inhibits the pancreozymin-induced Ca^{++} influx into the cell in a competitive manner.

Although reduction of the extracellular sodium concentration has been found to stimulate many Ca^{++}-dependent secretory systems (COCHRANE and DOUGLAS, 1976) the findings on the effect of Na^{+}-free media on pancreatic enzyme secretion are conflicting. In the intact mouse pancreas, PETERSEN and UEDA

(1976b) observed an inhibition of secretagogue-induced enzyme release if Na^+ was replaced by Tris in the incubation medium. Since readdition of Na^+ caused enzyme secretion, the authors referred the observed inhibition on enzyme release to an impaired washout phenomenon due to inhibition of fluid secretion in the absence of Na^+. In slices and isolated cells, WILLIAMS (1975) and SCHULZ et al. (1977) observed an augmented resting enzyme secretion when extracellular Na^+ was reduced or omitted, but no further stimulatory effect of secretagogues under these conditions. In parallel to this, the last-named authors observed that the in- and outflux of $^{45}Ca^{++}$ was augmented in isolated pancreas cells incubated in Na^+-free Tris, choline, or mannitol solutions, but that the secretagogues had no additional effect on the Ca^{++} fluxes. If, however, Na^+ was replaced by Li^+ or K^+, no change was observed with regard to amylase secretion and Ca^{++} fluxes as compared to the Na^+-containing controls (KONDO and SCHULZ, 1976b). Thus it appears that the effect of low Na^+ concentrations in the extracellular medium on amylase release and Ca^{++} fluxes is not an effect of Na^+ *per se*, but rather due to reduced ionic strength within the membrane, since $Tris^+$ and $choline^+$ cannot enter the membrane as freely as the smaller K^+ or Li^+.

Here a long known fact should be mentioned. Acetylcholine and pancreozymin or caerulein augment the phosphatidylinositol labelling and breakdown in pancreatic acinar cell membranes (HOKIN and HOKIN, 1955, 1956; BAUDUIN et al., 1971; BAUDUIN and CANTRAINE, 1972; HOKIN, 1974). Phosphatidylinositols have a high affinity for Ca^{++} ions (HAUSER and DAWSON, 1967) and are therefore possible candidates for membrane calcium storage and release. Since the phosphatidyl breakdown is only dependent on extracellular Ca^{++} to a small extent (HOKIN, 1966), its role in stimulus-secretion coupling is proposed to be either the intracellular liberation of membrane-bound Ca^{++} or in the opening of Ca^{++} gates (MICHELL, 1975). The increased ^{32}P phosphatidylinositol turnover seems to correlate with Ca^{++}-mediated secretory steps. Thus SELINGER (1974) observed that in the rat parotid gland only the Ca^{++}-mediated α-adrenergic or cholinergic response was associated with an increased incorporation of ^{32}P into the phosphatidylinositols, while isoproterenol which stimulates the adenylate cyclase did not lead to increased ^{32}P turnover. However, the exact role of the phosphoinositols in stimulus-secretion coupling remains to be clarified.

Recently evidence has been provided for a Ca^{++}-mediated increase in Na^+ permeability by three different methods. In the intact superfused mouse pancreas, the ionophore A 23187 caused a Ca^{++}-dependent acinar cell membrane depolarization only in the presence of Na^+. Substitution of Na^+ by $Tris^+$ abolished depolarization (POULSEN and WILLIAMS, 1977a). Micro-iontophoretic Ca^{++} injection into acinar cells of the superfused mouse pancreas resulted in membrane depolarization similar to those evoked by acetylcholine. This suggests that the acetylcholine effect is caused by a rise in cytosol intracellular free Ca^{++} which in turn influences the Na^+ permeability (IWATSUKI and PETERSEN, 1977b). Further support that intracellular Ca^{++} directly increases the Na^+ permeability of the cell membrane came from studies on isolated membrane vesicles (SCHULZ and HEIL, 1978). Na^+ efflux from these vesicles was increased when Ca^{++} was present only at the vesicle inside as compared to Ca^{++} being at both sides or in the absence of Ca^{++}. Ca^{++} only present at the outside decreased Na^+ efflux as compared to Ca^{++} being at both sides or in the absence of Ca^{++} (SCHULZ and HEIL, 1978).

Furthermore, the question arises as to how the Ca^{++} ions that enter the acinar cells during excitation are removed from the cell at the end of stimulation. A low-affinity Ca^{++}-ATPase has been found in pancreatic plasma membranes (FORGET and HEISLER, 1976; LÖFFLER et al., 1977), which could be distinguished from the mitochondrial Ca^{++}-Mg^{++}-ATPase by its insensitivity to azide and oligomycin (ARGENT et al., 1975). The lack of evidence for a Na^+-Ca^{++} countertransport system of the type described by BLAUSTEIN (1974) has led to the suggestion that this Ca^{++}-ATPase might be involved in pumping Ca^{++} out of the cell. The pancreatic enzyme secretion is not affected by ouabain (RIDDERSTAP and BONTING, 1969b). Furthermore, $^{45}Ca^{++}$ efflux from the cell is not augmented by an increase of Na^+ in the incubation medium, but rather inhibited (SCHULZ et al., 1977). Both facts speak against the possibility that Ca^{++} is pumped out of the cell in countertransport for Na^+ ions.

It should be mentioned that calcitonin inhibits secretagogue-evoked, as well as calcium shot-evoked enzyme secretion (SCHULZ and MANNIGEL, 1974). Since calcitonin affected neither the adenylate cyclase nor the $^{45}Ca^{++}$ fluxes and the Ca^{++}-ATPase (SCHULZ, I. unpublished observations), its mode of action is unknown.

II. Role of Cyclic AMP and Cyclic GMP in Stimulus-Secretion Coupling

It was shown by RUTTEN et al. (1972) and by SCHULZ et al. (1974) that pancreozymin activates a pancreatic adenylate cyclase. Experiments with isolated acinar cells revealed that these cells carry the pancreozymin-sensitive adenylate cyclase (MILUTINOVIĆ et al., 1977a), but that cAMP levels are not (BENZ et al., 1972; ROBBERECHT et al., 1974; HAYMOVITZ and SCHEELE, 1976) or only slightly increased by this hormone in the pancreas (DESCHODT-LANCKMAN et al., 1975; KEMPEN et al., 1977a). In the latter studies high concentrations of phosphodiesterase inhibitors were necessary to see the effect. In broken cells the acinar adenylate cyclase, however, could also be stimulated by secretin (MILUTINOVIĆ et al., 1976) and cAMP levels in intact cells are increased by secretin (KEMPEN, 1976; ROBBERECHT et al., 1976). Since the effects of both secretin and pancreozymin are not additive it is suggested that both hormones stimulate the same adenylate cyclase (KEMPEN et al., 1974; dePONT et al., 1977).

Thus the difference in all known actions of pancreozymin and secretin on the acini resides in the fact that pancreozymin, which evokes enzyme secretion and in some animals also fluid secretion, stimulates the activity of adenylate cyclase, but Ca^{++} and other ion fluxes as well, without raising cAMP levels in the cells; while secretin, which is able to potentiate the effect of pancreozymin on enzyme secretion, stimulates adenylate cyclase in broken membranes and raises cAMP levels in the acini. Acetylcholine on the other hand, does not stimulate acinar adenylate cyclase (RUTTEN et al., 1972) but it increases the cellular cyclic GMP level (ROBBERECHT et al., 1974; HAYMOVITZ and SCHEELE, 1976; CHRISTOPHE et al., 1976b), probably *via* an increased intracellular Ca^{++} concentration, which is known to stimulate the guanylate cyclase (SCHULTZ et al., 1973). The same holds also true for pancreozymin (ALBANO et al., 1976a and b; KAPOOR and KRISHNA, 1976; CHRISTOPHE et al., 1976b). In pancreas slices, in dispersed acinar cells and in the isolated perfused preparation of the cat pancreas, enzyme secretion could be evoked by the analogs of cGMP, dibutyryl-cyclic GMP and cyclic 8-bromo GMP (ROBBERECHT, 1974; SCHULZ, 1975a and b; ALBANO et al., 1976b; GARDNER and JACKSON, 1977) and by cyclic AMP or dibutyryl-cyclic AMP (KNODELL et al., 1970; KULKA and STERNLICHT, 1968; MORISSET and WEBSTER, 1971; BAUDUIN et al., 1971; SCHULZ, 1975a and b; KEMPEN et al., 1977a; BAUDUIN et al., 1977; GARDNER and JACKSON, 1977). Others, however, did not find such an effect of cyclic nucleotides on amylase release (BENZ et al., 1972; HEISLER and GRONDIN, 1975). Both enzyme secretagogues, pancreozymin and acetylcholine, but not cyclic nucelotides per se, augment Ca^{++} influx (KONDO and SCHULZ, 1976a) and both Ca^{++} and cyclic nucleotides were shown to potentiate each other in their effect on enzyme secretion (HEISLER and GRONDIN, 1975; SCHULZ, 1975a and b). Nevertheless it appears unlikely that both cyclic nucleotides play a major role in enzyme secretion from the exocrine pancreas, since the ability of secretagogues of enzyme secretion to increase intracellular cGMP levels is lost if extracellular calcium concentration is lowered to 0.05 mM. At this concentration, however, a marked secretory effect is elicited by secretagogues (HEISLER and LAMBERT, 1978) while the cAMP levels are not or only slightly increased (see above). Pentagastrin, which also stimulates enzyme secretion, was shown to compete with pancreozymin for the same binding site (MILUTINOVIĆ et al., 1977a). But in contrast to pancreozymin, it stimulates only the Ca^{++} influx into the acinar cells (KONDO and SCHULZ, 1976a), but not the adenylate cyclase (KLAEVEMAN et al., 1975) or only at relative high concentrations (KEMPEN, 1976). However, pentagastrin inhibits the effect of pancreozymin on the adenylate cyclase competitively. In addition, the amphibian peptides bombesin and litorin, showing similar action on enzyme secretion and calcium fluxes as pancreozymin and caerulein (ERSPARMER and MELCHIORRI, 1975; DESCHODT-LANCKMAN et al., 1976) do not stimulate adenylate cyclase (DESCHODT-LANCKMAN et al., 1976). Thus the mode of action of different hormones can be used as a model to study the principal ways of stimulus-secretion coupling, the influx of Ca^{++} and stimulation of adenylate cyclase.

The scheme depicted in Fig. 8 summarizes the events involved in stimulus-enzyme secretion coupling of acinar cells. Acetylcholine and pancreozymin act on specific receptors on the contraluminal cell side. Both liberate Ca^{++} from the membrane and increase the Ca^{++} permeability, resulting in increased Ca^{++} influx. Ca^{++} originating from the cell membrane and from the extracellular fluid produces an increase in cytoplasmic Ca^{++} concentration and triggers the fusion between zymogen granules and the luminal plasma membrane. The rise in cytoplasmic Ca^{++} concentration also stimulates a cytoplasmic guanylate cyclase. Furthermore, both pancreozymin and secretin act on a contraluminal

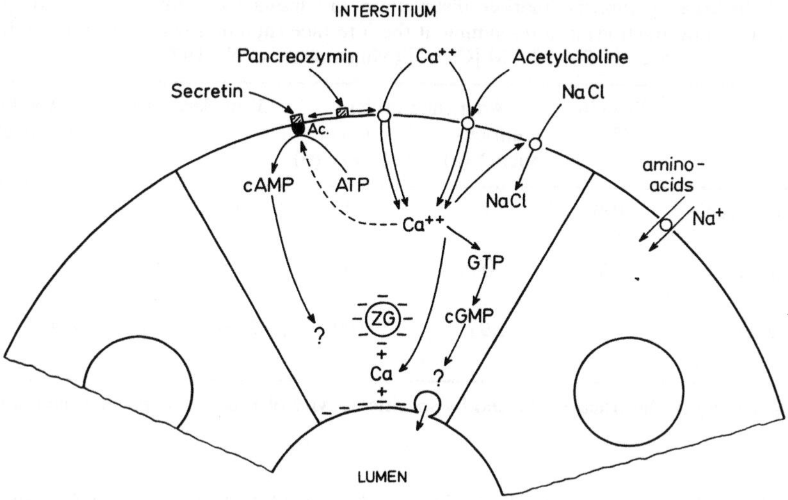

Fig. 8. Schematic representation of proposed cellular mechanism for enzyme secretion in pancreatic acinar cell (for explanation see text)

adenylate cyclase, but the cAMP level in the cell increases only in response to secretin and not to pancreozymin. An inhibition of adenylate cyclase activity by the influx of Ca^{++}, which is also stimulated by pancreozymin, could be a possible explanation.

III. Ca^{++} and Exocytosis of Secretory Granules

Exocytosis is a special case of membrane fusion. Thus, a general consideration on membrane fusion might be pertinent to the events that could be involved in fusion of zymogen granules with the luminal acinar cell membrane in the pancreas. This fusion is a fast event and hitherto it has been seen very rarely in electron microscopic pictures (PALADE, 1959, 1975; ICHIKAWA, 1965). Close membrane aggregation is the first step in fusion, which was recently studied with isolated zymogen granules and plasma membranes (MILUTINOVIĆ et al., 1977c). When 10^{-6} M Ca^{++} was added to the incubation medium it was observed that the zymogen granules aggregated with a fraction of pancreatic membranes that contains alkaline phosphatase but did not react with membranes containing other plasma membrane markers such as adenylate cyclase or 5'-nucleotidase (Table 2). The interaction between membranes and granules was specific for cat pancreatic membranes as opposed to cat liver membranes, and for pancreatic zymogen granules as opposed to pancreatic mitochondria. The aggregation was only partially reversible by EGTA. To prevent spontaneous lysis of the zymogen granules the experiments had to be performed at room temperature, and thus only the aggregation step could be studied. Similar behavior aggregation at low temperature but fusion and eventually lysis at a higher temperature, was also

Table 2. Difference in enzyme marker distribution in plasma membranes bound to zymogen granules (interacting fraction) and remaining at the interface (noninteracting fraction) in the presence of $5 \cdot 10^{-4}$ M [Ca^{++}] (MILUTINOVIĆ et al., 1977c)

	Protein (%)	Adenylate cyclase enrichment* (a)	Adenylate cyclase (%) recovery (b)	5'-nucleotidase a	5'-nucleotidase b	Alkaline phosphatase a	Alkaline phosphatase b
Plasma membranes (Fraction I)	100	1	100	1	100	1	100
Interacting fraction	33	0	0	0.15	11	1.56	58
Noninteracting fraction	67	1.91	100	2.10	89	0.90	42

* Specific activity of the fraction divided by specific activity of Fraction I (plasma membranes)

observed in other systems, such as virus-induced, Ca^{++}-dependent fusion of red cells (NEURATH, 1965; TOISTER and LOYTER, 1973) or Ca^{++}-induced fusion of negatively charged liposomes (PAPAHADJOPOULOS et al., 1975; POSTE and PAPAHADJOPOULOS, 1976; LANSMAN and HAYNES, 1975; PAGANO and HUANG, 1975). In all these studies it became evident that Ca^{++} ions play a predominant role in membrane fusion. Depending on the system, Ca^{++} interacts only with the membrane lipids and/or the membrane proteins and/or filamentous submembrane structures.

In liposomes Ca^{++} reduces or abolishes the negative surface charge, thereby diminishing the repulsive electrostatic forces (VAN DER KLOOT and KITA, 1973; DEAN, 1974) and increasing the transition temperature of the phospholipids (TRÄUBLE and EIBL, 1974; VAN DIJK et al., 1975). Since Ca^{++} ions induce lateral phase separation in membranes that are composed of mixed lipids (ITOH and OHNISHI, 1974; PAPAHADJOPOULOS et al., 1974; JACOBSEN and PAPAHADJOPOULOS, 1975) solid and fluid domains coexist in the same membrane. In the domain boundaries hydrophobic groups might be exposed and act as sites for the initiation of membrane fusion (VAN DER BOSCH and MCCONNEL, 1975). In agreement with this concept is the finding that an increase in temperature, or the addition of lipids and other substances that increase membrane fluidity, facilitates membrane fusion (BREISBLATT and OHKI, 1975; AHKONG et al., 1974; LUCY, 1975; KOSOWER et al., 1975; VAN DER BOSCH et al., 1973). The effect of Ca^{++} on lipid bilayers is completely different, according to whether Ca^{++} is applied from both membrane sides or from one side only. Thus Ca^{++} increased the permeability of phosphatidylserine membranes for K^{+} if it was applied from one side only, but it did not change the permeability if it was applied from both sides (PAPAHADJOPOULOS und OHKI, 1969). A more difficult question to answer is that concerning the behavior of membrane proteins during the fusion process. Ca^{++} was shown to induce clustering of intramembraneous particles (BÄCHI et al., 1973; SATIR, 1974; DAHL and GRATZL, 1976; GRATZL and DAHL, 1976b; SCHUDT et al., 1976; VOLSKY and LOYTER, 1977; GRATZL et al., 1977) which was also observed with other fusogenic substances as for instance glycerol (Mc

INTYRE et al., 1974). It was suggested that clustering of membrane particles simply leads to regions where the phospholipids are exposed and ready to fuse with another region of exposed phospholipids (AHKONG et al., 1975; LAWSON et al., 1977). It was, however, also observed that the clustered particles form the first site of contact which then changes to a site of membrane instability (KOSOWER et al., 1977). Aggregation of intramembranous particles of isolated insulin-containing secretory vesicles, of isolated Golgi-derived secretory vesicles from rat liver or of isolated neurohypophyseal membranes, could be induced when Ca^{++} in micromolar concentration was added to the incubation medium (DAHL and GRATZL, 1976; GRATZL and DAHL, 1976a and b; GRATZL et al., 1977). Under these conditions the vesicles fuse with each other and the intramembranous particles were found in the remaining waist region. Experiments on isolated pancreatic islets stimulated by glucose revealed that intramembranous particles of secretory vesicles aggregate at the site of contact with the plasma membrane (BERGER et al., 1975). When islets were exposed to a hypertonic sucrose solution the granules in contact with the plasma membrane swelled. From this observation it was suggested that membrane junctions formed during the attachment of secretory granules to the plasma membrane are permeable for molecules at least up to the size of sucrose, the compound with the highest molecular weight of those tested. The formation of these junctions is thought to precede the exocytotic process. In pancreatic acinar cells enzyme granules fusing with the luminal cell membrane could be demonstrated when the tissue was fixed during stimulation at low temperature. However, aggregation of intramembranous particles at the site of fusion could not be seen (HAASE, pers. comm.). DAVIS and LAZARUS (1976) have shown that secretory granules from the pituitary, adrenal medulla and the pancreatic β cells interact and fuse specifically with the plasma membranes of the same cell but not with the plasma membrane of other cells. As shown with mouse islet granules and cod islet plasma membranes this process shows an absolute requirement for physiological amounts of Ca^{++} (2 μM) in order for insulin release to proceed. These results indicate the presence of special recognition sites at the granules and plasma membranes as well which interact with each other when Ca^{++} is present. In the studies of MILUTINOVIĆ et al. (1977c), it appears, too, that both zymogen granules and luminal cell membranes of exocrine pancreas have specific protein recognition sites which are necessary for the initial step in fusion to occur. In this study the Ca^{++}-dependent interaction of zymogen granules and luminal plasma membranes was prevented when the latter was briefly trypsinized. Similarly the removal of sialic acid by neuraminidase inhibits the Sendai virus induced fusion of avian erythrocytes (DEN et al., 1974) and the Ca^{++}-induced fusion of Golgi-derived vesicles from the rat liver (GRATZL and DAHL, 1976b). Since the major part of the pancreatic zymogen granule membrane protein consists of a glycoprotein (MACDONALD and RONZIO, 1972; HOFBAUER, pers. comm.) it is possible that this glycoprotein plays a key role in membrane fusion during exocytosis. In this connection, one should keep in mind that the cell fusion inducing substances of viruses are glycoproteins (MAEDA et al., 1977; MANSERVIGI et al., 1977). Furthermore in herpes simplex virus beside a fusion-inducing glycoprotein another glycoprotein which is able to suppress fusion was found (MANSERVIGI et al., 1977). Hopefully more insight into the role of membrane proteins in the process of exocytosis will be gained in the near future.

It remains undecided however, whether cyclic nucleotides, such as cAMP and cGMP participate directly in fusion. It was suggested that phosphorylation of the fusing membranes may facilitate the exocytosis. Indeed phsophorylation of zymogen granule membranes by an endogenous protein kinase has been demonstrated. However, its activity was stimulated poorly or not at all by cAMP or cGMP (LAMBERT et al., 1974; MacDONALD and RONZIO, 1974). It is also not yet clear whether the luminal cell membrane can be phosphorylated or dephosphorylated. In chicken erythrocytes, at least, Sendai virus which induces membrane fusion causes a rapid and specific dephosphorylation of phosphorylated membrane proteins (BENZAQUEN et al., 1978). In this connection the question arises as to whether the Ca^{++}-ATPase found in the zymogen granule membrane of the parotid gland (WATSON et al., 1974) and of the exocrine pancreas (MILUTINOVIĆ et al., 1977b) acts by producing different Ca^{++} concentrations at the two sides of the zymogen granule membrane. For instance, fusion could be induced by translocation of Ca^{++} into the zymogen granules in a subsequent step following Ca^{++}-induced aggregation between luminal plasma and zymogen granule membranes. Another possible function of Ca^{++}-ATPase could be the removal of Ca^{++} from acinar cells, when after fusion the zymogen granule membrane had become part of the luminal plasma membrane. It is also possible that the Ca^{++}-ATPase of the granule membrane is associated with a contractile protein, as proposed by STORMORKEN (1969).

IV. Role of Microfilaments and Microtubules

The presence of microfilaments and microtubules in acinar cells of the exocrine pancreas has been recognized by GEUZE and POORT (1973) and SEYBOLD et al. (1975). Since their role in secretion has been assessed in a number of secretory tissues (LACY et al., 1968; ORCI et al., 1972; NEVE et al., 1970; WILLIAMS and WOLFF, 1971; WILLIAMS and LEE, 1976; BUTCHER and GOLDMAN, 1972) it has been suggested that they could play a role in the last step in the sequence of cellular events, leading to exocytosis (ORCI et al., 1972). According to BENZ et al. (1972) the antimicrotubular agent colchicine at 10^{-4} M concentrations had no effect on the discharge of amylase, whereas studies by JAMIESON (1972) showed that both colchicine and vinblastine inhibit discharge of proteins by only 25 to 30 percent, whereas they had a marked effect on protein synthesis. In extensive studies on protein synthesis, intracellular transport and discharge of secretory proteins from isolated lobules of rat pancreas SEYBOLD et al. (1975) have found that antimicrotubular agents inhibited intracellular transport of newly synthetized proteins rather than discharge of enzymes or protein synthesis. The effect of these agents was accompanied with deposition of secretory material in the cisternal space of the rough endoplasmic reticulum and the formation of paracrystals which could be demonstrated ultrastructurally in form of long needle-like inclusions. Similar paracrystals had also been observed in the same system after blockade of intracellular transport by high extracellular concentrations of aromatic amino acids (BIEGER and KERN, 1975). Subfractionation studies of rough and smooth microsomes which correspond ultrastructurally to elements of the endoplasmic reticulum and the Golgi complex also revealed a major block in transport between two organelles, which led the authors to conclude that microtubules not only participate in the translocation of secretory granules to the plasma membrane but more generally in the directional movement of newly synthetized protein to their site of packaging – the Golgi complex (BIEGER and KERN, 1975). These results agree with the studies of BAUDUIN et al. (1975) and WILLIAMS and LEE (1976) who observed inhibition of stimulated enzyme secretion by vinblastine, cholchicine or high concentrations of cytochalasin B, respectively, which was accompanied with disappearance of the thin microfilaments present in the apical region of acinar cells. At low concentrations (up to 1 $\mu g\ ml^{-1}$), however, the latter authors observed an increase in unstimulated secretion, which led them to assume a second, more sensitive site of cytochalasin B action located at the cell membrane.

Although vinblastine, colchicine and cytochalasin B were found not to alter $^{45}Ca^{++}$ exchange between the extracellular medium and pancreatic tissue (WILLIAMS and LEE, 1976; BAUDUIN et al., 1975), the elaborated meshwork of intramembrane fibrils forming the tight junctions appeared disordered and fragmented into small strands when pancreatic tissue lobules were incubated in vitro in Ca^{++}-free Krebs-Ringer bicarbonate medium containing EGTA. The readmission of Ca^{++} to the incubation medium resulted in rapid reformation of the tight junction (MELDOLESI et al. 1977). Although disassembly and reassembly of microfilaments in the formation of tight junctions could not be correlated to any functional alteration of these cell membranes which are segregated by the tight junctions (baso-lateral and luminal regions), the authors feel that the specificities of these cell membranes are probably maintained by the tight junctions (MELDOLESI et al., 1977). So the involvement of microfilaments in the secretory cycle could be at least twofold: one at the intracellular transport of newly synthesized proteins, the other at the formation of tight junction which would maintain the specificity of baso-lateral and luminal cell membranes and thereby influence secretory processes located at these cell sides.

D. Uptake of Amino Acids Used for Enzyme Synthesis

Following the observation, many years ago, that the mouse pancreas takes up some amino acids more extensively than any other tissue (WHEELER et al., 1949; BLAU and MANSKE, 1961; BERLINGUET et al., 1962), SCHOLEFIELD and JOHNSTONE and their associates have performed a series of important studies on amino acid transport. Thus BÉGIN and SCHOLEFIELD (1965a), using pancreatic slices, found that in mouse pancreas the uptake of L-valine, glycine, L-methionine, L-ethionine and the two unmetabolized amino acids 1-amino-cyclopentane carboxylic acid and 1-amino-2-methylcyclopentane carboxylic acid

(ACA, McACA) follows Michaelis-Menten kinetics. From this it was concluded that the transport of each amino acid investigated is mediated by only one transport carrier in each case. The K_m values of these six amino acids are very close to each other, ranging between 2.5 to 5.9 mM. Nevertheless, when the ABC test[2] was applied it was assumed that at least three sites must be involved in the transport of these amino acids, one for ACA, McACA, and L-valine, one for L-methionine and L-ethionine, and a third for glycine. L-proline, however, behaves differently (BÉGIN and SCHOLEFIELD, 1965b). In contrast to other amino acids, not the reciprocity of the intra-extracellular concentration difference, but the square root of this value, plotted against the reciprocity of the extracellular concentration, gives a straight line. From this it was concluded that two molecules of proline are transported with equal affinity by the same carrier. Glycine, L-methionine, and ACA, which inhibit proline uptake, were also considered to have affinities for both proline binding sites and to inhibit proline uptake competitively. The authors furthermore stated that although an amino acid may compete, and indeed have a high affinity, for the binding site responsible for the transport of a second amino acid, it is not necessarily transported. At the moment, however, it is dubious whether this interpretation of the data is pertinent if the amino acids are taken up in cotransport with Na^+. The same authors (BÉGIN and SCHOLEFIELD, 1964) have observed that the 60-min uptake of L-glycine, L-proline, and L-1-aminocyclopentane carboxylic acid by the mouse pancreas is reduced by 50 to 60 percent if the incubation medium is free of sodium or contained $4 \cdot 10^{-5}$ M ouabain. That glycine uptake by the *in-vitro* rat pancreas is reduced by 75 percent when ouabain is added to the incubation medium was also reported by CHENEVAL and JOHNSTONE (1976). LIN and JOHNSTONE (1971) also showed that the initial rate of glycine uptake is directly proportional to the extracellular Na^+ concentration. Application of Michaelis-Menten kinetics led to the conclusion that extracellular Na^+ increases the V_{max} of glycine uptake without a significant effect on the K_m. This approach, however, does not describe the Na^+ amino acid cotransport adequately, because the effect of the electrical potential difference is not considered (LAGARDE, 1976; GECK and HEINZ, 1976). An analysis of fluxes and conjugated driving forces with membrane vesicles was recently performed (TYRAKOWSKI et al., 1978). By analogy with brush-border vesicles derived from intestine (MURER et al., 1975) and from kidney (EVERS et al., 1976) it was shown that in the presence of a Na^+ gradient L-alanine was taken up by the pancreatic membrane vesicles against a concentration gradient of L-alanine (Fig. 9). This overshoot phenomenon, *i. e.* a transient intravesicular L-alanine concentration higher than the equilibrium value, was especially pronounced when the easily permeable thiocyanate was the accompanying anion, whereas it was abolished if the less permeable SO_4^{--} was used. Furthermore, among the cations tested (Na^+, Li^+, K^+), this overshoot phenomenon was shown only in the presence of Na^+, and it was abolished when vesicles were preincubated in the presence of the Na^+-specific ionophore

2 Two amino acids A and B must inhibit each other competitively, K_m and K_i for each being the same. Furthermore, an amino acid C must inhibit A and B with the same K_i.

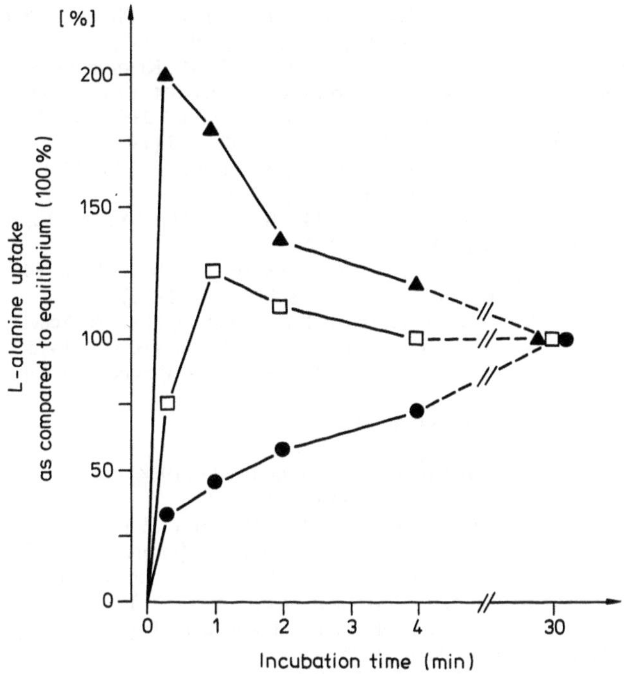

Fig. 9. Time-course of L-alanine uptake by isolated plasma membrane vesicles from cat pancreas at different L-alanine concentrations in incubation medium, 0.1 mM ▲, 1.0 mM □, or 10 mM ●. (TYRAKOWSKI et al., 1978)

monensin (TYRAKOWSKI et al., 1978). Thus, the amino acid uptake by the pancreas seems to occur *via* an electrogenic Na^+ gradient-driven cotransport system, and another energy source for amino acid uptake, *e. g.* ATP does not need to be postulated, although it cannot be excluded.

Acknowledgements

The authors wish to thank Dr. *J. J. H. H. M. DePont,* Prof. *S. L. Bonting* and Prof. *O. H. Petersen* for valuable discussions and criticism of the manuscript.

References

AGUS, Z. S., PUSCHETT, J. B., SENESKY, D., GOLDBERG, M.: J. clin. Invest. **50**, 617 (1971).
AHKONG, Q. F., HOWELL, J. I., TAMPION, W., LUCY, J. A.: FEBS Letters **41**, 206 (1974).
AHKONG, Q. F., FISHER, D., TAMPION, W., LUCY, J. A.: Nature **253**, 194–195 (1975).
ALBANO, J. D. M., BHOOLA, K. D., HARVEY, R. F.: In: Stimulus-Secretion Coupling in the Gastrointestinal Tract (R. M. Case and H. Goebell, Eds.). Lancaster, England: MTP 1976a, p. 227.

ALBANO, J., BHOOLA, K. D., CROKER, M., HARVEY, R. F., HEAP, P. F.: J. Physiol. **258**, P 87 (1976b).
ANASTASI, A., ERSPARMER, V., ENDEAN, R.: Experientia **23**, 699 (1967).
ANDERSON, J. C., BARTON, M. A., GREGORY, R. A., HARDY, P. M., KENNER, G. W., MCLEOD, J. K., PRESTON, J., SHEPPARD, R. C., MORLEY, J. S.: Nature (Lond.) **204**, 933 (1964).
ARGENT, B. E., CASE, R. M., SCRATCHERD, T.: J. Physiol. (London) **216**, 611 (1971).
ARGENT, B. E., CASE, R. M., SCRATCHERD, T.: J. Physiol. **230**, 575 (1973).
ARGENT, B. E., SMITH, R. K., CASE, R. M.: Biochem. Soc. Trans. **3**, 713 (1975).
BÄCHI, T., AGUET, M., HOWE, C.: J. Virol. **11**, 1004 (1973).
BALDAMUS, C. A., HIERHOLZER, K., RUMRICH, G., STOLTE, H., UHLICH, E., ULLRICH, K. J., WIEDERHOLT, M.: Pflügers Arch. **310**, 354 (1969).
BALL, E. G., TUCKER, H. F., SOLOMON, A. K., VENNESLAND, B.: J. biol. Chem. **140**, 119 (1941).
BAUDUIN, H., CANTRAINE, F.: Biochim. biophys. Acta **270**, 248 (1972).
BAUDUIN, H., ROCHUS, L., VINCENT, D., DUMONT, J. E.: Biochim. biophys. Acta **252**, 171 (1971).
BAUDUIN, H., STOCK, C., VINCENT, D., GRENIER, J. F.: J. cell Biol. **66**, 165 (1975).
BAUDUIN, H., STOCK, C., LAUNAY, J. F., VINCENT, D., POTVLIEGE, P., GRENIER, J. F.: Pflügers Arch. ges. Physiol. **372**, 69 (1977).
BAXTER, S. G.: Amer. J. Physiol. **96**, 343 (1931).
BAYLISS, W. M., STARLING, E. H.: J. Physiol. **28**, 325 (1902).
BÉGIN, N., SCHOLEFIELD, P. G.: Biochim. biophys. Acta **90**, 82 (1964).
BÉGIN, N., SCHOLEFIELD, P. G.: J. biol. Chem. **240**, 332 (1965a).
BÉGIN, N., SCHOLEFIELD, P. G.: Biochim. biophys. Acta **104**, 566 (1965b).
BEN-ZAQUEN, R., LOYTER, A., MILLNER, Y.: Israel J. Med. Sci. (in press).
BENZ, L., ECKSTEIN, B., MATTHEWS, E. K., WILLIAMS, J. A.: Brit. J. Pharmacol. **46**, 66 (1972).
BERGER, W., DAHL, G., MEISSNER, H. P.: Cytobiology **12**, 119 (1975).
BERLINGUET, L., BÉGIN, N., BABINEAU, L. M.: Canad. J. Biochem. Physiol. **40**, 1111 (1962).
BIEGER, W., KERN, H. F.: Virchows Arch. (path. Anat.) **367**, 289 (1975).
BLAU, M., MANSKE, R. F.: J. nucl. Med. **2**, 102 (1961).
BLAUSTEIN, M. P.: Rev. Physiol. Biochem. and Pharmacology **70**, 33 (1974).
BLOOM, S. R., JOFFE, S. N., POLAK, J. M.: European Pancreatic Club, VIIIth Symposium, Toulouse, 1975, p. 122.
BODANSZKY, M., ONDETTI, M. A., LEVINE, S. D., WILLIAMS, N. J.: J. Amer. Chem. Soc. **89**, 6753 (1967).
BODEN, G., SIVITZ, M. C., OWEN, O. E.: Science **190**, 163 (1976).
BONTING, S. L., DEPONT, J. J. H. H. M.: In: Secretory Mechanisms of Exocrine Glands (N. A. Thorn and O. H. Petersen, Eds). Munksgaard, Copenhagen 1974, p. 363.
BONTING, S. L., CASE, R. M., DEPONT, J. J. H. H. M., KEMPEN, M. J. M., SCRATCHERD, T.: J. Physiol. **240**, 34P (1974).
BREISBLATT, W., OHKI, S.: J. Membrane Biol. **23**, 385 (1975).
BRO-RASMUSSEN, F., KILLMANN, S. A., THAYSEN, J. H.: Acta physiol. scand. **37**, 97 (1956).
BROWN, J. C., HARPER, A. A., SCRATCHERD, T.: J. Physiol. **190**, 519 (1967).
BUTCHER, F. R., GOLDMAN, R. H.: Biochem. biophys. Res. Comm. **48**, 23 (1972).
CAPLE, I., HEATH, T.: Aust. J. biol. Sci. **25**, 155 (1972).
CARO, L. G., PALÁDE, G. E.: J. Cell Biol. **20**, 473 (1964).
CASE, R. M., CLAUSEN, T.: J. Physiol. **235**, 75 (1973).
CASE, R. M., SCRATCHERD, T.: Biochim. biophys. Acta **219**, 493 (1970).
CASE, R. M., SCRATCHERD, T.: J. Physiol. **223**, 649 (1972).
CASE, R. M., SCRATCHERD, T.: J. Physiol. **242**, 415 (1974).
CASE, R. M., HARPER, A. A., SCRATCHERD, T.: Proc. Int. Union Physiol. Sciences Washington, Vol. 7, Abstr. No. 230, 1968a.
CASE, R. M., HARPER, A. A., SCRATCHERD, T.: J. Physiol. **196**, 133 (1968b).
CASE, R. M., HARPER, A. A., SCRATCHERD, T.: In: Exocrine Glands (S. Y. Botelho, F. P. Brooks and W. B. Shelley, Eds). Philadelphia, Pa: Univ. of Pennsylvania Press 1969, p. 39.
CASE, R. M., SCRATCHERD, T., WYNNE, R. D'A.: J. Physiol. **210**, 1 (1970).
CECCARELLI, B., CLEMENTE, F., MELDOLESI, J.: J. Physiol. **245**, 617 (1975).
CHANG, K.-J., TRIGGLE, D. J.: J. theor. Biol. **40**, 125 (1973).
CHENEVAL, J. P., JOHNSTONE, R. M.: Biochim. biophys. Acta. **433**, 630 (1976).
CHRISTOPHE, J. P., CONLON, T. P., GARDNER, J. D.: J. biol. Chem. **251**, 4629 (1976a).

CHRISTOPHE, J. P., FRANDSEN, E. K., CONLON, T. P., KRISHNA, G., GARDNER, J. D.: J. biol. Chem. **251**, 4640 (1976b).
CLARK, D. G., GREENWELL, J. R., HARPER, A. A., SANKEY, A. M., SCRATCHERD, T.: J. Physiol. **189**, 247 (1967).
COCHRANE, D. E., DOUGLAS, W. W.: J. Physiol. **257**, 433 (1976).
COHEN, L. H., MUELLER, A., STEINMETZ, P. R.: J. clin. Invert **61**, 981 (1978).
COMLINE, R. S., HALL, L. W., HICKSON, J. C. D., MURILLO, A., WALKER, R. G.: J. Physiol. **204**, 10 (1969).
CREUTZFELDT, W., LANKISCH, P. G., FÖLSCH, U. R.: Dtsch. med. Wschr. **100**, 1135 (1975).
DAHL, G., GRATZL, M.: Cytobiology **12**, 345 (1976).
DAVIS, B., LAZARUS, N. R.: J. Physiol. **256**, 709 (1976).
DEAN, P. M.: In: Secretory Mechanisms of Exocrine Glands (N. A. Thorn and O. H. Petersen, Eds). New York: Academic 1974, p. 152.
DEBAS, H. T., GROSSMAN, M. I.: Digestion **9**, 469 (1973).
DÉLÈZE, J., LOEWENSTEIN, W. R.: J. Membrane Biol. **28**, 71 (1976).
DEN, H., LASTER, J., LOYTER, A.: Fed. Proc. **33**, 1296 (1974).
DEWHURST, D. G., the late HADI, N. A., HUTSON, D., SCRATCHERD, T.: J. Physiol. **277**, 103 (1978).
DEPONT, J. J. H. H. M., JANSEN, J. W. C. M., BONTING, S. L.: Acta Physiol. scand. Suppl. p. 373 (1978).
DEPONT, J. J. H. M., KEMPEN, H. J. M., BONTING, S. L.: In Hormonal Receptors in Digestive Tract Physiology (S. Bonfils, P. Fromageot and G. Rosselin, Eds.). Amsterdam: North-Holland Publ. Comp. 1977, p. 289.
DESCHODT-LANCKMAN, M., ROBBERECHT, P., De NEEF, P., LABRIE, F., CHRISTOPHE, J.: Gastroenterology **68**, 318 (1975).
DESCHODT-LANCKMAN, M., ROBBERECHT, P., De NEEF, P., LAMMENS, M., CHRISTOPHE, J.: J. clin. Invest. **58**, 891 (1976).
DIAMOND, J. M.: J. gen. Physiol. **48**, 15 (1965).
DOUGLAS, W. W.: Br. J. Pharmacol. **34**, 451 (1968).
EHRENSPECK, G., BRODSKY, W. A.: Biochim. biophys. Acta **419**, 555 (1976).
EIMERL, S., SAVION, N., HEICHAL, O., SELINGER, Z.: J. biol. Chem. **249**, 3991 (1974).
ERSPARMER, V., BERTACCINI, G., CHELI, R., CARO, G. de, ENDEAN, R., IMPICCIATORE, M., ROSEGHINI, M. L.: In: Advances in Experimental Medicine and Biology, Vol. 2 (N. Black, L. Martini and R. Paoletti, Eds). New York: Plenum 1968, p. 533.
ERSPARMER, V., MELCHIORRI, P.: In: Gastrointestinal Hormones (J. C. Thompson, Ed.). Austin, Tex.: University of Texas Press 1975, p. 575.
ERSPARMER, V., IMPROTA, G., MELCHIORRI, P., SOPRANZI, N.: Brit. J. Pharmacol. **52**, 227 (1974).
EVERS, J., MURER, H., KINNE, R.: Biochim. biophys. Acta **426**, 598 (1976).
EVERS, C., MURER, H., KINNE, R.: Biochim. J. **172**, 49 (1978).
FÖLSCH, U. R., CREUTZFELDT, W.: In: Stimulus-Secretion Coupling in the Gastrointestinal Tract (R. M. Case and H. Goebell, Eds). Lancaster: MTP 1975, p. 381.
FÖLSCH, U. R., WORMSLEY, K. G.: J. Physiol. **234**, 79 (1973).
FORGET, G., HEISLER, S.: Clin. exp. Pharmacol. Physiol. **3**, 67 (1976).
FRÖMTER, E., SATO, K., GESSNER, K.: In: Proceedings of the 6th International Congress of Nephrology (S. Giovannetti, V. Bonomini and G. D'Amico, Eds). Basel: Karger 1976, p. 108.
GANSER, A. L., FORTE, J. G.: Biochim. biophys. Acta **307**, 169 (1973).
GARDNER, J. D., CONLON, T. P., KLAEVEMAN, H. L., ADAMS, T. D., ONDETTI, M. A.: J. clin. Invest. **56**, 366 (1975).
GARDNER, J. D., CONLON, T. P., FINK, M. L., BODANSZKY, M.: Gastroenterology **71**, 965 (1976).
GARDNER, J. D., HAHNE, W. F.: Biochim. biophys. Acta **471**, 466 (1977).
GARDNER, J. D., JACKSON, M. J.: J. Physiol. **270**, 439 (1977).
GECK, P., HEINZ, E.: Biochim. biophys. Acta **443**, 49 (1976).
GEUZE, J. J., POORT, C.: J. cell. Biol. **57**, 159 (1973).
GIEBISCH, G.: Yale. J. Biol. Med. **45**, 191 (1972).
GOEBELL, H., STEFFEN, C., BODE, C.: Gut **13**, 477 (1972).
GRATZL, M., DAHL, G.: J. Cell Biol. **70**, 241a (1976a).
GRATZL, M., DAHL, G.: FEBS Letters **62**, 142 (1976b).
GRATZL, M., DAHL, G., RUSSELL, J. T., THORN, N. A.: Biochim. biophys. Acta **470**, 45 (1977).

GREENWELL, J. R.: Pflügers Arch. **353**, 159 (1975).
GREGORY, R. A., TRACY, H. J.: Gut **5**, 103 (1964).
HADI, N., HOTZ, J., SCRATCHERD, T., WYNNE, R. D. A.: J. Physiol. **259**, 56P (1976).
HART, W. M., THOMAS, J. E.: Gastroenterology **4**, 409 (1945).
HASHIMOTO, K., SATOH, S., TAKEUCHI, O.: Brit. J. Pharmacol. **43**, 739 (1971).
HAUSER, H., DAWSON, R. M. C.: Europ. J. Biochem. **1**, 61 (1967).
HAYMOVITZ, A., SCHEELE, G. A.: Proc. nat. Acad. Sci. (USA) **73**, 156 (1976).
HEIDBREDER, E., ECKLE, A., HENNEMANN, H., HEIDLAND, A.: Verh. dtsch. Ges. inn. Med. **78**, 1453 (1972).
HEISLER, S.: Life Sciences **19**, 233 (1976).
HEISLER, S., GRONDIN, G.: Experientia **31**, 936 (1975).
HEISLER, S., LAMBERT, M.: Can. J. Physiol. Pharmacol. **56**, 395 (1978).
HEISLER, S., FAST, D., TENENHOUSE, A.: Biochim. biophys. Acta **279**, 561 (1972).
HOKIN, L. E.: Biochim. biophys. Acta **115**, 219 (1966).
HOKIN, L. E., HOKIN, M. R.: Biochim. biophys. Acta **18**, 102 (1955).
HOKIN, L. E., HOKIN, M. R.: J. Physiol. **132**, 442 (1956).
HOKIN, M. R.: In: Secretory Mechanisms of Exocrine Glands, (N. A. Thorn and O. H. Petersen, Eds.). New York: Academic 1974, p. 101.
HOLTZMAN, E., SCHACHER, S., EVANS, J., TEICHBERG, S.: In: The Synthesis, Assembly and Turnover of Cell Surface Components (G. Poste and G. L. Nicolson, Eds.) Elsevier/North-Holland Biomedical Press, 1977, p. 165.
HUBEL, K. A.: Amer. J. Physiol. **212**, 101 (1967).
HUBEL, K. A.: J. clin. Invest. **53**, 964 (1974).
ICHIKAWA, A.: J. Cell Biol. **24**, 369 (1965).
ITOH, T., OHNISHI, S.: Biochim. biophys. Acta **352**, 29 (1974).
IWATSUKI, K., FURUTA, V., HASHIMOTO, K.: Experientia **29**, 319 (1973).
IWATSUKI, N., PETERSEN, O. H.: Nature **263**, 784 (1976).
IWATSUKI, N., PETERSEN, O. H.: J. Physiol. **269**, 735 (1977a).
IWATSUKI, N., PETERSEN, O. H.: Nature, **268**, 147 (1977b).
IWATSUKI, N., PETERSEN, O. H.: J. clin. Invest., **61**, 41 (1978).
JACOBSON, K., PAPAHADJOPOULOS, D.: Biochemistry **14**, 152 (1975).
JAMIESON, J. D.: In: Current topics in membranes and transport, Vol. 3 (F. Bronner and A. Kleinzeller, Eds). New York-London: Academic 1972, p. 273.
JAMIESON, J. D.: In: Secretin, Cholecystokinin, Pancreozymin and Gastrin. (J. E. Jorpes and V. Mutt, Eds). Handbook expfl. Pharma col. Vol. 34. Berlin-Heidelberg-New York: Springer 1973, p. 195.
JAMIESON, J. D., PALADE, G. E.: J. Cell Biol. **34**, 577 (1967a).
JAMIESON, J. D., PALADE, G. E.: J. Cell Biol. **34**, 597 (1967b).
JORPES, J. E., MUTT, V.: Acta chem. scand. **15**, 1790 (1961).
JORPES, J. E., MUTT, V.: In: The exocrine Pancreas (Ciba Foundation Symposium, A. V. S. de Reuck and M. P. Cameron, Eds). London: Churchill 1962, p. 150.
JORPES, J. E., MUTT, V.: In: Secretin, Cholecystokinin, Pancreozymin, and Gastrin. (J. E. Jorpes and V. Mutt, Eds). Handbook experimental Pharmacol. Vol. 34. Berlin-Heidelberg-New York: Springer 1973, p. 1.
JORPES, E., MUTT, V., TOCZKO, K.: Acta chem. scand. **18**, 2408 (1964).
KANNO, T.: J. Physiol. **226**, 353 (1972).
KANNO, T., NISHIMURA, O.: J. Physiol. **257**, 309 (1976).
KANNO, T., YAMAMOTO, M.: J. Physiol. **264**, 787 (1977).
KANNO, T., COCHRANE, D. E., DOUGLAS, W. W.: Canad. J. Physiol. Pharmacol. **51**, 1001 (1973).
KAPOOR, C. L., KRISHNA, G.: Fed. Proc. **35**, 295 (1976).
KEMPEN, H. J. M.: Thesis, Catholic University, Nijmegen, 1976.
KEMPEN, H. J. M., DEPONT, J. J. H. H. M., BONTING, S. L.: Biochim. biophys. Acta **370**, 573 (1974).
KEMPEN, H. J. M., DEPONT, J. J. H. H. M., BONTING, S. L.: Biochim. biophys. Acta **392**, 276 (1975).
KEMPEN, H. J. M., DEPONT, J. J. H. H. M., BONTING, S. L.: Biochim. biophys. Acta **496**, 65 (1977a).
KEMPEN, H. J. M., DEPONT, J. J. H. H. M., BONTING, S. L.: Biochim. biophys. Acta **496**, 521 (1977b).
KINNE, R., SHLATZ, L. J., KINNE-SAFFRAN, E., SCHWARTZ, I. L.: J. Membrane Biol. **24**, 145 (1975).

KINNE-SAFFRAN, E., KINNE, R.: Proc. Soc. exp. Biol. **146**, 751 (1974).
KLAEVEMAN, H. L., CONLON, T. P., GARDNER, J. D.: In: Gastrointestinal Hormones. (J. C. Thompson, Ed.). Austin and London: University of Texas Press, 1975, p. 321.
KNODELL, R. G., TOSKES, P. P., REBER, H. A., BROOKS, F. P.: Experientia **26**, 515 (1970).
KONDO, S., SCHULZ, I.: Biochim. biophys. Acta **419**, 76 (1976a).
KONDO, S., SCHULZ, I.: J. Membrane Biol. **29**, 185 (1976b).
KONTUREK, S. J., TASLER, J., OBTULOWICZ, W., COY, D. H., SCHALLY, A. V.: J. clin. Invest. **58**, 1 (1976).
KOSOWER, E. M., KOSOWER, N. S., WEGMAN, P.: Biochim. biophys. Acta **471**, 311 (1977).
KOSOWER, N. S., KOSOWER, E. M., WEGMAN, P.: Biochim. biophys. Acta **401**, 530 (1975).
KULKA, R. G., STERNLICHT, E.: Proc. nat. Acad. Sci. USA **61**, 1123 (1968).
LACY, P. E., HOWELL, S. L., YOUNG, D. A., FINK, C. J.: Nature (Lond.) **219**, 1177 (1968).
LAGARDE, A. E.: Biochim. biophys. Acta **426**, 198 (1976).
LAGUNOFF, D., CHI, E. Y.: J. cell. Biol. **75**, 367a (1977).
LAMBERT, M., CAMUS, J., CHRISTOPHE, J.: Abstract of the VII Meeting of the European Pancreatic Club, Dundee, 1974.
LANSMAN, J., HAYNES, D. H.: Biochim. biophys. Acta **394**, 335 (1975).
LAWSON, D., RAFF, M. C., GOMPERTS, B., FEWTRELL, C., GILULA, N. B.: J. cell. Biol. **72**, 242 (1977).
LEAF, A., DEMPSEY, E.: J. biol. Chem. **235**, 2160 (1960).
LIANG, C. T., SACKTOR, B.: Arch. Biochem. Biophys. **176**, 285 (1976).
LIGHTWOOD, R., REBER, H. A.: Gastroenterology **72**, 61 (1977).
LIN, K. T., JOHNSTONE, R. M.: Biochim. biophys. Acta **249**, 144 (1971).
LÖFFLER, G., LUCAS, M., SCHMID, G.: Hoppe-Seyler's Zeitschr. Physiol. Chem. **358**, 271 (1977).
LOEWENSTEIN, W. R., NAKAS, M., SOCOLAR, S. J.: J. gen. Physiol. **50**, 1865 (1967).
LUCY, J. A.: J. Reprod. Fert. **44**, 193 (1975).
MAC DONALD, R. J., RONZIO, R. A.: Biochem. biophys. Res. Comm. **49**, 377 (1972).
MACDONALD, R. J., RONZIO, R. A.: FEBS Letters **40**, 203 (1974).
MCINTYRE, J. A., GILULA, N. B., KARNOVSKY, M. J.: J. Cell Biol. **60**, 192 (1974).
MAEDA, Y., KIM, J., KOSEKI, I., MEKADA, E., SHIOKAWA, Y., OKADA, Y.: Exptl. Cell Res. **108**, 95 (1977).
MANGOS, J. A., MCSHERRY, N. R.: Amer. J. Physiol. **221**, 496 (1971).
MANGOS, J. A., MCSHERRY, N. R., NAUSIA-ARVANITAKIS, S., IRWIN, K.: Amer. J. Physiol. **225**, 18 (1973).
MANGOS, J. A., MCSHERRY, N. R., NOUSIA-ARVANITAKIS, S., SCHILLING, R. F.: Proc. Soc. exp. Biol. **146**, 321 (1974).
MANSERVIGI, R., SPEAR, P. G., BUCHAN, A.: Proc. nat. Acad. Sci. **74**, 3913 (1977).
MARTINEZ-CARRION, M., MIHOVILOVIC, M., RAFTERY, M. A., SATOR, V.: Fed. Proc. **35**, 1364 (1976).
MATTHEWS, E. K., PETERSEN, O. H.: J. Physiol. **231**, 283 (1973).
MATTHEWS, E. K., PETERSEN, O. H., WILLIAMS, J. A.: J. Physiol. **234**, 689 (1973).
MELDOLESI, J., DECAMILLI, P., BRENNA, A.: In: Hormonal Receptors in Digestive Tract Physiology (S. Bonfils, P. Fromageot and G. Rosselin, Eds.). Amsterdam: North-Holland Publ. Comp. 1977, p. 203.
MICHELL, R. H.: Biochim. biophys. Acta **415**, 81 (1975).
MILEDI, R.: Proc. roy. Soc. London B **183**, 421 (1973).
MILUTINOVIĆ, S., SCHULZ, I., ROSSELIN, G.: Biochim. biophys. Acta **436**, 113 (1976).
MILUTINOVIĆ, S., SCHULZ, I., ROSSELIN, G., FASOLD, H.: In: Hormonal Receptors in Digestive Tract Physiology (S. Bonfils, P. Fromageot and G. Rosselin, Eds.). Amsterdam: North-Holland Publ. Comp. 1977a, p. 213.
MILUTINOVIĆ, S., SACHS, G., HAASE, W., SCHULZ, I.: J. Membrane Biol. **36**, 253 (1977b).
MILUTINOVIĆ, S., ARGENT, B. E., SCHULZ, I., SACHS, G.: J. Membrane Biol. **36**, 281 (1977c).
MOQTADERI, F., HIMAL, H. S., RUDICK, J., DREILING, D. A.: Amer. J. Gastroent. **58**, 117 (1972).
MORISSET, J. A., WEBSTER, P. D.: Amer. J. Physiol. **220**, 202 (1971).
MURER, H., SIGRIST-NELSON, K., HOPFER, U.: J. biol. Chem. **250**, 7392 (1975).
MUTT, V.: In: Origin, Chemistry, Physiology and Pathophysiology of Gastrointestinal Hormones (W. Creutzfeld, Ed.). Stuttgart-New York: Schattauer 1970, p. 83.
MUTT, V., MAGNUSSON, S., JORPES, J. E., DAHL, E.: Biochemistry **4**, 2358 (1965).
MUTT, V., JORPES, J. E., MAGNUSSON, S.: Europ. J. Biochem. **15**, 513 (1970).

NEURATH, A. R.: Acta Virol. **9**, 34 (1965).
NÈVE, P., WILLEMS, C., DUMONT, J. E.: Exp. Cell Res. **63**, 457 (1970).
NISHIYAMA, A., PETERSEN, O. H.: In: Secretory Mechanisms of Exocrine Glands (N. A. Thorn and O. H. Petersen, Eds.). New York: Academic 1974a, p. 216.
NISHIYAMA, A., PETERSEN, O. H.: J. Physiol. **238**, 145 (1974b).
NISHIYAMA, A., PETERSEN, O. H.: J. Physiol. **244**, 431 (1975a).
NISHIYAMA, A., PETERSEN, O. H.: Proc. roy. Soc. London B. **191**, 549 (1975b).
ONDETTI, M. A., PLUŠČEĆ, J., SABO, E. F., SHEEHAN, J. T., WILLIAMS, N.: J. Amer. Chem. Soc. **92**, 195 (1970).
ORCI, L. GABBAY, K. H., MALAISSE, W. J.: Science **175**, 1128 (1972).
PAGANO, R. E., HUANG, L.: J. Cell Biol. **67**, 49 (1975).
PALADE, G. E.: In: Subcellular Particles (T. Hayashi, Ed.). New York: Ronald 1959, p. 64.
PALADE, G.: Science **189**, 347 (1975).
PALADE, G. E., SIEKEVITZ, P., CARO, L. G.: In: The Exocrine Pancreas (Ciba Foundation Symposium, A. V. S. de Reuck and M. P. Cameron, Eds.). London: Churchill 1962, p. 23.
PAPAHADJOPOULOS, D., OHKI, S.: Science **164**, 1075 (1969).
PAPAHADJOPOULOS, D., POSTE, G., SCHAEFFER, B. E., VAIL, W. J.: Biochim. biophys. Acta **352**, 10 (1974).
PAPAHADJOPOULOS, D., VAIL, W. J., JACOBSON, K., POSTE, G.: Biochim. biophys. Acta **394**, 483 (1975).
PETERSEN, O. H.: Physiol. Rev. **56**, 535 (1976).
PETERSEN, O. H., MATTHEWS, E. K.: Experientia **28**, 1037 (1972).
PETERSEN, O. H., UEDA, N.: In: Calcium Transport in Contraction and Secretion (E. Carafoli, F. Clementi, W. Drabikowski and A. Margreth, Eds). Amsterdam: North-Holland 1975a, p. 147.
PETERSEN, O. H., UEDA, N.: J. Physiol. **247**, 461 (1975b).
PETERSEN, O. H., UEDA, N.: J. Physiol. **263**, 223 (1976a).
PETERSEN, O. H., UEDA, N.: J. Physiol. **254**, 583 (1976b).
PETERSEN, O. H., UEDA, N.: J. Physiol. (Lond.) **264**, 819 (1977).
POSTE, G., PAPAHADJOPOULOS, D.: Proc. nat. Acad. Sci. (Wash.) **73**, 1603 (1976).
POULSEN, J. H., WILLIAMS, J. A.: J. Physiol. (Lond.) **264**, 323 (1977a).
POULSEN, J. H., WILLIAMS, J. A.: Pflügers Arch. **370**, 173 (1977b).
POULSEN, J. H., WILLIAMS, J. A.: Irish J. med. Science. Supplement to the October, 1977 Issue. Abstracts of papers presented at the 10th Symp. of the European Pancreatic Club. Dundee, Ireland, 1977c.
RAWLS, J. A., Jr, WISTRAND, P. J., MAREN, T. H.: Amer. J. Physiol. **205**, 651 (1963).
REBER, H. A., LIGHTWOOD, R.: J. appl. Physiol. **40**, 984 (1976).
REBER, H. A., WOLF, C. J.: Amer. J. Physiol. **215**, 34 (1968).
REDMAN, C. M., SIEKEVITZ, P., PALADE, G. E.: J. biol. Chem. **241**, 1150 (1966).
REGAL, H., FINK, M., LUGER, Th., OGRIS, E., FISCHER, M., IRSIGLER, K.: Wien. Klin. Wochenschr. **89**, 580 (1977).
RIDDERSTAP, A. S.: Pflügers Arch. ges. Physiol. **311**, 205 (1969a).
RIDDERSTAP, A. S., BONTING, S. L.: Amer. J. Physiol. **217**, 1721 (1969a).
RIDDERSTAP, A. S., BONTING, S. L.: Pflügers Arch. **313**, 53 (1969b).
RIDDERSTAP, A. S., BONTING, S. L.: Amer. J. Physiol. **216**, 547 (1969c).
ROBBERECHT, P.: Thesis, Brussels, 1974.
ROBBERECHT, P., CHRISTOPHE, J.: Amer. J. Physiol. **220**, 911 (1971).
ROBBERECHT, P., DESCHODT-LANCKMAN, M., DE NEEF, P., BORGEAT, P., CHRISTOPHE, J.: FEBS Letters **43**, 139 (1974).
ROBBERECHT, P., CONLON, T. P., GARDNER, J. D.: J. biol. Chem. **251**, 4635 (1976).
ROBBERECHT, P., DESCHODT-LANCKMAN, M., CAMUS, J., CHRISTOPHE, J.: In: Hormonal Receptors in Digestive Tract Physiology (S. Bonfils, P. Fromageot and G. Rosselin, Eds.). Amsterdam: North-Holland Publ. Comp. 1977, p. 261.
ROSE, B., LOEWENSTEIN, W. R.: J. Membrane Biol. **28**, 87 (1976).
ROTHMAN, S. S.: Nature (Lond.) **204**, 84 (1964).
ROTHMAN, S. S., BROOKS, F. P.: Amer. J. Physiol. **208**, 1171 (1965a).
ROTHMAN, S. S., BROOKS, F. P.: Amer. J. Physiol. **209**, 790 (1965b).

RÜBSAMEN, H., HESS, G. P., ELDEFRAWI, A. T., ELDEFRAWI, M. E.: Biochem. biophys. Res. Commun. **68**, 56 (1976).
RUTTEN, W. J., de PONT, J. J. H. H. M., BONTING, S. L.: Biochim. biophys. Acta **274**, 201 (1972).
SACHS, G., CHANG, H. H., RABON, E., SCHACKMAN, R., LEWIN, M., SACCOMANI, G.: J. biol. Chem. **251**, 7690 (1976).
SATIR, B.: J. Supramol. Structure **2**, 529 (1974).
SAUER, F. A.: Nonequilibrium thermodynamics of isotope flow through membranes. In: Vol. I, chapter 4 of this series.
SCHREURS, V. V. A. M., SWARTS, H. G. P., DEPONT, J. J. H. H. M., BONTING, S. L.: Biochim. biophys. Acta **404**, 257 (1975).
SCHREURS, V. V. A. M., SWARTS, H. G. P., DEPONT, J. J. H. H. M., BONTING, S. L.: Biochim. biophys. Acta **419**, 320 (1976a).
SCHREURS, V. V. A. M., SWARTS, H. G. P., DEPONT, J. J. H. H. M., BONTING, S. L.: Biochim. biophys. Acta **436**, 664 (1976b).
SCHUDT, C., DAHL, G., GRATZL, M.: Cytobiology **13**, 211 (1976).
SCHULTZ, G., HARDMAN, J. G., SCHULTZ, K., BAIRD, C. E., SUTHERLAND, E. W.: Proc. nat. Acad. Sci. (USA) **70**, 3889 (1973).
SCHULZ, I.: Pflügers Arch. **329**, 283 (1971).
SCHULZ, I.: Pflügers Arch. **360**, 165 (1975a).
SCHULZ, I.: Thesis, Johann Wolfgang Goethe University, Frankfurt/Main, 1975b.
SCHULZ, I., MANNIGEL, H.: Acta gastro-enterol. belg. **37**, 402 (1974).
SCHULZ, I., MILUTINOVIĆ, S.: In: Membrane Toxicity (M. W. Miller & A. E. Shamoo, Eds.). New York: Plenum 1977, p. 209.
SCHULZ, I., YAMAGATA, A., WESKE, M.: Pflügers Arch. **308**, 277 (1969).
SCHULZ, I., STRÖVER, F., ULLRICH, K. J.: Pflügers Arch. **323**, 121 (1971).
SCHULZ, I., PEDERSON, R., WIZEMANN, V., KONDO, S.: In: Secretory Mechanisms of Exocrine Glands (N. A. Thorn and O. H. Petersen, Eds.). New York: Academic 1974, p. 88.
SCHULZ, I., KONDO, S., SACHS, G., MILUTINOVIĆ, S.: In: Receptors in Digestive Tract Physiology, (S. Bonfils, P. Fromageot and G. Rosselin, Eds.). Amsterdam: North-Holland Publ. Comp. 1977, p. 275.
SCHULZ, I., HEIL, K.: Pflügers Arch. **373**, R49 (1978).
SCRATCHERD, T.: In: Secretory Mechanisms of Exocrine Glands (N. A. Thorn and O. H. Petersen, Eds). New York: Academic 1974, p. 379.
SCRATCHERD, T., CASE, R. M., SMITH, P. A.: Scand J. Gastroenterol. **10**, 821 (1975).
SELINGER, Z.: In: Secretory Mechanisms of Exocrine Glands, (N. A. Thorn and O. H. Petersen, Eds.). New York: Academic 1974, p. 114.
SEWELL, W. A., YOUNG, J. A.: J. Physiol. **252**, 379 (1975).
SEYBOLD, J., BIEGER, W., KERN, H. F.: Virchows Arch. A. Path. Anat. and Histol. **368**, 309 (1975).
SIEKEVITZ, P., PALADE, G. E.: J. Biophys. Biochem. Cytol. **7**, 619 (1960).
SIMON, B., THOMAS, L.: Biochim. biophys. Acta **288**, 434 (1972).
SIMON, B., KINNE, R., SACHS, G.: Biochim. biophys. Acta **282**, 293 (1972).
SMITH, P. A., CASE, R. M.: Biochim. biophys. Acta **399**, 277 (1975).
STENING, G. F., GROSSMAN, M. I.: Amer. J. Physiol. **217**, 262 (1969).
STORMORKEN, H.: Scand. J. Haematol. Suppl. 9 (1969).
SWANSON, C. H.: Ph. D. Thesis, Harvard University, Cambridge, Mass. 1971.
SWANSON, C. H., SOLOMON, A. K.: Nature New Biol. **236**, 183 (1972).
SWANSON, C. H., SOLOMON, A. K.: J. gen. Physiol. **62**, 407 (1973).
SWANSON, C. H., SOLOMON, A. K.: J. gen. Physiol. **65**, 22 (1975).
TERNOUTH, J. H., BUTTLE, H. L.: Brit. J. Nutr. **29**, 387 (1973).
TOISTER, Z., LOYTER, A.: J. biol. Chem. **248**, 422 (1973).
TRÄUBLE, H., EIBL, H.: Proc. nat. Acad. Sci. (Wash.) **71**, 214 (1974).
TYRAKOWSKI, T., MILUTINOVIĆ, S., SCHULZ, I.: J. Membrane Biol. **38**, 333 (1978).
UEDA, N., PETERSEN, O. H.: Pflügers Arch. **370**, 179 (1977).
ULLRICH, K. J.: In: Secretory Mechanisms of Exocrine Glands (N. A. Thorn and O. H. Petersen, Eds). New York: Academic 1974, p. 423.
ULLRICH, K. J., CAPASSO, G., RUMRICH, G., PAPAVASSILIOU, F., KLÖSS, S.: Pflügers Arch. **368**, 245 (1977).

VAN DER BOSCH, J., SCHUDT, C., PETTE, D.: Exp. Cell Res. **82**, 433 (1973).
VAN DER BOSCH, J., MCCONNELL, H.: Proc. nat. Acad. Sci. (Wash.) **72**, 4409 (1975).
VAN DER KLOOT, W., KITA, H.: J. Membrane Biol. **14**, 365 (1973).
VAN DIJK, P. W. M., VERVERGAERT, P. H. J. Th., VERKLEIJ, A. J., VAN DEENEN, L. L. M., DE GIER, J.: Biochim. biophys. Acta **406**, 465 (1975).
VOLSKY, D., LOYTER, A.: Biochim. biophys. Acta **471**, 243 (1977).
WALTON, K. G., DE LORENZO, R. J., CURRAN, P. F., GREENGARD, P.: J. gen. Physiol. **65**, 153 (1975).
WATSON, E. L., SIEGEL, I. A., ROBINOVITCH, M. R.: Experientia **30**, 876 (1974).
WAY, L. W., DIAMOND, J. M.: Biochim. biophys. Acta **203**, 298 (1970).
WHEELER, J. E., LUKENS, F. D. W., GYÖRGY, P.: Proc. Soc. exp. Biol. and Medicine **70**, 187 (1949).
WILLIAMS, J. A., LEE, M.: Biochem. biophys. Res. Commun. **60**, 542 (1974).
WILLIAMS, J. A.: Amer. J. Physiol. **229**, 1023 (1975).
WILLIAMS, J. A., CHANDLER, D.: Amer. J. Physiol. **228**, 1729 (1975).
WILLIAMS, J. A., LEE, M.: J. Cell. Biol. **71**, 795 (1976).
WILLIAMS, J. A., WOLFF, J.: Biochem. biophys. Res. Comm. **44**, 422 (1971).
WILLIAMS, J. A., CHANDLER, D. E., POULSEN, J. H.: J. Cell Biol. **70**, 221a (1976a).
WILLIAMS, J. A., CARY, P., MOFFAT, B.: Amer. J. Physiol. **231**, 1562 (1976b).
WIZEMANN, V., SCHULZ, I.: Pflügers Arch. **339**, 317 (1973).
WIZEMANN, V., SCHULZ, I., SIMON, B.: Biochim. biophys. Acta **307**, 366 (1973).
WIZEMANN, V., CHRISTIAN, A. L., WIECHMANN, J., SCHULZ, I.: Pflügers Arch. **347**, 39 (1974).
WORMSLEY, K. G.: Gastroenterology **54**, 197 (1968).
WÜNSCH, E.: Naturwissenschaften **59**, 239 (1972).
WÜNSCH, E., JAEGER, E., MORODER, L., SCHULZ, I.: In: Hormonal Receptors in Digestive Tract Physiology (S. Bonfils, P. Fromageot and G. Rosselin, Eds.). Amsterdam: North-Holland Publ. Comp. 1977, p. 19.
YOUNG, J. A., SCHÖGEL, E.: Pflügers Arch. **291**, 85 (1966).
ZIMMERMANN, M. J., MOORE, E. W., DREILING, D. A., JANEWITZ, H. D.: J. clin. Invest. **50**, 103a (1971).

Chapter 17

Transport in Gallbladder

L. REUSS

A. Introduction

The gallbladder was the first tissue in which transepithelial transport was shown to occur in the absence of a sizable spontaneous transepithelial electrical potential difference (DIAMOND, 1962b). In the 15 years that have passed since that observation, the experimental evidence accumulated from studies in this and other tissues has increasingly indicated that salt transporting epithelia possess widely different physiological properties, and can no longer be considered or modeled in a uniform fashion. The distinction between "tight" and "leaky" epithelia, first proposed by FRÖMTER and DIAMOND (1972), although perhaps an over-simplification, is a convenient way of pointing out this physiological heterogeneity.

According to all the presently accepted criteria (low transepithelial potential difference, high electrical conductance, high rate of transport, isosmotic transport, high hydraulic conductivity), the epithelium of the gallbladder is typically leaky, although not as leaky as the proximal tubule of the nephron.

It is beyond the purpose of this chapter to review the literature exhaustively. The information produced before 1971 has been thoroughly discussed by DIAMOND (1968) and WHEELER (1971). Recent articles of a general scope to which the reader is also referred are the ones by DIAMOND et al. (1971), DIAMOND (1974), and FRÖMTER and DIAMOND (1972). An excellent review of the passive properties of the gallbladder has been recently published (MORENO and DIAMOND, 1976).

The purpose of this chapter is to emphasize the recent experimental results that have increased our knowledge about the mechanisms of salt and water transport by this tissue. In some cases, fairly detailed descriptions of experimental procedures and theoretical analyses are given in order to make the chapter as nearly self-contained as possible. The author has tried to follow more or less historically a course defined by the most important contributions in the field; this may well be a biased approach.

B. General Transport Properties

I. Morphology of the Gallbladder

The gallbladder is a sac lined by a monolayered epithelium lying on a richly vascular *lamina propria.* The mucosa is surrounded by a layer of smooth muscle fibers, a subserosal layer of connective tissue, and the serosa (see Fig. 1).

Folds (100 to 500 μm high) are prominent in gallbladders of mammals. In some lower species, such as *Necturus maculosus,* the mucosa is flat. The overall thickness of the wall of the rabbit gallbladder is about 350 μm, two-thirds of which is comprised of muscular and subserosal layers.

Electron-microscopic observations (*e. g.,* KAYE et al., 1966; TORMEY and DIAMOND, 1967), reveal in the gallbladder some morphological characteristics shared by other transporting epithelia: (1) the epithelial cells have a PAS-positive microvillous apical border (less prominent than in the renal proximal tubule), and under it the cytoplasm is fibrillar and contains only a few vesicles. The nuclei are central and most mitochondria are apical. (2) At the apical end, the lateral plasma membranes of adjacent cells are associated to form junctional complexes, or limiting junctions. Two distinct regions are evident: the *zonula occludens,* closest to the lumen, is 200 to 300 nm long, and according to evidence obtained in other tissues, it is probably not a complete seal around the cell but consists of discrete contacts between the two membranes, leaving a mesh-

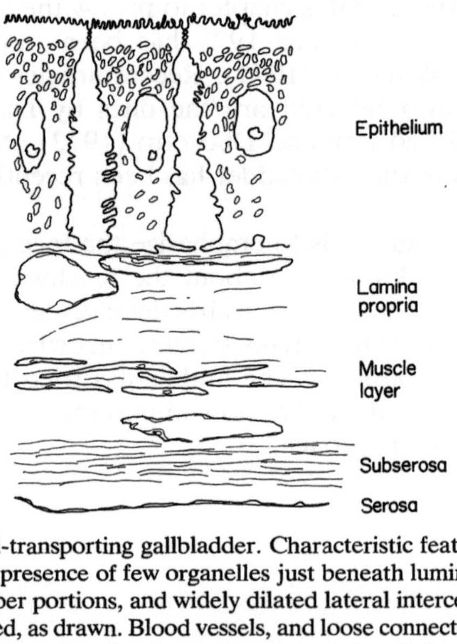

Fig. 1. Morphology of fluid-transporting gallbladder. Characteristic features of epithelial layer are microvillous apical border, presence of few organelles just beneath luminal cell membrane, abundance of mitochondria in upper portions, and widely dilated lateral intercellular spaces. Basal end of spaces is frequently narrowed, as drawn. Blood vessels, and loose connective tissue constitute *lamina propria.* Muscle layer is thin and rather loose in most species. Subserosa is made up of dense connective tissue. Diagram not drawn to scale. Height of the epithelium is approx. 15 percent of overall wall thickness

work of aqueous channels in the transepithelial direction. After the *zonula occludens* comes a narrow, more or less straight segment with no cellular contacts (*zonula adhearens*). Finally, most of the lateral surface of the epithelial cell is separated from adjacent cells by long, convoluted intercellular spaces. Their width varies, probably in relation to the transporting rate. Desmosomes are occassionally found along the lateral intercellular spaces.

The main barrier to ion flow is the epithelial layer itself (accounting for most of the transmural electrical resistance). However, it is clear that the epithelium is not the only barrier for osmotic water flow, since the whole wall of the organ shrinks when the mucosal solution is made hyperosmotic by addition of impermeant solutes, indicating the existence of a subserosal barrier to water flow (SMULDERS et al., 1972).

II. Composition of Gallbladder Bile

In situ, the gallbladder concentrates the impermeant solutes contained in hepatic bile by a factor of 5 to 10 (RAVDIN et al., 1932). When absorption stops, bile composition is characterized by a very high concentration of sodium salts of bile acids, increased concentrations of potassium and calcium and decreased concentrations of chloride and bicarbonate ions (Fig. 2). At this time the volume of the luminal fluid has been reduced by 80 to 90 percent and large net lumen-to-blood fluxes of sodium chloride and bicarbonate can be shown to have occurred.

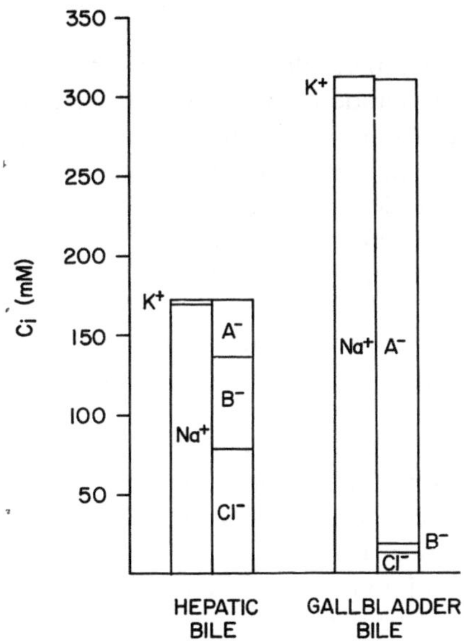

Fig. 2. Composition of hepatic and concentrated gallbladder bile. A^- = bile acid anions; B^- = bicarbonate ion. Concentration changes in gallbladder include an increase in Ca^{++}, not shown in Figure. Values are approximations from several species, quoted by DIAMOND (1968)

A calculation of concentrated bile osmolarity from molar ion concentrations yields a value almost twice of that of plasma (Fig. 2). However, bile has been shown to be isosmotic to plasma (DIAMOND, 1962a; KANG and HONG, 1962). The apparent discrepancy exists because sodium salts of bile acids have low osmotic coefficients compared to similar concentrations of sodium chloride (MOORE and DIETSCHY, 1964). DIAMOND (1962a) measured the sodium activity in bile and concluded that this cation was bound to the micelles produced by aggregation of bile acid anions, with a Na-to-anion ratio of 1.

Resorption of the gallbladder luminal fluid generates ion concentration gradients across the epithelium, and these in turn produce a significant transepithelial electrical potential difference (ca. 18 mV. lumen negative, in rabbit gallbladder *in vivo,* DIETSCHY and MOORE, 1964). Salt transport from lumen to blood takes place against an increasingly larger electrochemical potential gradient: Na is transported down its activity gradient but against the electrical potential gradient, while Cl and HCO_3 are transported down the electrical potential but against large activity gradients. At steady state, net transport ceases because active lumen-to-blood salt transport is exactly balanced by the increased backflux produced by the favorable electrochemical potential gradient for salt diffusion from blood to lumen.

Permeant ions that are not actively transported distribute passively, according to their charge and valence: thus, K and Ca concentrations in bile are higher than in plasma.

III. General Characteristics of Salt and Water Transport

Several procedures have been employed to measure net fluid transport (J_v) in a variety of species. These include (1) intermittent weighing of sacs suspended in Ringer solution in either the normal (DIAMOND, 1962a) or the everted (DIETSCHY, 1964) configuration, (2) measurement of the progressive concentration of impermeant indicators added to the mucosal solution (DIAMOND, 1962a, WHEELER, 1963), and (3) direct measurement of the transported fluid, by observation of the displacement of the meniscus in a horizontal pipette (DIAMOND, 1962a), collection of the fluid as it drips off a gallbladder filled with solution and maintained in a container filled with humidified air (DIAMOND, 1964b) or intermittent measurement of the volume of the serosal solution in a hermetic double perfusion system (WHEELER, 1963).

Net fluxes of ions (J_i) have been calculated in bag preparations, from J_v and the changes in composition of the luminal solution by the use of a simple balance equation. Since J_v is in general large, this method is accurate enough to measure the net rate of salt transport (DIAMOND, 1962a; DIETSCHY, 1964).

Unidirectional fluxes of ions, water, and other non-electrolytes have been determined with conventional isotopic techniques in gallbladders of several species, mounted as bags or flat sheets in Ussing-type chambers (DIAMOND, 1962b; WHEELER, 1963; DIETSCHY, 1964; ONSTAD et al., 1967). These determinations allow one to calculate the passive permeability coefficient of the species under study. In addition, a comparison of these results with those obtained from electrical measurements can provide information on the nature of the ion permeation process (see discussion on anion exchange diffusion below). Recently, the short-circuit current (SCC) technique has been employed to measure the rate of sodium transport in gallbladders from several vertebrate species, either by calculation from the open circuit transepithelial potential (V_{ms}) and the transepithelial resistance (R_t) (MORENO, 1974) or by measurements with the tissue clamped at $V_{ms} = 0$ (GELARDEN and ROSE, 1974). The appropriateness of this technique is not clear. First, the error produced by the corrections required for the electrical resistance of the bathing solutions (since the tissues have rather low resistances) can be significant. Second, this technique is valid only if equality of SCC and net Na flux can be demonstrated, and this has been done only in human gallbladder (ROSE et al., 1973). Among other possibilities, it is conceivable that a fraction of Na transport could be coupled to Cl transport in a neutral mechanism that would not be expressed as a transepithelial current (DIAMOND, 1962b). Furthermore, if $V_{ms} = 0$ in transporting gallbladders, the SCC technique measures no transport, whereas J_v measurements indicate that transport is taking place (DIAMOND, 1968). Finally, gallbladders that have a mucosa-positive potential would denote active Cl transport. In fact, as we will show later, the mechanism of generation of the transepithelial potential difference in this tissue is far more complicated than is implicitly assumed when the SCC technique is used to measure net Na transport.

Common transport features of gallbladders from all species (from fish to man) can be summarized as follows: (a) high rate of volume transport, similar to flows measured across other leaky epithelia; (b) isosmotic transported fluid (with respect to the luminal solution) under a wide variety of experimental conditions; (c) transport requires mucosal fluids with specific ion compositions; (d) net transport of salt occurs in the absence of or against a transepithelial electrochemical potential gradient; (e) water flow in the absence of or against bulk solution osmotic gradients (but in the same direction as net solute flux); (f) inhibition of fluid transport by cooling, lack of oxygen, ouabain, metabolic inhibitors, and other pharmacologic agents.

The experimental data that support this characterization will be described in the next paragraphs. For further details, the reader is referred to the excellent review articles by DIAMOND (1968) and WHEELER (1971).

1. Salt and Water Transport: Rate, Ion Dependency, Composition of the Transported Fluid

Typical results obtained *in vivo* and *in vitro* with bathing solutions of ionic composition similar to that of plasma on both sides are shown in Table 1. *In vivo*, the dog gallbladder has been shown to absorb about 16 percent of its luminal fluid per hour (RAVDIN et al., 1932), while the *in vitro* human gallbladder absorbs Ringer solution at a rate of about 13 percent per hour (ONSTAD et al., 1967). The corresponding figures for the *in vitro* preparations of rabbit and fish gallbladders are 35 percent (DIAMOND, 1964a) and 10 percent (DIAMOND,

Table 1. Rate of fluid transport (J_v) by gallbladders of several animal species

Species	Condition	J_v ($\mu l \cdot h^{-1} \cdot cm^{-2}$)	Reference
Fish (roach)	*in vitro*	15	DIAMOND (1962a)
Frog	*in vitro*	3.9	MORENO (1974)
Guinea-pig	*in vitro*	23	DIAMOND (1964a)
Rabbit	*in vitro*	55	WHEELER (1963)
Rabbit	*in vitro*	43	DIETSCHY (1964)
Rabbit	*in vitro*	50	DIAMOND (1964a)
Dog	*in vivo*	66	GRIM and SMITH (1957)
Man	*in vitro*	13%/h	ONSTAD et al. (1967)

1962a), respectively. Although when expressed per unit of serosal surface area these rates of volume flow are smaller than those measured in mammalian renal proximal tubule or upper intestine, they are nevertheless very high when compared to tighter epithelia such as distal segments of the nephron, urinary bladder, colon, or amphibian skin.

Net fluxes of Na, Cl and HCO_3 in the mucosa-to-serosa direction have been calculated for gallbladders of several of the above mentioned species from measurements of rates of volume flow and ion concentrations in the luminal fluid (Table 2).

Isotopically measured unidirectional Na, Br and Cl fluxes in fish and rabbit gallbladder (DIAMOND, 1962b; WHEELER, 1963; FRIZZELL et al., 1975) are very high when compared with tight epithelia (Table 3), indicating that the overall ionic permeability of the gallbladder is high. In addition, P_{Cl} and P_{Br} calculated

Table 2. Rates of ion transport by gallbladder

Species	${}_aJ_{Na}^{(ms)}$	${}_aJ_{Cl}^{(ms)}$		Reference
Fish (roach)	1.6		$\mu M \cdot h^{-1} \cdot cm^{-2}$	DIAMOND (1962a)
Rabbit	30.2	24.9	$\mu M \cdot h^{-1} \cdot 100\,mg^{-1}$ [a]	WHEELER (1963)
Rabbit	36.2	33.8	$\mu M \cdot h^{-1} \cdot 20\,mg^{-1}$ [b]	DIETSCHY (1964)

${}_aJ_i^{(ms)}$ = active flux of the ion *i* in the mucosa to serosa direction
[a] Wet weight
[b] Dry weight

Table 3. Tracer unidirectional Na and Cl fluxes in rabbit gallbladder

	${}^*J_i^{(ms)}$	${}^*J_i^{(sm)}$	${}^*_nJ_i^{(ms)}$
Na	35.1	22.1	12.9
Cl	28.1	13.3	14.8

Fluxes in $\mu Eq \cdot cm^{-2} \cdot h^{-1}$. First and second columns, fluxes from *m* to *s* and *s* to *m*, respectively. Third column, net *m*-to-*s* fluxes. Results are means from 5 paired gallbladders. Ratio of backfluxes = 1.66. (FRIZZELL et al., 1975)

from the fluxes are far greater than those computed from electrical measurements (see Section C). The most likely explanation for these discrepancies is that a large proportion of the passive anion fluxes takes place by exchange diffusion.

Fluid transport across the gallbladder almost ceases when Na is completely replaced with tetraethylammonium (in fish: DIAMOND, 1962a; in rabbit: KAYE et al., 1966), or choline (in rabbit: WHEELER, 1963; SULLIVAN and BERNDT, 1973). The addition of relatively small amounts of Na to Na-free mucosal solutions rapidly restores fluid transport (WHEELER, 1963; KAYE et al., 1966).

Under physiological conditions, Cl and HCO_3 are the anions transported with Na. However, Br (DIAMOND, 1962a), HCO_3 alone, DMO, propionate, acetate, lactate, and pivalate (trimethylacetate) can support vigorous fluid transport (WHITLOCK and WHEELER, 1967), while isethionate is less effective. When Cl is replaced by SO_4, fluid transport is virtually abolished (DIAMOND, 1962a; WHITLOCK and WHEELER, 1967).

If the mechanism of transport were *via* a carrier with cation and anion-binding sites, these observations would indicate a highly Na-selective site and a far less selective anion-binding site. However, only K and a few quarternary amines have been substituted for Na in studies of volume transport. Thus, the selectivity of the Na-binding site rests on less solid experimental evidence than the obviously low selectivity of the anion-binding site.

With salt solutions similar to plasma on both sides of the preparation, the osmolarity of the transported fluid, calculated from J_v and the composition of the mucosal fluid, is essentially identical to that of the bathing solution, regardless of the spontaneous transport rate (DIAMOND, 1962a, 1964a; WHITLOCK and WHEELER, 1964; DIETSCHY, 1964; ONSTAD et al., 1967). For example, the osmolarity of the transported fluid does not change when salt transport is reduced by lowering the sodium chloride concentration (C_{NaCl}) or by metabolic inhibitors. In other words, reductions in J_v parallel decreases in salt net flux.

Direct measurements of the absorbate composition were obtained by DIAMOND (1964b) from rabbit gallbladders suspended in the absence of serosal bathing solution. The fluid collected was shown to be essentially isotonic with the mucosal solution when C_{NaCl} was altered or when the osmolarity was increased by the addition of sucrose. In the first condition, C_{NaCl} in the absorbate was equal to C_{NaCl} in the lumen in the range of 50 to 300 mM. In the second, C_{NaCl} in the absorbate increased by 0.5 mM per mM of sugar.

When identical salt solutions bathe both sides of the gallbladder, salt transport from mucosa to serosa takes place in the absence of any passive net driving force, provided there is no significant hydrostatic pressure across the tissue. Furthermore, it has been shown that salt transport can be maintained against an electrochemical potential gradient (DIAMOND, 1962a; DIETSCHY, 1964). In both studies, C_{NaCl} in the mucosal solution was reduced by isotonic replacement with sucrose or mannitol, keeping the serosal solution composition constant. In DIAMOND's experiments with fish gallbladder, net volume flow saturated when mucosal C_{NaCl} was about 80 mmol $\cdot$ l^{-1}, was half-maximal at about 30 mmol $\cdot$ l^{-1}, and reversed below 10 mmol $\cdot$ l^{-1}. DIETSCHY's results in the rabbit gallbladder were very similar, but in order to reverse the direction of volume flow both

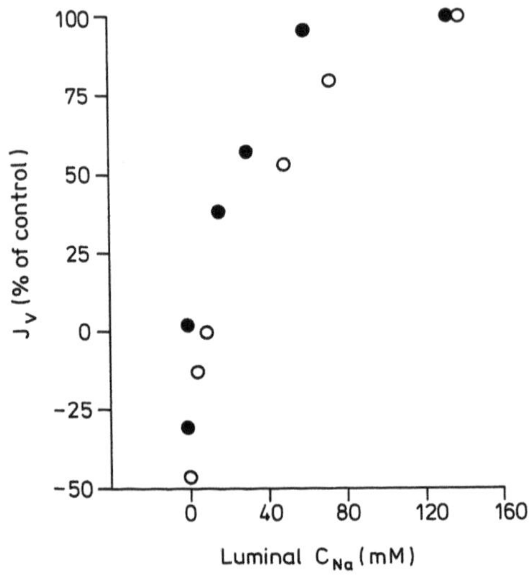

Fig. 3. Effect of luminal Na concentration on rate of fluid absorption by fish (○) and rabbit gallbladder (•). Lowest filled circle corresponds to removal of $NaHCO_3$ from luminal fluid in addition to NaCl substitution. Osmolalities of luminal solutions were kept constant by addition of sucrose or mannitol. Note that J_v reverses at low salt concentration, is 50 percent at about 30 mM Na, and saturates at about 80 mM. Values replotted from DIAMOND (1962a) for fish and from DIETSCHY (1964) for rabbit. Exclusive replacement of Cl with SO_4 gave essentially same results in rabbit gallbladder

NaCl and $NaHCO_3$ had to be reduced in concentration. The salt concentrations at which half maximal transport and saturation were observed were very similar to the values reported by DIAMOND (Fig. 3).

In DIAMOND's experiments, when $J_v = 0$, a steady state had been reached when active mucosa-to-serosa salt transport precisely balanced the increased passive backflux produced by the concentration gradient. The ratios of serosal-to-mucosal concentrations were: for Na, 14.3 and for Cl, 9.5. Both indicate active transport, especially the Cl concentration ratio, since in this experimental situation the mucosal solution becomes electrically positive. Thus, Cl transport takes place against both chemical and electrical potential gradients. Very similar results were obtained by KANG and HONG (1962) in dog gallbladders *in situ*. After the achievement of steady-state Na and Cl concentration gradients were about 20 and 10 $mEq \cdot l^{-1}$, respectively. Metabolic inhibition by the addition of iodacetate to the luminal fluid produced an immediate increase in mucosal solution volume and Cl concentration.

WHEELER (1963) measured unidirectional Na and Cl fluxes after reversing water flow in rabbit gallbladder by increasing the osmolarity of the mucosal solution with mannitol. Some of his results are summarized in Table 4. Note that the experimentally determined flux ratios were, for both ions, significantly larger than the values predicted from the transepithelial potential and the concentration in both solutions. These deviations are consistent with active transport of both ions.

Table 4. Flux ratios during osmotic inhibition of water transport in rabbit gallbladder

Ion	J_v μl · h^{-1} · 100 mg^{-1}	C_S mM	C_M mM	$_nJ_i^{(ms)}$ μEq · h^{-1} · 100 mg^{-1}	V_{ms} mV	$\frac{J_i^{(ms)}}{J_i^{(sm)}}$ obs	$\frac{J_i^{(ms)}}{J_i^{(sm)}}$ pred
Na	−13.4	152	88	13.3	17.6	1.47	1.07*
Cl	−5.5	137	92	18.2	14.6	2.0	0.4 *
K	12	6.0	6.1	0.9	10.5	1.24	1.45

J_v (volume flow) was reduced or reversed (negative sign) by addition of mannitol to mucosal solution. C_M and C_S are the ion concentrations in the mucosal and serosal solution, respectively. $_nJ_i^{(ms)}$ = net flux. V_{ms} = transepithelial potential (reference: serosal solution). Last two columns: observed and predicted (from electrochemical potential gradient) flux ratios; asterisks indicate statistically significant differences. Values for J_v and J_i normalized to 100 mg wet weight. (WHEELER, 1963)

2. Water Transport and Transmural Osmotic Gradients

When the osmolarity of the mucosal solution is progressively increased by the addition of a nonpermeant, nonionic solute, the rate of fluid transport decreases *pari passu* with the osmotic gradient, but J_v drops to 0 only when the osmotic gradient is 40 to 100 mOsm · l^{-1} (DIAMOND, 1962c; WHEELER, 1963; DIAMOND, 1964a; DIETSCHY, 1964; ONSTAD et al., 1967). The method and results of these experiments are shown in Fig. 4.

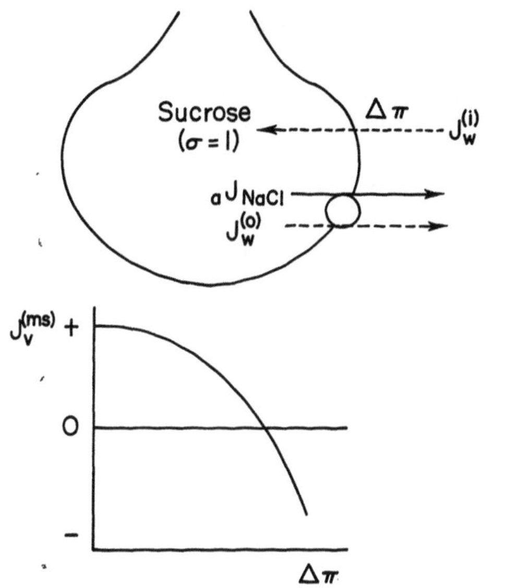

Fig. 4. Effect of transmural osmotic gradients on rate of fluid transport. As shown in upper diagram, addition of impermeant solute drives water into lumen down osmotic pressure gradient. Net transmural water flux is zero when osmotically produced flux (upper) equals water flux coupled to salt transport (lower). J_w = 0 (lower diagram) at sizable osmotic gradients: 40 in fish (DIAMOND, 1962c), 80 in rabbit (WHEELER, 1963; DIETSCHY, 1964; DIAMOND, 1964a), 30 to 80 mOsm · l^{-1} in man (ONSTAD et al., 1967). Thus, water can move "uphill". Presence of transportable salt in lumen is necessary for this phenomenon

These demonstrations of uphill water flow do not necessarily indicate active water transport. For one thing, water transport occurs only when solute is being transported. In addition, if water were the species actively transported, the solute flux, computed from the permeability characteristics of the tissue, would be at least one order of magnitude lower than the experimentally determined rate (DIAMOND, 1962c).

3. Metabolic and Pharmacologic Aspects of Transport

MARTIN and DIAMOND (1966) were able to divide gallbladder oxygen consumption into three fractions: one linked to fluid transport (Cl-dependent), one that is Na-dependent and ouabain-sensitive, interpreted as corresponding to a Na-K exchange pump, and a basal one that does not require Na or Cl and is ouabain-insensitive (Table 5). The molar ratio of transported NaCl to O_2 consumption was 24.6. On the assumption that 6 moles of ATP are generated per mole of O_2, the molar ratio of NaCl transported to ATP utilized was about 4. The available energy for NaCl translocation can be calculated to be eight times greather than the requirement.

Van Os and SLEGERS (1971) were able to obtain Na-K-ATPase activity from rabbit and guinea-pig gallbladders, and found a linear correlation with the transport rate. In rabbit gallbladder, the enzyme is located along the lateral cell membrane, mainly in the apical two-thirds (KAYE et al., 1966). Low concentrations of ouabain stimulate slightly, whereas high concentrations inhibit, both the activity of the enzyme and the rate of fluid transport. Cl substitution has no effects on ATPase activity (Van Os and SLEGERS, 1971).

The mechanism of the stimulatory effect of bicarbonate on fluid absorption has been studied by MARTIN (1974) and MARTIN and MURPHY (1974). Carbamyl phosphate may mediate the effect, since HCO_3 is a precursor of its synthesis, and since carbamyl phosphate has been shown to stimulate fluid absorption. Carbonic anhydrase inhibitors have no effect in fish gallbaldder (DIAMOND, 1962a), but diminish transport in rabbit gallbladder without altering the composition of the absorbed fluid (WHEELER et al., 1969; SULLIVAN and BERNDT, 1973).

Fluid transport can be inhibited by cooling, lack of oxygen supply, and metabolic inhibition (particularly of glycolysis). These results add circumstantial evidence to the conclusion that fluid transport is an active process, and strongly suggest that cellular metabolic energy for transport is provided, at least in part, by anaerobic glycosis (DIAMOND, 1968).

Other early studies of the effects of hormones and diverse pharmacological

Table 5. Fractions of oxygen consumption in rabbit gallbladder. (After MARTIN and DIAMOND, 1966)

9%	Cl-dependent	Fluid Transport
36%	Na-dependent Ouabain-sensitive	Na-K Pump
55%	Na-, Cl-, and ouabain-insensitive	Basal

agents on gallbladder transport have been reviewed by DIAMOND (1968). More recently, LEYSSAC et al. (1974a) and MORTON et al. (1974) have shown inhibitory effects of prostaglandins in rabbit and guinea-pig preparations, respectively. Angiotensin also inhibits absorption in rabbit gallbladder (LEYSSAC et al., 1974b).

Cyclic AMP reduces fluid transport in rabbit and guinea-pig gallbladder (HEINTZE et al., 1974; FRIZZELL et al., 1975). MERTENS (1974) showed that cholera enterotoxin also inhibits transport. In dog gallbladder, *in vivo*, the toxin tends in addition to elicit secretion (SCHAFER et al., 1969). These effects are very similar to those observed in intestine. Since in the latter organ the inhibitory effect of cAMP on absorption is exerted *via* inhibition of a neutral NaCl transport mechanism (NELLANS et al., 1973), an analogous observation in the gallbladder adds circumstantial evidence to the hypothesis of a similar uptake mechanism operating in this tissue (see Section C. IV).

C. Mechanisms of Ion Transport

The results summarized in the previous section conclusively demonstrate that the gallbladder actively transports salt (usually NaCl and $NaHCO_3$) in the mucosa-to-serosa direction. This process is tightly coupled to net water flow, in the same direction and in isotonic proportions. Both water and salt transport seem to be intimately related to the structural organization of the tissue, and recent experimental efforts have been directed at the identification and characterization of the pathways involved.

In this section, our discussion will include the evidence for the existence of at least two transepithelial ion pathways (cellular and paracellular), the properties of these pathways, and our present knowledge about the mechanisms of ion transport.

I. Black-Box Electrical Properties of the Gallbladder

1. The Transepithelial Potential Difference

In his studies with fish gallbladder, DIAMOND (1962b) found that the transepithelial potential was very close to zero. We know now that a low transmural potential with both tissue surfaces bathed by identical solutions is a general property of leaky epithelia. Although some gallbladders exhibit a V_{ms} of up to 8 mV, the potential is, in general, very small when compared to that developed by tight epithelia (Table 6).

A negligible V_{ms} across the gallbladder could be explained in several ways, represented in Figure 5.

Table 6. "Black-box" electrical properties of the gallbladder: Transepithelial potential (V_{ms}) and resistance (R_t)

		V_{ms}(mV)	R_t ($\Omega \cdot cm^2$)
Fish			
Roach	(1)	−0.8	113
Bream	(1)	−1.0	
Pike	(1)	−0.7	
Amphibian			
Frog	(2)	−2.2	129
Necturus	(3, 4, 5)	−1.4 to −2.2	267–450
Bird			
Goose	(6)	−3.6	30
Mammal			
Rabbit	(6, 7, 8)	0 to +1.4	23
Dog	(6)	−0.2	46
Monkey	(6)	−1.5 to −2.7	27 to 35
Man	(6)	−7.6	52

V_{ms}, referred to serosal solution. References: (1) Diamond (1962b), (2) Moreno (1974), (3) Frömter (1972), (4) Reuss and Finn (1975a), (5) Van Os and Slegers (1975), (6) Gelarden and Rose (1974), (7) Dietschy (1964), (8) Machen and Diamond (1969)

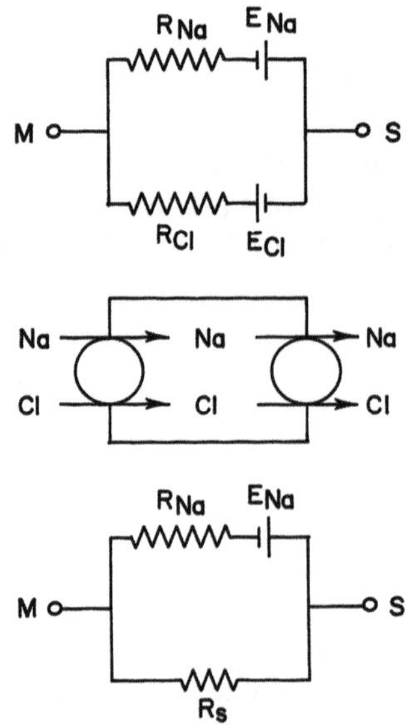

Fig. 5. Possible mechanisms of transepithelial active electrolyte transport without generation of a sizable transmural electrical potential difference. M and S, mucosal and serosal solutions, respectively. Top: cation and anion "electrogenic" transport mechanisms in parallel operate fortuitously at same rate. Middle: "electrically silent", coupled anion-cation transport. Bottom: active "electrogenic" ion pathway tends to generate a transmural pd, but is shunted by a parallel leak resistance (R_s)

a) There are two active transport pathways in parallel, one for Na and one for Cl (R_{Na} and E_{Na}; R_{Cl} and E_{Cl}), both oriented to transport the ion from mucosa to serosa. The potentials that the pumps generate across the epithelium are equal (because the rates of the pumps are equal) and opposite in polarity (because of the opposite charges of the ions). Thus, they cancel each other out and V_{ms} is zero.

b) Na and Cl are transported by a neutral mechanism, which requires simultaneous translocation of the two species in an uncharged form, *e. g.*, bound to a neutral carrier. This process is electrically silent. Different selectivities of the two sites in the carrier, for the cation and anion, could explain the specific requirement for Na, and the ability of several anions to substitute for Cl.

c) Only one of the ions is actively transported, as in the frog skin and the toad urinary bladder, and this mechanism tends to generate a transepithelial potential. However, the EMF of the pump is dissipated *via* a high conductance parallel pathway (shunt, R_s), and thus V_{ms} is very small.

The first two of these three possibilities were initially considered by DIAMOND, and the first one was experimentally discarded on the following basis: if the pumps are independent of each other but transport at the same rate, complete replacement of one of the transported ions by a non-transported ion should leave the other pathway intact, and its contribution to V_{ms} should be unmasked. For instance, bilateral SO_4 substitution for Cl should generate a mucosa-negative transepithelial potential, while bilateral substitution of Na with choline or tetraethylammonium should make the mucosal solution positive. The potential differences predicted from the measured transport rates are rather large in both Na-free and Cl-free bathing solutions. However, no change in transepithelial potential was observed when either of the ions was substituted (DIAMOND, 1962b). It was concluded that the salt transport mechanism in the gallbladder is a neutral, electrically silent pump.

The possibility of a parallel high-conductance pathway was raised a few years later in tight epithelia (USSING and WINDHAGER, 1964; CIVAN et al., 1966). The first direct demonstration of a shunt resulted from an intraepithelial cable analysis of the proximal tubule of *Necturus* (WINDHAGER et al., 1967). More recently, a large number of experimental results in gallbladders of several species have provided indirect or direct evidence of the existence of a high conductance pathway in parallel with the transcellular pathway. This concept is discussed in detail in Section C. II. The observation of a small or zero transepithelial potential difference is perhaps the most important experimental clue about the mechanism of ion transport by the gallbladder. By itself, this observation rules out the two-membrane hypothesis (KOEFOED-JOHNSEN and USSING, 1952) as an explanation of transport by the gallbladder, and narrows down the alternatives to a neutral salt pump mechanism, a high-conductance pathway in parallel with the active transport pathway, or a combination of both.

2. The Transepithelial Electrical Resistance

Like all leaky epithelia, the gallbladder has a very low transepithelial electrical resistance (R_t). Representative values from several species are shown in Table 6.

Note that the values range from 20 to about 400 $\Omega \cdot cm^2$. Tight epithelia have resistances in the order of several thousand $\Omega \cdot cm^2$. Most of the electrical resistance of the gallbladder wall has been shown to be associated with the epithelium itself (*e. g.*, WRIGHT and DIAMOND, 1968).

In addition to the low value of R_t, the current-voltage realtionship of the rabbit gallbladder is linear up to 800 mV, both with identical bathing solutions on both sides, or in the presence of a single salt concentration gradient. These findings are incompatible with the idea that one or more thin membranes constitute the sole pathway for transepithelial ion fluxes, since such membranes (*i. e.*, the cell membranes), would be expected to rectify when bathed by asymmetric solutions (if ion permeation is through polar channels) and also by identical solutions if permeation is carrier-mediated.

3. Overall Transepithelial Ionic Permeabilities

In addition to conventional measurements of unidirectional fluxes by isotopic techniques, electrical methods have been employed to determine permeability properties of epithelia. The techniques frequently employed are the measurements of dilution and biionic potentials across the whole epithelium (BARRY and DIAMOND, 1970; BARRY et al., 1971). In the papers referred to above, transport

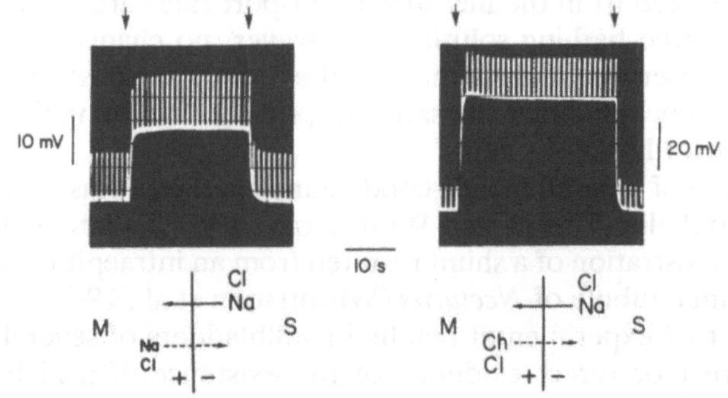

Fig. 6. Typical records of transepithelial diffusion potentials in frog gallbladder (REUSS and FINN, unpubl.). Continuous tracing of both records is transepithelial potential (V_{ms}). Ringer solution bathes both sides of tissue before and after mucosal solution replacement indicated by arrows. Vertical deflections are V_{ms} changes (ΔV_{ms}) produced by transepithelial d. c. pulses. Current density was kept constant in each record. Thus, changes in magnitude of ΔV_{ms} indicate directly proportional changes in transepithelial resistance (R_t). Left: 2:1 NaCl dilution potential. At first arrow, the mucosal solution was suddenly replaced with one containing half control amount of C_{NaCl}, identical concentrations of other ions, and sucrose to isotonicity. Upward (mucosa-positive) change in V_{ms} is accompanied by a slight increase in R_t. Changes are reversible upon return to control solution (second arrow). Since Na and Cl are the only ions whose concentrations differ across tissue, polarity of the change in V_{ms} indicates that $P_{Na} > P_{Cl}$. Diagram below record is a simplified explanation of V_{ms} change. Right: choline-Na bi-ionic potential. Replacement of all Na in mucosal solution with choline (first arrow) is followed by a large V_{ms} change (mucosa-positive) and a pronounced increase in R_t. Polarity of V_{ms} change indicates that $P_{Na} > P_{choline}$. After appropriate corrections for liquid junction potentials, from experiments such as these relative permeability coefficients can be derived. (See BARRY and DIAMOND, 1971; BARRY et al., 1971)

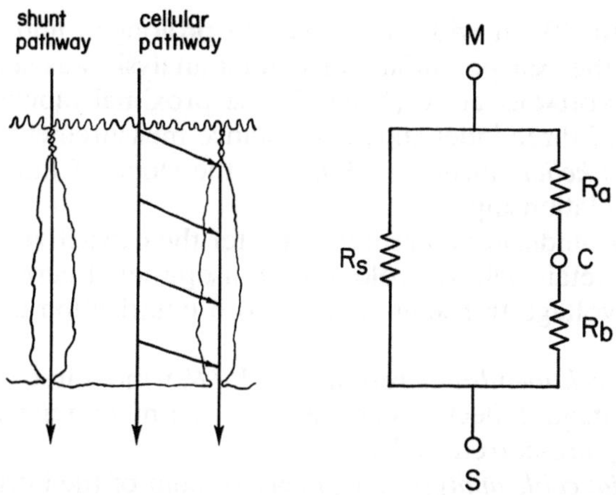

Fig. 7. Passive equivalent electrical circuit for gallbladder epithelium. As shown on left, ions can permeate through limiting junction and lateral intercellular space (shunt pathway) or through cells (cellular pathway), either across apical and basal membrane or across apical membrane and lateral membrane, in series with lateral space. On right, R_a represents equivalent electrical resistance of apical membrane, R_b equivalent resistance of basolateral membrane and lateral intercellular space, and R_s shunt resistance. M, C, and S stand for mucosal solution, cell, and serosal solution, respectively. Damaged areas could contribute to shunt conductance

was inhibited (*e. g.*, by low temperature) to avoid the contribution of active transport to the potential difference. It will be shown later that, given the relative conductances of the different pathways, this contribution is likely to be insignificant.

Typical records of dilution and biionic potentials are shown in Fig. 6. The alkali cation permeability ratios calculated from dilution and biionic potentials in rabbit gallbladder are: $P_K : P_{Rb} : P_{Na} : P_{Cs} : P_{Li} = 2.4 : 1.7 : 1.0 : 0.74 : 0.65$ (BARRY et al., 1971). It is striking that the whole range encompasses a factor of only about 5 while the ion permeability ratios across membranes of nonpolar cells are in the order of 50. Again, this observation is unlikely to be compatible with a transcellular pathway alone. Calculation of cation permeability ratios from measurements of transepithelial conductance yield the same values. These results strongly suggest the presence of a high-conductance, low-selectivity pathway in parallel with the cell membranes, as represented in the equivalent circuit shown in Fig. 7.

II. Demonstration of the Existence of a Paracellular Shunt Pathway

1. Electrophysiological Evidence

FRÖMTER (1972; see also FRÖMTER and DIAMOND, 1972) has recently provided convincing measurements of the total ionic conductances ot the cell membranes and the parallel shunt pathway in the gallbladder of *Necturus maculosus*. This preparation is particularly suitable for microelectrode work, since the cells are fairly large, about 30 μm wide, and tolerate prolonged impalement with no deterioration of the cell potentials. Frömter's analysis was similar to the one employed by WINDHAGER et al. (1967) in the proximal tubule of *Necturus,* in that it consisted of three independent resistance measurements:

a) *The transepithelial resistance* (R_t), *i. e.,* the slope of the "instantaneous" voltage-current relationship.

"Instantaneous" indicates measurement after the capacitive transient is over. As discussed in detail below, prolonged transepithelial current pulses induce time-dependent voltage transients due to both polarization effects and resistance changes.

b) *The ratio of cell membrane resistances* (R_a/R_b), measured as the ratio of the instantaneous voltage deflections across each cell membrane in response to a transepithelial square-current pulse.

c) *Intraepithelial cable analysis, i. e.,* measurement of the intracellular voltage change as a function of distance (V_x) when current is injected into a cell with a second microelectrode. V_x follows the Bessel function K_o (EISENBERG and JOHNSON, 1970), and from the best fit of the experimental data to this function it is possible to calculate the equivalent resistance ot the two cell membranes in parallel (Fig. 8).

From these three measurements, the resistances of the cell membranes and the shunt pathway can be easily obtained. FRÖMTER found the cell membranes to have resistances of 4500 (R_a) and 3000 (R_b) $\Omega \cdot cm^2$, while the shunt resistance was only about 300 $\Omega \cdot cm^2$. This means that about 95 percent of the current-carrying transepithelial ion fluxes take place through the shunt and not through the cell membranes.

In addition, FRÖMTER ingeniously obtained evidence for the intercellular loca-

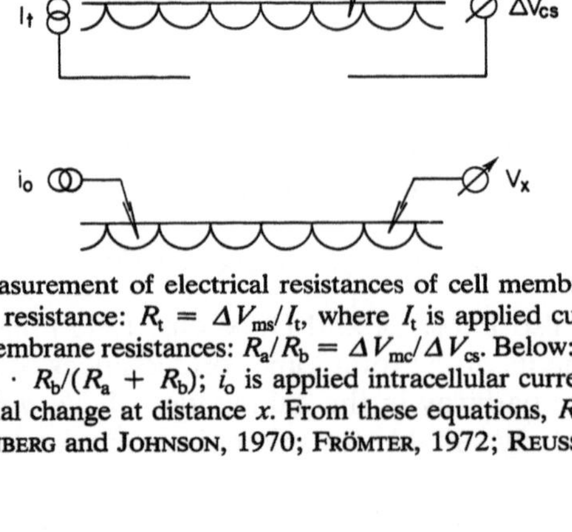

Fig. 8. Method for measurement of electrical resistances of cell membranes and shunt pathway. Above: Transepithelial resistance: $R_t = \Delta V_{ms}/I_t$, where I_t is applied current (per cm² of tissue). Middle: Ratio of cell membrane resistances: $R_a/R_b = \Delta V_{mc}/\Delta V_{cs}$. Below: Cable analysis that allows calculation of $R_x = R_a \cdot R_b/(R_a + R_b)$; i_o is applied intracellular current, and V_x is intracellular current-induced potential change at distance x. From these equations, R_a, R_b and R_s can be calculated. (See EISENBERG and JOHNSON, 1970; FRÖMTER, 1972; REUSS and FINN, 1975a)

tion of the shunt pathway. During transepithelial current flow, when a microelectrode was moved parallel and very close to the cell surface, a change in potential (current sink) was recorded each time the tip was placed above an intercellular space. This experiment required very high transepithelial currents (in the order of 1–2 mA · cm^{-2}), and such mucosa-to-serosa currents can themselves produce reductions of R_t (BINDSLEV et al., 1974), presumably by increasing the conductance of the limiting junctions. In spite of the drawback, the results of this experiment remain the most direct physiological evidence for the intercellular location of the shunt pathway.

2. Morphological Evidence

Another line of evidence for the intercellular location of the shunt pathway in gallbladder and intestine was provided by the experiments of MACHEN et al. (1972). Ionic lanthanum was added to the mucosal solution of rabbit gallbladder and rabbit ileum. Measurements of $^*J_{La}^{(ms)}$ demonstrated a finite transepithelial permeability coefficient. In addition, electron-microscopic observations showed precipitates in the lateral intercellular spaces and within the junctional complexe, filling almost the entire length of the *zonula occludens*. No lanthanum was observed inside the cells, the cell structure was well preserved, and the rate of fluid transport remained unaltered during exposure to the ion.

III. Properties of the Shunt Pathway

It seems clear that under physiological conditions the junctions are rate-limiting for passive ion permeation, and the lateral intercellular spaces represent a relatively small series resistance to ion fluxes. However, the significance of the spaces as a barrier could be larger when their width is experimentally reduced. The available experimental information on the permeation characteristics of the paracellular pathway in gallbladder is summarized below.

1. Physicochemical Characteristics of the Paracellular Pathway

DIAMOND and his co-workers have characterized the pathway responsible for ion permeation and have, thus, provided important clues about the mechanism of passive anion and cation transport. These results have been reviewed in detail by MORENO and DIAMOND (1976).

a) Charge of the Pathway

The addition of an impermeant nonelectrolyte to one of the solutions bathing the gallbladder produces a change in V_{ms} proportional to the osmotic gradient, and to the produced osmotic water flow (J_w^{osm}). The hyperosmolal side becomes electrically positive when NaCl is the main salt on both solutions, suggesting that the gallbladder is cation permselective. This so-called "streaming potential" was initially considered to be a true electrokinetic phenomenon due to drag of the most permeant species (cations) by J_w^{osm} (PIDOT and DIAMOND, 1964; DIAMOND and HARRISON, 1966). However, later experiments (WEDNER and DIAMOND, 1969) showed that it is essentially caused by the generation of a boundary diffusion potential, in the unstirred layers, by water flow: C_{NaCl} decreases in the hyperosmolal side and increases in the opposite side, generating an ion concentration gradient and, since the membrane is selective, a potential similar to the dilution potentials discussed above. The initial interpretation of the observation of streaming potentials was that the membrane behaves as an ion-exchanger, negatively charged, and containing higher concentrations of mobile counterions than coions (DIAMOND and HARRISON, 1966). However, this interpretation has since been modified in the light of the following experimental observations: (i) An ion-exchanger should rectify when the current is such that the carrier is saturated (mobile sites) or if the ion concentration in the membrane approaches the concentration of sites in the membrane (fixed sites). However, the current voltage relationship is linear (Section C. I). (ii) The total conductance should be independent of the concentration of salt in the bathing solutions until the latter becomes larger than the concentration of membrane sites. The experimental observation is that the conductance increases linearly with external salt concentration from 25 to 400 mM (WRIGHT et al., 1971). (iii) The ratio of cation to anion permeability should also be a function of the salt concentration, dropping at high concentration towards the free solution ratios. The measured decrease at high salt concentration is very small with RbCl and no change is observed with KCl (WRIGHT et al., 1971). (iv) The sequence of cation permeability coefficients should be close to the free solution series, since in ion-exchangers the main factor controlling ion permeation is the mobility and not the binding constant (WRIGHT et al., 1971). However, the cation permeability sequences deviate significantly from the free solution sequences in gallbladders of several species. (v) The ultrastructure of the limiting junctions is more consistent with a single thin barrier, or a series of thin barriers, to ion permeation than with the possibility of a thick membrane with the properties of an ion exchanger.

b) Mobility of the Sites

Carrier-mediated transport across the limiting junctions is unlikely because of the following properties of the gallbladder, which would not be expected for such a model (CIANI et al., 1969): (i) The IV plot is linear. (ii) Cation-selectivity ratios are too low. (iii) The effects of Ca^{+2}, Th^{+4}, and pH (see 1. d), below) would be ascribable to surface charge effects and not to effects on the carrier, but the different effects of Ca^{+2} on permeability coefficients of alkali cations are inconsistent with a surface charge effect.

c) Evidence for a Free-Solution Parallel Pathway

The total electrical conductance of the *in-vitro* rabbit gallbladder increases rapidly after mounting, reaching a stable value after 1–2 hours. In addition to this rise in conductance during this period the single salt dilution potentials decrease (*e. g.*, P_{Na}/P_{Cl} decreases), and the cation permeability ratios drop towards the free solution values (*e. g.*, P_{Cs}/P_{Li} increases). The increase in total conductance can be explained by a parallel increase in cation and anion conductances (g_+, g_-) in their free solution ratios, but not by specific increases in either g_- or g_+ alone. These observations are consistent with the "opening" of a free-solution shunt across the rabbit gallbladder after dissection (BARRY et al., 1971). However, the conductance of *Necturus* gallbladder decreases slightly after mounting (FRÖMTER, 1972; REUSS and FINN, unpubl.) and the dilution potentials remain approximately constant (REUSS and FINN, unpubl.). Moreover, the time-dependent changes in conductance and selectivity of rabbit gallbladder do not take place if the bathing solutions contain appropriate concentrations of potassium and phosphate (MORENO and DIAMOND, 1974b; WIEDNER and WRIGHT, 1975). Thus the original observation of a time-dependent change can be explained by tissue alterations secondary to the ionic composition of the bathing solutions. When these artifacts are avoided, the characteristics of nonelectrolyte permeation and the effect of triaminopyrimidinium (see below) seem to indicate that Na and Cl permeate through different pathways (MORENO, 1974, 1975a and b).

d) *Effects of pH and Polyvalent Cations*

Reduction of pH, addition of thorium, lanthanum, or other polyvalent cations, or increases in calcium concentration in the mucosal solution decrease or even reverse the polarity of potentials produced by unilateral reductions in salt concentration or unilateral hyperosmolarity in gallbladders of several species. These effects are due to both a decrease in g_+ and an increase in g_- (DIAMOND and HARRISON, 1966; WRIGHT and DIAMOND, 1968; MACHEN et al., 1972; MACHEN and DIAMOND, 1972; Moreno and Diamond, 1974b, 1975b). In addition, the permselectivity ratios of alkali metals fall towards the free solution values. The pH effect suggests that g_{Na} is controlled by dipoles with their negative poles facing the pathway. These anionic sites have pK_a's of about 4.4 in frog gallbladder and 4.6 in rabbit gallbladder. Cl conductance increases at lower pH values, indicating a pK_a of less than 3.0 (MORENO and DIAMOND, 1974b).

Triaminopyrimidinium (TAP) at high concentrations (*e. g.*, 10 mM) has been shown to inhibit g_{Na} in gallbladders of several species and in other leaky epithelia, with no change in g_{Cl} (MORENO, 1974, 1975a). The K_i for this effect is 2.6 mM. The lack of effect on g_{Cl} is consistent with Na and Cl permeation through different pathways. The structure of the TAP molecule suggests tight binding, by hydrogen bonds, to anionic sites in the limiting junction channels. However, the effects on the cell membranes have not been directly studied, and TAP has an amiloride-like effect when applied to the outer surface of the frog skin (ZEISKE, 1975). This observation suggests that a part of the overall effect is mediated by an action on the luminal cell membrane.

e) Permeation of Nitrogenous Cations

A large series of nitrogenous cations was employed by MORENO and DIAMOND (1974a, 1975a) to study the effect of molecular size and number of donor protons available for hydrogen-bond formation on permeability ratios. Permeability coefficients were shown to increase with the number of donor protons (up to 4), and to decrease with molecular size. Cation permeation appears to be mainly controlled by the electrostatic field strength. Differences between animal species may be related to steric restriction.

In summary, the main paracellular permeation pathway across the gallbladder epithelium is the junctional complex in series with the lateral intercellular space. The limiting junctions are highly cation-selective and it is possible to conclude tentatively that they have the physical properties of one or a series of barriers, probably of heterogeneous properties, with negatively charged sites. These results would account for the cation selectivity, and the effects of pH, calcium, thorium and lanthanum on ion permeation, and would be consistent with the ultrastructure of the *zonulae occludentes.* Anion permeation appears to follow a different pathway with the characteristics of a free solution shunt. The time-dependent opening of this pathway *in vitro* seems to be artifactual. The anatomical correlates of the paracellular pathways in the gallbladder will be discussed in detail below (nonelectrolyte permeation). For a detailed discussion on the passive permeation properties of the tissue, see the excellent review article by MORENO and DIAMOND (1976).

2. Role of the Lateral Intercellular Spaces in Passive Ion Permeation

Recently, there has been great interest in the role of the lateral intercellular spaces in solute and water transport across epithelia. The role of the spaces in passive permeation of ions and nonelectrolytes was examined by SMULDERS et al. (1972), who measured transepithelial electrical conductance and $P_{sucrose}$ under control conditions and after unilateral increases of the osmolarity of the mucosal or the serosal solution bathing the rabbit gallbladder. A hyperosmotic mucosal solution induces a J_w in the serosa-to-mucosa direction that is accompanied by a mucosa-positive potential, and marked reduction of the width of the lateral spaces and of the thickness of the subepithelial tissue. Under these circumstances, R_t increases (in proportion to the magnitude of the osmotic gradient), and $P_{sucrose}$ drops.

Osmotic gradients in the opposite direction (*i. e.,* addition of sucrose to the serosal solution) had no effect on transepithelial conductance or on nonelectrolyte permeabilities. The lateral intercellular spaces widened by one order of magnitude (at a gradient of 300 mOsm $\cdot$ l^{-1}) and the subepithelial thickness of the tissue increased.

SMULDERS et al. calculated that the collapse of the spaces could account for the observed effects of mucosal hyperosmolarity, while the fact that widening of the spaces (serosal hyperosmolarity) had no effect on permeabilities indicated that under control conditions the spaces do not limit significantly passive permeation. Dilution and biionic potentials in rabbit gallbladder change, when the mucosal solution is made hyperosmotic, towards free solution values. This observation is consistent with the lateral intercellular spaces becoming the rate-limiting step for passive ion permeation (WIEDNER and WRIGHT, 1975). In addition, the ultrastructural observations of SMULDERS et al. indicate the presence of a subserosal barrier for water flow, since the osmotic gradients produced large changes in subepithelial volume.

The effects of electrical and osmotic gradients on transepithelial conductance and dimensions of the lateral intercellular spaces in the frog gallbladder were studied by BINDSLEV et al. (1974). Mucosal hyperosmolarity had essentially the same effects on conductance and width of the spaces as in rabbit gallbladder. In addition, serosal hyperosmolarity produced a reduction in R_t, with widening of the lateral spaces. Externally applied transepithelial currents mimicked the effects of osmotic gradients on both the width of the spaces and R_t. Current flowing from serosa to mucosa, which generates salt and water flow in the same direction, closed the spaces and increased the resistance, while current flowing from mucosa to serosa had the opposite effects. These effects were reversed, as predicted, when the tissue was made anion-selective by the addition of La^{+++} to the mucosal solution. In contrast to observations on tight epithelia, where localized dilations ("blisters") of the limiting junctions have been observed upon the establishment of osmotic gradients (e. g., DiBONA and CIVAN, 1973) or the passage of large transepithelial currents (BINDSLEV et al., 1974), no changes in the ultrastructure of the junctional complexes were observed. At relatively high current densities, or when the serosal solution is slightly hyperosmotic, the increase in resistance during the current pulse is preceded by a fall in resistance. If the osmolarity of the serosal solution is increased further, the resistance drops monotonically during serosa to mucosa current pulses. Since in this condition the lateral intercellular spaces are widened, these observations seem to indicate two mechanisms of current-induced resistance changes. For both directions of transepithelial current, the resistance falls if the current density is high enough. When current is applied in the serosa-to-mucosa direction, the resistance increases, probably because of narrowing of the lateral spaces. The drop in resistance could occur at the limiting junctions themselves. That the shunt pathway is the site at which the change in electrical resistance due to mucosal hyperosmolarity takes place was confirmed directly by REUSS and FINN (1977), who showed by cable analysis in *Necturus* gallbladder that after the addition of sucrose to the mucosal solution the shunt resistance increases. Mucosal hyperosmolarity has also marked effects on the cell membranes (see Section CIV).

Doubts concerning the meaning of ultrastructural estimates of the width of the lateral intercellular spaces have been raised by FREDERIKSEN and ROSTGAARD (1974). These authors showed that the lateral spaces appear to be closed in the frog gallbladder *in vitro* when observed with Nomarski optics. However, the spaces become apparent during the preparation of the tissue for electron microscopy. Although these observations could be irrelevant to the changes observed in the presence of large transepithelial osmotic gradients, a note of caution seems appropriate at this time. Transport-dependent changes of width of the lateral intercellular spaces could take place only under *in-vitro* conditions, where transport occurs across the whole of the gallbladder wall, as opposed to the *in-vivo* situation, where the vascular compartment provides an effective fluid sink or reservoir that could be expected to prevent or limit changes in tissue fluid content. Finally, it is possible that electron-microscopic estimations of the width of the lateral spaces are artifactual, even though the studies described above appear to be internally consistent, in both the direction and the magnitude of the changes observed.

IV. Properties of the Epithelial Cell Membranes

Very recently, the electrical properties of the epithelial cells in *Necturus* and rabbit gallbladder have been successfully measured with microelectrode techniques. HÉNIN and CREMASCHI (1975) studied gallbladders from rabbit, REUSS and FINN (1975a and b) from *Necturus,* and van Os and SLEGERS (1975) gallbladders from both species. The results of these three independent studies are strikingly similar. Both the luminal and the basolateral cell membranes of the gallbladder epithelial cells are mainly potassium permeable; in other words, the cell membranes exhibit a lack of polarity in their ion permeability characteristics. This observation rules out the possibility of a two-membrane mechanism (as proposed by USSING) to explain salt transport in this tissue.

Concomitant with these electrophysiologic studies, FRIZZELL et al. (1975) and CREMASCHI and HÉNIN (1975) measured Na and Cl transepithelial fluxes, unidirectional fluxes of both ions from the lumen into the epithelium, and intracellular ionic concentrations in rabbit gallbladder. Both groups concluded that NaCl transport must be the consequence of a neutral coupled carrier-mediated mechanism operating at the luminal membrane of the cell.

1. Electrophysiological Studies

a) Equivalent Circuit for Gallbladder Epithelium

A Thevenin steady-state equivalent circuit for the epithelium, omitting the cell-to-cell coupling pathway, is shown in Fig. 9. Such an equivalent circuit is difficult to analyze, because the potentials measured across the cell membranes (V_{mc}, V_{cs}) or across the whole wall (V_{ms}) are not only functions of the batteries at the corresponding site (*e. g.*, $V_{mc} = f(V_a)$), but also of the other two batteries and all resistances in the circuit. In other words, a negative potential across the luminal membrane (cell negative to mucosal solution) would not necessarily indicate that E_a has that orientation, but could correspond to a small value of E_a with the opposite orientation, and a larger E_b oriented as indicated in the Figure (BOULPAEP, 1971; SCHULTZ, 1972).

Thus, in order to calculate E_a and E_b it is necessary to measure all resistances and all potentials in the circuit. The number of unknown elements is larger than the number of possible independent measurements (since V_{mc}, V_{cs}, and V_{ms} are

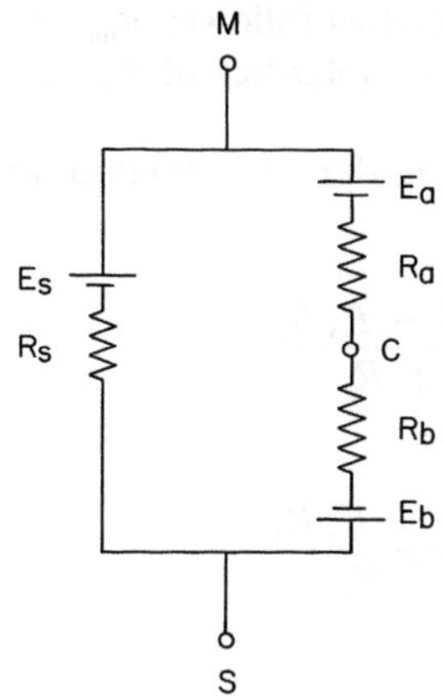

Fig. 9. Active steady-state equivalent circuit for gallbladder. According to Thevenin's theorem, each membrane and the shunt are represented by their *equivalent* electromotive forces in series with respective *equivalent* resistances (see text)

not independent, but $V_{ms} = V_{cs} - V_{mc}$). Thus, at least one of the values of the three active elements in the circuit has to be assumed.

R_a, R_b, and R_s represent, respectively, the total electrical resistances of the luminal membrane, the basolateral membrane (plus the lumped intercellular space), and the shunt pathway. The latter includes any pathways in parallel with the transcellular route (limiting junctions plus lateral intercellular space, damaged or absent cells, edge damage). E_a, E_b, and E_s represent the electromotive forces (EMFs) generated at the apical membrane, basolateral membrane, and shunt pathway, respectively. The cell membrane EMFs can result from diffusion potentials or rheogenic (electrogenic) mechanisms; the shunt EMF represents diffusion potentials across this pathway when the compositions of the bathing solutions are different, since the shunt is ion-permselective. Even when the bulk solutions are identical, diffusion potentials are predicted if salt is more concentrated in the lateral intercellular spaces.

The measured potentials across the cell membranes and the whole epithelium depend on all circuit parameters and are described by three equations:

$$V_{mc} = \frac{E_a (R_b + R_s) + R_a (E_b - E_s)}{R_a + R_b + R_s} \tag{1}$$

$$V_{cs} = \frac{E_b (R_a + R_s) + R_b (E_a + E_s)}{R_a + R_b + R_s} \tag{2}$$

$$V_{ms} = \frac{R_s (E_b - E_a) + E_s (R_a + R_b)}{R_a + R_b + R_s} \tag{3}$$

were the polarities are oriented as follows: $V_{mc} = \Psi^{(m)} - \Psi^{(c)}$; $V_{cs} = \Psi^{(s)} - \Psi^{(c)}$; $V_{ms} = \Psi^{(s)} - \Psi^{(m)}$. The polarities of E_a, E_b, and E_s are oriented as V_{mc}, V_{cs}, and V_{ms}, respectively.

If E_s is assumed to be zero under control conditions (Ringer solution on both sides), Eqs. 1–3 simplify to:

$$V_{mc} = \frac{E_a(R_b + R_s) + E_b R_a}{R_a + R_b + R_s} \tag{4}$$

$$V_{cs} = \frac{E_b(R_a + R_s) + E_a R_b}{R_a + R_b + R_s} \tag{5}$$

$$V_{ms} = \frac{(E_b - E_a) R_s}{R_a + R_b + R_s} \tag{6}$$

If the values of the three resistances are known, two equations of this set can be solved simultaneously, inserting the measured potentials, to calculate E_a and E_b. The resistances can be calculated (as described above) from R_t, R_a/R_b and intraepithelial cable analysis. The resultant EMFs can be compared with the results obtained when E_s differs from 0.

The results of these experiments in *Necturus* gallbladder by two groups are shown in Table 7. Van Os and SLEGERS did not measure resistances in their experiments, but used the mean values reported by FRÖMTER (1972). These are included for comparison. The studies demonstrate that E_a is oriented in such a way as to make the cell negative to the mucosal solution. Since incorrect estimates could result from the assumption that $E_s = 0$, the effect of this assumption on the values of all EMFs was appraised by the assumption of different NaCl concentration gradients across the limiting junction and calculation of E_s at each gradient from P_{Na}/P_{Cl} across the shunt ($P_{Na}/P_{Cl} = 3.18$ according to REUSS and FINN (1975a), or 3.12 according to van Os and SLEGERS (1975); see also Table 8. It can be seen that NaCl gradients greater than the maximum estimated by MACHEN and DIAMOND (1969) in rabbit gallbladder from the

Table 7. Circuit parameters in *Necturus* gallbladder epithelium

	V_{mc} (mV)	V_{cs} (mV)	V_{ms} (mV)	R_a ($\Omega \cdot cm^2$)	R_b ($\Omega \cdot cm^2$)	R_s ($\Omega \cdot cm^2$)	E_a (mV)	E_b (mV)
(1)	57.2	59.4	2.2	3350	2750	480	39.9	69.4
(2)	53.2	54.6	1.4	4470	2880	324	15	81

V_{mc}, V_{cs}, V_{ms}: apical, basolateral, and transmural potentials
R_a, R_b, R_s: apical, basolateral, and shunt resistances
E_a, E_b: apical and basolateral equivalent EMFs
(1) REUSS and FINN (1975a), (2) Potentials and EMFs: VAN OS and SLEGERS (1975); Resistances: FRÖMTER (1972)

Table 8. Effect of assumed NaCl concentration gradients across the limiting junctions on the calculated values of E_s, E_a, and E_b. (From REUSS and FINN, 1975a; with the kind permission of Journal of Membrane Biology)

NaCl concentration gradient (mM)	E_s	E_a	E_b
0	0	39.9	69.4
5	−0.6	33.2	75.0
10	−1.2	29.3	78.2
20	−2.2	21.9	84.3

From the assumed C_{NaCl} gradient, E_s was calculated from P_{Na}/P_{Cl} (= 3.18 across the shunt). E_a and E_b were computed by using mean values of potentials and resistances to solve Eqs. 1–3. Values for E are expressed in mV.

assumption of a standing osmotic gradient mechanism do not reverse the polarity of E_a. Thus, the luminal membrane cannot be preferentially Na selective, since C_{Na} in the solution is larger than in the cells (see below). The most likely explanation for the observed orientation is high K permeability.

The polarity of E_b is consistent with K permselectivity as postulated in most epithelia.

b) *Ionic Permeability of the Apical Membrane*

Equations 4–6 allow one to calculate the changes in E_a and E_s when the mucosal solution concentration of a given ion is altered, as long as the potentials are measured before significant changes in cell composition take place, and the properties of the contralateral cell membrane are unaltered by the experimental perturbation. With a microelectrode in a cell, it is possible to follow simultaneously changes in the potentials across both cell membranes, along with changes in the ratio of membrane resistances.

The effects of a single ion replacement in the mucosal solution are complicated, because both the resistance and EMF of the ipsilateral cell membrane and the shunt can change. For instance, if P_K of the luminal membrane exceeds P_{Na}, and since P_K/P_{Na} is about 1.8 across the shunt (REUSS and FINN, 1975a; van Os and SLEGERS, 1975), isomolar replacement of Na and K should generate a change in E_s (mucosa-negative), a reduction in E_a (less-cell-negative), and drops in R_a and R_s.

An example of this type of experiment is shown in Fig. 10. Note that the reduction of V_{mc} and V_{cs} and the drop in ratio of cell membrane resistances are proportional to the final K concentration in the mucosal solution. If one assumes that there are no rheogenic mechanisms involved in the generation of the membrane potentials, the measured changes of potentials and resistances allow one to compute ion permeability coefficients. The results of these calculations were $P_K = 12.2$, $P_{Cl} = 4.6$, and $P_{Na} = 1.1 \cdot 10^{-7}\,cm \cdot s^{-1}$ (REUSS and FINN, 1975b). Van Os and SLEGERS (1975) estimated P_K/P_{Na} to be about 2.7. A somewhat higher P_K/P_{Na} would have been obtained by van Os and SLEGERS if they had taken into account changes in membrane resistance during the ionic substitu-

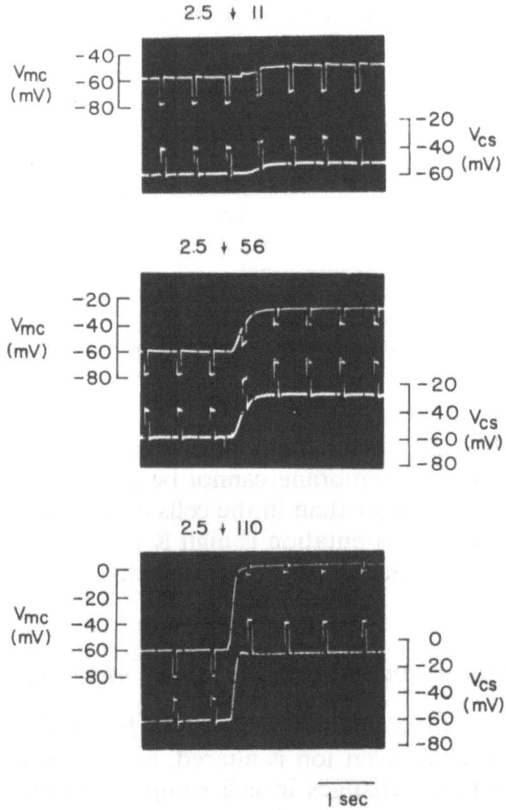

Fig. 10. Effects of increases in C_K in mucosal solution (given in mM above each tracing) on cell membrane potentials and ratio of cell membrane resistances in *Necturus* gallbladder. Potentials referred to serosal solution. Voltage deflections are due to externally applied identical transepithelial d.c. pulses. (From REUSS and FINN, 1975b)

tion. In addition, P_{Cl} was calculated to be neglibible by van Os and SLEGERS, and about 30 percent of P_K by REUSS and FINN. The reason for this discrepancy is not knowm. The anions used for Cl replacement were SO_4 in van Os and SLEGERS' experiments, and methylsulfate or gluconate in REUSS and FINN's. Finally, HÉNIN and CREMASCHI and van Os and SLEGERS, working with rabbit gallbladder, obtained results qualitatively similar to those described for *Necturus*. However, complete circuit analysis was not feasible in this tissue. Hence, interpretation of these results is difficult, since the contribution of the shunt EMF to changes in V_{mc} was neglected.

Thus, although quantitative differences have not been resolved, the luminal membrane of both *Necturus* and rabbit gallbladder is K-selective, with a P_K/P_{Na} ratio of at least 3 and probably larger than 10. This is the first instance of a Na-transporting epithelium in which a high K selectivity has been demonstrated for the luminal cell membrane.

The absolute value of P_{Na}, the electrical driving force (V_{mc}), and the external Na concentration can be used to predict the diffusional $J_{Na}^{(mc)}$, assuming a linear

C_{Na} profile in the membrane. The calculation yields a value of $J_{Na}^{(mc)}$ three orders of magnitude smaller than the net transepithelial Na flux, determined from the unidirectional isotopic fluxes (REUSS and FINN, 1975b). Although van Os and SLEGERS (1975) computed a larger P_{Na}, their predicted diffusional $J_{Na}^{(mc)}$ is also too small to account for the transepithelial flux. Therefore, the mechanism of Na transport across the apical membrane of gallbladder epithelial cells cannot be attributed to simple diffusion, because P_{Na} across this membrane is too low. The mechanism of Na uptake in this tissue is therefore qualitatively different from that presumed to exist in tight epithelia such as frog skin and toad urinary bladder.

c) Ionic Permeability of the Basolateral Membrane

The study of the permeability properties of the basolateral membrane is further complicated by the thickness of the unstirred layer on the serosal side of the epithelium. Thus, membrane potentials and resistances can be measured only after prolonged exposure of the serosal side to altered bathing solutions. During this period cell composition is likely to change significantly. With this limitation in mind, the results of HÉNIN and CREMASCHI (1975) in rabbit gallbladder, van Os and SLEGERS (1975) in rabbit and *Necturus*, and REUSS and FINN (unpubl.) in *Necturus* indicate that the basolateral membrane, as postulated for many other epithelia, behaves as a K-selective barrier. Na and Cl conductances across this barrier seem to be very small or negligible.

2. Tracer Analysis of Electrolyte Transport

FRIZZELL et al. (1975), in a series of careful experiments with rabbit gallbladder, used radioactive tracers to measure transmural unidirectional fluxes and influxes from the mucosal solution into the cells. Their results indicate that a sizable portion of Na or Cl influx depends on the presence of the counterion. The magnitude of the reductions in influx upon removal of the counterion is practically equal to the net active transepithelial flux. In addition, both Na and Cl backfluxes are large, about 60 percent of the m-to-s flux. $J_{Na}^{(sm)}/J_{Cl}^{(sm)}$ is about 1.7, even though electrical measurements yield P_{Na}/P_{Cl} ~3. Furthermore, Cl influx, measured from Cl media after preincubation of the tissue in a Cl-free, Na-rich solution is reduced. These two experiments are consistent with Cl exchange diffusion, as proposed by DIAMOND (1962b) and WHEELER (1963). A mean cellular potential of 45 mV was measured in gallbladders exposed to 12 mM K. This electrical potential and the intracellular ion concentrations (see Table 9) indicate that the electrochemical potential gradients favor Na uptake and Cl extrusion (FRIZZELL et al., 1975).

As seems clear from the electrophysiological experiments described above, FRIZZELL et al. concluded that most if not all Na entry takes place as a coupled, electrically silent, NaCl carrier-mediated transport. However, CREMASCHI and HÉNIN (1975) estimated a fraction of Na entry that could be Cl-independent, and thus diffusional. Na transport from cells to serosal solution was postulated

Table 9. Intracellular ion concentrations in rabbit gallbladder epithelium

Na	Cl	K
(1) 66	84	85
(2) 69	80	101

(1) Means of 4 tissues (FRIZZELL et al., 1975).
(2) Means of 8 tissues, 35-min incubation (HÉNIN and CREMASCHI, 1975).
Concentrations are expressed in mEq · l^{-1} cell water

by FRIZZELL et al. to be active, coupled to K uptake, as in a classic Na-K pump. Some circumstantial arguments favor this view. First, the demonstration of Na-K-ATPase in the tissue, and its correlation with the rate of fluid transport (van Os and SLEGERS, 1971). Second, the reduction in fluid transport by ouabain or potassium-free solutions on the serosal side.

In addition, it was proposed that outward Cl transport across the basolateral membrane could be passive and electrically coupled to Na transport. This hypothesis, also discussed by MORENO (1975a) and MORENO and DIAMOND (1975b), is based on the following observations: (i) the recent demonstration of values of V_{ms} significantly different from zero in gallbladders of several species, including values as high as 8 mV (mucosa-negative) in man (ROSE et al., 1973); (ii) the ability of several monovalent anions to maintain salt and water transport in the absence of Cl (see Section B); (iii) the large increase in Cl tracer backflux ($^*J_{Cl}^{(sm)}$) when the tissue is clamped at mucosa-positive potentials (FRIZZELL et al., 1975), suggesting a large diffusional component of J_{Cl}; (iv) the increase in V_{ms} (mucosa-negative) produced in gallbladders of several species by TAP, that could be interpreted as an indication of reduced paracellular "shunting" of a Na transport potential that could drive Cl (MORENO, 1975a).

On the other hand, objections to this view include:

i) the measured values of g_{Cl} both at the cell membranes and at the shunt are too small. Thus, the measured electrochemical potential gradients are not large enough to drive the ion through any of these pathways (REUSS and FINN, 1975b; van Os and SLEGERS, 1975; HÉNIN and CREMASCHI, 1975; CREMASCHI and HÉNIN, 1975). It might be argued that the "true" potential difference across the limiting junction would be larger than V_{ms} (*i. e.*, this potential is dissipated in the lateral intercellular spaces). However, only a potential in the order of 40 mV would account for the transepithelial Cl flux (CREMASCHI and HÉNIN, 1975).

ii) If Cl is a shunting ion, as in some tight epithelia, one would expect the transepithelial potential to increase after replacement of Cl with a nonpermeant ion (*e. g.*, sulfate), but the opposite effect is observed. Alternatively, it is conceivable that Na entry in this situation could be reduced (if its uptake requires the presence of a monovalent ion in the mucosal solution) and the rate of Na extrusion decreased.

iii) The results in human gallbladder are difficult to interpret, since the magnitude of the measured transepithelial potential does not seem consistent with the low resistance of the tissue (approx. 50 Ω cm^2). Since these results have been obtained by two groups of investigators (ROSE et al., 1973; ONSTAD et al., 1967)

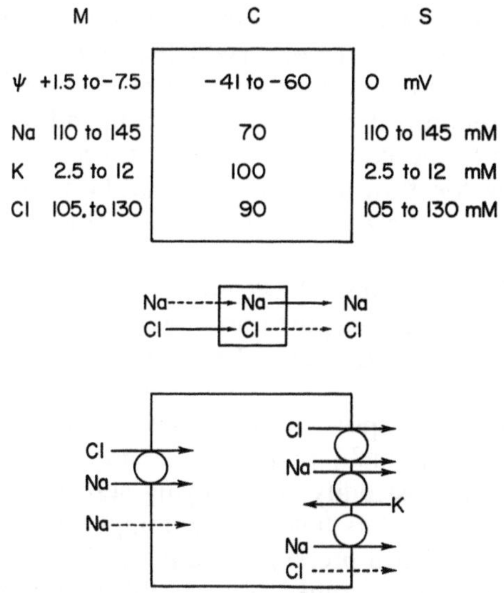

Fig. 11. Possible ion transport mechanisms across gallbladder epithelial cell membranes. Top panel: Summary of electrical potential profile (Ψ) across epithelium (cell potentials from rabbit and *Necturus*); extra- and intracellular ion *concentrations* in *in-vitro* conditions. Middle panel: Solid arrows, fluxes against electrochemical potential gradient; segmented arrows, fluxes down electrochemical potential gradient. Bottom panel: Left, mechanisms of apical ion uptake: a) neutral NaCl entry, b) possible small-Na entry by simple diffusion. Right, possible mechanisms of basolateral ion extrusion: a) coupled (neutral) NaCl transport, b) Na-K exchange pump, c) rheogenic Na extrusion, with electrically coupled passive Cl flux (see text for discussion)

and since the transepithelial potential can be abolished by reducing Na concentration in the bathing solutions or by the addition of ouabain, it is unlikely that they are artifactual. It is possible that the true electrical resistance of the tissue is much larger than estimated, given the extensive mucosal folds.

iv) The dependence of $^*J_{Cl}^{(sm)}$ on the transepithelial clamp potential, observed by FRIZZELL et al., is also difficult to interpret quantitatively, since externally applied mucosa-to-serosa currents are known to reduce R_t (BINDSLEV et al., 1974; REUSS and FINN, 1977), presumably by increasing the ion permeability of the limiting junctions. Thus, the change in Cl flux could be partly due to a current or voltage-dependent conductance change, and not to the altered electrical driving force.

In conclusion, the mechanisms of ion transport across the basolateral membrane are unclear at the present time. The possibilities discussed above are summarized in Figure 11.

3. Role of the Lateral Intercellular Spaces in Ion Transport

KAYE et al. (1966) and DIAMOND and TORMEY (1966b) observed that the lateral intercellular spaces of the rabbit gallbladder *in vitro* are distended during transepithelial fluid transport. When transport is inhibited by low temperature, ouabain, metabolic inhibitors, or ion replacements in mucosal solution, the spaces collapse. In addition, KAYE et al. observed that the ATPase activity of the epithelial cells is located in the lateral membrane, precisely along the intercellular spaces. These observations have been taken to indicate that salt is transported from the cells to the lateral intercellular spaces, tending to make them hypertonic and that water, flowing passively from the cells down along osmotic gradients makes the fluid contained in the space isotonic towards its basal end (see Section E).

On the other hand, TORMEY and DIAMOND (1967) point out that the observations of open spaces during transport and closed spaces in its absence do not prove fluid transport through this path. These investigators offer two alternative explanations:

First, accumulation of transport fluid in subepithelial spaces might increase the hydrostatic pressure, which, in turn, distends the spaces. TORMEY and DIAMOND examined this possibility by applying serosa to mucosa hydrostatic pressure to everted gallbladders held at low temperature. Distention became marked only at unphysiological pressures similar to those that irreversibly block transport (DIETSCHY, 1964). Of course, it is possible that pressure exerted on the spaces from the subepithelial space might not be equivalent to pressure across the entire wall, since the experiments of SMULDERS et al. (1972) have demonstrated a subserosal barrier to water flow. Thus, the application of hydrostatic pressure across the whole thickness of the organ, could result in the greatest pressure drop across the subserosal barrier and not across the epithelium. Correlation between width of the spaces and fluid transport has mostly been performed *in vitro,* when blood supply is absent and transport takes place across the whole wall. Fluid accumulation in the lamina propria (due to the subserosal barrier) could widen the spaces *in vitro,* but this effect would not take place *in vivo,* because of the sink role of blood circulation. However, both KAYE et al. (1966) and DIAMOND and TORMEY (1966) observed dilated spaces in gallbladder biopsies.

Second, it is conceivable that the distention of the lateral intercellular spaces during transport is related to a reduction of cell volume, which increases after inhibition of transport and, thus, obliterates the spaces. However, TORMEY and DIAMOND observed collapsed spaces in a variety of experimental situations in which cell volume was increased, unchanged, or decreased by the inhibitory agent. On the other hand, FREDERIKSEN and ROSTGAARD (1974) have recently proposed that the widening of lateral intercellular spaces is an artifact (see above).

4. Relative Contributions of Cell Membranes and Shunt Pathway to Transepithelial Potential and Transepithelial Resistance Changes

Since solid evidence supports the view that the intercellular shunt is the main pathway for passive ion permeation across the gallbladder and other leaky epithelia, it is reasonable to assume that experimentally induced changes in transepithelial electrical parameters of the tissue will be mainly due to changes in shunt parameters. This has been shown to be the case in several situations, but recent studies employing intracellular microelectrodes have shown that changes in the properties of the cell membranes cannot be disregarded. This section summarizes some of these results.

a) Estimation of Shunt Ion Permeability Ratios from Transepithelial Measurements

Employing the methods described earlier, REUSS and FINN (1975a and b) compared the changes in transepithelial potential (ΔV_{ms}) and the calculated changes in shunt EMF (ΔE_s) after single ion replacement in the mucosal solution bathing *Necturus* gallbladder. The composition of the serosal solution was kept constant. As expected from the relative resistances of the shunt pathway and the cell membranes, the values of ΔE_s and ΔV_{ms} were very close (Fig. 12).

From Eq. 6, ignoring changes in resistance, a change in transepithelial potential produced by an ion substitution in the mucosal solution is given by:

$$\Delta V_{ms} = \frac{\Delta E_a \cdot R_s + \Delta E_s \cdot R_c}{R_c + R_s} \tag{7}$$

where $R_c = R_a + R_b$, and all other expressions have the meanings described previously.

Since $R_c = 6.100\ \Omega \cdot cm^2$ and $R_s = 450\ \Omega \cdot cm^2$ (REUSS and FINN, 1975a),

$$\Delta V_{ms} = 0.07\ \Delta E_a + 0.93\ \Delta E_s \tag{8}$$

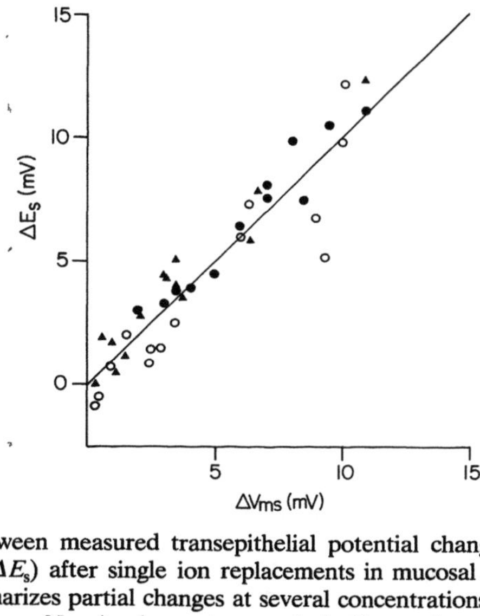

Fig. 12. Correlation between measured transepithelial potential change (ΔV_{ms}) and calculated change in shunt EMF (ΔE_s) after single ion replacements in mucosal solution bathing *Necturus* gallbladder. Figure summarizes partial changes at several concentrations. ●: K replaces Na; ○: N-methyl-D-glucamine replaces Na; ▲: gluconate or methylsulfate replace Cl. Straight line = identity line. (REUSS and FINN, unpubl.)

Thus, changes in E_s would be about 13 times more effective in altering V_{ms} than changes in E_a. Therefore, ΔV_{ms} values are an excellent index of ΔE_s. The contribution of changes in E_a to ΔV_{ms} (during ionic substitution in the mucosal bathing solution) is for practical purposes negligible.

b) Estimation of Cell Membrane Ion Permeability Ratios from Changes in Cell Membrane Potential

Large deviations from identity can be demonstrated when, after a similar analysis, ΔE_a is plotted as a function of ΔV_{mc} (Fig. 13; Reuss and Finn, 1975b).

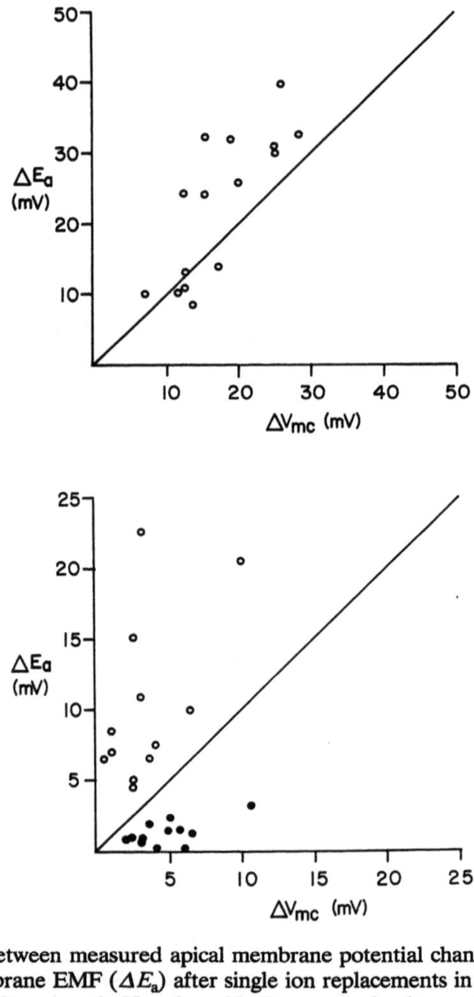

Fig. 13. Correlation between measured apical membrane potential change (ΔV_{mc}) and calculated change in apical membrane EMF (ΔE_a) after single ion replacements in mucosal solution bathing *Necturus* gallbladder. Upper graph: K replaces Na. Lower graph: gluconate or methylsulfate replace Cl (○); N-methyl-D-glucamine replaces Na (●). Several concentrations were employed in both graphs. Note large deviations from identity when ions less permeant across luminal membrane (Cl, Na) were substituted. (From Reuss and Finn, 1975b)

Assuming no changes in E_b or resistances, ΔV_{mc} can be calculated from Eq. 4:

$$\Delta V_{mc} = \frac{\Delta E_a (R_b + R_s) - \Delta E_s R_a}{R_a + R_b + R_s} \tag{9}$$

And from the measured values of the resistances in *Necturus* gallbladder (REUSS and FINN, 1975a), Eq. 9 can be written:

$$\Delta V_{mc} = 0.49\, \Delta E_a - 0.51\, \Delta E_s. \tag{10}$$

Thus, if ΔE_s is large enough (*i. e.*, the shunt permeability ratio for the two ions is high), its contribution to the measured potential across the luminal membrane will be significant, and the change in V_{mc} cannot be ascribed to a change in E_a. The error will be proportional to the shunt permeability ratio for the substituted ions: *e. g.*, Na compared with a large cation, or Cl compared with a large anion.

In summary, unless a complete circuit analysis is performed, changes in cell membrane potential produced by ionic substitutions in the bathing solutions do not allow accurate calculations of cell membrane permeability coefficients. The errors are largest if the shunt permeability ratio for the two ions is high.

c) Effects of Transepithelial Osmotic Gradients on Potentials and Resistances

The addition of a nonpermeant solute (*e. g.*, sucrose) to the mucosal bathing solution produces an increase in transepithelial resistance of the gallbladder and the development of a mucosa-positive potential. In addition, electron-microscopic observations show that the lateral intercellular spaces collapse (SMULDERS et al., 1972). The interpretation of these results is that the increase in transepithelial resistance is due to closing of the lateral intercellular spaces, adding a series resistance to the limiting junction. The change in transepithelial potential results from a diffusion potential produced across the epithelium by the osmotic water flow (increasing C_{NaCl} in the serosal unstirred layer, and decreasing it in the mucosal unstirred layer), with perhaps a small true electrokinetic component. In a few observations on *Necturus* gallbladder, FRÖMTER (1972) observed that exposure to hyperosmotic mucosal solutions (sucrose addition) decreased the ratio of resistances of the cell membranes (R_a/R_b). His interpretation was that the resistance of the basolateral membrane increased, because the lateral membrane is in series with a portion of the lateral intercellular space. The spaces collapse in this situation, and therefore their resistance rises.

These interpretations have been re-examined by REUSS and FINN (1977a). Cell potentials and resistances were measured before and after the addition of sucrose to the mucosal bathing solution of *Necturus* gallbladder. As shown in Table 10, the resistance of the shunt increases, and the mucosal solution becomes less negative, but, in addition, there are large reductions in the potentials across both cell membranes and the electrical resistance of the luminal membrane decreases to less than 50 percent of its control value. Thus, the change in R_a/R_b observed by FRÖMTER, and confirmed by REUSS and FINN, is not due to an increase in R_b, but to a decrease in R_a.

Table 10. Effects of mucosal solution hyperosmolarity on potentials and resistances in *Necturus* gallbladder

	V_{ms}	V_{mc}	V_{cs}	R_t	R_a	R_b	R_s
Control	3.2	56.2	59.4	450	3880	4900	480
Ringer + Sucrose	1.5	32.7	34.2	580	1810	4580	680
P	<0.02	<0.01	<0.01	<0.02	<0.02	NS	<0.05

Results are means of 6 gallbladders. V: mV; $R : \Omega \cdot cm^2$

Moreover, closing of the lateral intercellular spaces would be an unlikely explanation for an increase in R_b in this situation, on the following grounds: if one assigns all of the change in R_s to the collapse of the spaces, the increase is about 200 $\Omega \cdot cm^2$. If the lateral membrane consists of evenly distributed, parallel homogeneous resistive elements, the mean added series resistance by the collapse of the lateral intercellular spaces will be half of the increase in R_s, *i. e.,* 100 $\Omega \cdot cm^2$. Such a slight increase in resistance could not be detected by cable analysis and could not significantly change R_a/R_b, since the resistances of the membranes are in the order of 3000 to 4000 $\Omega \cdot cm^2$.

Although the mechanism of cell depolarization is unclear, ion substitutions in the mucosal solution indicate that the selectivity of the apical membrane is reduced (Fig. 14) by increases in both Na and Cl conductances. Even though g_K

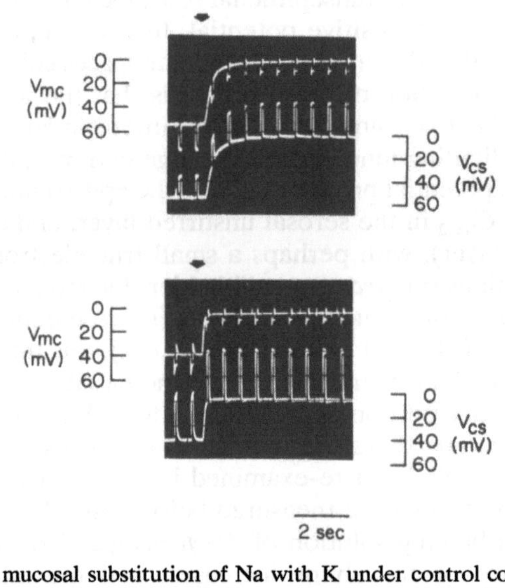

Fig. 14. Effects of mucosal substitution of Na with K under control conditions (upper panel) and after addition of sucrose to mucosal solution to double osmolality (lower panel). Scales indicate negative cell potentials referred to serosal solution. Arrows indicate solution changes. Both records obtained in same preparation. Note a) lower cell membrane potentials after sucrose addition (compare sections before arrows in both records); b) lower value of R_a/R_b in sucrose; c) marked reduction in changes of both V_{mc} and V_{cs} upon addition of K in presence of sucrose (lower panel). Cf. Table 10. (From Reuss and Finn, 1977a)

also increases, this change is proportionally less than the increase in Na and Cl conductance, and thus the cell potential falls.

Large changes in cell membrane EMFs could contribute to or oppose the change in transepithelial potential generated by the effects of the osmotic gradient on the shunt pathway. For instance, if the change in cell potential is secondary to a reduction of E_a alone, this effect would tend to make the mucosal solution more negative ($E_b - E_a$ would increase) and thus oppose the change in E_s, which tends to make the mucosal solution positive. Thus, the osmotically induced change of V_{ms} would give an erroneous estimation of the relative permeabilities of the shunt.

d) Effects of Transepithelial Current Pulses on Potentials and Resistances

Even though the instantaneous transepithelial current-voltage relationship is linear, in gallbladders of several species transepithelial constant current pulses of relatively long duration do not produce square-voltage deflections, but transient responses: when current is applied in the mucosa-to-serosa direction (inward), the potential falls with time, while outward pulses produce the opposite effect (FRÖMTER, 1972; BINDSLEV et al., 1974). BINDSLEV et al. have attributed these rather slow transients to "polarization" and resistance changes in the shunt pathway. All current pulses probably increase the conductance of the limiting junctions, and outward current pulses produce, in addition, a collapse of the intercellular spaces and thus increase the transepithelial resistance when this effect dominates over the limiting junction effect. REUSS and FINN (1977b) have studied the effects of current clamps on transepithelial and cell membrane resistances and potentials in *Necturus* gallbladder. A typical example of the effect of current on transepithelial parameters is shown in Fig. 15. Although the

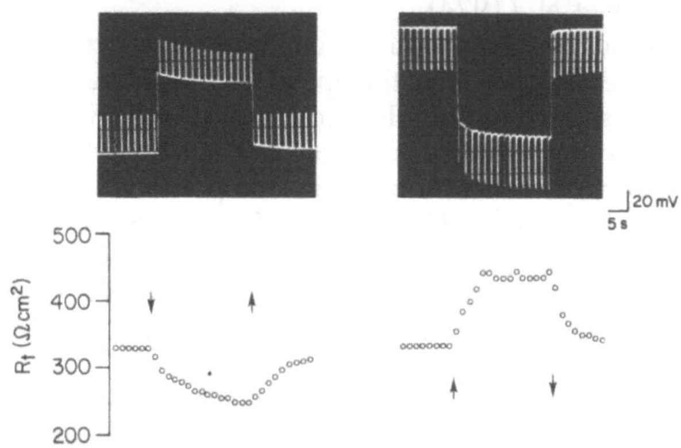

Fig. 15. Transient effects of transepithelial current clamps on transepithelial resistance (R_t) and potential in *Necturus* gallbladder. Upper tracings are V_{ms} records. Start and end of clamp indicated by lower arrows. Brief voltage deflections superimposed on continuous tracings were produced by short identical d.c. pulses, to measure transepithelial resistances at about 1.5-s intervals (lower plots). Left: *m*-to-*s* (inward) current. Right: *s*-to-*m* (outward) current. Note opposite changes in R_t. Calculations of predicted change in V_{ms}, from R_t alone, indicate that polarization phenomenon occurs in addition to resistance change. (REUSS and FINN, unpubl.)

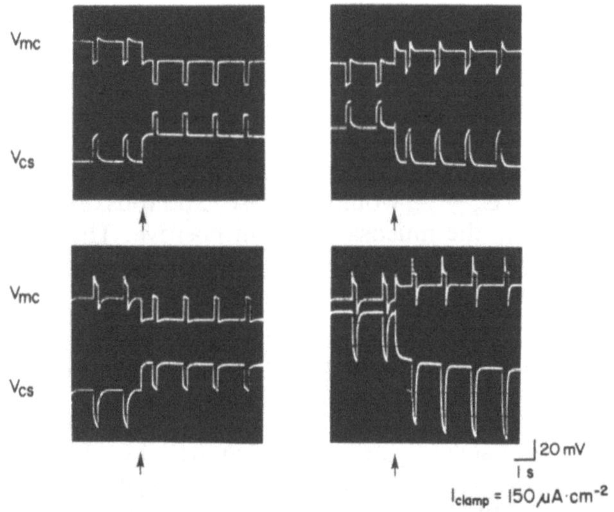

Fig. 16. Effects of transepithelial current clamps on cell membrane potential transient in *Necturus* gallbladder. Clamp current flow is inward in two records on left and outward in two records on right. Upper and lower records differ in direction of short superimposed pulses: inward and outward, respectively. Note that whenever transient voltage changes are observed (due to clamp or to short pulses) both V_{mc} and V_{cs} change in same direction (depolarization [upward]: *i. e.,* short pulses, upper right; hyperpolarization [downward]: *i. e.,* clamp and short pulses, lower right). This observation indicates that shunt is not main site of current-induced polarization (see text). (REUSS and FINN, unpubl.)

change in R_t is evident, this change does not account entirely for the change in V_{ms}. Thus, a change in tissue EMF (polarization) is necessary, as first demonstrated by BINDSLEV et al. (1974).

The effects of current clamps on the cell potentials are shown in Figure 16. The surprising observation is that the apical and basolateral membrane potentials change in the same direction. Inspection of the equivalent circuit shown in Figure 10 indicates that if only E_s changed during current passage, V_{mc} and V_{cs} should change in opposite directions. Thus, the polarization effect is *mainly* due to a change in E_a, E_b, or both. V_{ms}, V_{mc}, and V_{cs} *increase* during outward current pulses, whereas the three potentials *decrease* with opposite currents. Thus, the respective effects of transepithelial current are to increase or decrease E_b. The most likely mechanism of this effect is current-induced K depletion or accumulation in the lateral intercellular spaces.

Measurements of R_a/R_b and intraepithelial voltage spread during the current clamps indicate that neither R_a nor R_b changes significantly. Therefore, the changes in resistance take place in the shunt.

The results described above prove that changes in cell membrane parameters can contribute significantly to experimentally induced changes in transepithelial potential. The main conclusion to be derived from these studies is that even in leaky epithelia, transepithelial changes demonstrated by black-box measurements cannot be simply ascribed to the shunt pathway. Conversely, the shunt has to be taken into consideration when studying the cell membranes: analyses conventionally applied to nonpolar cells are simply inadequate for electrophysiologic studies in epithelia.

e) *Effects of Amphotericin B on Potentials and Resistances*

CREMASCHI et al. (1971) first showed that exposure of the luminal side of the rabbit gallbladder to the polyene antibiotic amphotericin B results in a change of V_{ms} (in the mucosa negative direction) of several mV. At the time, these authors attributed this V_{ms} change to an increase of g_{Na} at the apical membrane. More recently, ROSE and NAHRWOLD (1976) extended the original observations in rabbit gallbladder and concluded that the antibiotic reduces the cation selectivity of the limiting junctions. This would result in a decrease of the (mucosa positive) NaCl diffusion potential brought about by the higher NaCl concentration in the intercellular spaces (as compared to the mucosal medium) caused by the operation of the salt transport mechanism (see Section IV.1). ROSE and NAHRWOLD speculate that the obliteration of this potential would make evident the operation of a rheogenic basolateral Na pump.

These two conflicting hypotheses have been tested with microelectrodes techniques by CREMASCHI et al. (1977) in rabbit gallbladder, HÉNIN et al. (1977) in gallbladders of five other species and REUSS (1978) in *Necturus* gallbladder. In all species it was shown that amphotericin B causes a mucosa-negative change of V_{ms} which is accompanied by large depolarization of both cell membranes within a few minutes of its addition: V_{mc} depolarized by about 26 mV in rabbit (CREMASCHI et al., 1977) and 40 mV in *Necturus* (REUSS, 1978). Concomitantly, the total resistance of the luminal membrane largely drops, and its K selectivity is decreased by an increase of small cation conductance ($g_K > g_{Na} > g_{Li}$ in *Necturus*). Luminal membrane permeability to large cations (choline, tetraethylammonium, methylglucamine) changes only after prolonged exposure to the drug. In rabbit and *Necturus*, the Na:Cl selectivity ratio of the shunt pathway remained unchanged during the reversible period of action of the antibiotic. In *Necturus*, g_K across the shunt increased by about 50 percent. These results are, on a whole, consistent with a principal effect of amphotericin B on the luminal membrane, characterized by an increase in small monovalent cation conductance. When the tissue is exposed to Na bathing media, the increase in g_{Na} causes a fall of E_a which in turn produces both the large depolarization of the cell membranes and the relatively smaller change of V_{ms} (see Eqs. 1–3, Section IV.1.).

During the action of amphotericin B, net fluid transport decreases, and net Na transport at most increases transiently, falling below control values at a time when V_{ms} is still markedly negative (CREMASCHI et al., 1971; ROSE and NAHRWOLD, 1976; CREMASCHI et al., 1977). This observation constitutes a strong argument against the use of the ratio V_{ms}/R_t as an "open circuit current" indicative of the rate of active Na transport (MORENO, 1974), because during the action of amphotericin B V_{ms}/R_t can be increased several-fold at a time when net Na transport is reduced.

D. Transport of Nonelectrolytes

Nonelectrolytes may permeate epithelia *via* lipid membranes (i. e., cell membranes) or *via* polar channels (cellular or extracellular), with or without interaction with components of the pathway. The channels might be within cell membranes, limiting junctions and lateral intercellular spaces, or damaged portions of the tissue (edge damage, missing or altered cells). The state of our knowledge on permeation pathways across rabbit gallbladder is summarized in Table 11. The pertinent experimental evidence will be discussed below.

WRIGHT and DIAMOND (1969a and b) measured the permeability coefficients *(P)* of a series of nonelectrolytes in rabbit gallbladder by a technique which was based on the magnitude of the potential generated by addition of the solute, as compared to the addition of an equimolar concentration of sucrose. The permeability coefficients of most lipid-soluble solutes correlate directly with their lipid/water partition. However, branched molecules permeate less well than predicted from their partition coefficients.

Studies with rabbit gallbladder revealed anomalous behavior for smaller, polar molecules. These substances were more permeant than expected, and did not follow Overton's rules. In addition, when compared to lipid soluble solutes, permeation was less temperature dependent (WRIGHT and DIAMOND, 1969a and b; SMULDERS and WRIGHT, 1971; HINGSON and DIAMOND, 1972; VAN OS et al.,

Table 11. Permeation pathways in gallbladder epithelium. (Modified from MORENO, 1975b)

Pathway	Location	Transport	Properties
Lipid membranes	Cell membranes	*Passive* Hydrophobic nonelectrolytes (*e. g.*, 1,7-heptanediol, antipyrine) Water *Facilitated diffusion* Small polar nonelectrolytes (*e. g.*, urea; a small fraction) *Active* Na and Cl	Low electrical conductance
Cation channels	Limiting junctions	*Passive* Cations (largest fraction) Small polar nonelectrolytes (largest fraction) Water (small fraction?)	High electrical conductance Pores: $r = 0.4$ nm, numerous
Free solution shunt	Limiting junctions	*Passive* Anions Cations (small fraction) Large polar nonelectrolytes (*e. g.*, sucrose, inulin) Small polar nonelectrolytes (small fraction)	Low electrical conductance Pores: $r = 4$ nm, few

1974). However, significant interspecies variations in the rates of permeation of small, polar solutes were noted by HINGSON and DIAMOND (1972). In general, these rates followed the sequence rabbit > guinea pig > bullfrog ∼ goldfish. Restriction to diffusion of small nonelectrolytes across rabbit gallbladder yields an equivalent pore radius of 0.4 nm (VAN OS et al., 1974), in excellent agreement with the estimations of MORENO and DIAMOND (1974b) for the size of the cation selective channel in the limiting junctions.

On the other hand, some large hydrophilic solutes, such as sucrose and inulin, permeate through rabbit gallbladder epithelium (SMULDERS and WRIGHT, 1971; VAN OS et al., 1974; MORENO, 1975b), while albumin does not (VAN OS et al., 1974). Furthermore, the ratio of inulin and sucrose permeability coefficients is equal to the ratio of free-solution diffusion coefficients, and the apparent activation energy of sucrose permeability is equal to the activation energy for its diffusion in aqueous solution (SMULDERS and WRIGHT, 1971). These observations suggest the presence of a transepithelial pathway of polar nature, consisting of a small number of rather wide channels. From the restriction to albumin permeation, VAN OS and co-workers have calculated an equivalent pore radius no greater than 4 nm.

These lines of argument strongly suggest the existence of three pathways for nonelectrolyte permeation across the epithelium of the rabbit and, perhaps, other gallbladders: (i) a lipid phase, (ii) aqueous channels of $r = 0.4$ nm that account for small polar solute permeation, and (iii) aqueous channels of $r =$ 4 nm that account for the finite, but limited permeability to large polar molecules.

The fluxes of some small polar molecules such as urea exhibit saturation kinetics and can be inhibited by phloretin (VAN OS et al., 1974; WRIGHT and PIETRAS, 1974; CURCI et al., 1975), suggesting a mechanism other than simple diffusion. Thus, urea could permeate through both cell membranes (*via* a carrier) and through the narrow polar pathway. Conceivably, lipid-soluble solutes permeate mainly through cell membranes, but a small fraction could diffuse *via* the aqueous channels. Large molecules such as sucrose presumably permeate only through the large polar pathway.

MORENO (1975b) has taken advantage of the apparently specific blockade of cation-selective channels in the limiting junctions, by TAP, to search for the existence of more than one paracellular transepithelial pathway. Since TAP reduces the permeability coefficient of small polar solutes such as urea but does not alter $P_{sucrose}$ or the coefficients of lipid-soluble substances, it is likely that sucrose shares the permeation pathway used by Cl (free-solution shunt), and that the pathway for small molecules is the same as the one for cation permeation (limiting junctions and lateral intercellular spaces). Although the evidence is not conclusive, the most likely location of the free-solution shunt is in the limiting junction as well. These channels are thought to lie in parallel with the more numerous cation-selective channels.

E. Water Transport

Water transport across gallbladder epithelium can take place in the absence of or against an osmotic gradient between the bulk solutions bathing the tissue. In addition, it is always associated with net salt transport, so that the absorbate is isosmotic with the mucosal bathing solution (see Section D). Active water transport, classic osmosis, filtration, electroosmosis, pinocytosis and codiffusion have been ruled out as possible mechanisms to account for the flow of water that accompanies transport (DIAMOND, 1962c, 1968; DIETSCHY, 1966).

1. Local Osmosis Hypotheses

In recent years, several hypotheses have been proposed as possible explanations for the coupling of water flow to salt transport in epithelia. These have in common the concept of local osmosis, *i. e.,* that a compartment within the epithelium is rendered hyperosmotic by active salt transport so that water flow follows passively.

a) Three-Compartment Hypothesis

CURRAN and MCINTOSH (1962), using an artificial two-membrane, three-compartment system proved that water could move against its chemical potential if the membranes limiting the compartment are appropriately asymmetric in their water and solute permeabilities. PATLAK et al. (1963) have treated this model quantitatively. Using their equations, WHITLOCK and WHEELER (1964) tested the hypothesis in rabbit gallbladder, suggesting that the lateral intercellular spaces may be the compartment (see also DIETSCHY, 1966). The system predicts isotonicity of the transported fluid when both solutions contain only permeant solute, and hypertonicity when a nonpermeant solute (*e. g.,* sucrose) partially replaces the permeant solute (NaCl) in the mucosal compartment. WHITLOCK and WHEELER found the absorbate to be slightly hyperosmotic in the latter situation, whereas DIAMOND (1964) found the solution dripping from the "unilateral" rabbit gallbladder preparations (see Section B), to be isotonic. This hypothesis, in spite of minor differences in the quantitative measurements, offers a qualitative explanation for the mechanism of the coupling of water flux to the solute transport.

b) Standing-Gradient Osmotic Flow Hypothesis

DIAMOND and his co-workers (DIAMOND and TORMEY, 1966a and b; TORMEY and DIAMOND, 1967; DIAMOND and BOSSERT, 1967) have described and treated mathematically a somewhat different mechanism that would account for the isotonicity of the absorbate in all circumstances. Crucial to their hypothesis is the notion that the lateral intercellular spaces are an unstirred path for solute and water flow, and that local osmotic gradients can be generated and maintained in these spaces by active salt transport. Each space is considered to be a long channel, closed at one end (the limiting junction). Active solute transport takes place near the closed end, making the luminal fluid of the channel hypertonic. Water flows from the cells into the space down its chemical potential gradient, and NaCl moves toward the open end both by the sweeping effect of the bulk water flow and by diffusion. Under steady-state conditions, a standing osmotic gradient is established in the channel. The gradient is maximal at the closed end and zero at the open end. DIAMOND and BOSSERT (1967) treated the hypothesis quantitatively, and showed that with a number of assumptions the operation of such a system could in fact yield isotonic transport, if the channel is narrow and long, the osmotic water permeability of the walls high, and the diffusion coefficient of the solute low.

An important assumption of DIAMOND and BOSSERT's analysis is that the channels are effectively closed by the limiting junctions. However, as discussed in detail in Section C, there is electrophysiological and morphological evidence to indicate that the limiting junctions, in gallbladder and other leaky epithelia, are the main pathway for ion permeation. Thus, their osmotic water permeability cannot be assumed to be zero. In addition, for the emerging solution to be isotonic the pumping sites have to be restricted to the more apical segments of the lateral membranes. There are circumstantial arguments in favor of and against this assumption. Whereas ATPase is homogeneously distributed in the basolateral membranes of the epithelium of rabbit intestine (STIRLING, 1972) and frog skin (FARQUHAR and PALADE, 1954), KAYE et al. (1966) found most of the enzyme in the apical and middle portions of the lateral membranes in rabbit gallbladder.

c) Hypertonic Interspace Mechanisms with Leaky Junctions and Distributed Solute Input

SACKIN and BOULPAEP (1975) re-examined the problem in a more general way. They assumed that the limiting junctions are permeable to both salt and water and that active salt transport occurs uniformly along the lateral spaces. Two systems were considered: a continuous model similar to the one described by DIAMOND and BOSSERT, and a compartmental model similar to the one described by CURRAN and MSINTOSH. Both models explain quantitatively the movement of water across the epithelium bathed with identical solutions. Since the salt concentration profile along the lateral intercellular spaces turned out to be uniform, the channel could be considered to be a single compartment with no longitudinal osmotic gradient and the emerging fluid was hyperosmotic. The magnitude of this hyperosmolality depended on the values of the transport parameters, especially the osmotic water permeability of the limiting junction and the lateral cell membrane.

Further modifications of the composition of the absorbate could take place at the basement membrane (or subsequent barriers) and could also arise from volume transport across the basal side of the cell. First, when the solute reflection coefficient of the basement membrane (σ_s) is assumed to be zero, the fluid is hyperosmotic. Second, water could flow out of the cell, across the basal membrane, driven by osmotic and colloid-osmotic forces (*e. g.*, if albumin permeates the basement membrane), tending to dilute the fluid emerging from the lateral intercellular spaces.

d) Influence of Transport-Dependent Asymmetries in Fluid Composition

Recent studies in rat proximal tubule (FRÖMTER, 1974) strongly suggest that local osmosis in this epithelium could occur in the absence of total hyperosmolarity of the intercellular spaces, but because of a high *effective* osmolarity of the spaces as compared to the intratubular fluid. Under steady-state conditions, the intercellular spaces have a higher HCO_3 concentration than the luminal fluid, and the latter has a higher Cl concentration than the fluid in the spaces. Even though the total osmolarities are the same, the *effective* osmolarity of the intercellular spaces is larger on account of the difference in reflection coefficients ($\sigma_{HCO_3} > \sigma_{Cl}$), and water can be driven into the space by this osmotic gradient. In the proximal tubule, glucose and aminoacids also contribute to the gradient, since they are completely reabsorbed. Although interesting, this hypothesis does not seem applicable to the gallbladder, where a large difference in composition of the transported fluid (as compared with the mucosal solution) has not been observed. However, as discussed below, the osmotic gradient required to account for the measured water fluxes depends on the osmotic water permeability of the epithelium, and this has probably been underestimated by one order of magnitude when measured under steady-state conditions. Thus, if the osmotic water permeability is in fact 10 times larger than previously thought, very small osmotic gradients (total or effective) would provide enough driving force for solute-coupled water absorption. The difference in osmolarity or the asymmetries in fluid composition could conceivably be too small to be detected.

Proof of any of the mechanisms described above requires direct measurement of the concentration profiles, and perhaps composition profiles, of the lateral intercellular spaces. The only experimental information directly relevant to this issue is the measurement of hyperosmolarity in intercellular fluid (obtained by micropuncture) of insect rectal pads under transporting conditions *in vitro* (WALL et al., 1970). Other results that are consistent with the possibility of a hyperosmotic absorbate are summarized by SACKIN and BOULPAEP (1975). The availability of electron-microprobe analytical techniques may solve the problem of osmolarity and composition of the intercellular fluid directly.

2. Magnitude of the Hydraulic Conductivity of the Gallbladder

WRIGHT et al. (1972) have shown that the establishment of osmotic gradients across the tissue produces large transient responses in water fluxes. For instance, the osmotic water flow (J_W^{osm}) measured 5 min after adding sucrose to the mucosal solution was ten times larger than after a steady state had been reached. Since these changes in J_W^{osm} did not correlate with the measured changes of transepithelial conductance or nonelectrolyte permeability, WRIGHT et al. concluded that the observations cannot be ascribed to the collapse of the lateral intercellular spaces. The most likely explanation of this observation is an unstirred layer effect. The instantaneous transepithelial water flux both dilutes the hyperosmotic side and concentrates the hyposmotic side. Since the unstirred layer is thicker on the serosal side (approx. 800 μm), its concentration will exert the dominant effect. Thus the actual osmotic gradient across the tissue will be considerably reduced.

The main conclusion to be derived from these experiments is that the L_P of the gallbladder is probably underestimated by at least one order of magnitude when calculated from steady-state osmotic water fluxes. This means that only instantaneous changes in water flux yield accurate estimations of this important parameter. VAN OS (pers. comm.) using a method that allows continuous accurate measurement of J_w, has shown that the L_P can be even higher, since the sudden imposition of an osmotic gradient in rabbit gallbladder resulted in changes in tissue water content for up to 3 min before any change in transepithelial flow could be detected. In other words, the transepithelial osmotic gradient can be considerably reduced in the shortest time at which J_w can be measured.

If the L_p of the gallbladder is one order of magnitude larger than hitherto estimated, the required osmotic gradient to explain solute-coupled water transport is one order of magnitude smaller. Thus, a gradient of 2 instead of 20 mOsm l^{-1} may be enough, and the likelihood of detecting such a slight concentration difference in the intercellular spaces is very small indeed.

3. Route of Water Flow

Nonlinear osmosis was observed in gallbladder by DIAMOND (1966) and confirmed by WRIGHT et al. (1972). The main deviation from linear behavior was the reduction in steady-state water fluxes with large osmotic gradients (*e. g.*, 300 mM sucrose). When the solute was added to the mucosal solution, the steady-state flux was only one-third of the flow produced by serosal hyperosmolarity. Addition of solute to the mucosal medium collapses the lateral spaces (electron-microscopic studies by SMULDERS et al., 1972), but the contribution of the change in dimensions of the spaces to the steady-state apparent L_p was calculated to be only 5 to 10 percent. However, since the effective gradient was much reduced, and the L_p is probably at least 10-fold larger, the spaces could contribute 25 to 60 percent of the resistance to water flow in this situation. When the serosal bathing medium is hyperosmotic, the lateral intercellular spaces dilate. Values measured for L_p in this experimental condition are independent of the gradient, even though the spaces widen further at large serosal hyperosmolarity. This observation indicates that open lateral spaces do not contribute significantly to the total resistance to water flow.

These results, and the ones discussed in previous sections, suggest, but do not prove, that the lateral intercellular spaces are a pathway for transepithelial water flux, whether coupled to salt transport or osmotically induced.

Water flow from the mucosal medium to the lateral intercellular spaces can be either cellular or extracellular (*i. e.*, across the limiting junction). WRIGHT et al. (1972) calculated the possible contribution of the limiting junction route from the total ionic conductance, the length of the junction, the conductivity of Ringer solution, and the assumption of pores of 1.2 nm radius. Their result was an L_p of about $5 \cdot 10^{-4}$ cm $\cdot$ s^{-1}. If these calculations are correct, since the L_p of the tissue is at least $5 \cdot 10^{-3}$ cm $\cdot$ s^{-1}, only 10 percent of the J_w could follow this pathway. If, as is very likely, L_p is one order of magnitude higher, only 1 percent of the flux could be through the limiting junctions.

VAN OS and SLEGERS (1973) measured the ratio of osmotic and diffusional water permeabilities in rabbit gallbladder, corrected for unstirred layers, and calculated the equivalent pore radius to be about 0.45 nm (similar to the one calculated for red cell membranes). However, if the osmotic permeability coefficient is underestimated by a factor of 10, the pore radius is also underestimated, by a factor of about 3. Since molecules as large as inulin permeate the gallbladder, VAN OS and SLEGERS ascribed to the paracellular channels a radius of at least 1.2 nm, and concluded that a smaller pore is necessarily transcellular. However, there is strong evidence for two groups of channels in the extracellular pathway (see Section D) and, furthermore, an appropriate correction for the L_p yields an equivalent pore radius larger than 1.2 nm.

In conclusion, little is known about the pathways for water flow across the gallbladder. It is likely that the lateral intercellular spaces are involved as a common final path for both active fluid transport and osmotic water flow (some recent criticisms of the morphological evidence for this conclusion have been discussed in previous sections). On the basis of relative surface areas and calculated radii of the hypothetical channels in the limiting junctions, only a small contribution to the transepithelial L_p can be ascribed to these structures, especially if the actual L_p is much larger than initially calculated. Thus, the major proportion of water flux can be transcellular. It is obvious that much work needs to be done to end these uncertainties. However, present techniques appear to be insufficient to allow separate determinations of the L_p's of the limiting junctions and the cell membranes.

Acknowledgements

I wish to thank Drs. *P. J. DeWeer, A. L. Finn,* and *J. T. Gatzy* for constructive criticism of a preliminary version of this Chapter. I am also grateful to *G. Jerman* and *A. L. Guinn* for typing the manuscript. The personal work described in the text was supported by grants AM-17854 and AM-19580 from the National Institute of Arthritis, Metabolism, and Digestive Diseases.

References

BARRY, P. H., DIAMOND, J. M.: J. Membrane Biol. **3**, 93 (1970).
BARRY, P. H., DIAMOND, J. M., WRIGHT, E. M.: J. Membrane Biol. **4**, 358 (1971).
BINDSLEV, N., TORMEY, J. M., WRIGHT, E. M.: J. Membrane Biol. **19**, 357 (1974).

BOULPAEP, E. L.: In: Electrophysiology of Epithelial Cells (G. Giebisch, Ed.). Stuttgart: Schattauer 1971, p. 91.
CIANI, S. M., RISENMAN, G., SZABO, G.: J. Membrane Biol. **1**, 1 (1969).
CIVAN, M. M., KEDEM, O., LEAF, A.: Amer. J. Physiol. **211**, 569 (1966).
CREMASCHI, D., HÉNIN, S.: Pflügers Arch. ges. Physiol. **361**, 33 (1975).
CREMASCHI, D., HÉNIN, S., MEYER, G., BACCIOLA, T.: J. Membrane Biol. **34**, 55 (1977).
CREMASCHI, D., MONTANARI, C., SIMONIC, T., LIPPE, C.: Arch. Int. Physiol. Biochim. **79**, 33 (1971).
CURCI, S., CASAROLA, V., LIPPE, C.: Pflügers Arch. ges. Physiol. **355**, 267 (1975).
CURRAN, P. F., MACINTOSH, J. R.: Nature (Lond.) **193**, 347 (1962).
DIAMOND, J. M.: J. Physiol. (Lond.) **161**, 442 (1962a).
DIAMOND, J. M.: J. Physiol. (Lond.) **161**, 474 (1962b).
DIAMOND, J. M.: J. Physiol. (Lond.) **161**, 503 (1962c).
DIAMOND, J. M.: J. gen. Physiol. **48**, 1 (1964a).
DIAMOND, J. M.: J. gen. Physiol. **48**, 15 (1964b).
DIAMOND, J. M.: J. Physiol. (Lond.) **183**, 58 (1966).
DIAMOND, J. M.: In: Handbook of Physiology, Section 6, Vol. V (C. F. Code, Section Ed.). Baltimore, Md: Williams and Wilkins 1968.
DIAMOND, J. M.: Fed. Proc. **33**, 2220 (1974).
DIAMOND, J. M., BOSSERT, W. H.: J. gen. Physiol. **50**, 2061 (1967).
DIAMOND, J. M., HARRISON, S. C.: J. Physiol. (Lond.) **183**, 37 (1966).
DIAMOND, J. M., TORMEY, J. McD.: Nature (Lond.) **210**, 817 (1966a).
DIAMOND, J. M., TORMEY, J. McD.: Fed. Proc. **25**, 1458 (1966b).
DIAMOND, J. M., BARRY, P. H., WRIGHT, E. M.: In: Electrophysiology of Epithelial Cells (G. Giebisch, Ed.). Stuttgart: Schattauer 1971.
DIBONA, D. R., CIVAN, M. M.: J. Membrane Biol. **12**, 101 (1973).
DIETSCHY, J. M.: Gastroenterology **47**, 395 (1964).
DIETSCHY, J. M.: Gastroenterology **50**, 692 (1966).
DIETSCHY, J. M., MOORE, E. W.: J. Clin. Invest. **43**, 1551 (1964).
EISENBERG, R. S., JOHNSON, E. A.: Prog. Biophys. molec. Biol. **20**, 1 (1970).
FARQUHAR, M. G., PALADE, G. E.: Proc. nat. Acad. Sci. (Wash.) **51**, 569 (1964).
FREDERIKSEN, O., ROSTGAARD, J.: J. Cell Biol. **61**, 830 (1974).
FRIZZELL, R. A., DUGAS, M. C., SCHULTZ, S. G.: J. gen. Physiol. **65**, 769 (1975).
FRÖMTER, E.: J. Membrane Biol. **8**, 259 (1972).
FRÖMTER, E.: In: MTP International Review of Science. Physiology, Series 1, Vol. 6 (K. Thurau, Ed.). London: Butterworths; Baltimore, Md: University Park Press 1974, p. 1.
FRÖMTER, E., DIAMOND, J. M.: Nature New Biol. **235**, 9 (1972).
GELARDEN, R. T., ROSE, R. C.: J. Membrane Biol. **19**, 37 (1974).
GRIM, E., SMITH, G. A.: Amer. J. Physiol. **191**, 555 (1957).
HEINTZE, K., PETERSEN, K. U., HEIDENREICH, O.: Naunyn-Schmiedebergs Arch. Pharmak. exp. Path. **285**, 151 (1974).
HÉNIN, S., CREMASCHI, D.: Pflügers Arch. ges. Physiol. **355**, 125 (1975).
HÉNIN, S., CREMASCHI, D., SCHETTINO, T., MEYER, G., DONIN, C. L. L., COTELLI, F.: J. Membrane Biol. **34**, 73 (1977).
HINGSON, D. J., DIAMOND, J. M.: J. Membrane Biol. **10**, 93 (1972).
KANG, D. H., HONG, S. K.: Am. J. Physiol. **203**, 1015 (1962).
KAYE, G. I., WHEELER, H. O., WHITLOCK, R. T., LANE, N.: J. Cell Biol. **30**, 237 (1966).
KOEFOED-JOHNSEN, V., USSING, H. H.: Acta physiol scand. **42**, 298 (1958).
LEYSSAC, P. P., BUKHAVE, K., FREDERIKSEN, O.: Acta physiol. scand. **92**, 496 (1974a).
LEYSSAC, P. P., KRISTENSEN, L. Ø., CHRISTENSEN, P., FREDERIKSEN, O.: Acto physiol. scand. **92**, 508 (1974b).
MACHEN, T. E., DIAMOND, J. M.: J. Membrane Biol. **1**, 194 (1969).
MACHEN, T. E., DIAMOND, J. M.: J. Membrane Biol. **8**, 63 (1972).
MACHEN, T. E., ERLIJ, D., WOODING, F. B. P.: J. Cell Biol. **54**, 302 (1972).
MARTIN, D. W.: J. Membrane Biol. **18**, 219 (1974).
MARTIN, D. W., DIAMOND, J. M.: J. gen. Physiol. **50**, 295 (1966).
MARTIN, D. W., MURPHY, B.: J. Membrane Biol. **18**, 231 (1974).
MERTENS, R. B., WHEELER, H. O., MAYER, S. E.: Gastroenterology **67**, 898 (1974).

MOORE, E. W., DIETSCHY, J. M.: Am. J. Physiol. **206**, 1111 (1964).
MORENO, J. H.: Nature (Lond.) **251**, 150 (1974).
MORENO, J. H.: J. gen. Physiol. **66**, 97 (1975a).
MORENO, J. H.: J. gen. Physiol. **66**, 117 (1975b).
MORENO, J. H., DIAMOND, J. M.: Nature (Lond.) **247**, 368 (1974a).
MORENO, J. H., DIAMOND, J. M.: J. Membrane Biol. **15**, 277 (1974b).
MORENO, J. H., DIAMOND, J. M.: J. Membrane Biol. **21**, 197 (1975).
MORENO, J. H., DIAMOND, J. M.: In: Membranes – A Series of Advances (G. Eisenman, Ed.). New York: Dekker 1976, p. 383.
MORTON, I. K. M., SAVERYMUTTU, S. M., WOOD, J. R.: Brit. J. Pharmacol. **50**, 460P (1974).
NELLANS, H. N., FRIZZELL, R. A., SCHULTZ, S. G.: Amer. J. Physiol. **225**, 467 (1973).
ONSTAD, G. R., SCHOENFIELD, L. J., HIGGINS, J. A.: J. clin. Invest. **46**, 606 (1967).
PATLAK, C. S., GOLDSTEIN, D. A., HOFFMAN, J. P.: J. theoret. Biol. **5**, 426 (1963).
PIDOT, A. L., DIAMOND, J. M.: Nature (Lond.) **201**, 701 (1964).
RAVDIN, I. S., JOHNSTON, C. G., RIEGEL, C., WRIGHT, S. L., Jr.: Am. J. Physiol. **100**, 317 (1932).
REUSS, L.: J. Membrane Biol. **41**, 65 (1978).
REUSS, L., FINN, A. L.: J. Membrane Biol. **25**, 115 (1975a).
REUSS, L., FINN, A. L.: J. Membrane Biol. **25**, 141 (1975b).
REUSS, L., FINN, A. L.: Amer. J. Physiol. **232**, c99 (1977a).
REUSS, L., FINN, A. L.: J. Membrane Biol. **37**, 299 (1977b).
ROSE, R. C., GELARDEN, R. T., NAHRWOLD, D. L.: Amer. J. Physiol. **224**, 1320 (1973).
ROSE, R. C., Nahrwold, D. L.: J. Membrane Biol. **29**, 1 (1976).
SACKIN, H., BOULPAEP, E. L.: J. gen. Physiol. **66**, 671 (1975).
SCHAFER, D. E., NICOLOFF, D. M., GLEASON, D. F., CARLSON, T. J.: Gastroenterology **56**, 1195 (1969).
SMULDERS, A. P., TORMEY, J. McD., WRIGHT, E. M.: J. Membrane Biol. **7**, 164 (1972).
SMULDERS, A. P., WRIGHT, E. M.: J. Membrane Biol. **5**, 297 (1971).
STIRLING, E. S.: J. Cell Biol. **53**, 704 (1972).
SULLIVAN, B., BERNDT, W. O.: Amer. J. Physiol. **225**, 845 (1973).
TORMEY, J. M., DIAMOND, J. M.: J. gen. Physiol. **50**, 2031 (1967).
USSING, H. H., WINDHAGER, E. E.: Acta physiol scand. **61**, 484 (1964).
VAN OS, C. H., SLEGERS, J. F. G.: Biochim. biophys. Acta **241**, 89 (1971).
VAN OS, C. H., SLEGERS, J. F. G.: Biochim. biophys. Acta **291**, 197 (1973).
VAN OS, C. H., SLEGERS, J. F. G.: J. Membrane Biol. **24**, 341 (1975).
VAN OS, C. H., DE JONG, M. D., SLEGERS, J. F. G.: J. Membrane Biol. **15**, 363 (1974).
WALL, B. J., OSCHMANN, J. L., SCHMIDT-NIELSEN, B.: Science **167**, 1497 (1970).
WEDNER, H. J., DIAMOND, J. M.: J. Membrane Biol. **1**, 92 (1969).
WHEELER, H. O.: Amer. J. Physiol. **205**, 427 (1963).
WHEELER, H. O.: Amer. J. Med. **51**, 588 (1971).
WHEELER, H. O., ROSS, E. D., KING, K. K.: Amer. J. Physiol. **216**, 175 (1969).
WHITLOCK, R. T., WHEELER, H. O.: J. clin. Invest. **43**, 2249 (1964).
WHITLOCK, R. T., WHEELER, H. O.: Amer. J. Physiol. **213**, 1199 (1967).
WIEDNER, G., WRIGHT, E. M.: Pflügers Arch. ges. Physiol. **358**, 27 (1975).
WINDHAGER, E. E., BOULPAEP, E. L., GIEBISCH, G.: In: Proceedings of the 3rd International Congress on Nephrology, Washington, 1966. Karger: Basle–New York 1967, p. 35.
WRIGHT, E. M., DIAMOND, J. M.: Biochim. biophys. Acta **163**, 57 (1968).
WRIGHT, E. M., DIAMOND, J. M.: Proc. roy. Soc. B **172**, 203 (1969a).
WRIGHT, E. M., DIAMOND, J. M.: Proc. roy. Soc. B **172**, 227 (1969b).
WRIGHT, E. M., PIETRAS, R. J.: J. Membrane Biol. **17**, 293 (1974).
WRIGHT, E. M., BARRY, P. H., DIAMOND, J. M.: J. Membrane Biol. **4**, 331 (1971).
WRIGHT, E. M., SMULDERS, A. P., TORMEY, J. McD.: J. Membrane Biol. **7**, 198 (1972).
ZEISKE, W.: Pflügers Arch. ges. Physiol. **359**, R127 (1975).

Chapter 18

Transport of Ions in Liver Cells

M. CLARET

A. Introduction

The general characteristics of ionic distribution in hepatocytes are similar to those in other cells. Potassium concentration is higher and sodium and chloride concentrations are lower than in plasma, and most of the evidence suggests that the fundamental processes implicated in this distribution are the same as in other tissues. The membranes of hepatocytes, however, exhibit unique biophysical properties, especially with regard to the passive permeabilities, facilitated diffusion, and the electrogenic activity of the sodium pump. The characteristics of these different transport processes will be described in the present chapter, with a particular emphasis on potassium, sodium, chloride and calcium ions, since these have been studied most intensively.

Most of the data reported deal with *in-vitro* studies. Isolation of the liver from its normal physiological environment causes changes in the membrane properties, which may be of some importance. This is particularly apparent with respect to the membrane potential, which, for unknown reasons, is invariably lower *in vitro* than *in vivo.* Nethertheless, ionic concentrations as well as activities of major transport systems are maintained *in vitro,* and experiments may be carried out under steady-state conditions. This is not a negligible advantage, and may compensate for the disadvantages inherent in *in-vitro* techniques.

B. Ionic Concentrations and Activities

In vivo, the potassium, sodium and chloride concentrations range from 140 to 170, 20–40 and 20–35 mmol $\cdot$ l^{-1} of intracellular water, respectively (WILLIAMS and WOODBURY, 1971). The variability of reported values depends largely upon which extracellular marker is employed. *In vitro,* following a preliminary incubation period, intracellular electrolyte concentrations differ somewhat from those measured immediately after sacrifice of the animal. The potassium concentration is seldom altered but the sodium and chloride values are significantly higher and show much wider ranges of variation than observed *in vivo* (20–80 mmol $\cdot$ l^{-1} for sodium and 30–70 mmol $\cdot$ l^{-1} for chloride). As these concentrations do not change significantly over periods of several hours (BRIS-

TOW and KERLY, 1964; HAYLETT and JENKINSON, 1972a; CLARET and MAZET, 1972; LAMBOTTE, 1973a; MCLAUGHLIN, 1973; DAMBACH and FRIEDMANN, 1974; MACKNIGHT et al., 1974), it seems likely that the discrepancies between *in-vivo* and *in-vitro* values result from the establishment of a new steady state, and that the variability of values reflects the wide range of *in-vitro* conditions used by different investigators.

The diffusion and activity coefficients of sodium, potassium, and chloride in cytoplasm are not known. Indirect evidence from electrophysiological studies and kinetic analyses of fluxes suggests that at 38° C potassium and chloride are in true solution, without extensive binding to other components of the cytoplasm (HUMPHREY, 1961; HAYLETT and JENKINSON, 1972a; CLARET and MAZET, 1972). The activity of sodium, in contrast to that of potassium and chloride, is certainly much lower than the total intracellular sodium concentration (HOOPER and DICK, 1976). Mitochondria (SCARPA and AZZI, 1968), nuclei (LANGENDORF et al., 1966), and endoplasmic reticulum (SANUI and PACE, 1959) bind sodium, and this "trapping" may well contribute to the removal of sodium from cytoplasmic intracellular water. This hypothesis is reinforced by the existence of a multiexponential loss of ^{24}Na from the cell (VAN ROSSUM, 1966a).

C. Membrane Potential

The membrane potential of mammalian liver cells *in vivo* has been reported to be approximately -50 mV in the various species studied (TOIDA et al., 1958; LIMBERGER, 1963; BEIGELMAN et al., 1964; SCHANNE and CORABOEUF, 1966; PENN, 1966; BIEDERMAN, 1968; WILLIAMS et al., 1971a and b; CAILLÉ and SCHANNE, 1972; LAMBOTTE, 1970). It is quite sensitive to modifications of the plasma composition and, especially, is dependent on the animal's hormonal state (see BEIGELMAN and THOMAS, 1972). Little is known of the ionic dependence of the potential, primarily due to difficulties in controlling the extracellular ion concentration *in vivo*. The study of passive properties of the membrane is also hindered by the participation of an electrogenic component of the sodium pump (WILLIAMS et al., 1971a and b; FOLKE, 1971). This situation is illustrated by the variable effects of external potassium on the hepatic membrane potential: an increase in its plasma concentration may either hyperpolarize or depolarize the membrane, depending upon whether the preponderant effect is stimulation of the pump or a decrease in the driving force of potassium ions (SCHANNE and CORABOEUF, 1966; WILLIAMS et al., 1971a and b; CAILLÉ and SCHANNE, 1972).

Except for dog liver, which maintains a potential of about -50 mV *in vitro* (LAMBOTTE, 1973a), the absolute value of the liver membrane potential for all species tested declines to a stable level of about 30–35 mV in the isolated liver perfused with electrolyte solutions or blood (CLARET and CORABOEUF, 1970; SOMLYO et al., 1971; FOLKE, 1971; FRIEDMANN and DAMBACH, 1973), and in slices equilibrated in electrolyte solutions (LI and MCILWAIN, 1957; HAYLETT and JENKINSON, 1972a; BEIGELMAN et al., 1964; PENN, 1966; HUMPHREY and

MAENO, 1969; HELLER and VAN DER KLOOT, 1974). Isolated cells exhibit similar values (GREEN et al., 1972).

The cause of this sustained *in-vitro* depolarization in unknown. It is unlikely that alteration of ionic gradients constitutes a major factor, since these are roughly equivalent *in vitro* and *in vivo,* nor anoxia, as was suggested by LAMBOTTE (1973a); livers perfused with blood also exhibit a mean potential of about −35 mV (FOLKE, 1972). The depolarization can be counteracted by diverse experimental conditions, apparently unrelated, which can be summarized as follows:

i) In segments of guinea-pig liver much thicker than the usual slices, depolarization is unexpectedly reversible (HELLER and VAN DER KLOOT, 1974). For unknown reasons, this spontaneous reversibility depends upon the integrity of the fibrous capsule covering the liver. If the capsule is removed, the membrane potential remains at a value similar to that measured in thin slices. HELLER and VAN DER KLOOT never observed this effect in thick segments of rat liver, which invariably depolarized *in vitro,* whatever the conditions.

ii) Additions to perfusates of certain substrates of liver metabolism, such as pyruvate, fructose, alanine or palmitate, elicit a sustained hyperpolarization in perfused rat or dog liver (DAMBACH and FRIEDMANN, 1974; FOLKE, 1974; LAMBOTTE and PONTEGNIE-ISTACE, 1975). The origin of the hyperpolarization may vary according to the added substrate. The pyruvate- or alanine-induced hyperpolarization probably involves an activation of the electrogenic sodium pump (DAMBACH and FRIEDMANN, 1974; FOLKE and PALOHEIMO, 1975), while that generated by fructose is attributed to a change in potassium permeability (LAMBOTTE and PONTEGNIE-ISTACE, 1975).

iii) In isolated rat liver, the substitution of freshly taken blood for that recirculating in the perfusion system causes a transient hyperpolarization. After 10–20 minutes the membrane potential invariably returns to its original value. In contrast, if the liver, after initial perfusion *in vitro* with blood, is connected to the vascular system of a donor rat, the absolute value of the potential increases to about 50 mV and this new value is maintained (FOLKE, 1972).

These last experiments are of interest because they indicate that the membrane depolarization following the isolation procedure may not be caused by irreversible liver injury. They suggest that the maintenance of normal membrane potential depends on factors present in fresh blood, which the perfused liver rapidly exhausts and the intact living animal continuously provides (FOLKE, 1974). These factors may include hormones, which are known to alter hepatic membrane potential (SOMLYO et al., 1971; HAYLETT and JENKINSON, 1972a and b; LAMBOTTE, 1973a; FRIEDMANN, 1974; PETERSEN, 1974).

D. Passive Fluxes

I. Distribution of Ions

Analysis of ionic movements through the cell membrane requires the determination of relative contributions by different transport processes to each unidirectional flux. The average calculated *in-vitro* equilibrium potentials for potassium ($E_K = -90$ mV) and sodium ions ($E_{Na} = +30$ mV) are quite different from the measured membrane potential ($E_m = -35$ mV) (HAYLETT and JENKINSON, 1972a; CLARET and MAZET, 1972; DAMBACH and FRIEDMANN, 1974). In the steady-state, maintenance of this discrepancy implies that active transport must contribute to potassium influx and sodium efflux, and must be dependent on cell energy sources.

This expectation is substantiated by experimental evidence. Thus, potassium influx and sodium efflux are selectively reduced by pump inhibitors, and the uptake of potassium and extrusion of sodium are dependent on active cell metabolism (FLINK et al., 1950; HECKMANN and PARSONS, 1959; MCLEAN, 1963; ELSHOVE and VAN ROSSUM, 1963; CLARET and MAZET, 1972; MACKNIGHT et al., 1974; KOLB and ADAM, 1976). Conversely, the efflux of potassium seems to be a passive flux. Sodium influx may also be passive in the energetic sense, but it appears that inward movement of sodium cannot be attributed solely to simple diffusion (CLARET and MAZET, 1972). Influx and efflux of sodium are interdependent. A decrease in the external sodium concentration produces a decrease of sodium efflux, independent of changes in membrane potential. Such an interaction suggests that, at least in part, sodium movements are coupled through an exchange-diffusion system (see below).

The distribution of chloride across the membrane is more intricate. WILLIAMS et al. (1971a and b) reported that the calculated equilibrium potential (E_{Cl}) of rat liver cells was slightly more negative than the membrane potential (E_m) measured *in vivo*. They postulated that differences between E_{Cl} and E_m were the consequence of active chloride extrusion from the cells. These results have not been confirmed *in vitro* (CLARET and MAZET, 1972). The large variations of internal chloride concentration observed consequent to changes in membrane potential, and/or external chloride concentration are in close agreement with those calculated from the Nernst equation, so that E_{Cl} and E_m are always identical. A possible explanation of the discrepancy between *in-vivo* and *in-vitro* results might be that rat liver, when isolated, loses its capacity to extrude chloride.

II. Diffusion Fluxes

Despite uncertainties concerning chloride ion distribution, total chloride influx and efflux will be attributed to passive diffusion, since almost all experimental data presented in this review are derived from *in-vitro* studies. The diffusional

Table 1. Biophysical properties of hepatic cell membrane of rat (1) or guinea-pig (2)

	Rate constant min^{-1}	Flux pmoles cm^{-2} s^{-1}		Permeability coefficient 10^{-8} cm · s^{-1}	Conductance μmho · cm^{-2}	Relative permeability coefficients	Relative permeability coefficients determined from electro-physiologic studies (3)
		in	out				
K	0.0140–0.0134 0.0187 (2)	6.6	6.7	7.6	10	1	1
Na	0.17	13.4	11.9	4.0	12	0.53	0.15–0.45
Cl	0.077	7.7	8.3	12.3	30	1.62	1.0 –2.00

1) From HUMPHREY (1961); CLARET and MAZET (1972).
2) From HAYLETT and JENKINSON (1972a).
3) From SCHANNE and CORABOEUF (1966); BEIGELMAN and SCHLOSSER (1969); HUMPHREY and MAENO (1969); WILLIAMS et al. (1971a and b); CAILLE and SCHANNE (1972).

flux of sodium can be estimated by subtracting the magnitude of exchange-diffusion from measured influx. Table 1 shows the rate constants and fluxes of potassium, sodium, and chloride measured in rat liver. Permeabilities and conductances have been calculated according to the constant field theory (GOLDMAN, 1943; HODGKIN and KATZ, 1949).

Permeability coefficients to potassium, sodium, and chloride are in the same order of magnitude, in good agreement with the relative permeabilities indirectly deduced from *in-vivo* or *in-vitro* electrophysiological studies (Table 1). The value of the P_{Na}/P_K ratio may largely explain the low absolute value of membrane potentials measured in liver. *In vitro,* this situation is illustrated by utilizing the equation of MULLINS and NODA (1973). This equation is essentially equivalent to the constant field equation (GOLDMAN, 1943; HODGKIN and KATZ, 1949), but it takes into account a pump coupling ratio *(r),* as this may differ from unity (electrogenic pump). Calculation of the membrane potential from the permeability coefficients determined by radioactive tracers, and the coupling ratio measured in rat liver, demonstrate identical calculated and observed potentials. This equation predicts that rat liver membrane potential *in vitro* is mainly a diffusional potential (−30 mV), with a small contribution from the sodium pump (−5 mV).

A completely different set of values for relative permeability coefficients (P_{Na}/P_K = 0.04 and P_{Cl}/P_K = 0.3) has been proposed recently in guinea-pig (HELLER and VAN DER KLOOT, 1974). To account for the membrane potential measured in this tissue, mechanisms other than diffusional potentials and sodium electrogenic pump activity are assumed.

III. Facilitated Diffusion

1. Ouabain-Insensitive Sodium Exchange

Under physiological conditions, the observed ratio of ouabain-insensitive sodium efflux to influx (0.57) (CLARET and MAZET, 1972) is much nearer to unity than the value predicted (0.06) from the independence principle (USSING, 1949). This suggests that some interaction occurs between sodium fluxes. In its simplest form, this type of interaction predicts that the sodium efflux must be dependent on external sodium concentration. In the presence of 10^{-3} M ouabain, which is used to suppress any direct stimulating effect of lithium on the sodium pump, replacement of the total extracellular sodium by lithium reduces sodium efflux. Such a coupling of sodium transport may be the result of a sodium-for-sodium exchange that is insensitive to glycosides, as proposed by USSING (1949). Its contribution to the total flux is about 30 percent. Another factor that theoretically contributes to this deviation from the predicted ratio is incomplete inhibition of active sodium efflux by ouabain. It is of minimal importance quantitatively, however.

2. Ouabain-Sensitive Sodium Exchange

In slices recovering from lowered temperature and equilibrated at 30° C, the removal of potassium suppresses net sodium transport and reduces sodium efflux by about 40 percent (VAN ROSSUM, 1966b). Under these conditions, with no change in sodium content the addition of cardiac glycosides results in a further decrease of the residual efflux. VAN ROSSUM (1966a) suggested that this ouabain-sensitive sodium efflux, which persists in potassium-free solutions, is not explained by an exchange of sodium for traces of potassium leaking from the cells, since it is incapable of achieving net sodium flux. Accordingly, it has been proposed that a small component of active sodium extrusion is independent of inward potassium transport (VAN ROSSUM, 1966a). This transport system could represent another function of the sodium pump, which could exchange sodium for sodium instead of potassium for sodium, without expenditure of energy. In the presence of potassium, liver cell membranes do not exhibit such an exchange, since ouabain under these conditions does not alter sodium influx (CLARET et al., 1973). However, removal of potassium might activate this mechanism, which is not normally operative, as shown in other cells (GARRAHAN and GLYNN, 1967; KEYNES and STEINHARDT, 1968; BAKER et al., 1969).

3. Potassium-Facilitated Diffusion

In rat liver, a high concentration of external potassium induces marked acceleration of potassium fluxes. This type of potassium transport presents the following features (MAZET et al., 1974).

i) It occurs without any change in internal potassium concentration and does not result from membrane depolarization.

ii) It is selective. It occurs with no concomitant changes in sodium and chloride permeabilities, suggesting that potassium ions cross the membrane through a specific channel.

iii) It obeys the independence principle (USSING, 1949).

iv) It requires 10 minutes to reach maximum activity and exhibits great sensitivity to temperature and to metabolic inhibitors.

These properties have been explained by a large increase in potassium permeability selectively induced by potassium. It is assumed that potassium ions control the membrane concentration of a passive carrier responsible for potassium transport. However, great sensitivity to temperature and to metabolic inhibitors suggests that this potassium-dependent permeability is not a process limited to the membrane but also relates to cell metabolism.

Much longer periods of equilibration (from 60 to 300 min) in high-potassium solutions induce a net loss of cell potassium from isolated cells or perfused rat liver (FRIMMER and KROKER, 1973). This net flux is drastically accelerated if extracellular potassium concentration is increased above the intracellular level, and is completely suppressed by ouabain. The precise interpretation of this outward transport of potassium, which apparently occurs against a concentration gradient and is sensitive to glycosides, is not as yet completely clear.

IV. Factors Affecting Sodium, Potassium, and Chloride Permeabilities

A considerable amount of data has accumulated concerning factors that alter the ionic permeability of the liver cell membrane, especially hormones. It is beyond the scope of the present review to summarize all of these studies, and the discussion will therefore be restricted to the effects of ions on permeability. Reviews of hormonal effects are, at present, available (FRIEDMANN, 1974; RASMUSSEN et al., 1975; PETERSEN, 1976).

1. Effect of Calcium on Permeability

a) Presumed Effect of Calcium on Permeability of Junctional Membrane

If directly incubated in electrolyte solution at 38° C, liver slices spontaneously lose about 90 percent of their potassium and gain sodium, chloride, and water during the initial 10–20 minutes, then recover to nearly baseline *in-vivo* concentrations at 60–90 min (FLINK et al., 1950; HUMPHREY, 1961; KLEINZELLER and KNOTKOVA, 1964; HAYLETT and JENKINSON, 1972a; MCLAUGHLIN, 1973; MACKNIGHT et al., 1974). As suggested by HAYLETT and JENKINSON, these transitory biphasic movements of ions may be related to the existence of the numerous gap-junctions connecting liver cells. Normally insulated from external calcium, these junctions exhibit a high permeability to ions, which can diffuse freely from one hepatocyte to the next (PENN, 1966; LOEWENSTEIN and KANNO, 1967). As a result, cutting liver slices presumably alters not only the ionic composition of injured cells on each surface, but also that of intact deeper cells, and subsequently, as incubation proceeds, parenchymal cells will progressively seal off on contact with external calcium, as observed in salivary glands (LOEWENSTEIN et al., 1967; OLEIVEIRA-CASTRO and LOEWENSTEIN, 1971). This hypothesis is supported by the following observations:

i) From the detailed kinetic analysis performed by MCLAUGHLIN (1973), it can be calculated that both the initial loss of potassium and gain of sodium occurred at approximately the same rate of 16 mmol · kg^{-1} wet weight · min^{-1}. This figure is about 15 times greater than the magnitude of fluxes as measured at steady-state by radioactive tracers (VAN ROSSUM, 1966a; HAYLETT and JENKINSON, 1972a; CLARET and MAZET, 1972), suggesting that these rapid movements of ions do not result from temporary pump arrest.

ii) Isolated cells exhibit similar transitory changes in ionic concentrations KOLB and ADAM, 1976) while no net ion movements occur in intact tissue, such as perfused liver (D'SILVA and NEIL, 1954; BRISTOW and KERLY, 1964; CLARET and MAZET, 1972; DAMBACH and FRIEDMANN, 1974).

iii) The initial loss of potassium and gain of sodium are not dependent on external calcium, while recovery is achieved only in the presence of this cation (MCLAUGHLIN, 1973; HOWARD et al., 1973; BARNABEI et al., 1974; BAUR et al., 1975; KOLB and ADAM, 1976). This observation suggests that, in calcium-free solutions, the elevated junctional membrane permeability of damaged cells remains unaltered and prevents the sodium pump from restoring the initial ionic composition (OLEIVEIRA-CASTRO and LOEWENSTEIN, 1971).

b) Effect of External Calcium on Permeability

In addition to its probable effect on the junctional membrane, external calcium is necessary for the maintenance of low membrane permeability. After the drastic changes in ionic concentrations caused by cutting slices or isolating cells have been fully reversed, the removal of calcium induces a decline in potassium content and a corresponding increase in sodium content (MCLAUGHLIN, 1973; HOWARD et al., 1973; KOLB and ADAM, 1976). Both of these changes occur slowly (about 2.1 mmol · kg^{-1} · min^{-1}) and commence without any time-lag. Converse effects are observed when slices previously equilibrated without calcium are placed in calcium solution. As suggested by MCLAUGHLIN, the instantaneous response indicates that calcium acts at the external cell membrane, probably to reduce permeability to potassium and sodium. The observations of JUDAH and MCLEAN (1962), that liver slices lose more potassium in low- than in high-calcium media, probably reflect the same mechanism.

A detailed dose-response study has suggested that K permeability is a sigmoid function of external Ca concentration, which depends on temperature (KOLB and ADAM, 1976). At 37° C, P_K exhibits a sharp decrease at about 10^{-4} M $Ca^{++} \cdot l^{-1}$. This regulation of P_K by external concentration of divalent cations is described by a mechanism of cooperative change of binding site states depending on external Ca and K.

c) *Effect of Internal Calcium on Permeability*

The role of intracellular calcium on membrane permeability cannot be easily studied in hepatocytes. It seems, however, that a high intracellular calcium concentration, as induced by equilibration of slices at 1° C or in low-sodium media (AHMED and JUDAH, 1962; MCLEAN, 1963; JUDAH and AHMED, 1964; VAN ROSSUM, 1970a, c), prevents restoration of intracellular potassium concentration when such slices are subsequently incubated at 38° C in normal media. After this transfer, extrusion of sodium and calcium from cells begins immediately, while potassium uptake is starts about 10 minutes later. There is no clear-cut interpretation of the apparent dissociation of these transport processes. It is clear, however, that the calcium content must be lowered by its own transport mechanism before net potassium uptake begins. This delay of potassium accumulation and the dissociation effect can be prevented by previous lowering of tissue calcium (VAN ROSSUM, 1970a).

As sodium extrusion proceeds without any delay, it is believed that the failure of potassium retention does not result from a transitory inhibition of the sodium pump, but is largely due to increased permeability to potassium selectively induced by an elevated internal calcium concentration (VAN ROSSUM, 1970a). An increase in cytoplasmic calcium could also explain the elevated potassium permeability of liver cell membranes treated with various hormones (see FRIEDMANN, 1974; PETERSEN, 1974).

2. Effect of Foreign Anions on Membrane Permeability

The effect of foreign anions on membrane permeability may depend on animal species. Isethionate and methylsulfate do not change the membrane potential of guinea-pig or rat liver (HAYLETT and JENKINSON, 1972a; DAMBACH and FRIEDMANN, 1974; HELLER and VAN DER KLOOT, 1974; CLARET and CORABOEUF, 1970), whereas methylsulfate induces slight hyperpolarization of the mouse liver cell membrane (GRAF and PETERSEN, 1974). Pyroglutamate causes a slow hyperpolarization of rat liver membrane though it has no effect on guinea-pig liver (HELLER and VAN DER KLOOT, 1974). Other large foreign anions such as benzene sulfonate drastically modify the membrane potential. Measurement of diffusion fluxes indicates that the permeability to potassium, but not to sodium, is increased 2.6 times, which may partly explain the marked and sustained hyperpolarization observed.

3. Effect of Temperature on Permeability

Low temperatures (3–5° C) induce a net loss of potassium by blocking the sodium pump in a variety of isolated liver preparations. However, a lesser reduction of temperature (from 38° C to 25° C) causes a net accumulation of potassium in perfused dog or rat liver (LAMBOTTE et al., 1966; FRIMMER and KREH, 1971b; LAMBOTTE, 1973a). This paradoxical movement of potassium has been attributed to differential effects of temperature on passive and active transports. The permeability to potassium would be more reduced in this range of temperature than would active transport. Accordingly, the (Na-K)-activated ATPase from isolated plasma membranes, or from homogenates of rat liver exhibits a rather low sensitivity to temperature changes in the 30–40° C range (EMMELOT and BOS, 1968; BAKKEREN and BONTING, 1968).

E. Active Transport of Sodium and Potassium

When equilibrated under a variety of metabolically unfavorable conditions, liver slices gain sodium and lose potassium (FLINK et al., 1950; LEAF, 1956; PARSONS and VAN ROSSUM, 1962). With subsequent incubation under metabolically favorable conditions, these cells are capable of rapid extrusion of sodium and accumulation of potassium against their respective electrochemical gradients (MCLEAN, 1963; ELSHOVE and VAN ROSSUM, 1963; CASCARANO and SEIDMAN, 1966). In most experiments, the final content of potassium is 90 percent of the original value, and it can be maintained at this level for hours. Selective inhibitors of the sodium-potassium transport, such as cardiac glycosides, suppress the ability of liver cells to reverse these changes, suggesting that these movements of electrolytes are related to the sodium-potassium pump (ELSHOVE and VAN ROSSUM, 1963; JUDAH and AHMED, 1963; KLEINZELLER and KNOTKOVA, 1964).

I. Ouabain-Sensitive Fluxes

Under physiological conditions, the net active transport of sodium and potassium by liver slices is almost completely inhibited by cardiac glycosides (ELSHOVE and VAN ROSSUM, 1963, Fig. 1; KLEINZELLER and KNOTKOVA, 1964). However, very high concentrations must be used to inhibit the transport. When present at a concentration of 10^{-5} g · ml^{-1}, strophanthin-K has no effect on net movement of sodium and potassium. Maximal inhibition requires 10^{-3} g · ml^{-1} (Fig. 1). This result is in good agreement with the observation that (Na-K)-activated ATPase of rat liver plasma membrane has a very low affinity for cardiac glycosides (BAKKEREN and BONTING, 1968). The inhibition may be reversed by increasing potassium concentration (JUDAH and AHMED, 1964).

Ouabain at these high concentrations causes a reversible inhibition of sodium efflux from rat liver slices (VAN ROSSUM, 1966a and b). Kinetic analyses of the effect obtained in perfused liver show maximum inhibition after only 5 min (CLARET et al., 1973). The mean value of ouabain-sensitive sodium efflux is about 7.3 pmol · cm^{-2} s^{-1}, which represents 72 percent of the total sodium efflux. These effects occur without any significant alteration of sodium influx. Ouabain does not completely block sodium-potassium exchange. Probably about 16 percent of the "active flux" is resistant to ouabain at whatever concentration is used.

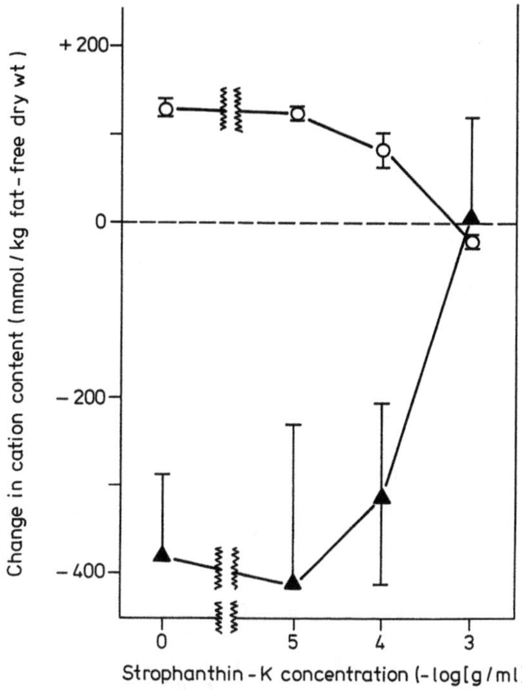

Fig. 1. The effect of strophanthin-K on net cation movements in rat liver slices recovering from cooled media. Each point represents the mean ± SEM of 3–6 observations. ▲ sodium; ○ potassium. (From ELSHOVE and VAN ROSSUM, 1963)

II. Coupling Between Sodium and Potassium Movements

The active movements of sodium and potassium are linked in the sense that net outward sodium transport occurs only if external potassium is transported inward. Such a coupling was first suggested by FLINK et al. (1950). Since then, a considerable amount of evidence has accumulated to confirm this hypothesis. The most important observation is that external potassium specifically activates net extrusion of sodium from slices recovering from cooled media, and concurrently augments the unidirectional sodium efflux (JUDAH and AHMED, 1963; ELSHOVE and VAN ROSSUM, 1963; VAN ROSSUM, 1966b).

The increment of sodium efflux generated by external potassium is shown in Fig. 2. Clearly, experimental points deviate at low potassium concentrations from a theoretical rectangular hyperbola. At zero -K, sodium efflux is reduced by 46 percent, which is significantly lower than the inhibition observed with ouabain. The incomplete inhibitory effect of potassium-free solutions does not contradict evidence of coupling between sodium and potassium transport mechanisms. This observation and the obvious sigmoid shape of potassium-induced sodium efflux may be explained by the following alternative hypotheses, which are not necessarily mutually exclusive:

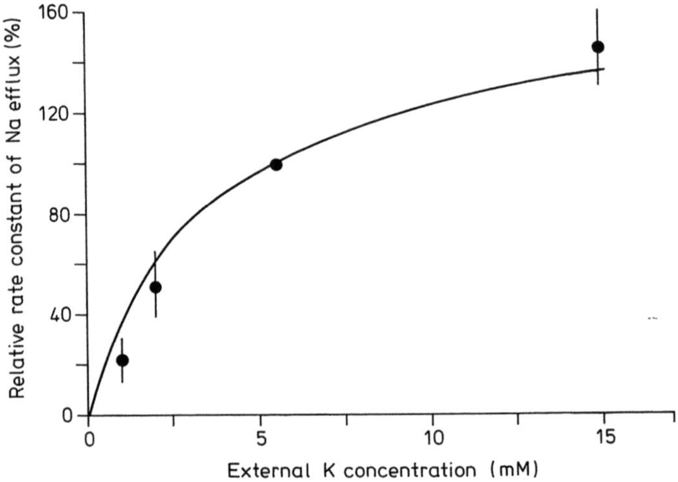

Fig. 2. Dependence of the relative rate constant of sodium efflux on external potassium concentration. The curve is plotted according to the equation of a rectangular hyperbola:

$$B = B_{max}\,[K]\;/\;(K_m + [K])$$

where B is the relative rate constant of sodium efflux, B_{max} is the maximal rate constant. The constant K_m corresponds to the value of [K] at which the binding sites are half-filled with potassium. B_{max} was taken as 160% and $K_m = 3.0$ mM. At low potassium concentration, experimental points seem to deviate from the theoretical curve. Each point represents the mean ± S.D. of an observation. (From CLARET et al., 1973)

i) A residual activity of the sodium pump as a result of incomplete removal of K from the interstitium. This hypothesis is supported by calculations indicating external potassium concentrations of 0.3–0.6 mM at the pumping sites in liver perfused with potassium-free solutions (CLARET et al., 1973).

ii) An ouabain-sensitive sodium exchange induced by potassium removal, as first suggested by VAN ROSSUM (1966a).

iii) A multisite carrier for ion transport, having more than one activation site sensitive to potassium. This assumption seems to fit with observations on the stoichiometry of the sodium-potassium pump.

III. Stoichiometry of the Sodium-Potassium Pump

1. Coupling Ratio Between Sodium and Potassium Fluxes

Questions about the coupling ratio, r (sodium-to-potassium flux ratio), of the sodium pump concern its normal value and whether it varies under different experimental conditions. The earliest studies suggested that the ratio was 1.0, on the basis of the observation that the ratio of net sodium lost to potassium gained in slices recovering from cooled media attained a constant value close to unity after the first 15 minutes of incubation (McLEAN, 1963; ELSHOVE and VAN ROSSUM, 1963). Under steady-state conditions, however, ouabain-sensitive sodium efflux is systematically greater than ouabain-sensitive potassium influx. The calculated ratio is 1.6, suggesting that three sodium ions are extruded for each two potassium ions pumped into a cell during one cycle of transport (CLARET et al., 1973). These results are not contradictory, but suggest that coupling between sodium and potassium transports is not rigid, depending rather upon conditions imposed on the pump. The following experimental data support this point of view:

i) As noted above, when external potassium is removed from incubating media, the pump may change its function and, instead of exchanging sodium for potassium, exchange sodium for sodium.

ii) A coupling ratio of 2.7 has been proposed to explain an increased contribution of the electrogenic pump to membrane potential, indirectly induced by external chloride removal (CLARET et al., 1973). Foreign anions like benzene sulfonate induce an increase in potassium permeability and a hyperpolarization of the membrane. In the steady state, the net potassium efflux is little changed while the net sodium influx is greatly increased. Under these conditions, it seems that the steady state can be maintained only if the pump changes its coupling ratio from 1.5 to 2.7.

2. Cation/ATP Ratio

A cation/ATP ratio (number of sodium ions that move across the membrane for each molecule of ATP that is split by the ATPase system) of about 3 has been determined in many tissues, usually on the basis of (Na-K)-activated ATPase (BONTING, 1970). A similar estimate is not feasible in hepatocytes, because the (Na-K)-actived ATPase is not confined exclusively to the sinusoidal portion of the cell membrane. A not inconsiderable fraction of the total enzyme activity found in isolated plasma membrane originates from bile canaliculi (SONG et al., 1969; TODA et al., 1975; BOYER and RENO, 1975). This fraction varies according to the isolation procedure used and probably causes an unpredictable overestimation of the ATPase that is tightly coupled to ionic transport characteristic of sinusoidal cell membrane.

The stoichiometric ratio may alternatively be deduced from the minimum energy required by the pump to extrude one mole of sodium. The equation for the calculation is given in the next chapter. Under steady-state conditions, the electrochemical work done by the sodium-potassium pump (with $r = 1.6$) can be estimated as 2.6 kcal $\cdot$ mol^{-1} of sodium extruded. Assuming that the free energy for hydrolysis of ATP in the cells is 11 kcal (VAN ROSSUM, 1970b), one molecule of ATP is probably sufficient to drive one cycle of the pump, *i. e.* to extrude three sodium ions from the cell. This value is compatible with those determined in many other cells. It is interesting to note that the stoichiometric ratio may remain unchanged in hepatocytes with a modified coupling ratio. In low-chloride solutions (CLARET et al., 1973), the minimum energy needed for extruding one mole of sodium *via* the sodium-potassium pump (with $r = 2.7$) is 2.5 kcal. Again, this value is compatible with a ratio of 3 Na/l ATP. Assuming that this ratio is fixed (GARRAHAN and GARAY, 1973; LINDENMAYER et al., 1974), the variation of the coupling ratio must reflect a change in the number of potassium ions bound to extracellular sites of the transport system. Under the present conditions, a coupling ratio of 2.7 may be best explained by a 3:1 coupling ratio for the majority of pump units and a 3:2 ratio for the remaining minority.

IV. Contribution of Electrogenic Sodium Pump to Membrane Potential

If the coupling ratio *(r)* is greater than one, and only sodium and potassium ions are transported, the pump contributes directly to the potential by generating a current across the cell membrane. Direct evidence has been accumulated in the last few years to show that the sodium pump of liver cells is effectively electrogenic.

The initial observation was the unexpected "ouabain effect" of *in-vivo* rat liver. WILLIAMS et al. (1971a) noted that treatment of the rat with 10 mg kg^{-1} ouabain induces elevation of plasma potassium and hyperpolarization of the liver cell membrane, whereas the Goldman equation predicts depolarization. The effect on membrane potential appeared to be mediated by the rise in plasma potassium induced by ouabain, which inhibits the muscle sodium pump but not, in the dose given, the liver sodium pump. Hyperkalemia induced by nephrectomy or by KCl injection resulted in similar hyperpolarization of the liver cell membranes (WILLIAMS et al., 1971b). Other studies have confirmed the electrogenic nature of the pump *in vitro:* in guinea-pig and mouse liver slices previously equilibrated in potassium-free solutions, the addition of potassium causes immediate hyperpolarization of 5 to 15 mV (HAYLETT and JENKINSON, 1972a; GRAF and PETERSEN, 1974), even if the resulting potassium equilibrium potential is less negative than the membrane potential. This hyperpolarization can be abolished by ouabain.

Under *in-vitro* physiological conditions, the contribution of the pump to membrane potential is small. The net active flux calculated from ouabain-sensitive fluxes is 3.1 pmol $\cdot$ cm^{-2} $\cdot$ s^{-1}, which corresponds to a current of 0.3 μA cm^{-2} (CLARET et al., 1973). Assuming the total membrane current is the sum of potassium, sodium, and chloride currents, the magnitude of the generated potential, which is inversely proportional to membrane conductance (G_m = 52 μmho cm^{-2}, Table 1) is 6 mV. This last calculated value is in reasonable agreement with measured depolarization values (2 to 7 mV) obtained with cardiac glycosides (HAYLETT and JENKINSON, 1972a; BEIGELMAN and THOMAS, 1972; CLARET et al., 1973; PETERSEN, 1974). However, the electrogenic component of the membrane potential may be greatly augmented by:

i) Simultaneous internal sodium loading and external potassium stimulation of the pump, as noted above.

ii) Reduction of the short-circuiting effect of chloride ion on membrane resistance, and changing of the coupling ratio by removal of chloride ions. Under these conditions, the theoretical pump current is increased to 0.70 $\mu A \cdot cm^{-2}$, the membrane conductance is decreased to 31 $\mu mho \cdot cm^{-2}$, and the resulting potential generated by the pump is increased to about 20 mV.

In contrast to these conclusions concerning mouse, rat, or guinea-pig liver, the sodium pump of dog liver does not appear to be electrogenic (LAMBOTTE, 1973a).

V. Energy Requirement for Sodium-Potassium Transport

1. Measured Energy Requirement

As the term implies, active transport of cation requires metabolic energy, which is directly supplied by ATP to the pump system. The origin and amount of energy necessary for the transport of ions has been extensively studied in liver. Results appear to differ somewhat according to the type of preparation used, but all show that sodium transport is quite resistant to energy limitation and that ATP content must be substantially decreased before it is altered.

In slices, most data indicate that energy is provided by oxidative phosphorylation, active cation transport being drastically reduced if slices are directly incubated anaerobically or with various respiratory inhibitors (FLINK et al., 1950; ELSHOVE and VAN ROSSUM, 1963; SEIDMAN and CASCARANO, 1966; VAN ROSSUM, 1972; MACKNIGHT, 1974). In a detailed study, VAN ROSSUM (1972) determined transport activity at different levels of mitochondrial respiration induced by varying the concentrations of diverse respiratory inhibitors. Results demonstrated that the rate of mitochondrial respiration just sufficient to maintain full transport activity is about 500 $\mu l\ O_2\ g^{-1} \cdot h^{-1}$ [1]. As stressed by VAN ROSSUM, this value is certainly an overestimation of the respiration tightly coupled to cation transport, since when respiration is partially inhibited, other processes with equal or higher affinity may efficiently compete with the (Na-K)-ATPase system for the limited supply of ATP.

The oxygen uptake required for ionic transport has also been estimated in slices by assuming that respiration is dependent on the pump activity. This value is not easy to define in liver, because the reduction of respiration observed when active transport is inhibited varies somewhat with experimental conditions. In chilled slices initially loaded with sodium and then incubated at 30° C, ouabain or potassium-free solutions lower the rate of respiration (ELSHOVE and VAN ROSSUM, 1963; KLEINZELLER and KNOTKOVA, 1964). The respiration tightly coupled to cation transport is about 200–250 $\mu l \cdot g^{-1} \cdot h^{-1}$ (about 10% inhibition) (VAN ROSSUM, 1970b, 1972). Similar figures are obtained from slices equili-

1 The original results have been recalculated. They are expressed as µl per g wet weight per hour, the tacit assumption being that 1 kg of liver contains 200 g of protein.

brated in calcium-free solutions, a situation which leads also to high sodium and low potassium within the cells. Ouabain causes an immediate decrease in oxygen consumption of about 200 μl · g^{-1} · h^{-1} (McLAUGHLIN, 1973). However, if slices are not preloaded with sodium, but are directly incubated in Ringer solution at 38°, ouabain has no significant immediate effect (McLAUGHLIN, 1973; AHMED and JUDAH, 1962). There is only a slow steady decline in respiration, which is attributable to the indirect effect of the changing intracellular ion composition (VAN ROSSUM, 1970a and b).

On the other hand, in perfused rat liver there is no clear evidence of cation transport dependence on oxygen uptake. In this preparation, the replacement of oxygen with nitrogen for 30 minutes, or the addition of 2–4 DNP does not reduce potassium transport activity, suggesting that perfused rat liver is much more resistant to anoxia than slices (FRIMMER and KREH, 1971a). This effect of anoxia might also depend on the species of animal, since in dog liver, unlike rat liver, it induces a net release of K (LAMBOTTE, 1970).

These variable data do not contradict the experimental evidence for coupling between respiration and cation transport. Possible contradictions can be resolved by the following observations:

i) The conditions imposed upon the system of ion transport define the energy requirement. In slices directly incubated at 38° C pump activity is probably weak, so that the fraction of respiration coupled to cation transport may be too low to be readily detected (steady state). Equilibration of slices at 1° C or in Ca-free solutions greatly increases internal sodium concentration and, consequently, the load on the pump system (non-steady state). Pump activity is proportionally augmented, with an equivalent increase of oxygen consumption (McLAUGHLIN, 1973). Under these conditions, the fall in respiration following addition of ouabain or removal of potassium represents the maximum estimated oxygen required for ion transport.

ii) Cell metabolism may differ according to the type of preparation used. The rate of anaerobic glycolysis in perfused rat liver is much higher than that in slices (WOODS and KREBS, 1971). In anoxic perfused liver, this process could effectively compensate for the decrease in ATP production by oxidative phosphorylation and provide sufficient energy to maintain cation transport (VAN ROSSUM, 1972). In slices, the inability to sustain anoxic cation transport might be due to insufficient anaerobic glycolysis. This point of view is supported by the fact that stimulation of glycolysis in anoxic slices by incubation in pyruvate, or preincubation in oxygen, stimulates cation transport (CASCARANO and SEIDMAN, 1965; SEIDMAN and CASCARANO, 1966; MACKNIGHT et al., 1974).

2. Theoretical Energy Requirement

A quantitative estimate of the energy required to support the transport mechanism may be obtained by calculating the work done by the pump under steady-state and non-steady-state conditions and comparing this with the energy requirements measured above.

Under steady-state conditions, the electrochemical work (W) performed by the electrogenic sodium-potassium pump to extrude one mole of sodium can be calculated from the following equation (CHRISTINAZ and SCHATZMAN, 1972):

$$W = F\{(E_{Na} - E_m) + 1/r(E_m - E_K)\} \quad (1)$$

where r is the coupling ratio of the pump, and F the Faraday constant. With values of membrane and equilibrium potentials measured in isolated preparations ($E_m = -35$ mV; $E_K = -90$ mV; $E_{Na} = +30$ mV) W equals 2.3 kcal. Since ouabain-sensitive sodium flux is 74 μm · g^{-1} · h^{-1} (CLARET et al., 1973), the electrochemical work of the pump is estimated to be 0.17 cal · g^{-1} · h^{-1}. Assuming that all ATP is supplied by oxidative phosphorylations and that the system is about 50 percent efficient in conversion of energy to ion transport (VAN ROSSUM, 1972), this work requires 120 μl oxygen · g^{-1} · h^{-1}.

Under non-steady-state conditions, the energy required for net observed sodium extrusion and potassium uptake can be calculated from an expression equivalent to Eq. (1). Such a calculation has been made for slices recovering from incubation in cold media (VAN ROSSUM, 1970b, 1972). The initial rate of net sodium extrusion during the first 10 min at 38° C required about 0.70 cal · g^{-1} · h^{-1}, corresponding to oxygen consumption of 470 μl · g^{-1} · h^{-1}. Clearly, this calculated value is in the same order of magnitude as the amount of energy estimated from ouabain-sensitive respiration (200–250 μl · g^{-1} · h^{-1}), or from titration with respiratory inhibitors (500 μl · g^{-1} · h^{-1}), measured in sodium-loaded slices. As suggested by VAN ROSSUM (1972), the mean of these values (300 μl · g^{-1} · h^{-1}) may represent a reasonable approximation of the energy requirement for maximum transport activity.

The rate of oxygen uptake by *in-vivo* rat liver has been estimated to be about 6–8 ml · g^{-1} · h^{-1} (FOLKE, 1974). The oxygen consumption of liver perfused with blood or artificial media, or of slices equilibrated in electrolyte solution, ranges from 2.0 to 6 ml · g^{-1} · h^{-1} (D'SILVA and NEIL, 1954; VAN ROSSUM, 1970b; SEGLEN, 1972; see also FOLKE, 1974). Similar values have been reported by FOLKE and SESTOFT (1977). Thus, cation transport should require about 10 percent of cell metabolic energy of isolated tissue when sodium-potassium pump activity is maximal, and only about 2–4 percent under steady-state conditions. These figures clearly indicate that the sodium-potassium pump may continue to work at maximum efficiency *in vitro* without making undue demands on cell metabolism, whatever the conditions. They also emphasize the requirement that sodium transport be maximal to demonstrate readily the contribution of the pump system to liver respiration. Failure to do this may partially explain apparent discrepancies in data reported in the present review.

F. Regulation of Cell Volume

It is generally admitted that cell volume is determined by the balance between passive movements and active transport of ions. If the sodium pump is inhibited, ionic concentrations tend to shift towards the Donnan equilibrium with an associated net influx of electrolytes and water, and a net efflux of potassium. In liver, this hypothesis has been substantiated by experimental data: specific inhibitors of the sodium pump (potassium-free solutions, ouabain) cause a reversible gain of cell water and electrolytes (KLEINZELLER and KNOTKOVA, 1964; VAN ROSSUM, 1966b, 1970).

Recently, the relationship between cell volume and pump activity has been re-examined experimentally by MACKNIGHT et al. (1974), and the validity of the model has been questioned. Their results negate previous interpretations and can be summarized as follows:

i) Rat or rabbit liver slices did not swell when incubated with ouabain or K-free solution, provided that they had previously been incubated in oxygenated media. Under these conditions, there was an unexpected dissociation between electrolyte and water transport, liver slices gaining sodium and losing potassium without any change of water content.

ii) Without such prior incubation, freshly cut liver slices exposed directly to oxygenated medium containing ouabain swelled, gained NaCl, and lost potassium as observed in earlier studies.

iii) Slices in which only aerobic metabolism was inhibited (N_2 replaced O_2) swelled much less than those in which anaerobic glycolysis was also blocked (iodoacetamide + N_2).

These results demonstrate clearly that liver possesses a mechanism for cell volume regulation, which is independent of ouabain and directly dependent on anaerobic metabolism. This hypothesis is further supported by the observation that preincubation in an oxygenated medium, which prevents cell swelling, stimulates liver glycolysis (WOODS and KREBS, 1971).

The ability of the rat liver cell to control its volume appears thus, like that of the kidney cell (MACKNIGHT & LEAF, 1977) to consist of two energy dependent components. The former appears to depend on the coupled Na-K transport. The precise nature of the latter remains unknown. It has been suggested that it depends on the secretion of Na and Cl into vesicles which release their contents into the bile canaliculi by an oligomycin-sensitive mechanism (RUSSO et al., 1977).

G. Calcium Transport

The total rat liver calcium concentration ranges from 0.5–2.5 mmol · kg^{-1} wet weight (WALLACH et al., 1966; VAN ROSSUM, 1970c; CHAMBAULT et al., 1974; CLARET et al., 1975). The concentration of cytoplasmic ionized calcium is unknown but it seems likely that hepatocytes, as the few large cells in which direct measurement has proved possible, maintain calcium activity lower than 10^{-6} M (CHAMBAULT et al., 1974; RASMUSSEN et al., 1976). Indeed, experimental data demonstrated intracellular calcium to be present in all subcellular fractions (THIERS and VALLÉE, 1957; CARAFOLI, 1967). These subcellular components reversibly bind or actively transport calcium ions, equally well in the intact cell or when isolated. They include endoplasmic reticulum, plasma membrane (Rossi et al., 1966; Chambault et al., 1974; Garnett and Kemp, 1975; Moore et al., 1975), and mitochondria (Carafoli, 1967; Lehninger, 1970; Kleineke and Stratman, 1974; Dubinsky and Cockrell, 1975). Chemical and tracer kinetic analyses reveal the presence of these exchangeable calcium pools (Wallach et al., 1966; Van Rossum, 1970c; Claret et al., 1975; Van Rossum et al., 1976; Claret-Berthon et al., 1977). In experiments on calcium efflux or influx, it is possible to detect three exponential components (Claret-Berthon, 1977). The rapidly exchangeable Ca and most of the slowly exchangeable Ca are located at the external surface of the cell membrane and constitute about 60% of the total exchangeable tissue Ca. The remaining compartment could represent the cytoplasmic Ca pool limited by the plasma membrane and could include the mitochondrial exchangeable Ca. The results are in agreement with the hypothesis that mitochondria are the major buffers of rapid changes in ionized ca while transport across the plasma membrane may contribute to long term regulation of cellular Ca.

If ionized calcium were distributed by simple diffusion across the cell membrane, there would presumably be intracellular accumulation. For a membrane potential of -50 mV and an external calcium concentration of 2.5 mM, the expected intracellular calcium concentration would be about 40 mM. The discrepancy between this value and the assumed intracellular ionized calcium activity lower than 10^{-6} M implies that calcium enters the cell passively and is extruded by an energy-dependent system. Wallach et al. (1966) have shown that influx is not explained by simple diffusion but requires a complex transport system involving a specific membrane carrier. The rate of calcium influx is not closely linked to a metabolic source of energy. Conversely, the extrusion of calcium does depend upon the metabolic state of the tissue. Under a variety of unfavorable conditions, calcium ions accumulate within the cells (Dawkins et al., 1959; Judah and Ahmed, 1963; Wallach et al., 1966; Van Rossum, 1970a), but can be subsequently extruded when normal conditions are restored. Calcium extrusion is maintained in the absence of sodium gradients across the plasma membrane, but is totally inhibited by cyanide (Van Rossum, 1970c), indicating that energy for this process is not provided by the sodium, electrochemical gradient but is directly derived from hydrolysis of ATP.

This type of calcium transport should be mediated by membrane-bound ATPase activity sensitive to changes in free calcium concentration. However, unlike the ATPase activities of plasma membranes isolated from mouse liver (Garnett and Kemp, 1975), the Mg-ATPase activity of rat liver plasma membranes isolated according to Neville's procedure (Chambaut et al., 1974) or with an isotonic medium technique (unpublished results) was not stimulated by varying Ca concentrations in the incubation medium. These results led Chambaut et al. (1974) to conclude that Ca extruding activity is mediated by a mechanism not requiring ATP hydrolysis at the plasma membrane level, and Van Rossum et al. (1976) to suggest that the Ca transport system exists in a region of endoplasmic reticulum communicating with the extracellular fluid. This apparent contradiction may be also explained by the fact that the specific ATPase activity linked to a Ca pump could be masked by the basal activity of the Mg-ATPase which is very high in rat liver (CLARET-BERTHON et al., 1977). Indeed, the Ca efflux across the plasma membrane of rat liver cells should require only 0.2 to 0.4 μmole ATP h^{-1} mg^{-1} proteins which constitutes a very small fraction (less than 1%) of the basal Mg-ATPase activity found in the plasma membranes isolated from rat liver.

Acknowledgements

I thank *P. M. Beigelman, D. H. Jenkinson, L. Lambotte,* and *G. D. V. van Rossum* for their criticism of the manuscript.

References

AHMED, K., JUDAH, J. D.: Biochim. biophys. Acta **57**, 245 (1962).
BAKKEREN, J. A. J. M., BONTING, S. L.: Biochim. biophys. Acta **150**, 460 (1968).
BAKER, P. F., BLAUSTEIN, M. P., KEYNES, R. D., MANIL, J., SHAW, J. I., STEINHARDT, R. A.: J. Physiol. **200**, 459 (1969).
BARNABEI, O., LEGHISSA, G., TOMASI, V.: Biochim. biophys. Acta **362**, 316 (1974).
BAUR, H., KASPEREK, S., PFAFF, E.: Hoppe-Seyler's Z. Physiol. Chem. **356**, 827 (1975).
BEIGELMAN, P. M., CORABOEUF, E., GUILBAULT, P.: Life Sci. **3**, 1343 (1964).
BEIGELMAN, P. M., SCHLOSSER, G. H.: Biochem. Med. **3**, 73 (1969).
BEIGELMAN, P. M., THOMAS, L. J.: J. Membrane Biol. **8**, 181 (1972).
BIEDERMAN, M.: Acta biol. med. germ. **21**, 827 (1968).
BONTING, S. L.: In: Membrane Ion Transport. (E. E. Bittar, Ed.). London: Wiley-Interscience 1970, p. 257.
BOYER, J. L., RENO, D.: Biochim. biophys. Acta **401**, 59 (1975).
BRISTOW, D. A., KERLY, M.: J. Physiol. **170**, 318 (1964).
CAILLE, J. P., SCHÂNNE, O. F.: Canad. J. Physiol. Pharmacol. **50**, 416 (1972).
CASCARANO, J., SEIDMAN, I.: Biochim. biophys. Acta **100**, 301 (1965).
CARAFOLI, E.: J. gen. Physiol. **50**, 1849 (1967).
CHAMBAULT, A. M., LERAY-PECKER, F., FELDMAN, G., HANOUNE, J.: J. gen. Physiol. **64**, 104 (1974).
CHRISTINAZ, P., SCHATZMANN, H. J.: J. Physiol. **224**, 391 (1972).
CLARET, M., CORABOEUF, E., EHRHART, J. C.: C. R. Soc. Biol. **160**, 476 (1966).
CLARET, M., CORABOEUF, E.: J. Physiol. **210**, 137 (1970).
CLARET, M., MAZET, J.-L.: J. Physiol. **223**, 279 (1972).
CLARET, B., CLARET, M., MAZET, J.-L.: J. Physiol. **230**, 87 (1973).
CLARET, B., CLARET, M., MAZET, J.-L., GUILLON, G.: C. R. Acad. Sci. Paris **281**, 1855 (1975).
CLARET-BERTHON, B., CLARET, M., MAZET, J. L.: J. Physiol. **272**, 529 (1977).
DAMBACH, G., FRIEDMANN, N.: Biochim. biophys. Acta **367**, 366 (1974).
DAWKINS, M. J. R., JUDAH, J. D., REES, K. R.: J. Pathol. Bact. **77**, 257 (1959).
D'SILVA, J. L., NEIL, M. W.: J. Physiol. **124**, 515 (1954).
DUBINSKY, W. P., COKRELL, R. S.: FEBS Letters **59**, 39 (1975).
ELSHOVE, A., VAN ROSSUM, G. D. V.: J. Physiol. **168**, 531 (1963).
EMMELOT, P., BOS, C. J.: Biochim. biophys. Acta **150**, 354 (1968).
FLINK, E. B., HASTINGS, A. B., LOWRY, J. K.: Amer. J. Physiol. **163**, 598 (1950).
FOLKE, M.: Proc. Int. Un. physiol. Sci. **9**, 181 (1971).
FOLKE, M.: Acta physiol. scand. **84**, 43A (1972).

FOLKE, M.: In: Regulation of Hepatic Metabolism (Alfred Benzon Symposium VI) (F. Lundquist and N. Tystrup, Eds.). Copenhagen: Munksgaard 1974, p. 596.
FOLKE, M., PALOHEIMO, M.: Acta physiol. scand. **95**, 44A (1975).
FOLKE, M., SESTOFT, L.: J. Physiol. **269**, 407 (1977).
FRIEDMANN, N., DAMBACH, G.: Biochim. biophys. Acta **307**, 399 (1973).
FRIEDMANN, N.: In: Regulation of Hepatic Metabolism (Alfred Benzon Symposium VI) (F. Lundquist and N. Tystrup, Eds.). Copenhagen: Munksgaard 1974, p. 32.
FRIMMER, M., KREH, W.: Naunyn-Schmiedebergs Arch. Pharmak. **270**, 313 (1971a).
FRIMMER, M., KREH, W.: Naunyn-Schmiedebergs Arch. Pharmak. **268**, 119 (1971b).
FRIMMER, M., KROKER, R.: Naunyn-Schmiedebergs Arch. Pharmak. **276**, 99 (1973).
GARNETT, H. M., KEMP, R. B.: Biochim. biophys. Acta **382**, 526 (1975).
GARRAHAN, P. J., GLYNN, I. M.: J. Physiol. **192**, 159 (1967).
GARRAHAN, P. J., GARAY, R. P.: J. Physiol. **231**, 297 (1973).
GOLDMAN, D. E.: J. gen. Physiol. **27**, 37 (1943).
GRAF, J., PETERSEN, O. H.: Proc. roy. Soc. **187**, 363 (1974).
GREEN, R. D., DALE, M. M., HAYLETT, D. G.: Experientia **28**, 1073 (1972).
HAYLETT, D. G., JENKINSON, D. H.: J. Physiol. **225**, 721 (1972a).
HAYLETT, D. G., JENKINSON, D. H.: J. Physiol. **225**, 751 (1972b).
HECKMANN, K. D., PARSONS, D. S.: Biochim. biophys. Acta **36**, 203 (1959).
HELLER, P., VAN DER KLOOT, W.: J. Physiol. **243**, 577 (1974).
HEMS, R., LUND, P., KREBS, H. A.: Biochem. J. **150**, 47 (1975).
HODGKIN, A. L., KATZ, B.: J. Physiol. **108**, 37 (1949).
HOOPER, G., DICK, D. A. T.: J. Gen. Physiol. **67**, 469 (1976).
HOWARD, R. B., LEE, J. C., PESCH, L. A.: J. Cell. Biol. **57**, 642 (1973).
HUMPHREY, E. W.: Cancer Res. **21**, 1566 (1961).
HUMPHREY, E. W.: Surgery (St. Louis) **60**, 411 (1966).
HUMPHREY, E. W., MAENO, T.: Biochim. biophys. Acta **183**, 236 (1969).
JUDAH, J. D., MCLEAN, A. E. M.: Biochem. Pharmac. **11**, 593 (1962).
JUDAH, J. D., AHMED, K.: Biochim. biophys. Acta **71**, 34 (1963).
JUDAH, J. D., AHMED, K.: J. cell. comp. Physiol. **64**, 355 (1964).
KLEINEKE, J., STRATMAN, F. W.: FEBS Letters **43**, 75 (1974).
KLEINZELLER, A., KNOTKOVA, A.: J. Physiol. **175**, 172 (1964).
KEYNES, R. D., STEINHARDT, R. A.: J. Physiol. **198**, 581 (1968).
KOLB, H. A., ADAM, G.: J. Memb. Biol. **26**, 121 (1976).
LAMBOTTE, L., KESYENS, P. J., HAXHE, J. J.: Arch. int. Physiol. Biochim. **74**, 541 (1966).
LAMBOTTE, L.: Europ. Surg. Res. **2**, 241 (1970).
LAMBOTTE, L.: J. Physiol. **232**, 181 (1973a).
LAMBOTTE, L.: Surgery (St. Louis) **73**, 8 (1973b).
LAMBOTTE, L., PONTEGNIE-ISTACE, S.: Arch. int. Physiol. Biochim. **83**, 623 (1975).
LANGENDORF, H. G., SIEBERT, G., KESSELRING, K., HANNOVER, R.: Nature **209**, 1130 (1966).
LEAF, A.: Biochem. J. **62**, 241 (1956).
LEHNINGER, A. E.: Biochem. J. **119**, 129 (1970).
LI, C., MCILWAIN, H.: J. Physiol. **139**, 178 (1957).
LIMBERGER, J.: Z. Krebsforsch. **65**, 590 (1963).
LINDENMAYER, G. E., SCHWARTZ, A., THOMPSON, H. K.: J. Physiol. **236**, 1 (1974).
LOEWENSTEIN, W. R., NAKAS, M., SOCOLAR, S. J.: J. gen. Physiol. **50**, 1865 (1967).
LOEWENSTEIN, W. R., KANNO, Y.: J. Cell Biol. **33**, 225 (1967).
MACKNIGHT, A. D. C., PILGRIM, J. P., ROBINSON, B. A.: J. Physiol. **238**, 279 (1974).
MCLAUGHLIN, C. W.: Biochim. biophys. Acta **323**, 285 (1973).
MCLEAN, A. E. M.: Biochem. J. **87**, 161 (1963).
MAZET, J.-L., CLARET, M., CLARET, B.: J. Membrane Biol. **18**, 335 (1974).
MOORE, L., CHEN, T., KNAPP, H. R., LANDON, E. J.: J. biol. Chem. **250**, 4562 (1975).
MULLINS, L. J., NODA, K.: J. gen. Physiol. **47**, 117 (1963).
OLEIVEIRA-CASTRO, G. M., LOEWENSTEIN, W. R.: J. Membrane Biol. **5**, 51 (1971).
PARSONS, D. S., VAN ROSSUM, G. D. V.: J. Physiol. **164**, 116 (1962).
PENN, R. D.: J. Cell Biol. **29**, 171 (1966).
PETERSEN, O. H.: J. Physiol. **239**, 647 (1974).

PETERSEN, O. H.: Physiol. Rev. **56**, 535 (1976).
RASMUSSEN, H., JENSEN, P., LAKE, W., FRIEDMANN, N., GOODMAN, D. B. P.: Advan. Cyclic nucleotide Res. **5**, 375 (1975).
ROSSI, C. S., CARAFOLI, E., DRAHOTA, Z., LEHNINGER, A. L.: In: Regulation of Metabolic Process in Mitochondria (S. McTager, S. Papa, E. Quagliariello, and E. C. Slater, Eds.). Amsterdam: Elsevier 1966, p. 317.
RUSSO, M. A., VAN ROSSUM, G. D. V., GALEOTTI, T.: J. Memb. Biol. **31**, 267 (1977).
SANUI, H., PACE, N.: J. gen. Physiol. **42**, 1325 (1959).
SCARPA, A., AZZI, A.: Biochim. biophys. Acta **150**, 473 (1968).
SCHANNE, O., CORABOEUF, E.: Nature **210**, 1390 (1966).
SEGLEN, P. O.: Biochim. biophys. Acta **264**, 398 (1972).
SEIDMAN, I., CASCARANO, J.: Amer. J. Physiol. **211**, 1165 (1966).
SOMLYO, A. P., SOMLYO, A. V., FRIEDMANN, N.: Ann. N.Y. Acad. Sci. **185**, 108 (1971).
SONG, L. S., RUBIN, W., RIFKIND, A. B., KAPPAS, A.: J. Cell Biol. **41**, 124 (1969).
STEINHARDT, R. A.: J. Physiol. **200**, 459 (1969).
THIERS, R. E., VALLEE, B. L.: J. biol. Chem. **226**, 911 (1957).
TOIDA, N., TAMAI, T., TAKAGISI, T.: Kyushu J. med. Sci. **9**, 163 (1958).
TODA, G., OKA, H., ODA, T., IKEDA, Y.: Biochim. biophys. Acta **413**, 52 (1975).
USSING, H. H.: Physiol. Rev. **29**, 127 (1949).
VAN ROSSUM, G. D. V.: Biochim. biophys. Acta **122**, 312 (1966a).
VAN ROSSUM, G. D. V.: Biochim. biophys. Acta **122**, 323 (1966b).
VAN ROSSUM, G. D. V.: Nature **225**, 638 (1970a).
VAN ROSSUM, G. D. V.: Biochim. biophys. Acta **205**, 7 (1970b).
VAN ROSSUM, G. D. V.: J. gen. Physiol. **55**, 18 (1970c).
VAN ROSSUM, G. D. V.: Biochem. J. **129**, 427 (1972).
VAN ROSSUM, G. D. V., SMITH, K. P., BEETON, P.: Nature, Lond. **260**, 335 (1976).
WALLACH, S., REIZENSTEIN, D. L., BELLAVIA, J. V.: J. gen. Physiol. **49**, 743 (1966).
WILLIAMS, J. A., WITHROW, C. D., WOODBURY, D. M.: J. Physiol. **212**, 101 (1971a).
WILLIAMS, J. A., WITHROW, C. D., WOODBURY, D. M.: J. Physiol. **212**, 117 (1971b).
WILLIAMS, J. A., WOODBURY, D. M.: J. Physiol. **212**, 85 (1971).
WOODS, H. F., KREBS, H. A.: Biochem. J. **125**, 129 (1971).

Subject Index

Acetate 503
Acetazolamide 329, 330, 332, 335, 373
–, effect on gastric HCl secretion 706
–, effect on large intestine 789
–, effect on renal tubules 329
–, –, effect on distal tubular calcium transport 373
–, effect on small intestine 761, 775
Acetylcholine 605, 610, 644, 821, 831, 839
Acidosis 199, 281, 312, 313, 335, 375, 384
Actinomycin-D 196
Activation energy 28, 523
– –, water permeation 28, 523
– –, –, capillary 28
Activity 48
–, bicarbonate 82, 83, 84
–, –, muscle fiber 83
–, chloride 77
–, –, muscle 77
– coefficient 56, 76
–, intracellular ions 56
–, ions 48
–, potassium 63, 64
–, –, distal tubule 64
–, –, proximal tubule 66
–, sodium 76
–, –, distal tubule 76
–, –, proximal tubule 71
Aldosterone 63
–, distal renal tubule 63, 134
–, –, cell potassium activity 63
–, large intestine 793, 794
–, tubular acidification 350
Alkalosis 281, 312, 384
Amiloride 134, 136, 199
–, effect on cortical collecting tubule 136
–, effect on distal tubular Na^+ and Ca^{++} transport 199
–, effect on distal tubular potassium transport 240, 274
–, effect on distal tubular transepithalial potential difference 135
–, effect on gastric mucosa 709, 710
–, effect on large intestine 783
–, effects on salivary glands 636, 640
Amino acids 150, 158, 429
–, effects on cell potentials in renal tubule 423, 433
–, γ-glutamyl cycle 435
–, genetic transport defects 429
–, sodium dependence of active renal transport 431
–, –, proximal renal tubule 431
–, specificity of renal transport mechanisms 429, 430
–, transport studies in renal tubular membrane vesicles 434, 435
–, proximal renal tubule 150
–, uptake in pancreas 843
–, uptake in salivary glands 659
Ammonia 286, 782
–, large intestine 782, 795
Amphiuma intestine 65
– –, potassium activity 65
Amphiuma urinary bladder 65
– – –, potassium activity 65
Amphotericin-B 899
–, effect on gallbladder epithelium 899
–, effect on large intestine 790
Amylase 569, 603, 814
Angiotensin 653, 794
Antidiuretic hormone (ADH) 203, 450, 457, 517, 794
– –, distal renal tubule 203
– –, isolated cortical collecting tubule 517, 518, 519
Arsenate 395
–, effect on tubular phosphate transport 395
Association-Induction Hypothesis 48
ATP 147, 156, 167, 169, 534, 535, 539, 546, 549, 551, 607, 636, 637, 733, 734, 735, 736, 741, 742, 763, 830
ATP-ase 364, 395, 636, 787, 790, 862, 911, 147, 156, 167, 169, 197, 200, 218
–, Ca^{++}-sensitive 550, 551
–, Cl^--sensitive 553
–, HCO_3^--sensitive 552, 734
–, K^+-sensitive 735, 830
–, Na^+-sensitive 549, 550, 735
–, Na^+-H^+-exchange 828

Barium 706
–, effect on gastric mucosa 702, 706, 722
Benzolamide 329, 332
Bicarbonate 82, 150
–, distribution across tubule cell membranes 87
–, effect on gallbladder transport 862
–, intracellular activity 82, 83, 84
–, large intestine 785, 786, 787, 788
–, proximal renal tubule 150, 173, 174
–, –, cell activity 85

Bicarbonate
–, role in renal tubular acidification 299, 311, 312, 328
Blood-brain barrier 4
– –, permeability 19
Bradykinin 35
Brain 19
–, capillary 19
–, –, facilitated glucose transport 19
–, –, hydraulic conductivity 19
Buffer capacity 84
– –, muscle 84

Cable analysis
– –, gallbladder epithelium 868
– –, Necturus proximal convoluted tubule 115
Calcium 357
–, clearance studies 358
–, complex formation with anions 357
–, distal tubule and collecting duct 370
–, –, factors affecting reabsorption 371
–, –, –, acetazolamide 373
–, –, –, acidosis 375
–, –, –, alkalosis 375
–, –, –, hypercalemia 371
–, –, –, miscellaneous factors 376
–, –, –, parathyroid hormone 372
–, –, –, phosphate deficiency 375
–, –, –, phosphate loading 375
–, –, –, sites of calcium reabsorption 370
–, –, –, thiazide diuretics 372
–, –, –, vitamin D 374
–, glomerulus 358
–, –, variability of ultrafilterable calcium 358
–, –, –, effect of calcium infusion 358, 359
–, interaction between renal sodium and calcium transport 198, 199
–, liver cells 905, 916, 917, 918
–, loop of Henle 366
–, –, ascending limb 367, 368, 369
–, –, pars recta 367
–, –, –, studies of isolated thick ascending limb segments 369
–, –, –, PTH-sensitive adenylcyclase 370
–, measurement of calcium 358
–, –, ^{45}Ca 358
–, –, electron-probe analysis 358
–, –, helium-glow photometry 358
–, micropuncture studies 358
–, protein binding 357
–, proximal tubule 359, 360
–, –, absence of PTH effects 361
–, –, bidirectional fluxes 361
–, –, calcium permeability 361
–, –, complexed calcium 362
–, –, factors affecting reabsorption 362, 363
–, –, mechanism of transport 364
–, –, –, active nature of transport 364
–, –, –, calcium-activated ATP-ase 364
–, –, –, countertransport against sodium 364
–, –, model of cellular phosphate transport 365
–, –, –, rabbit gallbladder 366
–, –, role of mitochondria 365
–, –, microperfusion and microinjection studies 361
–, reabsorption pattern along nephron 357, 359, 360
–, renal transport 357
–, –, transport modulation at post-proximal nephron sites 377
–, ultrafiltrability 357, 358, 359
–, role in gastric HCl-secretion 704
–, role in pancreas
–, role in salivary glands 600, 602, 604
Calcitonin 388, 403
Capillary 1
–, bradykinin 35
–, brain 19
–, –, hydraulic conductivity 19
–, "cell"pathway 19, 20, 27
–, cerebral 27
–, –, hydraulic conductivity 27
–, convective flow 39
–, cylindrical pores 38
–, diffusion 6, 8, 9, 12, 33
–, –, limitation 6
–, –, path length 12
–, dog lung 20
–, endothelium 1
–, –, cell permeability 27
–, –, transport of macromolecules 28, 30, 31, 35
–, fenestrated 2, 4, 28
–, histamine effects 35
–, hydraulic conductivity 2, 4, 18
–, hydrochynamic models 37
–, hydrostatic pressure 36
–, macromolecular sieving 31
–, mammalian intestine 2
–, mammalian muscle 2, 9, 10, 25, 27
–, –, hydraulic conductivity 27
–, –, isovolumetric osmotic transient 25
–, membrane channels 39
–, –, frictional interaction 39
–, membrane solvent drag coefficient 40
–, multiple pathway hypothesis 18, 19
–, osmotic reflection coefficient 10
–, osmotic transients 10, 17
–, parabolic flow profile 39
–, parallel pathway model 27
–, permeability 1, 9, 10, 12, 14, 16, 28, 32
–, –, charged dextrans 33
–, –, dog brain 17
–, –, dog heart 17
–, –, dog paw 32
–, –, equilibration kinetics 17
–, –, human forearm 17
–, –, neutral dextrans 33
–, –, rat cremaster 32
–, –, temperature coefficient 28
–, rabbit placenta 31
–, single capillary measurements 21

–, –, osmotic reflection coefficient 23
–, –, surface measurements 16, 33
–, pore area 12
–, "pore" pathway 9, 19, 20, 30
–, pore radius 20
–, pore theory 5
–, reflection coefficient 5, 9, 10, 13, 18, 19, 20, 23
–, renal glomerulus 2
–, slit half-width 17
–, solvent-solute interactions 29, 37
–, steric restriction 12
–, "third-pathway" transport 30, 35
–, "two-pathway" model 26
–, ultrafiltration 5, 8, 12, 29
–, transport 1
–, vesicles 4, 34, 36
–, visceral 4
–, water-filled channels 12, 13
–, water permeation 28
–, –, activation energy 28
Carbonic acid 323, 325, 329, 334, 347
– –, renal tubule 323, 325, 329, 334, 347
– anhydrase 301, 321, 322, 329, 332, 333, 334, 398, 700
– –, gastric mucosa 700
– –, proximal renal tubule 441, 442, 443
Cardiac glycosides 134, 218, 219
– –, effect on cortical collecting tubule 136
– –, effect on distal tubular potassium transport 256, 278, 279, 280, 281
– –, effect on distal tubular potential difference 134
– –, effect on gastric mucosa 709
– –, effect on large intestine 782
– –, effect on loop of Henle 189
– –, effect on proximal renal tubule 162, 172, 173
– –, effect on renal potassium transport 218, 219
Chloride 77, 150
–, activity in distal rat tubule cells 80, 81
–, activity in muscle fiber 77
–, activity in proximal *Necturus* tubule cells 78, 79, 80
–, transport 80, 145, 149, 161, 166, 188, 191, 197, 341, 503, 704, 785, 858, 899, 903
Concentration polarization 8
Constant field equation 116
Cortical collecting tubule 135
–, active chloride transport 137, 138
–, active hydrogen ion secretion 137
–, active sodium transport 137
–, active transepithelial ionic membrane parameters 138
–, passive transepithelial ionic membrane parameters 137
–, transepithelial conductance 137
–, –, partial ionic conductances 137, 138
–, transepithelial potential difference 136
–, –, effect of amiloride 136
–, –, effect of DOCA pretreatment 136, 138
–, –, effect of perfusion rate 136
–, –, effect of ouabain 136
–, –, effect of pH changes 136
–, –, effect of temperature 136
–, –, effect of triamterene 136
–, –, effect of spironolactone 137
–, single membrane properties 139
Cotransport 146, 155
–, primary active transport 147
–, renal tubule 146, 155
–, secondary active transport 146
Cyclic AMP 397, 647, 761, 775, 791, 792, 812, 813, 814, 837, 863
dbcAMP (dibutyryl adenosine 3′5′-cyclic monophosphate) 812, 814
Cytopempsis 4, 33

Diamox 134, 174, 274, 313, 329, 330, 332, 334, 335, 373, 441, 517, 702
Diffusion 9, 33
–, aqueous channels 9
–, capillary 6, 8, 9, 33
2.4-dinitrophenol 707, 782
Distal convoluted renal tubule 130, 190
–, active ionic membrane parameters of single cell barriers 135
–, amphibian and reptilian tubules 131, 132
–, antidiuretic hormone 203
–, cell heterogeneity in distal renal tubule 191
–, effect of mineralocorticoids 196
–, –, actinomycin-D 196
–, effect of renal vein occlusion 196
–, effect of sodium load on transport rate 194, 195, 196
–, effect of ureteral clamping 196
–, electrophysiology 193
–, –, peritubulare membrane potential 193
–, –, transepithelial potential difference 192, 193, 194
–, interaction between sodium and calcium 198, 199
–, –, adrenocortical insufficiency 199
–, –, amiloride 199
–, –, metabolic acidosis 199
–, –, parathyroid hormone 199
–, –, peritubular calcium-sodium exchange 200
–, –, thiazide diuretics 199
–, –, triamterene 199
–, –, Vitamin D 199
–, intercellular ionic shunt 202
–, –, lanthanum 202
–, limiting sodium concentration gradient 192
–, luminal transport mechanism 201
–, –, active potassium reabsorption 202

Distal, luminal
–, –, active chloride transport 201
–, macula densa 190
–, Na-K-activated ATPase 197, 200
–, net chloride movement 196
–, net salt movement 191
–, passive ionic membrane paramters of single barriers 135
–, passive transepithelial ionic membrane parameters 132, 133, 134, 135
–, peritubular transport mechanism 200
–, –, electrogenic sodium extrusion 200
–, –, Na-K-activated ATPase 200
–, transepithelial electrical conductance 132
–, transepithelial potential difference 131
–, –, change of potential with tubular length 131
–, –, effects of adrenalectomy 134
–, –, effects of aldosterone 134
–, –, effects of amiloride 134
–, –, effects of changes in diet 134
–, –, effects of chloride 134
–, –, effects of osmolality changes 134
–, –, effects of ouabain 134
–, –, effects of potassium 133
–, –, effects of sodium 133
–, –, mammalian tubules 131
–, water permeability 198
–, water transport 198
DOCA (desoxycorticosterone acetate) 136, 138, 196, 241

Electrical conductance 98
– –, renal tubule 98
Electrochemical driving forces 90, 215
– – –, across proximal tubule cell membranes 90, 91, 92
– electrical potential 49, 88, 90, 91, 92
– –, proximal tubule 88
– –, –, luminal cell membrane 88
– –, –, peritubular cell membrane 88, 89
Electrochemistry 47, 49, 51, 57
–, nephron 47, 49, 51, 92
Electrodes 50
–, ion-exchanger-filled 50, 52, 53, 69
–, –, double-barreled 50, 53
–, –, response time 54
–, –, single-barreled 52
–, ion-selective 50
–, potassium sensitive 60
–, –, selectivity to ions 60
–, sodium sensitive 69
–, –, glass 69
Electrophysiology 97
–, electrode techniques 98
–, –, electrolyte-filled microelectrodes 98
–, –, metal microelectrodes 98
–, –, microelectrodes 98
–, –, tip potentials 99
–, isolated kidney tubule 99
–, kidney 97
Endothelium 1, 35
–, capillary 1
–, cell pathway 36
–, changes in injury 35
–, cell membranes 35
–, –, negatively charged surface 35
–, interendothelial cell gaps 35
–, –, morphology 1
–, intercellular cement 4
–, intercellular pathway 36
–, transport of macromolecules 28
Ethacrynic acid 167
– –, proximal renal tubular sodium transport 167
– –, thick ascending limb of Henle's loop 189
Exchange diffusion 704
– –, gastric mucosa chloride transport 704, 706, 707
– –, of chloride in large intestine 787, 788

Fenestrae 1, 2, 4, 28
–, transcellular capillary 1
Ferry-Faxén equation 13, 14
Ferry-Happel-Brenner formula 14
Fluid balance 5
Furosemide 189, 274, 371
–, thick ascending limb of Henle's loop 189

Gallbladder 853
–, composition of gallbladder bile 855
–, –, osmotic coefficient 856
–, mechanism of ion transport 863, 864
–, –, active transport pathways in parallel 865
–, –, cell membrane resistance 868
–, –, cell models of ion transport 881
–, –, electrical circuit parameters 876
–, –, electrical equivalent circuit 864, 867, 874, 875
–, –, effect of transepithelial current on potentials and resistances 887, 888
–, –, effect of amphotericin B on potentials and resistances 889
–, –, effect of transepithelial osmotic gradients on potentials and resistances 885, 886
–, –, estimates of cell membrane permeabilities 884
–, –, intracellular ion concentrations 880
–, –, intraepithelial cable analysis 868
–, –, ionic permeabilities of apical cell membrane 877, 878
–, –, paracellular shunt 865, 867
–, –, –, „blister" formation 873
–, –, –, charge of pathway 870
–, –, –, effects of electrical and osmotic gradients 873
–, –, –, effects of pH and polyvalent cations 871
–, –, –, free-solution parallel pathway 870
–, –, –, ion permeabilities 883
–, –, –, mobility of charge sites 870
–, –, –, nitrogenous cations 872

–, –, –, physicochemical characteristics 869
–, –, –, role in ion transport 881, 882
–, –, –, role in passive ion permeation 872
–, –, –, triamino-pyrimidinium (TAP) 871, 891
–, –, properties of epithelial cell membranes 874
–, –, relationship between chloride and sodium transport 880
–, –, tracer analysis of electrolyte transport 879
–, –, transepithelial diffusion potentials 866, 871
–, –, transepithelial electrical resistance 864, 865, 868, 882
–, –, transepithelial ionic permeabilities 866
–, –, transepithelial potential difference 863, 864, 882
–, –, unidirectinal tracer fluxes 858, 861
–, metabolic aspects of transport 862
–, –, ATP-ase 862
–, –, components of oxygen consumption 862
–, –, role of bicarbonate
–, methods of studying salt and water transport 856
–, morphology 854
–, rate of fluid transport 857, 858
–, –, cyclic AMP 863
–, –, ionic dependence 859, 860
–, transport 853
–, –, general transport properties 854, 857
–, –, nonelectrolytes 890
–, –, –, permeation pathways 890, 891
–, water transport 861, 892
–, –, hydraulic conductivity 894
–, –, local osmosis 892
–, –, –, hypertonic interspace mechanisms 893
–, –, –, standing-gradient osmotic flow hypothesis 892
–, –, –, three-compartment hypothesis 892
–, –, –, transport-dependent asymmetries in fluid composition 894
–, –, route of water flow 895, 896
Gastric secretion 693
– –, anion dependence of H^+secretion 702
– –, –, Cl^--requirement for H^+-secretion 702, 703
– –, –, selectivity 703
– –, biochemical basis of HCl secretion 729
– –, –, ATP utilization hypotheses 733
– –, gastric microsomal vesicles 737, 738
– –, –, HCO_3^--stimulated ATPase 734
– –, –, K^+-stimulated ATPase 735, 736, 741, 742
– –, –, K^+-H^+-exchange 737, 742
– –, –, metachromic dye studies 739, 740
– –, –, metabolic requirements 729
– –, –, Na^+-K^+-ATPase 735
– –, –, redox hypotheses 731, 732
– –, –, thermodynamic work requirements 730
– –, black box model of secretion 727, 728
– –, –, components of ion fluxes 727, 728
– –, cation dependence of H^+-secretion 703
– –, –, compartments of K^+ 704
– –, –, removal of K^+ 703, 704
– –, –, role of Ca^{++} 704
– –, –, role of Na^+ 704
– –, chloride transport 704
– –, –, diffusional transport 704
– –, –, effects of acetazolamide 706
– –, –, effects of Ba^{++} 706
– –, –, exchange diffusion 704, 706, 707
– –, –, –, effects of 2.4-dinitrophenol 707
– –, –, –, halide selectivity 707
– –, –, selectivity of active anion transport mechanism 705
– –, –, short-circuit current technique 705
– –, –, unidirectional ion fluxes 705
– –, effect of Ba^{++} on tissue conductance 702
– –, electrophysiological analysis 716
– –, –, conductance pathways 722
– –, –, electrical coupling of epithelium 722, 725
– –, –, electrogenic hypothesis of HCl secretion 716, 717, 718
– –, –, –, equivalent circuit 717
– –, –, electroneutral hypothesis of HCl secretion 719
– –, –, intracellular potential differences 721
– –, –, permeability pathways 722
– –, –, cell membrane permeability 725
– –, –, –, effects of Cl^--variations 726
– –, –, –, effects of K^+-variations 726
– –, –, –, effects of metabolic inhibitors 727
– –, –, transcellular and paracellular conductances 722, 723, 724
– –, –, –, effect of Ba^{++} 722
– –, –, –. K^+ 722
– –, exchange of HCO_3^- for Cl^- 702
– –, general aspects 693
– –, ion transport 698
– –, –, chloride transport 699
– –, hydrogen ion secretion 699
– –, –, acid-base balance of oxyntic cells 700
– –, –, active 704
– –, –, active nature of transport 699
– –, –, acetazolamide 702
– –, –. carbonic anhydrase 700
– –, –, –, residual secretory rates 702

Gastric secretion, hydrogen
– –, –, role of CO_2 700, 701
– –, –, site of H^+-secretion 699
– –, models for H^+ and Cl^--transport 701, 715, 716, 720, 727, 741
– –, morphology of gastric mucosa 694
– –, morphology of oxyntic gland cells 695, 696, 697
– –, –, nonsecreting 696
– –, –, secreting 697
– –, –, –, tubulovesicular system 698
– –, Na^+-absorption 708
– –, –, amiloride effect 709, 710
– –, –, cardiac glycosides 709
– –, –, histamine effect 708
– –, –, pH-sensitivity 709, 710
– –, water transport 710, 712
– –, –, diffusional permeability 711
– –, –, endogenous HCl gradients 714, 715
– –, –, –, double membrane model 714
– –, –, during HCl secretion 714, 715
– –, –, –, endogenous osmotic gradients 715
– –, –, –, hydrostatie pressure 713
– –, –, –, ultrafiltration 713
– –, –, hydraulic conductivity 711
– –, –, unstirred layers 711
Gastrin 793, 831, 839
Glucagon 793
Glomerulo-tubular balance 151
– –, renal tubule 151
Glucose transport 19
– –, brain capillaries 19
Glycodiazine 150, 316, 317, 441, 442, 824
–, proximal renal tubule 150, 441
–, tubular ancidification 316, 317, 321

Histamine 35, 708
Hydraulic conductivity 6, 27
– –, cell membranes 6
– –, cerebral capillaries 27
– –, epithelial 6
– –, isolated renal tubule 483, 486, 487, 501
– –, mammalian skeletal muscle 27
Hydrogen ion transport 299
– – –, bicarbonate-CO_2 buffer system 300
– – –, renal tubule 299, 300, 321, 325, 328

Insulin 403
Intercellular pathways 1
Intracellular ions 47, 92
– –, physical state 47, 92
Ionic electromotive forces 115
– – –, renal tubule 115
– selectivity 109
– –, gallbladder epithelium 877, 878, 883
– –, Necturus proximal tubule 109
Ion-selective microelectrodes 50
Ions 47, 92
–, activity 48
–, bound 48
–, compartmentalized 48
–, exchangeability 48
–, free fraction in cells 48
–, intracellular 47, 51, 92
–, –, activity 47, 56, 92, 93
–, –, physical state 47
–, intracellular binding 48
–, surface adsorption 48
Isolated mammalian renal tubules 473
–, cortical collecting tubule 516
–, collecting duct system 513
–, –, mechanism of ADH action 517, 518, 519
–, –, –, activation energy 523
–, –, –, diffusional permeabilities of lipophilic solutes 520, 521, 522
–, –, pathways for water diffusion 524
–, –, role of parallel diffusion pathways 525
–, –, water and nonelectrolyte permeation 517
–, dimensions of isolated rabbit renal tubule segments 477
–, distal convoluted tubule 513
–, electrical measurements 488
–, –, Donnan voltages 489, 490, 491
–, –, electrical conductance pathways 489
–, –, ionic dilution potentials 492
–, –, liquid junction potentials 489, 490, 491
–, –, proximal straight tubule 491
–, –, transepithelial resistance 493
–, experimental techniques 475
–, –, isolation of renal tubule segments 475
–, –, tubular perfusion 476, 477, 478
–, heterogeneity of proximal tubules 509
–, hydraulic conductivity 483, 486, 487
–, loop of Henle 510
–, –, passive permeability characteristics 511, 512
–, measurement of net volume absorption 481
–, –, crimped end method 482
–, –, collection method 481
–, mechanism of salt and water transport in superficial pars recta 505, 506
–, –, effects of luminal hypotonicity 507, 508
–, –, isotonic fluid absorption 507
–, –, role of anion composition 508
–, perfusing and bathing solutions 479, 480
–, perfusion 473
–, –, availability of all nephron segments 474
–, –, unstirred layer effects 474, 483, 484
–, –, –, solute polarization 485
–, tracer flux measurements 494
–, –, bath-to-lumen fluxes 496
–, –, lumen-to-bath fluxes 495

–, –, tracer permeabilities 496, 497
–, –, transmembrane fluxes 498
–, transport properties of distal nephron segments 514
–, –, diffusional water permeability 514
–, –, electrical resistance 515
–, –, ion permeabilities 515
–, –, net ion fluxes 515
–, –, nonelectrolyte permeabilities 514
–, –, osmotic water permeability 514
–, –, reflection coefficients 515
–, –, transepithelial voltage 514
–, transport properties of isolated nephron segments 498
–, –, cell volume regulation 499
–, –, proximal tubule 499
–, –, –, acetate fluxes 503
–, –, –, active transport processes 502, 503
–, –, –, Cl^--fluxes 503
–, –, –, dissipative transport properties 500
–, –, –, electrical resistance properties 501
–, –, –, glucose fluxes 503
–, –, –, hydraulic conductivity 501
–, –, –, ionic permeabilities 501
–, –, –, Na^+-flux 503
–, –, –, net volume transport 503
–, –, –, nonelectrolyte permeabilities 501
–, –, –, PAH fluxes 503
–, –, –, permeability properties 501
–, –, –, phosphate fluxes 503
–, –, –, reflection coefficients 501
–, –, –, transepithelial voltage 503, 504

Junction potentials 99
– –, biionic potential 99, 107
– –, dilution potential 99
– –, renal tubular electrical potential measurements 99

Kidney 47
–, electrophysiology 97
–, tubule 97
–, –, electrical potential difference 97
–, –, –, amphibian proximal convoluted tubule 105
–, intracellular environment 47, 92

Large intestine 781
– –, acetazolamide 789
– –, amiloride 783
– –, ammonia 782, 795
– –, –, nonionic diffusion 796, 797
– –, –, –, coupled non-ionic diffusion 798, 799
– –, –, –, transport model 799, 800
– –, bicarbonate transport 785, 786, 787, 788
– –, cell transport model 786
– –, chloride-chloride exchange diffusion 787, 788
– –, chloride transport 785, 786, 787
– –, coupling of ion transport 788
– –, electrical parameters 783
– –, electrical resistance 782
– –, electrical potential difference 782
– –, electrolyte fluxes 785
– –, epithelial properties 782
– –, –, nonuniform transport capacity 782
– –, general transport properties 781
– –, –, ammonia 782, 795
– –, –, organic anions 781
– –, –, –, three-compartment model 801, 802
– –, –, oxalate 781
– –, –, volatile fatty acids 782, 795
– –, hormone effects 793
– –, –, aldosterone 793, 794
– –, –, angiotensin 794
– –, –, antidiuretic hormone 794
– –, –, gastrin 793
– –, –, glucagon 793
– –, metabolic inhibitors 782
– –, –, dinitrophenol 782
– –, –, ouabain 782
– –, Na^+-K^+-ATPase 787, 790
– –, organic anion transport 789
– –, osmotic permeability 783
– –, potassium secretion 785, 789, 794
– –, –, aldosterone 790, 793
– –, –, amphotericin B 790
– –, –, glucocorticoids B 790
– –, –, inflammatory bowel disease 790
– –, –, mechanism of secretion 790
– –, –, –, active and passive components 790, 791
– –, –, role of goblet cells 791
– –, –, –, β-adrenergic control 791
– –, –, –, cholinergic control 791
– –, reflection coefficient 784
– –, relative ion permeabilities 784
– –, secretion of fluid and electrolytes 791
– –, –, mechanism 791
– –, –, –, cyclic AMP 791, 792
– –, –, –, stimuli of colonic water and electrolyte secretion 792
– –, –, –, theophylline 792
– –, short-circuit current 782
– –, sodium transport 781, 785, 786
– –, –, mechanism of sodium entry 786, 787
– –, species differences of transport 787
– –, tight epithelium 782, 783
– –, volatile fatty acid transport 789
– –, water permeability 784
– –, –, hypertonic transport 784
Lanthanum 169, 202, 203, 204
–, distal renal tubule 202
–, proximal renal tubule 169
Leaky epithelium 148, 750
– –, gallbladder 148
– –, proximal renal tubule 148

Leaky epithelium
– –, small intestine 148
Liver cells 899
– –, calcium 905
– –, –, effects on ion permeabilities 905
– –, –, –, effects on junctional membrane 905
– –, –, –, effects on membrane permeability 906
– –, calcium transport 916, 917, 918
– –, –, role of ATP-ase 917, 918
– –, cation transport/ATP ratio 911
– –, coupling between sodium and potassium movements 909, 910, 911
– –, –, coupling ratio of sodium and potassium fluxes 910, 911
– –, –, multiple site carrier 910
– –, energy requirement for sodium-potassium transport 913
– –, –, oxidative phosphorylation 914
– –, –, oxygen requirement 913
– –, –, theoretical energy requirement 914
– –, ionic concentrations 899
– –, –, chloride 899
– –, –, potassium 899
– –, –, sodium 899
– –, membrane potential 900
– –, –, contribution of electrogenic pump 912
– –, –, *in vitro* depolarization 901
– –, –, –, factors 901
– –, passive ion fluxes 902
– –, –, diffusional fluxes 902
– –, –, facilitated 904
– –, –, –, ouabain-insensitive sodium exchange 904
– –, –, –, ouabain-sensitive sodium exchange 904
– –, –, –, potassium 904
– –, –, ionic permeability coefficients 903
– –, potassium 909
– –, –, active transport 909
– –, properties of cell membranes 903
– –, regulation of cell volume 915, 916
– –, sodium transport 908
– –, –, active transport 908
– –, –, –, ouabain-sensitive fluxes 908, 909
– –, transport properties 899
– –, –, temperature effects 906
Loop of Henle 185, 381
–, magnesium transport 381, 382
–, thick ascending limb 188, 189
–, –, effect of diuretics 189
–, –, effect of ethacrynic acid 189
–, –, effect of furosemide 189
–, –, effect of mersalyl 189
–, –, effect of ouabain 189
–, –, permeability to ions 189
–, –, transepithelial potential difference 188
–, thin ascending limb 188
–, salt transport 185, 186, 187, 188
–, –, active chloride transport 188
–, –, pars recta of proximal tubule 188
–, –, response to renal vein occlusion 187
–, –, response to sodium load 186
–, –, studies by tubular microperfusion 187

Macromolecules 28, 30, 31, 35
–, endothelial transport 28, 30, 35
–, sieving 31
Macula densa 190
Magnesium 357, 377
–, glomerulus 378
–, distal tubule and collecting duct 383, 384
–, –, effects of magnesium loading 384
–, factors regulating tubular reabsorption 385
–, –, acidosis 388
–, –, alcohol 388
–, –, alkalosis 388
–, –, diuretics 389
–, –, hormonal effects 387
–, –, –, adrenocortical steroids 387
–, –, –, calcitonin 388
–, –, –, growth hormone 388
–, –, –, parathyroid hormone 387
–, –, relationship to other cations 386
–, –, –, Vitamin D 388
–, loop of Henle 381
–, –, magnesium loading 382
–, –, mechanism of reabsorption 382
–, –, permeability 382
–, –, reabsorption pattern 382
–, proximal tubule 378, 379, 380, 381
–, –, mechanism of reabsorption 381
–, –, –, complexing with anions 381
–, –, permeability to magnesium 380
–, renal transport 357, 377
–, –, tubular pattern of reabsorption 378, 379
Medullary collecting duct 139
– – –, passive transepithelial ionic membrane parameters 140
– – –, transepithelial potential difference 139, 140
– – –, transepithelial electrical conductance 140
Mersalyl 189, 274, 278, 281
Mucoproteins 2
Multiple pathway hypothesis 18, 19
Muscle fiber 59
– –, potassium activity 59, 60, 61
– –, capillary 18, 19

Necturus proximal tubule 65
–, potassium activity 65
–, –, luminal cell membrane 66
–, –, peritubular cell membrane 66, 67, 68
Nephron 47, 49, 51
–, electrochemistry 47, 49, 51, 57

Organic buffers 441
– –, electrical effects on proximal tubule cells 442, 443
– –, glycodiazine 441

– –, ouabain-insensitivity of H^+-secretion 442
– –, proximal tubular transport 441
– –, –, dependence on carbonic anhydrase 441, 442, 443
– –, –, transport by non-ionic diffusion 441
– –, –, transport inhibition by SITS 441
– –, transport across specific cell membranes 437
– acids 440
– –, relationship between renal and red blood cell transport 440
– –, renal transport of citrate 436
– –, renal transport of lactate 436
– –, renal transport of N′-methylnicotinamide (NMN) 436
– –, renal transport of tetraaethylammonium (TEA) 436
– –, renal transport mechanisms 438
– –, –, electrical potential dependence 438
– –, –, Na^+-dependence 438
– –, properties of phenolsulfophtalein dyes 439
– renal transport sites 436, 437
– –, substrate-induced transport stimulation 440
– solutes 413
– –, mechanism of renal reabsorption 413
– –, –, amino acids 413
– –, –, carrier-mediated transport 416
– –, –, –, reconstituted liposomes 417
– –, –, cotransport with sodium 413
– –, –, countertransport 413
– –, –, sugars 413
– –, –, theoretical considerations of Na^+-coupled transport 417
– –, –, –, affinity-type 417
– –, –, –, mixed-type 417
– –, –, –, velocity-type 417
– –, –, weak acid buffers 413
– –, nephron site of transport 414
– –, proximal renal tubule 414
– –, –, methods of study 414
– –, –, –, electrical measurements 415
– –, –, –, membrane vesicles 416
– –, –, –, steady-state electrochemical potential difference 415
– –, –, –, transtubular measurements 414, 415
– –, –, transport inhomogeneity 414
– –, secretion of organic acids 413
– –, secretion of organic bases 413
Osmotic transients 10
– –, capillary 10
Oxalate 781
Oxygen consumption 146
– –, gallbladder 862
– –, proximal renal tubule 146

Papillary collecting duct 203
–, contribution of deep (juxtamedullary) nephrons to sodium excretion 204, 205
–, lanthanum 203, 204
–, sodium reabsorption 203, 204
–, –, active nature of transport 204
Pancreas 811
–, amino acid uptake 843
–, –, conjugate driving forces 844
–, –, Na^+-driven cotransport 845
–, –, plasma membrane vesicles 845
–, amylase release 814
–, –, CCK-OP (cholecystokinin-octa-peptide) 814
–, –, inhibition by neotilin 814
–, electrical cell-to-cell coupling 834
–, electrical potential differences 822
–, electrolyte and water secretion 812
–, –, enzyme secretagoge-stimulated secretion 821
–, –, flow dependence of concentration pattern 817
–, –, –, Ca^{++} 818
–, –, –, K^+ 816
–, –, –, inorganic phosphate 818
–, –, –, Mg^{++} 818
–, –, –, Na^+ 815, 816, 818
–, –, ionic requirements 815
–, –, mechanism of ion secretion 823
–, –, micropuncture and microperfusion techniques 819
–, –, model of secretion 821
–, –, –, H^+/OH^--exchange 824
–, –, –, re-equilibration 822
–, –, –, role of buffers 815, 823, 824
–, –, –, sidedness of buffer transport 825
–, –, secretin stimulated secretion 821
–, –, –, sulfonamide buffers 824
–, –, primary secretion 820
–, –, –, ionic concentrations 820, 821
–, –, spontaneous secretion 821
–, –, stimulatory processes 812
–, –, –, cAMP 812, 813, 814
–, –, –, dbc AMP 812, 814
–, –, –, localization of hormone receptors 813
–, –, –, secretin 812
–, –, –, species differences 814
–, –, –, VIP (vaso-active intestinal polypeptide) 813, 814
–, exocytosis of secretory granules 839
–, –, role of Ca^{++} 839, 840
–, –, role of microfilaments and microtubules 842, 843
–, –, secretory vesicles 841
–, secretion of enzymes 831
–, –, acetylcholine 831, 836
–, –, electrophysiological measurements 832
–, –, gastrin 831
–, –, model of enzyme secretion 839
–, –, pancreozymin (cholecystokinin) 831
–, –, role of Ca^{++} 833, 835, 836

Pankreas, exocytosis
–, –, role of cyclic AMP 837
–, –, role of cyclic GMP 837
–, secretion of Na^+-ions 826, 827, 828
–, transport processes 811
–, –, ATP 830
–, –, cell model 828
–, –, general aspects 811
–, –, –, electrolyte secretion 811
–, –, –, enzyme secretion 811
–, –, H^+-K^+-exchange 830
–, –, Na^+-H^+-exchange 830
–, –, mechanisms of buffer secretion 829, 830
–, –, paracellular pathway 828
Pancreocymin 821, 831
Para-aminohippuric acid transport 503
Paracellular transport 58, 148, 151, 865, 867
Parathyroid hormone 184, 199, 350, 372, 387, 390, 396, 397, 398
– –, effect on proximal renal tubule 184
– –, renal action 372, 405, 406
Permeability 1, 9, 10, 12, 14, 16, 23
–, capillary 23
–, –, micropuncture measurements 23
–, –, –, frog mesentery 24
–, renal glomerular capillaries 32
pH 83
–, intracellular 83
–, –, muscle fiber 83
–, –, –, effects of hypercapnia 83
–, renal control 299
Phloretin 124
Phosphate 318, 357
–, distal tubule and collecting duct 405, 406
–, –, effects of parathyroid hormone 405, 406
–, –, micropuncture studies 406
–, factors controlling reabsorption 393
–, –, calcitonin 403
–, –, calcium 401, 402
–, –, carbonic anhydrase inhibitors 398
–, –, glucose 403
–, –, growth hormone 403
–, –, insulin 403
–, –, metabolic alkalosis 398
–, –, phosphate depletion 400
–, –, phosphate infusions 390, 398, 399
–, –, renal failure 401
–, –, Vitamin D 402
–, glomerular filtration 391
–, –, ultrafilterability 391
–, –, –, relationship to plasma calcium 392
–, isolated renal tubules 503
–, loop of Henle 404, 405
–, nephron heterogeneity 406
–, proximal tubule 392, 393
–, –, effect of huminal pH 396, 397, 398
–, –, effect of parathyroid hormone 390, 396, 397, 398
–, –, –, cAMP 397
–, –, –, role of bicarbonate 397
–, –, heterogeneity of transport rate 393
–, –, mechanism of transport 395, 396
–, –, –, membrane vesicle studies 395
–, –, –, transport model 396
–, –, phosphate permeability 393
–, –, relationship to sodium reabsorption 394
–, –, role of mitochondria 395
–, –, role of Na-K-ATPase 395
–, –, sodium dependence of transport 395
–, –, transport inhibition by arsenate 395
–, reabsorption pattern along nephron 357, 389, 390, 391, 405, 406
–, role in renal tubular acidification 318, 319, 320, 342
–, secretion of phosphate 391
–, –, amphibians 391
–, –, birds 391
–, –, possibility in mammals 391, 394
–, –, reptiles 391
Pinocytosis 4, 33
Plasma colloids 5
Polyvinylpyrrolidone (PVP) 30
Pores 12, 15, 17, 20, 38
–, distributed population 15
–, radius 20
–, –, capillary 20
Potassium 59
–, activity 59
–, –, Amphiuma bladder 65
–, –, Amphiuma distal tubule 63
–, –, Amphiuma intestine 65
–, –, frog skeletal muscle 60
–, –, muscle fiber 59
–, –, Necturus proximal tubule 65, 66, 67, 68
–, –, rabbit cardiac muscle 60
–, –, rat muscle 60, 61
–, –, rat distal tubule 62, 63
–, –, –, changes in acid base balance 63
–, –, –, changes in potassium balance 62
–, –, –, effect of aldosterone 63
–, colon 215
–, renal tubule 62
–, –, distal 62
–, –, secretion 62
– tolerance 265, 266
– transport 215
– –, active cell uptake of potassium 220
– –, cardiac glycosides 218, 219
– –, cellular compartmentalization 221
– –, cellular mechanisms of transport 243
– –, –, amphibian species 245
– –, –, collecting duct 245
– –, –, collecting tubule 245, 255, 258
– –, –, –, cellular transport model 255
– –, –, distal tubule 245
– –, –, –, active potassium transport across luminal membrane 257
– –, –, –, cellular transport model 254, 255, 256, 257, 262
– –, –, –, compartmental model 259, 260, 261

– –, –, effects of ouabain 256
– –, –, –, effect of voltage clamping on luminal potassium 256, 259
– –, –, –, paracellular pathway 263
– –, –, –, potassium concentration of cells 266
– –, –, –, potassium content of cells 166
– –, –, –, potassium permeability 258
– –, –, –, properties of luminal cell membrane 251, 252, 253, 255, 257
– –, –, –, properties of peritubular cell membrane 249, 250, 251, 255
– –, –, –, transcellular pathway 261, 262, 263
– –, –, –, transepithelial electrical potential profile 245, 246, 247, 248
– –, –, proximal tubular epithelium 243
– –, –, –, active potassium transport 244
– –, –, –, cellular potassium activity 244
– –, –, –, intercellular shunt pathway 245
– –, –, –, net potassium reabsorption 244
– –, cell potassium concentration 216
– –, –, renal papilla 216, 217
– –, collecting ducts 215
– –, colonic epithelium 215
– –, dissociation of filtered from excreted potassium 224, 225
– –, electrochemical driving forces 215
– –, exchange for sodium 226
– –, factors regulating potassium excretion 264
– –, –, changes in acid-base balance 281
– –, –, –, acidosis 281
– –, –, –, alkalosis 281
– –, –, –, changes in cell potassium 285
– –, –, –, distribution of sodium reabsorption along nephron 287
– –, –, effect of changes in ammonium excretion 286
– –, –, nephron site of potassium changes 282, 283
– –, –, –, relationship to intratubular pH 283, 284
– –, –, –, relationship to plasma potassium concentrations 284
– –, –, delivery of sodium 271, 274, 275, 276
– –, –, –, effect on collecting duct Na^+-K^+-exchange 273
– –, –, –, potassium reabsorption 272
– –, –, –, transepithelial electrical potential difference 273
– –, –, dietary potassium deprivation 265
– –, –, distal tubular flow rate 270, 271, 274, 275, 276
– –, –, –, changes in distal tubular sodium concentration 277
– –, –, –, stop-flow experiments 270
– –, –, diuretics 274
– –, –, –, amiloride 274
– –, –, –, carbonic anhydrase inhibitors 274
– –, –, –, chlorthiazide 274
– –, –, –, ethacrynic acid 274
– –, –, –, furosemide 274
– –, –, –, mercurial diuretics 274, 278, 281
– –, –, –, ouabain 278, 279, 280, 281
– –, –, –, spironolactone 274
– –, –, –, triampterene 274
– –, –, mineralocorticoids 288
– –, –, –, changes in potassium permeability 291, 292
– –, –, –, changes in cellular potassium activity 291
– –, –, effects on potassium secretion 288
– –, –, –, effects on sodium reabsorption 288
– –, –, –, electrophysiological effects 290
– –, –, –, "escape" phenomenon 289
– –, –, –, isolated cortical rabbit collecting ducts 289
– –, –, –, mechanism of action 290, 291
– –, –, –, morphological effects of DOCA-treatment 289
– –, –, –, nephron site of steroid action 289
– –, –, variations in hydration 264
– –, –, variations in potassium intake 264
– –, general aspects 216, 223, 224
– –, –, acid-base balance 216
– –, –, cell excitability 216
– –, –, cell growth 216
– –, –, cell volume control 216
– –, –, enzyme functions 216
– –, homocellular transport 218
– –, microperfusion 215
– –, Na-K-ATP-ase 218
– –, micropuncture 215
– –, potassium pool 222, 260, 261, 263
– –, –, exchangeable potassium in distal tubule 222
– –, potassium secretion 215, 220
– –, –, cortical collecting tubule 223
– –, –, distal tubule 215, 220, 223, 225, 231
– –, –, –, cell heterogeneity 222, 231, 232, 233, 234
– –, –, loop of Henle 229
– –, –, papillary collecting duct 223
– –, –, saturation of secretory mechanism 265
– –, potassium tolerance (potassium adaptation) 265, 268
– –, –, cellular potassium uptake 268
– –, –, Na-K-ATP-ase 268, 269
– –, –, "physiological" 267
– –, –, role of mineralocorticoids 269
– –, relationship between potassium and sodium transport 218, 220
– –, renal clearance 215
– –, sodium-potassium exchange pump 218, 219

Potassium, renal, sodium
– –, –, rheogenic mode of action 220, 249, 250
– –, tubular sites of transport 226, 238
– –, –, collecting duct 238, 239, 240, 241
– –, –, –, effect of amiloride 240
– –, –, –, effect of changes in luminal contact time 240
– –, –, –, effect of dietary potassium deprivation 241
– –, –, –, effect of DOCA 241
– –, –, –, effect of low sodium intake 241
– –, –, –, effect of luminal acidification 240
– –, –, microcatheterization method 243
– –, –, –, potassium reabsorption 240, 241
– –, –, distal tubule 231, 235, 236
– –, –, –, cell heterogeneity 232, 233, 234
– –, –, –, effect of changes in acid-base balance 237
– –, –, –, effect of adrenal steroids 237
– –, –, –, effect of changes in fluid and sodium delivery 238
– –, –, –, effect of dietary potassium deprivation 237
– –, –, –, effect of diuretics 237
– –, –, –, effect of potassium loading 237
– –, –, –, main site of regulation 235, 236
– –, –, –, potassium concentration changes 237
– –, –, loop of Henle 228
– –, –, –, benzolamide 230
– –, –, –, concentration changes of potassium 228
– –, –, –, differences between superficial and juxtamedullary nephrons 230
– –, –, –, furosemide 230
– –, –, –, secretion of potassium into descending limb of Henle's loop 229
– –, –, proximal tubule 227
– –, –, –, potassium concentrations 227
– –, –, –, potassium reabsorption 228
– –, –, –, straight segment of proximal tubule 228
– –, –, thin ascending limb of Henle's loop 230
– –, transcellular transport 218
Proximal convoluted tubule 102, 111, 146, 148, 149, 150, 154
–, acid-base balance 146
–, active ionic membrane parameters of single barriers 120, 122
–, amino acid reabsorption 150
–, bicarbonate reabsorption 149
–, brush-border vesicles 155, 156
–, –, cotranport 155, 156
–, –, –, effects on transepithelial potential difference 157, 158
–, –, –, "overshoot" phenomenon 157
–, cable analysis 115
–, chloride permeability 158
–, chloride reabsorption 149
–, cotransport of sodium with sugars and aminoacids 146, 155, 156, 157
–, –, amphibian tubule 155
–, –, electrophysiology 155
–, –, mammalian tubule 155, 156
–, –, primary active transport 147
–, –, secondary active transport 146
–, –, specificity 155
–, coupling of sodium and water reabsorption 169
–, –, differential reflection coefficients 169, 171
–, –, isotonic fluid transport 169
–, –, osmotic linkage 169
–, –, standing osmotic gradient hypothesis 169, 170
–, dependence of water reabsorption on sodium chloride transport 160, 161
–, effect of diamox 174
–, effect of renal nerves 184, 185
–, electrical conductance 147
–, electrophysiology 152
–, epithelial inhomogeneity 148
–, –, functional 148, 150
–, –, morphological 148, 149
–, equivalent circuit model 158
–, fluid reabsorption 161
–, –, effect of anion substitutions 161, 162
–, –, effects of bicarbonate 173, 174
–, –, effect of cation substitutions 161, 162
–, –, effect of cooling 162
–, –, effect of metabolic inhibitors 172
–, –, effect of ouabain 162, 172, 173
–, –, effect of transepithelial chloride gradients 172, 173, 174
–, –, hydraulic conductance 170
–, –, solute entrainment 170
–, general properties 146
–, glomerulo-tubular balance 151
–, glucose reabsorption 150
–, glycodiazine 150
–, inorganic phosphate 150
–, intercellular shunt 148, 151
–, isolated perfused tubules 151
–, –, convoluted tubules 152
–, –, straight segment 152
–, isosmotic fluid reabsorption 149
–, juxtamedullary nephrons 104
–, laetic acid 150
–, lanthanum 169
–, leaky epithelium 148
–, model of sodium transport 163, 166, 177
–, Na-H countertransport 165
–, Na-K-activated ATPase 147, 156, 167, 169
–, net transport rates of ions 151, 152
–, oxygen consumption 146
–, p-Aminohippurate (PAH) 150

–, passive sodium chloride reabsorption 171, 172, 173
–, pH 150
–, regulation of salt transport 175
–, –, changes in extracellular fluid volume 175
–, –, changes in glomerular filtration rate 175
–, –, colloid osmotic pressure 176, 177, 179, 180, 181, 182
–, –, electrical conductivity changes 178
–, –, filtration fraction 183
–, –, flow rate 176, 182
–, –, hydrostatic pressure 176, 179
–, –, luminal diameter 176
–, –, natriuretic hormone 176, 184
–, –, parathyroid hormone 184
–, –, peritubular capillary control mechanism 177
–, –, peritubular capillary flow rate 183
–, –, rectification of fluid movement 177
–, –, role of passive backflux 175
–, seasonal variations in organic acid transport 112
–, seasonal variations of electrical potential differences 112
–, –, superficial nephrons 104
–, sodium permeability 158
–, sodium chloride entry across luminal cell membrane 163, 164, 165
–, –, relationship to hydrogen ion secretion 165, 166
–, sodium chloride transport across peritubular cell membrane 167
–, –, Na-K-exchange pump 167
–, –, rheogenic mode of operation 168
–, –, sensitivity to ethacrynic acid 167
–, –, "uncoupled" transport 167
–, –, active 147, 160
–, –, equivalent circuit model 158
–, –, sodium-calcium countertransport 168
–, sodium transport 147
–, sodium transport pool 168
–, stop-flow microperfusion 160
–, transepithelial ion conductance 147
–, transepithelial potential difference 102, 147, 150, 152, 153, 162
–, –, Ambystoma 104, 105, 106, 107, 112
–, –, amphibian proximal convuluted tubule 104, 105
–, –, changes along the tubule 153
–, –, chloride diffusion potential 118
–, –, effect of amino acids 102
–, –, effect of bicarbonate 103
–, –, effect of chloride 102
–, –, effect of glucose 102, 110
–, –, effects of luminal composition 102
–, –, equivalent electrical circuit 113, 115
–, –, heterogeneity of proximal renal tubule 103, 104
–, –, ionic selectivity 108, 109
–, –, luminal cell membrane potential difference 112, 113
–, –, luminal cell membrane properties 119
–, –, mammalian proximal convoluted tubule 105
–, –, Necturus 104, 105, 107, 112, 114
–, –, –, anion selectivity 107
–, –, –, short-circuit current 111
–, –, paracellular membrane properties 117
–, –, peritubular cell membrane properties 118, 119
–, –, permeability coefficients 116
–, –, permselectivity 117, 118
–, –, –, Necturus 116, 117
–, –, potassium diffusion potential 118
–, –, rabbit isolated proximal convoluted tubule 103
–, –, rat proximal tubule 103, 108
–, –, rheogenic ion pumping 121, 122, 123
–, –, sensitivity to amino acids 153, 154
–, –, sensitivity to anions (chloride and bicarbonate) 153, 154
–, –, sensitivity to glucose 153, 154
–, –, sensitivity to organic solutes 153, 154
–, –, Thamnophis 105
–, –, transference number sequence 108
–, –, Triturus 104, 105, 112
–, transepithelial resistance 114
–, voltage divider ratio 114
Proximal straight tubule 123, 148, 149, 150, 154
–, transepithelial electrical conductance 124
–, transepithelial ionic membrane parameters 124
–, transepithelial potential difference 123, 124
–, –, superficial proximal straight tubules 124
–, –, effect of bicarbonate 125
–, –, effect of chloride 125
–, –, effect of cyclamate 125
–, –, effect of osmotic gradients 125
–, –, effect of potassium removal 125
–, –, effect of sodium removal 125
–, –, juxtamedullary proximal straight tubule 125
Psammomys 452, 468, 469
–, urinary concentrating mechanism 452, 468

Rana pipiens 128
– –, diluting segment of renal tubule 128
Reflection coefficient 5, 13, 169, 171, 501, 515, 784
– –, capillary 5, 13
Renal acidification 87

Renal concentrating mechanism 449
–, active transport of water 449
–, collecting ducts 456, 457
–, –, effects of ADH 457
–, –, water permeability 457
–, countercurrent hypothesis 449
–, countercurrent mechanism 451, 458, 460
–, –, central vascular core model 462, 463, 464, 469
–, –, –, permeability requirements 466, 467
–, –, countercurrent multiplier 449
–, –, –, three-channel model 459, 460, 461
–, –, function of thin limb 461
–, –, –, transport characteristics 462
–, –, mathematical model 459
–, –, passive concentrating model for loop of Henle 465, 466, 467, 469
–, –, "single effect" 458, 459, 460
–, distal convolution 454
–, –, effects of ADH 455
–, –, osmolality 455
–, –, osmotic equilibration 456
–, –, –, strain differences in rats 456
–, effects of urea 465, 466, 467
–, effects of vasopressin 450
–, hypothalamic diabetes insipidus 468
–, juxtamedullary nephrons 452
–, localization of concentrating process in the nephron 451
–, –, proximal convoluted tubule 451, 452
–, –, –, hydraulic conductivity 451
–, –, –, isosmotic fluid transport 451, 452
–, –, osmolality changes along the nephron 451
–, –, –, loop of Henle 452, 453
–, –, –, hydraulic permeability 454
–, –, –, thick ascending limb 454
–, medullary hypertonicity 450
–, passive diffusion of water 449
–, Psammomys 452, 468, 469
–, regulation of osmolality 449
–, relationship between length of loop of Henle and maximum urinary concentrating ability 450
–, relationship of osmolality in papillary structures 454
–, solute-free water 449
–, vasa recta 457
–, vascular pattern 453
Renal dilution mechanism 449
Renal transport of calcium 357
Renal transport of magnesium 357, 377
Renal transport of inorganic phosphate 357
Renal tubular acidification 299
–, bicarbonate reabsorption along the nephron 311, 312
–, –, bicarbonate transport during potassium deficiency 340
–, –, carbonic anhydrase inhibition 313, 319
–, –, distal tubule 313
–, –, hypochloremic alkalosis 341, 342
–, –, loop of Henle 312
–, –, metabolic acidosis 313
–, –, metabolic alkalosis 313
–, –, proximal tubule 312
–, –, redistribution of sodium to distal nephron sites 342
–, –, role of extracellular volume changes 342
–, bicarbonate secretion along the nephron 314
–, –, isolated rabbit cortical collecting tubule 314
–, carbonic anhydrase 301, 321, 332, 334
–, cell mechanism of acidification 300
–, –, acetazolamide 329, 330, 332, 335
–, –, –, effect on anion exchange 332
–, –, cell bicarbonate concentration 327
–, –, cellular pH estimates 326
–, –, components of distal tubular bicarbonate reabsorption 333
–, –, –, acidosis 335
–, –, components of proximal tubular bicarbonate reabsorption 333
–, –, –, acidosis 335
–, –, coupling of H transport to other ions 337, 341, 342, 343, 344
–, –, effects of anion substitutions 341, 342
–, –, effects of diuretics 341
–, –, H/Cl coupling 341, 342, 343
–, –, K/H coupling 339, 340
–, –, –, potassium depletion 340
–, –, –, potassium loading 340
–, –, Na/H exchange 336, 337, 338
–, –, –, brush border vesicles 338
–, –, –, effects of cardiac glycosides 339
–, –, –, results of microperfusion studies 338, 339
–, –, rate of carbonic acid generation 329
–, –, rate of uncatalyzed H-ion generation 330
–, –, recirculation of carbonic acid 334
–, –, role of buffer base reabsorption 328, 329, 331
–, –, role of proton transport 328, 329
–, –, supply of peritubular H-ions 331
–, –, studies with bromcresol purple 331
–, cellular permeability to CO_2 304
–, CO_2 in final urine 325
–, –, ampholytic nature of bicarbonate 325
–, –, role of urinary concentration impairement 325
–, cellular CO_2 production 304, 324

–, disequilibrium pH 303, 304, 321, 322, 332
–, –, acid 303, 304
–, –, alkaline 305
–, –, role of H_2CO_3 303
–, –, Role of pCO_2 303, 304, 323
–, evidence for active transport of hydrogen ions 325
–, factors controlling renal acidification 344
–, –, dependence of bicarbonate reabsorption upon luminal bicarbonate concentration 345, 346
–, –, humoral regulation 349
–, –, –, aldosterone 350
–, –, –, parathyroid hormone 350
–, –, modification of bicarbonate reabsorption by extracellular volume changes 344, 345, 347
–, –, role of extracellular volume changes 344, 345, 347
–, –, role of peritubular base transfer 348, 349
–, –, role of peritubular pH and pCO_2 346, 347
–, –, segmental buffer load 344
–, HCO_3/CO_2 buffer system 300
–, –, equilibrium conditions 302
–, –, kinetic considerations 300, 301
–, –, –, dehydration reaction 301
–, –, –, hydration reaction 301
–, kinetic aspects of acidification 314, 415, 316, 317, 318
–, –, color changes of indicators 316
–, –, pump leak system 316
–, –, stop-flow studies with luminal HCO_3-perfusion 315, 316, 318
–, –, stop flow studies with luminal phosphate perfusions 315, 316, 317, 318, 319
–, –, use of H^3-glycodiazine 316, 317, 321
–, –, titration characteristics of bicarbonate 320
–, –, titration characteristics of phosphate 320
–, mechanism of acidification 321
–, –, carbonic anhydrase 322, 333, 329
–, –, –, histological localization 322
–, –, –, membrane localization 323
–, –, generation of CO_2 in turtle bladder 324
–, measurement of tubular bicarbonate concentrations 309
–, –, osmometric method 309
–, –, picapnotherm method 310
–, –, quinhydrone electrode 309
–, measurement of titratable acidity 311
–, measurement of tubular pCO_2 310
–, permeability of tubule to HCO_3^- 327
–, permeability of tubule to H^+-ions 328
–, role of bicarbonate reabsorption 299, 331, 335
–, role of hydrogen ion secretion 299
–, –, water splitting redox pump 302
–, tubular pH measurements 305
–, –, antimony microelectrode 308, 309, 317
–, –, –, bicarbonate error 309
–, –, glass electrode 306, 307, 308
–, –, quinhydrone electrode 305, 306, 309
Renal tubular transport 529
–, active and passive components 529
–, ATP content and turnover 546
–, cell models of transport components 558, 559
–, effect of metabolic inhibitors on transepithelial sodium transport 534, 535, 536, 537, 538, 539
–, enzyme activities of renal tubule 545
–, energy requirements for active sodium transport 534
–, metabolic correlates 529
–, –, α-aminoisobutyric acid transport 547
–, –, –, brush-border vesicles 547
–, –, –, kidney slice 547
–, –, carbon dioxide production 531
–, –, general aspects 530
–, –, heat production in the kidney 530
–, –, –, effect of transport inhibitors and diuretics 531
–, –, oxygen consumption 531
–, –, –, relationship to sodium reabsorption 531, 532, 533
–, –, intrarenal compensation 533
–, –, proximal tubule 533
–, –, role of ATP 534, 535, 539, 546, 549
–, –, uncoupling of Na^+-transport and O_2 consumption 533
–, –, tricarboxylic acid cycle 539
–, renal substrate metabolism 540
–, –, relationship to sodium transport 540, 542, 543
–, –, –, free fatty acids 540, 543
–, –, –, glucose oxidation 541, 544
–, –, –, lactate uptake 540
–, –, –, pyruvate oxidation 541
–, –, studies in isolated perfused kidney 541
–, sodium-solute cotransport systems 554
–, –, transport by renal vesicles 554
–, –, –, electrical potential difference as driving force 557
–, –, –, importance of sodium gradient as driving force 555
–, –, –, overshoot phenomenon 556
–, –, –, sodium-glucose coupling 556
–, thermodynamic considerations 529

Renal tubular transport
–, transport-related ATP-hydrolases 549
–, –, bicarbonate and protons 552
–, –, –, electrogenic proton transfer 552
–, –, –, gastric mucosa 552
–, –, –, mitochondrial contamination 552
–, –, calcium 550, 551
–, –, –, localization 551
–, –, –, relationship to transepithelial Ca^{++}-transport 551
–, –, chloride 553
–, –, sodium 549
–, –, –, adaptive changes 549
–, –, –, effect of inhibitors 550
–, –, –, localization 549
–, –, –, species differences 550
Renal tubule 97
–, sea also isolated mammalian renal tubule 473
–, sea also proximal convoluted tubule 102, 103, 104
–, active transepithelial ionic membrane parameters 110
–, compartmental model 101
–, electrical conductance 97, 98, 105, 109
–, –, methodological problems 99
–, –, –, axial electrode system 100
–, –, –, distribution of current 100
–, –, –, point current source 99
–, electrical potential difference 97, 101, 104
–, –, intracellular (single cell) measurements 98, 100
–, –, ion diffusion barrier 101
–, –, ionic electromotive forces 115
–, –, junction potentials 99
–, –, methodological problems 99
–, passive transepithelial ionic membrane parameters 106
–, –, antoperfused dog kidney 107
–, –, juxtamedullary convoluted tubule 107
–, micropuncture methodology 473
–, –, limitations 473
–, single barrier electrical conductances 112
–, single membrane electrical properties 111
–, sodium transport 145
–, –, superficial proximal convoluted tubule 107
–, –, transepithelial (transmural) measurements 98, 104, 110
–, electroosmotic coefficient 97
–, electrophysiological methods 97
–, ionic transference number 97, 106, 107, 108, 116
–, Na^+-H^+ exchange 87
–, paracellular pathway 115
–, paracellular potential difference 111
–, passive ionic parameters of single barriers 115
–, peritubular cell membrane potential difference 111, 113
–, potassium activity 65, 66, 67, 68
–, rheogenic sodium pumping 110
–, –, effect of carbonic anhydrase inhibitors 111
–, –, effect of cooling 110
–, –, effect of ethacrynic acid 110
–, –, effect of mercurial diuretics 110
–, –, effect of ouabain 110
–, sodium activity 69, 71, 72, 73, 75
–, sodium transport 146
–, transepithelial resistance 115
–, water transport 145, 146
Rheogenic ion pumping 121, 122, 761, 762, 764, 765, 770
–, renal tubular epithelium 121, 122, 123

Salivary glands 563
–, anatomy 564
–, –, duct system 565
–, –, –, excretory 565, 566
–, –, –, intercalated 565
–, –, –, striated 565, 571
–, –, mucous cells 565
–, –, myoepithelial cells 566, 571
–, –, secretory endpieces 564
–, –, –, heterocrine 564
–, –, –, homocrine 564
–, –, serous cells 565
–, functions 563
–, heterogeneity 563
–, –, functional 563, 564
–, –, structural 563
–, mammalian vertebrates 563
–, –, mandibular 563
–, –, parotid 563
–, –, sublingual 563
–, transport of proteins 655, 656
–, –, albumin 650
–, –, condensing vacuoles 658
–, –, control of protein secretion 662
–, –, –, amylase 663
–, –, –, mucoproteins 663
–, –, discharge of secretion granules 661
–, –, exocytosis of secretion granules 662
–, –, immunoglobulin A (IgA) 656
–, –, intracellular polypeptide transport 659, 660
–, –, mechanism of transport 658, 659
–, –, secretory proteins 657
–, –, transitional vesicles 658
–, –, uptake of amino-acids 659
–, transport of water and electrolytes 567
–, –, active sodium transport 635, 636
–, –, –, substrate requirements of transport 633
–, –, adrenergic secretomotor receptors 571, 603
–, –, amylase secretion 569, 603
–, –, composition of primary secretion 575
–, –, –, bicarbonate 578
–, –, –, chloride 580
–, –, –, electrolyte composition

in different species 576, 577, 578
–, –, –, organic solutes 580
–, –, –, osmotic activity 575
–, –, –, potassium 579
–, –, –, sodium 579
–, –, control of secretion 572
–, –, –, innervation of endpieces 572, 573
–, –, ductal modification of primary secretion 608
–, –, –, active and passive transport components 633
–, –, –, angiotensin 653
–, –, –, autonomic control of duct transport 643, 644, 645
–, –, –, bicarbonate 616
–, –, –, calcium 620
–, –, –, chloride permeability 631, 632
–, –, –, effects of acetylcholine 645, 646, 647
–, –, –, effects of amiloride 636, 640
–, –, –, effects of cyclic AMP 647
–, –, –, effects of cyclic GMP 647
–, –, –, effects of triamterene 636, 640
–, –, –, endocrine control of ductal transport 648
–, –, –, gastro-intestinal polypeptide hormones 653
–, –, –, magnesium 621
–, –, –, micropuncture and perfusion studies 622, 623, 624
–, –, –, mechanism of mineralocorticoid action 649, 650, 651, 652
–, –, –, mineralocorticoids 648, 649, 650, 651, 652
–, –, –, permeability properties of salivary ducts to ions 625, 626, 632
–, –, –, phosphate 621, 622
–, –, –, potassium 615, 616
–, –, –, potassium fluxes 627, 628, 629
–, –, –, potassium permeability 627, 632
–, –, –, sodium permeability 626, 630, 632
–, –, –, water permeability 632
–, –, –, effects of changes in flow rate 608, 610, 612
–, –, –, parasympathetic stimulation 608, 610, 611, 612, 644, 645
–, –, –, sodium 609, 612, 613
–, –, –, sympathetic stimulation 610, 611, 617, 618, 619, 644
–, –, –, transport model for duct epithelium 640, 641, 642
–, –, –, variability of secretory response 614
–, –, electrical properties of resting endpiece cell 581, 582, 583, 584
–, –, –, effects of nerve stimulation 586
–, –, –, effect of potassium ions 589
–, –, –, resting potentials across basal plasma membrane 585, 587
–, –, electrical properties of stimulated endpiece cell 590, 596
–, –, –, adrenergic stimulation 597, 599, 601, 603
–, –, –, changes in cell membrane resistance 596
–, –, –, cholinergic stimulation 592, 593, 594, 601
–, –, –, mechanism of secretory potentials 595, 596
–, –, –, parasympathetie innervation 590
–, –, –, receptors 590
–, –, –, secretory ionic fluxes 592
–, –, –, secretory potentials 592, 597, 598, 599
–, –, –, species differences 591
–, –, –, sympathetie innervation 591
–, –, isotonic fluid transport 606
–, –, –, electrochemical potential difference for ions 606, 607
–, –, –, Na^+-K^+-ATPase 607, 636, 637
–, –, poly-L-lysine 569, 570
–, –, potassium secretion 637
–, –, –, interrelationship with bicarbonate or protons 639
–, –, salivary cell ionic content 587, 588
–, –, sites of fluid secretion 568, 571
–, –, –, retrograde injection of toxins 569
–, –, stimulus-secretion coupling 600
–, –, –, cell model of noradrenalin and acetylcholine action 605
–, –, –, role of calcium 600, 602, 604
–, –, –, role of cyclic nucleotides 600, 610, 602
–, –, theories of secretion 567
–, –, –, isotonic primary secretion 568
–, –, –, parasympathetie nerve activity 567, 574
–, –, –, sympathetie nerve activity 567, 574
–, –, –, two-stage secretory hypothesis 567
Salt glands 563
–, adaptation to salt loading 680
–, elasmobranchs 675
–, electrolyte composition 682, 683
–, hormonal control 681
–, –, adrenal glucocorticoids 681
–, –, cholinergic innervation 681
–, lacrimal glands 678
–, marine birds 675
–, marine vertebrates 675
–, mechanism of salt secretion by tubular endpiece 683
–, –, behavior of anions 688
–, –, intracellular electrolyte concentrations 686
–, –, ion pumps 686, 687
–, –, Na^+-K^+-ATPase 688
–, –, retrograde ouabain injection 685
–, –, secretory cell potenteials 684
–, –, two-stage process 684
–, microanatomical structure 677
–, neural control 681
–, occurrence in vertebrates 676
–, rectal glands 678, 682

Salt glands
–, reptiles 675
–, role of duct system 689, 670
–, teleosts 676
–, ultrastructure 678
–, zonula occludens 679, 684
Salt transport 145, 146
–, renal tubule 145, 146
–, –, equivalent electrical circuit 121
Small intestine 749
–, active sodium transport 758
–, electrophysiology 755
–, –, electrical coupling between cell membranes 757, 758
–, –, equivalent circuit model 755, 756
–, –, intracellular potential 757
–, functional properties 751
–, paracellular pathway 749, 750, 751, 756
–, –, influence on transcellular sodium transport 769, 770
–, –, multiple parallel pathways 753
–, –, permeability to ions 750, 752, 756
–, –, –, cation selectivity 753
–, –, –, effect of triaminopyrimidinium (TAP) 753
–, –, –, role of fixed dipoles 753
–, –, permeability to nonelectrolytes 754, 755
–, –, –, reflection coefficients 754
–, paracellular sodium transport 759
–, –, role in transepithelial water movement 754, 755
–, –, sodium permeability 759, 760
–, transport of sodium 749
–, –, carrier-mediated exchange with H^+-ions 761
–, –, coupled NaCl transport 761, 775
–, –, –, effect of acetazolamide 761, 775
–, –, –, effect of cAMP 761, 775
–, –, coupling to sugars and L-amino acids 761
–, –, exit step from cell 763, 765
–, –, –, coupling to cell potassium uptake 766
–, –, –, localization of ouabain binding sites 763
–, –, –, Na^+-K^+-ATPase 763
–, –, relationship to cell potassium 766
–, –, –, sodium transport pool 767, 768
–, –, intracellular sodium activity 760
–, –, relationship to other solutes 749, 771
–, –, –, amino acids 771, 772
–, –, –, bullfrog small intestine 775
–, –, –, human ileum 775
–, –, –, ileum 771
–, –, –, jejunum 771
–, –, –, mammalian ileum 775
–, –, –, mutual inhibition between sugar and amino acid transport 774
–, –, –, role of electrical potential 774
–, –, –, role of ion gradients 772, 773
–, –, –, sugars 771, 772
–, –, role of intercellular spaces in sodium-solute coupling 771
–, –, rheogenic nature of transport 761, 762, 764, 765, 770
–, –, transcellular transport 760
Sodium 69, 70, 71, 72, 145, 146, 151, 191, 196, 503, 708, 760, 785, 858, 899, 903
–, coupled sodium transport 155, 417, 554, 772, 775
–, metabolic requirements 529, 536, 542
–, passive 11
– acivity 69
– –, muscle fiber 69, 70
– –, Necturus proximal tubule cell 71, 72
– –, –, effects of glucose 72
– –, –, effects of luminal [Na^+] changes 72
– –, –, effect of luminal pH changes 73
– –, rat distal tubule 75, 77
– –, renal tubule cell 70
– transport 130
– –, renal tubule 145
– –, –, sodium bicarbonate 145
– –, –, sodium chloride 145
– –, thick ascending limb of Henle's loop 130
Solute-solvent interaction 29, 37, 41
– –, capillary 29, 37, 41
Solutes 8
–, lipid-insoluble 8, 15
–, lipidsoluble 8
Solvent drag 40
– –, capillary 40
Steric restriction 12
– –, capillary 12
Sugar transport 418
– –, across contraluminal cell side 427
– –, requirements for active transport across proximal renal tubule 419
– –, sodium-dependence of active transport 421, 422
– –, specificity of renal transport 418
– –, sugar and phlorhizin binding to brushborder membrane 423
– –, sugar and sodium transport across brush-border 423, 424
– –, –, electrical equivalent circuit 425
– –, –, electrical measurements 424
– –, –, electrogenic mode of Na^+-glucose transfer 427
– –, –, glucose uptake in brush-border vesicles 426
– –, –, measurements in brush-border vesicles 426
– –, –, specificity requirements 427, 428
– –, sugars not actively transported 420, 421

TAP (triaminopyrimidinium) 753, 871, 891
Theophylline 792
Thiazide diuretics 274, 372

Thick ascending limb of Henle 127
–, active transepithelial ionic membrane parameters 129
–, –, chloride transport 130
–, passive transepithelial ionic membrane parameters 128, 129
–, sodium transport 130
–, transepithelial conductance 128
–, transepithelial potential difference 128, 129
–, –, *Ambystoma tigrinum* 128
–, –, amphibian diluting segment 128
–, –, *Bufo marinus* 128
–, –, effect of acetazolamide 130
–, –, effect of amiloride 130
–, –, effect of diuretics 130
–, –, effect of furosemide 130
–, –, effect of mersalyl 130
–, –, mammalian tubule 128
–, –, *Rana pipiens* 128
Thin ascending limb of *Henle* 126
–, transepithelial ionic membrane parameters 127
–, –, isolated perfused tubules 127
–, –, –, anion selectivity 127
–, –, –, cation selectivity 127
Thin descending limb of *Henle* 125
–, electrical potential difference 125, 126
Tissue conductance of epithelia 751, 752
–, relative ionic permeabilities of shunt pathway in epithelia 752
–, transepithelial potential difference 126
Transcapillary exchange 8
Transcellular fenestrae 1
– transport 58
Transport 58
–, isolated renal tubule 503
–, hydrogen 299
–, paracellular 58, 158, 263, 502, 749, 828, 865
–, potassium 215
–, rheogenic 121, 159, 168, 257
–, sodium 145
Triamterene 136, 199, 274
Tubule 57
–, chloride cell activity 78, 79, 80
–, renal 57
–, –, cell potassium 62
–, –, ion activity 57, 58

Ultrafiltration 5, 8, 9, 29, 33
–, capillary 5, 8, 29
Unstirred layer effects 474, 483, 484, 802

Vitamin D 199, 374, 388, 402

Water transport 5, 145, 146, 169, 449, 507, 567, 711, 714, 782, 812, 856

Zonula adhaerens 1, 855
– occludens 1, 100, 111, 750, 854, 872

Membrane Transport in Biology

Editors: G. Giebisch, D. C. Tosteson, H. H. Ussing
Associate Editor: M. T. Tosteson

Volume I

Concepts and Models

Editor: D. C. Tosteson

With contributions by O. S. Andersen, J. E. Hall, D. J. Hanahan, U. V. Lassen, P. K. Lauf, R. J. Lefkowitz, E. Racker, B. E. Rasmussen, S. A. Rudolph, F. A. Sauer, C. W. Slayman, G. Stark, O. Sten-Knudsen, H. H. Ussing

1978. 108 figures, 36 tables. XX, 357 pages
ISBN 3-540-08687-0

Contents: Membrane Transport Biology. – Passive Transport Processes. – Interpretation of Tracer Fluxes. – Nonequilibrium Thermodynamics of Isotope Flow Through Membranes. – Use of Microelectrodes for Measurement of Membrane Potentials. – Chemical Composition of Membranes. – Genetic Determination of Membrane Transport. – Mechanisms of Ion Transport and ATP Formation. – Membrane Immunological Reactions and Transport. – Membrane Receptors, Cyclic Nucleotides, and Transport. – Permeability Properties of Unmodified Lipid Bilayer Membranes. – Carrier-Mediated Ion Transport Across Thin Lipid Membranes. – Channels in Black Lipid Films. – Subject Index.

Volume II

Transport Across Single Biological Membranes

Editor: D. C. Tosteson

With contributions by L. Beaugé, R. B. Gunn, A. Herold, G. Inesi, V. L. Lew, R. I. Macey, L. J. Mullins, B. Sarkadi, A. Scarpa, D. C. Tosteson, R. A. Venosa, D. Walker, G. Weissmann

1979. 103 figures, 41 tables. XX, 443 pages
ISBN 3-540-08780-X

Contents: Transport of Water and Non-Electrolytes Across Red Cell Membranes. – Transport of Anions Across Red Cell Membranes. – Passive Cation Fluxes in Red Cell Membranes. – Active Cation Transport in Human Red Cells. – Transport Across Axon Membranes. – Ionic Movements Across the Plasma Membrane of Skeletal Muscle Fibers. – Transport Across Mitochondrial Membranes. – Transport Across Sarcoplasmic Reticulum in Skeletal and Cardiac Muscle. – Transport Across the Lysosomal Membrane. – Transport Across Chloroplast Envelopes – The Role of Phosphates. – Subject Index.

This volume describes transport across single biological membranes, both plasma membranes separating the inside and outside of cells, and intracellular membranes separating the inside and outside of organelles such as mitochondria. Emphasis is on fairly detailed treatment of well-characterized systems such as mammalian red cells and squid axons.

Volume III

Transport Across Multi-Membrane Systems

Editor: G. Giebisch

With contributions by M. A. Bisson, D. Erlij, A. L. Finn, J. Gutknecht, D. F. Hastings, J. Hess Thaysen, A. Leaf, E. A. C. MacRobbie, S. H. P. Maddrell, A. Martinez-Palomo, J. L. Oschmann, H. H. Ussing, E. M. Wright, J. A. Zadunaisky, K. Zerahn

1978. 97 figures, 26 tables. XVIII, 459 pages
ISBN 3-540-08596-3

Contents: Transport Across Multimembrane Systems. – Role of Tight Junctions in Epithelial Function. – Morphological Correlates of Transport. – Transport of Giant Plant Cells II – Freshwater Cells. – Ion Transport and Turgor Pressure Regulation in Giant Algal Cells. – Transport Across Amphibian Skin. – Transport Across Amphibian Urinary Bladder. – Transport Across Insect Excretory Epithelia. Transport Across Insect Gut Epithelium. – Transport in Eye Epithelia: The Cornea and Crystalline Lens. – Transport in Eye Epithelia: Ciliary Epithelium and Retina Pigment Epithelium. – Ion Transport Across the Choroid Plexus. – The Sweat Glands. – The Lacrimal Gland. – Subject Index.

Springer-Verlag
Berlin
Heidelberg
New York

Biochemistry of Membrane Transport

FEBS-Symposium No. 42
Editors: G. Semenza, E. Carafoli
1977. 392 figures, 76 tables. XIX, 669 pages
(Proceedings in Life Sciences)
ISBN 3-540-08082-1

This book contains 47 articles, dealing with various aspects of the topic "Biochemistry of Membrane Transport". The area covered spans the molecular architecture of biological membranes and the interaction of membrane lipids and proteins, the chemistry of ionophoric molecules and various theoretical aspects of transport processes, with a detailed discussion of the properties of several natural and reconstituted transport processes. Distinctive for the book is its interdisciplinary character, a quality frequently found in membrane transport research. A further characteristic is the balance between established knowledge, offered in some review-type articles and the numerous contributions presenting the most recent advances in the field.

Molecular Aspects of Membrane Phenomena

International Symposium held at the Battelle Seattle Research Center
Seattle, WA, USA, November 4–6, 1974
Editors: H. R. Kaback, H. Neurath, G. K. Radda, R. Schwyzer, W. R. Wiley
1975. 144 figures, 31 tables. XIII, 338 pages
ISBN 3-540-07448-1

Molecular studies of biological membrane phenomea have progressed over the past decade to the point where it is now realistic to expect future resolution of the physico-chemical processes or forces governing the organization, function, and dynamic properties of membranes. This book is a compilation of formal lectures made during a three-day invitational conference at the Battelle Research Center in Seattle, Washington. The purpose of organizing and publishing the proceeding of the conference was to provide a comprehensive survey of present knowledge on the determinants of membrane structure, the molecular specificity of membrane function, and the dynamic properties of membranes. The overriding emphasis of the invited contributions is on the molecular aspects of three key membrane phenomena: membrane dynamics, recognition, and energy coupling. The lectures describe recent progress in defining the nature of the biochemical information which specifies membrane formation and the manner in which information, encoded in membranes, is functionally implemented at the molecular level. New research approaches are presented which delineate the dynamic properties of membranes and the significance of these properties to the molecular membrane energy-coupling processes.

Structural and Kinetic Approach to Plasma Membrane Functions

Proceedings of a Meeting Held on September 6–9, 1976, in Grignon (France)
Editors: C. Nicolau, A. Paraf
1977. 119 figures, 31 tables. XIV, 204 pages
(Proceedings in Life Sciences)
ISBN 3-540-08265-4

This book is an up-to-date synthesis of advances in the field of membrane research. It covers the structural aspects e.g., lipid protein interactions, domain formation, and cooperative phase transitions, freeze facture and spin-label studies of membranes, lipid organization, structure and shape of membrane proteins. Functional studies are also reported, regarding enzymic activities associated with membranes, their modulation by different ligands, changes occurring in the membranes upon tumor-virus transformation, correlation of the density-dependent cell-growth control with physical parameters, correlation of the membrane structure with optimal enzymic function etc. The aim of this book is to give an insight into the structure-function relationship in membranes as seen both from the physical and from the biological standpoint.

M. H. Saier Jr., C. D. Stiles

Molecular Dynamics in Biological Membranes

1975. 47 figures, 8 tables. XIII, 129 pages
(Heidelberg Science Library, Volume 22)
ISBN 3-540-90142-6

Contents: Introduction: Cell Structure and Function. – Constituents of Biological Membranes. – Structure of Membranes and Serum Lipoprotein Complexes. – Biological Consequences of Membrane Fluidity and Fusion. – Transmembrane Solute Transport Mechanisms. – Sensory Perception I: Chemoreception. – Sensory Perception II: Transmission Mechanisms. – Hormonal Regulation of Cellular Metabolism. – Cell Recognition. – Role of the Plasma Membrane in Growth Regulation and Neoplasia.

Springer-Verlag
Berlin Heidelberg New York

MIX
Papier aus verantwortungsvollen Quellen
Paper from responsible sources
FSC® C105338

If you have any concerns about our products,
you can contact us on
ProductSafety@springernature.com

In case Publisher is established outside the EU,
the EU authorized representative is:
Springer Nature Customer Service Center GmbH
Europaplatz 3, 69115 Heidelberg, Germany

Printed by Libri Plureos GmbH
in Hamburg, Germany